IRON TRANSPORT
IN BACTERIA

IRON TRANSPORT IN BACTERIA

Edited by
Jorge H. Crosa
Alexandra R. Mey
Shelley M. Payne

Washington, D.C.

Copyright © 2004   ASM Press
American Society for Microbiology
1752 N St., N.W.
Washington, DC 20036–2904

Library of Congress Cataloging-in-Publication Data

Iron transport in bacteria/edited by Jorge H. Crosa, Alexandra R.
  Mey, and Shelley M. Payne.
    p.; cm.
  Includes bibliographical references and index.
  ISBN 1-55581-292-9 (hardcover)
  1. Microbial metabolism. 2. Iron—Physiological transport.
  3. Siderophores. I. Crosa, Jorge H. II. Mey, Alexandra R.
  III. Payne, Shelley M.
  [DNLM: 1. Bacteria—metabolism. 2. Iron—metabolism. 3. Biological
  Transport—physiology. 4. Iron-Binding
  Proteins—physiology.

QW 52 171 2004]
QR92.171755 2004
572'.517293–dc22

2004007714

10 9 8 7 6 5 4 3 2 1

All rights reserved
Printed in the United States of America

Address editorial correspondence to ASM Press, 1752 N St., N.W., Washington, DC
20036–2904, USA

Send orders to ASM Press, P.O. Box 605, Herndon, VA 20172, USA
Phone: 800–546–2416; 703–661–1593
Fax: 703–661–1501
E-mail: books@asmusa.org
Online: www.asmpress.org

DEDICATION

This book is dedicated to the memory of Igor Stojiljkovic, M.D., Ph.D., who passed away on 10 October 2003 after a heroic 2-year battle with cancer. Igor was one of the bright young stars in the fields of iron transport and bacterial infection. However, he was not just a highly creative and imaginative scientist but was, foremost, a first-class gentleman, a kind human being, and, for those of us who had the fortune of being close to him, an outstanding and warm friend, always ready to give positive advice and constructive criticism. He was an indefatigable worker—until not long before his untimely passing, Igor managed to write the first drafts and give directions for the completion of his contribution to this book, the two chapters that were finished by his postdoctoral fellows.

Igor received his M.D. and Ph.D. degrees from the University of Zagreb, Croatia, and carried out postdoctoral research at the University of Tübingen, Germany, and at the Oregon Health and Science University in Portland. He became a tenured Associate Professor at Emory University in 2002.

(Continues)

One of his major scientific interests was to understand how microorganisms obtain and use iron for their metabolic purposes. He focused on the ability of bacteria to assimilate heme, a ubiquitous source of iron. This research included identifying and characterizing bacterial components that are involved in the assimilation of heme (iron), understanding the mechanisms of function of heme assimilation systems, and studying genetic mechanisms causing the expression of these systems to vary. He also determined the virulence potential of these systems in different pathogens, the importance of these mechanisms for the population biology of pathogens, and novel approaches that would exploit heme assimilation as a target against pathogenic bacteria.

Igor's great joy for life and his incredible capacity to be imaginative and creative will always be remembered by all of us. We know that his scientific legacy will be perpetuated by those students and postdoctoral fellows who had the fortune of training with him.

CONTENTS

Contributors xi
Preface xvii

I. SIDEROPHORES AND HEMOPHORES: PROPERTIES AND BIOSYNTHESIS OF BACTERIAL IRON AND HEME CARRIERS 1

1. Biochemical and Physical Properties of Siderophores
Kenneth N. Raymond and Emily A. Dertz
3

2. Siderophore Biosynthesis in Bacteria
Christopher T. Walsh and C. Gary Marshall
18

3. Hemophore-Dependent Heme Acquisition Systems
Laurent Debarbieux and Cécile Wandersman
38

II. IRON TRANSPORT PROTEINS: STRUCTURAL STUDIES 49

4. Structure of Outer Membrane Receptor Proteins
Dick van der Helm
51

5. Bacterial Heme and Hemoprotein Receptors
Donna Perkins-Balding, Andrew Rasmussen, and Igor Stojiljkovic
66

6. Bacterial Heme Oxygenases
Melanie Ratliff-Griffin, Angela Wilks, and Igor Stojiljkovic
86

7. The TonB, ExbB, and ExbD Proteins
Kathleen Postle and Ray A. Larsen
96

8. Periplasmic Binding Proteins Involved in Bacterial Iron Uptake
Karla D. Krewulak, R. Sean Peacock, and Hans J. Vogel
113

III. IRON TRANSPORT, ENERGETICS, AND REGULATION IN *ESCHERICHIA COLI* K-12: A PROTOTYPE FOR IRON TRANSPORT SYSTEMS IN GRAM-NEGATIVE BACTERIA 131

9. Iron Uptake via the Enterobactin System
Charles F. Earhart
133

10. Transport Biochemistry of FepA
Phillip E. Klebba
147

11. Ferrichrome- and Citrate-Mediated Iron Transport
Volkmar Braun, Michael Braun, and Helmut Killmann
158

12. Ferrous Iron Transport
Klaus Hantke
178

13. Mode of Binding of the Fur Protein to Target DNA: Negative Regulation of Iron-Controlled Gene Expression
Víctor de Lorenzo, José Perez-Martín, Lucía Escolar, Graziano Pesole, and Giovanni Bertoni
185

IV. IRON TRANSPORT SYSTEMS IN PATHOGENIC BACTERIA 197

14. Pathogenic *Escherichia coli*, *Shigella*, and *Salmonella*
Shelley M. Payne and Alexandra R. Mey
199

15. *Yersinia*
Robert D. Perry
219

16. *Vibrio*
Manuela Di Lorenzo, Michiel Stork, Alejandro F. Alice, Claudia S. López, and Jorge H. Crosa
241

17. *Neisseria*
Cynthia Nau Cornelissen and P. Frederick Sparling
256

18. *Haemophilus*
Daniel J. Morton and Terrence L. Stull
273

19. *Pseudomonas*
Keith Poole
293

20. *Bordetella*
Timothy J. Brickman, Carin K. Vanderpool, and Sandra K. Armstrong
311

21. *Porphyromonas gingivalis*
Caroline Attardo Genco, Waltena Simpson, and Teresa Olczak
329

22. *Corynebacterium diphtheriae*
Michael P. Schmitt
344

23. Pathogenic Mycobacteria
G. Marcela Rodriguez and Issar Smith
360

24. *Legionella*
Nicholas P. Cianciotto
372

25. *Staphylococcus*, *Streptococcus*, and *Bacillus*
David E. Heinrichs, Andrea Rahn, Suzanne E. Dale, and Michael Tom Sebulsky
387

26. *Erwinia*, a Plant Pathogen
Dominique Expert, Lise Rauscher, and Thierry Franza
402

27. Therapeutic Uses of Iron(III) Chelators and Their Antimicrobial Conjugates
Vinay Girijavallabhan and Marvin J. Miller
413

V. IRON TRANSPORT AND ECOLOGY 435

28. Ecology of Siderophores
Günther Winkelmann
437

29. Environmental Fluorescent *Pseudomonas* and Pyoverdine Diversity: How Siderophores Could Help Microbiologists in Bacterial Identification and Taxonomy
Jean-Marie Meyer and Valérie A. Geoffroy
451

30. Mechanisms and Regulation of Iron Uptake in the Rhizobia
Andrew W. B. Johnston
469

Index   489

CONTRIBUTORS

Alejandro F. Alice
Department of Molecular Microbiology L220, Oregon Health and Science University,
3181 SW Sam Jackson Park Road, Portland, OR 97239

Sandra K. Armstrong
Department of Microbiology, University of Minnesota, MMC 196,
420 Delaware Street S.E., Minneapolis, MN 55455-0312

Giovanni Bertoni
Dipartimenti di Genetica e Fisiología, via Celoria, 26,
20133 Milano, Italy

Michael Braun
Mikrobiologie/Membranphysiologie, Universität Tübingen, Auf der Morgenstelle 28,
D-72076 Tübingen, Germany

Volkmar Braun
Mikrobiologie/Membranphysiologie, Universität Tübingen, Auf der Morgenstelle 28,
D-72076 Tübingen, Germany

Timothy J. Brickman
Department of Microbiology, University of Minnesota, MMC 196,
420 Delaware Street S.E., Minneapolis, MN 55455-0312

Nicholas P. Cianciotto
Department of Microbiology and Immunology, Northwestern University Medical School,
320 East Superior St., Chicago, IL 60611

Cynthia Nau Cornelissen
Department of Microbiology and Immunology, Medical College of Virginia,
Virginia Commonwealth University, Richmond, VA 23298-0678

Jorge H. Crosa
Department of Molecular Microbiology L220, Oregon Health and Science University,
3181 SW Sam Jackson Park Road, Portland, OR 97239

Suzanne E. Dale
Department of Microbiology and Immunology, University of Western Ontario,
London, Ontario N6A 5C1, Canada

Laurent Debarbieux
Unité des Membranes Bactériennes, Institut Pasteur (CNRS URA 2172),
25 rue du Dr. Roux, 75724 Paris Cedex 15, France

Victor de Lorenzo
Centro Nacional de Biotecnología del CSIC, Campus Universidad Autónoma,
Madrid 28043, Spain

Emily A. Dertz
Department of Chemistry, University of California, Berkeley, CA 94720–1460

Manuela Di Lorenzo
Department of Molecular Microbiology L220, Oregon Health and Science University,
3181 SW Sam Jackson Park Road, Portland, OR 97239

Charles F. Earhart
Section of Molecular Genetics and Microbiology, The University of Texas at Austin,
Austin, Texas 78712–1095

Lucía Escolar
Centro Nacional de Biotecnología del CSIC, Campus Universidad Autónoma,
Madrid 28043, Spain

Dominique Expert
Laboratory of Plant Pathology, UMR 217 INRA/INA P-G, Université Paris 6,
16 rue Claude Bernard, F-75005 Paris, France

Thierry Franza
Laboratory of Plant Pathology, UMR 217 INRA/INA P-G, Université Paris 6,
16 rue Claude Bernard, F-75005 Paris, France

Caroline Attardo Genco
Department of Medicine, Section of Infectious Diseases, Boston University
School of Medicine, Boston, MA 02118

Valérie A. Geoffroy
Laboratoire de Microbiologie et Génétique, Université Louis-Pasteur–CNRS FRE 2326,
F-67000 Strasbourg, France

Vinay Girijavallabhan
Department of Chemistry and Biochemistry, 251 Nieuwland Science Center,
University of Notre Dame, Notre Dame, IN 46556

Klaus Hantke
Mikrobiologie/Membranphysiologie, Universität Tübingen, Auf der Morgenstelle 28,
D-72076 Tübingen, Germany

David E. Heinrichs
Department of Microbiology and Immunology, University of Western Ontario,
London, Ontario N6A 5C1, Canada

Andrew W. B. Johnston
School of Biological Sciences, University of East Anglia, Norwich NR4 7TJ,
United Kingdom

Helmut Killmann
Mikrobiologie/Membranphysiologie, Universität Tübingen, Auf der Morgenstelle 28,
D-72076 Tübingen, Germany

Phillip E. Klebba
Department of Chemistry & Biochemistry, University of Oklahoma,
620 Parrington Oval, Norman, OK 73019

Karla D. Krewulak
Structural Biology Research Group, Department of Biological Sciences, University of
Calgary, Calgary, Alberta T2N 1N4, Canada

Ray A. Larsen
Department of Biological Science, Bowling Green State University,
Bowling Green, OH 43403–0212

Claudia S. López
Department of Molecular Microbiology L220, Oregon Health and Science University,
3181 SW Sam Jackson Park Road, Portland, OR 97239

C. Gary Marshall
Department of Biological Chemistry and Molecular Pharmacology, Harvard Medical School,
Boston, MA 02115

Alexandra R. Mey
Department of Molecular Genetics and Microbiology, The University of Texas at Austin,
Austin, TX 78712

Jean-Marie Meyer
Laboratoire de Microbiologie et Génétique, Université Louis-Pasteur—CNRS FRE 2326,
F-67000 Strasbourg, France

Marvin J. Miller
Department of Chemistry and Biochemistry, 251 Nieuwland Science Center,
University of Notre Dame, Notre Dame, IN 46556

Daniel J. Morton
Department of Pediatrics, University of Oklahoma Health Sciences Center,
Oklahoma City, OK 73104

Teresa Olczak
Institute of Biochemistry and Molecular Biology, Wrocław University, Przybyszewskiego
63/77, 51–148 Wrocław, Poland

Shelley M. Payne
Department of Molecular Genetics and Microbiology, The University of Texas at Austin,
Austin, TX 78712

R. Sean Peacock
Structural Biology Research Group, Department of Biological Sciences, University of Calgary,
Calgary, Alberta T2N 1N4, Canada

José Perez-Martín
Centro Nacional de Biotecnología del CSIC, Campus Universidad Autónoma,
Madrid 28043, Spain

Donna Perkins-Balding
Department of Microbiology and Immunology, Emory School of Medicine,
Rollins Research Center, 1510 Clifton Rd., Rm. 3152, Atlanta, GA 30322

Robert D. Perry
Department of Microbiology, Immunology, and Molecular Genetics, MS415, Medical
Center, University of Kentucky, Lexington, KY 40536–0298

Graziano Pesole
Dipartimenti di Genetica e Fisiología, via Celoria, 26,
20133 Milano, Italy

Keith Poole
Department of Microbiology and Immunology, Queen's University,
Kingston, Ontario K7L 3N6, Canada

Kathleen Postle
School of Molecular Biosciences, Washington State University,
Pullman, WA 99164–4234

Andrea Rahn
Department of Microbiology and Immunology, University of Western Ontario,
London, Ontario N6A 5C1, Canada

Andrew Rasmussen
Department of Microbiology and Immunology, Emory School of Medicine,
Rollins Research Center, 1510 Clifton Rd., Rm. 3152, Atlanta, GA 30322

Melanie Ratliff-Griffin
Department of Microbiology and Immunology, Emory School of Medicine,
3152 Rollins Research Center, Atlanta, GA 30322

Lise Rauscher
Laboratory of Plant Pathology, UMR 217 INRA/INA P-G, Université Paris 6,
16 rue Claude Bernard, F-75005 Paris, France

Kenneth N. Raymond
Department of Chemistry, University of California, Berkeley, CA 94720–1460

G. Marcela Rodriguez
TB Center, Public Health Research Institute, International Center for Public Health,
225 Warren Street, Newark, NJ 07103–3535

Michael P. Schmitt
Center for Biologics Evaluation and Research, Food and Drug Administration,
8800 Rockville Pike, Building 29, Room 108, HFM-437, Bethesda, MD 20892

Michael Tom Sebulsky
Department of Microbiology and Immunology, University of Western Ontario,
London, Ontario N6A 5C1, Canada

Waltena Simpson
Department of Biological Sciences, South Carolina State University,
Orangeburg, SC 29117

Issar Smith
TB Center, Public Health Research Institute, International Center for Public Health,
225 Warren Street, Newark, NJ 07103–3535

P. Frederick Sparling
Department of Medicine, University of North Carolina at Chapel Hill,
Chapel Hill, NC 27599–7031

Igor Stojiljkovic
Department of Microbiology and Immunology, Emory School of Medicine,
Rollins Research Center, 1510 Clifton Rd., Rm. 3152, Atlanta, GA 30322

Michiel Stork
Department of Molecular Microbiology L220, Oregon Health and Science University,
3181 SW Sam Jackson Park Road, Portland, OR 97239

Terrence L. Stull
Department of Microbiology/Immunology, University of Oklahoma Health Sciences Center,
Oklahoma City, OK 73104

Dick van der Helm
Department of Biochemistry and Microbiology, University of Victoria,
P.O. Box 3055, Victoria, British Columbia V8W 3P6, Canada

Carin K. Vanderpool
Department of Microbiology, University of Minnesota, MMC 196,
420 Delaware Street S.E., Minneapolis, MN 55455–0312

Hans J. Vogel
Structural Biology Research Group, Department of Biological Sciences,
University of Calgary, Calgary, Alberta T2N 1N4, Canada

Christopher T. Walsh
Department of Biological Chemistry and Molecular Pharmacology, Harvard Medical School,
Boston, MA 02115

Cécile Wandersman
Unité des Membranes Bactériennes, Institut Pasteur (CNRS URA 2172),
25 rue du Dr. Roux, 75724 Paris Cedex 15, France

Angela Wilks
Department of Pharmaceutical Sciences, School of Pharmacy,
University of Maryland, Baltimore, MD 21201

Günther Winkelmann
Institut für Mikrobiologie, Universität Tübingen, Auf der Morgenstelle 28,
D-72076 Tübingen, Germany

PREFACE

Iron is one of the most abundant elements in the Earth's crust. It also displays remarkable chemical properties; it has two stable valences and an extremely wide range of oxidation-reduction potentials. Given its abundance and properties, it is not surprising that most organisms evolved iron-dependent enzymes to perform many essential functions.

The importance of iron in microbial growth, metabolism, and interactions with the host has been recognized for many years. Studies such as those by Waring and Workman in the 1940s defined the iron requirements of several microorganisms. Subsequent studies showed that iron played a critical role in electron transport, metabolism, protection against oxidative stress, DNA metabolism, and regulation of gene expression.

Despite its abundance, iron acquisition represents a major challenge for many organisms. In the presence of oxygen, iron forms insoluble ferric hydroxides and the level of free iron in aerobic environments is below that required for microbial growth. For microbes that colonize or invade mammalian hosts, iron limitation is exacerbated by the presence of host high-affinity iron-binding proteins. Studies begun in the 1940s by Schade and Caroline demonstrated the presence of iron-binding proteins in host fluids and showed that these proteins inhibited microbial growth. Numerous studies over the ensuing 60 years have shown that competition for iron within the host is a critical factor in host-pathogen interactions.

Microbes have evolved an impressive array of systems designed to solubilize and capture iron from their environments. These include siderophores, low-molecular-weight iron-chelating compounds that are secreted into the environment, and cell surface receptors for host iron proteins. The growth-promoting effects of siderophores were detected as early as 1912, when Twort and Ingram reported a cell-associated growth factor in mycobacteria, but it was not until the 1950s that the isolation and structural characterizations of siderophores were first reported. Ferrichrome, a reddish-brown iron-binding compound produced by the fungus *Ustilago*, was the first to be isolated and characterized. This work by J. B. Neilands was one of many pioneering studies from his laboratory that made enormous contributions to understanding not only the structure but also the biosynthesis, genetics, and regulation of expression of microbial siderophores. Neilands, along with Charles Lankford, put forward the hypothesis that these iron-binding compounds were involved in microbial iron transport and helped supply essential iron to the microorganism. Following the identification of ferrichrome, a large number of other siderophores have been described, and many details of their biosynthesis and transport have been elucidated. A number of these are described in this volume, although this description is by no means complete.

It was also recognized that iron could be extremely toxic to cells. Iron catalyzes the formation of potentially lethal reactive oxygen species. Therefore, the uptake of iron into cells must be tightly regulated, and negative regulation of genes by iron is a recurring theme in this book. This regulation extends beyond control of iron uptake to regulation of genes associated with virulence in a number of pathogens. Studies by Clarke of diphtheria toxin production and by van Heyningen of Shiga toxin synthesis showed that iron concentration was the environmental factor that determined the amount of toxin produced by the bacteria.

This book is an attempt to survey and consolidate the research on microbial iron transport that has taken place over the past 50 years. It is not intended to be an exhaustive catalog of everything that has been done. Rather, we hope to give the reader an overview of microbial iron transport and insight into where the field has been and where it is going. The authors were asked to provide a brief list of selected readings instead of a comprehensive bibliography of all the papers in their field. Any omission of important references should be blamed on the editors rather than on the authors of the chapters.

The availability of complete sequences of microbial genomes and increasingly sophisticated technology has led to an explosion of information in recent years, and many of the chapters focus on these recent advances. New iron transport systems have been identified based on analysis of genomes, and new models for transport have been suggested by crystallography and structural determinations of the membrane transport proteins. Some of these studies have raised as many questions as have been answered. One question that remains unresolved is the mechanism for transduction of energy from the inner membrane to the outer membrane receptors for transport of siderophores and other iron complexes through the outer membrane in gram-negative bacteria. It is well established that the cytoplasmic membrane protein TonB is required for this process, but precisely how it interacts with the receptors and how it transfers energy is still an open question. This question and possible models for TonB-mediated transport are described in several chapters of this book.

Overall, the book is organized into five sections. The first is an overview of the structures, chemical properties, and biosynthesis of the microbial products, siderophores and hemophores, used by these organisms to acquire iron. The second section describes the transport of these compounds into gram-negative bacteria and covers the structure of the receptors, their initial interaction with the ligand, the movement of the ligand through the outer membrane and the role of the TonB system in this process, and the structure and function of the periplasmic binding proteins. A chapter on heme oxygenases, which remove the iron from heme following transport, completes this section. The third section gives a relatively complete picture of iron transport in the gram-negative prototype, *Escherichia coli* K-12. Because this is the most extensively studied of the gram-negative bacteria, there is a wealth of information on the genetics, regulation, and structure of iron transport systems in *E. coli* K-12. The fourth section is a collection of descriptions of iron transport systems in selected pathogenic microorganisms. In many cases, the choice of organisms was arbitrary. We have tried to include a representative selection of gram-negative, gram-positive, and acid-fast bacteria and chose ones for which a relatively large amount of information was available. However, many equally important pathogens were omitted for the sake of brevity. Many of the iron transport systems presented in this section are described relative to the prototype, and the similarities to and differences from *E. coli* systems are noted. A chapter on the potential exploitation of microbial iron transport systems for antibiotic design completes this section. The book concludes with a section on iron transport and ecology. This is an extremely important area but one that is often overlooked in the emphasis on pathogens.

We are indebted to the many authors who gave generously of their time in writing the chapters. It is our hope that this book will inform and stimulate discussion of the importance of iron in biological systems.

JORGE H. CROSA
ALEXANDRA R. MEY
SHELLEY M. PAYNE

SIDEROPHORES AND HEMOPHORES: PROPERTIES AND BIOSYNTHESIS OF BACTERIAL IRON AND HEME CARRIERS

I

BIOCHEMICAL AND PHYSICAL PROPERTIES OF SIDEROPHORES

Kenneth N. Raymond and Emily A. Dertz

1

As discussed throughout this volume, the significance of iron as a nutrient and its role in biological processes have major consequences in areas of science ranging from geochemistry to human disease. Iron is generally required for an enormous variety of metabolic processes in virtually all organisms, with the possible exception of some lactobacilli. However, in an aerobic, neutral-pH environment, the concentration of free Fe^{3+} is limited to 10^{-18} M by the insolubility of $Fe(OH)_3$; this concentration is well below that generally required by cells. The concentration of iron available to mammalian bacterial pathogens is even lower because Fe^{3+} is sequestered by host proteins such as serum transferrin and lactoferrin. Consequently, serum is bacteriostatic. Many microorganisms circumvent this nutritional limitation by producing siderophores, low-molecular-weight compounds secreted under iron-limited conditions. These chelating agents strongly and specifically bind, solubilize, and deliver iron to microbial cells via specific cell surface receptors.

In addition to their major role in iron solubilization and transport, siderophores have been identified as germination or sporulation factors within selected organisms. Invading microorganisms exposed to circulating blood produce siderophores to compete for iron with the human transport protein transferrin; this constitutes one aspect of virulence and pathogenicity. Pathways for iron uptake in plants are not as well understood, but it is fairly well established that root-colonizing nonpathogenic microorganisms can prevent the invasion of pathogenic strains by secreting siderophores, which bind iron strongly and make it unavailable for potentially harmful microorganisms. Additionally, siderophores may be used clinically to treat some human diseases; ferrioxamine B is employed clinically for iron removal from the body after acute iron poisoning and hence in the treatment of diseases such as β-thalassemia, and the use of antibiotic-siderophore complexes may represent a new approach to treating bacterial and fungal infections (see chapter 27).

This introduction to the biochemical and physical properties of the siderophores focuses on the coordination chemistry of these remarkable compounds, where a lot of function is packed into small molecules.

WHY IRON?

The Fe^{3+} ion has a d^5 electronic configuration and is typically six coordinate. The variation of its redox chemistry and spin state with ligand

---

Kenneth N. Raymond and Emily A. Dertz, Department of Chemistry, University of California, Berkeley, CA 94720–1460.

environment causes enormous changes in its chemical properties. The aqueous ion, $Fe(H_2O)_6^{3+}$, is very acidic—hence the insolubility as $Fe(OH)_3$. As in the aqueous ion, the siderophores generally surround Fe^{3+} in an octahedral arrangement of oxygen atoms. This relatively weak ligand field makes all of the ferric siderophore complexes high-spin (five unpaired d electrons) with corresponding fast ligand exchange. The relatively small size and high charge of Fe^{3+} account for its strong complexation by charged or polar ligands, as can be seen in the large hydration enthalpy of about 1,000 kcal/mol. Most of this energy is due to the formation of the inner FeO_6 coordination octahedron. The high charge/radius ratio of $Fe(H_2O)_6^{3+}$ causes the proton of a coordinated water molecule to be quite acidic (pK_a values are shown in Table 1). While divalent cations have lower charge/radius ratios and have water protons which are correspondingly less acidic, Fe^{3+} must be complexed to be in aqueous solution.

SIDEROPHORE TYPES

The three broad groups of siderophores (catecholates, hydroxamates, and hydroxycarboxylates) are distinguished by the chemical structure of the metal-binding functionality (Fig. 1). However, with a few hundred different siderophores now characterized, the variety of resolved siderophore structures also includes oxazoline, thiazoline, hydroxypyridinone, α- and β-hydroxy acids, and α-keto acid components. The ligand denticity (number of coordinating atoms per molecule) constitutes an important feature of siderophores and ranges from bidentate

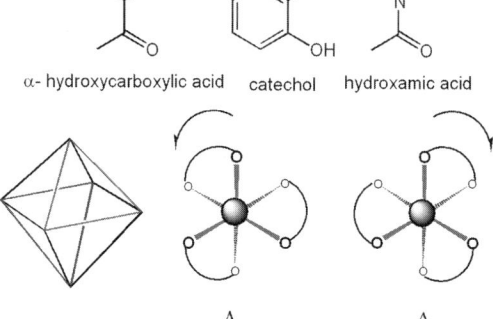

FIGURE 1 (Top) Principal functional groups found in siderophores for Fe^{3+} coordination. The three functional groups are bidentate because each has two oxygen atoms that coordinate iron. (Bottom) The octahedral coordination of Fe^{3+} occurs by having six ligand atoms in a close-packed geometry around the metal ion. A schematic view of the pseudo-octahedral geometry of Fe^{3+} in siderophores composed of bidentate chelating units (such as hydroxamate and catecholate) is shown. Chirality at the metal centers gives Δ or Λ stereochemistry, and the sequence and orientation of the rings can give a number of geometric numbered isomers. For the Δ complex, the chelates from a right-handed screw around the metal center. A left-handed screw is found for the Λ complex.

to hexadentate. Although structurally diverse, most siderophores have some common features, including hard donor atoms (usually oxygen but occasionally nitrogen or sulfur) and the formation of thermodynamically stable, high-spin Fe^{3+} species. Although a high-spin d^5 metal has no ligand field stabilization energy, the resulting complex is very stable due to the very strong interaction between the negatively charged oxygen atoms and the positively charged iron center. The most common geometry for high-spin d^5 iron is octahedral, allowing six ligands to be arranged around the metal center with a minimal amount of ligand repulsion. Depending on the nature of the ligand, the octahedral geometry can distort to a trigonal prisomatic geometry, as seen in some siderophore complexes (Color Plate 1). Some representative siderophores discussed in this chapter are shown in Fig. 2. Additional siderophores are discussed in more detail in later chapters of this volume.

TABLE 1 Comparison of Fe^{3+} with physiologically relevant cations

Cation	Ionic radius (Å)	pK_a	Heat of solvation (kcal/mol)
Fe^{3+}	0.65	2.2	1,035
Mg^{2+}	0.72	11.4	455
Ca^{2+}	1.00	12.7	380
Cu^{2+}	0.57	7.7	498
Zn^{2+}	0.74	9.0	484

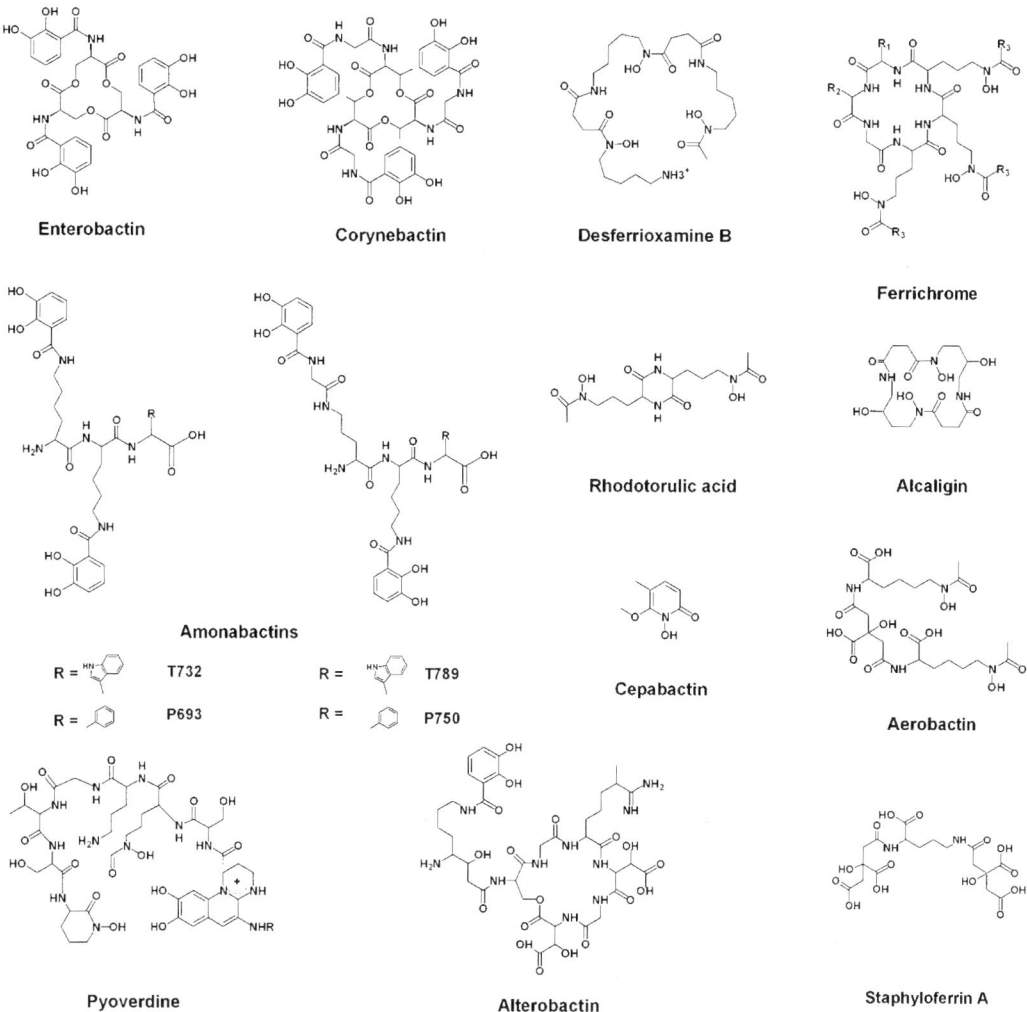

FIGURE 2 Some representative siderophore structures. The hexadentate siderophores include enterobactin, corynebactin, desferrioxamine B, ferrichrome, pyoverdine, alterobactin, aerobactin, and staphyloferrin A. The tetradentate siderophores include the amonabactins, rhodotorulic acid, and alcaligin. Cepabactin is a bidentate siderophore.

While some bacteria produce exclusively one siderophore, many secrete multiple types, since specific features may be more efficient than others under particular circumstances. The production of several siderophores may permit the microbe to colonize different environments. For example, some *Escherichia coli* strains produce both enterobactin and aerobactin (Fig. 2). Aerobactin may be more effective than enterobactin in competing for iron with the host iron-binding proteins in vivo, because enterobactin binds to and is inactivated by serum albumin. Further, it has recently been found that NGAL (neutrophil gelatinase-associated lipocalin), a protein associated with the human immune response, specifically binds enterobactin and reduces its effectiveness in supplying iron to the bacterium. Aerobactin, however, is not bound by serum albumin (and presumably also not by NGAL) and is considered a virulence factor.

Enterobactin and aerobactin are the primary siderophores of the members of the gram-negative family *Enterobacteriaceae*, such as *E. coli*, *Shigella*, *Salmonella*, *Yersinia*, and *Klebsiella* species. Recently, enterobactin has also been isolated from gram-positive *Streptomyces* species. The ferrioxamine siderophores are produced by both gram-negative and gram-positive bacteria, including *Streptomyces* species, *Erwinia herbicola*, and *Pseudomonas stutzeri*. Fungal siderophores commonly use hydroxamate groups as their iron-chelating functionality. Recently, other functional group types have been reported. Hydroxycarboxylate-containing siderophores, such as rhizoferrin, are produced by fungi of the Zygomycete group (such as *Rhizopus* strains), and phenolate-catecholate siderophores have been identified in wood-decaying basidiomycetes. The catecholate siderophore corynebactin (Fig. 2) has been found in two species of gram-positive bacteria, *Corynebacterium glutamicum* and *Bacillus subtilis*.

Catecholate Siderophores

Of all the catecholate siderophores, enterobactin has been the most intensively studied with regard to solution thermodynamics, microbial iron transport, chemical synthesis, and biosynthetic pathways, and several of these aspects are covered in the following chapters. The enterobactin molecule (Fig. 2) possesses threefold symmetry and is composed of three 2,3-dihydroxybenzoic acid groups, each appended to an L-serine group. Three serines form the trilactone macrocycle backbone. Metal coordination at neutral pH occurs through the six catecholate oxygens (Color Plate 1, left). The closely related corynebactin (Color Plate 1, right) is the only other example of a siderophore containing a trilactone backbone, and it is composed of three identical units consisting of 2,3-dihydroxybenzoic acid, glycine, and L-threonine.

Several linear catecholate siderophores have also been identified, including protochelin, cepaciachelin, azotobactin, and aminochelin. All of these, except cepaciachelin, are produced by *Azotobacter vinelandii*. Protochelin contains three units of 2,3-dihydroxybenzoic acid attached to a linear backbone formed by the amide of lysine with diaminobutane. Other catecholate siderophores containing triamine spermidine as a backbone include parabactin, agrobactin, vibriobactin, and fluvibactin. The four types of bis(catecholate) amonabactins, produced by *Aeromonas hydrophila*, are composed of an amino acid backbone containing two lysines, one (L configuration) phenylalanine or tryptophan, and an optional glycine.

Hydroxamate Siderophores

The first compounds recognized as microbial iron transport agents were the ferrioxamines and the ferrichromes. Hydroxamate siderophores are produced by both bacteria and fungi. Structures of representative hydroxamate siderophores are presented in Fig. 2. Hydroxamate is a bidentate ligand formed by acylation of a hydroxylamine; all natural hydroxamates originate from L-ornithine, which is N^δ hydroxylated and N^δ acylated. In nature, hydroxamate siderophores are predominantly hexadentate. At low concentration, a hexadentate ligand is a much more powerful chelator than lower-denticity ligands (see below). Hydroxamate groups have two major resonance forms, one with a negative charge on a single oxygen atom, and one with a distributed negative charge over the two oxygen atoms. The nature of the carbonyl and nitrogen substituents plays an essential role in the charge distribution. The stability of the metal complex is enhanced when these substituents promote the donation of electron density from the nitrogen lone pair to the carbonyl oxygen, so that both oxygen atoms are negatively charged. Siderophores from the ferrichrome family are cyclic hexapeptides in which one tripeptide (of glycine, alanine, or serine) is linked to a second tripeptide of N^δ-acyl-N^δ-hydroxyornithine. These are produced exclusively by fungi, such as basidiomycetes and ascomycetes. Although only tetradentate, rhodotorulic acid is the primary siderophore of basidiomycetes. It is a dihydroxamate composed of the diketopiperazine of N^δ-acetyl-L-N^δ-hydroxyornithine.

Dimerumic acid is a derivative of rhodotorulic acid. Both siderophores form dimer complexes composed of two iron ions and three siderophore molecules.

All of the above hydroxamate siderophores, and others related to them, are linear or exocyclic, in that the bidentate hydroxamate groups are focused toward a metal center that is away from any molecular ring (as in the ferrichromes). Endocyclic ligands focus toward a center that is both a macrocyclic ring and metal center, a distinctive structural family. Alcaligin is a cyclic bis(hydroxamate) siderophore produced by an aquatic alga, *Alcaligenes xylosoxidans* subsp. *xylosoxidans*, as well as two mammalian pathogens, *Bordetella pertussis*, which causes whooping cough in humans, and *Bordetella bronchiseptica*, which is the agent of swine atrophic rhinitis and kennel cough in dogs. Bisucaberin and putrebactin, two structural analogs of alcaligin, are produced by the marine bacterium *Alteromonas haloplanktis* and *Shewanella putrefaciens*, respectively. Alcaligin and putrebactin are 20-member macrocyclic dihydroxamate siderophores (Fig. 2), while bisucaberin is a 22-atom-ring dihydroxamic acid. The tetradentate nature of these ligands is particularly interesting, since they need to form a binuclear metal complex with a stoichiometry of Fe_2L_3 (where L represents one dihydroxamate ligand). Another endocyclic example is from *Streptomyces pilosus*, which produces ferrioxamine E, a cyclic trihydroxamate siderophore, along with other ferrioxamine siderophores, and forms an FeL complex. Most other ferrioxamine siderophores are linear trihydroxamates, except for ferrioxamine H, which is an unusual pentadentate siderophore. Cepabactin (Fig. 2) can be considered an aromatic hydroxamate with a unique bidentate 1,2-hydroxypyridonate. The siderophore has been isolated from *Burkholderia cepacia* (*Pseudomonas cepacia*) and *Pseudomonas alcaligenes* strains. It is the only hydroxypyridinone known to be produced by a microorganism.

Carboxylate Siderophores

A few siderophores chelate Fe^{3+} only through carboxylate and α-hydroxy donor groups. Examples include staphyloferrin A and B, rhizobactin, and rhizoferrin (Fig. 2).

Other Siderophore Types

Plants produce some aminocarboxylates that are related in function to siderophores and in structure to the common synthetic chelator EDTA (e.g., mugineic acid). Several carboxylate-hydroxamate mixed siderophores have been isolated, including aerobactin, arthrobactin, shizokinen, acinetoferrin, and nannochelin. Aerobactin (Fig. 2) has been isolated from *Aerobacter aerogenes*, *Escherichia*, *Salmonella*, and *Shigella* species, and many other bacteria. Aerobactin has also been isolated from *Alteromonas luteoviolacea* and is the first siderophore from an open-ocean bacterium to be structurally characterized.

A number of mixed hydroxamate-catecholate siderophores and other structural types are described elsewhere. We close this section by mentioning the remarkable fluorescent siderophores, whose optical properties have not been well related to any functional use. Fluorescent pseudomonad strains produce pyoverdines and pseudobactins. Pyoverdine was the first siderophore to be characterized from the *Pseudomonas* species. Siderophores belonging to the pyoverdine family are heteropeptides composed of a linear or cyclic peptide, a small dicarboxylic acid (or its monoamide), and a quinolinic chromophore (Fig. 2). The chromophore is identical for all pyoverdines characterized except for the one produced by *Pseudomonas putida*. The chromophore structure is one of the binding sites for Fe^{3+} and confers the characteristic UV-visible spectrum and strong fluorescence to the molecule. Pyoverdine structures differ only in the amino acid composition of the peptide component.

FERRIC SIDEROPHORE STRUCTURES

Enterobactin and Corynebactin

The only crystal structure of enterobactin (Color Plate 1) is with vanadium (IV) and was reported in 1992; $[V(ent)]^{2-}$ is Δ and has approximate C_3 symmetry. Comparison of the triserine backbone of the metal complex to a related trilactone reveals very little structural perturbation on metal complexation, suggesting that the trilactone provides an optimal cavity size. Corynebactin (Color Plate 1) is structurally similar to enterobactin but incorporates a glycine spacer in the chelating arms and L-threonine instead of L-serine in its trilactone ring. Circular dichroism spectra indicate that corynebactin assumes the Λ configuration while enterobactin forms the Δ complex. Molecular modeling (including MM3 and density functional theory [pBP86/DN★] calculations) confirmed these experimental observations and indicated that the addition of the glycine spacer contributes to the Λ conformation of corynebactin. However, this trend is opposed by methylation of the trilactone ring (serine to threonine) in corynebactin. Since the specific chirality of the ferric siderophore complex is often required for recognition, incorporation, and utilization, investigation of the effect of the different chiralities of the two similar siderophores will be interesting.

Alcaligin

Tetradentate siderophores, such as alcaligin and rhodotorulic acid (Fig. 2), need to form Fe_2L_3 complexes to completely satisfy the coordination shell of iron. The stoichiometry of these types of complexes yields two different coordination geometries: triple helicates and monobridged dimers. The ferric alcaligin structure (Color Plate 2, right) is the first example of the latter geometry and possesses several remarkable features. The two terminal bis(bidentate) fragments are structurally similar in both free and ferric alcaligin, revealing a siderophore preorganized for metal binding. This structural arrangement renders alcaligin ready to form the 1:1 FeL^+ complex, resulting in a 32-fold increase in the stability of the complex compared to that of more flexible siderophores such as rhodotorulic acid. However, this organization is costly for bridge formation, since significant deformation of the ligand is required to span both iron centers, resulting in a lower stability constant for this reaction.

Ferrioxamine B (or Desferrioxamine B)

Ferrioxamine B (Fig. 2) from *Streptomyces pilosus* is used as a therapeutic iron chelator in the treatment of diseases such as β-thalassemia. Although ferrioxamine B is the most intensively studied member of the ferrioxamine family, the crystal structure of the ferric complex was solved only recently and contains a racemic mixture of Λ- and Δ-N-*cis*, *cis* coordination isomers (the Δ chirality is shown in Color Plate 3). As a linear ligand, ferrioxamine B is much more flexible and far less organized for ferric ion binding than is the cyclic ferrioxamine E. The lower iron formation constant for ferrioxamine B than for ferrioxamine E reflects this greater flexibility. Ferrioxamine B forms a ferric complex with two closed loops and one open chain, which is directed away from the metal center. The pendant amine arm may play a role in receptor recognition.

Ferrichrome A

Many fungi produce ferrichromes, all trihydroxamates (Fig. 2). While many structural varieties exist, all but one include a cyclic hexapeptide ring containing a tripeptide sequence of N^δ-acyl-N^δ-hydroxyornithine. The other tripeptide in the backbone yields the diversity of this class of siderophores. The crystal structure of ferrichrome A (Color Plate 4) verified the structure proposed on the basis of chemical methods. The configuration for ferrichrome A (and all ferrichromes subsequently crystallized) is Λ,-C-*cis*, *cis*.

FERRIC SIDEROPHORE SOLUTION THERMODYNAMICS

The formation constants of siderophore complexes define the thermodynamic limits for the producing organisms to compete for iron, thus eventually providing optimal propagation conditions for the organism producing the superior complexing agent. The strength of iron binding may determine the mechanism of iron release (such as ligand exchange, reductive removal, or ligand destruction). The solution thermodynamics and iron exchange kinetics of siderophores are reviewed extensively elsewhere. Only some illustrative features are discussed here.

The stability of an iron-siderophore complex must minimally be greater than that of iron hydroxide. Siderophores involved in virulence (which remove iron from transferrin or other iron transport/storage proteins) must have a greater affinity for iron than does the protein in order to be effective. Additionally, the stability of an iron-siderophore complex influences the Fe^{3+}/Fe^{2+} reduction potential and is a determinant of the possible mechanism of iron release from a siderophore.

The stepwise association equilibria for a triprotic acid are given in equations 1–3:

$$K_1 \quad H^+ + L^{3-} \rightleftharpoons HL^{2-}$$
$$K_1 = [HL^{2-}]/[H^+][L^{3-}] \tag{1}$$

$$K_2 \quad H^+ + HL^{2-} \rightleftharpoons H_2L^-$$
$$K_2 = [H_2L^-]/[H^+][HL^{2-}] \tag{2}$$

$$K_3 \quad H^+ + H_2L^- \rightleftharpoons H_3L$$
$$K_3 = [H_3L]/[H^+][H_2L^-] \tag{3}$$

These reactions are the reverse of the proton dissociation equilibria, and their corresponding association constants (K) are given in the opposite sequence as the proton dissociation constants (pKa) (i.e., K_1 refers to the most basic protonation reaction whereas pKa$_1$ is the most acidic). This protocol follows the format for the stepwise equilibria involving a metal and a ligand.

$$L + M \rightleftharpoons ML \quad K_1 = [ML]/[L][M] \tag{4}$$

$$L + ML \rightleftharpoons ML_2 \quad K_2 = [ML_2]/[L][ML] \tag{5}$$

pM Values

By a standard convention, overall equilibria are expressed as β_{mlh} values for the reaction $mM + lL + hH = M_mL_lH_h$. Hence, $\beta_{110} = K_1$ and $\beta_{120} = K_1K_2$ (Table 2). The very large differences in the acidity of various siderophores at the pH experienced by the organism do not account for the fact that most siderophores are protonated. Hence, H^+ and Fe^{3+} compete with each other for binding to the siderophore. Thus, the formal stability constant of a complex is not, by itself, a relative measure of the ability of a ligand to bind a metal. Variation in the protonation constants and the concentration dependence can lead to large differences in the magnitude of the overall formation constant among ligands, which differ in pH dependence. To compare the true relative ability of different ligands to bind a metal some measure of the metal-ion free energy in the complex must be used. We have used the pM value, which has now become widely used as a standard convention for comparison. The pM value, analogous to pH, is defined as $-\log[M]$ where [M] represents the (usually very low) concentration of free aqueous metal ion. This is a direct measure of the chemical free energy of the metal ion at equilibrium with the chelating siderophore, which is why this is so direct and simple in application. The pM value is reported for a defined set of experimental conditions, usually pH = 7.4, [M]$_{tot}$ = 1 µmol liter^{-1}, [L]$_{tot}$ = 10 µmol liter^{-1}. The following example demonstrates the calculation of the pM value for the siderophore aerobactin.

Aerobactin has five dissociable protons from two hydroxamate groups (K_1 and K_2) and three carboxylate groups (K_3, K_4, and K_5). The protonation constants are log K_1 = 9.44, log K_2 = 8.93, log K_3 = 4.31, log K_4 = 3.48, and log K_5 = 3.11. The hydroxy proton is not included, nor does it need to be, since it cannot be removed from the free ligand below pH 14. At pH 7.4, the fraction of the ligand in increasing protonation states (usually called the α function) can be calculated. Hence, L^{5-} represents the fully deprotonated ligand (except for the hydroxy proton). In the following, the charges are not always included for simplicity of the expressions. Rearrangement of equations 1, 2, and 3 yields

$$[HL]/[L] = K_1[H^+] \quad [HL] = K_1[H^+][L] \tag{6}$$

$$[H_2L]/[HL] = K_2[H^+]$$
$$[H_2L] = K_2[H^+][HL] = K_1K_2[H^+]^2[L] \tag{7}$$

$$[H_3L]/[H_2L] = K_3[H^+] \tag{8}$$

TABLE 2 Protonation and iron formation constants of representative siderophores

Siderophore	Log K_{a6}	Log K_{a5}	Log K_{a4}	Log K_{a3}	Log K_{a2}	Log K_{a1}	Log β_{110}	Log β_{120}	Log β_{130}	Log β_{230}	pM
Enterobactin	6.0	7.5	8.6	12.1	12.1	12.1	49				35.5
Amonabactin		6.7	7.0	12.1	12.1	12.1	34.3			86.3	25.1
Ethyl-2,3-DHB[a]					7.3	12.1	18.7	33.6	43.7		19.7
Ferrioxamine B					8.39	9.03	9.70	30.5			26.6
Alcaligin					8.61	9.42	23.5			64.7	23.0
N-Methyl-acetohydroxamic acid						9.36	11.4	21.1	28.3		13.3

[a] Ethyl-2,3-DHB, ethyl-2,3-dihydroxybenzoic acid.

Describing [HL] and [H$_2$L] in terms of only [H$^+$] and [L] allows the calculation of α_L (protonation states relative to the fully deprotonated ligand) of the three major species:

$$[HL]/[L] = K_1[H^+] = 10^{9.44} \times 10^{-7.4} = 109.6 \quad (9)$$

$$[H_2L]/[L] = K_1 K_2 [H^+]^2 = 10^{9.44} \times 10^{8.93} \times (10^{-7.4})^2 = 3{,}715.3 \quad (10)$$

$$[H_3L]/[L] = K_1 K_2 K_3 [H^+]^3 = 10^{9.44} \times 10^{8.93} \times 10^{4.31} \times (10^{-7.4})^3 = 3.0 \quad (11)$$

$$\alpha_L = 1 + 109.6 + 3{,}715.3 + 3.0 = 3{,}828.9 \quad (12)$$

The concentrations of other species (H$_4$L$^-$ and H$_5$L) are so low that they can be neglected. The fractions (as percentages) of the protonated forms of the ligand at pH 7.4 are thus [L] = 0.03%, [HL] = 2.86%, [H$_2$L] = 97.0%, and [H$_3$L] = 0.08%.

The corresponding metal complex equilibria (and their implicitly defined equilibrium constants) are

$$M + L \rightleftharpoons ML \quad \log K_f = 22.93 \quad (13)$$

$$MLH_{-1} + H^+ \rightleftharpoons ML \quad \log K_{ML} = 4.27 \quad (14)$$

$$ML + H^+ \rightleftharpoons MLH \quad \log K_{MLH} = 3.48 \quad (15)$$

$$MLH + H^+ \rightleftharpoons MLH_2 \quad \log K_{MLH_2} = 3.10 \quad (16)$$

The last two reactions account for the two carboxylates, which are not involved in metal binding, and the second reaction corresponds to the loss of the hydroxy proton of the coordinated bidentate α-hydroxycarboxylate group. The acidity of this proton is increased by about 12 orders of magnitude (from a log K$_a$ of about 16 to 4.27), owing to the hydroxyl coordination to Fe^{3+}. (If this were included in defining the fully deprotonated ligand, the formal formation constant for aerobactin would be log K$_f$ = 22.93 + 16 − 4.27 = 34.7. However, this would be a meaningless description because a functional group with a pK$_a$ of 16 cannot be significant in equilibria in any aqueous solution.)

Describing the major species relative to [ML], where α_{ML} is the protonation states relative to ML, yields

$$\frac{[MLH_{-1}]}{[ML]} = \frac{1}{K_{MLH_{-1}}[H^+]} = 1{,}348.96 \quad (17)$$

$$[MHL]/[ML] = K_{MLH}[H^+] = 10^{3.48} \times 10^{-7.4} = 1.2 \times 10^{-4} \quad (18)$$

$$\alpha_{ML} = 1 + 1{,}348.96 + 1.2 \times 10^{-4} = 1{,}349.96 \quad (19)$$

Again, the other forms of the protonated metal complex are present in insignificant concentrations. The percentages of deprotonated and monoprotonated complex (these are the only forms considered here) at pH 7.4 are [MLH$_{-1}$] = 99.93% and [ML] = 0.07%. Addition of the terms that describe the doubly and triply protonated metal complex and also the terms that describe the formation of hydroxide metal complex species are too small to affect the pM value (as shown below).

If we follow the frequently used convention that the pM value is calculated at $[L_{total}] = 10^{-5}$ mol liter^{-1} and $[M_{total}] = 10^{-6}$ mol liter^{-1} then $[L] = 10^{-5} - 10^{-6} = 9 \times 10^{-6}$ mol liter^{-1} since complex formation is quantitative (as confirmed by the final results). The concentration of the fully deprotonated ligand is given by

$$[L] = 9 \times 10^{-6}/\alpha_L = 2.35 \times 10^{-9} \text{ mol liter}^{-1} \quad (20)$$

The equilibrium constant for equations 13 and 14 can be written in terms of [M] and [ML] as seen in equations 21 and 22, respectively. Substitution of equation 22 into equation 21 allows the calculation of [M] as described in equation 23 for the standard pM conditions. Calculating the pM for pH 7.4 and setting $[Fe]_{total} = 1$ μM and $[L]_{total} = 10$ μM gives a pM value for aerobactin of 23.43 (equation 24).

$$[M] = [ML]/K_f [L] \quad (21)$$

$$[ML] = K_{ML}[MHL_{-1}][H^+] \quad (22)$$

$$[M] = \frac{K_{ML}[MHL_{-1}][H^+]}{K_f [L]}$$
$$= \frac{(10^{4.27})(0.9993 \times 10^{-6})(10^{-7.4})}{(10^{22.93})(2.35 \times 10^{-9})} \quad (23)$$
$$= 3.68 \times 10^{-24} \text{ mol liter}^{-1}$$

$$pM = -\log [M] = -\log (3.68 \times 10^{-24}) = 23.43 \quad (24)$$

LIGAND COMPETITION STUDIES

The overall ferric ion complex formation constants usually cannot be determined directly at neutral pH because the extremely high stability of siderophore complexes precludes direct measurement of the equilibrium of interest. Often, even at very low pH, the iron-siderophore complex does not fully dissociate and so the Fe + L ⇌ FeL equilibrium cannot be observed. One method of circumventing this problem is the spectrophotometric measurement of competition for the metal by another thermodynamically well characterized ligand (equation 25), typically EDTA:

$$FeL^{3-n} + EDTA^{-4} + nH^+ \rightleftharpoons Fe(EDTA)^- + H_nL \quad (25)$$

Using the known value for the formation constant of ferric EDTA, a value of the proton-dependent equilibrium constant can be calculated. Enterobactin (ent) employs three catechol groups to tightly encapsulate the ferric ion in a hexadentate coordination sphere. The overall reaction that defines the formal stability constant (β_{110}) between enterobactin and trivalent metal ions, such as Fe^{3+}, is M^{3+} + $ent^{6-} \rightleftharpoons [M(ent)]^{3-}$.

To determine the equilibrium concentration of the fully deprotonated (and quite hypothetical) ent^{6-} species, the following protonation constants are needed:

$$ent^{6-} + H^+ \rightleftharpoons Hent^{5-} \quad K_1 \quad (26)$$

$$Hent^{5-} + H^+ \rightleftharpoons H_2ent^{4-} \quad K_2 \quad (27)$$

$$H_2ent^{4-} + H^+ \rightleftharpoons H_3ent^{3-} \quad K_3 \quad (28)$$

$$H_3ent^{3-} + H^+ \rightleftharpoons H_4ent^{2-} \quad K_4 \quad (29)$$

$$H_4ent^{2-} + H^+ \rightleftharpoons H_5ent^- \quad K_5 \quad (30)$$

$$H_5ent^- + H^+ \rightleftharpoons H_6ent \quad K_6 \quad (31)$$

Since the triserine ring of ent is susceptible to base-catalyzed hydrolysis in aqueous solution, the first three (high-pH) protonation constants cannot be measured. Values for those protonation constants (Table 2) have been estimated from model compounds to give the overall protonation constant for H_6ent, estimated to be $\beta_{016} = 10^{58.5}$. The formation constant for iron binding is then estimated as $\beta_{110} = 10^{49}$. The protonation of the ferric enterobactin complex, $Fe(ent)^{3-} + H^+ \rightleftharpoons Fe(Hent)^{2-}$, has log K = 4.95, with other protonation reactions at lower pH. Hence, near pH 5.5 the primary equilibrium reaction for the competition between EDTA and enterobactin for Fe^{3+} is given by

$$Fe(EDTA)^- + H_6ent = Fe(ent)^{3-} + H_2EDTA^{2-} + 4H^+ \quad (32)$$

The strong proton concentration dependence of this reaction means that at low pH, EDTA can compete with enterobactin for iron if used in excess. At this pH and under these conditions, the equilibrium free [Fe^{3+}] concentration is the same for both ligands. Since the EDTA constants are known, the pM value (for these conditions) can be calculated—and this is the same as for enterobactin. Calculating the pM for pH 7.4 (using the experimental protonation constants from Table 2 that are relevant) and setting [Fe]$_{total}$ = 1 µM and [ent]$_{total}$ = 10 µM gives a pM value for enterobactin of 35.5. (Note that this is directly determined from experimental data and not based on estimated values for the high pH protonation constants.)

A free Fe^{3+} concentration of $10^{-35.5}$ M for enterobactin is much lower than the inverse of Avogadro's number. This would imply only one atom of free ferric ion in $\sim 10^{13}$ liters of water! However as noted earlier, the pM value is still meaningful since it is proportional to the chemical free energy (ΔG) released by metal-ligand binding and remains a useful thermodynamic parameter for the comparison of siderophores.

SIDEROPHORE DENTICITY

The change of pM values as a function of siderophore concentration is a good way to illustrate the dramatic effect of siderophore denticity on the stability of the complex at low siderophore concentrations. Table 2 presents some representative ferric siderophore formation constants and pM values. Reduction of Fe^{3+} to Fe^{2+} releases siderophore-transported Fe^{3+} inside the cell and allows for recycling of the siderophore. Although reduction of Fe^{3+} to Fe^{2+} is the primary transport mechanism of plants, such a reduction has not been shown to be the rate-determining step in any microbial iron transport system. The shift in potential for a reversible couple is Nernstean:

$$E = E^0 - 0.059 \log([Fe^{2+}]/[Fe^{3+}])$$

Hence, the shift of potential from the standard ferric/ferrous potential is just given by 0.059 (pM$_{Fe3+}$ − pM$_{Fe2+}$). Since both the ferrous and ferric pM values change from one ligand to another, both the ferrous and ferric stability constants must be known in order for the redox potential to be calculated. Conversely, the ferrous stability constant and the redox potential must both be known before the ferric siderophore stability constant can be determined. Figure 3 shows a plot of the pM values and redox potentials for a series of various siderophores and related species.

A fully formed iron complex corresponds to the following equilibria (charges are suppressed for simplicity):

Bidentate: $\quad Fe + 3L \rightleftharpoons FeL_3$
$\beta_{130} = [FeL_3]/[Fe][L]^3$

Tetradentate: $\quad 2Fe + 3L \rightleftharpoons Fe_2L_3$
$\beta_{230} = [Fe_2L_3]/[Fe]^2[L]^3$

Hexadentate: $\quad Fe + L \rightleftharpoons FeL$
$\beta_{110} = [FeL]/[Fe][L]$

The pM value for such complexes has a dependence on log [L] that is 3, 1.5, and 1, respectively. The slopes of the plots in Fig. 4 correspond to these values.

Another feature of the stability constants in Table 2 is the difference between, for example, catecholate and hydroxamate siderophores. The formation constant of a tris(bidentate) complex, FeL$_3$, forms the same fully formed iron complex as an FeL hexadentate complex. For the hydroxamates, these numbers are nearly the same—always about 10^{28} for a tris(hydroxamate) complex and about 10^{21} and 10^{11} for bis- and mono(hydroxamate) complexes, respectively. This is an accident of the use of 1 M concentrations as the thermodynamic standard state. As can be seen, these numbers are not comparable for corresponding catecholate siderophores. As shown in Fig. 4, a change of standard state to some other concentration would find a point where the pM values for ligands of different denticity cross

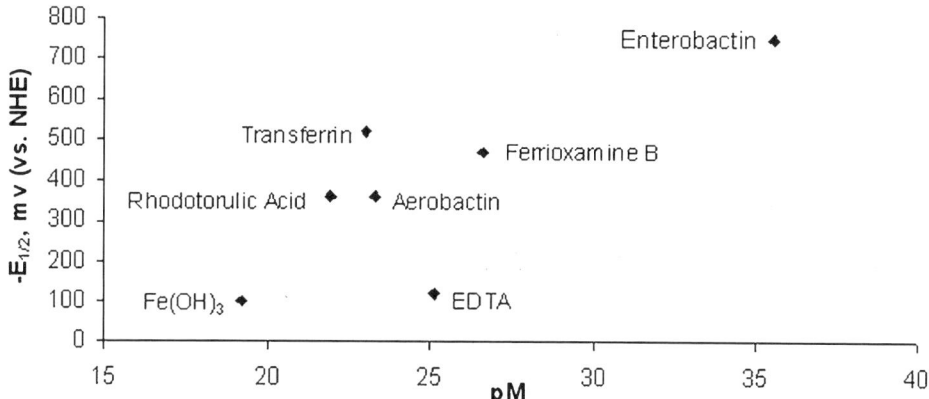

FIGURE 3 The pM and standard reduction potentials ($E_{1/2}$) for representative ferric siderophores and comparison species. Note that the scale is $-E$, such that the reduction potential decreases from bottom to top. While there is a general correlation of $E_{1/2}$ with pM, the several large exceptions show that the pM for both ferric and ferrous complexes must be known to predict the potential accurately. The hydroxide potential is calculated using Fe(OH)$_2$ $K_{sp} = 10^{-16}$ and Fe(OH)$_3$ $K_{sp} = 10^{-39}$.

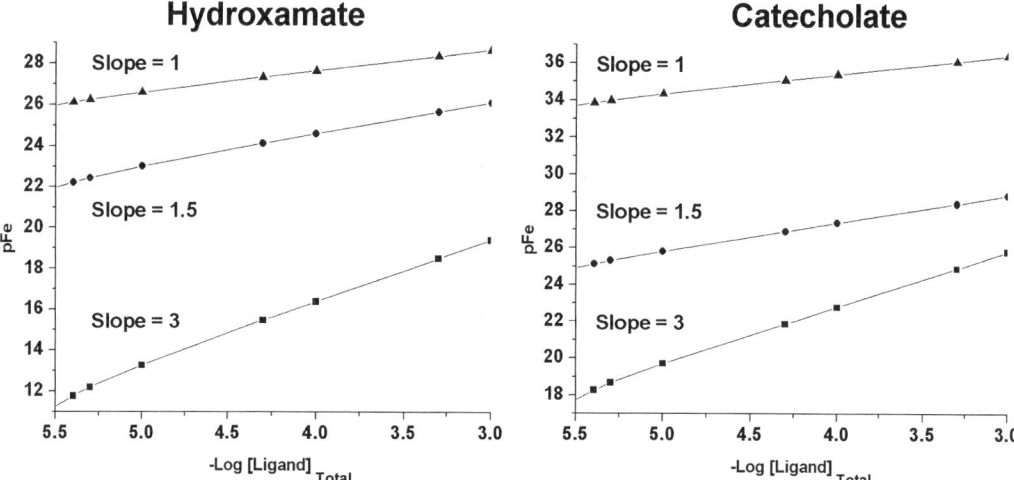

FIGURE 4 The change in pM (pH 7.4; total metal ion concentration, 1 μM) as the total ligand concentration is changed for bidentate (■), tetradentate (●), and hexadentate (▲) complexes of hydroxamate and catecholate siderophores (or analogs). Note the marked increase in sensitivity to concentration for the lower-denticity ligands.

and where, for this chosen standard, the thermodynamic formation constants for fully formed catecholate complexes would be nearly equal.

CONFORMATIONAL EFFECTS ON STABILITY

Deviations from the expected stability of a siderophore complex (based on comparison compounds) is usually an indication of an impact of a specific feature of the ligand. Enterobactin is about 10^6-fold more stable than hexadentate synthetic analogs, but this extra stability disappears when the triserine ring is hydrolyzed. The special conformational stability of enterobactin and its role in the stability of the iron complex have been extensively reviewed recently and are not reported here.

Another dramatic example of siderophore conformation and structure can be found in alcaligin (Fig. 2; Color Plate 2). Alcaligin is a cyclic dihydroxamic acid that forms 1:1 ferric complexes (FeL$^+$) under acidic conditions and 2:3 ferric complexes (Fe$_2$L$_3$) at and above neutral pH. The stability constants of these macrocyclic dihydroxamate siderophores differ significantly from that of rhodotorulic acid (Fig. 2), a linear dihydroxamate siderophore. Notably, K_{FeL} of alcaligin is 32 times greater than that of rhodotorulic acid, while the subsequent stepwise formation constant for Fe$_2$L$_3$ is 3 times less. The structure of the Fe$_2$L$_3$ alcaligin complex (Color Plate 2) is a topological alternative to the triple-helicate structure of the rhodotorulic complex Fe$_2$(RA)$_3$. The structures of the free ligand and the bis(bidentate) ligand in the FeL complex are essentially identical, indicating that alcaligin is highly preorganized for metal ion binding. This explains the difference in K_{FeL} between alcaligin and rhodotorulic acid, as well as explaining the monobridged topology of the Fe$_2$L$_3$ alcaligin complex. As illustrated in Color Plate 2 (and with the reminder that the word "chelate" comes from the Greek "chelos," for a crab's claw), the extra stability gained in the 1:1 complex (left, top) is partially offset when one of the alcaligin molecules (the crab at center) has to open up to bridge two Fe^{3+} centers (left, bottom).

FERRIC SIDEROPHORE ELECTRONIC STRUCTURE AND SPECTRA

The electron paramagnetic resonance, Mössbauer, and nuclear magnetic resonance spectroscopies of hydroxamate and catecholate siderophores were extensively discussed by Matzanke and coworkers. Although that review was written some time ago, there has been relatively little information to add here about these topics. However, there has been a substantial change in understanding the electronic structure and corresponding visible-UV and circular dichroism spectra of ferric siderophore complexes, particularly the catecholates.

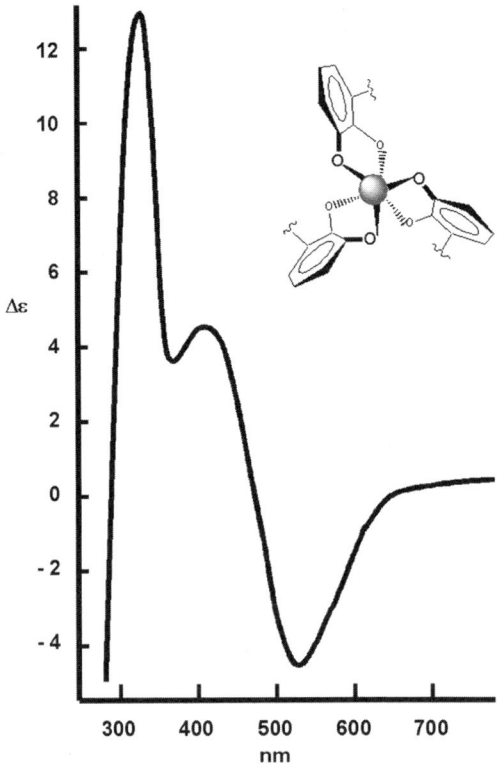

FIGURE 5 Circular dichroism spectrum of ferric enterobactin in aqueous solution at pH 7.5 in 0.1 M HEPES buffer. The cartoon shows the Δ chirality of the metal center of ferric enterobactin.

Ferric EDTA and similar complexes are essentially colorless, because the Fe^{3+} high-spin d^5 electronic configuration has no spin-allowed d-d electronic transitions. However, the siderophores (hydroxamate and catecholate) are all intensely red! Indeed, the early name given to this class of compounds, "siderochromes," referred to the fact that these were highly colored ferric complexes. The intense and rather broad visible-UV absorption spectra were described as "ligand to metal charge transfer," with the implicit assumption that this was too complicated to pursue further. This changed with the analysis by Karpishin et al. of the ferric tris (catecholate) complex and, by extension, ferric enterobactin.

The chirality of ferric siderophores and its role in recognition and transport have been reviewed elsewhere and are not discussed here. However the chirality, as presented in Fig. 5, is the origin of the circular dichroism spectrum and, in turn, is a powerful tool in understanding the electronic structure of the ferric siderophore complex. In addition to the spectra of the native siderophore iron complexes, siderophore analogs were used to assign the electronic structure and transitions. Single-crystal polarized absorption and magnetic circular dichroism were used to elucidate the electronic structure of the iron(III) tris(catecholate) complex $[Fe(cat)_3]^{3-}$ and so to investigate the bonding in ferric enterobactin and similar catecholate siderophores (Fig. 6). Two related complexes were studied by magnetic circular dichroism to provide a perturbation of the electronic structure of $[Fe(cat)_3]^{3-}$ and to determine differences in bonding between the three complexes. Similar studies were performed on a series of bicapped macrocycle analogs, which allowed for variation of the geometric, as well as electronic, structures of the complexes. The ligand-to-metal charge transfer band characteristic of the catecholate siderophores is found for $[Fe(cat)_3]^{3-}$ to be composed of two overlapping x,y-polarized transitions at 18,414 and 22,018 cm^{-1}. These transitions are ligand π to metal d in nature; assignments of these and four other transitions and an experimental energy order for the molecular orbitals of the complexes were made in

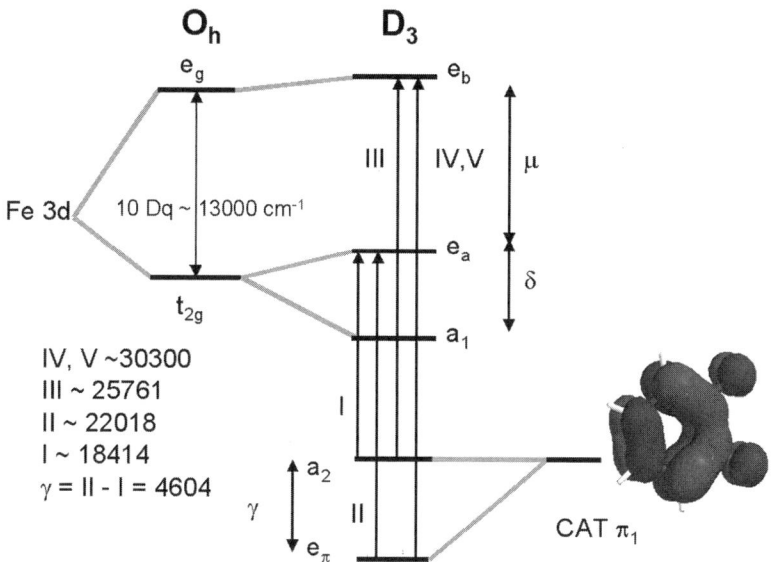

FIGURE 6 One-electron molecular orbital energy level diagram for iron(III) tris (catecholates). The observed charge-transfer transitions are indicated by arrows and are labeled with their corresponding energies in reciprocal centimeters.

D_3 symmetry. The siderophore charge transfer excited state is best described as a semiquinone Fe^{2+} species.

As shown in Fig. 6 (left) and following the published analysis, the octahedral coordination field splits the d orbitals into the triply degenerate t_{2g} and doubly degenerate e_g sets. As the symmetry is lowered to D_3 [corresponding to a tris(catecholate) complex], the t_{2g} set splits into a_1 and e (labeled e^a) symmetry orbitals. The e_g set goes to e symmetry (labeled e^b to distinguish it from the orbitals of t_{2g} origin). The corresponding highest occupied molecular orbital of a catecholate dianion fragment is sketched on the right. Three of these combine to give linear combinations of a_2 and e (labeled e_π) symmetry. The a_2 combination is rigorously nonbonding, since it does not match the symmetry of any of the Fe d orbitals. The e_π is lowered in energy by (γ) and the e_a is raised by (∂), relative to their isolated energies, due to ligand and metal π bonding. This is a significant contribution to the Fe—O interaction and accounts for the high stabilities of these complexes relative to similar Fe(III) complexes with other ligands (e.g., oxalate).

SUMMARY

This chapter has summarized our knowledge of siderophores, a remarkable group of microbial iron-binding molecules. The emphasis has been on the function, stability, structure, and spectroscopy of these compounds. The basic biochemical and physical properties do much to explain the growing body of biomedical and biochemical science surrounding the siderophores and their iron transport functions.

ACKNOWLEDGMENTS

We gratefully acknowledge the continuous support of NIH grant AI11744.

We thank our past coworkers, whose names appear in the referenced papers, and the other cited colleagues who have continued to expand and enlighten this fascinating field. We also thank Al Crumbliss and Suraj Dhungana for their help with the ferrioxamine B structure.

SUGGESTED READING

Boukhalfa, H., and A. L. Crumbliss. 2002. Chemical aspects of siderophore mediated iron transport. *Biometals* **15:**325–339.

Braun, V. 1998. Pumping iron through cell membranes. *Science* **282:**2202–2203.

Braun, V., and K. Hantke. 1997. Receptor-mediated bacterial iron transport, p. 81–116. *In* G. Winkelmann and C. J. Carrano (ed.), *Transition Metals in Microbial Metabolism.* Harwood Academic Publishers, Amsterdam, The Netherlands.

Budzikievicz, H., A. Bossenkamp, K. Taraz, A. Pandey, and J. M. Meyer. 1997. Bacterial constituents. 72. Corynebactin, a cyclic catecholate siderophore from *Corynebacterium glutamicum* ATCC 14067 (*Brevibacterium* sp. DSM 20411). *Z. Naturforsch. Ser. C* **52:**551–554.

Budzikiewicz, H. 1997. Siderophores of fluorescent pseudomonads. *Z. Naturforsch. Ser. C* **52:**713–720.

Cotton, F. A., and G. Wilkinson. 1999. *Advanced Inorgainc Chemistry,* 6th ed. John Wiley & Sons, Inc., New York, N.Y.

Crosa, J. H. 1989. Genetics and molecular biology of siderophore mediated iron transport in bacteria. *Microbiol. Rev.* **53:**517–530.

Dhungana, S., P. S. White, and A. L. Crumbliss. 2001. Crystal structure of ferrioxamine B: a comparative analysis and implications for molecular recognition. *J. Biol. Inorg. Chem.* **6:**810–818.

Earhart, C. 1987. Ferrienterobactin transport in *Escherichia coli,* p. 67–84. *In* G. Winkelmann, D. van der Helm, and J. Neilands (ed.), *Iron Transport in Microbes, Plants and Animals.* VCH Publishers, Weinheim, Germany.

Goetz, D. H., M. A. Holmes, N. Borregaard, M. E. Bluhm, K. N. Raymond, and R. K. Strong. 2002. The neutrophil lipocalin NGAL is a bacteriostatic agent that interferes with siderophore-mediated iron acquisition. *Mol. Cell* **10:**1033–1043.

Konopka, K., and J. B. Neilands. 1984. Effect of serum-albumin on siderophore-mediated utilization of transferrin iron. *Biochemistry* **23:**2122–2127.

Leggett, D. J. 1985. *Computational Methods for the Determination of Formation Constants.* Plenum Press, New York, N.Y.

Martell, A. E. 1988. *The Determination and Use of Stability Constants.* VCH Publishers, New York, N.Y.

Matzanke, B. F., G. Müller-Matzanke, and K. N. Raymond. 1989. Siderophore-mediated iron transport, p. 1–121. *In* T. M. Loehr (ed.), *Iron Carriers and Iron Proteins.* VCH Publishers, New York, N.Y.

Moeck, G. S., and J. W. Coulton. 1998. TonB-dependent iron acquisition: mechanisms of siderophore-mediated active transport. *Mol. Microbiol.* **28:**675–681.

Neilands, J. B. 1991. A brief history of iron metabolism. *Biol. Metab.* **4:**1–6.

Neilands, J. B. 1952. A crystalline organo-iron pigment from the smut fungus *Ustilago sphaerogena*. *J. Am. Chem. Soc.* **74:**4846–4847.

Neilands, J. B. 1995. Siderophores: structure and function of microbial iron transport compounds. *J. Biol. Chem.* **270:**26723–26726.

Otto, B. R., A. M. J. J. Verweij-van Vught, and D. M. MacLaren. 1992. Transferrins and heme-compounds as iron sources for pathogenic bacteria. *Crit. Rev. Microbiol.* **18:**217–233.

Postle, K. 1999. Active transport by customized beta-barrels. *Nat. Struct. Biol.* **6:**3–6.

Raymond, K. N., E. A. Dertz, and S. S. Kim. 2003. Enterobactin—an archetype for microbial iron transport. *Proc. Natl. Acad. Sci. USA* **100:**3584–3588.

Raymond, K. N., G. Müller, and B. F. Matzanke. 1984. Complexation of iron by siderophores. A review of their solution and structural chemistry and biological function, p. 50–102. *In* F. L. Boschk (ed.), *Topics in Current Chemistry*. Springer-Verlag KG, Berlin, Germany.

Renshaw, J. C., G. D. Robson, A. P. J. Trinci, M. G. Wiebe, F. R. Livens, D. Collison, and R. J. Taylor. 2002. Fungal siderophores: structures, functions and applications. *Mycol. Res.* **106:**1123–1142.

Rusnak, F., M. Sakaitani, D. Drueckhammer, J. Reichert, and C. T. Walsh. 1991. Biosynthesis of the *Escherichia coli* siderphore enterobactin. *Biochemistry* **30:**2916–2927.

Stintzi, A., and K. N. Raymond. 2002. Siderophore chemistry, p. 273–319. *In* D. M. Templeton (ed.), *Molecular and Cellular Iron Transport*. Marcel Dekker, Inc., New York, N.Y.

Weinberg, E. D. 1990. Cellular iron metabolism in health and disease. *Drug Metab. Health* **22:**531–579.

SIDEROPHORE BIOSYNTHESIS IN BACTERIA

Christopher T. Walsh and C. Gary Marshall

2

The evolution of the biosynthetic routes to bacterial siderophores has been driven by the chemical demands imposed by the task of coordinating (and therefore making soluble) individual atoms of ferric iron (Fe^{3+}). Ferric iron has 5 valence electrons and requires 12 donor electrons (e.g., hexadentate coordination by atoms with a lone pair) to reach an energetically stable thermodynamic state. The lowest-energy geometrical configuration for these electrons to coordinate the iron atom is the octahedral configuration, in which lines drawn to connect the donors from two pyramids that share a common base with the iron atom at its center (Fig. 1). Thus, the task required of the bacteria which are in need of assembling a biological molecule capable of coordinating ferric iron is to meet these chemical and geometrical requirements. This summary of our current knowledge of these assembly schemes will accordingly focus on functional groups capable of providing these donor electrons and the flexible scaffolds on which they are arranged.

To serve as a chemically competent ligand to ferric iron, a functional group must possess a lone pair of electrons that has good electron donor properties. Bacteria have marshalled a diverse array of such functional groups for use in their siderophores, including phenolic hydroxyls, *N*-hydroxamates, the nitrogen constituents of five-member heterocyclic rings, and carboxylates. Each functional group defines a class of siderophores, and while some are the sole contingent of electron donors in a given siderophore, each of them has been found in combination with another. Siderophore scaffolds require flexibility and a certain amount of spatial extension, given that the molecule must wrap around an iron atom such that coordinating functional groups can be positioned at or near the optimal distances and angles. To make effective scaffolds, bacteria have chosen readily available materials that have the appropriate molecular handles to attach the required functional groups. The carbon backbones assembled are derived from shorter monomeric units and are linked via amide bonds, often catalyzed by nonribosomal peptide synthetases. These large enzyme systems are composed of specialized domains, usually tethered to each other, responsible for aminoacyl group activation and condensation (see the following section). Intermediates are sequestered from bulk solvent by the assembly machinery. Alternatively, carbon skeletons can

---

Christopher T. Walsh and C. Gary Marshall, Department of Biological Chemistry and Molecular Pharmacology, Harvard Medical School, Boston, MA 02115.

Iron Transport in Bacteria, Edited by Jorge H. Crosa, Alexandra R. Mey, and Shelley M. Payne
© 2004 ASM Press, Washington, D.C.

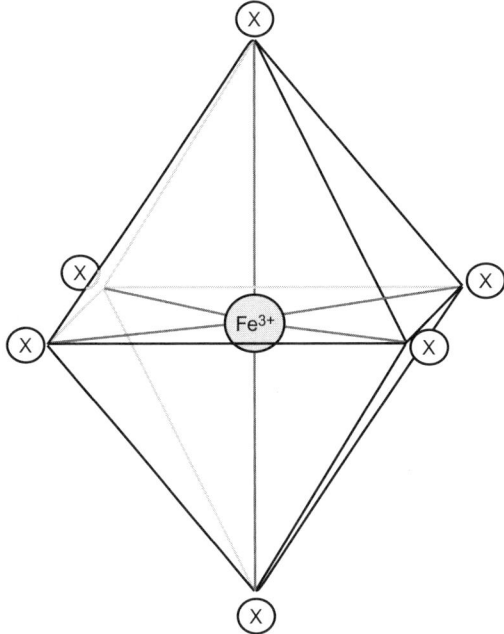

FIGURE 1 Octahedral configuration of ferric iron coordination by ligands (X).

be assembled by individual enzymes (which may form noncovalent complexes) producing freely diffusible intermediates. Either way, the amide bonds formed are kinetically stable but thermodynamically labile, allowing easy dismantling of the siderophore scaffold after ferric iron capture and retrieval into the cell.

Recent advances in whole-genome sequencing of bacterial species, combined with improved bioinformatic technology, have provided a bolus of information on putative siderophore biosynthetic pathways. However, there is still only limited biochemical information, and mechanistic details of many siderophore assembly pathways are still wanting. This treatment seeks to serve not as an encyclopedic reference to siderophore biosynthesis but, rather, as a representative sampling of distinct biosynthetic routes to a common functional goal.

PRIMER ON NONRIBOSOMAL PEPTIDE BIOSYNTHESIS

Nonribosomal peptide synthetases (NRPSs) are assembly lines of specialized domains that link amino acids via a thioester intermediate. Three types of domains are required to effect peptide bond formation—adenylation (A), peptidyl carrier protein (PCP), and condensation (C) domains. While there is considerable variation in the organization of these domains, the following describes the "typical" NRPS. A, PCP, and C domains are physically tethered to each other in repeats of A-PCP-C, each repeat being designated a module, and each module being responsible for the incorporation of one amino acid into the peptide product. Modules are colinear with the amino acids they link in the peptide, suggesting a role for domain organization in programming of the system. The A domains activate amino acids as aminoacyl adenylates and then thioesterify them to a phosphopantetheine (cofactor that is posttranslationally added to PCP domains (Fig. 2A). C domains catalyze peptide bond formation between two activated amino acids (the upstream electrophile, or donor, and the downstream nucleophile, or acceptor) resulting in transfer of the growing peptide to the PCP domain of the downstream module (Fig. 2B). This "elongation" occurs iteratively down the length of the NRPS system, with the peptide being translocated with each transfer in an N-to-C direction. The final module employs a thioesterase (TE) domain in place of the C domain, which transfers the fully grown peptide to a water molecule and off the PCP of the last module (Fig. 2C). When a nucleophile from the peptide itself is used in place of water, a cyclic product is released instead.

In addition to these four domains, auxiliary domains can be found in a given module that modify the corresponding amino acid in a particular way (Fig. 3). Cyclization (Cy) domains replace C domains and form five-member heterocyclic rings using amino acids that possess a β-carbon nucleophile. Epimerization domains sit between the PCP and C domains and racemize the α-carbon. Methylation domains methylate N or C atoms in amino acids, changing ionization properties and adding important hydrophobicity to the final peptide. Oxidase (Ox) and reductase (Red) domains are similarly located and change the redox state of an amino

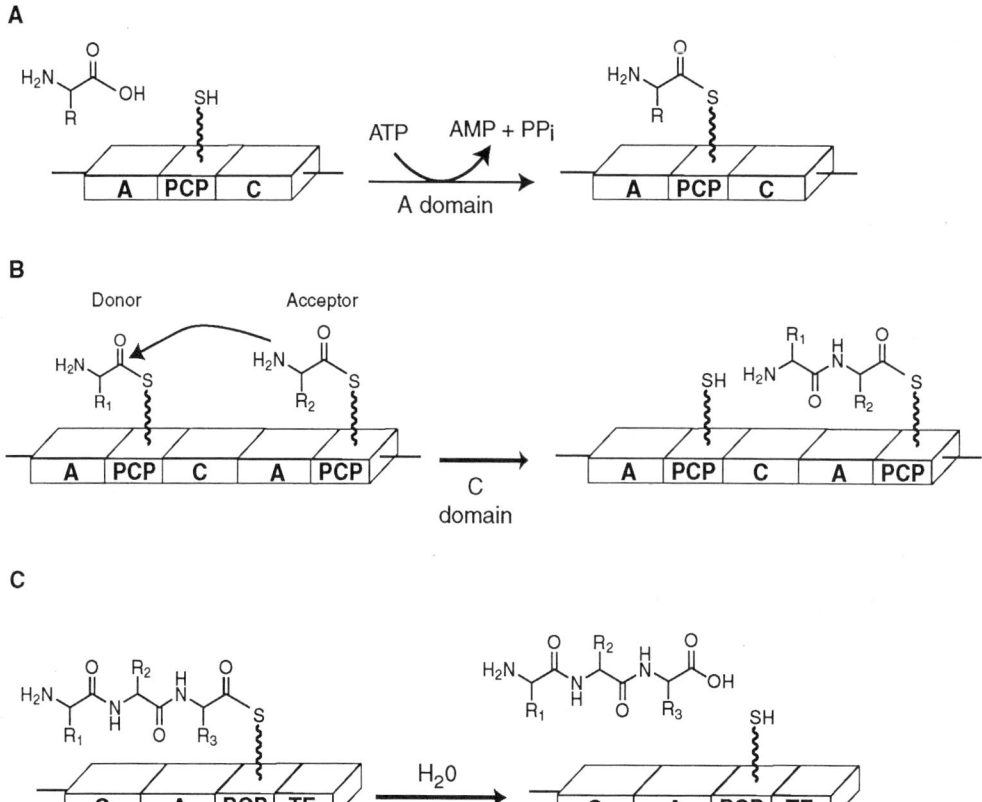

FIGURE 2 Required activities in NRPS modules. (A) A domains use ATP to activate amino acids and thioesterify them to the phosphopantetheine cofactor (wavy line) of a PCP, (B) C domains couple upstream donor and downstream acceptor amino acids, elongating the peptide and translocating it down the NRPS. (C) TE domains release the finished peptide via hydrolysis.

acid. The functional diversity added by these auxiliary domains is often critical to the biological activity of the peptide.

While some NRPS systems incorporate all modules onto a single polypeptide, the majority of NRPSs distribute modules onto a number of proteins, keeping protein sizes more manageable. Therefore, some transfers of the growing peptide between modules are intramolecular (or in *cis*) while others are intermolecular (or in *trans*). In some highly fragmented systems, divisions occur even within modules, forcing in *trans* interactions between the PCP with the A or the C domain of its own module. This is relatively common in siderophore systems. Another exception to the simplified scheme outlined above is that monomers of assembly are not always amino acids. Monomers capable of both upstream and downstream attachment must, of course, be amino acids, but often "peptides" manufactured in this way are N capped with carboxylic acids or C capped with amines. Several siderophore peptides are N or C capped in this way.

CATECHOLATE AND PHENOLATE SIDEROPHORES

Siderophores are known with one, two, or three catecholate or phenolate chelating groups attached to a peptide backbone. The *Escherichia*

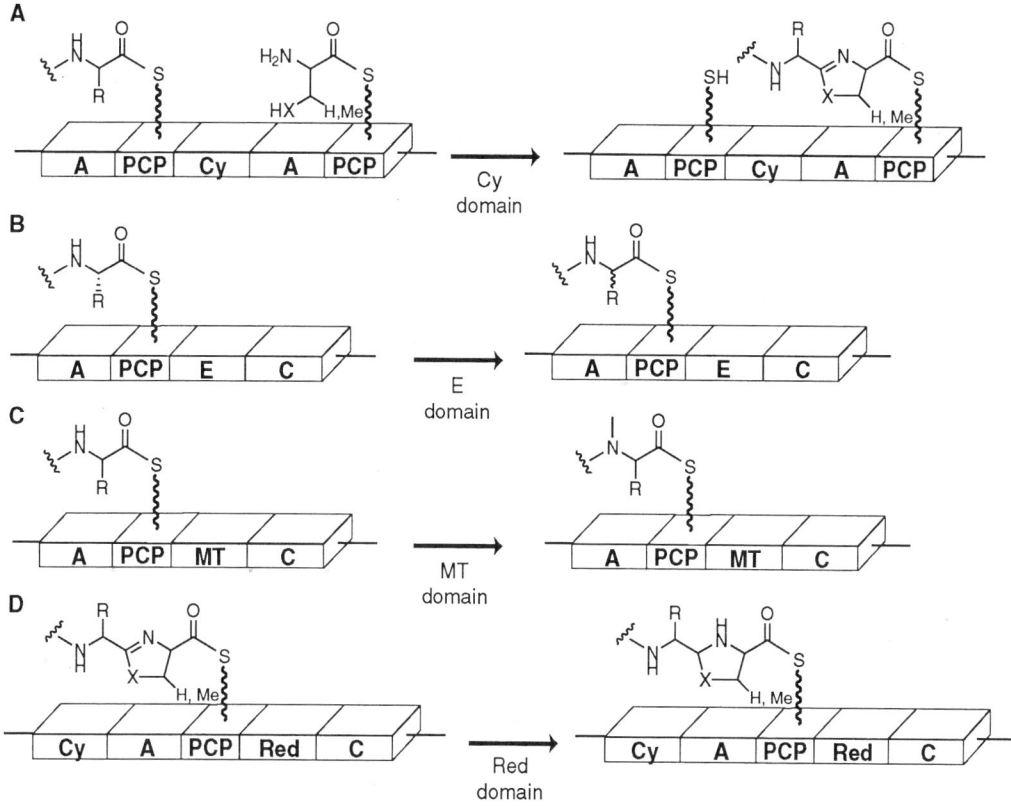

FIGURE 3 Auxilliary activities in NRPS modules. Commonly observed domains catalyze cyclization-dehydration (A), epimerization (B), methyl transfer (C), and oxidation or reduction (D). Oxidation is not shown. X indicates O or S; R indicates an amino acid side chain.

coli siderophore enterobactin is perhaps the best-characterized catecholate siderophore in terms of the logic of the NRPS assembly line and the demonstrated biochemical and enzymatic function of the modules. As shown in Fig. 4, enterobactin has three catechol rings presented as 2,3-dihydroxybenzoate (DHB) moieties in amide linkage to the amino groups of three serine residues, themselves linked in a 12-member ring as a cyclic trilactone. The trilactone serves as a scaffold for the three DHB catecholate groups to swing up and function as tridentate ligands for ferric iron with an impressively low K_d value. An extended version of a tris (catechol) on a trilactone platform was recently detected in *Bacillus subtilis*, with the elucidation of the bacillibactin structure. The DHB moieties are not directly linked to the threonyl trilactone scaffold but have an interposed glycine, perhaps for additional flexibility to allow the three catechols in one siderophore to provide all six ligands to ferric iron. The siderophore myxochelin, from myxobacteria, utilizes two DHB groups as chelators, tethered in amide linkage to a modified lysine residue as the scaffold (Fig. 4). Pyochelin and pseudomonine are nonribosomal peptide siderophores with a single aryl chelating group, in these cases a phenolate rather than catecholate, and are both isolated from pseudomonads.

The DHB and the 2-hydroxybenzoate (salicylate) moieties are made as dedicated monomers for assembly of these siderophores by enzymes encoded by genes that are switched on

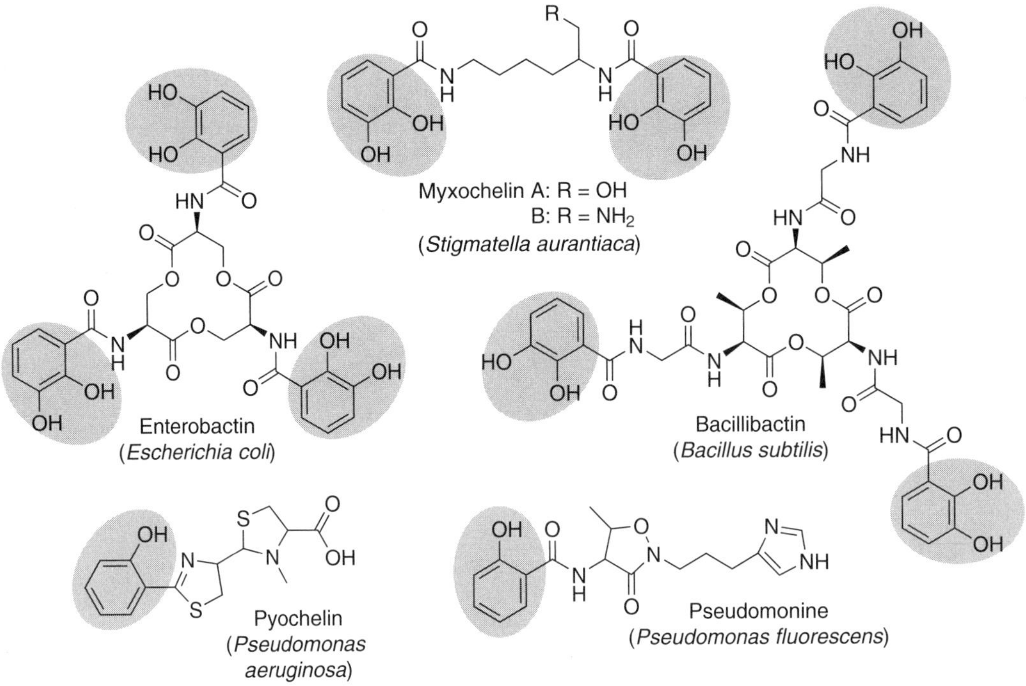

FIGURE 4 Some representative catecholate or phenolate siderophores.

at low iron concentrations. Both DHB and salicylate represent specialized monomers of secondary metabolism which arise by rerouting the central intermediate of aromatic amino acid biosynthesis, chorismate. Under iron starvation conditions, the flux of chorismate is directed away from phenylalanine, tyrosine, and tryptophan production and shunted to the siderophore precursors. As shown in Fig. 5, chorismate is derived via the shikimate pathway. When flux to DHB is desired, bacteria synthesize isochorismate synthase (EntC), isochorismate lyase (EntB), and dihydrodihydroxybenzoate dehydrogenase (EntA) and the three-enzyme pathway converts the dihydroaromatic chorismate into the catecholic DHB metabolite. Analogously, in pseudomonads making pyochelin or pseudomonine and in *Yersinia* species making yersiniabactin, a two-enzyme pathway (PchAB in *Pseudomonas aeruginosa*, PmsCB in *Pseudomonas* sp. [Table 1]) reroutes chorismate to isochorismate and then directs the elimination of the enolpyruvyl substituent during aromatization to yield salicylate.

With the aryl acid monomers available as N-capping substrates for the NRPS, the assembly lines can begin the initiation, elongation, and termination steps of siderophore chain elaboration. Figure 6 summarizes the pathway for enterobactin synthetase, composed of the Ent B, EntE, and EntF proteins that act together as a two-module NRPS assembly line. DHB has to be activated for amide bond formation, this is the task of EntE, an A domain, to make DHB-AMP, hold it in its active site, and transfer it to the partner HS-pantetheinyl-PCP domain. This PCP domain is on EntB, which is a bifunctional enzyme. The N-terminal two-thirds is the isochorismate lyase activity, noted in Fig. 5, while the C-terminal 100 amino acid residues comprises the PCP of the first NRPS module, albeit in *trans* with respect to both EntE and EntF. Thus, EntE transfers the DHB-AMP to the EntB PCP to yield the DHB-*S*-pantetheinyl thioester covalently tethered. Because the EntB PCP functions as an aryl carrier protein rather than a PCP, it has been designated an ArCP domain.

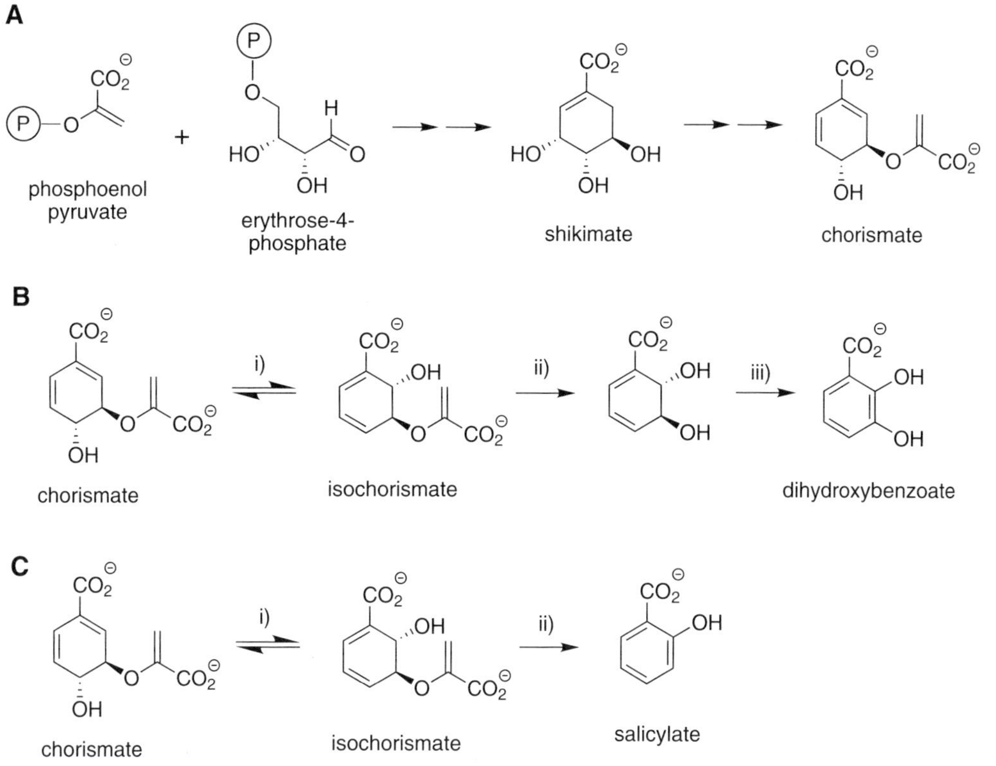

FIGURE 5 Biosynthesis of catechols and phenolates. Chorismate is generated from primary metabolites (A) and then used to manufacture DHB (B), while salicylate has a related but distinct biosynthetic pathway (C). Roman numerals refer to enzymes listed in Table 1.

The second module of the enterobactin synthetase assembly line is wholly contained within EntF, a four-domain (C-A-PCP-TE) 140-kDa protein. The A domain is specific for L-serine, activating it as the seryl-AMP mixed anhydride and then transferring it to the PCP domain to make the seryl-S-pantetheinyl PCP thioester. Now the C domain of EntF can carry out its condensation catalysis, utilizing DHB-S-EntB as the donor and the seryl-S-PCP as the intramolecular acceptor, forming the amide bond to accumulate the DHB-seryl-S-EntF thioester.

How do the subsequent steps proceed in this and the bacillibactin assembly line? Inspection of the enterobactin structure shows three

TABLE 1 Biosynthesis of catechols and phenolates

		Proteins used to synthesize:			
	Enzyme	DHB		Salicylate	
		Enterobactin	Myxochelin	Pyochelin	Pseudomonine
(i)	Isochorismate synthase	EntC	MxcD	PchA	PmsC
(ii)	Isochorismate lyase	EntB	MxcF	PchB	PmsB
(iii)	2,3-Dihydro-2,3-DHB dehydrogenase	EntA	MxcE		

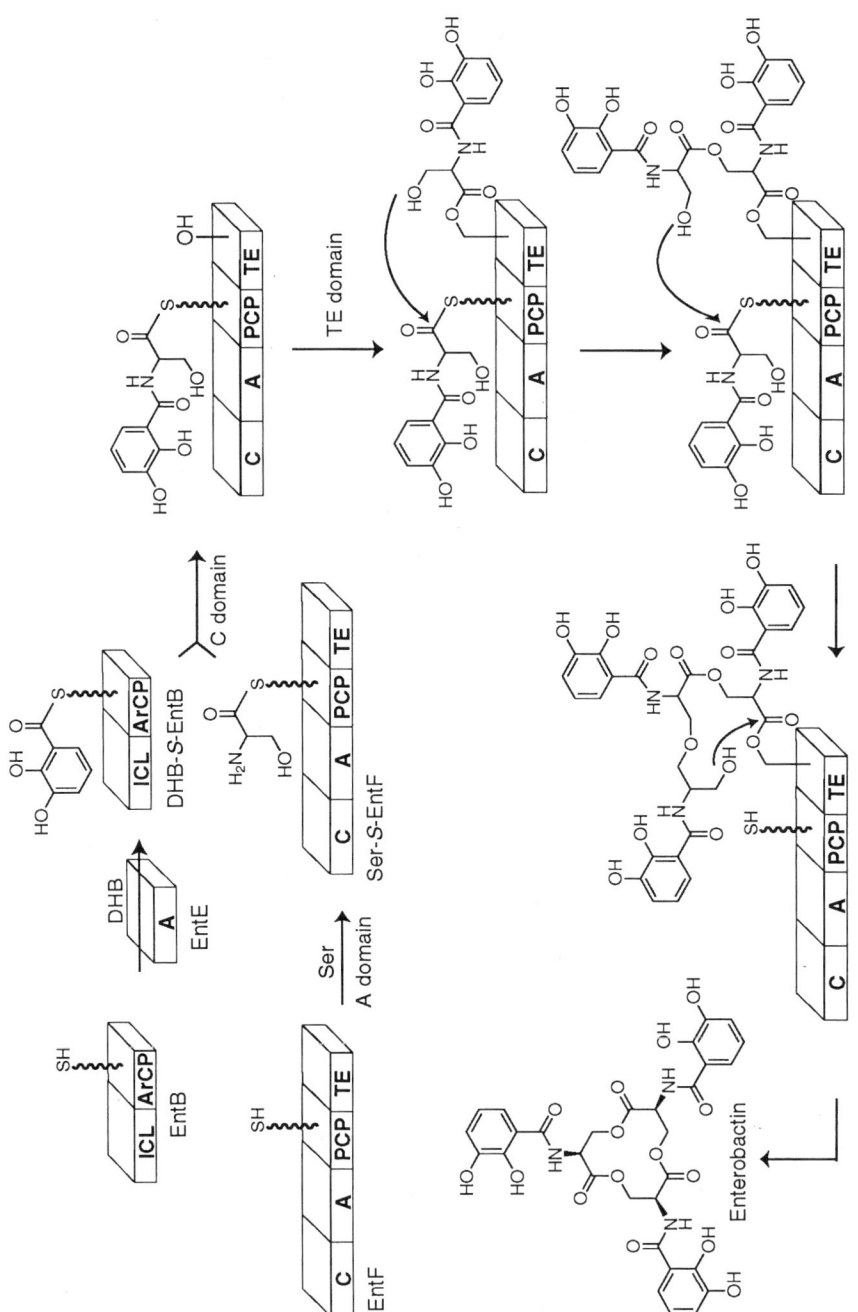

FIGURE 6 Biosynthesis of enterobactin from DHB and Ser by EntB, EntE, and EntF.

amide linkages (DHB-Ser) and three ester bonds (in the Ser trilactone scaffold). While one of the six bonds, the first amide linkage, has been generated to this point, there is only one domain, the TE domain of EntF, left for catalysis. The DHB-Ser chain is transferred to the active-site serine of the the TE domain, freeing up the EntF PCP for another round of DHB and Ser activation and for the second amide bond formation, to create an assembly line with one DHB-Ser in oxoester linkage at the TE domain and the second DHB-Ser in thioester linkage on the PCP domain of EntF (Fig. 6). The first ester bond can form by attack of the nucleophilic-OH of one DHB-Ser moiety on the activated carbonyl of the second, here shown as producing a DHB-Ser−O−DHB−Ser−O-TE intermediate. This frees up the PCP domain again for a third round of DHB and Ser activation and DHB-Ser−S-PCP formation. Iteration of the ester bond formation yields a linear trimer (DHB−Ser−O−DHB-Ser−O−DHB−Ser) tethered in the active site of the TE domain. Now, if the linear trimer can fold up so that the remaining seryl-OH can approach the carbonyl of DHB-Ser$_1$, then intramolecular cyclization to the 12-ring trilactone will be effected and the completed siderophore will be released into solution. For bacillibactin, the aryldipeptidyl enzyme intermediate is built up by a three-module system, activating DHB, glycine, and threonine, and a comparable scheme for elongation to a liner trimeric acyl-O-TE and TE-driven cyclization is envisioned.

The underlying logic of myxochelin assembly is similar to enterobactin assembly in the first part of the assembly line. DHB is activated by MxcE (Fig. 7) as DHB-AMP and then installed on the ArCP domain of MxcF, again a bifunctional isochorismate lyase/aryl carrier protein. The MxcG subunit is a four-domain enzyme, C-A-PCP-Red, differing from EntF in that the C-terminal TE of EntF is replaced by the C-terminal Red domain of MxcG. The A domain of MxcG recognizes and activates Lys, producing the tethered thioester Lys-S-PCP. The C domain of MxcG catalyzes amide bond formation, with DHB transfer to yield DHB-Lys−S-MxcG. It is unclear if the first DHB transfer is to the C-2 or C-6 amino group, but iteration of another round of DHB activation and condensation yields two DHB groups anchored in amide linkage at C-2 and C-6 of the MxcG aminoacyl-S-PCP enzyme intermediate (Fig. 7). At this point, the full-length chain is to be released. In EntF, the chain is transferred covalently to the TE active site. Chain release in MxcG is thought to be reductive, under the catalytic aegis of the C-terminal domain. The reductase (Red) domain uses NAD(P)H to reduce the thioester, presumably first to the aldehyde and then to the alcohol, to produce myxochelin A. An alternative siderophore, myxochelin B, has a carboxamide functionality in place of the alcohol group of myxochelin A. A fourth enzyme, MxcL, is thought to function as an amidotransferase, utilizing NH_3 rather than H_2O as the nucleophile to release the siderophore chain from covalent attachment to the PCP domain. The arylamide bond-forming steps that effect chain initiation in pyochelin synthetase, to create a salicyl-cysteinyl−S-PCP intermediate, have also been characterized and are essentially identical to those described here.

Thus, catecholate siderophore assembly utilizes aryl acids with hydroxyl groups capable of iron chelation as chain-initiating acyl moieties in what amounts to peptidyl N capping. The iterative addition of one or two additional catecholate moieties to generate the bis- and tris (catechol) siderophores requires kinetic control of the acyl-S and acyl-O enzyme intermediate lifetimes on the assembly line, in particular the avoidance of adventitious hydrolysis. We have also noted distinct chain termination catalytic domains to tailor the acyl chain released from the assembly line to yield C-terminal alcohols, amides, and cyclic lactones.

HYDROXAMATE SIDEROPHORES

Several siderophores, including exochelin, aerobactin, rhizobactin 1021, ferrichrome, and alcaligin, employ hydroxamate groups as a means of chelating iron (Fig. 8). Hydroxamate sidero-

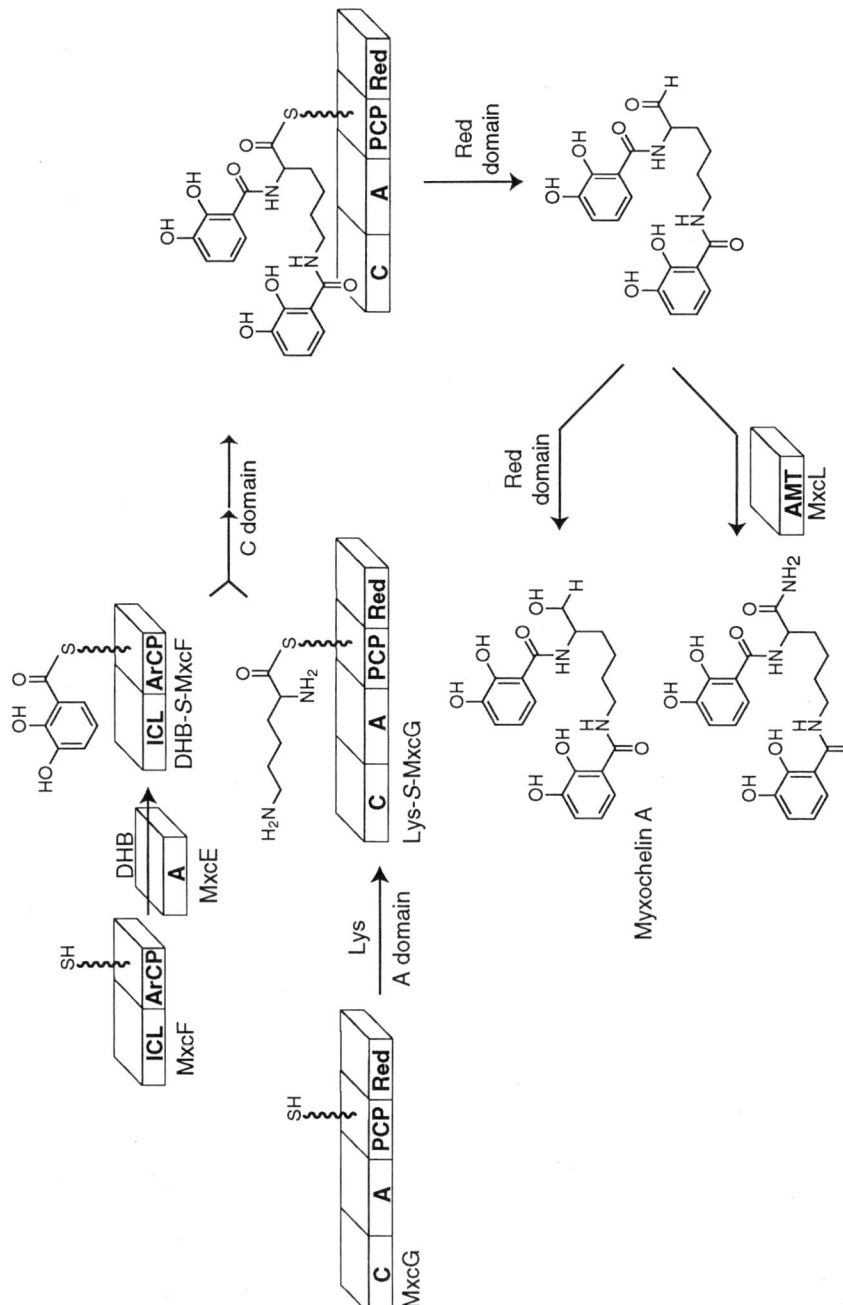

FIGURE 7 Biosynthesis of myxochelins A and B from DHB and Lys by MxcF, MxcE, MxcG, and MxcL.

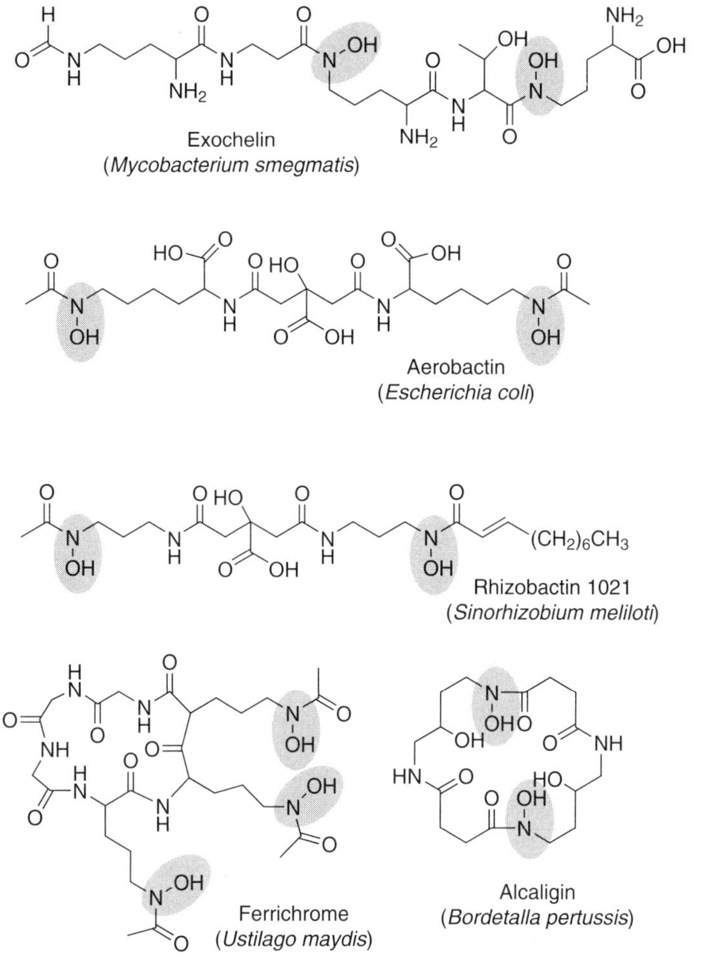

FIGURE 8 Some representative hydroxamate siderophores.

phores are produced by both gram-positive (exochelin) and gram-negative (alcaligin) bacteria and, in the case of ferrichrome, by a corn smut fungi. They are produced by both opportunistic pathogens of mammalian hosts (aerobactin) and soil-dwelling plant symbionts (rhizobactin 1021). Hydroxamate siderophores are observed with both linear and cyclic backbone configurations, and typically two hydroxamate groups are found per molecule. Hydroxamates are often used in conjunction with other iron-chelating functional groups; however, they can exist as the lone chelating group type in a given siderophore.

The lone-pair electrons of hydroxamate oxygens are known to be particularly nucleophilic due to a phenomenon known as the alpha effect. This increases their ability to participate as ligands to ferric iron. The hydroxyl is derived from molecular oxygen, which must be reduced by two electrons (to the level of a peroxide) to generate the hydroxamate functionality. This amine oxygenation reaction is catalyzed by a monooxygenase, the best characterized of which is IucD, or lysine:N^6-hydroxylase, which is used in aerobactin biosynthesis. Due to its tendency to associate with membranes, in vitro studies of IucD have been performed with recombinant versions (rIucD) that either have hydrophilic N-terminal additions or deletions of the native N-terminal hydrophobic sequence. rIucD functions as a tetramer and uses

NADPH as the preferred source of reducing equivalents (Fig. 9). FAD serves as a cofactor, accepting electrons from NADPH that are used to generate the characteristic 4a-peroxyflavin typically formed in this class of enzymes. This electrophilic species is attacked by N-6 of lysine with monooxygen transfer to give the final products, N^6-hydroxylysine, water, and NADP$^+$, with FAD regenerated in the enzyme active site.

While lysine serves as the amine substrate in aerobactin biosynthesis, a variety of amines are used to generate hydroxamate groups in siderophore biosynthetic schemes. One of the most common is ornithine (exochelin, ornibactin), which can be readily generated from glutamate via a semialdehyde intermediate. Diaminopropane, found in rhizobactin 1021, is thought to be manufactured from glutamate and aspartate. In alcaligin, putrescine is the target and is produced by decarboxylation of ornithine. In anguibactin, histamine is generated from the common amino acid histidine, also via a simple decarboxylation. Thus, the microorganisms that make these molecules have developed a variety of solutions to a common need to make diamine scaffolds.

Hydroxamate scaffolds are assembled by using both NRPS and non-NRPS methods, and probably the best-characterized example of the latter brings us again to the biosynthesis of aerobactin. The aerobactin backbone is assembled from two molecules of lysine, two acetyl groups derived from acetyl coenzyme A (acetyl-CoA) and one molecule of citrate that serves as a central linker (Fig. 10A). While several organisms can produce aerobactin, including species of *Shigella, Yersinia,* and *Salmonella,* most work has been done on the five-gene cluster encoded by the ColV-K30 plasmid of *E. coli* (Fig. 10B). *iutA* encodes a 77-kDa outer membrane receptor, while the remaining four genes, *iucA* to *iucD* are responsible for aerobactin biosynthesis. Both in vivo gene deletion analysis and in vitro kinetic analysis have been used to establish the sequence of events of aerobactin assembly (Fig. 10C). IucD, a 49-kDa monooxygenase, uses molecular oxygen to generate N^6-hydroxy lysine, as discussed above. This intermediate is then substrate for the 33-kDa IucB, which N acetylates the hydroxamate by using acetyl-CoA. The final coupling of two molecules of N^6-acetyl-N^6-hydroxylysine to citrate occurs in two discrete steps and is catalyzed by a synthetase composed of α and β subunits, encoded by *iucA* and *iucC*, respectively. Amide bond formation probably proceeds via a high-energy intermediate, such as an acyl-phosphate or acyl-adenylate, although no in vitro characterization of this synthetase has been reported.

The gene cluster for the biosynthesis of rhizobactin 1021, a siderophore produced by *Sinorhizobium meliloti,* parallels that of aerobactin and is thus predicted to follow a very similar biosynthetic scheme. In rhizobactin 1021, lysine is replaced by 1,3-diaminopropane and RhbE, RhbD, RhbC, and RhbF function homologously to IucD, IucB, IucA, and IucC, respectively. The identity of the N-acylase catalyzing the last step of this pathway, addition of the lipid moiety, is not known.

Of the NRPS-assembled hydroxamate siderophores, exochelin of *Mycobacterium smeg-*

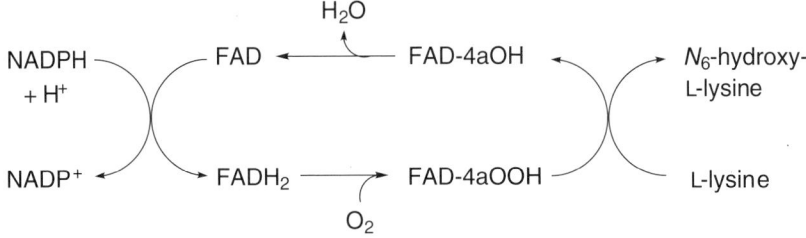

FIGURE 9 Proposed scheme for lysine-N^6-hydroxylation by the monooxygenase IucD of the aerobactin biosynthetic pathway.

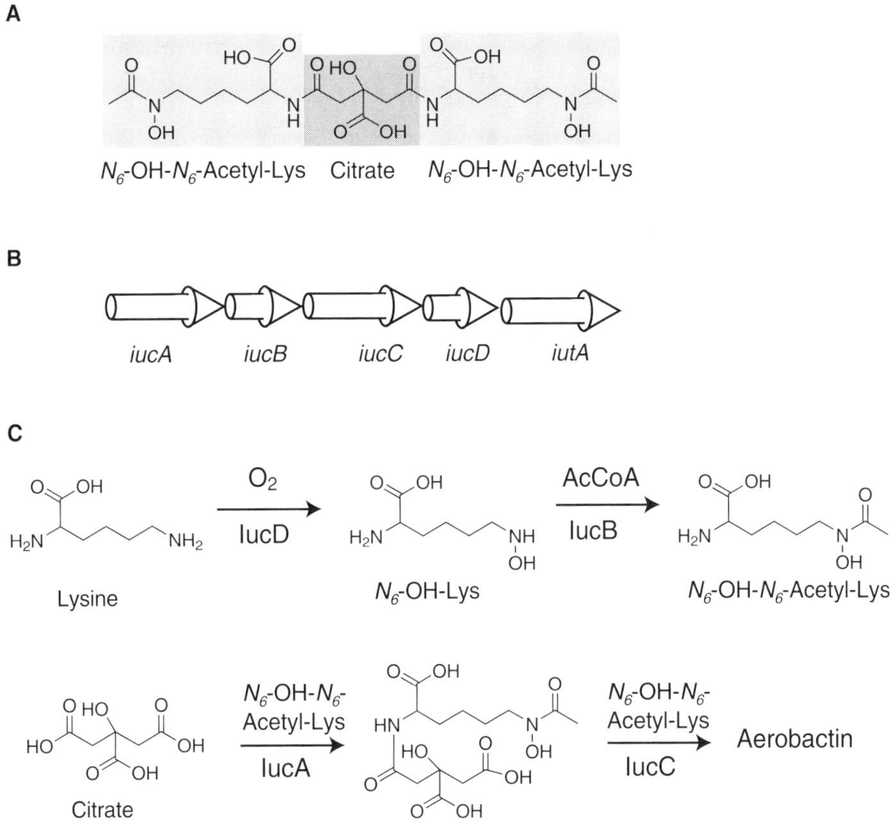

FIGURE 10 Biosynthesis of aerobactin. (A) Structure of aerobactin. (B) Structure of aerobactin biosynthetic gene cluster. (C) Scheme for aerobactin synthesis from lysine, citrate, acetyl-CoA (AcCoA), and O_2 by IucD, IucB, IucA, and IucC.

matis is one of the best characterized. Exochelin is composed of N-formyl-D-ornithine, β-alanine, N^5-hydroxy-D-ornithine, D-threonine, and N^5-hydroxy-ornithine (Fig. 11A). Part of the gene cluster for its biosynthesis has been identified and is shown in Fig. 11B. Several of the genes, including *fxuA* to *fxyC ORF1*, and *ORF2*, are thought to be involved in iron permeation of the cell, while *fxuD* is a putative membrane-associated receptor for ferric exochelin. The remaining *fxb* genes are components of the biosynthetic pathway. Bioinformatic analysis suggests that *fxbA* is probably involved in N formylation of ornithine. FxbB and FxbC are large modular proteins and clearly are components of an NRPS system (Fig. 11C). FxbB has two modules, and FxbC has four; the placement of epimerase domains corresponds to the placement of *R* amino acids in the exochelin peptide. That exochelin has only five monomeric units suggests that one of the two final modules of FxbC is inactive, or, less likely, that a hexamer is manufactured which is then cleaved to give the final product. The gene for N-hydroxylation has not been identified. Furthermore, the timing of hydroxamate formation is not known. It could conceivably occur prior to, during, or after peptide assembly. The timing of N formylation is similarly unknown, although this N capping may be the chain initiation step.

The cyclic siderophore ferrichrome, composed of three molecules of glycine and three molecules of N-acyl-N-hydroxyornithine (Fig. 8), also employs an NRPS to build its scaffold. The ferrichrome biosynthetic scheme

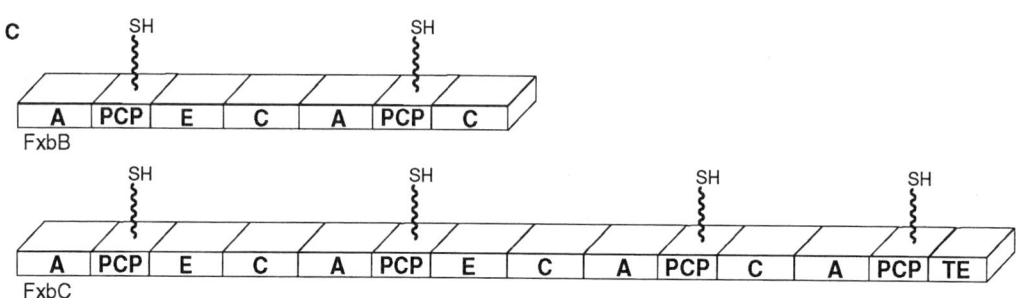

FIGURE 11 Biosynthesis of exochelin. (A) Structure of exochelin. (B) Structure of exochelin biosynthesis gene cluster. (C) NRPS proteins FxbB and FxbC shown with posttranslationally modified (but unacylated) PCPs.

utilizes a six-module NRPS designated Sid2; unlike for exochelin, no epimerization domains are required nor found. However, as in the exochelin system, the timing of acetylation and N-hydroxylation in ferrichrome assembly has yet to be determined.

FIVE-MEMBER HETEROCYCLIC RING-CONTAINING SIDEROPHORES

A third ferric iron-chelating functional group found in siderophores is a five-member heterocycle that arises by cyclization of cysteine, serine, or threonine side chains during siderophore chain elongation on NRPS assembly lines. Enzymatic cyclization of cysteinyl side chains onto the preceding amide carbonyl yields a thiazoline ring, found in such siderophores as anguibactin, pyochelin, and yersiniabactin (Fig. 12). The corresponding cyclization of seryl side chains yields oxazoline rings in mycobactin, the mycobacterial siderophore, while threonine produces β-methyloxazolines, for example in vibriobactin from *Vibrio cholerae*. Validation that the five-ring heterocycle is a ligand to chelated metal cation is provided by the X-ray structure of micacocidin in complex with zinc (Fig. 13).

Redox adjustment of the thiazoline and oxazoline rings can alter both their electronic and geometric properties. For example, the second heterocyclic rings in both yersiniabactin and pyochelin have undergone a two-electron reduction at the C=N imine linkage, using NADPH to create a thiazolidine ring. The pucker of the thiazoline and thiazolidine rings is distinct, and this may be important for the geometry of chelation of ferric iron by the thiazoline and/or thiazolidine nitrogens. The thiazoline and oxazoline rings can also be enzymatically oxidized by two electrons to the fully aromatic thiazole and oxazole rings, as happens in epothilone and bleomycin biosynthesis, but

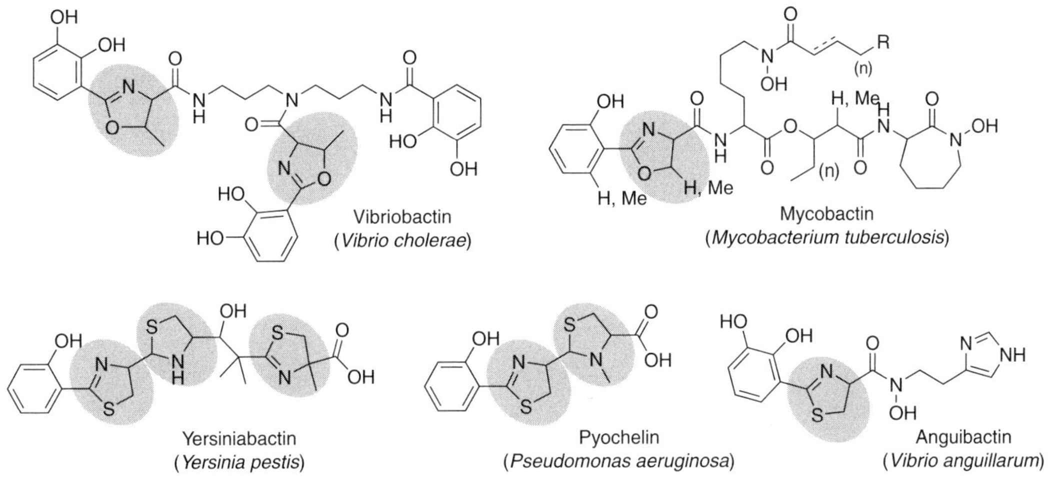

FIGURE 12 Some representative five-member heterocyclic ring-containing siderophores.

in this oxidation state the nitrogen is less basic and less effective at iron chelation. The five-ring heterocycles in the siderophores of Fig. 12 are found in different contexts: connected to a catechol or phenol ring (vibriobactin, yersiniabactin, and mycobactin); in a tandem array of thiazoline-thiazolidine (yersiniabactin and pyochelin), or nonadjacent to a ring structure (yersiniabactin, third heterocyclic ring). The first context suggests the use of salicylate or 2,3-dihydroxybenzoate monomers in the NRPS chain initiation steps, as discussed above for the catechol/phenol siderophores. The second context suggests tandem NRPS modules activating cysteine and double cyclizations. The isolation of the protein subunits of the yersiniabactin, pyochelin and vibriobactin synthetase assembly lines and the full reconstitution of these synthetases in vitro have validated these predictions.

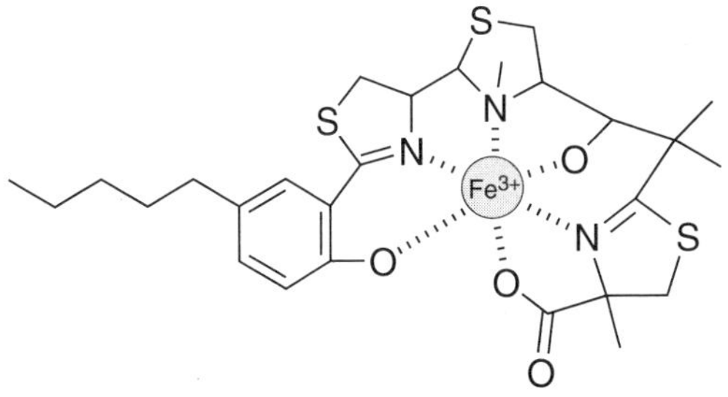

The most distinctive feature of NRPS assembly lines that produce five-ring heterocycles during chain elongation is the replacement of the 50-kDa C domains by a 50-kDa Cy domain upstream of modules activating Cys, Ser, or Thr. A variety of studies have provided evidence that Cy domains carry out the initial condensation typical of amide bond-forming C domains (Fig. 14) but then go on to drive cyclodehydration before the chain moves downstream to the next module. The cyclodehydration catalysis is proposed to have two steps, with the nucleophilic thiol of Cys, or the β-OH of Ser or Thr, directed by the Cy domain active site to attack the carbonyl group of the newly formed amide to create a tetrahedral adduct. Loss of H_2O generates the C=N imine double bond, drives the decomposition in the forward direction, and allows accumulation of the cyclic thiazoline/oxazoline rings. This cyclodehydration enzymology radically alters the backbone connectivity of the nonribosomal peptide chain during elongation.

The vibriobactin synthetase from *V. cholerae* has been reconstituted from purified components and the molecular logic of monomer se-

FIGURE 13 Scheme of the chelation of ferric iron by the siderophore micacocidin, as determined by X-ray crystallography.

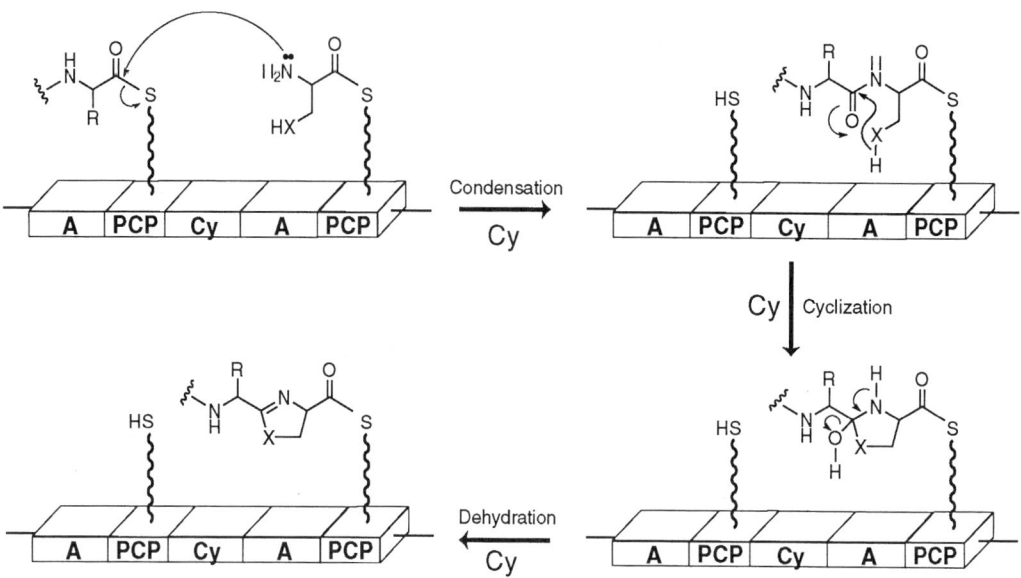

FIGURE 14 Predicted mechanism for the formation of heterocyclic rings by Cy domains.

lection and assembly has been deciphered. As shown in Fig. 15, the first module is typical for a DHB-activating system: a free-standing A domain (VibE) activates DHB and transfers it to the HS-pantetheinyl arm of the ArCP domain of a bifunctional ICL/ArCP protein (VibB). The scaffold for attachment of three such activated DHB residues from DHB-S-VibB is a triamine norspermidine (NSPD), a dedicated metabolite made in *V. cholerae* for siderophore production. One of the three DHB groups is transferred directly to a primary amine of NSPD, to yield DHB-NSPD, and this is catalyzed by a free-standing C domain, VibH, whose crystal structure has been determined. The remaining two DHB groups are transferred first to threonyl moieties, which are cyclodehydrated to methyloxazolinyl groups, and then to the remaining two amine groups of the NSPD backbone. This chemistry is accomplished by the VibF subunit, a six-domain NRPS protein complete with a PCP, an A domain, two Cy domains, and two C domains. The threonine-specific A domain and PCP function to make threonyl-S-PCP. Unusually, the two Cy domains of VibF split the tasks of condensation (Cy2) and cyclodehydra-tion (Cy1), to yield the dihydroxyphenyl-methyloxazolinyl-S-PCP intermediate. The final two transfers to DHB-NSPD are both carried out by C2, with no enzymatic role for Cl in assembly. Unlike for other NRPS assembly lines, several of the vibriobactin intermediates are freely diffusible in bulk solvent, a property that facilitates kinetic analysis. The vibriobactin assembly line thus combines several operations to build a complex iron chelator from three DHB molecules, two threonines, and the triamine scaffold norspermidine, using both the catechol and five-ring oxazoline chelator moieties. Anguibactin assembly by *V. anguillarum* uses an equivalent strategy for construction of a dihydroxyphenyl-thiazoline ring from DHB and cysteine and transfers the dihydroxyphenyl-thiazolinyl-S-PCP donor to histamine as the acceptor nucleophile in place of norspermidine. Then N hydroxylation creates the hydroxamate linkage and efficiently introduces all three types of iron-chelating groups in a compact siderophore structure.

Yersinia pestis in iron-depleted microenvironments turns on the genes for making the siderophore yersiniabactin, enabling growth and enhancing virulence. Yersiniabactin (Ybt)

2. SIDEROPHORE BIOSYNTHESIS IN BACTERIA ■ 33

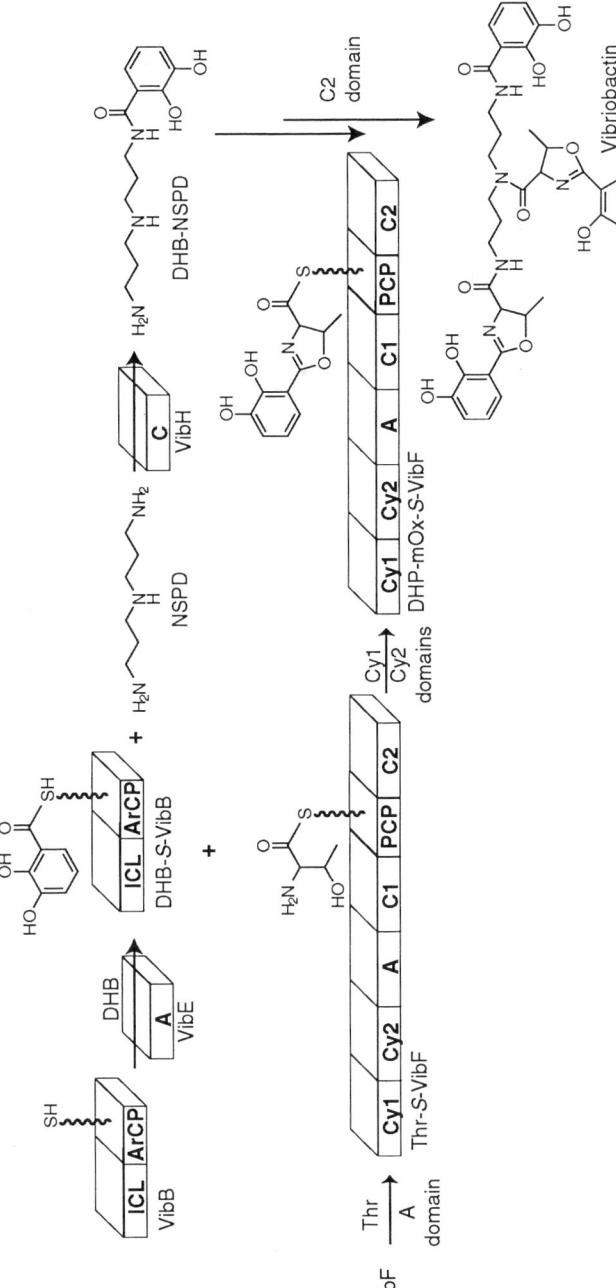

FIGURE 15 Biosynthesis of vibriobactin from DHB, Thr, and NSPD by VibB, VibE, VibH, and VibF. DHP-mOx, dihydroxyphenyl-methyloxazolinyl.

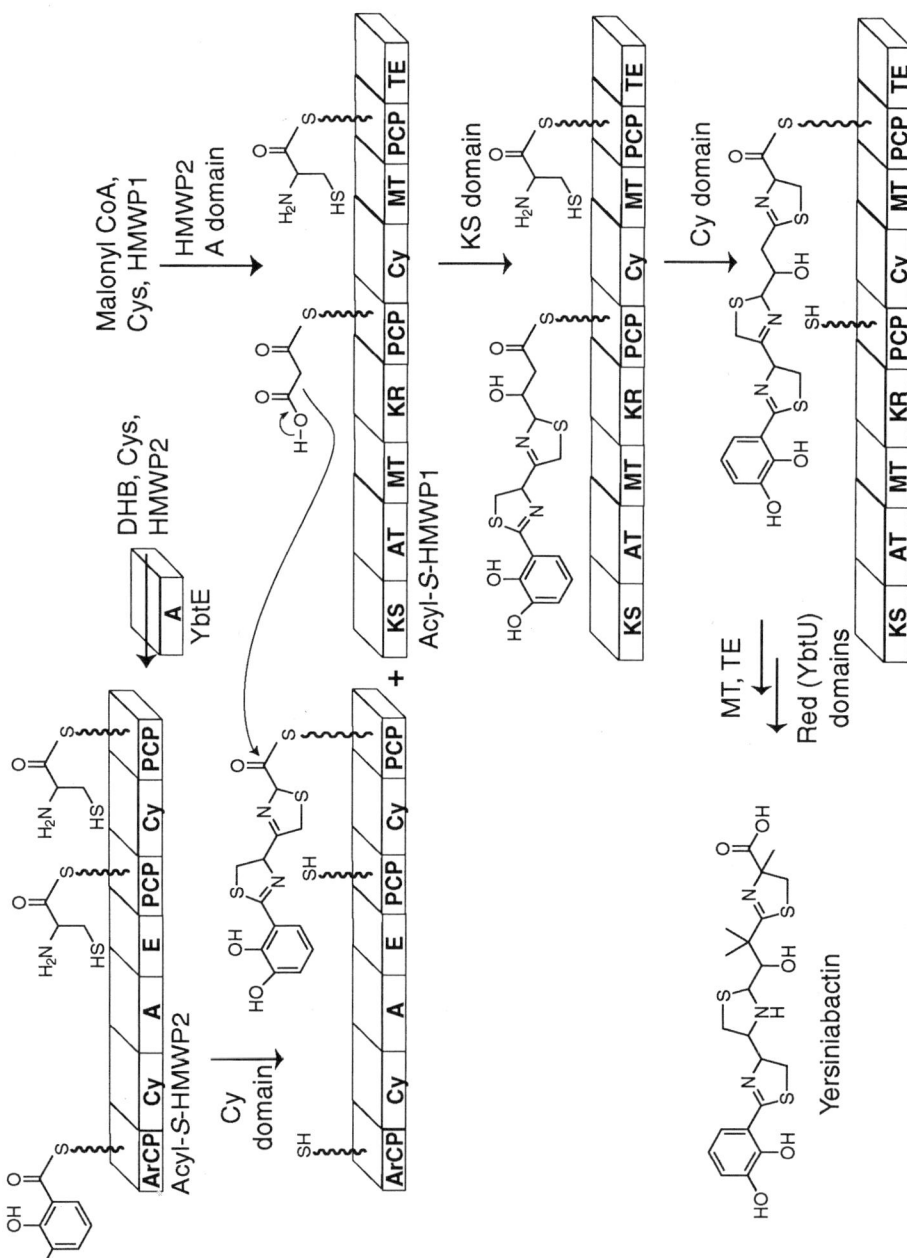

FIGURE 16 Biosynthesis of yersiniabactin by YbtE, HMWP2, the hybrid NRPS-PKS HMWP1, and the reductase YbtU.

offers three additional complexities of biosynthetic logic. First, while most of its origin can be parsed into DHB and three cysteine monomers as the source of the three sulfur-nitrogen heterocycles (thiazoline, thiazolildine and methythiazoline), there are an extra two carbons in the backbone between the thiazolidine and the methylthiazoline that do not come from an amino acid. Indeed, the Ybt structure predicts a mixed nonribosomal peptide-polyketide origin, with the extra carbons arising from decarboxylative insertion of malonyl-CoA, a typical monomer for polyketide synthase elongation steps. The second anomaly is the C methylation of the last thiazoline ring and of the bridging malonyl unit, typical of S-adenosylmethionine-derived methylations.

The third is the reduction of the middle thiazoline ring to the tetrahydrothiazolidine oxidation state.

The Ybt synthetase assembly line has all the domains and modules necessary to explain these three additional complexities. YbtE is a salicylate-activating A domain for the N-capping chain initiation step. Its partner, ArCP, is the first domain in a seven-domain protein called high-molecular-weight protein 2 (HMWP2). HMWP2 has two other PCP domains and one A domain such that cysteinyl groups can be installed as thioesters on PCP_1 and PCP_2, as shown in Fig. 16. There are two Cy domains, one to condense salicyl onto the first Cys-S-PCP and cyclohydrate it and one to analogously translocate the growing chain by condensation and cyclodehydration, to build up a hydroxyphenyl-thiazolinyl-thiazolinyl-S-PCP_2 intermediate at the most downstream domain of HMWP2.

Inspection of the next protein HMWP1 reveals nine predicted domains the first five as a polyketide synthase module and the last four in an NRPS module. The PKS module has a ketosynthase, acyl transferase, methyltransferase (MT), ketoreductase, and acyl carrier protein (ACP) domains. These five domains function to (i) transfer (acyl transferase) a malonyl group from malonyl-CoA onto the ACP domain; (ii) condense (ketosynthase) the hydroxyphenyl-thiazolinyl-thiazolinyl donor onto the decarboxylating malonyl-S-ACP to make a β-keto acyl-S-ACP; (iii) doubly C methylate (MT) with two molecules of cosubstrate S-adenosylmethionine as methyl donor; and then (iv) reduce (ketoreductase) the β-keto group to the hydroxy group found in Ybt. At this point, the NRPS module in the last four domains, with its activated Cys-S-PCP_3 installed, translocates the acyl chain from the ACP domain via the Cy domain, to the acyl-thiazolinyl–S-PCP_3. The third domain of this NRPS module is another methyl transferase (MT) and carries out a final C methylation, this time at C-2 of the acyl-thiazolinyl–S-PCP_3 to produce the carbon skeleton of Ybt. Two events remain. The chain has to be released from its covalent thioester attachment on PCP_3, and this is the job of the C-terminal TE domain of HMWP1, catalyzing hydrolytic release. The other chemical step that has to occur is reduction of the middle thiazoline ring by two electrons to the thiazolidinyl ring. This is the function of YbtU in the Ybt synthetase assembly line (and by the cognate PchG in the pyochelin synthetase assembly line). It appears that the imine reduction by NADPH occurs while the growing chain is still tethered to the NRPS assembly line.

The Ybt siderophore is a small molecule bristling with functional group complexity, admirably engineered for scavenging ferric iron, with a K_d of 10^{-35} M, and is put together via 18 autonomous protein domains distributed over four proteins (YbtE, YbtU, HMWP1, and HMWP2) with more than 20 chemical steps. The logic of the assembly strategy can be predicted from the placement and identity of the domains and suggests that siderophore assembly lines represent ongoing experiments in combinatorial biosynthesis of natural products, of both peptide and polyketide origin.

CONCLUDING THOUGHTS

Siderophores are assembled from monomer units on a just-in-time basis by biosynthetic enzymes whose synthesis is turned on by promoters sensitive to iron depletion in bacterial (and fungal) cells. The siderophores are small molecules composed of flexible scaffolds that bear electron-rich iron-chelating groups. Most often, the scaffolds have peptide backbones, although diamines generated via enzymatic decarboxylation steps are also used. The peptide backbones are usually constructed on nonribosomal peptide synthetase assembly lines. The peptide and diamine backbones are often N capped enzymatically with salicyl or dihydroxybenzoyl groups in amide linkage. These comprise the first category of phenol/catechol ligands for ferric iron. For scaffolds that contain side chain amines, such as lysine and ornithine residues, enzymatic N hydroxylation generates the hydroxamates that are the second class of ferric chelating groups. The *N*-hydroxyl moieties are almost always found in amide linkages where the acylation steps probably follow the N hydroxylations. The third type of ferric iron chelator, found in the five-ring thiazoline, thiazolidine, and oxazoline rings, is formed by siderophore synthetase assembly lines, in which Cy domains replace C domains. Cysteine, serine, and threonine residues are condensed and then cyclodehydrated to form the conformationally constrained five-ring systems with the imine or amine lone pair as the iron coordination ligand. It is not clear whether one type of chelating group is superior to or preferred over another, and there are examples such as anguibactin where all three types are found in this densely functionalized siderophore. The diversity of iron chelators probably reflects a balance of biosynthetic availability of dedicated monomers and shuffling of NRPS domains to create evolving assembly line organization.

ACKNOWLEDGMENTS

This work has been supported by the National Institutes of Health (grant AI042738 to C.T.W.). C.G.M. is a Fellow of the Canadian Institutes of Health Research.

SUGGESTED READINGS

Actis, L. A., W. Fish, J. H. Crosa, K. Kellerman, S. R. Ellenberger, F. M. Hauser, and J. Sanders-Loehr. 1986. Characterization of anguibactin, a novel siderophore from *Vibrio anguillarum* 775(pJM1). *J. Bacteriol.* **167:**57–65.

Bearden, S. W., J. D. Fetherston, and R. D. Perry. 1997. Genetic organization of the yersiniabactin biosynthetic region and construction of avirulent mutants in *Yersinia pestis*. *Infect. Immun.* **65:**1659–1668.

Budde, A. D., and S. A. Leong. 1989. Characterization of siderophores from *Ustilago maydis*. *Mycopathologia* **108:**125–133.

Cox, C. D., K. L. Rinehart, Jr., M. L. Moore, and J. C. Cook, Jr. 1981. Pyochelin: novel structure of an iron-chelating growth promoter for *Pseudomonas aeruginosa*. *Proc. Natl. Acad. Sci. USA* **78:**4256–4260.

Drechsel, H., H. Stephan, R. Lotz, H. Haag, H. Zahner, K. Hantke, and G. Jung. 1995. Structural elucidation of yersiniabactin, a siderophore from highly virulent *Yersinia* strains. *Liebigs Ann.* **1995:**1727–1733.

Gaitatzis, N., B. Kunze, and R. Muller. 2001. In vitro reconstitution of the myxochelin biosynthetic machinery of *Stigmatella aurantiaca* Sg a15: biochemical characterization of a reductive release mechanism from nonribosomal peptide synthetases. *Proc. Natl. Acad. Sci. USA* **98:**11136–11141.

Gehring, A. M., K. A. Bradley, and C. T. Walsh. 1997. Enterobactin biosynthesis in *Escherichia coli*: isochorismate lyase (EntB) is a bifunctional enzyme that is phosphopantetheinylated by EntD and then acylated by EntE using ATP and 2,3-dihydroxybenzoate. *Biochemistry* **36:**8495–8503.

Gehring, A. M., E. DeMoll, J. D. Fetherston, I. Mori, G. F. Mayhew, F. R. Blattner, C. T. Walsh, and R. D. Perry. 1998. Iron acquisition in plague: modular logic in enzymatic biogenesis of yersiniabactin by *Yersinia pestis*. *Chem. Biol.* **5:**573–586.

Harris, W. R., C. J. Carrano, S. R. Cooper, S. R. Sofen, A. E. Avdeef, J. V. McArdle, and K. N. Raymond. 1979. Coordination chemistry of microbial iron transport compounds. 19. Stability constants and electrochemical behavior of ferric enterobactin and model complexes. *J. Am. Chem. Soc.* **101:**6097–6104.

Keating, T. A., C. G. Marshall, and C. T. Walsh. 2000. Reconstitution and characterization of the *Vibrio cholerae* vibriobactin synthetase from VibB, VibE, VibF, and VibH. *Biochemistry* **39:**15522–15530.

Keating, T. A., C. G. Marshall, and C. T. Walsh. 2000. Vibriobactin biosynthesis in *Vibrio cholerae*: VibH is an amide synthase homologous to nonribosomal peptide synthetase condensation domains. *Biochemistry* **39:**15513–15521.

Keating, T. A., C. G. Marshall, C. T. Walsh, and A. E. Keating. 2002. The structure of VibH represents nonribosomal peptide synthetase condensation, cyclization and epimerization domains. *Nat. Struct. Biol.* **9:**522–526.

Keating, T. A., D. A. Miller, and C. T. Walsh. 2000. Expression, purification, and characterization of HMWP2, a 229 kDa, six domain protein subunit of yersiniabactin synthetase. *Biochemistry* **39:** 4729–4739.

Keating, T. A., and C. T. Walsh. 1999. Initiation, elongation, and termination strategies in polyketide and polypeptide antibiotic biosynthesis. *Curr. Opin. Chem. Biol.* **3:**598–606.

Kobayashi, S., H. Nakai, Y. Ikenishi, W. Y. Sun, M. Ozaki, Y. Hayase, and R. Takeda. 1998. Micacocidin A, B and C, novel antimycoplasma agents from *Pseudomonas* sp. II. Structure elucidation. *J. Antibiot.* (Tokyo) **51:**328–332.

Lynch, D., J. O'Brien, T. Welch, P. Clarke, P. O. Cuiv, J. H. Crosa, and M. O'Connell. 2001. Genetic organization of the region encoding regulation, biosynthesis, and transport of rhizobactin 1021, a siderophore produced by *Sinorhizobium meliloti. J. Bacteriol.* **183:**2576–2585.

Marshall, C. G., M. D. Burkart, T. A. Keating, and C. T. Walsh. 2001. Heterocycle formation in vibriobactin biosynthesis: alternative substrate utilization and identification of a condensed intermediate. *Biochemistry* **40:**10655–10663.

Marshall, C. G., N. J. Hillson, and C. T. Walsh. 2002. Catalytic mapping of the vibriobactin biosynthetic enzyme VibF. *Biochemistry* **41:**244–250.

May, J. J., T. M. Wendrich, and M. A. Marahiel. 2001. The *dhb* operon of *Bacillus subtilis* encodes the biosynthetic template for the catecholic siderophore 2,3-dihydroxybenzoate-glycine-threonine trimeric ester bacillibactin. *J. Biol. Chem.* **276:**7209–7217.

Miller, D. A., L. Luo, N. Hillson, T. A. Keating, and C. T. Walsh. 2002. Yersiniabactin synthetase: a four-protein assembly line producing the nonribosomal peptide/polyketide hybrid siderophore of *Yersinia pestis. Chem. Biol.* **9:**333–344.

Patel, H. M., and C. T. Walsh. 2001. In vitro reconstitution of the *Pseudomonas aeruginosa* nonribosomal peptide synthesis of pyochelin: characterization of backbone tailoring thiazoline reductase and N-methyltransferase activities. *Biochemistry* **40:** 9023–9031.

Perry, R. D., P. B. Balbo, H. A. Jones, J. D. Fetherston, and E. DeMoll. 1999. Yersiniabactin from *Yersinia pestis:* biochemical characterization of the siderophore and its role in iron transport and regulation. *Microbiology* **145:**1181–1190.

Quadri, L. E., T. A. Keating, H. M. Patel, and C. T. Walsh. 1999. Assembly of the *Pseudomonas aeruginosa* nonribosomal peptide siderophore pyochelin: in vitro reconstitution of aryl-4, 2-bisthiazoline synthetase activity from PchD, PchE, and PchF. *Biochemistry* **38:**14941–14954.

Reimmann, C., H. M. Patel, L. Serino, M. Barone, C. T. Walsh, and D. Haas. 2001. Essential PchG-dependent reduction in pyochelin biosynthesis of *Pseudomonas aeruginosa. J. Bacteriol.* **183:** 813–820.

Sharman, G. J., D. H. Williams, D. F. Ewing, and C. Ratledge. 1995. Isolation, purification and structure of exochelin MS, the extracellular siderophore from *Mycobacterium smegmatis. Biochem. J.* **305:** 187–196.

Tolmasky, M. E., L. A. Actis, and J. H. Crosa. 1995. A histidine decarboxylase gene encoded by the *Vibrio anguillarum* plasmid pJM1 is essential for virulence: histamine is a precursor in the biosynthesis of anguibactin. *Mol. Microbiol.* **15:**87–95.

Walsh, C. T., J. Liu, F. Rusnak, and M. Sakaitani. 1990. Molecular studies on enzymes in chorismate metabolism and the enterobactin biosynthetic pathway. *Chem. Rev.* **90:**1105–1129.

Yu, S., E. Fiss, and W. R. Jacobs, Jr. 1998. Analysis of the exochelin locus in *Mycobacterium smegmatis:* biosynthesis genes have homology with genes of the peptide synthetase family. *J. Bacteriol.* **180:** 4676–4685.

Yuan, W. M., G. D. Gentil, A. D. Budde, and S. A. Leong. 2001. Characterization of the *Ustilago maydis sid2* gene, encoding a multidomain peptide synthetase in the ferrichrome biosynthetic gene cluster. *J. Bacteriol.* **183:**4040–4051.

Zhu, W., J. E. Arceneaux, M. L. Beggs, B. R. Byers, K. D. Eisenach, and M. D. Lundrigan. 1998. Exochelin genes in *Mycobacterium smegmatis:* identification of an ABC transporter and two nonribosomal peptide synthetase genes. *Mol. Microbiol.* **29:**629–639.

HEMOPHORE-DEPENDENT HEME ACQUISITION SYSTEMS

Laurent Debarbieux and Cécile Wandersman

3

The poor solubility of iron (Fe^{3+}) salts at physiological pH in the presence of oxygen is a serious obstacle to the growth of microorganisms, since iron is required for many metabolic pathways. Microorganisms colonizing living hosts are confronted with another problem: the iron in their environment is not readily available. It is bound to iron carrier proteins or to heme in hemoproteins. Therefore, a microorganism unable to assimilate iron from these diverse sources would be an unsuccessful pathogen, since it would not be able to colonize the mammalian host. It is now clear that several iron-heme acquisition systems may be present in the same bacterium.

One way for microbes to acquire iron is to produce and secrete siderophores (see chapter 1). Siderophores chelate ferric ion with extremely high affinity, allowing its solubilization and its extraction from most mineral or organic complexes. In gram-negative bacteria, iron-siderophore complexes are recognized and are bound by specific outer membrane receptors at the cell surface and then transported through the inner membrane by periplasmic binding protein-dependent ABC permeases. In gram-positive bacteria, the iron-siderophore complexes are recognized by specific binding proteins anchored to the inner membrane and then transported by ABC permeases (Fig. 1).

Bacteria can also acquire iron from heme. Following the transport of heme into the cytoplasm, the iron is extracted by heme oxygenases that break the tetrapyrrole ring. Heme uptake has been identified in both gram-positive and gram-negative bacteria. Heme transport by a gram-positive bacterium, *Corynebacterium diphtheriae*, is described in chapter 22. Heme uptake in gram-negative bacteria has been studied in a number of organisms, and the molecular mechanisms of this transport have been substantially elucidated. Two distinct systems have been identified in gram-negative bacteria. One involves outer membrane receptors and periplasmic binding protein-dependent ABC permeases specific for heme (see chapter 5). The second involves the secretion of specialized bacterial proteins able to sequester heme from diverse sources and bring it to a specific outer membrane receptor. Due to the functional similarity of these proteins (acting like a heme carrier) with siderophores, we called them hemophores (Fig. 2).

A hemophore system has been described in *Haemophilus influenzae*. The hemophore HxuA

Laurent Debarbieux and Cécile Wandersman, Unité des Membranes Bactériennes Institut Pasteur (CNRS URA 2172), 25 rue du Dr. Roux, 75724 Paris Cedex 15, France.

Iron Transport in Bacteria, Edited by Jorge H. Crosa, Alexandra R. Mey, and Shelley M. Payne
© 2004 ASM Press, Washington, D.C.

3. HEMOPHORE-DEPENDENT HEME ACQUISITION SYSTEMS ■ 39

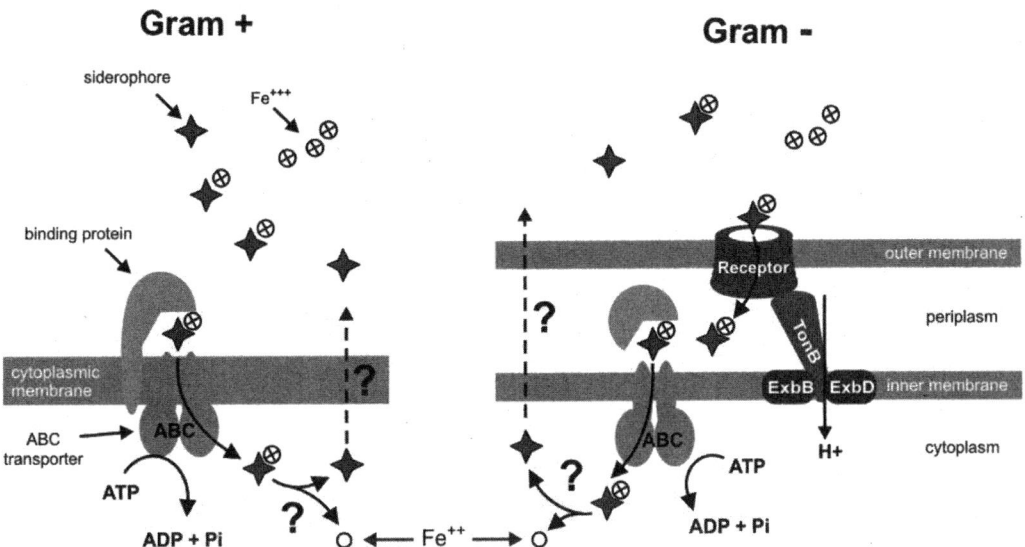

FIGURE 1 Siderophore uptake in gram-positive and gram-negative bacteria. Question marks indicate unelucidated steps. (Reprinted, with permission, from the *Annual Review of Microbiology* [volume 58, 2004] by Annual Reviews [www.annualreviews.org].)

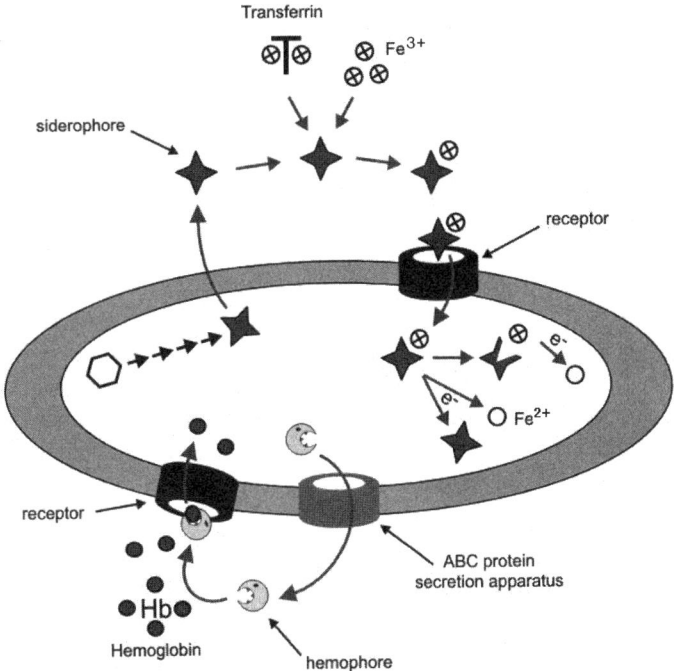

FIGURE 2 Schematic comparison between hemophore and siderophore iron acquisition pathways. (Reprinted, with permission, from the *Annual Review of Microbiology* [volume 58, 2004] by Annual Reviews [www.annualreviews.org].)

is secreted by a signal peptide-dependent pathway. It is able to acquire heme from hemopexin and to present it to HxuC, a specific receptor in the outer membrane.

A second hemophore system has been identified in several bacteria including *Serratia marcescens*, *Pseudomonas aeruginosa*, *Pseudomonas fluorescens*, and *Yersinia pestis*. These hemophores, called HasA (for "heme acquisition system"), form a new family of proteins without homology to any other known proteins. They have no signal peptide and are secreted by dedicated ABC transporters directly into the external medium. Once in the medium, hemophores capture free heme or extract heme from hemoglobin, owing to their higher affinity for heme, and present it to specific outer membrane receptors (Fig. 3A). These receptors are essential for heme acquisition, whereas the hemophores themselves are not essential but stimulate the process, reducing the minimal heme or hemoglobin concentration required for growth.

The proteins making up the secretion apparatus (HasD, HasE, and HasF), the hemophore (HasA), and the outer membrane receptor (HasR) are encoded in most strains by linked genes in an operon repressed by the iron-loaded Fur repressor (Fig. 3B). The secretion apparatus is composed of an ABC protein (HasD), a member of the membrane fusion protein family (HasE), and a TolC-like outer membrane protein (HasF). The *has* operons in the various species have similar genetic organization, although differences have been noted (Fig. 3B). In *S. marcescens* and *Y. pestis*, the last gene of the *has* operon is *hasB*, encoding a TonB homolog, and *hasF* is unlinked and is not regulated by iron. In *P. aeruginosa* and *P. fluorescens*, there is no *hasB* gene and the last gene encodes the outer membrane component of the transporter. In *S. marcescens*, *P. aeruginosa*, and *P. fluorescens*, there are two genes upstream of *hasR* that are also preceeded by a Fur operator. DNA sequence homologies indicate that these genes might code for an extracytoplasmic sigma factor (HasI) and its modulator (HasS).

In this chapter, we describe hemophore secretion and its interaction with heme and hemoproteins, specific receptors, mechanisms of heme transfer from hemophore to receptor, TonB function at various steps in the heme acquisition process, and the regulation of the *has* operon.

HEMOPHORE SECRETION

HasA hemophores are secreted by ABC transporters. These ABC pathways are widespread in microbes. For example, toxins, proteases, lipases, S-layer proteins, colicins, and hemophores in gram-negative bacteria all use ABC secretion systems, as do most peptides and bacteriocins in gram-positive bacteria. ABC systems are characterized by the presence of an inner membrane ATPase named the ABC protein. ABC proteins constitute one of the largest families of proteins involved in the vectorial movement of solutes across biological membranes. The typical feature of these proteins is the presence of a large (200-amino-acid) conserved cytoplasmic domain, which contains the two conserved ATP binding motifs that form the ATP binding pocket. Another characteristic feature of this family is a common basic structure, most often consisting of two energy-coupling domains and two membrane-embedded hydrophobic domains with six transmembrane spanners each.

In gram-negative bacteria, the ABC exporters involved in protein secretion have an ABC protein in the inner membrane and two helper proteins. One of these is a second inner membrane protein belonging to the membrane fusion protein family, and the other is an outer membrane component which belongs to the TolC family, a multifunctional protein involved in drug efflux and colicin import. Very recently, the TolC protein of *Escherichia coli* was crystallized, revealing a new fold for embedded outer membrane proteins, with a very large periplasmic domain forming a tunnel between the inner and outer membranes.

Taking advantage of the heme-binding properties of HasA, it was possible, by affinity chromatography on hemin agarose, to show an association between HasA and the HasD-HasE-HasF exporter. The association of the

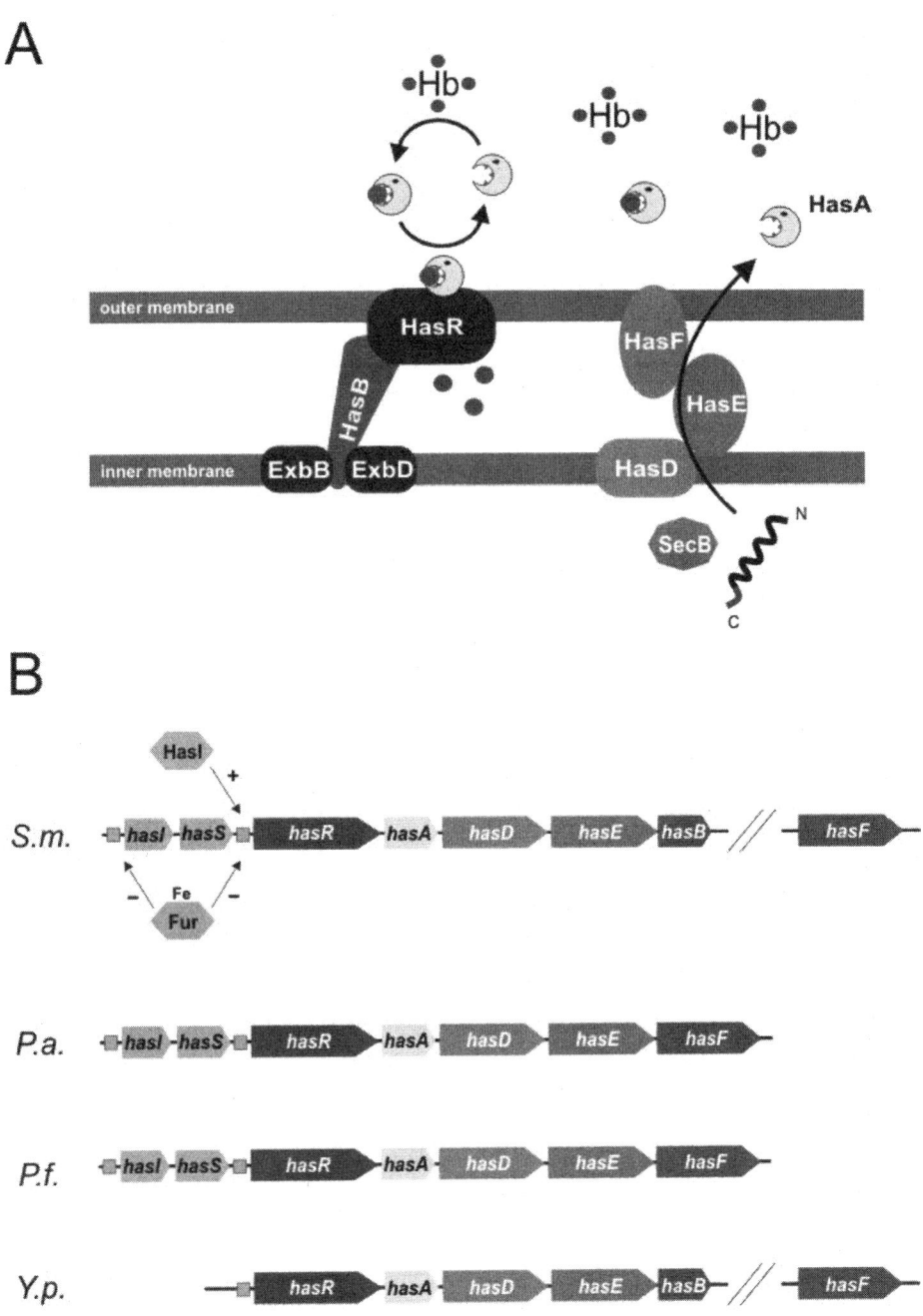

FIGURE 3 (A) HasA system of *S. marcescens*. (B) Comparison of the genetic organization of the *has* operons of *S. marcescens* (S.m.), *P. aeruginosa* (P.a.), *P. fluorescens* (P.f.), and *Y. pestis* (Y.p.). (Reprinted, with permission, from the *Annual Review of Microbiology* [volume 58, 2004] by Annual Reviews [www.annualreviews.org].)

three secretion components was ordered and induced by substrate binding to the ABC protein, HasD. By constructing hybrid exporters using components from various systems, we demonstrated that the ABC protein is responsible for the substrate specificity.

Hemophores, like most proteins secreted by an ABC exporter, have an α-helical C-terminal secretion signal. This signal is characterized by the presence of a negatively charged residue followed by several hydrophobic residues. The location of the ABC secretion signal at the C terminus of the secreted protein implies that the protein is fully synthesized before the signal is recognized. We found that HasA requires the SecB chaperone for efficient secretion. Chaperones help maintain newly synthesized proteins in a conformation competent for secretion. The nature of the chaperone needed depends on the protein secreted: *E. coli* alpha-hemolysin requires DnaJ, whereas HasA requires SecB. Recently, we showed that HasA mutant proteins carrying deletions or point mutations resulting in their being folded less rapidly than the wild type are secreted in a SecB-independent manner. Thus, the major contribution of SecB appears to be to maintain the protein in an unfolded state and not for specific targeting to its ABC transporter. SecB was characterized as the chaperone dedicated to the Sec general secretory pathway. However, it is now clear that it is involved in at least one other fundamentally distinct secretion mechanism.

In the absence of the Has secretion apparatus, the hemophore accumulates in the cytoplasm, folds, and is able to bind heme. This folded polypeptide is not competent for secretion, even if the secretion apparatus is expressed subsequently. Moreover, the folded protein (loaded with heme or unloaded) still interacts with the exporter and inhibits the secretion of newly synthesized hemophore. Only HasA expressed in the presence of a functional exporter was secreted. This points toward a paradoxical situation in which a protein with a C-terminal secretion signal is secreted cotranslationally.

HEME BINDING AND CAPTURE PROPERTIES OF HEMOPHORES

One of the four HasA hemophores identified, the *S. marcescens* HasA protein, has been biochemically characterized. It is a monomer of 19 kDa, and the crystal structure of the holo-HasA resolved at 1.9 Å revealed that it is a globular protein with a fold showing two faces: four α-helices on one face and seven β-sheets on the other. Iron in the heme-hemophore complex is in the oxidized form, with a very low redox potential (-550 mV). Nuclear magnetic resonance analysis of the apo-HasA revealed that the hemophore does not undergo major structural changes on heme binding.

The heme pocket was identified from the crystal structure and is highly exposed to the solvent. The amino acids histidine 32 (H32) and tyrosine 75 (Y75) are the axial iron ligands in this pocket, and the close proximity of histidine 83 (H83) to the heme pocket suggested that it is involved in heme binding as well (Fig. 4). Site-directed mutagenesis was used to construct proteins in which these three residues, H32, Y75, and H83, were replaced by alanine. The heme-binding constants were determined by absorption spectrometry and isothermal microcalorimetry. The wild-type protein has a K_d of 10^{-11} M for heme, and each of the single alanine mutants has a lower K_d. Thus, either of the iron axial ligands H32 or Y75 is sufficient

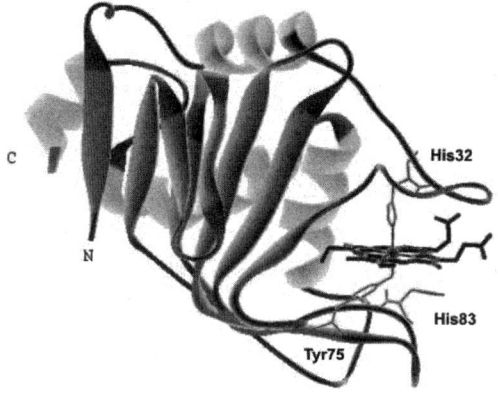

FIGURE 4 Crystal structure of *S. marcescens* hemophore (holo-HasA). Reprinted from Arnoux et al. (1999) with permission from *Nature*.

to bind heme, albeit less efficiently. H83 strengthens the bond between Y75 and iron. It may serve as an alternative iron ligand in the absence of Y75 or both H32 and Y75. The pKa of each of the various histidines of the hemophore has been measured using gallium instead of iron in the protoporphyrin IX as the bound substrate. These measurements indicate that the phenolate of Y75 is tightly hydrogen bonded to the Nδ atom of H83. It is therefore possible that local pH variations that can change the strength of the bond between Y75 and H83 might contribute to the mechanism involved in heme capture or release. For heme capture, this suggests that HasA and hemoglobin form a complex. However, analytical ultracentrifugation experiments could not detect a stable complex, which suggests that this complex would be transient.

Hemophores are also able to capture heme from hemopexin and myoglobin. This large range of substrates is consistent with passive transfer of heme from the heme carrier protein to the hemophore owing to its higher affinity.

In the case H. influenzae hemophore system, hemopexin is the only heme carrier protein of several tested that serves as a substrate for the hemophore. It is possible that this system has evolved to the point where the hemophore and its receptor recognize only hemopexin and have a specific molecular mechanism to capture heme from hemopexin.

OUTER MEMBRANE HEMOPHORE RECEPTORS

Hemophore receptors belong to the large family of TonB-dependent receptors. Their activity is dependent on a protein complex comprising TonB and the inner membrane proteins ExbB and ExbD (see chapter 7). TonB-dependent receptors have significant similarity (25% identity) despite the diversity of substrates transported (which include heme, siderophores, vitamins, and others) and the high specificity of substrate recognition. The crystal structures of several of these outer membrane receptors have been determined (see chapter 4): a series of β-sheets linked by loops form a pore closed by the N-terminal domain (the plug), also exposed to the periplasm, where it can make contact with TonB. The periplasmic C-terminal part of TonB interacts with a short conserved region of the receptor plug domain named the TonB box. Ligand binding to the cognate receptor is TonB independent but enhances the interaction between these receptors and TonB. Ligand binding sites on several outer membrane receptors have been characterized. They comprise residues mapping both to external loops and to residues in the N-terminal plug. Crystal structures of loaded and unloaded receptors reveal that on ligand binding, there are major structural changes in the plug consistent with the opening of a channel for the ligand.

The S. marcescens HasR hemophore receptor also functions as a heme receptor that allows the uptake of free and hemoglobin-bound heme, albeit less efficiently than of hemophore-bound heme. Thus, HasR has a function analogous to that of heme receptors such as the Yersinia enterocolitica HemR and the Y. pestis Hmu receptors. Two histidines in the Y. enterocolitica HemR are involved in heme uptake, one belonging to the predicted plug domain and the second belonging to the predicted β-barrel of the receptor. It has been proposed that the ligand is transferred from the external histidine to the other histidine by a mechanism that is TonB dependent. These two histidines are conserved in HasR and could be involved in heme uptake and/or heme transfer from the hemophore.

HEMOPHORE-RECEPTOR INTERACTIONS

The interaction between HasA and its receptor was first investigated by binding experiments with cells expressing HasR immobilized on nitrocellulose membranes. All the hemophore mutants affected in heme binding residues (single, double, and triple mutants) bound to the receptor with a similar affinity to the wild type, suggesting that the interaction between the hemophore and its receptor does not require an intact heme pocket region. Second, the binding of HasA to whole cells expressing HasR in liquid media was determined. Both heme-free and heme-loaded HasA bind to HasR with the same apparent K_d (5 nM), and thus the protein-protein interaction between the hemophore and its receptor does not require the presence of heme. Competition experiments between apo-hemophore (using the H32A Y75A double mutant, which does not bind heme at a concentration lower than 10 µM) and holo-hemophore showed that there are one possibly two overlapping sites on HasR for the apo and holo-hemophores.

HEME DELIVERY FROM THE HEMOPHORE TO THE RECEPTOR

Hemophores are not transported into cells; rather, the heme is stripped off at the cell surface and then taken up intact through heme receptors. The very-high-affinity bond between hemophore and heme must be broken, and this must be followed by heme transfer to the outer membrane receptor, which has a lower affinity for this ligand. It is not known how this step is achieved and whether TonB is required. Indeed, heme stripping and heme uptake processes have never been dissociated. Only the very first step of hemophore binding to the receptor has been shown to be TonB independent.

After binding to the receptor, a conformational change of the hemophore may move the heme away from the axial ligands, resulting either in iron reduction or formation of a new bond with a HasR residue. All the hemophore single and double alanine mutants that still bind heme with a significant affinity retained the ability to stimulate hemophore-dependent heme uptake in vivo. Thus, residues H32, Y75, and H83 are not individually necessary for heme transfer to the receptor.

MULTIPLE TonB PROTEINS

In *S. marcescens*, there are two TonB-like proteins, TonB and HasB, with 56 and 17.5% identity, respectively, to the *E. coli* TonB protein. A mutation in *tonB* prevents *S. marcescens* growth in iron-restricted medium but does not prevent the utilization of free heme or hemoglobin heme as iron sources. In contrast, *hasB* inactivation does not affect any iron source utilization. Inactivation of both genes prevents all TonB-dependent iron and heme utilization, indicating that this bacterium does not have additional TonB analogs that can perform these functions. The HasB protein is not more efficient than TonB for heme transport through the outer membrane in *S. marcescens*, implying that these two proteins have partially redundant functions with respect to heme utilization.

The presence of a specific TonB-like protein encoded by a dedicated operon may ensure that a sufficiently large pool of TonB is maintained for other physiologically important TonB-dependent functions. A similar redundancy of TonB proteins has been found in other species including *Vibrio cholerae* and *P. aeruginosa*, and sequence data indicate the presence of a *hasB* gene, in addition to *tonB*, in *Y. pestis*.

HasB does not complement any of the TonB functions in *E. coli*. However, variants of *hasB* that allow heme utilization by an *E. coli* heme auxotroph expressing the *has* system have been selected. One such mutant, named HasB6, has an insertion of 2 amino acids downstream from residue 24. This insertion is in the HasB transmembrane anchor and thus might affect the interaction with ExbB (see chapter 7). The HasB6 protein does not complement other *E. coli* TonB functions. Thus, the mutation does not change the specificity of HasB but makes it functional in *E. coli*.

has OPERON REGULATION

The *has* operon of *S. marcescens* is negatively regulated by the Fur repressor in the presence of iron and contains a Fur operator. The DNA sequence indicates that the two open reading frames upstream of *hasR* are also preceeded by a Fur operator. These two open reading frames are found upstream of the *has* operon in *P. aeruginosa* and *P. fluorescens* as well. DNA sequence homologies indicate that these two genes might code for an extracytoplasmic sigma factor (HasI) and its modulator (HasS).

This system appears to be very similar to the *E. coli* Fec system for uptake of ferric dicitrate (see chapter 11), which responds to the external ferric dicitrate concentration through a specific sigma factor (FecI) and a membrane-anchored sigma modulator (FecR) that sequesters the sigma factor. On binding of the ferric dicitrate to the outer membrane receptor FecA, there is a TonB-driven conformational change in the N-terminal extension preceding the plug domain in FecA. This change modifies the modulator and releases the sigma factor from the membrane, allowing *fec* operon induction. A similar N-terminal extension is found in the hemophore outer membrane receptor and in several outer membrane receptors subject to signal transduction.

Using a chromosomal *hasR-lacZ* transcriptional fusion and nonpolar mutants with mutations in the *S. marcescens hasI, hasS, hasR, hasA*, and *hasB* genes, we showed that HasI is necessary for *hasR* transcription and that the *hasS* mutation leads to constitutive expression. Overproduction of HasS from a plasmid resulted in complete repression of the *hasR-lacZ* gene fusion. With the *hasA* mutant, we showed that the *hasR-lacZ* fusion is induced by holo-HasA added exogenously to concentrations above 10 nM but not by apo-HasA or heme at any concentration. Holo-HasA protein induces the fusion in iron-depleted medium but not in iron-rich medium, showing that iron repression is epistatic on holo-HasA induction. Holo-HasA induces the fusion in wild-type strains and *hasA* and *hasB* mutants, but not in *hasR* and *hasI* mutants.

These results are consistent with a cascade of induction that requires HasI as a positive regulator, HasS as an anti-sigma factor, and HasR as the surface sensor which transmits the signal. The inducer of the *has* operon (holo-HasA) and the transported substrate (heme) in this system are different molecules. Like pyoverdin-mediated signaling, the hemophore signaling pathway involves an endogenously produced molecule which, after secretion, activates its own synthesis. This suggests that hemophores and siderophores may constitute a previously unrecognized class of signaling molecules. Since the inducer is a molecule secreted by the bacterium, the hemophore can be considered to be a new type of quorum-sensing molecule.

CONCLUSIONS AND QUESTIONS FOR THE FUTURE

Several aspects of the hemophore pathway for heme acquisition are still unclear. Which part of the hemophore is involved in the receptor recognition, and, reciprocally, which part of the receptor is required for this interaction? For example, it is not clear whether residues of the plug domain are involved. If they were, it would suggest that the hemophore is able to "enter" the receptor or, alternatively, that the plug can reach the membrane surface. We also do not know whether the HasR N-terminal extension predicted to be involved in regulation also plays a role in heme extraction from the hemophore in a TonB-dependent fashion. Finally, the molecular basis of the differences between HasB and TonB functions are not known.

In a wider perspective, it would be interesting to investigate whether hemophore-like proteins exist in gram-positive bacteria and whether hemophores are species specific or are used by nonproducing bacteria, similarly to certain siderophores. Finally, many bacterial species (including *S. marcescens, P. aeruginosa*, and *Y. pestis*) have multiple heme acquisition systems, with only one of these being hemophore dependent. It is not clear how these various systems are coordinated. Are the systems used for direct heme uptake induced at higher heme source concentrations? Or is heme sequestration a major hemophore function to avoid heme overload? Heme permeases have been found in several cases to be encoded by genes clustered close to the gene encoding the hemophore-independent outer membrane receptor. It is not known whether these heme permeases are common to the various outer membrane heme transporters or whether there is also a permease redundancy.

Hemophores enhance the efficiency of heme delivery to bacterial cells and enlarge the heme source range. Like siderophores, they are produced in the apo form and acquire their ligand in the extracellular medium. As for several siderophores, it has been demonstrated that unloaded and loaded hemophores bind to their cognate receptor with the same efficiency. This raises the puzzling question of how the change from empty to loaded siderophore or hemophore is achieved. Since siderophores are produced in large amounts when iron is absent, a high ratio of unloaded to loaded siderophores is expected, and this would inhibit uptake. Stintzi et al. proposed that there is exchange of iron without swapping of siderophores on the receptors. Other groups found that release of the empty siderophore is energy dependent, suggesting that there may be at least two different mechanisms for uptake of iron from siderophores.

In the hemophore, the *has* operon is positively regulated by holo-HasA through a sigma/anti-sigma cascade. Thus, hemophore synthesis and secretion occur only when heme is present, maintaining a high ratio of loaded to unloaded hemophores. The low HasA concentration in iron-chelated medium in the absence of heme is such that it is substantially below the K_d of HasA for HasR, thereby leaving the receptors unoccupied. In such a situation, the hemophore functions as a heme sniffer. As soon as heme is detected, the system is induced to use this newly available iron source.

ACKNOWLEDGMENT

We gratefully acknowledge P. Delepelaire for helpful discussions.

SUGGESTED READING

Arnoux, P., R. Haser, N. Izadi, A. Lecroisey, M. Delepierre, C. Wandersman, and M. Czjzek. 1999. The crystal structure of HasA, a hemophore secreted by *Serratia marcescens*. *Nat. Struct. Biol.* **6:**516–520.

Bracken, C. S., M. T. Baer, A. Abdur-Rashid, W. Helms, and I. Stojiljkovic. 1999. Use of hemeprotein complexes by the *Yersinia enterocolitica* HemR receptor: histidine residues are essential for receptor function. *J. Bacteriol.* **181:**6063–6072.

Braun, V., and M. Braun. 2002. Active transport of iron and siderophore antibiotics. *Curr. Opin. Microbiol.* **5:**194–201.

Cadieux, N., C. Bradbeer and R. J. Kadner. 2000. Sequence changes in the ton box region of BtuB affect its transport activities and interaction with TonB protein. *J. Bacteriol.* **182:**5954–5961.

Cope, L. D., R. Yogev, U. Muller-Eberhard, and E. J. Hansen. 1995. A gene cluster involved in the utilisation of both free heme and heme:hemopexine by *Haemophilus influenzae* type B. *J. Bacteriol.* **177:**2644–2653.

Debarbieux, L., and C. Wandersman. 2001. Folded HasA inhibits its own secretion through its ABC exporter. *EMBO J.* **20:**4657–4663.

Delepelaire, P., and C. Wandersman. 1998. The SecB chaperone is involved in the secretion of the *Serratia marcescens* HasA protein through an ABC transporter. *EMBO J.* **17:**936–944.

Enz, S., S. Mahren, U. H. Stroeher, and V. Braun. 2000. Surface signaling in ferric citrate transport gene induction: interaction of the FecA, FecR, and FecI regulatory proteins. *J. Bacteriol.* **182:**637–646.

Ferguson, A. D., R. Chakraborty, B. S. Smith, L. Esser, D. van der Helm, and J. Deisenhofer. 2002. Structural basis of gating by the outer membrane transporter FecA. *Science* **295:**1715–1719.

Ghigo, J. M., S. Létoffé, and C. Wandersman. 1997. A new type of hemophore-dependent heme acquisition system of *Serratia marcescens* reconstituted in *Escherichia coli*. *J. Bacteriol.* **179:**3572–3579.

Higgins, C. F. 1992. ABC transporters—from microorganisms to man. *Annu. Rev. Cell Biol.* **8:**67–113.

Izadi, N., Y. Henri, J. Haladjan, M. E. Goldberg, C. Wandersman, M. Delepierre, and A. Lecroisey. 1997. Purification and characterization of an extra cellular heme binding protein, HasA, involved in heme iron acquisition. *Biochemistry* **36:**7050–7057.

Lamont, I. L., P. A. Beare, U. Ochsner, A. I. Vasil, and M. L. Vasil. 2002. Siderophore-mediated signaling regulates virulence factor production in *Pseudomonas aeruginosa*. *Proc. Natl. Acad. Sci. USA* **99:**7072–7077.

Létoffé, S., C. Deniau, N. Wolff, E. Dassa, P. Delepelaire, A. Lecroisey, and C. Wandersman. 2001. Haemophore-mediated bacterial haem transport: evidence for a common or overlapping site for haem-free and haem-loaded haemophore on its specific outer membrane receptor. *Mol. Microbiol.* **41:**439–450.

Létoffé, S., F. Nato, M. E. Goldberg, and C. Wandersman. 1999. Interactions of HasA, a bacterial haemophore, with haemoglobin and with its outer membrane receptor HasR. *Mol. Microbiol.* **33:** 546–555.

Létoffé, S., V. Redeker, and C. Wandersman. 1998. Isolation and characterisation of an extracellular heme binding protein from *Pseudomonas aeruginosa* that shares function and sequence similarities with the *Seratia marcescens* HasA hemophore. *Mol. Microbiol.* **28:**1223–1234.

Paquelin, A., J. M., Ghigo, S. Bertin, and C. Wandersman. 2001. Characterization of HasB, a *Serratia marcescens* TonB-like protein specifically involved in the haemophore-dependent haem acquisition system. *Mol. Microbiol.* **42:**995–1005.

Rossi, M. S., A. paquelin, J. M. Ghigo, and C. Wandersman. 2003. Haemophore-mediated signal transduction across the bacterial cell envelope in *Serratia marcescens*: the inducer and the transported substrate are different molecules. *Mol. Microbiol.* **48:**1467–1480.

Sapriel, G., C. Wandersman, and P. Delepelaire. 2003. The SecB chaperone is bifunctional in *Serratia marcescens*: SecB is involved in the Sec pathway and required for HasA secretion by the ABC transporter. *J. Bacteriol.* **185:**80–88.

Seliger, S., A. Mey, A. Valle, and S. Payne. 2001. The two TonB systems of *Vibrio cholerae*: redundant and specific functions. *Mol. Microbiol.* **39:**801–812.

Stintzi, A., C. Barnes, J. Xu, and K. N. Raymond. 2000. Microbial iron transport via a siderophore shuttle: a membrane ion transport paradigm. *Proc. Natl. Acad. Sci. USA* **97:**10691–10696.

Vanderpool, C. K., and S. K. Armstrong. 2001. The *Bordetella bhu* locus is required for heme iron utilization. *J. Bacteriol.* **183:**4278–4287.

Wandersman, C., and P. Delepelaire, 1990. TolC, an *Escherichia coli* outer membrane protein required for hemolysin secretion. *Proc. Natl. Acad. Sci. USA* **87:**4776–4780.

Wandersman, C., and I. Stojiljkovic. 2000. bacterial heme sources: role of hemophores and receptors. *Curr. Opin. Microbiol.* **3:**215–220.

Wolff, N., C. Deniau, S. Letoffe, C. Simenel, V. Kumar, I. Stojiljkovic, C. Wandersman, M. Delepierre, and A. Lecroisey. 2002. Histidine pK(a) shifts and changes of tautomeric states induced by the binding of gallium-protoporphyrin IX in the hemophore HasA(SM). *Protein Sci.* **11:**757–765.

Zhao, Q., and K. Poole. 2000. A second *tonB* gene in *Pseudomonas aeruginosa* is linked to the *exbB* and *exbD* genes. *FEMS Microbiol. Lett.* **184:**127–132.

IRON TRANSPORT PROTEINS: STRUCTURAL STUDIES

II

STRUCTURE OF OUTER MEMBRANE RECEPTOR PROTEINS

Dick van der Helm

4

The outer membrane receptor proteins for ferric siderophores are physically similar to porins, being less hydrophobic than most other membrane proteins. This indicated that their basic structure would most probably consist of β-strands rather than helices, but confirmation of the structure awaited the results of crystal X-ray diffraction. Before 1980, it was commonly predicted that membrane proteins could not be crystallized, because detergents are required for their solubilization. This attitude changed once the first structure of a membrane protein, the photosynthetic reaction center, was solved in 1985. Even before that time, during a visit to the Biozentrum in Basel, Michael Garavito demonstrated to me the systematic methods to crystallize porins and thus gave rise to the idea that this should also be possible for the outer membrane receptor proteins of ferric siderophores. The first crystallization of one of the proteins in the ferrisiderophore receptor family, FepA, was subsequently reported in 1989. Another advance was the determination of the sequences for the proteins in this family. This was important for two reasons. First, it made plasmids available for overproduction, and second, structural determination by single-crystal X-ray diffraction is aided by knowing the primary structure of the protein.

The crystal structures of three of the ferrisiderophore receptor proteins have been determined so far, and the following sections describe the comparison of these structures and their analysis. This comparison allows the identification of various features that probably are important for the understanding of the function of the proteins and may form the basis for new experiments.

OVERVIEW OF THE STRUCTURES

The first structures of the ferric siderophore outer membrane receptor proteins (OMRP) FhuA and FepA were reported in 1998 and 1999; respectively. More recently, the structure of FecA was solved. All three proteins (Color Plate 5) are found in the outer membrane of *Escherichia coli*.

The structures each reveal two domains, a 22-strand β-barrel with short periplasmic loops and much longer extracellular loops, and a N-terminal globular domain nestled inside the barrel, which is covalently bonded to the first strand of the barrel. The presence of the globular or plug domain was an unexpected finding

---

Dick van der Helm, Department of Biochemistry and Microbiology, University of Victoria, Victoria, Canada, and Department of Chemistry and Biochemistry, Oklahoma University, Norman, OK 73069.

Present address: 5895 Bay Pine Court, Ferndale, WA 98248.

that was not predicted prior to the structure determinations. The plug domain has a novel fold and forms a cavity at the periplasmic side. The barrel domain, whose apices are involved in ligand binding, extends above the outer leaflet of the membrane surface by 13 Å (FhuA) to more than 20 Å (FepA). The β-barrels have 10 periplasmic loops (P1 to P10) and 11 extracellular loops (L1 to L11). The strands are at an angle of about 45° with respect to the surface. The shear number of the β-strands in the barrels is 24, which makes this angle, more accurately, 43°. Ten or eleven residues of each strand are embedded in the membrane, and this portion of the β-strands localized in the membrane is characterized by bands of aromatic residues on the exterior of the barrel. The residues of the barrel, pointing into the membrane, are not strictly hydrophobic. Between 20 and 25% are polar, and some are even charged residues, with half close to the surfaces of the membrane and the other half in the middle of the barrel. The cross section of the barrel is elliptical rather than circular, with the longest and shortest axes being 46 and 35 Å in FecA and FhuA and 44 and 37 Å in FepA. In all three proteins, the longest axis runs from the bottom of strand 8 to the bottom of strand 19. However, a least-squares fit of 220 equivalent residues from the barrels in the three structures is far from perfect, with a root mean square deviation of 1.4 Å and maximum differences of 5.5 Å. Many, but not all, of the β-strands extend up to 12 to 14 residues above the membrane surface. For loops L7 to L11 in FecA and FhuA, this gives the appearance of a wall; however, this feature is not apparent in FepA. For convenience, the term "extracellular loop" is assigned to the string of amino acids once it emerges from the membrane surface until it enters the membrane to form the adjacent strand and thus includes the β-strand residues extending above the membrane surface. There are about 40 to 50 fewer residues in the extracellular loops of FecA than in those of FhuA and FepA. This is apparent in the comparison of the three structures in Color Plate 5, where FecA appears shorter and stubbier. The extracellular loops of the three proteins differ greatly in length, and show, with one exception (see "Possible second site of interaction with TonB," below), no sequence similarities in a simultaneous comparison of 20 proteins (see "Simultaneous sequence alignment," below). The extracellular loops are hydrophilic but also contain a fairly large number of Tyr and Phe residues, primarily in FhuA and FepA, where these residues line the top of the loops corresponding to the entrance site of the ferric siderophore. This feature is not nearly as obvious in FecA; this may well be related to the fact that ligands for FhuA and FepA are amphipathic while the ligand for FecA is decidedly hydrophilic. The external loops slope inward, although they leave an opening for the entrance of the ligand. The external loops and the apices of the plug domain form another large cavity, providing adequate space for the binding of the ligand.

The availability of three different structures makes it possible to ascertain the common features of the proteins; these features are discussed in separate sections, at times together with the information available from simultaneous sequence comparisons of many proteins in this family.

BIPARTITE GATING

The structures of both FhuA and FecA were determined with and without their bound ligands, ferrichrome and diferric dicitrate, but this was not possible for FepA. Although the FepA crystals were soaked for a short period with ferric enterobactin before data collection, the structure that has been reported should be considered to describe the ligand-free protein. Electron density was observed close to the top of the extracellular loops in FepA, estimated to have an iron occupancy of 20%, but with no discernible density for the enterobactin ligand. The location of this partial density is far above (15 Å) the putative binding site for the ligand and may indicate an initial binding site. Soaking the crystals for a longer period never improved the electron density nor resulted in any indication of ligand at the binding site in contact with apices A, B, and C of the globular domain. The final result of soaking was always cracking of the crystals. This may be an indication that binding causes a conformational change in the extracellular loops, eliminating intramolecular packing interactions and thus destroying the crystals.

There is no doubt about the localization of the ferric siderophore from the electron density in the liganded FhuA structures. The ligand rests on apices A, B, and C of the globular domain and also interacts with several extracellular loops. Unexpectedly, however, no change in the conformation of the extracellular loops was observed on ligand binding, although many biochemical experiments with both FhuA and FepA indicate that conformational changes do occur in the extracellular loops on ligand binding. The lack of closure of the extracellular loops over the binding site, observed in the FhuA-ligand structure, poses a logical conundrum. For further transport from the binding site to the periplasm, the formation of a transient channel is required. This can be achieved by removal of the globular domain from the barrel or by conformational changes within the globular domain (see "Formation of a transient channel," below). However, once the channel is formed, if the extracellular loops are still open, as observed in the crystal structure, many foreign molecules could diffuse to the periplasm, along with the ferric siderophore; all selectivity would be lost. Failure to observe the closure may be an artifact caused by the crystal packing or may indicate that relatively little free energy (ΔG) is involved in the binding process, due to the large decrease in entropy (the equilibrium constant is determined by $\Delta G°$ rather than ΔG). The failure to observe the closure in FepA and FhuA should be addressed by properly designed biochemical experiments to observe the closure of the extracellular loops after binding of the substrate.

Possibly fortuitously, in FecA, the structures of the free protein and the protein complexed with the ligand show very obvious and large changes in the extracellular loops. In fact, L7 and L8 move by 11 and 15 Å, respectively, while other loops change by smaller amounts. The loops close over the ligand, forming one gate. The second gate is formed by the globular domain. While the first gate is closed, a transient channel is formed in the globular domain, represented by the opening of the second gate. Transport is therefore specific because only the bound ligand is now in a position to diffuse from the binding site to the periplasm. This system is properly described as bipartite gating, i.e., the use of two independent gates for specific transport. It is not likely that closure of the loops forces the transport of the ligand. A more likely effect of the ligand binding is the transduction of a signal (see "The switch helix and signaling," below).

TOPOLOGY AND STRUCTURE OF THE GLOBULAR DOMAINS

The structure of the N-terminal globular domain in FecA is shown in Color Plate 6. It exhibits a central β-sheet with four strands. The first residue of the globular domain that could be located from the electron density maps is Ala81, and the last residue of the globular domain is Phe223; the β-barrel domain starts at residue 224. Ala81 is part of the TonB box (residues 80 to 84), a conserved region characteristic of this group of outer membrane proteins that depend on TonB for function (see chapter 7). As with FepA and FhuA, the TonB box of FecA is on the periplasmic side of the domain. This region is followed by the switch helix (residues 94 to 99). The switch helix can partially unwind in response to substrate binding and is described in detail below (see "The switch helix and signaling," below). The three apices of the globular domain, indicated by residues 138, 155, and 176, are clearly observed well above the outer membrane surface (Color Plates 5 and 7). The structures of the globular domains in FecA, FhuA, and FepA are similar but not identical. The topologies, however, are identical in the three structures (Table 1). For instance, the apices occur in the same relationship with respect to the secondary-structure components in the three structures. The same parallel and antiparallel mixed-sheet structure occurs in all three proteins. It can be described, from the bottom in Color Plate 6, by the fol-

TABLE 1 Topologies of the globular domain in FecA, FhuA and FepA

Structure	Location in[a]:		
	FecA	FhuA	FepA
TonB box	80–84*	7–13**	12–18
Switch helix	94–99	24–29	18–23
β1 strand	103–109	49–55	27–33
α2 helix	110–113***	56–60	34–38
α3 helix	121–125	66–70***	46–50
β2 strand	129–131	75–77	54–58
Apex A	138	81	63
β3 strand	148–150	91–93	71–75
Apex B	155	100	78
β4 strand	162–165	104–107	83–87
β4 (cont.)	167–169	109–111	89–91
Apex C	176	116	101
β5 strand	188–196	125–134	118–127
β5-β6 loop	197–209	135–145	128–138
β6 strand	210–217	146–154	139–147
β-barrel	224	161	154

[a] The numbers are residue numbers in the three structures. Symbols: *, partially located; **, not located; ***, partially helical.

lowing notation: strand 1 −, strand 5 +, strand 6 −, strand 4 −. Other topological features that will prove to be important are the two small β-strands preceding and following apex A and forming a small antiparallel sheet (left side of Color Plate 6). This sheet contains residue R150, which is conserved in all the characterized proteins in this family. Finally the 10- to 12-residue coil, the β5-β6 loop, is maintained in the topology of the structures (front of Color Plate 6, Arg196-Gly210). It may play an important role in the formation of a transient channel for transport of the substrate from the binding site to the periplasm (see "Formation of a transient channel," below).

While the topologies of the globular domains are identical in FhuA, FepA, and FecA, there are differences in the structures. There are 20 residues between the switch helix and the first β-strand in FhuA, but there are only 4 residues in FecA and FepA. There are some 10 residues between the TonB box and the switch helix in FecA and FhuA, but in FepA the last residue in the TonB box is the first one for the switch helix. Also, the coil lengths between the β-strands and the apices are, more often than not, different. In other words, the structures of the globular domains are not superimposable.

To make objective comparisons, one needs to use a least-squares procedure to fit the structures. This has been done here in two different ways. One way was to select the 200 to 220 equivalent residues of the β-barrel embedded in the membrane and to use the least-squares fit of those residues as reference for comparisons. Another way was to use the same β-barrel residues along with the plug domain for the least-squares calculation. These calculations were performed with the SPDBV program, which is readily available without cost. The two calculations show small differences, of the order of 0.1 Å or less. These differences are not significant, considering the accuracies of the structural determinations of the proteins.

Both methods were used to compare the mixed β-sheets in the globular domains of the three structures. Equivalent residues, four in each strand, were compared in FepA, FecA, and FhuA. For the C-α atoms in these residues, most differences in location were less than 1 Å and the maximum difference was 2.0 Å. This is a remarkable result. Not only is the same mixed β-sheet present in the three structures and present with the same topology, it also is found at the same location and in the same orientation with respect to the β-barrel. The fact that the barrel is not circularly symmetrical but elliptical has already been pointed out. It is obvious from this discussion that the central mixed β-sheet must recognize the orientation of the elliptical barrel.

BINDING OF FERRIC SIDEROPHORES

The binding site of the ferric siderophore is located in the cavity formed by the top of the globular domain and the extracellular loops. The interactions with the ligand involve residues on several extracellular loops and residues of apices A, B, and C of the globular domain. The specific interactions of the ligands with the proteins were described for both FhuA and

FecA in the original publications, and the details are not repeated here. In general, there are more interactions with the extracellular loops than with the apices. Additionally, in both structures the interactions with the apices involve one purely electrostatic interaction, rather than a specific H bond. This site may initiate the signaling (see the next section). In FhuA, but not FecA, several aromatic residues line the binding pocket. This can be understood by remembering that the FhuA ligand, ferrichrome, is distinctly amphipathic while the FecA ligand is hydrophilic.

For FhuA, structures were determined for the protein bound to the siderophores ferricrocin, ferrichrome, and phenylferricrocin and bound to the structurally related antibiotic albomycin. Analysis and comparison of the structures of all these complexes indicate that binding of the siderophores to FhuA is the same in each case. Binding involves both the hydroxamate groups chelating the ferric ion and some parts of the cyclic hexapeptide found in this type of siderophore. Albomycin is built from a linear tripeptide, tri-N-hydroxy-N-acetyl-L-ornithine, the same grouping that occurs in the cyclic peptides. Attached covalently by an aminoacetyl linker is the antibiotic moiety thioribosylpyrimidine. The coordination geometry for the iron in albomycin, however, is the same as in the ferrichromes. Albomycin binds FhuA in the same orientation and manner as the other ligands. The antibiotic moiety in the crystal structure is observed in two different conformations due to the flexible linker group. The structural results allow insight into the design of other antibiotics that would be able to use FhuA for transport across the outer membrane

Apices A, B, and C of the globular domain (Table 1; Color Plate 7) are part of the binding site of the ferric siderophore. Topologically they are in the same position in the three structures, but after a least-squares fit, they do not superimpose. This is in contrast to the good fit described above for the mixed β-sheets. The apices are at different heights with respect to the β-barrel and also at different locations (Color Plate 7; apices A, B, and C are shown in blue for FepA, in green for FhuA, and in red for FecA). Apices A and C in FepA are by far the highest, rising approximately 20 Å above the surface of the outer leaflet of the membrane, whereas apex B of FepA is the lowest (5 Å). For FecA, apex A is the highest (14 Å), and for FhuA, apex C is the highest (11 Å). The relative locations are also different with respect to each other. Apices A are at distances of 9 to 12 Å from one another, and apices C are at distances of 8 to 13 Å. Apices B are somewhat closer and are found at distances of 5 to 7 Å from one another. The differences in location of the apices in the three structures are, in fact, understandable. The positions are adapted to the different molecular structure and chemical character of the particular ferric siderophore that fits into the binding site of the three proteins.

Recently, hybrid proteins of FepA and FhuA were constructed in which the globular domain of FepA was replaced by the one of FhuA and vice versa. The results show that the barrels in the hybrid proteins determine the ligand that is bound; for instance, the FepA (barrel)-FhuA (plug) construct binds ferric enterobactin but not ferrichrome. This shows that the binding sites on the extracellular loops are more important than those at the apices of the plug domain for ligand specificity. The K_d values for substrate binding by the hybrids were found to be normal. This remarkable finding may also be explained by the relatively greater contribution of the extracellular loops than the apices to ligand binding. Differences were observed in the hybrids with respect to the wild-type receptors; these may be attributable to interaction of the ligand with the apices. The binding capacity of the hybrids was only 3 to 7% of that of the wild type. The very low binding capacity may result from misplacement of the apices; in the FepA-FhuA hybrids, one expects the apices, whose relative locations in the two proteins are different, to be misplaced by 7 to 14 Å in height and location. Furthermore, the transport properties of the hybrid proteins are adversely affected. This also could be the result of the misplacement of the apices, which would, in turn, affect the signaling from the apices to the TonB box (see the next section) and thus affect the subsequent transport.

THE SWITCH HELIX AND SIGNALING

It is likely that two different and distinguishable conformational changes occur in the plug domain. The first is the one caused by the binding of the substrate, resulting in a conformational change in the domain, signaling the switch helix to unwind or partially unwind, and allowing the interaction between TonB (see chapter 7) and the TonB box of the outer membrane receptor protein to occur. The second conformational change in the domain would occur after the TonB protein had sensed the presence of the ligand by the change in the switch helix. Using the proton motive force, TonB would cause the second conformational or structural change in the outer membrane receptor protein to allow the ligand to be transported through to the periplasm (see "Formation of a transient channel," below). The first conformational change is discussed here; the TonB-dependent change is discussed in a later section.

Both independent structural determinations of FhuA showed that the switch helix, which follows the TonB box topologically, was unwound in the structures in which the protein was bound to the ligand. Unwinding causes drastic changes in the location of the residues forming the helix, as well as in those preceding the helix, and, by inference, the location of the TonB box. For example, Glu19 moves 17 Å, which is a considerable distance when one realizes that the barrel has a diameter of only about 40 Å. In the FhuA structures, the switch helix is formed by residues 24 to 29 (identified as residues 22 to 25 in one structure, but it appears that Ala25 and Ala26 were not properly located, probably due to the great thermal motion in that part of the structure). In FepA the switch helix is formed by residues 18 to 23, and in FecA it is formed by residues 94 to 99. Structures have also been determined for FecA with and without ligand. The switch helix in FecA is easily recognized, but in the liganded structure the unwinding of the helix is only partial and involves only the first two residues of the helix. The first residue that can be located in both structures, Asn95, moves 1.1 Å, a significant amount but still much less than in FhuA, where the equivalent residue, the second one in the helix, moves 6 Å. If one compares this last value with the one for Glu19 in FhuA, it is obvious that any unwinding of the helix is amplified in the preceding residues where the TonB box is located. This amplification is easily visualized in Color Plate 6 when one considers the relative locations of the switch helix and residue 81, which is part of the TonB box in FecA. Even partial unwinding in FecA may have a large effect on the location of the TonB box.

In an earlier section it was pointed out that the mixed β-sheet in the globular domain is observed to be in the same location and in the same orientation with respect to the β-barrel in all three structures. This, however, is not true for the location of the switch helices in the three structures. The differences in the location, orientation, and level of the switch helices within the β-barrels are large (Color Plate 7). When the helices are compared in pairs, the middles of the helices are 5, 14, and 15 Å apart, with the shortest distance being the one between FecA and FepA. The switch helices are in contact with the inner wall of the barrel. In FecA these contacts are centered on strands 12 and 13 of the β-barrel, in FepA they are centered on strand 11, and in FhuA they are centered on strand 15. Even though the switch helices are not positioned identically before unwinding, it is likely that on ligand binding, unwinding brings the TonB box of each protein to the same location with respect to the barrel in order for the TonB protein to be able to productively interact. Depending on the protein, the unwinding or partial unwinding of the switch helix could meet this requirement for the interaction.

The signal for the unwinding is a conformational change at one of the apices as a result of ligand binding, and this signal is transduced to the switch helix on the periplasmic side of the globular domain. In FhuA, Arg81 of apex A, involved in an electrostatic interaction with the ligand, moves up by a rather large amount, 2.5 Å, after binding the ferrichrome. This has an apparent consequence for all the residues between apex A and Thr54. This part of the structure moves up by about 1 Å, while no other changes of this magnitude are observed in the remainder of the plug domain. Thr54 is part of the first strand (β1) of the mixed β-sheet and is at a van der Waals' distance from Ala30, the first residue after the switch helix (Table 1). The upward motion thus destabilizes the helix; in fact, Ala30 follows the upward motion of Thr54, probably causing the unwinding.

In FecA, the structural changes that are observed after ligand binding are quite different from the ones observed in FhuA. There are no changes in apex A or in the residues preceding apex A, as occur in FhuA. Instead, apex B, also involved in an electrostatic interaction with the ligand, moves down by a small amount (0.5 Å). This motion is a concerted one for the residues following apex B (Table 1). Residues 159 and 160 are in contact with residues 206 and 209 of the β5-β6 loop. This loop (residues 197 to 209) shows a concerted downward motion of 0.7 to 1.1 Å. Residue 198 of the loop is in contact with residues 102 and 103, which follow the switch helix, moving these residues down. This contact may therefore be the cause of the partial unwinding of the switch helix in FecA and hence may be the structural component of signal transduction between the binding site and the switch helix. These observations cannot be considered to constitute a general proof of the signaling effect; confirmation would require the results for another pair of structures or directed biochemical experiments.

Another relationship between the binding site and the switch helix can be found in the structure of FhuA complexed with a derivative of rifamycin. This antibiotic is known to use FhuA for entry into the cell, but it is structurally unrelated to ferrichrome. In the complex with FhuA, the rifamycin derivative occupies the binding site for ferrichrome but does not interact at all with apex A, which is the initiator of the signal to the switch helix in the FhuA-ferrichrome structure. As a matter of fact, the switch helix in this complex is not unwound. The thermal parameters of the residues in the switch helix are high, however, indicating partial disorder and leading to the tentative conclusion that the switch helix is unwound only part of the time or only in a fraction of the molecules.

SIMULTANEOUS SEQUENCE ALIGNMENT

In 1997, before the crystal structures of any of the proteins were known, an attempt was made to obtain information from the known sequences of members of this family of proteins. A group of 20 proteins that transport a wide variety of ferric siderophores was chosen. It seemed most advantageous to discover general properties by comparing the sequences simultaneously rather than in small groups. Fortuitously, large gaps were allowed in the comparison. It was immediately clear that sequences of proteins transporting the same or related ferric siderophores were most similar. For instance, this was the case for FepA, BfeA, and PfeA, all transporting ferric enterobactin, or for proteins transporting pseudobactins (pyoverdines). This, in turn, suggested the idea that the homologies that were found among all 20 proteins reflect a property that is common to all the proteins and independent of the particular siderophore that the protein binds. It was concluded that all the proteins might have a common transport mechanism and that the remaining homologies, after simultaneous comparison of all 20 proteins, represented residues involved in this transport mechanism. Some 18 to 20 residues, identical in all proteins or strictly conserved (not including hydrophobic ones), were identified.

Once the first structure was solved, the possible function of many of the conserved residues could be recognized. A number of homologous residues were observed to interact with one another, and, when both the FepA and FhuA coordinates were used, they were all found to be structurally conserved and superimposable in the structures. The structural similarities that are found verified the value of the simultaneous comparison of sequences. No homologies were observed in the apices or the residues preceding and following the apices or at the binding sites on the extracellular loops. This was not unexpected, because the 20 proteins used for the alignment bind a diverse set of ferric siderophores. In fact, no homologies were found on any of the extracellular loops, with the exception of a portion of L11 described in more detail below. All homologies that were originally identified are positioned well below the apices and occur in both the globular domain and the β-barrel, embedded in the membrane.

It therefore seems a valid hypothesis that all 20 conserved residues are involved in various stages of the transport process and that these functions can be separated from the binding event. In the following sections, they are discussed separately in the context of the transport functions of the outer membrane receptor proteins. One set of residues form the lock region, which positions the globular domain properly with respect to the β-barrel. A second set of homologous residues is most probably involved in conformational changes within the globular domain, allowing the formation of a transient channel through which the substrate can diffuse to the periplasm. A third set appears to be a likely second site on the outer membrane receptor proteins that interacts with TonB.

Location of the Globular Domain and the Lock Region

It has been pointed out that the mixed β-sheet of the globular domains in FecA, FepA, and FhuA, involving strands 1, 4, 5, and 6 (Table 1) superimpose very well, while the switch helices and apices A, B, and C do not superimpose. One can therefore ask which parts of the globular domains superimpose and which do not. Least-squares fits using 220 equivalent residues in the barrels and the approximate 140 residues of the globular domains are used for this purpose, and the comparisons are done in pairs. Between 74 and 89 residues of the 140 superpose within 2.5 Å. The strands of the mixed β-sheet and the two remaining β-strands (strands β2 and β3, before and after apex A) fit very well. Approximate fits are obtained for the string of 20 amino acids between strands β1 and β2, for the loop between strands β5 and β6, and for the amino acids joining the end of strand β6 with the first residue of the barrel. The 4 to 6 residues before and after each apex and the switch helices do not fit.

One basic question is to ask what keeps the globular domain in its place and so well aligned with the β-barrel. The answer involves four strictly conserved residues, two in the globular domain and two located on the barrel embedded in the membrane. In the globular domain, the residues are Arg75 in FepA (Arg93 in FhuA and Arg150 in FecA) at the end of strand β3, which is identical in all 20 proteins in the simultaneous sequence alignment, and Arg126 in FepA (Arg133 in FhuA and Arg196 in FecA) at the end of strand β5, which is Lys in some proteins. The conserved residues on the barrel are Glu511 in FepA (Glu522 in FhuA and Glu541 in FecA) on strand β14, which is identical in all 20 proteins in the sequence alignment, and Glu567 in FepA (Glu571 in FhuA and Glu587 in FecA) on strand β16, which is either Glu or Asp in all proteins. When these four residues in the three crystal structures are compared pairwise, their locations only differ by an average of 1 Å. Obviously they are conserved not only in the sequence alignment but also in the structures of the proteins. These four residues form a quadrupole with a distinct hydrogen-bonding pattern that is maintained in the three crystal structures. They form two separate points of attachment between the globular domain and the barrel, as shown in Color Plate 8. The third point of attachment between the two domains is the covalent bond between the last residue of the plug domain and the first strand of the barrel (Color Plate 8). It is these three points of attachment that place the globular domain in identical positions with respect to the elliptical barrel in the three structures (Color Plate 8). It is interesting that when heme transporters are included in the simultaneous sequence alignment, the same four residues are conserved, indicating a similar positioning of the equivalent domain in these proteins and possibly a similar mechanism for these TonB-dependent transporters.

The structural region centered around the two Arg residues and two Glu residues has been named the "lock" region. It is supported and surrounded by another eight conserved residues. Using the numbering of FepA, Arg75 is always an RG sequence in the 20 proteins, Arg126 is strictly conserved as an RG or KG sequence, E511 is always identical in the sequence EXG, and E567 is strictly conserved as a GXE or GXD sequence. It is worth noting that the importance of Arg75 and Arg126 was also revealed by random mutagenesis. PCR-generated and spontaneous mutants were tested for binding of and sensitivity to colicin B, for which FepA is the receptor. Missense mutations at both Arg75 and Arg126 were identified among the mutants with decreased sensitivity to the colicin.

Exploratory experiments were carried out on single and double mutants of the four central residues of the lock region in FepA. All the mutants showed, quantitatively, normal binding of ferric enterobactin. A conservative E-to-Q mutation showed almost normal transport, but the E511Q E567Q double mutant and the more drastic E511A and R75A mutants showed significantly reduced transport in quantitative measurements. It was not possible to establish a clear mechanistic principle for the lock region. However, it seems most likely that in mutants with mutations that affect transport, the globular domain is not positioned properly with respect to the barrel and thus inhibits or at least diminishes the signaling process to the TonB box. More comprehensive mutational studies and associated transport and binding studies, as well as structural studies of mutants, will yield valuable information.

There are no other conserved residues in the globular domain that form hydrogen bonds with conserved residues in the barrel. The other conserved residues in the globular domain are in the β5, β6 loop and in the middle and lower end of strands 4, 5, and 6. The latter includes a number of hydrophobic residues. These are discussed in the following section.

Formation of a Transient Channel
Presumably the proton motive force of the cytoplasmic membrane activates TonB, which then interacts with the ORMP to form a transient channel through which the substrate can diffuse from the binding site to the periplasm. It is clear that in the crystal structures, the globular domain closes off the barrel. Therefore, the channel has to be formed either by removal of the plug domain from the barrel or by conformational changes within the plug domain. The former is certainly possible although many hydrogen bonds (more than 50) between the plug domain and the barrel have to be broken to remove the plug. It also requires the destruction of the genetically and structurally conserved quadrupole discussed above. This structural feature may, however, help to reassemble the plug with respect to the barrel after the substrate has diffused through the open channel. Alternatively, if the channel is formed by conformational changes within the globular domain, the quadrupole most probably stays intact. The topological identity and structural similarity of the plug domains in the three crystal structures favor the latter possibility.

If the channel is indeed formed by conformational changes within the globular domain, there is a likely channel candidate between the globular domain and the barrel, suggested by analysis of the three receptor structures. On the barrel, this potential channel is lined by Arg431 in FepA (Arg467 in FecA and Arg452 in FhuA). This residue is identical in the simultaneous sequence alignment of all 20 proteins, as indicated in Color Plate 8, which also shows the relative location of this residue with respect to the quadrupole. The formation of the channel would require small changes in the loops containing apices B and C, allowing the substrate to slide between the loops (Color Plate

6). The main obstruction of the channel in the three structures would be the 10 to 12 residues forming the β5–β6 loop between strands β5 and β6 (Table 1; Color Plate 6). It is absolutely necessary for this loop to be rotated out of the way before the channel can be formed. At the time when the sequence of FepA was determined and compared to other receptor sequences, a homology was noted for two Gly residues at i and i + 7 in the first 150 residues of the sequence (region IV). This homology is verified in the simultaneous sequence comparison of 20 proteins, where the pair of Gly residues at i and i + 7 line up in all 20 proteins. These Gly residues (Gly127 and Gly134 in FepA, Gly134 and Gly141 in FhuA, and Gly197 and Gly204 in FecA) are also structurally conserved. Gly (i) is the first residue in the β5-β6 loop, and Gly (i + 7) is in the middle of the loop. Glycine, in contrast to other amino acid residues, has a large amount of conformational freedom and a much larger range of allowed values for the phi/psi values in a Ramachandran plot than do other amino acids. Rotations around glycines are therefore preferred when conformational changes occur within a protein or between protein domains. The conformational angles of the two conserved Gly residues are similar in the three structures, with phi/psi values of approximately 70/140, far outside the allowed region for other amino acids. A Gly-to-Ala substitution would necessarily result in small local changes in the structure and would also impede conformational changes around the two residues. In preliminary experiments for FepA, it was found that the double mutation G127A G134A allowed normal binding of the substrate but impeded transport significantly. This result implies that rotation around these two Gly residues (and possibly others) is important for the formation of the channel and for transport to occur. If one inspects both the structure and the topology of the globular domain (Table 1; Color Plate 6), one might expect that this rotation would disrupt the mixed β-sheet. There are, in fact, homologies (primarily hydrophobic ones) in the top and bottom part of the sheet, but it is not possible to draw conclusions about their significance in the absence of additional data. The same general location for the channel was pointed out for FhuA; however, the 8 residues lining the putative channel, considered to be strictly conserved, are homologous only in ferrichrome transporters and are not apparent in the simultaneous sequence alignment of the 20 diverse outer membrane receptor proteins.

Possible Second Site of Interaction with TonB

After the structures of FepA and FhuA were reported, investigators explored the properties of variants of each protein lacking the globular domain. These mutants lacked the putative TonB box and hence the first site of interaction between the receptor and TonB. Unexpectedly, these variants both showed TonB-dependent uptake (more so for the FhuA deletion mutant than for the FepA barrel variant, for which the efficiency, as measured by V_{max}, was very low). The results implied that a second site for TonB interaction is present on the β-barrel of the outer membrane receptor proteins. More recently, another FepA construct lacking the plug domain was designed. This construct did not show TonB-dependent uptake, and the conflicting results are being investigated in several laboratories. It is generally accepted that in the intact protein after ligand binding, the TonB box is at a particular place with respect to the barrel to allow for the interaction with TonB to occur. The question that has not been adequately addressed is how TonB would recognize this location without having a reference point on the barrel. One answer is that this can be achieved only by another part of TonB interacting with a specific location on the barrel of the outer membrane receptor protein.

One of the homology regions originally identified in the alignment of FepA with three other receptor sequences was the region from residues 672 to 683. Analysis of this region, termed homology region II, does not seem to have been followed up on, although conservation of the region is very clear in the simultaneous comparison of 20 proteins, reinforcing its significance. The homology is centered on Asn677 in FepA (Asn707 in FecA and Asn682 in FhuA), and this residue is identical in all 20 proteins.

However, more residues are involved. In the 20 proteins, there are 5 conserved residues that form a common and superposable motif in the three crystal structures (shown for FecA in Color Plate 9). Four of these residues, located in the sequence from 707 to 710 for FecA, are Asn707, Ile708 (Ile, Leu, Val in all proteins), Phe709 (Phe, Tyr in 17 proteins), and Asp710 (Asp, Asn in all proteins). The fifth residue is Arg731 (identical in all proteins). The sequence from 707 to 710, and corresponding sequences in FepA and FhuA, is located at the beginning of loop L11 in, or just above, the outer leaflet of the membrane. Two of the four residues form a bulge (residues 708 and 709 and equivalent residues in FepA and FhuA), pointing away from the barrel and slanting toward the end of loop L11 just before strand 22 begins. The bulge is caused by the conformations of the Asn (i) and the Asp (i + 3) residues. Both are in the uncommon left-handed helix conformation in all structures. The Arg731 residue in FecA, the last residue in L11 (704 in FhuA and 714 in FepA), participates in this conserved structural region by lying on top of the bulge (Color Plate 9). In all three structures, the main-chain N of the Arg forms an H bond with the side chain carboxyl group of the Asp and the side chain of the Arg forms an H bond with the C=O group of the Ile residue (708 in FecA, and corresponding residues in FepA and FhuA). Also, the other H bonds indicated in Color Plate 9 are maintained in the other two structures. This conserved structure is called the strand 21 region, for lack of a better name at this time.

The fact that the strand 21 region is derived from sequence homology of many proteins and from the structural identity in the three crystal structures suggests that it has a common function in this family of proteins. This is reinforced by the fact that when heme transporters or transferrin and lactoferrin transporters, all of which are TonB dependent, are included in the simultaneous sequence alignment, these proteins show the same homology for the five residues discussed in the previous paragraph. It has therefore been suggested that this strand 21 region is the second site of interaction for the TonB protein

The strand 21 region is at an unusual location for an interaction with TonB, being in the outer leaflet of the membrane and pointing into the membrane rather than to the center of the barrel. On the other hand, there have been indications that TonB might be associated, at least temporarily, with the outer membrane. No experiments have been carried out to explore the hypothetical function of the strand 21 region and its possible interaction with TonB, although some approaches appear to be quite obvious. For instance, it is possible to change Arg731 in FecA to Asp or Glu or to change Asp710 to a hydrophobic or basic residue. Equivalent mutations could be made in any of the TonB-dependent receptor proteins. Biochemical and structural experiments with mutants could yield valuable results. Similarly, Asn, Gln, and Asp have a much higher tendency to assume a left-handed helical conformation than do the other residues, and therefore a change of the conserved Asn and/or Asp residue to any other amino acid may disturb the local bulge structure and interfere with a possible TonB interaction.

The least-squates fit is generally applied to the barrels, and one may therefore expect a fairly good overlap for the short periplasmic loops in the three structures. This is true for periplasmic loops P2, P3, P4, P5, and P6. However, none of these loops show sequence homology. The overlap of loop P4 in the three structures, in both their liganded and unliganded forms, is most noticeable. It shows a distinct structure pointing downward and away from the barrel. This might, however, be a coincidence rather than a significant feature of the structures.

BtuB STRUCTURES

Recently, the structures of apo-BtuB, BtuB with Ca^{2+}, and BtuB with Ca^{2+} and the substrate cobalamin were determined. In addition, the coordinates of the structures are available. Although this protein is not involved in iron transport, it belongs to the same family as the outer membrane receptor proteins for ferric siderophores. It is therefore interesting to discuss certain aspects of these BtuB structures in the context of the topics covered in this chapter, because at times the discussions imply predictions for new structures in this family of proteins.

The BtuB structure shows the N-terminal plug domain and the 22-strand β-barrel, as expected and as observed for the other three outer membrane receptors. The protein contains 60 to 100 fewer residues than the ferric siderophore receptor proteins; this is reflected in a slightly smaller plug domain and shorter external loops. The cross section of the barrel in BtuB is elliptical and fits nicely, using least-squares fits, with the barrels of the other three proteins. The largest axis runs from the bottom of strand 8 to the bottom of strand 19, the same as in FecA, FhuA, and FepA. The cross section for all proteins is somewhat deformed from an ellipse and has an egg shape. In all four structures, the largest curvature of the barrel occurs on a line perpendicular to the membrane starting at the bottom of strand 19. The shape of the β-barrel, therefore, seems to be strictly maintained in all these proteins.

Although the plug domain is smaller in BtuB (141 residues), the topology is the same as for FecA, FepA, and FhuA, as shown in Table 1. For instance, apex A is located between the β2 and β3 strands, apex B is located between the β3 and β4 strands, and apex C is located between the β4 and β5 strands. In the structure of BtuB, the β2 and β3 strands form a small antiparallel sheet and the β1, β5, β6, and β4 strands form the mixed β-sheet, as observed in the other three structures. Most difficult to recognize is the switch helix in BtuB, but, it most probably consists of residues 19 to 23. Four of the five residues have an α-helical conformation (also see below).

In contrast to the topology, only parts of the structures of the plug domain are the same in the proteins. The location and orientation of the mixed β-sheet is the same in FecA, FepA, and FhuA. This is true for the BtuB structure. The least-squares fit procedure, described above, shows that the differences in location of 16 Cα atoms in the mixed β-sheet of BtuB, compared to the one of FecA, are all less than 1.6 Å. The same is true for the small sheet formed by the β2 and β3 strands. The sheets in BtuB, therefore, also recognize the shape and orientation of the barrel.

The cobalamin substrate interacts with both extracellular loops and apices A, B, and C, as is the case with FecA and FhuA. The interactions of cobalamin with apex A (more correctly, the end of strand 2 and the beginning of strand 3) are H bonds. The same is the case for the interaction with apex B. The interaction with apex C, however, is neither an H bond nor an electrostatic interaction but, rather, an induced fit in which residues from apex C move by as much as 6 Å up toward the substrate (it involves, however, a van der Waals' contact between the protein and substrate, involving two nitrogens, which is too small [2.8 Å]). Chimento et al. observed an important and significant change in the structure of the TonB box for the protein bound to cobalamin compared to the ligand-free protein on the periplasmic side. However, the switch helix (residues 19–23), which unwinds partially in FecA and completely in FhuA, does not unwind in BtuB, nor does the analysis show an apparent concerted molecular motion from apex C to either the switch helix or the TonB box. This brings into question the significance and function of the switch helix, as discussed above. It is likely that the binding of the substrate in BtuB and the resulting changes in apex C are related to the structural changes in the TonB box. The molecular motions involved in this signal relating the two sites are not clear.

The substrate cobalamin is only partially covered by the extracellular loops. This involves loops L2, L3, L5 (if it is in a similar location as in the apo-BtuB structure; part of the loop is not observed in the substrate-bound form), L7, L8, L9, and L10. In BtuB, the extracellular loops are smaller than in the other three proteins, and the substrate cobalamin is much larger than the substrates of FecA and FhuA. The substrate is sufficiently covered that it cannot escape to the medium without changes in the loops. In addition, several extracellular loops are involved in intermolecular interactions in the crystal structure, possibly preventing changes in their location. These loops could close off the substrate-binding site more completely. The principle of bipartite gating is thus quite possible but would require additional proof.

If BtuB is included in the simultaneous sequence alignment of many proteins in this family, it fits the same pattern as observed for the ferric siderophore receptor proteins. The lock region involved in the plug domain in BtuB, Arg69 at the end of strand $\beta 3$ and Arg111 at the end of strand $\beta 5$, as well as Glu419 and Glu465 on strands $\beta 14$ and $\beta 16$ of the β-barrel, are apparent in the simultaneous alignment, but they are also superposable, when using a least-squares fit, with the equivalent residues in FecA, FepA, and FhuA structures. The four residues form a quadrupole. These interactions, together with the covalent bond between the last residue of the plug domain and the first residue of the barrel, form the three points of attachment which place the plug domain in the proper and same orientation with respect to the β-barrel in all structures.

The $\beta 5$–$\beta 6$ loop in BtuB is also apparent from the simultaneous sequence alignment, with Gly residues at i and i + 7. This loop in all structures, including BtuB, occludes a potential channel for the transport of the substrate from the binding site to the periplasm. In a previous section, it was suggested that a transient channel can be formed by conformational changes in this loop, primarily allowed by the great conformational freedom of glycines. In BtuB, this channel is lined on the barrel by Arg358 on strand $\beta 11$; this residue is identical in all proteins in the simultaneous alignment.

Finally, it is suggested that there exists a second site on the receptor proteins, in addition to the TonB box, which interacts with TonB. This site is located at the beginning and end of L11. Its existence is strongly implied by the simultaneous sequence alignment as well as the unusual structure of this site in the three proteins. In BtuB this involves Asn567 and Arg584, identical in all proteins, and the strictly conserved Asp570. Both Asn567 and Asp570 have the unusual left-handed helix conformation as observed in FecA, FhuA, and FepA. The interaction of Arg584 with residues 567 to 570 and the structure of this site in BtuB are identical to those observed in the other three structures (Color Plate 9).

SUMMARY AND CONCLUSION

In this review a number of topics were discussed that pertain to functions of outer membrane receptor proteins in the transport of ferric siderophores. The discussions were based on comparisons of the crystal structures of three of the proteins and on the simultaneous sequence alignment of many TonB-dependent proteins. The topics involved binding of substrate, signaling to the TonB box, interaction(s) with TonB, and effects of activated TonB on the structure of the outer membrane receptor protein. In addition, the organization of the globular domain was explored and the principle of bipartite gating was presented.

At various points, suggestions were put forward for new structural determinations. There are established methods to produce, purify, crystallize, and solve the structures of members of this family of proteins. This is applicable to newly identified members of this family and possibly to hybrid proteins and protein domains. Furthermore, the structural determination of a mutant protein or proteins with other small molecular changes is a much easier task than for new structures, because the time-consuming phase problem does not have to be solved. Instead, the intensity data of the mutant can be combined with the phases of the parent protein to obtain the initial structure, decreasing the required time from months to days for this task.

The most difficult problem will be to obtain an understanding of the function and the effect of the proton motive force transduced by the TonB protein. There is a distinct need for a structural determination of a complex between an outer membrane receptor protein and a TonB construct. The stability and folding of any construct and the nature of the interaction between the two proteins will make this a difficult but not impossible enterprise.

ACKNOWLEDGMENTS

I thank M. A. F. Jalal for innovative work at the beginning of the FepA project and R. Chakraborty for his contributions to the later stages. Thanks are extended to B. S. Smith for the systematic and successful crystallizations of many proteins and derivatives and to J. T. Buckley for constructive comments on the manuscript.

Part of this work was supported by NIH grant GM21822.

SUGGESTED READING

Barnard, T. J., M. E. Watson, Jr., and M. A. McIntosh. 2001. Mutations in the *Escherichia coli* receptor FepA reveal residues involved in ligand binding and transport. *Mol. Microbiol.* **41:** 527–536.

Braun, M., H. Killmann, and V. Braun. 1999. The beta-barrel domain of FhuA Δ5–160 is sufficient for TonB-dependent FhuA activities of *Escherichia coli*. *Mol. Microbiol.* **33:** 1037–1049.

Buchanan, S. K., B. S. Smith, L. Venkatramani, D. Xia, L. Esser, M. Palnitkar, R. Chakraborty, D. van der Helm, and J. Deisenhofer. 1999. Crystal structure of the outer membrane active transporter FepA from *Escherichia coli*. *Nat. Struct. Biol.* **6:**56–63.

Chakraborty, R., E. A. Lemke, Z. Cao, P. E. Klebba, and D. van der Helm. 2003. Identification and mutational studies of conserved amino acids in the outer membrane receptor protein, FepA, which affect transport but not binding of ferric enterobactin in *Escherichia coli*. *Biometals* **16:**507–518.

Chimento, D. P., A. K. Mohanty, R. J. Kadner and M. C. Wiener. 2003. Substrate-induced transmembrane signaling in the cobalamin transporter BtuB. *Nat. Struct. Biol.* **10:**394–401.

Coulton, J. W., P. Mason, D. R. Cameron, G. Carmel, R. Jean, and H. N. Rode. 1986. Protein fusions of β-galactosidase to the ferrichrome-iron receptor of *Escherichia coli* K- 12. *J. Bacteriol.* **165:** 181–192.

Ferguson, A. D., E. Hofmann, J. W. Coulton, K. Diederichs, and W. Welte. 1998. Siderophore-mediated iron transport: crystal structure of FhuA with bound lipopolysaccharide. *Science* **282:** 2215–2220.

Ferguson, A. D., J. Ködding, G. Walker, C. Bös, J. W. Coulton, K. Diederichs, V. Braun, and W. Welte. 2001. Active transport of an antibiotic rifamycin derivative by the outer membrane protein FhuA. *Structure* **9:**707–716.

Ferguson, A. D., J. W. Coulton, K. Diederichs, and W. Welte. 2001. The ferric hydroxamate uptake receptor FhuA and related TonB-dependent transporters in the outer membrane of gram-negative bacteria, p. 834–849. *In* A. Messerschmidt, R. Huber, T. Poulos, and K. Wieghardt (ed.), *Handbook of Metalloproteins*. John Wiley & Sons, Ltd., Chichester, United Kingdom.

Ferguson, A. D., R. Chakraborty, B. S. Smith, L. Esser, D. van der Helm, and J. Deisenhofer. 2002. Structural basis of gating by the outer membrane transporter FecA. *Science* **295:**1715–1719.

Ferguson, A. D., V. Braun, H.-P. Fiedler, J. W. Coulton. K. Diederichs, and W. Welte. 2000. Crystal structure of the antibiotic albomycin in complex with the outer membrane transporter FhuA. *Protein Sci.* **9:**956–963.

Jalal, M. A. F., and D. van der Helm. 1989. Purification and crystallization of ferric enterobactin receptor protein, FepA, from the outer membranes of *Escherichia coli* UT5600/pBB2. *FEBS Lett.* **243:** 366–370.

Locher, K. P., B. Rees, R. Koebnik, A. Mitschler, L. Moulinier, J. P. Rosenbusch, and D. Moras. 1998. Transmembrane signaling across the ligand-gated FhuA receptor: crystal structures of free and ferrichrome-bound states reveal allosteric changes. *Cell* **95:**771–778.

Lundrigan, M. D., and R. J. Kadner. 1986. Nucleotide sequence of the gene for the ferricenterochelin receptor FepA in *Escherichia coli*. *J. Biol. Chem.* **261:** 10797–10801.

Pressler, U., H. Staudenmaier, L. Zimmermann, and V. Braun. 1988. Genetics of the iron dicitrate transport system of *Escherichia coli*. *J. Bacteriol.* **170:** 2716–2724.

Scott, D. C., Z. Cao. Z. Qi, M. Bauler, J. D. Igo, S. M. C. Newton, and P. E. Klebba. 2001. Exchangeability of N termini in the ligand-gated porins of *Escherichia coli*. *J. Biol. Chem.* **276:** 13025–13033.

Smith, B. S., B. Kobe, R. Kurumbail, S. K. Buchanan, L. Venkatramani, D. van der Helm, and J. Deisenhofer. 1998. Crystallization and preliminary X-ray analysis of ferric enterobactin receptor FepA, an integral membrane protein from *Escherichia coli*. *Acta Crystallogr. Sar. D* **54:**697–699.

Vakharia, H. L., and K. Postle. 2002. FepA with globular domain deletions lacks activity. *J. Bacteriol.* **184:**5508–5512.

van der Helm, D., and R. Chakraborty. 2001. Structures of siderophore receptors, p. 261–287. *In* G. Winkelmann (ed.), *Microbial Transport Systems*, Wiley-VCH, Weinheim, Germany.

van der Helm, D., R. Chakraborty, A. D. Ferguson, B. S. Smith, L. Esser, and J. Deisenhofer. 2002. Bipartite gating in the outer membrane protein FecA. *Biochem. Soc. Trans.* **30:** 708–710.

van der Helm, D. 1998. The physical chemistry of bacterial outer membrane siderophore receptor proteins, p. 355–401. *In* A. Sigel and H. Sigel (ed.), *Metal Ions in Biological Systems,* vol. 35. Marcel Dekker, New York, N.Y.

BACTERIAL HEME AND HEMOPROTEIN RECEPTORS

Donna Perkins-Balding, Andrew Rasmussen, and Igor Stojiljkovic[†]

5

Heme, or iron-protoporphyrin IX, is an important nutrient for many bacteria (Fig. 1). It can serve as a source of essential iron, or it may function as a cofactor for oxygen transport processes, energy generation via electron transport systems, cellular oxidative reactions, or hydrogen peroxide catabolism. Heme belongs to the metalloporphyrin family of molecules, which have δ-aminolevulinic acid as a common precursor but have different central metal atoms complexed within a tetrapyrrole ring. The importance of heme as a biological catalyst stems from the ability of the central iron atom to undergo reversible oxidative change. When the porphyrin contains reduced iron (Fe^{2+}) it is referred to as heme, whereas the oxidized form (Fe^{3+}) is known as hemin. Both heme and hemin are commonly referred to as heme, and therefore the name "heme" is used in this chapter.

Scavenging heme from the environment is less energetically expensive to a bacterium than is synthesizing heme *de novo*. Clearly, many bacteria take advantage of environmental sources of heme, as demonstrated by the ever-increasing number of heme acquisition systems described (Table 1). However, numerous systems remain unidentified in bacteria that can either utilize or bind heme and hemoglobin (Table 2). A majority of the known heme acquisition systems belong to pathogenic bacteria, as indicated in Table 1; however, this probably reflects a bias in the organisms being studied rather than an unequal distribution of heme acquisition systems among bacteria. In fact, heme is probably an important source of environmental iron, such as that in in soil, because of its ubiquitous presence in living matter. Certain nitrogen-fixing bacteria found in soil have newly discovered heme acquisition systems. Most of our current knowledge of heme acquisition, however, reflects the bias toward studying pathogenic bacteria.

Pathogenic bacteria face serious obstacles in obtaining heme (and iron) for growth, because 99.9% of total body iron is localized intracellularly and is not readily available for use. Intracellular iron consists of heme tightly bound to the host hemoproteins, hemoglobin and myoglobin, as well as insoluble ferric salts stored in ferritin. Human plasma contains transferrin, small amounts of lactoferrin, and trace amounts of hemoglobin (from spontaneous hemolysis).

Donna Perkins-Balding, Andrew Rasmussen, and Igor Stojiljkovic, Department of Microbiology and Immunology, Emory School of Medicine, Rollins Research Center, 1510 Clifton Rd., Atlanta, GA 30322.

[†] Deceased.

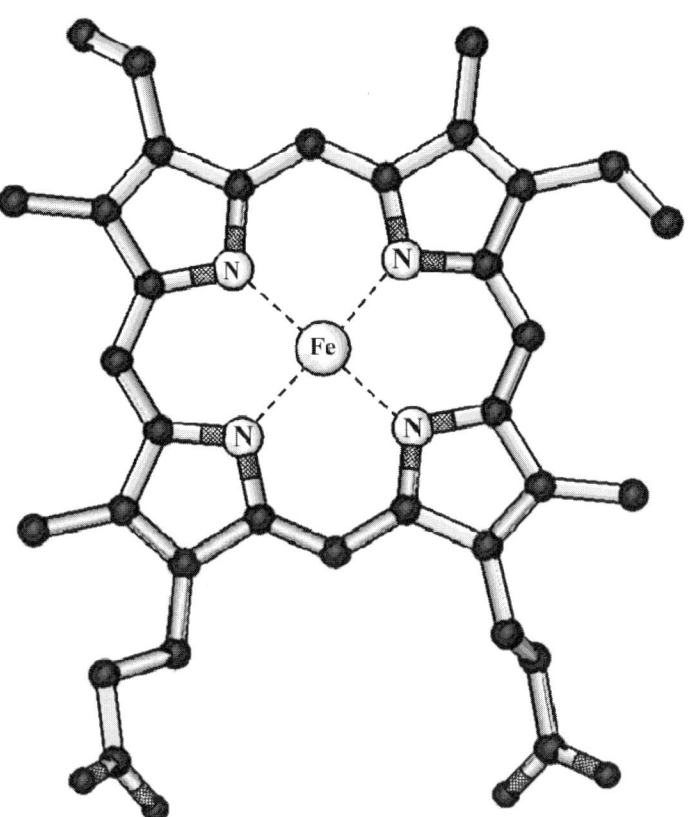

FIGURE 1 Structure of heme. A ball-and-stick model of iron-protoporphyrin IX, with propionic side chains located at the bottom. Four nitrogen atoms coordinate the central iron atom. Possible axial ligands of iron are not shown. Double bonds are indicated by gray shading.

When hemoglobin is found in the serum, it may acquire three structurally different forms, depending on whether oxygen is bound, or can bind, as the sixth ligand of the iron atom. These forms are deoxyhemoglobin (Fe^{2+}), oxyhemoglobin (Fe^{2+}), and methemoglobin (Fe^{3+}). Circulating hemoglobin in human plasma, however, is quickly bound by haptoglobin and removed by the liver. Similarly, free heme is bound and removed from the circulation by the host protein hemopexin, serum albumin, and some lipoproteins. Mucosal surfaces, on the other hand, have significant amounts of inorganic iron and heme resulting from dead epithelial cells, ingested food material, and menstruation in human females.

Bacteria have several mechanisms to overcome the limited access to heme and hemoglobin imposed by the host through sequestration mechanisms. Numerous pathogenic bacteria secrete exotoxins such as hemolysins and cytolysins, which can quickly increase the local concentration of free heme and hemoglobin. Both *Streptococcus pyogenes* and *Staphylococcus aureus*, which have well-characterized hemolysins, also have heme transport systems (see chapter 25). Some bacteria secrete proteases, which are capable of releasing heme from host carrier proteins. In addition, recent evidence indicates that *Haemophilus* species have a two-component regulator, consisting of a sensor kinase and a cognate response regulator that positively regulate the expression of components of the heme acquisition system when heme is available for use.

Heme has a molecular weight of over 600 Da, which is just above the size limit for passive diffusion across the outer membrane of gram-negative bacteria. This, along with the limited availability of heme, makes active heme transporters

TABLE 1 Characterized bacterial heme uptake systems

Species (disease)[a]	Receptor (mass)[b] (hemophore)[c]	Substrate[c,e]	ABC transport (heme utilization)[d,e]	Accessory proteins (function)[e]	Notable features[e]
Gram negative					
Actinobacillus actinomycetemcomitans (periodontitis)	HgpA (114)	Hm, Hb		AfuA, FbpA	
Bartonella henselae (cat scratch fever)	Pap31 (31)	Hm		HbpB, HbpC, HbpD, HbpE (paralogues)	High Hm requirement (20–40 µg/ml in vitro)
Bartonella quintana (trench fever)	HbpA (30)	Hm			Hm-responsive positive regulators
Bordetella avium (avian respiratory disease)	BhuR (91)	Hm, tHb, hHb, Mb	BhuT, BhuU, BhuV, (BhuS)	RhuI (ECF σ factor); RhuR (cytoplasmic membrane regulator)	Hm-responsive positive regulators
Bordetella bronchiseptica (mammalian respiratory disease)	BhuR (90)	Hm, Hb	BhuT, BhuU, BhuV (BhuS)	HurI (ECF σ factor); HurR (cytoplasmic membrane regulator)	Hm-responsive positive regulators
Bordetella pertussis (pertussis)	BhuR (90)	Hm, Hb, Hb-Hp, Hm-BSA	BhuT, BhuU, BhuV (BhuS)	HurI (ECF σ factor); HurR (cytoplasmic membrane regulator)	
Bradyrhizobium japonicum	HmuR (86)	Hm, Hb, legHb	HmuT, HmuU, HmuV		
Burkholderia cepacia (pneumonia)	97-kDa HBP	Hm		77 and 149 kDa (Hm-binding proteins)	Conversion of monomeric Hm into µ-oxo dimers
Campylobacter jejuni (gastroenteritis)	71 kDa	Hm, Hb, Hb-Hp, Hm-Hx		RfaH (antitermination factor)	
Escherichia coli O157:H7 (hemorrhagic colitis)	ChuA	Hm, Hb			
Haemophilus ducreyi (STD:[c] chancroid)	HgbA (110) TdhA (82)	Hb Hm			High Hm requirement (50 µg/ml in vitro)
Haemophilus influenzae (otitis media, meningitis, pneumonia)	HgpA, HgpB, HgpC (116, 122, 113)	Hb, Hb-Hp			Phase-variable Hgp system (CCAA repeats)
	HxuA, HxuC (100, 62) (HxuA)	Hm, Hm-Hx (Hb, HuxC only)		HxuB (may be required for HxuA secretion)	HxuC may be receptor for secreted HxuA
	HhuA (115)	Hb-Hp			
	57 kDa	Hm-Hx		HgbA, HgbB, HgbC P2 (porin protein) 29 kDa	
	Lipoprotein e (aka P4) (28)	Hm			P4 may serve dual function as an NMN 5' nucleosidase

Organism (disease)	Receptor	Substrate	Other components	Notes
Helicobacter pylori (gastritis and peptic ulcer disease)		Hm, Hb	77 kDa	
Neisseria gonorrhoeae (STD: gonorrhea)	HpuAB (35, 85)	Hb, Hb-Hp	(HO: HemO*)	50 and 48 kDa (outer membrane-associated Hm-binding proteins) Bipartite receptor phase-variable expression
Neisseria meningitidis (meningitis and meningococcemia)	HpuAB (35, 85) HmbR (99)	Hb, Hb-Hp Hm, Hb	(HO: HemO)	Hpu system: Bipartite receptor with single Hb-binding site; phase-variable expression of both Hm utilization systems
Plesiomonas shigelloides (gastroenteritis)	HugA (72)	Hm	HugB, HugC, HugD	
Porphyromonas gingivalis (periodontitis)	HmuR (75)	Hm, Hb, Hb-Hp, Hm-Hx, Hm-HSA	HugW, HugX, HugY (use of Hm for iron but not for porphyrin) Kgp, HRgpA, and HagA (proteases)	Requires gingipains for Hm transport
Pseudomonas aeruginosa (opportunistic infections)	PhuR (82) HasR (94) (HasA)	Hm, Hb Hm, Hb	PhuT, PhuU, PhuV (PhuS, HO: PigA) PhuW HasD, HasE, HasF (hemophore export)	Phu system may have interdependence on Has system
Pseudomonas fluorescens	HasR (101) (HasA)	Hm	(HO: PigA*) HmuT, HmuU, HmuV (HmuS, HmuP) HasD, HasE, HasF (hemophore export)	
Rhizobium leguminosarum	Unknown	Hm		
Shigella dysenteriae (bacillary dysentery)	ShuA (70)	Hm	ShuT, ShuU, ShuV (ShuS) ShuW, ShuX, ShuY; ShuS (sequesters heme)	
Serratia marcescens (opportunistic infections)	HasR (92) (HasA)	Hm, Hb	HasI, HasR (σ and anti-σ of has operon); HasD, HasE, HasF (hemophore export); HasB (TonB dedicated to Has system); SecB (HasA chaperone)	HasA removes Hm from Hb; HasA structure known; HasA-Hm and HasA-HasR interaction sites described; Holo-HasA binding induces has operon
Sinorhizobium meliloti	ShmR (91)	Hm		No homology to other Hm receptors
Treponema denticola (periodontitis)	HbpA, HbpB (43, 42)	Hm		
Vibrio anguillarum (hemorrhagic septicemia in marine fish)	HuvA (79)	Hm, Hb		36 kDa (Hm- and Hb-binding protein); 97 and 56 kDa (Hb-binding proteins)
Vibrio cholerae (cholera)	HutA (75) HutR (79)	Hm, Hm	HutB, HutC, HutD	HutA and HutR function with either of two Ton systems, but TonB1 more efficient
	HasR (88) (unknown)	Hm		HasR uses TonB2 system; triple mutant is Hm-utilization negative
Vibrio vulnificus (septicemia and wound infections)	HupA (77)	Hm, Hb		Protease required to use Hb

(Continues)

TABLE 1 Continued

Species (disease)[a]	Receptor (mass)[b] (hemophore)	Substrate[c,e]	ABC transport (heme utilization)[d,e]	Accessory proteins (function)[a]	Notable features[e]
Yersinia enterocolitica (gastroenteritis)	HemR (78)	Hm, Hb, Mb, Hb-Hp, Hm-Hx, Hm-HSA, catalase	HemT, HemU, HemV (HemS) HemP)		Essential histidine residues in HemR identified
Yersinia pestis (bubonic and pneumonic plague)	HmuR (77)	Hm, Hb, Mb, Hb-Hp, Hm-Hx, Hm-HSA	HmuT, HemU, HmuV (HmuS, HmuP')		
	HasR (90) (HasA)	Hm		HasD, HasE, HasF (hemophore export); HasB (TonB dedicated to Has system)	
Gram positive					
Corynebacterium diphtheriae (diphtheria)	HmuT (35)	Hm, Hb	HmuT, HmuU, HmuV (HO: HmuO)		Hmu system is the only system required for Hm usage
Corynebacterium ulcerans (zoonotic diphtheria)	HmuT (35)	Hm, Hb	HmuT, HmuU, HmuV (HO: HmuO*)		
Staphylococcus aureus (various types of infections)	IsdA, B (39, 71)	Hm, Hb		IsdC, IsdD, IsdE, IsdF (membrane-associated translocation factors); IsdG (cytoplasmic Hm-binding protein); SrtA, SrtB (Sortases)	Heme relay by several proteins; receptors anchored to cell membrane by sortases
Streptococcus pyogenes (various types of infections and sequelae)	Shr (144)	Hm, Hb, Mb, Hb-Hp, Hm-HSA			No homology to other heme receptors
	Shp (32)	Hm			
	SiaA (33)	Hm, Hb	SiaA, SiaB, SiaC		

[a] Bacteria that have heme uptake systems described in peer-reviewed journals.
[b] Heme/hemoprotein receptor with approximate molecular mass (in kilodaltons) given in parentheses; processed form is cited if reference was available.
[c] Substrates as cited in Suggested Reading.
[d] Cytoplasmic heme oxygenase; putative oxygenases are indicated by asterisks.
[e] Abbreviations: HO, heme oxygenase; Hb, hemoglobin; Hm, heme; Hp, haptoglobin; Hx, hemopexin; Mb, myglobin; legHb, Hb of leguminous plants; tHb, turkey hemoglobin; hHb, human hemoglobin; HSA, human serum albumin; BSA, bovine serum albumin; HBF, heme-binding protein; STD, sexually transmitted disease; NMN, nicotinamide mononucleotide.

TABLE 2 Bacteria that are able to bind or utilize heme or hemoproteins but have uncharacterized heme assimilation systems

Aeromonas spp.	Legionella pneumophila (15-kDa secreted HBP)[a]
Aeromonas salmonicida	Neisseria gonorrhoeae (TdfG, TdfH; 97- and 44-kDa HBPs)
Corynebacterium glutamicum	Neisseria meningitidis (TdfH; 97- and 50-kDa HBPs)
Corynebacterium jeikeium	Shigella flexneri (101-kDa HBP)
Corynebacterium pseudotuberculosis	Streptococcus pneumoniae (43-kDa HBP)
Corynebacterium pyogenes	Photobacterium damselae subsp. piscicida
Corynebacterium renale	Porphyromonas gingivalis (HemR, IhtA, Tlr)
Haemophilus influenzae (P4)	Prevotella intermedia
Helicobacter felis	Vibrio alginolyticus
Helicobacter acinonyx	Vibrio anguillarum (HupA)
Helicobacter mustelae	Vibrio fluvialis
Helicobacter fennelliae	Vibrio mimicus (MhuA)
Helicobacter cinaedi	Vibrio parahaemolyticus
Helicobacter muridarum	Vibrio vulnificus (HupA)
Helicobacter bilis	Yersinia enterocolitica (Has)
Helicobacter hepaticus	Yersinia pseudotuberculosis

[a] HBP, heme-binding protein.

an essential part of the heme acquisition system. Three common themes are emerging for the mechanism of heme transport by bacteria. (i) Analogous to the siderophore-dependent iron acquisition pathways (see chapters 1, 2, 9, and 11), in which small, secreted, high-affinity iron-binding molecules shuttle iron into the cell, some bacteria secrete hemophores that serve a similar function for heme (see chapter 3). The hemophores deliver heme or hemoproteins to an outer membrane transporter, which then delivers heme into the cell. (ii) Bacteria may also use non-hemophore-dependent outer membrane transporters, which recognize heme alone or host hemoproteins (hemoglobin, hemoglobin-haptoglobin, heme-hemopexin, myoglobin, and heme-albumin). (iii) A few of these transporters are high-affinity receptors, which are limited to only one or a few structurally similar substrates. Another aspect worthy of mentioning, but not typically used to differentiate between receptors, is that hemoprotein receptors have an additional function. This involves the removal of heme from its high-affinity association with the host carrier protein prior to transport.

Heme assimilation systems are often redundant in bacteria; for example, three systems have been found in *Vibrio cholerae*, and *Haemophilus influenzae* may have four or even five. Besides multiple heme assimilation systems, bacteria may, at the same time, have redundant iron assimilation systems, such as siderophore-dependent systems or high-affinity receptors for transferrin or lactoferrin. In addition, recent evidence suggests that some heme assimilation systems may be interdependent. *Pseudomonas aeruginosa* is unable to utilize heme when expressing a functional hemophore-dependent Has system if the strain also contains mutations in the Phu heme acquisition system.

Certain common aspects of heme transport in gram-negative bacteria exist, with only a few exceptions. These include (i) the active transport of heme across the outer cell membrane, (ii) the requirement for a proton motive force and the TonB system for transport, (iii) maximal expression of heme acquisition proteins under iron-poor growth conditions, (iv) the presence of a highly conserved amino acid motif (described below) in the transporter protein, and (v) structural homology to ferric siderophore receptors. This chapter focuses on describing the general features of heme acquisition systems by using the most extensively studied heme transport systems. Exceptions to common trends are noted, as well as notable current findings in the field.

HEMOPHORE-DEPENDENT RECEPTORS

Several bacterial species produce hemophores and have corresponding hemophore-dependent receptors; these species include *Serratia marcescens*, *H. influenzae*, *P. aeruginosa*, *Pseudomonas fluorescens*, and *Yersinia pestis* (see Fig. 3). A putative hemophore-dependent receptor is found in *V. cholerae*; however, no hemophore has yet been identified. Homologues of hemophore-dependent receptors have also been found in *Neisseria* and *Pasteurella* species, but these have not been shown to transport heme or interact with hemophores. The best-characterized hemophore-dependent heme uptake systems are the Has system of *S. marcescens* and the Hxu system of *H. influenzae*. These systems are described at length in chapter 3.

GENERAL HEME RECEPTORS AND HEMOPROTEIN-SPECIFIC RECEPTORS

General heme receptors have been identified in numerous bacteria. These receptors recognize either free heme or heme bound to host-carrier proteins. Interestingly, heme, and not the surrounding protein scaffolding, appears to be key to recognition by these receptors; the presence of iron is also crucial, since protoporphyrin IX is either not bound or weakly bound. Three examples of general heme receptors, which can recognize numerous heme-containing substrates, are HmuR of *Y. pestis*, HemR of *Yersinia enterocolitica*, and HmuR of *Porphyromonas gingivalis*. It remains unclear how these receptors recognize heme in the context of so many different host carrier proteins and often with limited surface exposure; for example, only two propionic side chains of heme are exposed on the surface of hemoglobin. However, comparison of the hemoglobin-binding affinity of the *Neisseria meningitidis* hemoglobin-specific receptor HmbR and the *Y. enterocolitica* heme receptor HemR indicated that HemR does not possess a high-affinity hemoglobin-binding domain.

Evolutionary relationships, based on amino acid sequence homology, suggest that a subgroup of heme receptors are probably high-affinity hemoprotein-specific receptors. Examples of this type of receptor are found among the *Neisseria*, *Haemophilus*, *Pasteurella*, and *Actinobacillus* genera of bacteria (Fig. 2). Members of this group include the hemoglobin-haptoglobin and hemoglobin receptors HpuAB and HmbR, respectively, of pathogenic *Neisseria* species and the redundant hemoglobin-haptoglobin and hemoglobin receptors, HgpA, HgpB and HgpC, of *H. influenzae*. Evidence indicates that these proteins have evolved to specifically recognize a host hemoprotein or a limited number of host hemoproteins with very high affinity and that this recognition involves the host protein itself. Many receptors listed in Table 1 have not been studied with respect to the variety of substrates they recognize or their substrate affinity; therefore, dividing the current list of receptors into separate categories is premature.

Unfortunately, no structural data for any of the heme or hemoprotein receptors are currently available. These receptors are related to ferric siderophore receptors; therefore, some predictions have been made about the structure of heme receptors based on this relationship. Modeling studies suggest that heme receptors have multiple membrane-spanning domains, which probably form β-strands and may fold into β-barrels in the outer membrane, like their close relatives. For example, the two-dimensional model of the hemoglobin receptor HmbR of *N. meningitidis* consists of 22 putative transmembrane β-strands and 11 surface exposed loops. The amino-terminal end of the protein is thought to extend into the pore of the predicted barrel structure from the periplasm, like the plug domains of the ferric siderophore receptors. Another common, but not exclusive, feature characteristic of this family of proteins is the presence of a highly conserved amino acid motif, $FRAPX_{10}HX_2NPNLX_2E$, which may correspond to some common mechanistic function. Only a few heme transporters have been characterized to any extent with regard to specificity or structure-function relationships; these are described below.

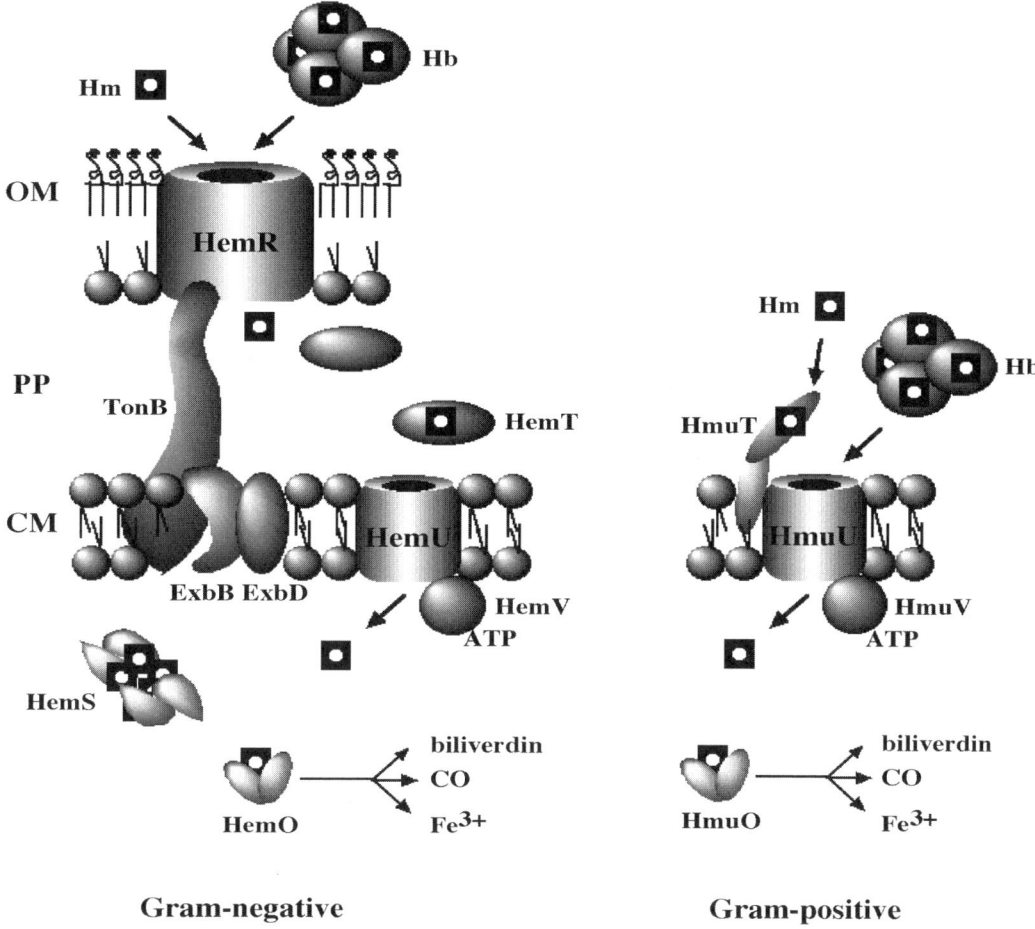

FIGURE 2 Heme assimilation systems of selected gram-negative and gram-positive bacteria. The Hem system of the gram-negative bacterium *Y. enterocolitica* (left) and the Hmu system of the gram-positive bacterium *C. diphtheriae* (right) are drawn schematically. The heme transport locus in *Y. enterocolitica*, *hemPRSTUV*, encodes HemP, a protein of unknown function (not shown), HemR, the outer membrane hemoprotein receptor, HemS, a cytoplasmic heme-sequestering protein, HemT, a heme PBP, HemU, a membrane permease, and HemV, an ATPase. HemO, a heme oxygenase found in *N. meningitidis*, is included because some gram-negative bacteria use heme oxygenases in the catabolism of heme to ferric iron, biliverdin, and CO, although a homologue has not been identified in *Y. enterocolitica*. The Hmu system of *C. diphtheriae* consists of HmuT, a membrane-associated heme-binding protein, HmuU, a membrane permease, and HmuV, an ATPase. HmuO of *C. diphtheriae* converts heme in the cytoplasm to biliverdin, CO, and ferric iron. Abbreviations: Hm, heme; Hb, hemoglobin; OM, outer membrane; PP, periplasm; CM, cellular membrane; CO, carbon monoxide.

HemR and HmuR of *Yersinia* Species

General heme receptors were first identified in *Yersinia* species. Many other bacteria, such as *Shigella dysenteriae* and *P. aeruginosa*, possess homologous heme acquisition systems. An operon of six genes, *hemPRSTUV*, is responsible for heme utilization in *Y. enterocolitica* (Fig. 3). When introduced into an *Escherichia coli* strain deficient for heme and enterobactin synthesis, these genes permitted the use of heme under iron-limiting conditions. The Hmu system of *Y. pestis* is essentially identical to the Hem system, except that HmuP is truncated and

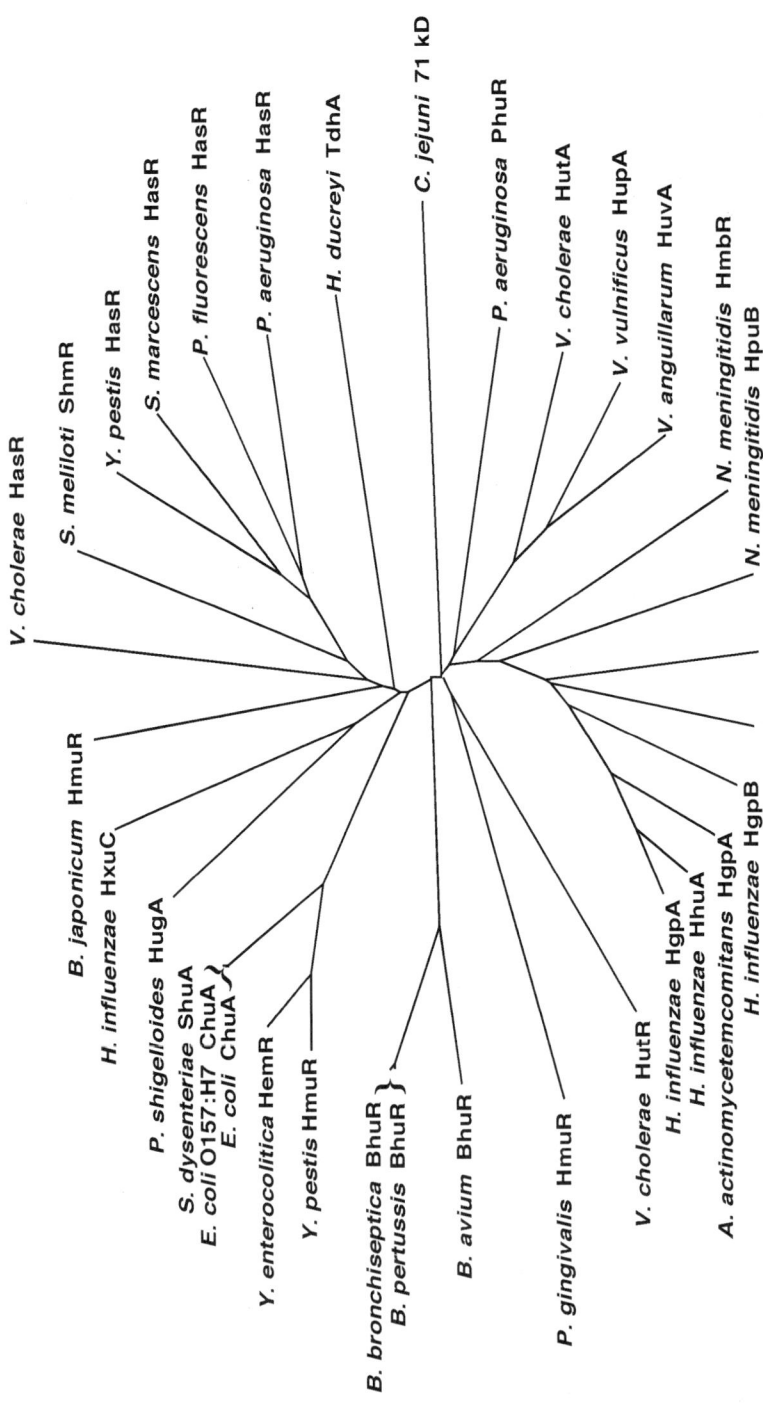

FIGURE 3 Evolutionary relationships of bacterial heme and hemoprotein receptors. An unrooted phylogenetic tree of heme and hemoprotein receptors from Table 1, excluding HbpA (*T. denticola*), Pap31 (*B. henselae*), HbpA (*B. quintana*), HxuA (*H. influenzae*), lipoprotein e (*H. influenzae*), and the gram-positive receptors, is shown. The evolutionary distance is represented schematically by branch length. ClustalX and Treeview software were used to align data, perform bootstrapping analysis, and construct the tree.

HemR and HmuR differ in their primary amino acid sequences within a small region near their carboxyl termini. HmuP may not be an essential part of the system, however, since the homologous HemP was not required for heme transport in *E. coli*. The HemTUV and HmuTUV proteins comprise a heme-specific periplasmic-binding protein dependent cytoplasmic transport system (described below). Although the role of HemS and HmuS has not been confirmed, their presence in the cell prevents heme toxicity and their purpose may be to sequester heme, a prediction based on studies involving the homologous ShuS of *S. dysenteriae*.

HemR and HmuR of *Yersinia* species recognize a wide variety of substrates, which include heme, hemoglobin, hemoglobin-haptoglobin, myoglobin, heme-hemopexin, and heme-albumin. Mutational analysis of HemR revealed that two highly conserved histidine residues play an essential role in the utilization of hemoproteins and free heme. Single amino acid changes in HemR, H461L and H461A, completely eliminate the utilization of heme from hemoprotein complexes and severely impair the use of free heme. Similarly, mutants with the HemR mutation H128K were unable to use hemoprotein complexes. Further, mutant receptors were unable to bind either heme or Hb agarose in vitro. Whole-cell labeling of different HemR cysteine derivatives with a lipid-impermeable, fluorescein-labeled biotin-maleimide indicated that both H128 and H461 are surface accessible, and they are likely candidates as axial ligands of heme-iron that bind and transport heme through the receptor pore

HmuR of *P. gingivalis*
The HmuR receptor of *P. gingivalis* enables the use of hemoglobin and heme. This receptor is a close homologue of HmuR and HemR of the yersiniae. HmuR is encoded in an operon with an upstream gene, *hmuY*, and this operon is predicted to be Fur -regulated; Fur and its capacity to regulate bacterial gene expression are discussed by in chapter 13. A separate locus, identified by genome searching, is predicted to encode a heme transport system homologous to the HemSTUV system of *Y. enterocolitica*. Biochemical analysis of different substrates in vitro indicated hemoglobin, hemoglobin-haptoglobin, heme-hemopexin, and heme-albumin were bound by HmuR whereas substrates without heme, such as apo-haptoglobin, albumin, apohemopexin, and transferrin, were not bound. Furthermore, HmuR preferentially recognized heme and other metalloporphyrins over hemoglobin. Metalloporphyrins containing iron, copper, and zinc bound more tightly to HmuR than did protoporphyrin IX, suggesting that the active site of HmuR interacts directly with the central metal ion. It is likely that a histidine residue within the active site of HmuR interacts directly as an axial ligand to the metal ion. In addition, HmuR demonstrated a preference for ethyl or vinyl side chains on the heme moiety.

HpuAB of Pathogenic *Neisseria* spp.
The HpuAB receptor serves as a high-affinity receptor for three substrates, hemoglobin, hemoglobin-haptoglobin, and apo-haptoglobin, and transports heme from hemoglobin and hemoglobin-haptoglobin. HpuAB is unusual among heme receptors because it consists of two outer membrane-associated proteins. HpuA is predicted to form a pore in the outer membrane, and HpuB is an accessory lipoprotein. Several features of HpuAB resemble the transferrin and lactoferrin iron transport systems of *Neisseria* species, including the genetic organization, iron regulation, the accessory lipoprotein, and sequence similarity between the pore protein HpuB and TbpA/LbpA. However, the HpuAB system has unique differences from the transferrin and lactoferrin utilization systems. The HpuA lipoprotein is much smaller than the TbpB and LbpB lipoproteins and has no sequence similarity. In addition, HpuA, unlike the other lipoproteins, is essential for function; HpuAB, together, form a single high-affinity binding site for hemoglobin ($K_d \approx 150$ nM). Evidence from heme- and hemoglobin-binding competition assays suggests that the HpuAB binding depends on the

whole hemoglobin molecule and that heme recognition is not essential as it is with heme receptors that use a variety of hemoproteins or heme alone.

HmbR of *N. meningitidis*

HmbR is a second high-affinity hemoglobin receptor found in *N. meningitidis*, and it appears to be restricted to human hemoglobin as its substrate. The binding kinetics for the interaction between hemoglobin and HmbR was measured with purified proteins in vitro, using surface plasmon resonance. HmbR has a binding affinity for hemoglobin of $K_d \approx 13$ nM. The association rate, or rate of HmbR-hemoglobin complex formation, is 4.4×10^5 M^{-1} s^{-1}, and the stability or dissociation rate of the complex is 6.8×10^{-3} s^{-1}.

No detailed studies of the structure-function relationships of heme receptors have yet been reported in the literature. Data from our laboratory have discerned functionally important domains of HmbR, including a region putatively involved in hemoglobin binding, in experiments using a series of HmbR deletions and site-directed mutations. Deletion analysis also revealed that the highly conserved protein motif (FRAP/NPNL motif, described above) of HmbR, found in most heme receptors, is essential for hemoglobin utilization but is not required for hemoglobin binding. This finding suggests a mechanistic involvement of this motif in heme removal from hemoglobin. In addition, an amino-terminal deletion in the putative cork-like domain of HmbR affected hemoglobin usage but not hemoglobin binding. This result supports a role of the cork domain in utilization steps that are subsequent to hemoglobin binding.

Nonhomologous Receptors

A few heme transporters have been identified that have no related proteins in the current databases. Among these are Shr and Shp of *S. pyogenes*. These proteins are part of a 10-gene operon, which contains an ABC transport system named *sia* (for "streptococcal iron acquisition"). Shr is both membrane associated and secreted, and it was found to bind hemoglobin, hemoglobin-haptoglobin, myoglobin, and heme-albumin. An *S. pyogenes* strain deficient in Shr was more resistant to both streptonigrin and hydrogen peroxide, indicating a decrease in heme uptake and reduced intracellular iron concentration. Shr does not have homology to hemoprotein receptors, but a motif was identified in Shr that can also be found in P4 and HbpA of *H. influenzae* as well as cytochrome c_3. Another heme-binding protein named Shp, encoded by a gene located just downstream of *shr*, is also membrane associated but is not secreted. Shp is expressed in vivo in human infections, and antibodies against it impair the growth of group A *S. pyogenes* strains in vitro. Two outer membrane proteins that are required for efficient heme utilization in *Treponema denticola* were reported to have no homologues in the database. These two proteins, which were 49% identical to each other, were named HbpAB.

STRATEGIC METHODS USED BY BACTERIA TO ACQUIRE HEME

Influencing Heme Availability

As stated above, bacteria use hemolysins and cytolysins to gain access to intracellular iron stores. However, several bacteria have also been identified that secrete extracellular proteases to assist in the release of heme from host hemoproteins. The production of proteases has been correlated with the ability of these bacteria to use heme-containing proteins as sources of iron. Protease-deficient *Vibrio vulnificus* strains were unable to grow on iron-limited medium containing various hemoproteins, yet supplementation of the medium with purified *V. vulnificus* protease enabled growth of the mutants. Results from further experiments concluded that free heme and hemoglobin can be used by *V. vulnificus* for growth without the need for extracellular proteases, but when the bacteria are faced with other heme sources, such as hemoglobin-haptoglobin, heme-hemopexin, or heme-albumin, proteases are essential.

A similar situation is observed for *P. gingivalis*, which secretes potent proteases known as gingipains (see chapter 21). The arginine-specific (RGP) and lysine-specific (KGP) proteases of *P. gingivalis* are known to degrade a number of host proteins and appear to be involved in heme acquisition. These proteins have hemoglobin-binding domains, and Kgp can degrade hemopexin, haptoglobin, and transferrin. Recent studies have shown that Kgp has hemophore as well as hemoproteinase activities. Kgp is essential for black-pigment formation, involving the deposition of heme on the bacterial surface, and a *kgp* mutant had reduced growth on hemin as the sole iron source. It is speculated that these proteases allow the use of hemoprotein complexes such as heme-hemopexin and hemoglobin-haptoglobin by these bacteria.

Regulated Expression of Heme Transport Systems

The regulation of heme assimilation systems mimics the regulation of other iron assimilation systems, in which iron and the global, negative transcriptional regulators Fur and DtxR are involved. For a summary of negative regulation by Fur, DtxR, and iron, see chapters 13 and 22.

In addition to these well-characterized negative regulators, investigations have uncovered a family of positive transcriptional activators of heme transport. Numerous genes associated with heme transport have been identified which either are positively regulated by the presence of heme or hemoglobin alone or are responsive to both heme-associated and classical iron regulation strategies. These include *hemR* of *P. gingivalis*, *hgbA* of *H. influenzae*, *shuA* of *S. dysenteriae*, and *hmuO* of *Corynebacterium diphtheriae*. *Bordetella* species have two regulatory genes located upstream of the Bhu/Rhu heme transport operons, designated *hurIR* and *rhuIR*, respectively (see chapter 20). HurI and RhuI are homologous extracytoplasmic function (ECF) sigma factors. A nucleotide sequence characteristic of ECF promoters was identified upstream of the *bhuRSTUV* operon. Furthermore, no Fur-binding activity was detected in the *bhuRSTUV* promoter region. The transcription of *rhuIR*, however, is probably Fur dependent. Both systems were induced by the presence of heme and hemoproteins, and transcriptional activation was dependent on the presence of the heme receptor. A model was proposed for the regulatory mechanism of these positive transcriptional activators where, under iron-starved conditions, there is derepression by Fur of the *hurIR* genes. There is subsequent production of HurI and HurR and low-level expression of the Bhu system components. HurI would remain inactive by interacting with the anti-sigma factor, HurR, until BhuR binds heme. This binding signal would be transferred through HurR, which would release HurI, allowing it to activate transcription of the *bhuRSTUV* operon.

Another heme-responsive positive activator has been identified in *V. vulnificus* and is homologous to the LysR family of transcriptional activators. This protein, HupR, exhibits Fur-dependent regulation, like HurI and RhuI, and positively regulates the expression of the heme receptor HupA.

In the symbiotic plant bacteria known collectively as 'rhizobia,' many species have iron-regulated genes, including heme transport system genes, that are not responsive to the Fur repressor. Recent studies have indicated that a novel global regulator of iron-responsive genes exists in these species and that this regulator, named RirA, is not a homologue to Fur or DtxR. In fact, the Fur homologue in these bacteria is Mur, a global regulator of manganese-responsive genes.

Sensing the Presence of Heme

The first two-component regulatory system capable of sensing heme has been found in *C. diphtheriae* (see chapter 22). It consists of the sensor kinase, ChrS, and the cognate response regulator, ChrA. This two-component system was shown to activate transcription from the *hmuO* (heme oxygenase) promoter. Site-directed mutagenesis of a conserved aspartic acid in ChrA abolished heme-dependent transcription of the *hmuO* promoter. Homologues of this system may be present in *P. aeruginosa*.

Phase-Variable Expression of Heme Transport Systems

Certain heme acquisition systems in *Neisseria* and *Haemophilus* species experience phase-variable expression, where they can be turned completely off. The obvious advantage associated with phase-varying surface proteins is evasion of the host immune response. The expression of the HmbR hemoglobin receptor of *N. meningitidis* undergoes phase variation due to slip strand mispairing of poly(G) tracts within the *hmbR* gene. Interestingly, among clinical isolates of serogroups A, B, and C, the phase variation frequency of hemoglobin receptors has been found to vary dramatically, ranging from approximately 10^{-6} to 10^{-2} CFU^{-1}. Frequencies of phase variation were shown to be the genetic trait of a particular strain, where two unlinked hemoglobin receptors phase vary with similar frequencies within a given isolate. The reason for elevated mutation rates in a study of epidemic serogroup A isolates was attributed to defects in mismatch repair pathways. These studies found a relatively high frequency of hypermutable strains among epidemic isolates, suggesting that the ability to mutate at high frequency may be advantageous to the meningococci.

Like the HmbR receptor and many other virulence-related proteins in *N. meningitidis*, expression of HpuAB varies due to translational frameshifting resulting from slip strand mispairing of a poly(G) tract within the coding sequence of *hpuA*. The rate of phase variation of *hpuAB* is similar to the rate for *hmbR* within any given strain, provided that the number of repeat sequences in the poly(G) tract is equal. Interestingly, the frameshifting mutation in *hpuA* correlates not only with a lack of HpuA expression but also with a lack of HpuB.

In *H. influenzae*, two heme transport proteins, HgpA and HgpB, are phase variable. Phase variation of the two genes encoding these proteins has been attributed to translational frameshifting caused by the slip strand mispairing of CCAA repeat sequences. Two other heme transport genes, *hgpC*, linked to *hgpA* and *hgpB*, and an unlinked heme receptor gene, *hhuA*, have also been identified as having repetitive CCAA sequences, but whether they undergo phase variation remains to be determined. In *P. gingivalis*, 12 CCAA repeats were found in the gene encoding HmuR, a newly identified heme receptor, but it is unknown whether this receptor is phase variable.

PERIPLASMIC AND CYTOPLASMIC MEMBRANE TRANSPORT

Once heme has crossed the outer membrane via its receptor, it must traverse the periplasm and be actively transported into the cytoplasm. ABC transport systems function in these activities. An ABC transport system typically consists of a periplasmic binding protein (PBP) and permease and ATPase components. Detailed descriptions of well-characterized nonheme ABC transport systems can be found in chapters 8 and 11.

While nonheme ABC transporter systems occasionally encode multiple permease and ATPase components within the same locus, heme-specific ABC transporters typically consist of only one PBP, permease, and ATPase. Mutants in heme-specific ABC transport systems do not always result in abrogation of the ability of the bacteria to use heme as an iron source. Mutations in the *Y. pestis* heme-specific ABC transporter system genes (*hmuTUV*) abrogated iron utilization from myoglobin, heme, and heme-albumin but did not affect the use of hemoglobin and hemopexin. In addition, other uncharacterized mechanisms exist for heme transport. For example, *E. coli* K-12 cannot normally utilize heme as an iron source, except when heme receptors from another bacterial species are expressed, which also occasionally requires a species-specific TonB protein (examples include HugA from *Plesiomonas shigelloides* and HmbR from *N. meningitidis*). The nature of these additional mechanisms of heme transport remains elusive and may not involve the typical ABC transport systems. At least one heme transport system reconstituted in *E. coli*, however, is more effective in the presence of its own species-specific ABC transporter; this is the Hut system of *V. cholerae*. The following briefly characterizes what is known about heme-specific ABC transport components.

Heme Periplasmic Binding Proteins

Heme PBPs must recognize heme in the periplasm and deliver it to the inner membrane transport machinery. The percent identity between described heme PBPs generally ranges from 31 to 40%, except for homologues of closely related species (HemT of *Y. enterocolitica* and HmuT of *Y. pestis* have 90% identity; ChuT of *E. coli* O 157:H7 and ShuT of *S. dysenteriae* have 97% identity). Homology is spread throughout the proteins. Potentially conserved heme-binding residues such as tyrosine and histidine have not been identified, with the exception of a single tyrosine residue. This tyrosine is located at amino acid 200 of HemT and is shared by HmuT (*Y. pestis*), BhuT (*Bordetella avium, Bordetella pertussis*, and *Bordetella bronchiseptica*), ChuT (*E. coli* O 157:H7), HmuT (*Rhizobium leguminosarum*), HutB (*V. cholerae*), HmuT (*Bradyrhizobium japonicum*), HugB (*P. shigelloides*), ShuT (*S. dysenteriae*), and PhuT (*P. aeruginosa*). Heme PBPs have ~25% identity to certain other PBPs from other ABC transporter systems that do not transport heme as their primary ligand, such as BhuF, a vitamin B_{12} PBP. It has been proposed that in the absence of a heme PBP, a homologous PBP may be able to compensate, although this has yet to be proven.

Heme Permeases and ATPases

Heme transport requires interaction between the heme PBP and its cognate permease. Functional analysis of heme-specific permeases and ATPases is still forthcoming. Current understanding of these proteins is based on the elucidated function of nonheme permeases and ATPases (see chapter 11). Heme permeases have between 37 and 68% identity to each other, with the exception of closely related species. The $EAAX_3GX_9IXLP$ motif, identified in a comparison of 61 permease proteins, is not entirely conserved in heme permeases; only $EXAX_3G$ is present. Several other motifs of unknown function are present in heme permeases. In HemU, these are (53) $NIRLPRX_2LX_3VGX_{11}FRNP$, (150) LXGI-AINAL, and (255) $GXIGFXGLX_2PH$; invariant W181 and H226 residues were also identified in the majority of heme permeases analyzed.

Heme-specific ATPases interact with their corresponding permease to facilitate the transport of heme across the cellular membrane. Heme-specific ATPases have 41 to 90% identity. ATPases perhaps contain some of the most highly conserved regions of ABC transporter systems, which can be attributed to their ATP-binding function. Typical conserved Walker A (GPNGAGKSTLL) and Walker B (hhhD, where h is a hydrophobic residue) motifs are present which bind ATP and Mg^{2+}, respectively. The linker motif $LSGGEX_2R$ is also found in all ATPases and is essential for transport function. Heme-specific ATPases contain an invariant histidine at residue 206 of HemV, along with a VLHDLNXAAXYXD motif in the switch region, which is required for transport but is not required for ATP binding.

Y. enterocolitica Heme-Specific ABC Transport

The Hem system of *Y. enterocolitica* was among the first heme transport systems to be characterized and remains one of the best-characterized systems. The products of the *hemSTUV* operon, in conjunction with the heme receptor, HemR, confer utilization of heme from a variety of hemoproteins. HemT is the PBP, HemU is a permease, and HemV is a ATP-binding protein. HemT, surprisingly, is not essential for heme uptake, whereas HemU and HemV are required for efficient uptake of heme into the cytoplasm. HemT is most probably required for optimal uptake of heme, but a redundant hPBP may exist that can compensate for the loss of HemT. The first gene in the operon codes for HemS, which was initially proposed to degrade heme in the cytoplasm once it had passed through the ABC transporter. A *hemS* mutation is lethal in *Y. enterocolitica*, but this is not true of an *hmuS* mutation in *Y. pestis*. ShuS from *S. dysenteriae*, a homologue of HemS, functions as a heme-binding protein (described below).

Notable Attributes of Other Heme-Specific ABC Transporters

Certain systems have unique features or uncharacterized components. For example, the genes encoding the *V. cholerae* heme-specific ABC transporter are not linked with the heme receptor gene *hutA*, as is typically the case for other systems, but instead are closely associated with one of the two *exbB*, *exbD*, and *tonB* loci in *V. cholerae*. In many bacteria, the loci for heme-specific ABC transporters also contain a variety of genes of unknown function. Many of these are homologous to *orfX* and *orfY* of *Yersinia* species which do not appear to play a role in heme uptake or utilization. *S. dysenteriae* has an additional gene, *shuW*, which is not present in *Yersinia* species, but also has no known function. A homologue, PhuW, from *P. aeruginosa*, was found not to be essential for heme utilization but was required for optimal function. Additional genes, named *hugWXZ* in *P. shigelloides*, were not required for utilization of heme as a porphyrin source but were required for the use of heme as an iron source. This suggests that the heme transport story is complex and that much remains to be discovered.

TRANSPORT OF HEME ACROSS THE OUTER MEMBRANE IN GRAM-POSITIVE BACTERIA

Much less is known about the heme and hemoprotein binding and uptake systems of gram-positive bacteria, although recent studies have begun to shed light on their mechanisms of heme transport. Heme uptake systems of *Corynebacterium* spp. (*C. diphtheriae* and *C. ulcerans*), *S. aureus*, and *S. pyogenes* have all been recently documented. The *hmu* system of *C. diphtheriae* (see chapter 22) has homology to the ABC heme transport system apparatus of gram-negative bacteria (Fig. 2). *S. pyogenes* and *S. aureus* hemoprotein binding and heme uptake systems also have certain aspects that resemble those of gram-negative heme ABC transporters (see chapter 25). One emerging theme in gram-positive heme transport is that the protein serving as the periplasmic heme-binding protein in gram-negative bacteria assumes the role of a heme receptor on the plasma membrane surface. This component of the ABC transport system is anchored to the cellular membrane instead of being free floating, but it serves the same function of ushering the heme substrate to the permease. Other accessory proteins, in addition to heme-specific ABC transporters, function in heme transport. In *S. pyogenes* open reading frames upstream of the *sia* genes designated *shr* and *shp* encode surface-exposed hemoproteins (discussed above). *S. aureus* produces a series of proteins (Isd) that function to relay heme through the cell wall and across the cell membrane. The newly identified heme utilization systems of these three bacteria have laid the foundation for our understanding of heme-iron utilization in gram-positive bacteria.

The *C. diphtheriae hmuTUV* genes were identified as additional genes linked to *hmuO*, a heme oxygenase gene that could complement heme and hemoglobin utilization mutants in *C. ulcerans*. The predicted products of *hmuTUV* were homologous to heme transport systems in gram-negative organisms. A major difference was recognized in HmuT of *C. diphtheriae*, which was proposed to be a heme- and hemoglobin-binding lipoprotein. It has ~25% identity to PBPs from *Y. pestis*, *Y. enterocolitica*, and *B. avium*. Whereas PBPs float free in the periplasm in gram-negative organisms, HmuT is proposed to be anchored to the cell membrane via its lipid motif and serves as the receptor portion of the heme uptake apparatus. It then extracts heme from hemoglobin and delivers it to the permease. *C. ulcerans* also contains the *hmuT* gene.

Similarly, *S. pyogenes* has been recently shown to contain ABC transporter system-like genes, *siaABC*, whose encoded proteins shuttle heme through the cellular membrane. SiaA has homology to a gram-positive bacterial ferrichrome-binding protein, to HmuT of the *C. diphtheriae* heme uptake system, and to HutB, the PBP in the gram-negative *V. cholerae*. SiaA, a putative lipoprotein, contains at its N terminus a signature sequence [(82)PDIELIASLKP-TWILSPNS] of binding proteins of the iron complex family. SiaB and SiaC, like HmuT and HmuV of the corynebacteriae, are the permease and ATPase components that are homologues of ABC transporters in gram-negative bacteria.

A Novel Heme Transport System in *S. aureus*

The Isd system has been recently described in *S. aureus* as a hemoglobin utilization system. It consists of three transcriptional units, each repressible by Fur. One transcriptional unit is composed of only *isdA*, which encodes a protein partially exposed on the bacterial surface and containing an LPXTG membrane-sorting motif, which targets polypeptides for linkage to the cell wall. Another transcriptional unit consists solely of *isdB*, whose protein product is surface exposed and also possesses the LPXTG membrane-sorting motif. The third transcriptional unit is composed of *isdC*, *isdD*, *isdE*, *isdF*, *srtB*, and *isdG*. The IsdC protein is located in the cell wall. IsdD appears to be inserted in the plasma membrane. IsdE is predicted to be a lipoprotein, while IsdF is membrane protein with an ABC motif, and IsdG is proposed to reside the cytoplasm. IsdA, IsdB, IsdC, IsdD, IsdE, and IsdF have been shown experimentally to bind heme. However, IsdB is the only Isd protein that bound to hemoglobin. Attachment of Isd proteins to the cell wall is the result of the action of two sortases. SrtB (sortase B) attaches IsdC to murein via the NPQTN motif. SrtA (sortase A), which is not encoded in the *isd* operon, targets IsdA and IsdB to the cell wall via the above-mentioned LPXTG motif. Mutations in SrtB or SrtA result in a significant reduction or a complete absence of staphylococcal growth, respectively.

The model for the function of the Isd heme acquisition system predicts that hemoglobin binds to IsdB on the staphylococcal cell surface via an amino-terminal domain. Heme is removed from hemoglobin by IsdA and/or IsdB and is subsequently forwarded to IsdC, IsdD, IsdE, and IsdF, which promote heme's crossing of the plasma membrane. IsdG, the cytoplasmic binding protein, may be involved in the removal of iron from heme for utilization.

CYTOPLASMIC FATE OF HEME

Very little is known about the fate of heme once it reaches the cytoplasm. For iron contained in heme to be utilized by bacteria, the heme must first be degraded. The only method of degradation that has been characterized is that by bacterial heme oxygenases. A detailed description of bacterial heme oxygenases can be found in chapter 6.

Heme oxygenases of *C. diphtheriae* (HmuO), *N. meningitidis* (HemO), and *P. aeruginosa* (PigA) have been described. However, the scanning of completed genomes for homologues did not uncover many additional candidates. This indicates that other mechanisms of heme degradation may exist in bacteria that do not require heme oxygenases or that heme iron is not used as an iron source in many bacteria. It is also possible that is there is independent utilization of heme iron and nonheme iron, thus eliminating the need for heme degradation. Heme iron may be directly shuttled to electron transport components and other heme-requiring reactions, whereas iron required for other metabolic processes comes from storage or the environment.

The presence of heme inside bacteria puts them in a precarious situation. Although they require heme as a biological catalyst and sometimes as a source of iron, heme is toxic. Thus, great care must be taken by bacteria not to accumulate free heme, which can cause oxidative damage. HemS of *Y. enterocolitica* was initially described as a cytoplasmic factor involved in the degradation of heme. However, when HemS is absent, *Y. enterocolitica* is still able to use heme and hemoglobin. ShuS, the HemS homologue in *S. dysenteriae*, has been recently shown to sequester heme. One monomer of

ShuS binds one molecule of heme, and ShuS-heme forms an oligomeric complex. This has been viewed to be a heme-ferritin like complex. ShuS was also shown to possibly protect against the toxic affects of heme by binding to DNA. Although this is speculative, a function of ShuS may be to deliver heme to an as yet unknown heme oxygenase for degradation. Homologues of ShuS are found in *P. aeruginosa*, *Bordetella* spp., *R. leguminosarum*, and *Y. pestis*.

ACTIVE TRANSPORT OF HEME REQUIRES TonB

Although some evidence suggests there may be a TonB-independent mechanism of heme transport in *Neisseria* and *Haemophilus* species, heme transport in currently characterized systems is TonB dependent. Energy for heme transport is delivered to heme receptors via the cytoplasmic membrane proteins TonB, ExbB, and ExbD and requires the proton motive force of the cell. More information about the Ton system and about TonB-dependent iron transport can be found in chapter 7.

Like iron and heme transport systems, redundant *ton* systems have been identified in bacteria. Multiple *ton* systems exist in *V. cholerae*, *S. marcescens*, and *P. aeruginosa*. In addition, the genomic sequence of *Helicobacter pylori* reveals three putative sets of *exbB*- and *exbD*-like genes. Some common features of these redundant *ton* systems are beginning to emerge. In both *S. marcescens* and *V. cholerae*, one of the *tonB* loci is found within heme utilization operons. HasB, a TonB homologue of *S. marcescens*, is a component of the *has* operon (for "hemophore-dependent heme acquisition system"), and TonB1 of *V. cholerae* is cotranscribed with part of the *hut* heme transport system. Moreover, it appears that even though there is some functional redundancy between TonB systems in the same bacteria, certain TonB systems are specific for the utilization of heme and hemoproteins. The TonB1 system, and not the TonB2 system, appears to be involved in heme transport in *P. aeruginosa*. *V. cholerae* has three heme transporters, HutA, HutR, and HasR, which exhibit a functional requirement for specific TonB systems. HutA and HutR function with either *V. cholerae* TonB1 or TonB2, but heme transport through either receptor was more efficient in strains carrying *tonB1* system genes. In contrast, heme uptake through HasR was TonB2 dependent. In *S. marcescens*, both TonB and HasB can be involved in heme transport, but HasB does not function in the transport of nonheme iron.

Several TonB-dependent nonheme receptors have been reported to interact directly with TonB. However, only limited information is available about the specific interactions between bacterial TonB homologues and heme transporters or the mechanism by which the TonB-ExbBD complex transduces energy from the proton motive force into the active transport of heme. A characteristic motif, known as the TonB box, which is a site of interaction between TonB and ferric siderophore receptors, is a common feature found near the amino-terminal domain of most heme receptors. At least three studies have shown interactions involving TonB and heme receptors. (i) In our laboratory, the direct interaction between a high-affinity hemoglobin receptor, HmbR of *N. meningitidis*, and TonB has been demonstrated with purified proteins in vitro. Surface plasmon resonance was used to measure the affinity and kinetics of the interaction. The binding kinetics of the HmbR-TonB interaction were examined by surface plasmon resonance, and the binding affinity for neisserial TonB was found to be high, with a K_d of 76 nM. The rate of HmbR-TonB complex formation (k_a) is $1.3 \times 10^5 \ M^{-1} \ s^{-1}$, and the rate of dissociation of the complex (k_d) is $9.7 \times 10^{-3} \ s^{-1}$. (ii) Studies examining the specificity of one of the two *V. cholerae* TonB proteins for certain receptors have also been reported. Chimeric TonB proteins, made from *V. cholerae* TonB1 and *E. coli* TonB, and single amino acid substitutions, indicated that the carboxy-terminal end of the protein confers receptor specificity. (iii) Additionally, studies involving the high-affinity hemoglobin-haptoglobin receptor, HpuAB, suggested a conformational change in the receptor as a result of an interaction with TonB. The binding kinetics of hemoglobin was dramatically altered in strains deficient in TonB and those treated with protonophores that disrupt the proton motive force. The affinity of hemoglobin for HpuAB in both situations was essentially the same as when TonB was present; however, both the binding capacity of the receptor and the release of hemoglobin from the receptor were diminished. Based on the altered binding kinetics of hemoglobin and changes in the protease accessibility of receptor components, a model was proposed in which HpuA and HpuB interact to form a single high-affinity hemoglobin-binding site and TonB serves to enhance this productive or "binding-competent" state.

ROLE OF HEME TRANSPORT IN VIRULENCE

Bacterial virulence has been enhanced in numerous animal models of infection by the addition of heme and hemoproteins. The value of heme as an in vivo source of iron or porphyrin appears to vary among bacterial pathogens. For some, heme acquisition is important for virulence. Examples include uropathogenic *E. coli*, *Haemophilus ducreyi*, *Vibrio anguillarum*, and *N. meningitidis*. Meningococci deficient in two hemoglobin-specific receptors, HmbR and HpuAB, were attenuated in an infant rat model of infection. In uropathogenic *E. coli*, disruption of the heme receptor, ChuA, also had an effect on virulence; mutants remained able to infect mouse kidneys but failed to compete with the wild-type strain in mixed infections. Mixed infections with a BhuR heme receptor mutant and the parental strain of *B. avium* demonstrated a reduced ability of the mutant to colonize the tracheas of birds by this avian respiratory pathogen. Similarly, for the fish pathogen *V. anguillarum*, wild-type strains and complemented strains were more virulent in fish than was a mutant with a mutation in the HuvA heme receptor. Studies have shown that heme receptor, TdhA, in *H. ducreyi* is expressed during a human volunteer model of infection; sera from patients with chancroid have elevated levels of antibodies to both TdhA and HgbA. HgbA was also shown to be an important virulence factor in the establishment of infection in the human model system.

In contrast, disruption of heme transport in other pathogenic bacteria seems to have no effect on virulence. For example, *Y. pestis* strains lacking the Hmu system are as virulent as the parental strain when injected into mice either subcutaneously or intravenously. In the disease model for *S. dysenteriae*, ShuA is not required for bacterial invasion or multiplication. Redundant heme systems may function in the absence of these heme transporters, but in one bacterium, *V. cholerae*, this is not the case. A *V. cholerae* triple mutant with mutations in the three known heme receptors is completely deficient in vitro for heme utilization. However, this mutant is virulent when examined in the infant-mouse model of cholera and competes efficiently with the wild type for colonization in the mouse intestine. This suggests that heme utilization is not essential and that other sources of iron are available in vivo. Caution should be used, however, when evaluating the relevance of heme transport in virulence, keeping in mind the stage of infection under assessment. For instance, studies of in vivo expression of the Hmu system of *Y. pestis* have not been done, but a second heme receptor, HemR, has been examined. In the mouse model, HemR is highly expressed in some tissues and minimally expressed in others, suggesting that the evaluation methods would be highly relevant in determining the contribution of HemR to virulence.

Several bacteria, including *Burkholderia cepacia* and *P. gingivalis*, exhibit an unusual phenotype that is not related to heme transport but is related to heme and virulence. In these bacteria, there is cell surface deposition of ferrihemes, which form a shell or armor of heme around the cell. Bacteria with this heme shell are more virulent than those without it. A number of putative functions for cell surface heme have been proposed, including extracellular heme storage and antioxidant defense. Ferrihemes may act defensively against attack by hydrogen peroxide by virtue of their inherent catalase activity.

OUTLOOK

These are exciting times for the field of bacterial heme transport. Numerous heme transport systems are just now being discovered in a diverse array of bacterial species, including both gram-negative and gram-positive bacteria. Genome sequencing will undoubtedly result in the discovery of more systems within the next few years. Systems that were discovered several years ago in bacteria such as *Yersinia*, *Porphyromonas*, *Neisseria*, and *Vibrio* are now yielding valuable information: information about structural domains and essential amino acids that are crucial to transporter function, information about receptor interactions with the energy transducer TonB, and information about substrate recognition based on detailed biochemical analysis. Structural information about receptors and other system components will undoubtedly follow, allowing us to investigate structure-function relationships and further understand the mechanism of heme transport. Questions such as how heme can be recognized by some receptors in the context of a variety of host hemoproteins and how heme is removed from tight associations with host proteins may be answered. In fact, structural data are already available for several related proteins, including the hemophore of *S. marcescens*, HasA, and *E. coli* TonB.

In addition, there remain a wealth of new questions to answer about the regulation of heme acquisition, the role of heme in virulence, and possible new heme acquisition systems, like the novel heme transporter system of *S. aureus*. A heme-sensitive two-component regulation system and positive regulators of heme transport have only recently been identified, and these may be only part of a larger signaling cascade that regulates heme and iron utilization.

Heme acquisition systems are highly redundant and, in some species, phase variable; both features attest to their importance. But what is the role of heme in virulence? Results ranging from heme transport being essential to virulence to being dispensable for virulence have been reported for different model systems of bacterial infection. Studies have also shown that a heme acquisition system can be virtually turned off in some tissues and highly expressed in others within the same animal model. Certainly, the heme requirements of some bacteria are higher than those of others, and some bacteria may not require heme in vivo to the extent that others do. However, additional investigations are required to determine the contribution of heme to bacterial pathogenesis.

Furthermore, many components of heme acquisition systems have not been identified or have, thus far, unknown functions. An excellent example is the *phuW* gene of *P. aeruginosa*, which is not essential for heme utilization but is required for optimal function. A majority of bacterial genomic sequences have no homologues of heme oxygenases; therefore, it is likely that another type of enzyme involved in heme catabolism exists. Certain gram-negative species have receptors but no identified heme-specific ABC transport systems. The mechanism of heme transport across the inner membrane remains a mystery in these organisms. Although genome searching may be useful for the identification of homologues of already identified systems, other methods are needed to find and characterize these undescribed components.

ACKNOWLEDGMENTS

We thank S. Pittard, K. Gernert, and L. Balding for their help with the computer software. We thank W. Shafer, G. Churchward, and H. Alexander for suggestions and for reading the manuscript.

SUGGESTED READING

Bates, C. S., G. E. Montanez, C. R. Woods, R. M. Vincent, and Z. Eichenbaum. 2003. Identification and characterization of a *Streptococcus pyogenes* operon involved in binding of hemoproteins and acquisition of iron. *Infect. Immun.* **71:**1042–1055.

Bracken, C. S., M. T. Baer, A. Abdur-Rashid, W. Helms, and I. Stojiljkovic. 1999. Use of hemeprotein complexes by the *Yersinia enterocolitica* HemR receptor: histidine residues are essential for receptor function. *J. Bacteriol.* **181:**6063–6072.

Braun, V., and H. Killmann. 1999. Bacterial solutions to the iron-supply problem. *Trends Biochem. Sci.* **24:**104–109.

Cornelissen, C. N., and P. F. Sparling. 1994. Iron piracy: acquisition of transferrin-bound iron by bacterial pathogens. *Mol. Microbiol.* **14:**843–850.

Crosa, J. H. 1989. Genetics and molecular biology of siderophore-mediated iron transport in bacteria. *Microbiol. Rev.* **53:**517–530.

Drazek, E. S., C. A. Hammack, and M. P. Schmitt. 2000. *Corynebacterium diphtheriae* genes required for acquisition of iron from haemin and haemoglobin are homologous to ABC haemin transporters. *Mol. Microbiol.* **36:**68–84.

Genco, C. A., and D. W. Dixon. 2001. Emerging strategies in microbial haem capture. *Mol. Microbiol.* **39:**1–11.

Gray-Owen, S. D., and A. B. Schryvers. 1996. Bacterial transferrin and lactoferrin receptors. *Trends Microbiol.* **4:**185–191.

Hantke, K. 2001. Iron and metal regulation in bacteria. *Curr. Opin. Microbiol.* **4:**172–177.

Henderson, D. P., E. E. Wyckoff, C. E. Rashidi, H. Verlei, and A. L. Oldham. 2001. Characterization of the *Plesiomonas shigelloides* genes encoding the heme iron utilization system. *J. Bacteriol.* **183:**2715–2723.

Kirby, A. E., D. J. Metzger, E. R. Murphy, and T. D. Connell. 2001. Heme utilization in *Bordetella avium* is regulated by RhuI, a heme-responsive extracytoplasmic function sigma factor. *Infect. Immun.* **69:**6951–6961.

Lee, B. C. 1995. Quelling the red menace: haem capture by bacteria. *Mol. Microbiol.* **18:**383–390.

Mazmanian, S. K., E. P. Skaar, A. H. Gaspar, M. Humayun, P. Gornicki, J. Jelenska, A. Joachmiak, D. M. Missiakas, and O. Schneewind. 2003. Passage of heme-iron across the envelope of *Staphylococcus aureus*. *Science* **299:**906–909.

Mey, A. R., and S. M. Payne. 2001. Haem utilization in *Vibrio cholerae* involves multiple TonB-dependent haem receptors. *Mol. Microbiol.* **42:**835–849.

Mey, A. R., and S. M. Payne. 2003. Analysis of residues determining specificity of *Vibrio cholerae* TonB1 for its receptors. *J. Bacteriol.* **185:**1195–1207.

Murphy, E. R., R. E. Sacco, A. Dickenson, D. J. Metzger, Y. Hu, P. E. Orndorff, and T. D. Connell. 2002. BhuR, a virulence-associated outer membrane protein of *Bordetella avium*, is required for the acquisition of iron from heme and hemoproteins. *Infect. Immun.* **70:**5390–5403.

Olczak, T., D. W. Dixon, and C. A. Genco. 2001. Binding specificity of the *Porphyromonas gingivalis* heme and hemoglobin receptor HmuR, gingipain K, and gingipain R1 for heme, porphyrins, and metalloporphyrins. *J. Bacteriol.* **183:**5599–5608.

Rohde, K. H., A. F. Gillaspy, M. D. Hatfield, L. A. Lewis, and D. W. Dyer. 2002. Interactions of haemoglobin with the *Neisseria meningitidis* receptor HpuAB: the role of TonB and an intact proton motive force. *Mol. Microbiol.* **43:**335–354.

Schmitt, M. P. 1999. Identification of a two-component signal transduction system from *Corynebacterium diphtheriae* that activates gene expression in response to the presence of heme and hemoglobin. *J. Bacteriol.* **181:**5330–5340.

Schryvers, A. B., and I. Stojiljkovic. 1999. Iron acquisition systems in the pathogenic *Neisseria*. *Mol. Microbiol.* **32:**1117–1123.

Stojiljkovic, I., and D. Perkins-Balding. 2002. Processing of heme and heme-containing proteins by bacteria. *DNA Cell Biol.* **21:**281–295.

Vanderpool, C. K., and S. K. Armstrong. 2003. Heme-responsive transcriptional activation of *Bordetella bhu* genes. *J. Bacteriol.* **185:**909–917.

Wandersman, C., and I. Stojiljkovic. 2000. Bacterial heme sources: the role of heme, hemoprotein receptors and hemophores. *Curr. Opin. Microbiol.* **3:**215–220.

BACTERIAL HEME OXYGENASES

Melanie Ratliff-Griffin, Angela Wilks, and Igor Stojiljkovic[†]

6

Most of the body's iron is found intracellularly, and two-thirds of that iron is found within the hemoglobin molecule (Table 1). Hemoglobin, a critical transport molecule for tissue-specific oxygen exchange, contains four heme moieties, each containing a single iron atom. Thus, hemoglobin may serve as a powerful source of iron for invading bacteria. Many microbial pathogens are known to possess heme acquisition systems that allow the organisms to utilize heme or heme-bound proteins as sources of iron. These systems have been demonstrated to procure iron from heme, hemoglobin, haptoglobin, or hemopexin and are identified as nonsiderophore heme iron acquisition systems. As with free iron, free heme is kept in a limited state by being bound by the high-affinity host protein hemopexin. Similarly, hemoglobin is complexed with haptoglobin. Therefore, the versatility of the bacterial heme acquisition systems enables them to circumvent the host defense of bacteriostasis by iron limitation. Pathogens that use heme or heme-bound proteins as sources of iron include *Neisseria* spp., *Yersinia* spp., *Pseudomonas* spp., *Serratia* spp., *Corynebacterium diphtheriae*, *Bacilllus fragilis*, *Vibrio* spp., *Haemophilus* spp., *Plesiomonas* spp., *Porphyromonas gingivalis*, *Helicobacter pylori*, and *Shigella* spp. As the systems of these various organisms and others are being investigated, the basic heme acquisition and assimilation pathways for both gram-negative and gram-positive pathways are becoming clearer.

HEME ACQUISITION PATHWAYS

Recent reviews have extensively and thoroughly covered the experimental data of the heme acquisition pathways, and therefore only the general mechanisms are discussed here.

Gram-negative organisms have the difficult challenge of transporting nutrients across a double membrane barrier, whereas gram-positive organisms have only a single cell membrane. Heme transport in gram-negative bacteria, as modeled by *Yersinia enterocolitica*, begins with surface heme receptors that recognize and bind the various heme species (i.e., heme, hemoglobin, and haptoglobin). Transportation across the outer membrane is generally TonB dependent and requires auxiliary proteins. Once in the periplasmic space, the heme molecule is thought to be bound by a heme-binding

---

Melanie Ratliff-Griffin and Igor Stojiljkovic, Department of Microbiology and Immunology, Emory School of Medicine, 3152 Rollins Research Center, Atlanta, GA 30322. *Angela Wilks*, Department of Pharmaceutical Sciences, School of Pharmacy, University of Maryland, Baltimore, MD 21201.

[†] Deceased.

Iron Transport in Bacteria, Edited by Jorge H. Crosa, Alexandra R. Mey, and Shelley M. Payne
© 2004 ASM Press, Washington, D.C.

TABLE 1 Steady-state distribution of iron in humans

Iron containing molecule	% of total body iron	Function
Hemoglobin	70	Tissue-specific oxygen transport and iron reutilization
Myoglobin	5	Oxygen transport to muscle
Ferritin	15	Iron storage
Catalases, hydroxylases, cytochromes, electron transport, etc.	10	"Housekeeping functions"

protein and then shuttled to the inner membrane, where it is translocated across the membrane for further processing. For gram-positive organisms, as demonstrated by *C. diphtheriae*, the process is similar, but it is thought that lipoproteins anchored in the cytoplasmic membrane act as the cell surface receptors. These lipoproteins bind heme or heme-bound molecules and shuttle them directly to cytoplasmic membrane proteins in a manner similar to that for their counterparts in the gram-negative organisms.

An additional pathway of heme assimilation has also been observed in several gram-negative bacteria, including *Serratia marcescens*, *Yersinia pestis*, *Pseudomonas aeruginosa*, and *Pseudomonas fluorescens*. With this system, the bacteria secrete extracellular proteins called hemophores, which function to scavenge heme or heme-bound proteins. The hemophores bind and deliver these molecules back to the cell surface receptor for internalization. Heme assimilation beyond this step is presumed to be similar to the mechanism previously described for heme uptake systems.

What happens once the heme molecule is delivered to the cytoplasm is the remaining focus of this review. Only recently, with the discovery of bacterial heme oxygenases, have we gained some insight into this question.

HEME OXYGENASES

Heme oxygenase activity from mammalian cells was first demonstrated in the 1960s. This enzyme is capable of degrading heme to α-biliverdin with the liberation of free iron (Fig. 1). Since then, HO-1, an inducible isoform of human heme oxygenases, has been identified and shown to respond to a diverse group of stimuli ranging from its cofactor, heme, to heat shock and heavy metals, suggesting that it may play a significant role in protection against oxidative stress. Heme, being a highly hydrophobic molecule, has the ability to intercalate within DNA strands, causing damage. Thus, it is also plausible that microbial pathogens capable of utilizing heme and hemoglobin as a source of iron may have heme oxygenase-like enzymes for the degradation of heme molecules, not only for the acquisition of iron but perhaps also for heme detoxification.

Bacterial heme oxygenase activity was first found in the gram-positive pathogen *C. diphtheriae*, the causative organism of the respiratory disease diphtheria (see also chapter 22). The gene encoding this enzyme, *hmuO*, was identified and characterized based on its homology (33%) to eukaryotic heme oxygenase HO-1 and its induction by heme. In vitro studies of the *hmuO* gene product revealed that the enzyme could oxidatively cleave heme to α-biliverdin with the release of carbon monoxide and iron (Fig. 1). This reaction, as in the mammalian system, also requires an NADPH-dependent reductase source. Two notable differences between HmuO and HO-1 are solubility and size. HmuO is a soluble cytoplasmic protein of 25 kDa, whereas HO-1 has a mass of 33 kDa

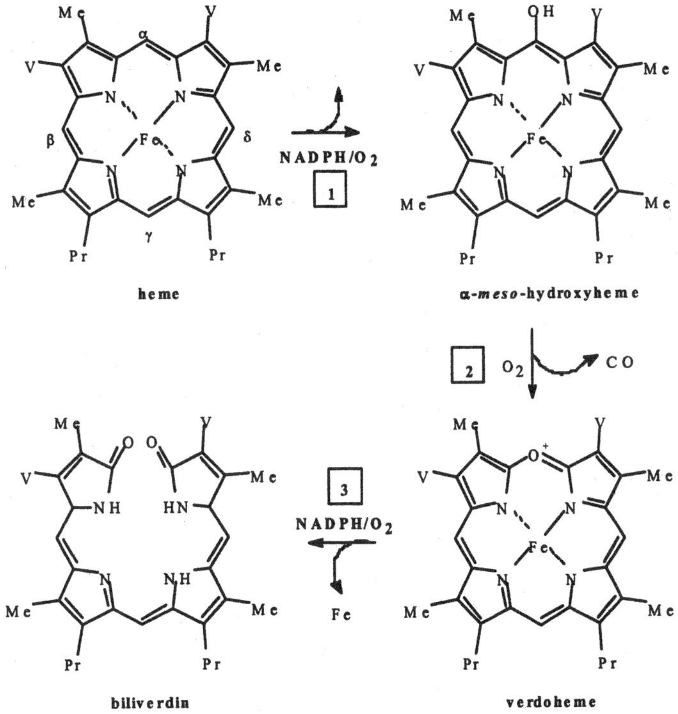

FIGURE 1 Chemical steps in heme degradation, as defined by the studies of eukaryotic heme oxygenases and *C. diphtheriae* HmuO. Me, methyl side chain; V, vinyl side chain; Pr, propionic side chain. All characterized heme oxygenases cleave the porphyrin ring at the α-carbon atom, with the exception of *P. aeruginosa* PigA, which cleaves at the β-carbon.

and is anchored within the membrane. However, it should be further noted that a truncated form of HO-1 remains soluble and retains partial wild-type activity. Wilks et al. have concluded that the role of HmuO, as it is induced under iron-limiting conditions, is to release the iron from heme for its subsequent use by the pathogen.

Structure-function relationships have been investigated by site-directed mutagenesis in the hopes of identifying crucial residues participating in the binding of heme and the enzymatic cleavage of the porphryin ring (see below). Overall, human heme oxygenase HO-1 and bacterial heme oxygenase HmuO are mechanistically the same, and they show only slight structural variations when investigated by such methods as electron paramagnetic resonance, resonance Raman spectroscopy, and nuclear magnetic resonance studies.

We have recently investigated two additional bacterial heme oxygenases, HemO of *Neisseria meningitidis* and PigA of the opportunistic pathogen *P. aeruginosa*, the first identified in gram-negative microorganisms. HemO is of extreme importance to the obligate human pathogen *N. meningitidis*, which, unlike many other pathogens, cannot scavenge free iron by the production and utilization of siderophores, a common bacterial system of high-affinity iron chelation. Due to spontaneous hemolysis and to the presence of heme-associated proteins in human body fluids and surfaces, sufficient levels of heme can be derived from these sources to support bacterial growth. HemO is essential to the utilization of heme and hemoglobin as the sole source of iron, since an *N. meningitidis hemO* knockout mutant is unable to utilize either. Mechanistically, HemO is identical to HmuO in that it converts heme to α-biliverdin with the liberation of iron and carbon monoxide. On the other hand, PigA, which is 33% identical to HemO, may represent a novel bacterial heme oxygenase. As seen with *N. meningitidis*, PigA of *P. aeruginosa* is essential to heme utilization under iron starvation conditions in minimal growth media. The knockout mutant also could not utilize hemoglobin. However, on in vitro characterization of the PigA protein, we discovered that PigA produced a structurally different biliverdin from that produced by all previously characterized heme oxygenases. Instead of α-biliverdin, PigA yields β-biliverdin (Fig. 1). Preliminary data suggest that PigA does not participate in particular proprionic bonding as seen with HmuO and HemO. The reason for this different isomer pattern is unknown, but, interestingly, *P. aeruginosa* also possesses a second heme oxygenase system that is most probably different from the PigA system. This system could not complement the *pigA* knockout mutation and may be expressed under different metabolic conditions. This second heme oxygenase system, termed BphO, has been identified in other bacteria. It will be of interest to determine whether the two systems produce different isomer patterns and, if so, whether the different biliverdin isomer patterns correlate with the use of a particular heme oxygenase system.

BACTERIAL PHYTOCHROME HEME OXYGENASES

Cyanobacteria have long been known to be able to carry out photosynthesis for the production of energy. To accomplish this, they use a photoreception system similar to that found in plants, in which chromophore proteins are required to harvest the energy from light. Functional phytochrome-like receptors, BphPs (for "bacterial phytochrome-like proteins") have been characterized by Davis and colleagues for *P. aeruginosa* and *Deinococcus radiodurans* and are most probably part of a functioning photosensory system. The phytochrome receptor protein acts as a light-activated kinase and requires a phytobilin cofactor. The synthesis of phytobilins requires the action of bacterial heme oxygenases, BphOs (for "bacterial phytochrome heme oxygenase"). The α-biliverdin product of BphO is converted to the final bilin cofactor by a series of enzymatic reactions. In general, it appears that the *bphO* genes lie just upstream from other known bacterial photoreception genes; this suggests that they collaborate with each other. We report here the widespread existence of genes with identity to either HmuO of *C. diphtheriae* or BphO of *P. aeruginosa* in both photosynthetic and nonphotosynthetic bacteria, further suggesting the presence and diversity of the heme oxygenases genes (Table 2).

STRUCTURAL SIMILARITIES IN THE OVERALL FOLD OF THE MAMMALIAN AND BACTERIAL HEME OXYGENASES

The recently determined crystal structures of a truncated soluble form of the human and rat HO-1 enzymes complexed with heme have provided significant insight into both the mechanism and regiospecificity of the heme oxygenase reaction. The structure of the human HO-1 lacking the C-terminal 55 amino acids, including a 23-residue hydrophobic membrane anchor, has been refined to 1.5Å. Despite the truncation of the protein, the activity and regiospecificity of the reaction were largely unaffected. Recently, the crystal structures of two of the bacterial heme oxygenases have been solved. The *C. diphtheriae* HmuO crystal structure has been solved, and, more recently, the crystal structure of the soluble 24-kDa HemO from *N. meningitidis* has been solved to 1.5 Å. The most striking feature of the HemO structure, given the limited sequence identity between HO-1 and HemO, is the high degree of structural conservation (Color Plate 10B).

All of the heme oxygenase proteins thus far crystallized have a novel α-helical fold. The δ-*meso* edge of the heme and the propionates are exposed at the molecular surface of the protein. Interestingly, the bacterial enzymes appear to have a more compact overall fold, with the distal helix more closely approaching the heme. It is unclear whether this is structurally or catalytically relevant, but

TABLE 2 Amino acid comparison of heme oxygenase-like proteins from diverse microorganisms

Organism	% Identity[a] to:		
	HemO[b]	HmuO[c]	BphO[d]
HemO homologues			
Pseudomonas aeruginosa	**33**	23	
Agrobacterium tumefaciens	**41**		29
Xanthomonas axonopodis	**33**		32
Rhodobacter sphaeroides	**32**		28
HmuO homologues			
Nostoc sp. strain 1		**31**	
Nostoc sp. strain 2		**34**	
Bradyrhizobium sp.		**33**	
Clostridium perfringens		**28**	
Streptomyces coelicolor		**41**	
Synechococcus sp. strain PCC		**34**	
Clostridium glutamicum		**58**	
Prochlorococcus marinus		**30**	
Synechocystis sp. strain 1		**31**	
Synechocystis sp. strain 2		**33**	
BphO homologues			
Deinococcus radiodurans	27		**40**
Pseudomonas fluorescens	20		**38**
Pseudomonas syringae	20		**40**
Rhizobium leguminosarum	24		**38**

[a] Percent identity between two proteins is presented in bold. Comparison was done using the Pairwise BLAST program at NCBI.
[b] HemO from *N. meningitidis*.
[c] HmuO from *C. diphtheriae*.
[d] BphO from *P. aeruginosa*.

it may be that the human HO-1 structure is more relaxed as a consequence of the missing C-terminal domain.

Heme Pocket

The heme is inserted between two helices, often referred to as the proximal and distal helices. The proximal helix donates the histidine ligand (His25 in HO-1, His20 in HmuO, and His23 in HemO [Color Plate 10A]), with Thr21 in HO-1 and the corresponding Thr19 in HemO contacting the heme through a water molecule. Additional contact residues with the heme include Val26 and Asp27 in HemO, corresponding to Ala28 and Glu29 of HO-1. In site-directed mutagenesis studies of the proximal ligand, the HO-1 (H25A) mutant was shown to be inactive, although activity could be rescued by the addition of exogenous imidazole. In contrast, the *C. diphtheriae* HmuO (H20A) mutant converted heme to verdoheme in the absence of exogenous ligand, and, on reconstitution with imidazole, reacted to give biliverdin as the final product. The reason why the HmuO (H20A) mutant, in contrast to the HO-1 (H25A) mutant, can catalyze the initial hydroxylation of the heme is not understood. Resonance Raman studies on the HmuO (H25A)-heme complex have shown that the ligand to the heme is similar in nature to that of the HO-1 (H25A)-heme complex and most probably arises from a water molecule or interaction with a Glu or Asp side chain. In addition, we have recently constructed the HemO (H23A) mutant and found that in contrast to both the HO-1 and HmuO proximal mutants, this mutation resulted in a heme complex that had no activity in the presence or absence of excess imidazole. The variation in the results obtained with the proximal His mutants of the respective proteins suggests that subtle changes within the active-site polarity, redox potential, and/or geometry may account for the different reactivity of the proximal mutants.

The distal pocket of the heme-HemO complex is significantly smaller, having a solvent-accessible volume of 7.5 Å, in contrast to 53.5 Å for the more closed form of the heme–HO-1 complex. The distal helix has a kink of approximately 50° directly over the heme, provided by the glycines of the highly conserved sequence ^{116}Gly-Ser-Asn-Leu-Gly-Ala^{121} in HemO, corresponding to ^{139}Gly-Asp-Leu-Ser-Gly-Gly^{144} in HO-1. This kink in the helix is proposed to provide flexibility to the helix, which is required for binding of the substrate (heme) and release of the product (biliverdin). The helix closely approaches the heme with direct backbone contacts from Gly139 and Gly143 in HO-1. Recently, it was shown that mutation of Gly139 or Gly143 resulted in a loss of oxygenase activity and an increase in peroxidase activity. This suggested that these residues are critical in maintaining an electrophilic environment that is conducive to oxygenase activity. The reversible conversion of Fe^{2+} to Fe^{3+} is essential for this activity; hence, the primary role of the distal helix may be to suppress the improper formation of iron species that interfere with the reaction.

Regiospecificity

The heme propionates and their interaction with the protein are critical for binding and for correctly orienting the α-*meso* carbon of the heme for hydroxylation. Residues possibly involved in interactions with the heme in HO-1 are Lys18, Lys22, Lys179, Arg183, and Tyr134. In the heme-HemO complex, only Lys16 and Tyr112, corresponding to Lys18 and Tyr134, are absolutely conserved in the structure. Lys22 is replaced by Thr20, and Trp53 largely fills the space occupied by Arg183 and Lys179. Recent mutagenesis studies of Arg183 have suggested that this residue is involved in orienting the heme within the active site through direct hydrogen bonding to the propionate group. The significance of the ionic/hydrogen-bonding interactions of the heme propionates with Lys16 and Tyr112 in the HemO protein is highlighted by the recent finding that the *Pseudomonas* heme oxygenase (PigA), in which these residues are replaced by Asn19 and Phe117, oxidatively cleaves heme at the δ- and β-*meso*-carbons. Replacement of both Asn19 and Phe117 of PigA with a Lys and Tyr, respectively, as is found in HemO, has demonstrated that the formation of the α-isomer as well as the β- and δ-isomer is a function of two alternate in-plane heme seatings, one representative of the wild-type PigA enzyme and one more clearly identified as similar to that of the wild-type HemO. Clearly, orientation of the heme within the active site, and hence regioselectivity, is controlled in part by interactions with the heme propionates. As a result of a rotation of the heme by 180° along the α/γ axis on binding the active site, PigA gives a 2.3:1 ratio of the δ- and β-isomers. This phenomenon has been observed previously for the human HO-1, in which the ratio is closer to 1:1. It is clear that such a rotation would have no effect on the α-selective enzymes, since the α-*meso*-carbon does not change spatially. However, a rotation along the α/γ-axis in the δ-selective enzyme would result in the β- and δ-*meso*-carbons switching positions and allowing cleavage at alternative positions in a ratio reflective of the isomer pattern observed by nuclear magnetic resonance spectroscopy. It should be noted that in a previous publication we reported the isomer ratio as 2.3:1 β/δ; however, this discrepancy arose as a consequence of misinterpretation of a published high-performance liquid chromatography method, and it is clear from our recent mutagenesis studies and further analysis of the high-performance liquid chromatography systems that the ratio is 70% δ-biliverdin and 30% β-biliverdin.

Further clarification of many of the unresolved questions arising from the crystal structures of both HO-1 and HemO require structural characterization of reaction intermediates in the conversion of heme to biliverdin, as well as heteronuclear nuclear magnetic resonance and molecular dynamic simulation studies directed at understanding the interplay between protein dynamics and heme electronics in the reaction of the heme oxygenase enzymes.

DISTRIBUTION OF HEME OXYGENASES AMONG BACTERIAL SPECIES

Previously, heme oxygenases were identified in only a small number of bacteria. This may have been due to the limited number of available sequenced genomes or to the extremely diverse nucleotide sequences between potential heme oxygenases. Here, we have used the amino acid sequences of HemO, HmuO, PigA, and BphO to conduct BLAST searches of the finished and unfinished whole-genome sequences. Only HmuO and BphO revealed significant identities, as seen in Table 2. Potential heme-degrading enzymes were found in nearly all branches of the eubacterial branch of the tree of life, including the purple bacteria, proteobacteria, cyanobacteria, and the *Thermococcales*. All of the cyanobacteria, which are characterized as being photosynthetic bacteria, contain open reading frames whose products have amino acid sequence identity in the range of 30 to 34% to HmuO of *C. diphtheriae*. While genes involved in photoreception have not currently been identified in *C. diphtheriae*, it is likely that the heme oxygenases of the cyanobacteria are involved in this process, as predicted for the BphO enzymes. For at least one of these organisms, *Synechocystis* sp. strain PCC 6803, the heme oxygenase gene has been characterized and determined to be able to functionally complement the heme oxygenase *hy1* mutant of the plant *Arabidopsis thaliana* which was defective in phytochrome responses, suggesting its role as a BphO-like enzyme. Similarly, the organisms of the phylum *Proteobacteria*, including plant pathogens such as the pseudomonads and *Xanthomonas axonopodis* and the plant-associated bacteria *Rhizobium leguminosarum* and *Agrobacterium tumefaciens*, most probably possess photoreception systems for which the heme by-products are consumed. The remaining organisms range from gram-positive human pathogens to the extremely radioresistant *D. radiodurans*. We speculate that these organisms may use these heme-degrading enzymes for heme iron consumption versus heme metabolism, since this would be the function most suited to their biological niches. As mentioned above, *P. aeruginosa* is the only organism identified thus far to possess both types of heme oxygenases; interestingly, this may correlate with its well-documented existence in diverse environments.

The various related heme oxygenases and bacterial phytochrome heme oxygenases can be aligned with some degree of conservation of particular amino acid residues (Fig. 2). Notably, the critical proximal-helix histidine residue of HemO (His23) is absolutely conserved among all heme oxygenases. Second, the specific flexible glycine-rich region of the distal helix is conserved among some related bacterial species (Fig. 2). As described above, the glycine residues are necessary for orientation of the heme molecule within the active site and for release of the biliverdin product. The GSNLGA consensus sequence observed in HemO and PigA is also present in the putative heme oxygenase of *A. tumefaciens*. With only slight variations, the cyanobacteria have the conserved sequence of GDLSGG, which is also seen in the first identified bacterial heme oxygenase, HmuO, of *C. diphtheriae*. Lastly, the BphO proteins all have the conserved sequence GATLGG in common with each other and with human HO-1. Although the glycine residues in these sequences are most probably critical, as shown for HO-1, each group of seemingly related proteins also contains conserved adjacent residues, as seen in the sequences. Interestingly, the

HemO-like proteins

```
             10        20        30        40        50        60    68
PA   ---MDTLAPESTRQNLRSQRLNLLTNEPHQRLESLVKSKEPFASRDNFARFVAAQYLFQHDLEPLYRN
AT   MTSAEPAAPVAEIEQSRVKRLKAATRGAHGDLDSFIMAAKPFESRENFGKFVETQYLFHRDLDVFFSN
NM   ------MSETENQALTFAKRLKADTTAVHDSVDNLVMSVQPFVSKENYIKFLKLQSVFHKAVDHIYKD
                  :.**:    *       *   ::.::  :.**  *::*: :*:    * :*::  ::  ::  :
             70        80        90       100       110       120      135
PA   EALARLFPGLASRARDDAARADLADLGHPVPEGDQS--VREADLSLA-EALGWLFVSEGSKLGAAFLF
AT   VTLDGLLPDLKGRRRLAMIEQDLADLGHAIPDTDEPRFTAEMPFDLP-EAMGWLYVVEGSNLGAAFLL
NM   AELNKAIPELEYMARYDAVTQDLKDLGE-----EPYKFDKELPYEAGNKAIGWLYCAEGSNLGAAFLF
         *  :.* *  *        **  ***.        :    *     :.*:***.  ***.******.
            140       150       160       170       180       190      200
PA   KKAAALELDENFGARHLAEPEGGRAQGWKSFVAILDGIELNEEEERLAAKGASDAFNRFGDLLERTFA
AT   KDAAKLGLGEEFGARHLAGAPEGRGLHWRTFTAALDEISLTVQEEERVVAGAEAAFRAVHAYAQQRMG
NM   KHAQKLDYNGEHGARHLAPHPDGRGKHWRAFVEHLNALNLTPEAEAEAIQGAREAFAFYKVVLRETFG
       *.*   *    .:.******   **. *::*.   *:  ::.*. :  *      **          .  ::
```

HmuO-like proteins

```
             10        20        30        40        50        60    67
SC   ----MDSFSTLIRTASHQQHVEAETSTFMSDLLGGGLGVDAYARYTEQLWFVYEALEAAAGR--LAA
CD   MTTATAGLAVELKQSTAQAHEKAEHSTFMSDLLKGRLGVAEFTRLQEQAWLFYTALEQAVDA--VRA
Bs   ----------MRERTKTLHVTAERTGVVAELLRGRGTVRAYALLLRNLLPVYEALEAELVR--HQA
Syn  ---MSVNLASQLREGTKKSHSMAENVGFVKCFLKGVVEKNSYRKLVGNLYFVYSAMEEEMAK--FKD
Ns   ---MSSNLANKLRVGTKKAHTMAENVGFVKCFLKGVVEKSSYRKLVANFYVVYSAMEEEMEK--HSQ
Ss   ---MGANLATKLREGTKKAHTMAENVGFVKCFLKGVVEKNSYRKLVANLYFVYSAMEEEELER--HRD
PM   ---MAVALAGQLREGTKKSHTMAENVGFVACFLKGVVEKKSYRKLISDLYFVYEAMEEEIERLVQEE
CP   ----MNSFMMDIKNNSNDLHAVAEKTGFLKRLLEGKASTESYAEYLYNLYEVYNAIEVNLEK--CKD
                      ::  *  **  .: :**    :        * :       :.* *.*:*
             70        80        90       100       110       120      130
SS   DPVAGPFVRPELLRLASLERDLAHLRG-ADWRTGLTALPATEAYAARVRECAEEWP-AGYVAHHYTR
CD   SGFAESLLDPALNRAEVLARDLDKLNGSSEWRSRITASPAVIDYVNRLEEIRDNVDGPALVAHHYVR
Bs   SPVVGLTVRPELHRCPAIKADLAALDA-----SDLPLLPEAIAYVRAIQEAGSGSG-HPLLAHAYTR
Syn  HPILSHIYFPELNRKQSLEQDLQFYYG-SNWRQEVKISAAGQAYVDRVRQVAATAP-ELLVAHSYTR
Ns   HPIVSKINFSQLNRKQTLEQDLSYYYG-ANWREQIQLSPAGEAYVQRIREISATEP-ELLIAHSYTR
Ss   NDKIAGIYFPVLNRKTSLERDLAFYYG-EDWRQQIQPSKAAQSYVARIREVSNTAP-ELLVGHAYTR
PM   HPVIKHIGFKSLFRKETLENDLKFYFG-DNWQKEINISKSAKEYVNRIHLVAKKSP-ELLVGHHYTR
CP   NKVVKDFVLPEIYRAEAILKDLKFLLE--ENLNTMKPLASTRAYVARINEIGETAP-ELLVAHAYTR
              :  *     :  **       :             :           *.  ::           :.* *.*
            140       150       160       170       180       190      200
SC   YLGDLSGGQIIRDKAERTWGFARKGDGVRFYVFEEISN---PAAFKREYRDLLDGIRADDLEKQRVV
CD   YLGDLSGGQVIARMMQRHYGVDP--EALGFYHFEGIAK---LKVYKDEYREKLNNLELSDEQREHLL
Bs   YLGDLSGGQIIKKILARSLELQP--EALSFYEFPAITD---IPRFKTEYREALEQAGSAMTEHDSVV
Syn  YLGDLSGGQILKKIAQNAMNLHG-GTAFYEFADIDD---EKAFKNTYRQAMNDLPIDQATAERIV
Ns   YLGDLSGGQILKNIAVTAMNLND-GQGTAFYEFADISD---EKAFKAKYRQTLDELAIDEATGDRIV
Ss   YLGDLSGGQILKTIAQRGMNL-D-GAGTAFYEFEAIED---EKAFKQTYRQAMDTLPVDEATADRIV
PM   YIGDLSGGQILKKIAKKALNLDG-DNGLNFYEFKNIDD---EKKFKEEYSKTLNQLPINQNIADQII
CP   YLADLFGGRTIYGMVKDLYKIDE--EGLNYYKYETLSDGSEMKGFVMNYHNKLNNIELNEEMKERFI
     *:.** **  :.                     .    : *: :       *  :. *          :  : .:
```

FIGURE 2 Amino acid sequence alignment of heme oxygenase proteins. Comparisons were made by using the ClustalW 1.8 program. (A) HemO-like proteins; (B) HmuO-like proteins; (C) BphO-like proteins. NM, *Neisseria meningitidis*; AT, *Agrobacterium tumefaciens*; PA, *Pseudomonas aeruginosa*; SC, *Streptomyces coelicolor*; CD, *Corynebacterium diphtheriae*; Bs, *Bradyrhizobium* sp.; Syn, *Synechocystis* sp.; Ns, *Nostoc* sp.; Ss, *Synechococcus* sp.; PM, *Prochlorococcus marinus*; CP, *Clostridium perfringens*; DR, *Deinococcus radiodurans*; PF, *Pseudomonas fluorescens*; PS, *Pseudomonas syringae*; RL, *Rhizobium leguminosarum*; RS, *Rhodobacter sphaeroides*; XA, *Xanthomonas axonopodis*. Bold type indicates residues that are critical to the coordination of the heme molecule within the proteins. Dots and asterisks indicate similar and identical amino acids, respectively, among all proteins.

R. sphaeroides enzyme contains a glycine box that is homologous to that of BphO, yet it has 32% identity to HemO, nearly the same degree of identity as that between HemO and PigA. It has only 28% identity to BphO. Preliminary data suggest that the *Rhodobacter* putative heme oxygenase cannot complement HemO in vitro, further suggesting the importance of this glycine box as a potential marker for specific heme oxygenases. Perhaps future investigations will demonstrate that these conserved residues, along with the conserved motifs described above, may correlate with mechanistic differences in these enzymes.

ACKNOWLEDGMENTS

We thank Donna Perkins-Balding and Heather Alexander for critical review and suggestions for the manuscript. We thank Rahul Deshmukh for providing pictures of the active site.

This work is supported by Public Health Service grant A142870 to I.S.

SUGGESTED READING

Beale, S. I., and J. Cornejo. 1991. Biosynthesis of phycobilins. 3(Z)-Phycoerythrobilin and 3(Z)-phycocyanobilin are intermediates in the formation of 3(E)-phycocyanobilin from biliverdin IX alpha. *J. Biol. Chem.* **266:**22333–22340.

Bhoo, S.-H., S. J. Davis, J. Walker, B. Karniol and R. D. Vierstra. 2001. Bacteriophytochromes are photochromic histidine kinases using a biliverdin chromophore. *Nature* **414:**776–779.

Caignan, G. A, R., Deshmukh, A. Wilks, Y. Zeng, H.-W. Huang, P. Moënne-Loccoz, R. A. Bunce, M. A. Eastman, and M. Rivera. 2001. The oxidation of heme to beta- and delta-biliverdin by *Pseudomonas aeruginosa* heme oxygenase is a consequence of an unusual seating of the heme. *J. Am. Chem. Soc.* **124:**14879–14892.

Chu, G. C., S.-Y. Park, Y. Shiro, T. Yoshida, and M. Ideka-Saito. 1999. Crystallization and preliminary X-ray diffraction analysis of a recombinant bacterial heme oxygenase (HmuO) from *Corynebacterium diphtheriae*. *J. Struct. Biol.* **126:**171–174.

Davis S. J., A. V. Vener, and R. D. Vierstra. 1999. Bacteriophytochromes: phytochromes from nonphotosynthetic eubacteria. *Science* **286:**2517–2520.

Gambetta, G. A., and J. C. Lagarias. 2001. Genetic engineering of phytochrome biosynthesis in bacteria. *Proc. Natl. Acad. Sci. USA* **98:**10566–10571.

Genco, C. A., and D. White Dixon. 2001. Emerging strategies in microbial haem capture. *Mol. Microbiol* **39:**1–11.

Létoffé, S., V. Redeker, and C. Wandersman. 1998. Isolation and characterization of an extracellular haem-binding protein from *Pseudomonas aeruginosa* that shares function and sequence similarities with the *Serratia marcescens* HasA haemophore. *Mol. Microbiol.* **28:**1223–1234.

Liu, Y., L. K. Lightning, H. Huang, P. Moenne-Loccoz, D. J. Schuller, T. L. Poulos, T. M. Loehr, and P. R. Ortiz de Montellano. 2000. Replacement of the distal glycine 139 transforms human heme oxygenase-1 into a peroxidase. *J. Biol. Chem.* **275:**34501–34507.

Maines, M. D. 1997. The heme oxygenase system: a regulator of second messenger gases. *Annu. Rev. Pharmacol. Toxicol.* **37:**517–554.

Ratliff, M., W. Zhu, R. Deshmukh, A. Wilks, and I. Stojiljkovic. 2001. Homologues of neisserial heme oxygenase in gram-negative bacteria: degradation of heme by the product of the *pigA* gene of *Pseudomonas aeruginosa*. *J. Bacteriol.* **183:**6394–6403.

Schluchter, W. M., and A. N. Glazer. 1997. Characterization of cyanobacterial biliverdin reductase. Conversion of biliverdin to bilirubin is important for normal phycobiliprotein biosynthesis. *J. Biol. Chem.* **272:**13562–13569.

Schuller, D. J., W. Zhu, A. Wilks, I. Stojiljkovic, and T. L. Poulos. 2001. Crystal structure of heme oxygenase from the gram-negative pathogen *Neisseria meningitidis* and a comparison with mammalian heme oxygenase-1. *Biochemistry* **40:**11552–11558.

Schuller, D. J., A. Wilks, and P. R. Ortiz de Montellano. 1999. Crystal structure of human heme oxygenase-1. *Nat. Struct. Biol.* **6:**860–867.

Stojiljkovic, I., and D. Perkins-Balding. 2002. Processing of heme and heme-containing proteins by bacteria. *DNA Cell Biol.* **21:**281–295.

Sugishima, M., Y. Omata, Y. Kakuta, H. Sakamoto, M. Noguchi, and M. K. Fukuyama. 2000. Crystal structure of rat heme oxygenase-1 in complex with heme. *FEBS Lett.* **471:**61–66.

Sun, J., T. M. Loehr, A. Wilks, and P. R. Ortiz de Montellano. 1994. Identification of histidine 25 as the heme ligand in human liver heme oxygenase. *Biochemistry* **33:**13734–13740.

Sun, J., A. Wilks, P. R. Ortiz de Montellano, and T. M. Loehr. 1993. Resonance Raman and EPR spectroscopic studies on heme-heme oxygenase complexes. *Biochemistry* **32:**14151–14157.

Tenhunen, R., H. S. Marver, and R. Schmid. 1969. Microsomal heme oxygenase: characterization of the enzyme. *J. Biol. Chem.* **244:**6388–6394.

Wilks, A., and P. Moenne-Loccoz. 2000. Identification of the proximal ligand His-20 in heme oxygenase (HmuO) from *Corynebacterium diphtheriae*: oxidative cleavage of the heme macrocycle does not require the proximal histidine. *J. Biol. Chem.* **275:**11686–11692.

Wilks, A., S. M. Black, W. L. Miller, and P. R. Ortiz de Montellano. 1995. Expression and characterization of truncated human heme oxygenase (hHO-1) and a fusion protein of hHO-1 with human cytochrome P450 reductase. *Biochemistry* **34:**4421–4427.

Wilks, A., J. Sun, T. M. Loehr, and P. R. Ortiz de Montellano. 1995. Heme oxygenase His25Ala mutant: replacement of the proximal iron ligand by exogenous bases restores catalytic activity. *J. Am. Chem. Soc.* **117:**2925–2926.

Willows, R. D., S. M. Mayer, M. S. Foulk, A. DeLong, K. Hanson, J. Chory, and S. I.

Beale. 2000. Phytobilin biosynthesis: the *Synechocystis* sp. PCC 6803 heme oxygenase-encoding *ho1* gene complements a phytochrome-deficient *Arabidopsis thaliana hy1* mutant. *Plant Mol. Biol.* **43:** 113–120.

Zhou, H., C. Taiko, M. Sato, D. Sun, X. Zhang, M. Ideka-Saito, H. Fujii, and T. Yoshida. 2000. Participation of carboxylate amino acid side chain in regiospecific oxidation of heme by heme oxygenase. *J. Am. Chem. Soc.* **122:**8311–8312.

Zhu, W., A. Wilks, and I. Stojiljkovic. 2000. Degradation of heme in gram-negative bacteria: the product of the *hemO* gene of neisseriae is a heme oxygenase. *J. Bacteriol.* **182:**6783–6790.

THE TonB, ExbB, AND ExbD PROTEINS

Kathleen Postle and Ray A. Larsen

7

The outer membrane gives gram-negative bacteria some distinct competitive advantages in less than ideal natural environments. This structure is a permeability barrier to a wide range of toxins, detergents, and degradative enzymes, but it still permits energy-independent diffusion of small (<600-Da) hydrophilic molecules. Unfortunately, the metabolic demands of gram-negative bacteria are not fully satisfied by the set of diffusible nutrients. Iron-bearing siderophores (discussed in chapters 1 and 2) and vitamin B_{12} molecules exceed the diffusion limit of the outer membrane and are sufficiently rare in the environment that high-affinity, energy-dependent transport systems are required for their efficient acquisition. The outer membrane lacks local access to the classic means of energizing transport—an established ion gradient or ATP. Instead, the proton motive force (PMF) of the cytoplasmic membrane is harnessed as the energy source. This spatial separation of the energy source from the energy sink necessitates an intermediate linkage between the two processes. That function is fulfilled by three proteins, TonB, ExbB, and ExbD, that couple the cytoplasmic membrane PMF to the active transport of nutrients through outer membrane transporters.

MUTANT PHENOTYPES

Luria and Delbrück isolated the first *tonB* mutations in 1943 during their classic fluctuation tests to study the origin of mutations conferring bacteriophage T1 resistance (hence the mnemonic "ton" for "T-one"). As the story goes, they obtained big colonies and little colonies from those selections, with the corresponding mutant alleles named *tonA* and *tonB*, respectively. The *tonA* mnemonic became *fhuA* when its role in ferric hydroxamate uptake was apparent. The pleiotropic phenotype of *tonB* mutants posed a puzzle for several years. With further experimentation on the part of many laboratories, it gradually became clear that mutations in *tonB* prevented all high-affinity siderophore-mediated iron transport (probably accounting for the small colony size) and vitamin B_{12} transport. In addition, they conferred protection against (i.e., tolerance to) bacteriophages T1 and φ80 and B-group colicins—B, D, G, H, Ia, Ib, M, Q, S1, and V. The secondary effects of iron limitation deepened the confusion; e.g., mutations in *tonB* could derepress the *trp* attenuator, due to lack of an iron-de-

---

Kathleen Postle, School of Molecular Biosciences, Washington State University, Pullman, WA 99164-4234. *Ray A. Larsen,* Department of Biological Science, Bowling Green State University, Bowling Green, OH 43403-0212.

Iron Transport in Bacteria, Edited by Jorge H. Crosa, Alexandra R. Mey, and Shelley M. Payne
© 2004 ASM Press, Washington, D.C.

pendent modification of all tRNAs that recognize codons beginning with uracil. More directly, iron limitation leads to higher levels of virtually all proteins involved in siderophore synthesis and transport, with a resultant hypersecretion of the siderophore enterochelin (also called enterobactin). This useful confirmatory phenotype manifests as slightly pink agar on a streak plate.

Mutations in *exb*, named for "excretion of inhibitor of colicin B," initially included the *exbA*, *exbB*, and *exbC* genes. The inhibitor turned out to be enterochelin, hypersecreted as a result of the iron limitation induced by these various mutations. As we now know, enterochelin inhibits colicin B killing of wild-type cells due to competition for binding to the FepA outer membrane transporter to enter *Escherichia coli*. The *exbA* gene was ultimately recognized as *tonB*, while *exbC* has just faded away, but not until after sequence analysis of the *exb* region revealed two genes in the operon, named *exbB* and *exbD*. Mutants with mutations in *exbB* or *exbD* appear to have a *tonB* leaky phenotype, with greatly reduced (but not absent) iron or vitamin B_{12} transport and decreased sensitivity to B-group colicins and ϕ80. The full effect of these mutations is revealed only when the homologous *tolR* and *tolR* genes are knocked out. Under that circumstance, ExbB and ExbD phenotypes are the same as TonB phenotypes. (Cross talk from TolQ and TolR supports a low level of TonB activity that must be considered in all experiments with the TonB system.)

So which protein is the energy transducer? Since TonB overexpression can compensate for the absence of ExbB and ExbD (leading to low-level sensitivity to colicins and phage) but ExbBD overexpression cannot compensate for the absence of TonB, it is evident that TonB must be the energy transducer. Subsequent data suggested that ExbB and ExbD harness the PMF to energize TonB. This chapter focuses on our current understanding of the means by which energy is harvested at the cytoplasmic membrane and delivered to the outer membrane transporters in *E. coli* and then reviews some of the diversity in this system that has recently become apparent.

TonB

E. coli TonB is a 239-amino-acid (aa) protein (~26 kDa), the bulk of which occupies the periplasmic space, where it can associate with components of both the cytoplasmic and outer membranes (Fig. 1). TonB can be cross-linked by monomeric formaldehyde in vivo to several different proteins in either the cytoplasmic membrane (ExbB and ExbD) or the outer membrane, (Lpp, OmpA, and the outer membrane transporter, FepA). *tonB* transcription is regulated aerobically over a three-fold range of expression by Fur protein and iron availability (see chapter 13). Under conditions of iron limitation, *tonB* expression is derepressed. Anaerobically the situation is similar, except that *tonB* expression is repressed 10-fold, this time requiring both Fur and an uncharacterized DNA binding protein. Based on evidence to date, it

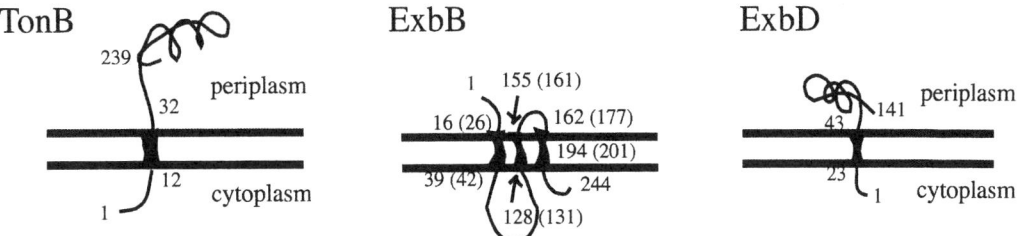

FIGURE 1 Predicted topologies of TonB, ExbB, and ExbD in the cytoplasmic membrane. The locations of the periplasm and cytoplasm are indicated. Parentheses indicate alternative designations for transmembrane domains. Reprinted from Postle and Kadner (2003) with permission from the publisher.

is reasonable to consider TonB to consist of three distinct functional domains: the amino-terminal domain, the central domain, and the carboxy-terminal domain.

AMINO-TERMINAL DOMAIN

The amino-terminal region (aa 1 to 32) appears to serve as the cytoplasmic membrane signal anchor. It resembles a classic signal sequence in that it has a hydrophobic (presumably) transmembrane domain of about 20 aa (aa 12 to 32) that is preceded by two positive charges. A Gly-to-Asp mutation in the hydrophobic region can inhibit export and is suppressible by mutation in the prlA gene of the sec system. The amino terminus is retained in the mature protein due to lack of a leader peptidase cleavage site. When a cleavage site is engineered, this domain is removed with a concomitant loss of activity, the first evidence for its direct importance in energy transduction.

The transmembrane domain contains a conserved S16, H20, L27, S31 motif that occupies one face of the predicted α-helical hydrophobic region. Of those four amino acids, the S16 and H20 residues, and the register between them, are clearly important for TonB activity as an energy transducer, although their precise roles are not known. Interestingly, overexpression of TonB proteins carrying mutations that eliminate S16, H20, or the register between them can move the slight amount of activity into a window where it can now be detected, suggesting that a small percentage of TonB can achieve an "active" conformation independent of those residues. Nonetheless, some feature of the transmembrane domain is crucial. Replacement of aa 1 to 32 with the first transmembrane domain of the TetA protein leads to complete loss of TonB activity, even though it is still properly exported to the cell envelope. The amino-terminal 32 aa (at least) are required for in vivo cross-linking with ExbB and for a conformational response to the presence or absence of a PMF in which carboxy-terminal aa 155 to 239 play no role. Recently, the transmembrane domain was implicated in the formation of TonB dimers in vivo, as well as in conformational changes at the carboxy terminus.

CENTRAL DOMAIN

Unlike the amino terminus, the central region (aa 33 to 100) appears not to play an essential role in energy transduction. Deletion of aa 33 to 64 results in a TonB with somewhat reduced specific activity, while deletion of the very interesting proline-rich domain, $_{70}$EPEPEPEPI-PEP-9 aa-KPKPKPKPKPKP$_{102}$ has no observable effect unless the periplasmic space is temporarily expanded. Under such conditions, TonBΔ(66–100) functions about half as well. A similar result has also been obtained for deletion of the proline-rich domain of one of the Vibrio cholerae TonB proteins (TonB 1). Although not essential, this region almost certainly makes contributions to survival following shifts to environmental extremes and is probably more important in natural (as opposed to benchtop) settings. The proline-rich region is also responsible for the aberrant decreased mobility of TonB in sodium dodecyl sulfate-polyacrylamide gels, where it has an apparent molecular mass of 36 kDa. Consistent with the lack of an essential participation in energy transduction, aa 33 to 102 are not highly conserved, except for a preponderance of proline residues variously arranged in most (but not all) TonBs.

CARBOXY-TERMINAL DOMAIN

Parts of the carboxy-terminal region (aa 103 to 239) are absolutely required for TonB activity, for interaction with the outer membrane, and for interaction with the outer membrane transporters. Amino acids from 199 to 216 are especially well conserved among enteric bacteria and are capable of forming an amphipathic helix, as predicted by sequence-gazing and then demonstrated by determination of the crystal structure of the carboxy-terminal 75 aa of TonB. The crystal structure of the TonB carboxy terminus does not appear to represent an active form of TonB, as explained more fully below.

ExbB and ExbD

TonB-dependent energy transduction requires the participation of additional proteins, ExbB and ExbD. ExbB is a ~26-kDa cytoplasmic membrane protein with a periplasmically localized amino terminus. The cytoplasmic domain of ~90 aa is bordered by two transmembrane domains, for a total of three. ExbD is a ~17-kDa cytoplasmic membrane protein with a topology identical to that of TonB and a periplasmic domain of ~90 aa (Fig. 1). These two proteins are expressed from an operon, with *exbB* being the first gene. ExbB has not yet been subjected to mutational analysis; however, three different suppressors of TonB transmembrane domain mutations have been isolated in ExbB. These are not allele-specific suppressors but, rather, make ExbB somehow less discriminating in its TonB interactions. Surprisingly, they are all substitutions of a glutamate or an aspartate for a nonpolar amino acid at position 35, 36, or 39 of ExbB. The meaning of these suppressors will have to await a thorough analysis of ExbB function. Mutant analysis has revealed aspartate 25 and leucine 132 to be essential for the function of ExbD in *E. coli*.

Even though *exbB* and *exbD* are transcribed as an operon, there is a significant difference in their levels of expression in the cell, with ExbB being present at about 3.5 times the level of ExbD, which is present at about twice the level of TonB (TonB is present at 340 ± 80 molecules per cell under iron-replete conditions). Since neither ExbB nor ExbD appears to have a function independent of the other and since these two proteins form homomultimers in vivo, it is possible to predict a theoretical molecular mass for a unit energy transduction complex. If TonB is indeed a dimer, as suggested by in vivo studies, a single energy transduction complex could be as large as 520 kDa.

COMPARISON OF TonB PHENOTYPIC ACTIVITY ASSAYS

The ability of cells to transport cobalamin and various iron-siderophore complexes and their susceptibility to killing by bacteriophages and colicins are routinely used in assays to characterize phenotypes of mutant proteins in the TonB-dependent energy transduction system. These assays vary greatly in sensitivity and are subject to perturbation by overexpression of TonB and, perhaps, other proteins that contribute to the process. Thus, the choice of assay and the means by which a potential mutant is expressed can greatly influence the interpretation, and even the identification, of a given mutant.

Recently, bacteria in which TonB was underexpressed to various degrees were evaluated quantitatively in a panel of assays. Since TonB is the limiting protein in the energy transduction system under normal conditions, it became clear that the fundamental difference among the available assays was in their ability to detect TonB activity. Based on the results obtained, it is useful to think in terms of sensitivity windows for phage and colicin sensitivity assays, transport, irreversible φ80 adsorption, enterochelin excretion, and cobalamin-dependent growth (Fig. 2). However, nutritional-disk assays generally gave the disturbing result that less TonB resulted in larger growth zones, most probably because decreased transport rates and correspondingly slow cell growth allow for greater diffusion of the nutrient on the plate. For these assays, the interplay between diffusion rate, growth rate, and absolute minimal concentrations sufficient to sustain growth render interpretation of the results unpredictable and problematic.

In addition to the choice of assay, another important consideration in determining TonB activity is the maintenance of stoichiometry within the TonB system. Obviously, if TonB or some other essential component is, for whatever reason, present at decreased levels, the amount of activity observed in a correctly chosen phenotypic assay is decreased. Interestingly, this is also the case when TonB is overexpressed. This dominant negative gene dosage effect was first noted in cobalamin transport studies, where the expression of *tonB* from a multicopy plasmid resulted in reduced transport. This observation has been widely con-

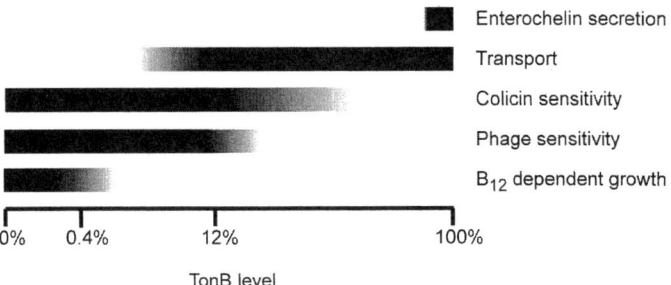

FIGURE 2 Relative sensitivity windows for commonly used assays of TonB function. Cells expressing functional TonB at either wild-type (100%) or lower (12% or 0.4% or absent) levels were examined, and the results are summarized here. The black region indicates the range of TonB activity that a given assay can clearly distinguish. For example, colicin sensitivity can clearly distinguish between TonB activity levels of 0, 0.4, and 12%. The transition from gray to white indicates the portion of the sensitivity window in which the end point is unclear; for example, colicin sensitivity can distinguish between 12 and 100%, but levels intermediate between these values have not been tested. A gray zone is not depicted for enterochelin hypersecretion due to the overall insensitivity of this assay. Note that the 0.4% level corresponds approximately to a single copy of TonB per cell. Enterochelin secretion was measured as a zone of clearing on CAS (Chrome azurol S) indicator plates. Transport includes both [^{35}Fe] derophore uptake and irreversible φ80 adsorption assays. Colicin sensitivity is based on spot titer assays with the TonB-dependent colicins B, D, Ia, and M. Phage sensitivity is based on spot titer assays with φ80. B$_{12}$-dependent growth was measured by the ability of *metE* strains to grow in the presence of vitamin B$_{12}$.

firmed in several laboratories and has recently been revisited in experiments with quantitated levels of TonB. When TonB was only modestly overexpressed (~2.5-fold), ferrichrome transport and irreversible φ80 adsorption rates were decreased to less than half of the wild-type values, similar to the values corresponding to the expression of TonB at a level only ~10% that of the wild-type level. These results clearly demonstrate the complexity of the TonB system and indicate the necessity of determining the relative expression levels of any TonB mutant protein.

SHUTTLING BETWEEN MEMBRANES, A NEW BIOLOGICAL PARADIGM

A diversity of models has been proposed over the years to explain the activity of the TonB protein. These include the production of a soluble messenger, molecular processing of permeases, a mechanical model where TonB serves as a propeller to crank open the transporters, and shuttling between the cytoplasmic and outer membranes to deliver conformationally stored potential energy. All except the shuttling model require or suggest that TonB remains anchored to the cytoplasmic membrane, whereas the shuttling model requires TonB to disengage from the cytoplasmic membrane entirely and associate with the outer membrane.

The first hint that TonB might shuttle to the outer membrane came from the detection of TonB in the outer membranes of maxicell preparations—a result that was never published but was alluded to in a review article. When highly specific anti-TonB monoclonal antibodies became available, the potential artifacts associated with the maxicell expression system could be circumvented by localization of chromosomally encoded TonB in wild-type *E. coli* K-12. This time, TonB was found to segregate in sucrose density gradients with both the cytoplasmic and outer membrane fractions, at approximately 60 and 40%, respectively. Several potential artifacts were excluded, first by the demonstration in a flotation experiment that TonB colocalized with the outer membrane and second by the observation that ExbB and ExbD (two proteins with which TonB can closely associate) remained in the cytoplasmic membrane under circumstances where TonB localized to both the cytoplasmic and outer membranes. It was further demonstrated that without its carboxy terminus, TonB could not associate with the outer membrane. In contrast, without ExbB and ExbD (and in the absence of cross talk from TolQ and TolR), all of the TonB localized to the outer membrane. Soluble periplasmic intermediates were not detected. These data were interpreted to suggest that TonB shuttles back and forth between the two membranes of gram-negative bacteria in the process of transducing energy to the outer membrane transporters.

The potential artifact that could not be excluded was one introduced by the in vitro fractionation process. Could the lysis and fractionation procedure have ripped TonB out of the cytoplasmic membrane at the point in the cycle when it was most strongly associated—perhaps even briefly covalently—with the outer membrane transporters? In that model, TonB would remain continuously associated with the cytoplasmic membrane, cyclically forming strong, transient associations with the outer membrane transporters at some point in the energy transduction cycle.

The mechanistic implications of the two models are different. In the case of shuttling, TonB must store the cytoplasmic membrane PMF as an energy-rich conformation. In the case where TonB remains tethered to the cytoplasmic membrane throughout the energy transduction cycle, models that can be considered include proton wires, changes in the amino terminus coupled to changes in the carboxy terminus, and mechanical models such as rotary motion.

The resolution of the question clearly required an in vivo approach in which the periplasmic accessibility of the extreme amino terminus of TonB could be assayed. If the amino terminus ever crossed into the periplasm, as required in the proposed shuttling mechanism, perhaps this event could be recorded by using a specific label. To facilitate such a study, two versions of TonB, each containing a sole cysteine residue at the extreme cytoplasmic amino terminus but with one version lacking a carboxy terminus, were engineered. The cytoplasmic membrane positioning of the two proteins was shown to be identical—an essential control—since they could both cross-link in vivo to the cytoplasmic membrane protein, ExbB, a reaction known to require the amino-terminal transmembrane domain of TonB. In addition, both full-length and truncated TonB could respond conformationally (and identically) to PMF, an assay known to require both an intact transmembrane domain, and the ExbB and ExbD proteins.

Cells expressing the individual proteins were incubated in vivo with the cysteine-biased reagent, Oregon Green maleimide (OGM). This probe is permeant to the outer membrane but largely impermeant to the cytoplasmic membrane. As predicted by the shuttling model, only the full-length TonB could be specifically labeled, and the labeling occurred at a rate and to an extent similar to that for a control TonB bearing a periplasmically localized cysteine substitution in the carboxy terminus of TonB (TonB lacking a cysteine residue did not label specifically). TonB lacking the carboxy terminus did not associate with the outer membrane and did not specifically label; this served not only to demonstrate the correlation between outer membrane association and ability to be labeled but also as an apparent control for the leakage of OGM across the cytoplasmic membrane. The in vivo OGM labeling data indicate that TonB does indeed shuttle between the cytoplasmic and outer membranes, confirming the predictions of the previous fractionation study.

MODEL FOR TonB-DEPENDENT ENERGY TRANSDUCTION

Transport of iron-siderophores and vitamin B_{12} from the external medium to the cytoplasm involves three distinct classes of ligand-specific components: (i) a high-affinity (0.3 to 10 nM) outer membrane transporter (see chapter 4), (ii) a moderate-affinity (micromolar) periplasmically localized binding protein required for transport across the cytoplasmic membrane (see chapter 8), and (iii) a standard cytoplasmic membrane ABC transporter (described for ferric citrate and hydroxamate transport in chapter 11). The moderate affinity of the periplasmic binding proteins is not sufficient to capture siderophores bound by the high-affinity transporters, hence the requirement for input of energy. TonB does not participate in the transport of nutrients across the cytoplasmic membrane.

Color Plate 11 depicts a model for the mechanism of energy transduction between membranes in gram-negative bacteria that is most consistent with available data (which have admittedly been gathered almost entirely from studies of *E. coli* K-12).

In step 1, TonB is in a large complex of ExbB and ExbD, since TonB can cross-link specifically in vivo to both proteins. These cross-linking interactions are conserved among several other gram-negative bacteria, suggesting that they represent widespread and meaningful interactions in other bacteria. It has been clearly demonstrated that TonB-dependent energy transduction requires the cytoplasmic membrane PMF, either directly or indirectly. Based on current data, the protons are most probably translocated through the ExbB-ExbD complex.

In step 2, TonB has been converted to an energized conformation by the combined actions of ExbB, ExbD, and PMF. Recent data have shed light on the energy transducing conformation of TonB. Mutational analysis of the carboxy terminus demonstrated that the carboxy terminus is a dynamic and flexible domain. Consistent with that flexibility, a double-mutant cycle analysis suggested that all the aromatic residues in the carboxy-terminal 75 aa were capable of interacting with one another to form an aromatic cluster—a conformation not represented by the recent crystal structure (Fig. 3). Although interactions among the aromatic residues are not required to form the cluster, they are clearly necessary for energy transduction to the outer membrane transporters. The existence of the aromatic cluster and a dimeric state for TonB were confirmed in vivo and, interestingly, shown to depend on ExbBD and the TonB amino-terminal transmembrane domain. The requirement for ExbBD to form this conformation at the carboxy terminus suggests that given the topologies of all the proteins involved, there is the possibility that signal transduction between ExbD and the carboxy terminus supports the formation of the aromatic cluster. The aromatic cluster is most likely to be a transition state on the way to an active conformation, since it is observed only in cytoplasmic membrane fractions. The data also indicate that the crystal structure of the carboxy terminus does not represent an active conformation of TonB, which is not surprising since it does not include the transmembrane domain.

In step 3, TonB associates with two proteins in the outer membrane, OmpA and Lpp. Mutations in these proteins do not have strong effects on TonB function, possibly due to a high degree of functional redundancy in *E. coli*. It is not clear whether this interaction occurs

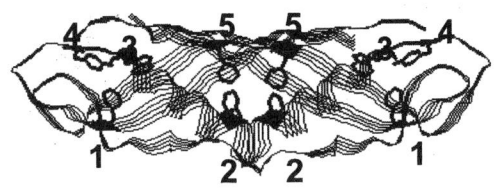

1:Phe^{180}, 2:Phe^{202}, 3:Trp^{213}, 4:Tyr^{215}, 5:Phe^{230}

FIGURE 3 Crystal structure of the C-terminal 75 residues of *E. coli* TonB, with positions of all aromatic residues indicated in the dimer. Modified from Chang et al. (2001) and reprinted from Ghosh and Postle (2004) with permission from the publisher.

prior to or following the energy transduction event. In this model, TonB does not release its association with the cytoplasmic membrane until some contact has been made with the outer membrane, consistent with the results of deleting the proline-rich region, where it appears to be capable of measuring the distance between cytoplasmic and outer membrane.

During step 3, as is evident from crystal structures and direct experimentation, the outer membrane transporters undergo a change in conformation on binding of their cognate ligands. It is likely that TonB-mediated release of conformationally stored potential energy occurs following the ligand-induced conformational change in the transporter. TonB does not appear to transduce energy to outer membrane transporters that are without bound ligand.

The crystal structures reveal the outer membrane transporter proteins to be large β-barrels completely occluded by an amino-terminal globular domain that signals ligand binding on the external face by undergoing large conformational changes at the periplasmic face. Recognition of outer membrane transporters by TonB is enhanced by the presence of ligand, suggesting that TonB has a higher affinity for ligand-bound transporter than for unoccupied transporter.

It is clear that TonB amino acids in the region of residue 160 directly interact with a region near the extreme amino terminus of the transporters called the TonB box. (Colicins also have TonB box sequences at their amino termini.) It is also clear that a subset of mutations in the TonB box renders the transporters inactive. Nonetheless, the role of the internal globular domain (also called the cork or the plug) has been controversial, with some evidence suggesting that empty FepA or FhuA barrels (lacking the internal globular domain and TonB box) were fully active and exhibited TonB-dependent transport of nutrients. It first became clear for FepA that the empty FepA barrels were actually inactive but that activity could be detected in strains chromosomally expressing what subsequently turned out to be a globular domain (aa 1 to 150) and eight β-strands of the barrel of FepA (M. McIntosh kindly supplied the DNA sequence). It was suggested that results showing activity of empty FepA and FhuA barrels were actually due to interprotein complementation by chromosomally encoded globular domains of the respective transporters. This idea was later rigorously experimentally confirmed and extended for FhuA.

To summarize, then, in step 3, energy-charged TonB directly contacts the ligand-bound outer membrane transporter and subsequently transduces PMF-induced, conformationally stored potential energy, allowing the transporter to actively transport ligand into the periplasmic space. In spite of all we do know, the actual nature of the TonB-dependent step in ligand transport is unknown. It could be that TonB somehow releases the cork from the globular domain to allow transport not only of the relatively small ($\sim$0.7 to 1.2-kDa) iron-siderophore or vitamin B_{12} molecules, but also of the much larger ($\sim$55- to 70-kDa) colicin molecules. Alternatively, a smaller conformational change might occur that allows a channel to form.

In step 4, TonB has discharged the conformationally stored potential energy and returns to the cytoplasmic membrane. ExbB and ExbD are also required for this reassociation with the cytoplasmic membrane. Once TonB is repositioned within the cytoplasmic membrane complex of ExbB and ExbD, the cycle can be repeated.

CYTOPLASMIC MEMBRANE PMF DOES NOT ENERGIZE SHUTTLING

An interesting facet of the model (not illustrated here) is that TonB shuttles back and forth between the cytoplasmic and outer membrane regardless of whether it is charged via the PMF. The protonophore carbonyl cyanide m-chlorophenyl hydratone (CCCP) does not affect either TonB distribution between the cytoplasmic and outer membranes or the ability to cross-link in vivo to FepA. Furthermore, mutations in the transmembrane domain that inactivate TonB still support an association with the outer membrane and cross-linking to FepA in vivo. Thus, the participation of an additional energy source seems to be required, most probably for recovery of TonB from the outer membrane but also possibly for shuttling to the outer membrane. It may well be that ATP hydrolysis serves to energize the shuttle and is mediated through the cytoplasmic domain of ExbB—the only part of the TonB system that has significant contact with the cytoplasm and one part of the system that has yet to be examined in any detail.

Involvement of a Mystery Protein(s)

A requirement for ATP hydrolysis would be consistent with the participation of an as yet uncharacterized cytoplasmic protein in TonB-dependent energy transduction. Years ago, it was demonstrated that TonB activity decayed rapidly in the presence of protein synthesis inhibitors. Chromosomally encoded TonB was later determined to be a stable protein in the presence of these inhibitors, pointing to the existence of a different short-lived protein in the process. Now that chromosomally encoded ExbB and ExbD are also known to be stable in the presence of protein synthesis inhibitors, it is likely that the short half-life of TonB activity under these conditions represents the contribution of one or more, as yet uncharacterized, proteins.

TonB IN OTHER SPECIES: VARIATIONS ON A THEME

All gram-negative bacteria face constraints imposed by their unique cell wall architecture. It is reasonable to suspect that they have all retained some sort of energy transduction system to support the active transport of large, important molecules across the unenergized outer membrane. This suspicion has been largely confirmed by the explosion of data emerging from the on going genomics revolution. As of June 2004, the National Center for Biotechnology Information (NCBI) sequence repository contained entries for putative *tonB* genes representing over 65 genera of gram-negative bacteria, a number no doubt obsolete well before this book goes to press. Similar numbers exist for putative *exbB* and *exbD* genes, with genes encoding what appear to be TonB-dependent outer membrane transporters numbering in the hundreds. Although only a handful of these assignments have been experimentally confirmed, similarities are of an order that misidentifications will probably prove rare.

The biggest surprise to stem from this wealth of data is the realization of how fortunate we have been to work with *E. coli* K-12. This domesticated bacterium carries a single *tonB* gene and a single *exbBD* operon, and although it does encode seven separate TonB-dependent outer membrane transporters, it normally synthesizes but one ligand—enterochelin—that is specifically recognized by a single transporter—FepA (see chapters 1 and 4). Even this apparently simple energy transduction system has proven complicated to unravel. Owing to their expression in an operon, it has taken successive efforts to demonstrate that ExbB and ExbD function as a unit both phenotypically and biochemically, while the true importance of ExbB and ExbD in energy transduction was initially masked by the phenomenon of cross talk with Tol system components. Nevertheless, it is now evident that *E. coli* K-12 offers the simplest version that one could have chosen for the dissection of energy transduction. As might be expected for a system of such importance, evolution has engendered variations and redundancies within this framework to an extent that we are only beginning to appreciate.

The first *tonB* gene to be sequenced was from *E. coli* K-12 and was predicted to encode a 244-, 241-, or 239-residue polypeptide with features ultimately determined to represent the three domains described above. The next two *tonB* genes to be sequenced, from *Salmonella enterica* serovar Typhimurium (NCBI accession no. NP_46096) and *Serratia marcescens* (accession no. P26185) predicted proteins that did not depart greatly from the *E. coli* benchmark (indeed, all *tonB* sequences to date from the family *Enterobacteriaceae* predict similar proteins). In particular, it was noted that the predicted transmembrane domains of these proteins contained a motif (S16, H20, L27, S31) shared with TolA (Fig. 4). Beyond indicating the most likely basis for cross talk between the TonB and the Tol systems, this homology misleadingly suggested that the TonB transmembrane domain might be widely conserved over the range of gram-negative bacteria. The sequence of the *Pseudomonas putida tonB* gene (accession no. X70139) predicted a protein that did not fit this mode, with only the histidine and the second serine retained. Unlike the *Salmonella* and *Serratia* genes, the *P. putida tonB* gene was not able to complement an *E. coli tonB* strain. As more sequences were determined, it became evident that the *P. putida* TonB transmembrane domain was not exceptional: while the family *Enterobacteriaceae* retains the SHLS motif intact, many other gram-negative bacteria do not; indeed, even the highly conserved histidine residue (required for activity in *E. coli* TonB) is not inviolate (Fig. 4). Nor, apparently, is the position of the transmembrane domain: a subset of *tonB* genes (including ones from *Pseudomonas aeruginosa* [NP_254218] and *Rhizobium leguminosarum* [CAC34389]) encode a hydrophilic, ~100-residue amino-terminal domain prior to the putative transmembrane domain. These amino-terminal domains have few sequence similarities, and their significance remains unclear. Their topologies relative to the cytoplasmic membrane are as yet unexamined. It will be especially interesting to determine the location of their extreme amino termini in light of the shuttling behavior of *E. coli* TonB.

The second surprise offered by the *P. putida tonB* sequence involved gene organization, with TonB encoded by the third gene in a Fur-regulated operon that also encoded ExbB and ExbD. This is counter to the arrangement in the *Enterobacteriaceae*, where the *tonB* gene is unlinked to *exbB* and *exbD* and shares a bidirectional rho-independent terminator with an open reading frame that, based on homology to the *E. coli yciA* gene (formerly known as P14), is predicted to encode an acyl coenzyme A hydrolase. While this relationship may have potential regulatory consequences for TonB, it would probably be specific to the *Enterobacteriaceae*, since few other groups examined have maintained that gene arrangement.

The *tonB* operon of *P. putida* parallels the Tol system, where the TonB analog, TolA, is the product of the third gene of an operon that encodes the ExbB and ExbD homologs TolQ and TolR (accession no. M16489). Similarly arranged *tonB* operons have now been identified in other genera including *Haemophilus* and *Leptospira*, which might suggest that this gene relationship represents the primordial operon from which the Tol and TonB systems diverged. This hypothesis is far from established,

```
TonB              12      TMD      32
    E. coli       WPTLLSVCIHGAVVAGLLYTSVH
    Y. pestis     WSLIFSIGLHGSVVAALLYVSVE
    B. melitensis GAGILVLLAHAAGAYVIHMAQED
    P. putida     GSLALVLGVHAVAVLLTLNWSVP
    B. pertussis  GAGFTVLALHAAVIGAVFLSKTE
    N. gonorrhoeae AVVFSVALLHLAIVALLWQAHKL
    P. multocida  IGFAISLLFHASFVSFLYWIVQK
    H. influenzae LGLLISLIAHGIVIGFILWNWNE
    P. shigelloides LASGLCMSAHAALLLFGLQPTQT
    V. cholerae #1 IAGGLSLAFHALLLITTDEAQVF
    V. cholerae #2 IALLVTLALFSLMAWMVDNGGKS
TolA
    E. coli       RAIIISAVLHVILFAALIWSSFD
    V. cholerae   KSITISLAMHGALVAILLWGADF
```

FIGURE 4 Conservation in the TonB transmembrane domain. Sequences are presented relative to residues 11 to 33 of *E. coli*. TonB, the predicted transmembrane domain (TMD) of which residues 12 to 32 are indicated by the bar at the top. The corresponding region from the TonB analog TolA is included for two species at the base of the figure. Shading indicates conserved aa in the SHLS motif. The NCBI accession numbers for the sequences presented are as follows: TonB sequences: *E. coli* (BVEC), *Yersinia pestis* (NP_405736), *Brucella melitensis* (AAK08071), *P. putida* (Q05613), *B. pertussis* (CAB53383), *Neisseria gonorrhoeae* (O06432), *Pasteurella multocida* (NP_246125), *Haemophilus influenzae* (P42872), *Plesiomonas shigelloides* (AAG23396), *V. cholerae* TonB1 (O052042) *V. cholerae* TonB2 (AAC69456); TolA sequences: *E. coli* (NP_415267), *V. cholerae* (NP_231471).

however, and the model is complicated by the occurrence of *tonB* in operons where the gene order differs. In some genera, including *Neisseria*, *tonB* is the first gene in the operon. An interesting variation occurs in *Xanthomonas campestris*, where a *tonB* operon contains a second *exbD* gene. Nor is operon composition restricted to these three genes—a *tonB exbB exbD* operon in *Vibrio cholerae* also includes *hutB* (encoding a periplasmic heme binding protein) and *hutCD* (encoding proteins with homology to cytoplasmic membrane ABC-type permeases involved in iron transport) (see chapter 11). More recently, *tonB* genes from *Rhizobium* and *Bradyrhizobium* have been described that either are associated with or are within operons encoding a variety of proteins involved in iron transport, including a TonB-dependent outer membrane transporter.

V. cholerae has been a bellwether organism for unfolding diversity in TonB systems. In addition to providing the initial example of cotranscribed TonB and heme transport system components, that same study marked the first identification of multiple *tonB* genes. Interestingly, the second distinct gene (designated *tonB2*) was also carried in an operon, here following a second *exbBD* pair (designated *exbB2* and *exbD2*). Multiple different *tonB* genes have also been experimentally confirmed to exist in *Pseudomonas aeruginosa*, in this case with one copy (*tonB2*) as the first gene in an operon that includes *exbB* and *exbD* and the other copy (*tonB1*) present as a single cistron (and potentially a lone *exbD* gene) (accession no. NC_002516). In this case, there do not appear to be extra copies of *exbB* and *exbD*. The presence of multiple *tonB* genes has also been indicated in the complete genomic sequences of several species. The complete *Campylobacter jejuni* sequence (NC_002163) reveals three *tonB* genes, two occurring as *exbB exbD tonB* operons and one monocistronic copy (with a third set of *exbBD* genes occurring elsewhere as a two-gene operon). Other in silico-identified multiple *tonB* genes are predicted in *Xylella fastidiosa* (NC_002488), with five copies; the current record of six copies goes to two species of *Xanthomonas* (NC_003919 and NC_003902).

The occurrence of multiple versions of TonB suggests the possibility of divergent function. This has been best studied with *V. cholerae*, where the two TonBs differ not only in their ability to support heme uptake but also in the pattern of siderophore usage (*tonB1* knockouts cannot use schizokinen, whereas *tonB2* knockouts cannot use enterochelin). This difference in siderophore transport suggests that the two TonB species recognize distinct outer membrane transporters. Interestingly, when expressed in *trans* with their cognate *exbB* and *exbD* genes, only *tonB2* could complement an *E. coli tonB* knockout strain. Since the inclusion of their own ExbB and ExbD should ensure that each TonB is capable of being energized (and, in fact, TonB1 retains the SH transmembrane motif essential in *E. coli* TonB for energization by ExbB and ExbD), this differential ability to function in *E. coli* suggests that the targets to which they deliver energy differ, with the target of *V. cholerae* TonB1 being absent from *E. coli*. Consistent with this notion, while the carboxy-terminal region (essential for interaction with the outer membrane transporters) of both *V. cholerae* TonB proteins bears features conserved in most (but not all) TonB proteins (the YP at residues 163 and 164 and the G's at residues 174 and 186 of the *E. coli* sequence), TonB1 lacks a number of motifs that TonB2 shares with the TonB of *E. coli* and other members of the *Enterobacteriaceae* (Fig. 5). In addition, single-amino-acid substitutions near the carboxy terminus of TonB1 have been identified that permit TonB1 to function with *E. coli* receptors, providing further evidence that this is the region of TonB1 that governs the specificity of the receptor interaction.

It now appears that the possession of multiple *tonB* genes may be the rule rather than the exception. The observation that an *S. marcescens tonB* strain retained the ability to utilize hemoglobin led to the identification of a protein called HasB that was shown by subsequent studies to be a TonB homologue. More recently, studies of a *Shigella* colicin (colicin Js)

```
                            12        TMD        32                                    160
E. coli              WPTLLSVCIHGAVVAGLLYTSVH.//.QPQYPARAQALRIEGQVKFDVTPDGRVDNVQILSA_KPANMFEREVKNAMRWRYEPGKPSGIVNLFKING
S. marcescens        VPFVLSVGLHSALVAGLLYASVK.//.NPLYPPRAQALQIEGNVRVQFDIDSDGRVSNVRILSA_EPRNMFEREVKQAMRKWRYEAKEAKDRTVTIR_FKLNG
V. cholerae #2       IALLVTLALFSLMAWMVDNGGKS.//.EPNYPAKALQRGVEGYVILRFTIDELGKTRDIEVVDAN_PKRYFEREAMLALRNWKYQSKIVDGQPVSQPGQTVRL
V. cholerae #1       IAGGLSLAFHALLLITTDEAQVF.//.QPRYPRIARKRGIEGTVMYEIWLDAQGNQI_KQQLLSSSGTEALDQSALEAIKQWKFSPHILDGVPVAHRIHIPIR
E. coli CjrB         SWFICSLLLHGLIFLTFIWRFSE.//.YKRYPGDARKRARTGTAVVTFTVNTEGTIVSS_FLEISSGTLSLDREAIAVLERAQPLPKPPEILEGG_LFKVKM
S. marcescens HasB   RCLVLVLALH_LLVAALLWPRRD.//.FKRYFKDALRLRKRQGVGQVRFTLDRQGHVLAVT_LVSSAGLPSLDREIQALVKRASPLPTPPADAYVNGTVELTLP
                                                                                                                    234
```

FIGURE 5 The C-terminal region of "extra" TonB proteins differs from that of "classic" TonB. Sequences are presented relative to residues 11 to 33 of *E. coli* TonB as in Fig. 4 and to residues 160 to 234 of the carboxy-terminal region of *E. coli* TonB. In the TMD region, only the conserved S and H residues are shaded. In the remainder, the shading indicates regions of significant residue homology. The NCBI accession numbers for the sequences presented are as follows: *E. coli* TonB (BVEC), *S. marcescens* TonB (P26185), *V. cholerae*. TonB1 (O052042), *V. cholerae* TonB2 (AAC69456), *E. coli* CjrB (AAK67306), *S. marcescens* HasB (sequence provided by C. Wandersman).

that specifically targets enteropathogenic strains of *Shigella* and *E. coli* determined that these strains carry a (presumably plasmid-borne) operon that encodes three products: CjrA, a putative lipoprotein; CjrC, the outer membrane transporter targeted by colicin Js; and CjrB, a TonB-like protein. The ability of CjrB to support the action of colicin Js appeared to be dependent on the presence of an *exbBD* operon. Examination of the carboxy-terminal regions of CjrB and HasB reveals these proteins to be similar to TonB1 of *V. cholerae* (Fig. 5). (The sequence of *S. marcescens hasB* is not yet available at NCBI but was kindly provided by C. Wandersman.)

Interestingly, both colicin Js and its transporter, CjrC, lack the TonB box characteristic of the classic TonB-dependent colicins and their transporters, strengthening the hypothesis that the unique departure of the CjrB and *V. cholerae* TonB1 carboxy termini from that of other TonB proteins indeed represents the machinery of transporter discrimination. In support of this, receptors that are energized by *V. cholerae* TonB1 also exhibit unique sequence features in the TonB box region, and these are essential for productive interactions with TonB1. Just to keep matters confused, HasR, the transporter serviced by HasB, does have a TonB box motif, but it is displaced from the traditional extreme N terminus of the protein.

TonB proteins are clearly much more diverse than once thought. In contrast, ExbB and ExbD still seem to be almost as conserved as we once thought, although here, too, some distinctions are now becoming evident. The degree of conservation among these proteins is most extensive in the transmembrane domains, although in the case of ExbD there also exist several highly conserved residues in the carboxy-terminal half of the periplasmic domain.

The membrane topology of ExbB (and its homolog, TolQ) is distinctive, since it has three transmembrane domains, an unusual periplasmically exposed N terminus, and almost half of the protein in the cytoplasmic domain. This design has remained common for the species examined to date, with the exception of several ExbB proteins that have a larger, presumably also periplasmically displayed N-terminal domain (for example, *P. putida* and *Bordetella pertussis*). There is little conservation within the cytoplasmic domain. Furthermore, in several genera, including *Pasteurella, Haemophilus*, and *Helicobacter* (NP_208131), the size of the cytoplasmic domain is greatly diminished (from ~100 to ~20 residues). The first transmembrane domain was initially noted for the high degree of similarity (38% identity) between ExbB and TolQ in *E. coli*. (Indeed, the similarities between ExbB and TolQ are such that in more divergent organisms one cannot assign a given sequence to either group. These proteins are therefore referred to as ExbB/TolQ-like in the databases, pending more rigorous characterization. The same can be said for ExbD/TolR.) As more data became available, the interspecies variability in this domain became more evident, to the degree that it now appears that these sequences can be segregated into two groups (Fig. 6). The relevance of this distinction, which is also evident in comparisons among ExbD proteins (see below), is unclear, since it is not paralleled by differences in the corresponding TonB sequences of the same organism. Perhaps these differences define contributions made by the one or more mystery proteins hypothesized above.

While the two-group theme is reiterated in the first half of the second ExbB transmembrane domain, the latter portion of this domain is highly conserved between groups, as well as with TolQ and the third transmembrane domain of MotA, a component of the flagellar motor. In particular, this conservation involves three glycine residues and a proline that is thought to participate in proton conductance in MotA. The third transmembrane domain similarly contains a conserved hydrophobic face that is also conserved in the fourth transmembrane domain of MotA. Of these residues, mutation of the alanine at position 177 in *E. coli* TolQ (corresponding to position 180 in the *E. coli* ExbB sequence) is known to eliminate function, suggesting an important functional role for that transmembrane domain.

	16	TMD#1	39		128		TMD#2	155		162		TMD#3	194

```
ExbB
    E. coli         MYQHADIVVKCVMIGLLILASVVTWAIF.//..MGRGNGYLATIGAISPFVGLFGTVWGIMNSFI_GIAQTQTTNLAVVAPGIAEBALLATAIGIVAAIPAVVIYN
    Y. pestis       MYQNADVVVKGVMIGLVLASIVTWTIL.//..MGKGNGFLATIGAISPFVGLFGTVWGIMNSFI_GIAHSQTTNLAVIAPGIAEBALLATAIGLVAAIPAVVIYN
    B. melitensis   MFMAADWVVKAVMIGLAIASLATWTVW.//..LAKGTGILATIGSVAPFVGLFGTVWGIMNSFI_NISEAQTTNLAVVAPGIAEBALLATAIGIVAAIPAVVIYN
    P. putida       MYKNADIVVKIVMIGLAIASIITWTIW.//..MSSGTGVLATIGSTAPFVGLFGTVWGIMNSFI_GIAKTQTTNLAVVAPGIAEBALLATAIGLVAAIPAVVIYN
    B. pertussis    FVAQSDFVGKTLFIILIVMSLVTWYLI.//..LENGLTVLASVGSTAPFVGLFGTVWGVYHALV_GIGLSDGVTINRIAGPVGEALIMTGLGLAVAIPAVLAYN
    N. gonorrhoeae  VFESGDPVLIGVFVLMLLMSTVTWCLV.//..FDYGMTALASIGATAPFIGLFGTVWGIYHALI_NIGQSQMSIAAVAGPIGEALVATAAGLFVAIPAVLAYN
    P. multocida    LQQYIDYIILGLLAFMSFIMVWLVIER.//..LQRHLTAISTIGSNAPYVGLLGTVIGILLTFVELGHSGGDIDAAAIMVHLSLALKATAVGILVAIPAMVCYN
    H. influenzae   LQQYSDYFIIGLILLMSIIMLAMVIER.//..LNRNMTVISTIGANAPYVGLLGTVIGILLTFYQIGHGGGDIDPSVIMLHLSLALKATALGILVAIPSMVFYN
    P. shigelloides QLEHQLGSMMWPLLTCSFLTGVILLDR.//..LSSGLKILNLIAAISPLIGLLGTIIGLIQMFQDIGASNSPVTPALLADGLGVAMYTTAAGLFIALPALVGSQ
    V. cholerae #1  QLQHQLGLMAWPLFICSALTVMLLAER.//..FNSGLRLLTLIGVISPLIGLLGTVIGLIEMFKGVAATTGSITPNVLADGLGVAMYTTAAGLLIAVPAVAGAQ
    V. cholerae #2  EFMAQGGAVLWWLAVVVLLCWLLVIER.//..LNQNLNLIKVLVAICPMLGLLGTVTGMISVF_DVMATQGSSDPKLMASGISLATLPTMAGMVAALAGLFVHA
TolQ
    E. coli         LFLKASLLVKLIMLILIGFSIASWAII.//..LETHIPFLGTVGSISPYIGLFGTVWGIMHAFI_ALGAVKQATLQMVAPGIAEBALIATAIGLFAAIPAVMAYN
    V. cholerae     LFLQASFLVKVMLTLLGMSVASWAAI.//..LETSLPFLATVGSISPYIGLFGTVWGIMHAFI_ALGEVKQATLAMVAPGIAEBALIATAIGLFAAIPAVMAYN
MotA
    B. subtilis                          ...HQAGAAIFTQAGTYAPTLGVLGAVIGLIAALSHMDNTDE..//..LGHAISAAFVATLLGIFTGYVLWHPFA
```

FIGURE 6 Conservation in the ExbB TMDs. Sequences are presented relative to residues 15 to 41 and 126 to 196 of E. coli ExbB, with the predicted TMDs indicated at the top. ExbB sequences are divided into two groups primarily on the basis of differences in TMD number 1. The corresponding region from the ExbB homolog TolQ is included for two species at the base of the figure, as is flagellar motor protein MotA. The shading indicates regions of significant residue homology. The NCBI accession numbers for the sequences presented are as follows: ExbB sequences: E. coli (NP_417479), Y. pestis (NP_404318), B. melitensis (AAK08069), P. putida (S28442), B. pertussis (T44782), N. gonorrhoeae (AAC45287), P. multocida (NP_246123), H. influenzae (NP_438422), P. shigelloides (AAG23397), V. cholerae ExbB1 (O025897), V. cholerae ExbB2 (AAC69454); TolQ sequences: E. coli (NP_308799), V. cholerae (NP_231473); MotA sequence: Bacillus subtilis (NP_389252).

Like ExbB, the highest degree of conservation among ExbD proteins occurs in the transmembrane domain, and it again appears that these sequences can be segregated into the same two groups (Fig. 7). Of particular interest is the aspartic acid residue at position 25 in the *E. coli* sequence, which is shared between groups in a conservation that extends to TolR and another flagellar motor component, MotB, in which it has been suggested to play a role in the proton channel. In ExbD, mutation of this aspartic acid results in loss of function. There is also some conservation in the carboxy-terminal region of ExbD, in a pattern consistent with that observed for the transmembrane domain and for ExbB. Of particular note is an aspartic acid residue at position 108 in the *E. coli* sequence, which is nearly as highly conserved as its transmembrane domain counterpart.

Significantly, there is no apparent conservation between the carboxy termini of ExbD and MotB. This is relevant because the corresponding region of MotB contains a highly conserved motif that is hypothesized to mediate anchorage to the peptidoglycan layer. This would allow MotB to function as a stator, which is obligatory for the function of a rotational system. It is here that parallels between the flagellar motor and ExbB and ExbD break down. While they probably have commonly derived ion channels for harvesting the proton gradient, the absence of any apparent anchorage for ExbD suggests that the systems diverged at an early point, and thus, consistent with the dem-

FIGURE 7 Conservation in the ExbD TMDs and C terminus. Sequences are presented relative to residues 18 to 44 and 100 to 136 of *E. coli* ExbD, with the predicted TMD indicated at the top. The corresponding region from the ExbD homolog TolR is included for two species at the base of the figure, as is flagellar motor protein MotB. The shading indicates regions of significant residue homoPgy. The NCBI accession numbers for the sequences presented are as follows: ExbD sequences: *E. coli* (NP_417478), *Y. pestis* (NP_404319), *B. melitensis* (AAK08070), *P. putida* (AAK70857), *B. pertussis* (T44783), *N. gonorrhoeae* (O06434), *P. multocida* (NP_246124), *H. influenzae* (NP_438421), *P. shigelloides* (AAG23398), *V. cholerae* ExbD1 (O52044), *V. cholerae* ExbD2 (Q9ZHV9); TolR sequences: *E. coli* (P05829), *V. cholerae* (NP_231472), MotB sequence: *B. subtilis* (P28612).

onstration that *E. coli* TonB shuttles between cytoplasmic and outer membranes, the transduction of energy is not a simple mechanical process.

In many respects, the availability of additional sequences has deepened the mysteries surrounding the mechanisms of energy transduction as mediated by TonB, ExbB, and ExbD. The paradigm that has been developed largely in *E. coli* might simply serve as a point of departure when we begin to consider the various alternatives that nature has selected to meet the diverse demands placed on gram-negative bacteria by their environments. In the end, it may well be that the unifying theme of the TonB system is that it serves to globally energize outer membrane processes, with iron transport being only the tip of the iceberg. Indeed, the next TonB epiphany could well come from beyond the *E. coli* realm, provided that other systems receive the molecular scrutiny which they clearly deserve.

SUGGESTED READING

Barnard, T. J., M. E. Watson, Jr., and M. A. McIntosh. 2001. Mutations in the *Escherichia coli* receptor FepA reveal residues involved in ligand binding and transport. *Mol. Microbiol.* **41:**527–536.

Bradbeer, C. 1993. The proton motive force drives the outer membrane transport of cobalamin in *Escherichia coli*. *J. Bacteriol.* **175:**146–3150.

Braun, M., F. Endriss, H. Killmann, and V. Braun. 2003. In vivo reconstitution of the FhuA transport protein of *Escherichia coli* K-12. *J. Bacteriol.* **185:**5508–5518.

Braun, M., H. Killmann, and V. Braun. 1999. The beta-barrel domain of FhuADelta5–160 is sufficient for TonB-dependent FhuA activities of *Escherichia coli*. *Mol. Microbiol.* **33:**1037–1049.

Braun, V., S. Gaisser, C. Herrman, K. Kampfenkel, H. Killman, and I. Traub. 1996. Energy-coupled transport across the outer membrane of *Escherichia coli*: ExbB binds ExbD and TonB in vitro, and leucine 132 in the periplasmic region and aspartate 25 in the transmembrane region are important for ExbD activity. *J. Bacteriol.* **178:**2836–2845.

Braun, V., and C. Herrmann. 1993. Evolutionary relationship of uptake systems for biopolymers in *Escherichia coli*: cross-complementation between the TonB-ExbB-ExbD and the TolA-TolQ-TolR proteins. *Mol. Microbiol.* **8:**261–268.

Braun, V., S. I. Patzer, and K. Hantke. 2002. Ton-dependent colicins and microcins: modular design and evolution. *Biochimie* **84:**365–380.

Cadieux, N., and R. J. Kadner. 1999. Site-directed disulfide bonding reveals an interaction site between energy-coupling protein TonB and BtuB, the outer membrane cobalamin transporter. *Proc. Natl. Acad. Sci. USA* **96:**10673–10678.

Cascales, E., R. Lloubes, and J. N. Sturgis. 2001. The TolQ-TolR proteins energize TolA and share homologies with the flagellar motor proteins MotA-MotB. *Mol. Microbiol.* **42:**795–807.

Chang, C., A. Mooser, A. Pluckthun, and A. Wlodawer. 2001. Crystal structure of the dimeric C-terminal domain of TonB reveals a novel fold. *J. Biol. Chem.* **276:**27535–27540.

Fischer, E., K. Günter, and V. Braun. 1989. Involvement of ExbB and TonB in transport across the outer membrane of *Escherichia coli*: phenotypic complementation of *exb* mutants by overexpressed *tonB* and physical stabilization of TonB by ExbB. *J. Bacteriol.* **171:**5127–5134.

Ghosh, J., and K. Postle. 2004. Evidence for dynamic clustering of carboxy-terminal aromatic amino acids in TonB-dependent energy transduction. *Mol. Microbiol.* **51:**203–213.

Ghosh, J., and K. Postle. Disulfide trapping of an in vivo energy transducing conformation of *Escherichia coli* TonB protein. Submitted for publication.

Higgs, P. I., R. A. Larsen, and K. Postle. 2002. Quantitation of known components of the *Escherichia coli* TonB-dependent energy transduction system: TonB, ExbB, ExbD, and FepA. *Mol. Microbiol.* **44:**271–281.

Higgs, P. I., T. E. Letain, K. K. Merriam, N. S. Burke, H. Park, C. Kang, and K. Postle. 2002. TonB interacts with nonreceptor proteins in the outer membrane of *Escherichia coli*. *J. Bacteriol.* **184:**1640–1648.

Kampfenkel, K., and V. Braun. 1993. Topology of the ExbB protein in the cytoplasmic membrane of *Escherichia coli*. *J. Biol. Chem.* **268:**6050–6057.

Karlsson, M., K. Hannavy, and C. F. Higgins. 1993. ExbB acts as a chaperone-like protein to stabilize TonB in the cytoplasm. *Mol. Microbiol.* **8:**389–396.

Larsen, R. A., G. J. Chen, and K. Postle. 2003. Performance of standard phenotypic assays for TonB activity, as evaluated by varying the level of functional, wild-type TonB. *J. Bacteriol.* **185:**4699–4706.

Larsen, R. A., T. E. Letain, and K. Postle. 2003. *In vivo* evidence of TonB shuttling between the cytoplasmic and outer membrane in *Escherichia coli*. *Mol. Microbiol.* **49:**211–218.

Larsen, R. A., M. G. Thomas, and K. Postle. 1999. Protonmotive force, ExbB and ligand-bound

FepA drive conformational changes in TonB. *Mol. Microbiol.* **31:**1809–1824.

Larsen, R. A., G. E. Wood, and K. Postle. 1993. The conserved proline-rich motif is not essential for energy transduction by *Escherichia coli* TonB protein. *Mol. Microbiol.* **10:**943–953.

Letain, T. E., and K. Postle. 1997. TonB protein appears to transduce energy by shuttling between the cytoplasmic membrane and the outer membrane in Gram-negative bacteria. *Mol. Microbiol.* **24:**271–283.

Liu, J., J. M. Rutz, P. E. Klebba, and J. B. Feix. 1994. Site-directed spin-labeling study of ligand-induced conformational change in the ferric enterobactin receptor, FepA. *Biochemistry* **33:**13274–13283.

Mann, B. J., C. D. Holroyd, C. Bradbeer, and R. J. Kadner. 1986. Reduced activity of TonB-dependent functions in strains of *Escherichia coli*. *FEMS Microbiol. Lett.* **33:**255–260.

Mey, A. R., and S. M. Payne. 2003. Analysis of residues determining specificity of *Vibrio cholerae* TonB1 for its receptors. *J. Bacteriol.* **185:**1195–1207.

Moeck, G. S., P. Tawa, H. Xiang, A. A. Ismail, J. L. Turnbull, and J. W. Coulton. 1996. Ligand-induced conformational change in the ferrichrome-iron receptor of *Escherichia coli* K-12. *Mol. Microbiol.* **22:**459–471.

Moeck, G. S., J. W. Coulton, and K. Postle. 1997. Cell envelope signaling in *Escherichia coli*. Ligand binding to the ferrichrome-iron receptor FhuA promotes interaction with the energy-transducing protein TonB. *J. Biol. Chem.* **272:**28391–28397.

Occhino, D. A., E. E. Wyckoff, D. P. Henderson, T. J. Wrona, and S. M. Payne. 1998. *Vibrio cholerae* iron transport: haem transport genes are linked to one of two sets of *tonB*, *exbB*, *exbD* genes. *Mol. Microbiol.* **29:**1493–1507.

Postle, K., and R. J. Kadner. 2003. Touch and go: tying TonB to transport. *Mol. Microbiol.* **49:**869–882.

Sauter, A., S. P. Howard, and V. Braun. 2003. In vivo evidence for TonB dimerization. *J. Bacteriol.* **185:**5747–5754.

Scott, D. C., Z. Cao, Z. Qi, M. Bauler, J. D. Igo, S. M. Newton, and P. E. Klebba. 2001. Exchangeability of N termini in the ligand-gated porins of *Escherichia coli*. *J. Biol. Chem.* **276:**13025–13033.

Seliger, S. S., A. R. Mey, A. M. Valle, and S. M. Payne. 2001. The two TonB systems of *Vibrio cholerae*: redundant and specific functions. *Mol. Microbiol.* **39:**801–812.

Smajs, D., and G. M. Weinstock. 2001. The iron- and temperature-regulated cjrBC genes of *Shigella* and enteroinvasive *Escherichia coli* strains code for colicin Js uptake. *J. Bacteriol.* **183:**3958–3966.

Vakharia, H. L., and K. Postle. 2002. FepA with globular domain deletions lacks activity. *J. Bacteriol.* **184:**5508–5512.

Young, G. M., and K. Postle. 1994. Repression of *tonB* transcription during anaerobic growth requires Fur binding at the promoter and a second factor binding upstream. *Mol. Microbiol.* **11:**943–954.

PERIPLASMIC BINDING PROTEINS INVOLVED IN BACTERIAL IRON UPTAKE

Karla D. Krewulak, R. Sean Peacock, and Hans J. Vogel

8

Bacteria use their surrounding membranes and peptidoglycan layers to protect themselves from their environment. Gram-positive bacteria possess only a single cytoplasmic membrane, whereas all gram-negative bacteria have both a cytoplasmic membrane and an outer membrane. The outer membranes contain porins, a group of trimeric proteins which create pores that allow the diffusion of nutrients required for bacterial growth into the periplasmic space, the area between the outer and cytoplasmic membranes. However, the pores created by the porins are of limited size, and, typically, compounds that are larger than 600 Da cannot gain entry into the periplasm in this manner. Thus, while sugars, amino acids, and phosphate or sulfate can pass through the porins, compounds such as ferric-siderophores or vitamin B_{12} are too big and have to rely on uptake through specialized outer membrane receptors to gain access to the periplasm. Irrespective of their size, once they have entered to the periplasmic compartment, all small nutrients as well as the larger molecules can be actively transported through the cytoplasmic membrane of gram-negative and gram-positive bacteria via integral membrane permeases and ATP binding cassette proteins (collectively called ABC transporters). The membrane transport proteins themselves do not necessarily have a strong binding affinity for the substances they transport into the cytoplasm. Therefore, a number of soluble periplasmic binding proteins (PBPs) in gram-negative bacteria and extracellular surface-bound lipoproteins in gram-positive bacteria have evolved. The PBPs serve as high-affinity carriers that escort nutrients or ligands to the membrane transporters for subsequent passage into the cytoplasm of the cell. PBPs are usually an absolute requirement for uptake, and they have also been implicated in chemotaxis and sensory transduction.

PBPs constitute a large and diverse family of proteins. Recent amino acid sequence alignment searches show hundreds of hits in numerous bacterial species, and this number will continue to grow with the completion of further bacterial genome-sequencing projects. The genes for the PBPs, permeases, and ATPase components of the transport systems for each particular substrate are often found in the same operon, in that respective order. This operon layout is thought to be evidence of a common origin of PBPs and transport systems in general,

---

Karla D. Krewulak, R. Sean Peacock, and Hans J. Vogel, Structural Biology Research Group, Department of Biological Sciences, University of Calgary, Calgary, Alberta, Canada T2N 1N4.

Iron Transport in Bacteria, Edited by Jorge H. Crosa, Alexandra R. Mey, and Shelley M. Payne
© 2004 ASM Press, Washington, D.C.

with variation arising from gene duplication events.

To better understand how PBPs work, it is necessary to divide this large family into smaller classes. The PBP family was first classified, on the basis of amino acid sequence homology, into eight different families in the early 1990s, and since then one additional family has been uncovered. In Table 1 we summarize all of the proteins from these nine clusters for which at least one three-dimensional structure has been solved. To date, the structures of at least 25 different PBPs have been determined, with every family represented except cluster 7, the organic polyanion binding proteins.

It has been noted that PBPs are homologous to a number of eukaryotic receptors such as glutamate-gated ion channel (NMDA) receptors, binding domains from G-protein-coupled receptors (GPCRs) and guanylate cyclase-atrial natriuretic peptide (ANP) receptors. PBPs are thought to be a good model for understanding the mechanism of ligand binding to these proteins and a good model for drug development.

The wealth of structural information currently available for PBPs means that they have served as an excellent model for understanding protein-ligand binding interactions, domain motions that occur on ligand binding, and domain conservation. The remainder of this chapter focuses on the structural biology of PBPs in general and, more specifically, on the iron binding PBPs whose structures have been solved: FbpA, FhuD, and BtuF.

GENERAL STRUCTURAL FEATURES

PBPs have very little sequence conservation. Within the cluster families of the PBPs, there is often less than 10% sequence identity between members. Consequently, even after exhaustive iterative sequence alignment searches, members of one family of PBPs do not necessarily return hits from other families. Nonetheless, the structures of all of the PBPs which have been solved to date are strikingly similar; in fact, all of the PBPs fall into one of three basic topological arrangements.

All the PBPs are bilobal, with the two lobes being joined either by two or three β-strands (clusters 1 to 6) or by a long backbone helix (clusters 8 and 9). Each of the lobes has a mixed αβ structure. A study of 15 structures representing five of the PBP clusters and several non-PBP but structurally related proteins showed that the sequences of these proteins could be grouped into one of two structural

TABLE 1 The nine PBP clusters[a]

Cluster	Specificity	Members with solved structures[b]
1	Malto-oligosaccharides, multiple sugars, phosphate, iron (Fe^{3+})	MalE (MBP), FbpA (FBP), PstS (PBP), AlgQ2
2	Hexoses and pentoses	AraF (ABP), MglB (GBP), RbsB (RBP), AlsB (ALBP)
3	Polar amino acids	ArgT (LAOBP), HisJ (HBP), GlnH (GlnBP)
4		
5	Aliphatic amino acids	LivJ (LIVBP), LivK (LBP)
6	Peptides and nickel	OppA, DppA, NiKA
7	Inorganic polyanions	PstS (PBP), Sbp (SBP), ModA
8	Organic polyanions	No structures solved
9	Iron complexes and vitamin B_{12}	FhuD, BtuF
	Zinc and manganese	PsaA, TroA

[a] All the PBPs for which at least one three-dimensional structure has been solved, sorted into the nine clusters described in the text, are shown.

[b] Proteins are given in the following form: protein product of the gene name as found in the NCBI GenBank entry (common acronym found in the literature). FBP, ferric ion binding protein; ABP, L-arabinose binding protein; GBP, galactose-binding protein; RBP, ribose-binding protein; ALBP, D-allose-binding protein; LAOBP, lysine/arginine/ornithine-binding protein; HBP, histidine-binding protein; GlnBP, glutamine-binding protein; LIVBP, leucine/isoleucine/valine-binding protein; LBP, leucine-binding protein; SBP, sulfate-binding protein.

groups and sequence groups depending on the arrangement of the β-strands in the N-terminal region and the number of strands (two or three) in the linking hinge region (Fig. 1). This was proposed to be a result of an evolutionary event that the authors termed a domain translocation. It was also proposed that the bilobal domain structure of the PBPs might have arisen from an αβ ancestor protein such as CheY, a protein involved in chemotaxis.

More recently, the structures of several proteins in groups 8 and 9 have been solved. These PBPs (PsaA, TroA, FhuD, and BtuF) have a long backbone helix linker and therefore represent a unique structural class of PBPs, as discussed in detail below.

The iron binding and siderophore binding PBPs belong to clusters 1 and 8 of the PBP family, with the iron binding ferric binding protein (FbpA, cluster 1), the siderophore binding PBP (FhuD, cluster 8), and the vitamin B_{12} binding protein (BtuF, cluster 8) being the representatives for which structures have been determined to date. In gram-negative bacteria, the transport of iron, whether via FBP or via siderophore binding proteins, is dependent on the outer membrane receptor-TonB-ExbB-ExbD system to internalize the iron and the

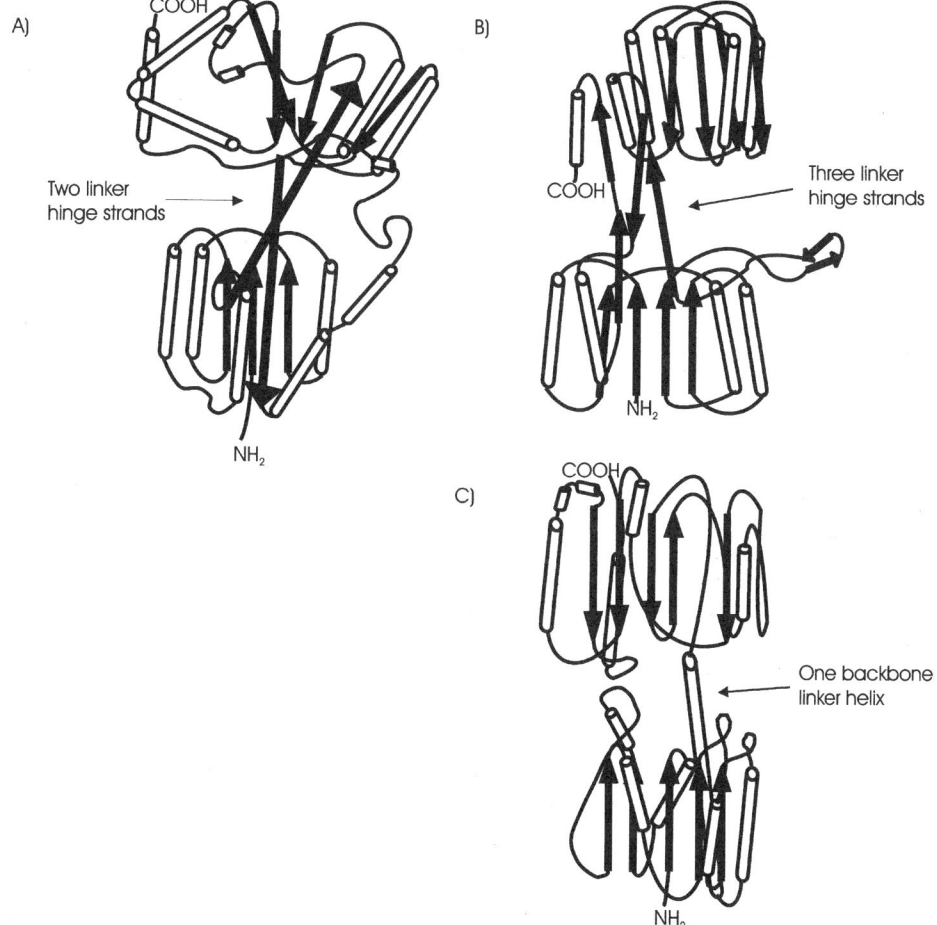

FIGURE 1 Topological arrangement of PBPs. (A) Two-strand linker as represented by ferric ion binding protein. (B) Three-strand linker as represented by L-arabinose binding protein. (C) Helix linker as represented by ferrihydroxamate uptake binding protein (FhuD).

siderophores into the periplasm. In gram-positive bacteria, the iron binding protein is tethered to the outer leaflet of the cytoplasmic membrane via a lipid anchor.

It appears that the mechanisms of ligand binding of FbpA and FhuD are quite different from each other. To better explain the ligand binding to PBPs, it is best to look at the extensive work done with maltose binding protein (MBP), which is structurally similar to the Ferric binding protein. We also discuss some studies of the TroA protein, which is structurally related to FhuD.

MALTOSE BINDING PROTEIN

Although not involved in iron transport or iron transport-related events, MBP is one of the best-characterized periplasmic binding proteins and is therefore a good mechanistic model of periplasmic transport. In particular, the structural biology of this protein has been extensively characterized in the presence and absence of ligands. It is worthwhile to devote a portion of this chapter to describing different experimental techniques which have been used to characterize the motions that MBP undergoes on ligand association or dissociation.

MBP is a 40.6-kDa periplasmic protein that serves as the high-affinity binding component of malto-oligosacchharides in bacteria, as well as being a mediator in chemotaxis toward malto-oligosaccharides. It is part of the *mal* operon in *Escherichia coli* (*malE* is the gene encoding MBP), and it passes the malto-oligosaccharides on to the cytoplasmic membrane transporter MalFGK$_2$ complex. In protein engineering, MBP is often used as a fusion tag for protein expression since its high solubility appears to enhance the yields of proteins fused to it.

A number of X-ray crystallographic structures of MBP have been solved, including the apo form and ligand-bound forms with maltose, β-cyclodextrin, maltotriose, maltotetraose, maltotetraitol, and maltotriitol. Most of these structures have been solved for the MBP from *E. coli*. One of the most striking features is that the apo structure of MBP is a great deal more "open" than the structure of MBP with maltose, maltotriose, or maltotetraose (Fig. 2). The domain closure observed when MBP binds its ligand is the origin of the term "Venus flytrap mechanism." This has also been referred to as a "PacMan motion" by analogy to a com-

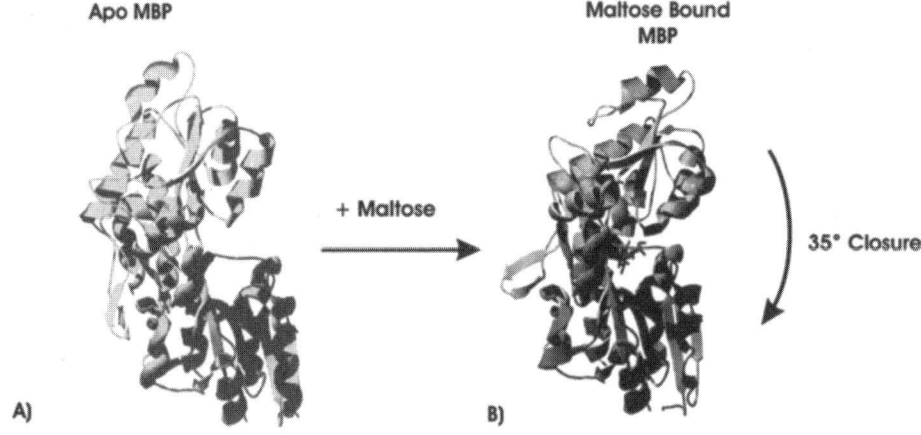

FIGURE 2 Conformational change of MBP on ligand binding. apo MBP (A) and MBP bound to maltose (B) are shown. Domain closure when MBP binds its ligand demonstrates the term "Venus flytrap mechanism." Note that although the structure of the N- and C-terminal domains remains unchanged on ligand binding, their orientation with respect to each other changes drastically. This hinge motion is very efficient since it requires changes in the orientation of only a few residues in the region connecting the two domains. As in all subsequent structural figures, the coordinates for the protein structures were obtained from the protein database.

puter game that was popular in the early 1980s. Binding of maltose to MBP is characterized by a hinge bending motion that involves the β-strands connecting the two domains of MBP and a slight twist of one domain with respect to the other. Notably, three ligand-bound structures have been shown to be in the open state: MBP complexed with β-cyclodextrin, maltotetraitol, and maltotriotol.

Crystal structures sometimes have artifacts that can arise from the crystal packing. Various other studies have been done to verify that there are two MBP conformations in solution—the ligand-bound closed form and the ligand-free open form—as opposed to a dynamic equilibrium between an open and closed apo-form that becomes trapped on ligand binding.

Small-angle X-ray scattering (SAXS) can be used to measure the radius of gyration of proteins, with a smaller radius of gyration (R_g) indicating a compaction of the protein—in this case via domain closure. Using MBP in the apo, maltose-bound, or maltotriose-bound form, SAXS showed a significant decrease in the R_g of MBP in the ligand-bound form in comparison to the apo form. This indicates that there are two forms in solution, an open ligand-free form and a closed holo form, and not a continuous exchange between open and closed ligand-bound form that is trapped in the closed conformation by the binding of ligand.

Extensive nuclear magnetic resonance (NMR) studies have been performed on MBP. MBP is a particularly interesting protein in terms of NMR because of its size (40.6 kDa), which is larger than proteins conventionally accessible by NMR, and because the two domains are difficult to orient correctly with respect to each other by using conventional proton-proton nuclear Overhauser enhancements. The group of Lewis Kay has shown in a number of different papers that it is possible to orient the domains of MBP by using residual dipolar couplings as well as anisotropy and carbonyl chemical shift change measurements. Surprisingly, these measurements, which have been done with apo and β-maltodextrin-bound forms of MBP, show that in contrast to the crystal structure, in which that β-maltodextrin binds to an open form of MBP, the NMR structures are considerably more closed (from 7 to 11° more closed) than the apo form. This is interesting because it shows that the crystal structure may not be entirely representative of the protein structure in solution, reminding us that some caution must be used in interpreting crystal structures.

A third alternative method to study domain reorientation in bilobal proteins is through computational molecular dynamic calculations, which can be complemented by various filtering methods, such as normal-mode analysis or essential dynamics, to focus exclusively on the large-scale motions that take place in proteins. Various computational methods (e.g., Dyndom) are available to extract the kind of motions typically seen in many PBPs.

FERRIC ION BINDING PROTEIN

FbpA proteins are 37-kDa periplasmic proteins found in the *Neisseriaceae* and *Pasteurellaceae* families; they are involved in the transport of iron from the bacterial transferrin (Tf) and lactoferrin (Lf) receptors across the periplasmic space to the respective cytoplasmic membrane ABC transporter. The iron-free and iron-loaded forms of *Haemophilus influenzae* FbpA have been solved to 1.6 and 1.75 Å, respectively. Although FbpA and Tf have less than 20% sequence identity, the FbpA structure is similar to a single lobe of the Tf and Lf structures. Like Tf, FbpA has two domains consisting of an alternating helix-sheet structure connected by antiparallel β-strands. Comparison of apo-FbpA and holo-FbpA reveal a Venus fly trap domain mechanism with a 21° rotation of the two structural domains about a central hinge (Fig. 3). By comparison, other members of the PBP family can exhibit larger domain movements between bound and unbound forms, ranging from a 35° domain swing between the liganded and unliganded forms of MBP to the 63° rotation found in Tf. Although the domain motions vary among these PBPs, the general position of the hinge is conserved between FbpA, PBPs, and the Tfs. Antiparallel

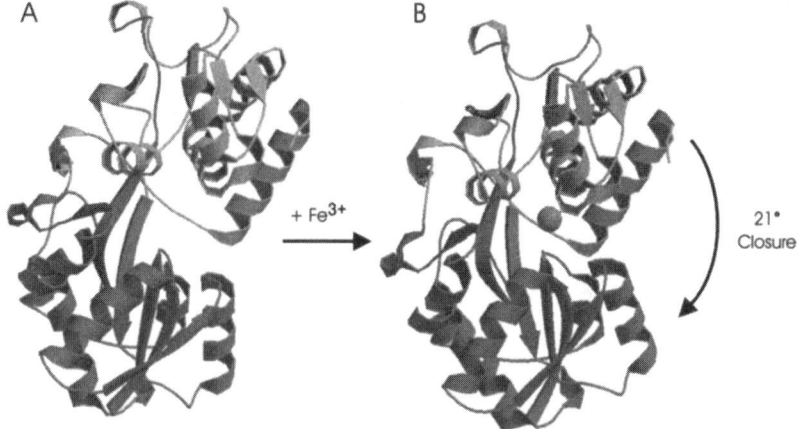

FIGURE 3 Conformational change of FbpA on ligand binding. Fe^{3+}-free FbpA, (A) and Fe^{3+}-bound FbpA (B) are shown. The most dramatic difference on binding of Fe^{3+} is a 21° rotation about a central β-sheet hinge that separates the two structural domains. The same structure is seen for the C-terminal domain in both the apo and holo forms of FbpA. Some reorganization of key residues in the N-terminal domain occurs on iron binding.

β-strands form the hinge between the two domains. Another similarity between Tf and FbpA is that Fe^{3+} binds in the cleft between the two domains. An additional loop in the Tfs covers the binding site; therefore, the Fe^{3+} in FbpA is more solvent exposed than that in Tf. Like Tf, FbpA has six ligands coordinating the Fe^{3+} with octahedral geometry (see Fig. 4). Both Tf and FbpA coordinate Fe^{3+} with two oxygens from two Tyr residues, an imidazole nitrogen from His, and a carboxylate oxygen from Glu. While Tf completes the octahedral coordination of Fe^{3+} with two oxygens from a carbonate anion, FbpA uses oxygens from phosphate and a water molecule. The carbonate ion acts as a "synergistic anion" which plays an important role in the release of iron from Tf and Lf. By analogy, it has been suggested that the phosphate, which was found to be bound to the apo structure of FBP, may play a similar role in FBP. Nevertheless, we have been able to crystallize apo-FBP without any bound phosphate and have found that other anions such as citrate can act as the FBP-bound anion as well. Clearly, the requirement for a bound anion for Fe^{3+} binding to FBP requires further study. Although FbpA and Tf exist in different environments (periplasm and serum, respectively), the homology between them hints at the possibility that the Tfs and the PBPs may have evolved from a common ancestor.

CLASS 9 PERIPLASMIC BINDING PROTEINS

One of the members of the newest class of PBPs is TroA, a 31-kDa protein found in *Treponema pallidum*, the spirochete that causes syphilis. TroA is the periplasmic component involved in shuttling Zn^{2+}, and possibly other transition metals, across the cytoplasmic membrane of this organism. TroA is attached to the outer leaflet of the cytoplasmic membrane via a 23-amino-acid hydrophobic leader peptide. The structure of Zn^{2+}-bound TroA was solved to a resolution of 1.8 Å. Like other PBPs, two independent globular domains interact with each other to create a Zn^{2+} binding cleft. However, in this case, the N and C domains are connected by a long backbone α-helix. This feature appears to be unique to the class 9 and class 8 family of PBPs, since it is also found in the structure of the zinc binding pneumococcal surface protein PsaA.

As discussed above, members of the PBP family that bind and release nonmetal ligands mechanism. A 4° tilt of the C-terminal domain about an axis parallel to the relatively inflexible backbone helix occurred on release of Zn^{2+}. Because the domain-domain interface is highly hydrophobic, exposure of this region to solvent is not favorable; therefore, on release of Zn^{2+}, TroA adopts a closed conformation in which the volume of the binding pocket shrinks from 97.2 to 59.3 Å^3. The bound and apo structures of TroA are presented in Fig. 5. The domain change is small compared to that in other PBPs but large enough to distort the correct geometry for Zn^{2+} coordination, thereby disallowing metal binding. The proposed mechanism involves three states: a nonpermissive unbound form, a hydrated form, and the Zn^{2+} bound form. Two water molecules are incorporated into the binding pocket of the nonpermissive form, thus restoring the geometry for the Zn^{2+} binding site, and a negatively charged Asp residue in the binding pocket is responsible for attracting the positively charged Zn^{2+} molecule.

SIDEROPHORE BINDING PROTEINS

It has been established in previous chapters in this book that siderophores can be used to chelate iron and transport it into a gram-negative bacterial cell through an iron uptake system which includes an outer membrane receptor, a PBP, and several inner membrane proteins (Fig. 6). A phylogenetic tree relating PBPs from the FbpA family of different bacteria (cluster 1) to those of cluster 8 and cluster 9 is shown in Fig. 7. The evolutionary relationship between different gram-negative bacteria shows that although the sequences between the PBPs in cluster 8 and cluster 9 are not similar (<20% identity), they are structurally similar; therefore, because FhuD is the only periplasmic siderophore binding protein (PSBP) to have its structure reported, it can be used as a representative of all PSBPs to understand the mechanism of siderophore binding. Furthermore, we note that the Hem T protein, which functions as the PBP in bacterial heme

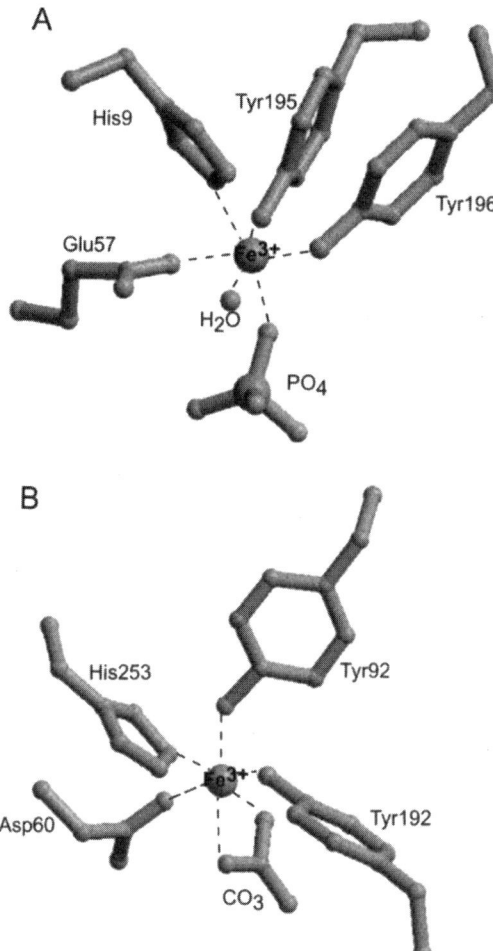

FIGURE 4 Comparison of Fe^{3+} binding sites of FbpA (A) and the N lobe of lactoferrin (B). Similar ligands coordinate the Fe^{3+} ion in each binding site. As in the transferrin superfamily, six ligands coordinate the Fe^{3+} ion in FbpA. FbpA contributes two oxygen ligands from Tyr195 and Tyr196, a carboxylate ion ligand from Glu57, and an imidazole nitrogen ligand from His9. A water molecule and a phosphate ion contribute the two remaining oxygen ligands, thus fulfilling the octahedral geometry preferred by Fe^{3+}.

such as sulfate, phosphate, amino acids, oligopeptides, monosaccharides, and oligosaccharides open and close in a so-called Venus flytrap mechanism and, on release of their respective ligand, adopt an open conformation. Recently, the structure of apo-TroA was solved. It revealed an exception to this

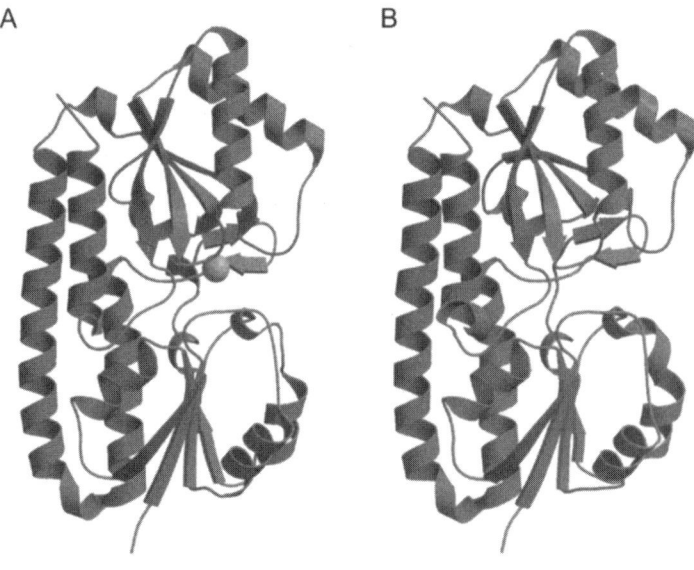

FIGURE 5 Ribbon representation of Zn(II)-bound (A) and Zn(II)-free (B) TroA. On binding Zn(II), TroA adopts a more open conformation due to a 4° tilt of the C-terminal domain. Two C-terminal domain loops collapse into the Zn(II) binding cavity, thus decreasing the cavity volume from 97.2 to 59.3 Å^3. The α-helix that connects the two domains of TroA is structurally different from the typical β-strands that connect two domains in non-metal-binding PBPs. The poorly flexible helix may explain why ligand binding and release by TroA differs from the Venus flytrap mechanism seen in other PBPs.

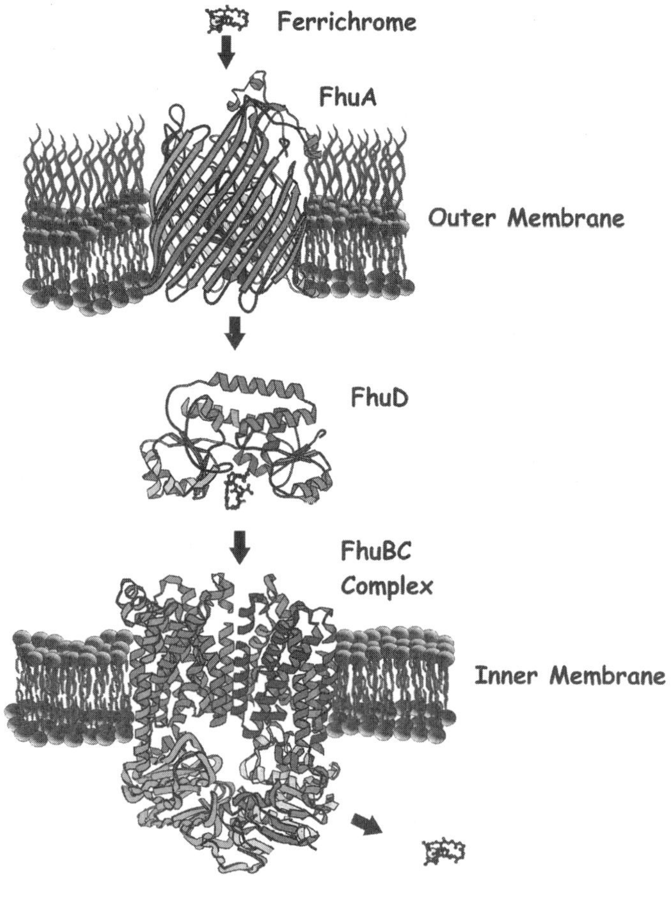

FIGURE 6 Overview of the ferric siderophore transport pathway for gram-negative bacteria. Siderophores secreted from bacteria chelate iron from host proteins (e.g., TF, LF, and heme proteins) or the environment (e.g., free iron or iron hydroxide precipitates). These ferric siderophores are transported through specific outer membrane receptors into the periplasmic space. A periplasmic protein carries the ferric siderophore across the periplasm to an inner membrane receptor. With the assistance of an associated ATPase protein, the intact ferric siderophore is transported through this receptor into the cytoplasm, and iron is released for bacterial survival.

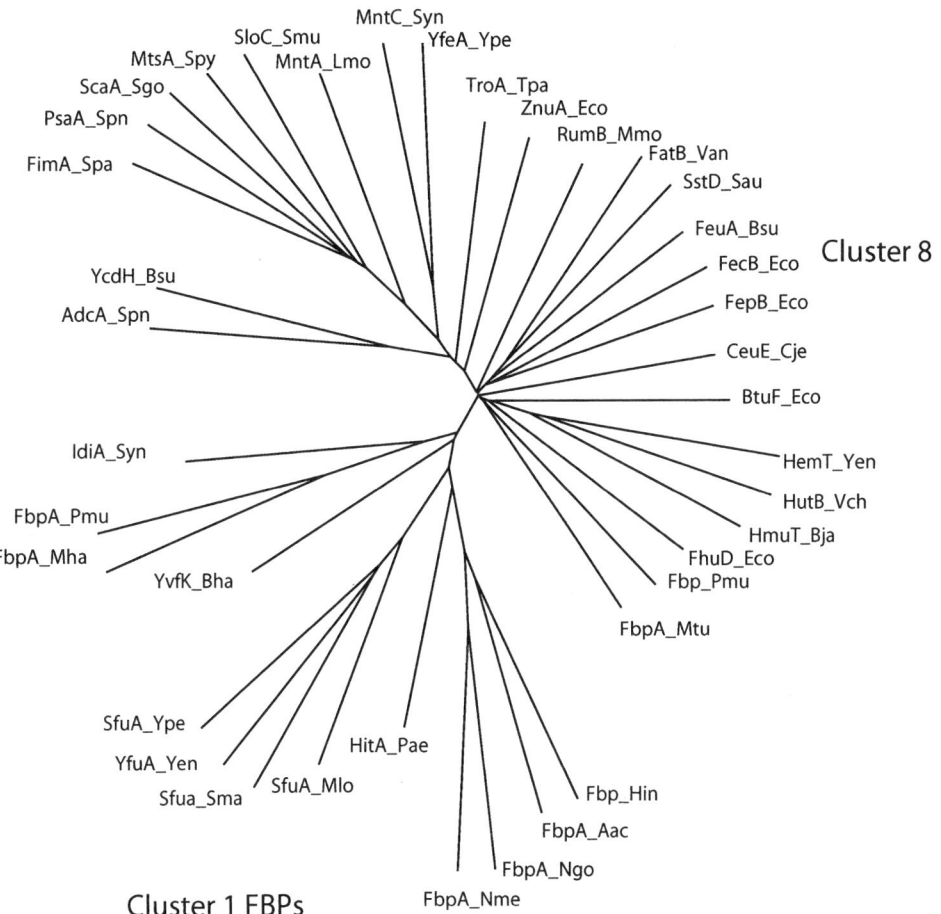

FIGURE 7 Clustering of selected PBPs from cluster 1 iron binding proteins and cluster 8 and cluster 9 proteins. A dendrogram of a selection of proteins thought to be members of the FBP family or of the cluster 8 or cluster 9 families of PBPs is shown. Within cluster 8 and cluster 9, the sequence identities between proteins is often very low—around 10%. In general, the proteins cluster into the families that they should, although ZnuA and TroA are comparatively homologous to some of the cluster 8 proteins. Notice that two FBPs cluster almost as well with the cluster 8 proteins as with the other FBPs in cluster 1. This may simply be a case of incorrect annotation in sequencing projects. Abbreviations: Aac, *Actinobacillus actinomycetemcomitans*; Bsu, *Bacillus subtilis*; Bha, *Bacillus halodurans*; Bja, *Bradyrhizobium japonicum*; Cje, *Campylobacter jejuni*; Eco, *Escherichia coli*; Hin, *Haemophilus influenzae*; Lmo, *Listeria monocytogenes*; Mha, *Mannheimia haemolytica*; Mlo, *Mesorhizobium loti*; Mmo, *Morganella morganii*; Mtu, *Mycobacterium tuberculosis*, Nme, *Neisseria meningitidis*; Ngo, *Neisseria gonorrhoeae*; Pae, *Pseudomonas aeruginosa*; Pmu, *Pasteurella multocida*; Sau, *Staphylococcus aureus*; Sgo, *Streptococcus gordonii*; Sma, *Serratia marcescens*; Smu, *Streptococcus mutans*; Spa, *Streptococcus parasanguis*; Spn, *Streptococcus pneumoniae*; Spy, *Streptococcus pyogenes*; Syn, *Synechocystis* sp.; Tpa, *Treponema pallidum*; Van, *Vibrio anguillarum*; Vch, *Vibrio cholerae*; Yen, *Yersinia enterocolitica*; Ype, *Yersinia pestis*.

uptake systems, also falls into PBP cluster 8, suggesting that it is very similar to FhuD.

Many microorganisms secrete siderophores (low-molecular-weight, highly specific iron chelators) to chelate iron from their environment when they are under low-iron stress (see chapter 1). These siderophores can generally be classified by their iron binding moieties as either hydroxamate, catecholate, polyhydroxycarboxylate, or citrate, type siderophores. FhuD binds a variety of hydroxamate-type siderophores: ferrichrome, coprogen, ferrioxamine B, and rhodotorulic acid (Fig. 8). The ferric hydroxamate PBP, FhuD, is structurally homologous, despite limited sequence identity, to TroA. The structure of gallichrome-bound FhuD was solved to 1.8 Å. Gallichrome was used rather than ferrichrome in order to assist with the phasing of the crystallographic data. However, subsequent studies showed that the structures of FhuD with bound ferrichrome and alumichrome were identical to that with gallichrome. FhuD is a kidney bean-shaped bilobate protein. Like TroA, the N- and C-terminal domains are connected by a 23-residue backbone helix (Fig. 9), with gallichrome binding in a 10-Å-deep groove between the N-terminal and C-terminal domains. Only 45% (246 of 398 Å^2) of the molecular surface of gallichrome

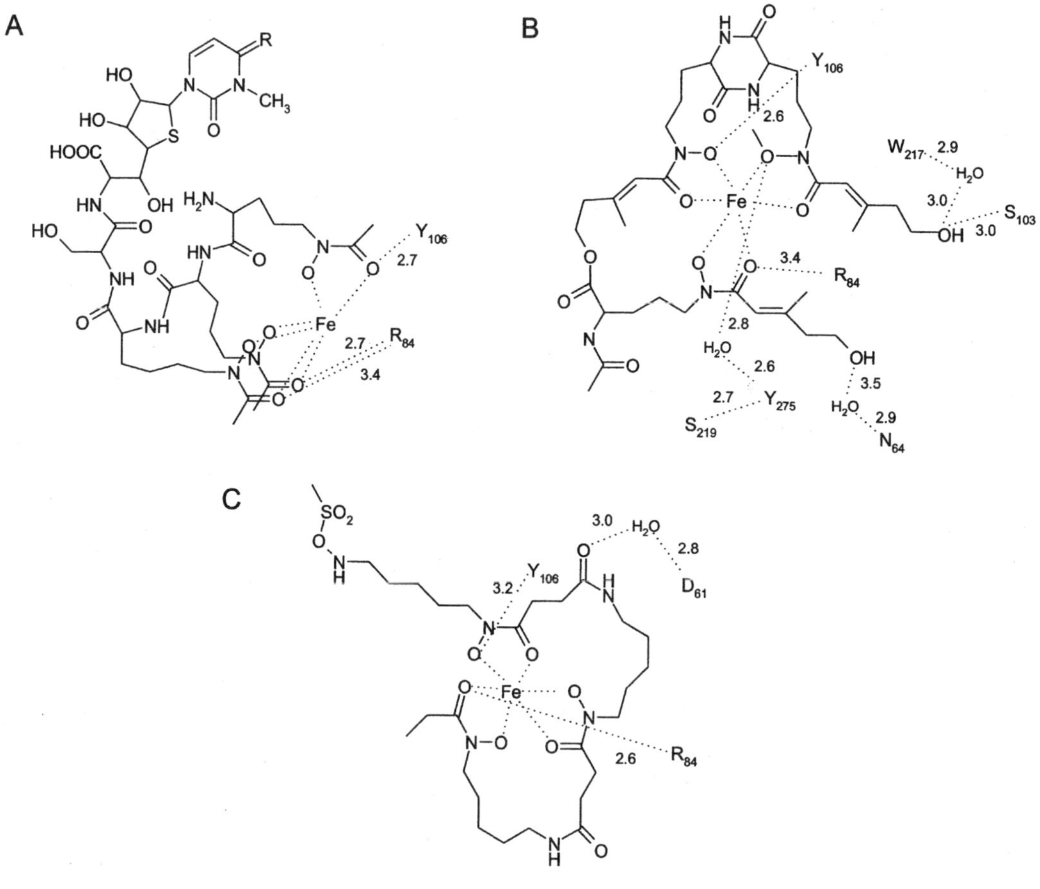

FIGURE 8 Chemical structures of various hydroxamate-type siderophores. The chemical structures of albomycin (a), coprogen (b), and Desferal (c) are shown with FhuD side chain residues. Interactions are indicated by dotted lines (distance indicated). Reprinted from Clarke, Braun, et al. (2002) with permission from the publisher.

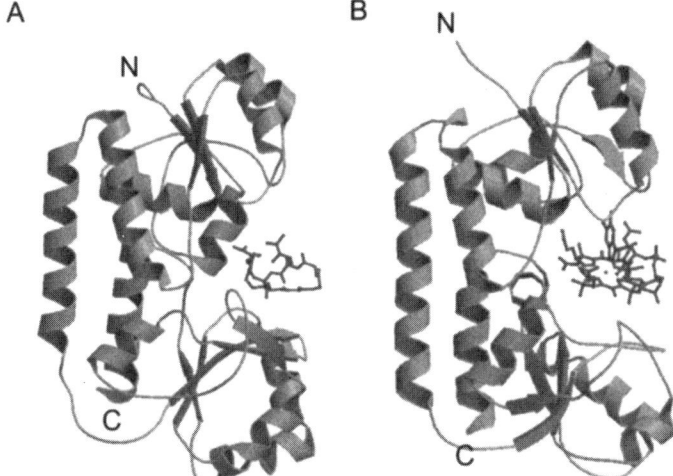

FIGURE 9 Structures of FhuD bound to gallichrome (A) and BtuF bound to vitamin B_{12} (B). Although these two proteins have little sequence homology, they have a surprisingly similar fold. Both have two topologically similar domains consisting of a central five-strand β-sheet surrounded by helices. The domains are connected by an α-helix that spans the length of the protein. A similar fold is seen in the TroA and PsaA proteins (discussed in the text) from cluster 9.

is buried in the complex (Fig. 10). Only four hydrogen bonds are involved in binding gallichrome; this is surprising, considering the 0.1 μM dissociation constant indicating a relatively high-affinity binding. The guanidinium group of Arg84 forms two hydrogen bonds to two of the carbonyl oxygen atoms of the hydroxamic acid moieties of gallichrome (Fig. 8 and 9). The hydroxyl group of Tyr106 forms the third hydrogen bond to the third carbonyl oxygen atom of the hydroxamic acid moiety. The fourth hydrogen bond is water mediated between residues Asn215 and Ser219 to the backbone of gallichrome. After mutation of Arg84 or Tyr106 to Ala, gallichrome no longer binds, suggesting that the hydrogen bonds are vital for siderophore binding. Hydrophobic residues (Trp, Tyr, Ile, and Leu) lining the binding pocket of FhuD interact with the hydrophobic regions of gallichrome (methylene carbon atoms), and these are also important for the strong binding of gallichrome to FhuD. An apo structure for FhuD will soon be reported. Because of the structural similarity and the similar hydrophobic residue-lined binding pocket found in both FhuD and TroA, the mechanism proposed for TroA binding to Zn(II) could be similar to that of FhuD binding to a hydroxamate-type siderophore.

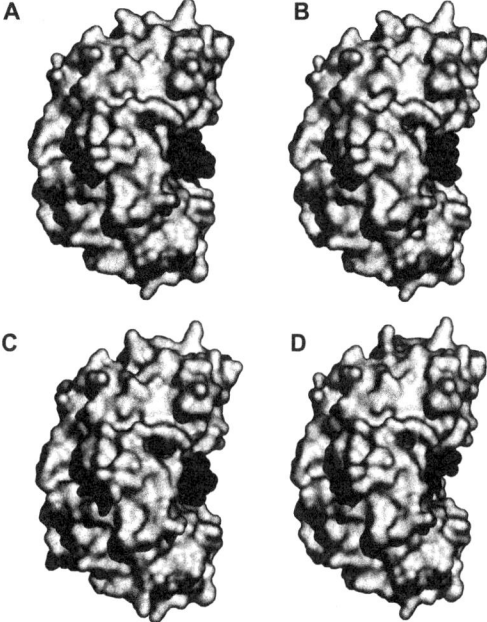

FIGURE 10 Space-filling models of FhuD bound to various hydroxamate siderophores. FhuD bound to gallichrome (A), albomycin (B), coprogen (C), and desferal (D) is shown. Although these hydroxamate-type siderophores are structurally different, all are transported by the same periplasmic protein, FhuD.

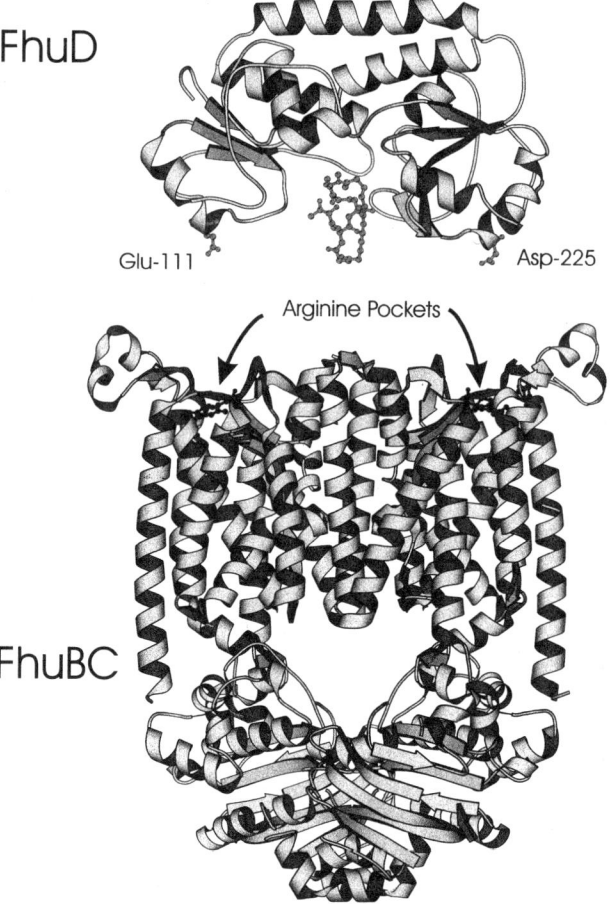

FIGURE 11 Proposed interaction of FhuD with its ABC transporter, FhuBC. The negatively charged Glu-111 and Asp-225 on the apex of each lobe of FhuD (shown in ball-and-stick) can potentially interact with the positively charged arginine pockets of FhuB (shown in ball-and-stick), revealing a path for transport of ferrichrome from FhuD to the cytoplasm. The FhuBC structure is a homology model made by using the BtuCD crystal structure as a template.

Transport across the cytoplasmic membrane is mediated by three proteins: FhuD in the periplasm, FhuB in the cytoplasmic membrane, and FhuC (ATPase) associated with the inside of the cytoplasmic membrane. It is unknown how FhuD acquires ferrichrome from the outer membrane receptor FhuA; from structural analysis, it is not clear that the two proteins interact. Indeed, FhuD may just bind ferric siderophores in the periplasm after their release from the outer membrane receptors. Competitive peptide-mapping studies provide some insight into how FhuD interacts with FhuB. Synthetic peptides derived from FhuB were shown to bind to FhuD and inhibit ferrichrome transport. It was found that the FhuD-ferrichrome complex interacts with the transmembrane region and cytoplasmic segment 7 of FhuB. The transmembrane region of FhuB may form a pore through which FhuD inserts itself into the cytoplasmic membrane to transfer ferrichrome to the cytoplasm in a process driven by ATP hydrolysis by FhuC.

Recently, the structure of BtuF, another cluster 8 protein, was reported. BtuF is a vitamin B_{12} PBP found in *E. coli* and *Salmonella*. Similar to FhuD and TroA, BtuF has two separate domains connected by a single backbone helix. The two domains are structurally similar, each composed of a five-strand β-sheet surrounded by helices. A vitamin B_{12} molecule is bound in a deep cleft between the domains (Fig. 9B). It is hydrogen bonded to the main-chain nitrogen of Ala32 and the side chains of

Asp242 and Arg246. Similar to FhuD, hydrophobic residues lining the binding pocket interact with the ligand. Because of these hydrophobic residues, it is likely that no large domain motion occurs on the release of vitamin B_{12}, as in TroA. Because the structure of the cytoplasmic membrane transporter-ATPase complex BtuCD was solved previously, it is possible to gain an understanding of the interactions of BtuF with BtuCD. This can serve as a model for other PBPs involved in iron transport. Complex formation between BtuF and BtuCD was confirmed by gel electrophoresis. How the two proteins interact can be hypothesized by examining the structures of BtuF and the periplasmic part of BtuC. BtuF has two negatively charged Glu residues (Glu72 and Glu202) on each lobe that are 48 Å apart. The periplasmic surface of BtuC has two positively charged pockets (Arg56 and Arg59 from transmembrane helix TM2, and Arg295 from TM9) that are 46 Å apart. Theoretically, BtuF could dock onto BtuC via charge-charge interactions between these regions. In this orientation, the binding pocket of BtuF would be aligned with BtuCD such that vitamin B_{12} could be transported through the center of the protein into the cytoplasm (Fig. 11). The two Glu residues and the three Arg residues are conserved in the periplasmic protein and the ABC-transporter complex, respectively, in iron transport proteins from various organisms, indicating a common mechanism. These studies are concurrent with the competitive-mapping study of FhuB, discussed above. The FhuB peptide that bound to FhuD corresponds to a loop on the periplasmic surface of BtuC located between the conserved arginine pockets.

Fe-SIDEROPHORE RECOGNITION

As mentioned above, FhuD also binds a number of other hydroxamate-type siderophores (Fig. 10; Color Plate 12). The uptake of different hydroxamate-type siderophores in *E. coli* requires several distinct receptors including FhuA for ferrichrome, FoxA for ferrioxamine B, and FhuE for coprogen. FhuD is the common periplasmic protein used to shuttle these hydroxamate siderophores to the inner membrane-associated proteins FhuB and FhuC. Obviously, the outer membrane receptors are quite specific for their respective hydroxamates whereas FhuD is not. The crystal structures determined for FhuD bound to coprogen and Desferal (desferrioxamine) confirm the ability of FhuD to accommodate structurally different hydroxamate-type siderophores and also confirm the large solvent exposure of the hydroxamate-type siderophores on binding to FhuD, as seen in Fig. 10. Many of the interactions between the siderophore and FhuD are mediated by a few hydrogen bonds with key amino acid residues: Arg84 and Tyr106 (Color Plate 12). Slight movements of these residues allow for a variety of different hydroxamate structures to bind. The ferrioxamine portion of Desferal binds to FhuD. The absence of a bulky functional group on the peptide backbone results in the enveloping of Desferal to a greater extent than for the other siderophores. One hydroxamate carbonyl oxygen is hydrogen bonded to Arg84, with the second hydroxamate carbonyl sharing a hydrogen bond with Tyr106. Unlike other siderophore-FhuD interactions, the third hydroxamate carbonyl oxygen is not hydrogen bonded to FhuD. The orientation and position of the hydrophobic residues lining the binding pocket are similar to those in the gallichrome and albomycin complexes.

Coprogen is another fungal siderophore. The binding mode of coprogen to FhuD is somewhat different from the binding modes of gallichrome and albomycin (Fig. 10; Color plate 12). One hydrogen bond is formed between the hydroxamate oxygen opposite the diketopiperazine ring and the terminal amine of Arg84. The hydroxyl group of Tyr106 makes a hydrogen bond to the other hydroxamate oxygens on the side of the ring system, and a water-mediated hydrogen bond to Tyr275 is formed with the remaining hydroxamate oxygen. The positions of the hydrophobic residues within the binding site shift slightly compared with those in the FhuD-gallichrome complex; the most dramatic change is the reorientation of Trp217. This movement allows the *trans*-anhydromevalonic acid group to insert into the protein. The overall backbone structure of the FhuD proteins complexed with gallichrome, albomycin, Desferal, and coprogen are virtually identical, confirming the ability of FhuD to acquire various hydroxamate-type siderophores with no conformational change.

Ferrichrome enters the periplasm via the outer membrane receptor, FhuA. Under most environmental conditions, the concentration of the Fe^{3+} siderophore is too low to meet the iron requirements of the cell (10^5 to 10^6 ions per bacterial cell per generation). Receptor proteins are presented to the environment on the surface of the cell wall. The low K_D of the receptors, on the order of 0.1 μM, favors binding of the Fe^{3+} siderophores. Exact complementary fit between the Fe^{3+} siderophore and the receptor causes high specificity and usually also high affinity, which is advantageous if the outer membrane proteins serve the purpose of extracting the rare Fe^{3+} siderophores from the culture medium and concentrating them at the bacterial cell surface. The 2.7-Å structure of ferrichrome bound to FhuA reveals a high-affinity ferrichrome binding site located slightly above the external outer membrane interface. Superposition of the ferrichrome ligand within the binding sites of FhuD and FhuA suggests why FhuA specifically binds ferrichrome and albomycin (discussed below) whereas FhuD is a more promiscuous binder of hydroxamate siderophores. Both binding pockets are lined with hydrophobic residues that interact with the ornithyl moieties of the siderophores. In both FhuA and FhuD, the terminal amino groups of an Arg residue forms hydrogen bonds with ferrichrome. FhuA has an additional interaction with Gln that is not present in FhuD. This could confer the ferrichrome specificity of FhuA. Another difference is the solvent exposure of the hydroxamate-type siderophore. When bound to FhuA, less of the hydroxamate-type siderophore is solvent exposed compared to the exposure when bound FhuD. Only the iron-hydroxamate portion of the siderophore interacts with the shallow binding pocket of FhuD.

TROJAN HORSE ANTIBIOTICS

Although the binding pocket of FhuA is deeper than that of FhuD, it still could accommodate a more bulky ligand such as albomycin. Albomycin is a naturally occurring siderophore-antibiotic similar to ferrichrome. It consists of an antibiotic group attached to a hydroxamate siderophore via an amino acyl linker. It displays bactericidal activity against the gram-negative bacteria *E. coli*, *Salmonella* spp., *Bordetella pertussis*, and *Spirochaeta* spp. A siderophore-antibiotic conjugate's bactericidal activity is dependent on its ability to utilize the bacterial siderophore uptake system for its uptake across both the outer and cytoplasmic membranes. Albomycin uses the ferric hydroxamate uptake system, first binding to the outer membrane receptor (FhuA), then binding to periplasmic ligand binding protein (FhuD), and finally internalizing within the cell through the inner membrane FhuBC transmembrane complex. Once in the cell, the antibiotic group is released by a peptidase. The intracellular target for albomycin is unknown. The structures of albomycin bound to FhuA and albomycin bound to FhuD have been determined. Conclusions can be drawn from each of these structures to understand the important interactions between this natural siderophore-antibiotic conjugate and the respective protein to synthesize a potent siderophore-antibiotic conjugate.

The structure of the antibiotic albomycin in complex with the outer membrane transporter FhuA was determined to 3.10 Å. The iron-chelating component of albomycin binds to FhuA in the same orientation as ferrichrome does. Also, the same hydrogen bonds and van der Waals' contacts are seen between FhuA and the iron-chelating component, as seen in Fig. 12. The amino acyl linker connecting the antibiotic group of albomycin to the hydroxamate siderophore is flexible; consequently, two conformers of albomycin are found in the FhuA

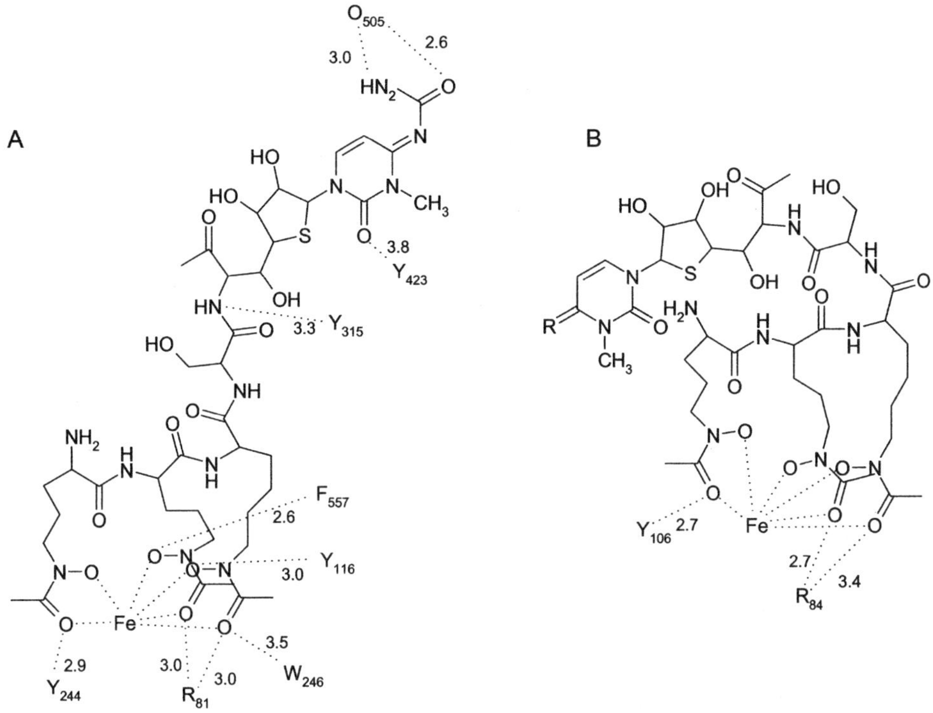

FIGURE 12 Comparison of the binding modes of albomycin in FhuA (A) and FhuD (B). The hydrogen bonds from the side chain residues to albomycin are shown. It can be seen that FhuA interacts with the antibiotic portion of albomycin but FhuD does not. Reprinted from Clarke, Braun, et al. (2002) with permission from the publisher.

structure (extended and compact), although in aqueous solution, many more conformations could exist. When superimposed, the structures of FhuA bound to albomycin and ferrichrome are identical. The increased bulkiness in albomycin versus ferrichrome makes no additional conformational change in FhuA. The extracellular binding pocket may potentially accommodate a larger ligand than albomycin. The structure of albomycin bound to FhuD was solved to 2.6 Å. Albomycin also binds to FhuD in a similar orientation to ferrichrome. The same hydrogen-bonding patterns are seen, except that there is no water-mediated hydrogen bond. When the structure of FhuD bound to ferrichrome and the structure of FhuD bound to albomycin are superimposed, they are again virtually identical. Intriguingly, no electron density is found for the antibiotic group; due to the flexible linker, the antibiotic portion of albomycin can freely move around.

Theoretically, a drug conjugated to a siderophore would be an ideal way to smuggle drugs in a "Trojan horse"-like manner into a pathogenic gram-negative bacterial cell (see chapter 27). Such siderophore-antibiotic conjugates, similar to albomycin, contain an iron-chelating group, a linker, and an antibiotic. From the structures of FhuD and FhuA bound to albomycin, it is evident that the iron-coordinating moieties are important for ligand recognition in both FhuA and FhuD. FhuA is more spatially restrictive than FhuD, thus limiting the size of the siderophore-antibiotic conjugate. FhuD is not as restricted because of the vast solvent-exposed regions, suggesting that this approach could be more successful for gram-positive bacteria. Although rational design of a novel, efficient antibiotic by chemical conjugation is challenging, this Trojan horse design may become increasingly important due to the growing problem of antibiotic resistance.

FINAL COMMENTS

In this chapter, we have summarized our present knowledge of PBPs involved in iron transport. Although a great deal of research has been done on the PBP families, the field involving PBPs and iron transport is still lacking in information in three key areas. First, it is not yet understood how the Fe^{3+} (free or bound to a siderophore) is transported from the outer membrane receptor to the PBP—is it simply released by the receptor into the periplasm, or do FhuA and FhuD interact? Studies must also be done to determine the path of the Fe^{3+} siderophore through the outer membrane transporter to the periplasm. Second, because PBP classes 8 and 9 are new, more research is required to understand their dynamic motions on binding and release of their ligand. This can be done largely by solving crystal structures for more PBPs from these classes. To further understand the range of motion occurring on binding and release of a ligand, solution NMR measurements together with SAXS studies could be done for several class 8 and 9 proteins to complement the information provided by the crystal structures. Computational molecular dynamic studies should be done to obtain a picture of the dynamic movements on binding and release of the ligand and the dynamic movements, if any, involving the backbone helix. Third, it has been hypothesized how the PBP may interact with the ABC transporter to transport bound Fe^{3+} to the cytoplasm; however, little experimental evidence currently supports this. Ultimately, a crystal structure with the PBP docked to the membrane transporter would remedy this void. Because crystallization is a limiting factor, a theoretical study involving molecular dynamics calculations and docking of the periplasmic protein onto the inner membrane transporter-ATPase complex should be done in the meantime.

ACKNOWLEDGMENTS

This work is supported by an operating grant from the Canadian Institutes of Health Research. K.D.K. and R.S.P. are the recipients of studentship awards from the Natural Sciences and Engineering Research Council of Canada and the Alberta Heritage Foundation for Medical Research (AHFMR). H.J.V. is an AHFMR Scientist.

SUGGESTED READING

Baker, E. N., H. M. Baker, and R. D. Kidd. 2002. Lactoferrin and transferrin: functional variations on a common structural framework. *Biochem. Cell Biol.* **80:**27–34.

Berendsen, H. J., and S. Hayward. 2000. Collective protein dynamics in relation to function. *Curr. Opin. Struct. Biol.* **10:**165–169.

Borths, E. L., K. P. Locher, A. T. Lee, and D. C. Rees. 2002. The structure of *Escherichia coli* BtuF and binding to its cognate ATP binding cassette transporter. *Proc. Natl. Acad. Sci. USA* **99:**16642–16647.

Bruns, C. M., D. S. Anderson, K. G. Vaughan, P. A. Williams, A. J. Nowalk, D. E. McRee, and T. A. Mietzner. 2001. Crystallographic and biochemical analyses of the metal-free *Haemophilus influenzae* Fe^{3+}-binding protein. *Biochemistry* **40:**15631–15637.

Bruns, C. M., A. J. Nowalk, A. S. Arvai, M. A. McTigue, K. G. Vaughan, T. A. Meitzner, and D. E. McRee. 1997. Structure of *Haemophilus influenzae* Fe^{+3}-binding protein reveals convergent evolution within a superfamily. *Nat. Struct. Biol.* **4:**919–924.

Clarke, T. E., V. Braun, G. Winkelmann, L. W. Tari, and H. J. Vogel. 2002. X-ray crystallographic structures of the *Escherichia coli* periplasmic protein FhuD bound to hydroxamate-type siderophores and the antibiotic albomycin. *J. Biol. Chem.* **277:**13966–13972.

Clarke, T. E., S. Y. Ku, D. R. Dougan, H. J. Vogel, and L. W. Tari. 2000. The structure of the ferric siderophore binding protein FhuD complexed with gallichrome. *Nat. Struct. Biol.* **7:**287–291.

Clarke, T. E., M. R. Rohrbach, L. W. Tari, H. J. Vogel, and W. Koster. 2002. Ferric hydroxamate binding protein FhuD from *Escherichia coli*: mutants in conserved and non-conserved regions. *Biometals* **15:**121–131.

Claverys, J. P. 2001. A new family of high-affinity ABC manganese and zinc permeases. *Res. Microbiol.* **152:**231–243.

Duan, X., J. A. Hall, H. Nikaido, and F. A. Quiocho. 2001. Crystal structures of the maltodextrin/maltose-binding protein complexed with reduced oligosaccharides: flexibility of tertiary structure and ligand binding. *J. Mol. Biol.* **306:** 1115–1126.

Evenas, J., V. Tugarinov, N. R. Skrynnikov, N. K. Goto, R. Muhandiram, and L. E. Kay. 2001. Ligand-induced structural changes to maltodextrin-binding protein as studied by solution NMR spectroscopy. *J. Mol. Biol.* **309:** 961–974.

Felder, C. B., R. C. Graul, A. Y. Lee, H. P. Merkle, and W. Sadee. 1999. The Venus flytrap of periplasmic binding proteins: an ancient protein module present in multiple drug receptors. *AAPS Pharm. Sci.* **1:**E2.

Ferguson, A. D., V. Braun, H. P. Fiedler, J. W. Coulton, K. Diederichs, and W. Welte. 2000. Crystal structure of the antibiotic albomycin in complex with the outer membrane transporter FhuA. *Protein Sci.* **9:**956–963.

Ferguson, A. D., E. Hofmann, J. W. Coulton, K. Diederichs, and W. Welte. 1998. Siderophore-mediated iron transport: crystal structure of FhuA with bound lipopolysaccharide. *Science* **282:** 2215–2220.

Fukami-Kobayashi, K., Y. Tateno, and K. Nishikawa. 1999. Domain dislocation: a change of core structure in periplasmic binding proteins in their evolutionary history. *J. Mol. Biol.* **286:**279–290.

Kapust, R. B., and D. S. Waugh. 1999. *Escherichia coli* maltose-binding protein is uncommonly effective at promoting the solubility of polypeptides to which it is fused. *Protein Sci.* **8:**1668–1674.

Koster, W. 2001. ABC transporter-mediated uptake of iron, siderophores, heme and vitamin B12. *Res. Microbiol.* **152:**291–301.

Lawrence, M. C., P. A. Pilling, V. C. Epa, A. M. Berry, A. D. Ogunniyi, and J. C. Paton. 1998. The crystal structure of pneumococcal surface antigen PsaA reveals a metal-binding site and a novel structure for a putative ABC-type binding protein. *Structure* **6:**1553–1561.

Lee, Y. H., R. K. Deka, M. V. Norgard, J. D. Radolf, and C. A. Hasemann. 1999. *Treponema pallidum* TroA is a periplasmic zinc-binding protein with a helical backbone. *Nat. Struct. Biol.* **6:** 628–633.

Lee, Y. H., M. R. Dorwart, K. R. Hazlett, R. K. Deka, M. V. Norgard, J. D. Radolf, and C. A. Hasemann. 2002. The crystal structure of Zn(II)-free *Treponema pallidum* TroA, a periplasmic metal-binding protein, reveals a closed conformation. *J. Bacteriol.* **184:**2300–2304.

Locher, K. P., A. T. Lee, and D. C. Rees. 2002. The *E. coli* BtuCD structure: a framework for ABC transporter architecture and mechanism. *Science* **296:** 1091–1098.

Mademidis, A., H. Killmann, W. Kraas, I. Flechsler, G. Jung, and V. Braun. 1997. ATP-dependent ferric hydroxamate transport system in *Escherichia coli*: periplasmic FhuD interacts with a periplasmic and with a transmembrane/cytoplasmic region of the integral membrane protein FhuB, as revealed by competitive peptide mapping. *Mol. Microbiol.***26:**1109–1123.

Quiocho, F. A., and P. S. Ledvina. 1996. Atomic structure and specificity of bacterial periplasmic receptors for active transport and chemotaxis: variation of common themes. *Mol. Microbiol.***20:**17–25.

Schneider, R., and K. Hantke. 1993. Iron-hydroxamate uptake systems in *Bacillus subtilis*: identification of a lipoprotein as part of a binding protein-dependent transport system. *Mol. Microbiol.***8:** 111–121.

Shilton, B. H., M. M. Flocco, M. Nilsson, and S. L. Mowbray. 1996. Conformational changes of three periplasmic receptors for bacterial chemotaxis and transport: the maltose-, glucose/galactose- and ribose-binding proteins. *J. Mol. Biol.***264:**350–363.

Tam, R., and M. H. Saier, Jr. 1993. Structural, functional, and evolutionary relationships among extracellular solute-binding receptors of bacteria. *Microbiol. Rev.***57:**320–346.

IRON TRANSPORT, ENERGETICS, AND REGULATION IN *ESCHERICHIA COLI* K-12: A PROTOTYPE FOR IRON TRANSPORT SYSTEMS IN GRAM-NEGATIVE BACTERIA

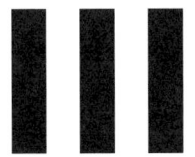

IRON UPTAKE VIA THE ENTEROBACTIN SYSTEM

Charles F. Earhart

9

Iron assimilation by means of the enterobactin (Ent) system of *Escherichia coli* has served as a paradigm for siderophore-dependent iron uptake in prokaryotes. The subject has a comparatively long history, beginning in the mid-1960s, when workers in Australia and the United States used extracts from *Aerobacter aerogenes* (now *Enterobacter aerogenes*) and *Salmonella typhimurium* (now *Salmonella enterica* serovar Typhimurium) as well as *E. coli* to determine the structure of enterobactin (enterochelin) (Fig. 1) and the initial reactions of the Ent biosynthetic pathway. The field blossomed in the 1970s, when microbiologists with varied interests, including colicin uptake, outer membrane proteins, growth lags, and bacteriophage adsorption, all recognized the relevance of iron assimilation to their work. In retrospect, the work progressed slowly despite the convergence of interests. In 1972, the *E. coli* linkage map had seven Ent cluster genes; with time, the number gradually increased, although some genes were eliminated (*fesB* and *entG*) and one was renamed (*fepE* is now *wzz*). The most recent traditional linkage map (1998), which includes 14-Ent cluster genes, is not current because it lists *fepE* rather than *wzz* and has an open reading frame (ORF) between *fepD* and *fepB* instead of *entS*. Also, *ybdB*, a gene not shown on the 1998 map, may yet be found to play a role in this iron uptake system. Work on basic aspects of the system continues.

It is an interesting exercise to consider what bioinformatics would contribute if the Ent gene cluster were discovered today. The DNA sequence would disclose a transport system consisting of five genes (*fepA*, *fepB*, *fepC*, *fepD*, and *fepG*) including a gene for a large outer membrane protein with a TonB box (*fepA*), which could be surmised to function as a TonB-dependent gated pore, a periplasmic protein that is a family member of those involved in siderophore uptake (*fepB*), and a cytoplasmic membrane ABC transporter complex (*fepC*, *fepD*, and *fepG*) of the subfamily concerned with iron uptake. The presence of multiple binding sites for the Fur repressor protein would confirm the relationship of the system to iron. Other genes would encode biosynthetic enzymes (*entA* to *entF*). One set would encode a gene that recognizes chorismic acid as a substrate (*entC*) and another set (*entB*, *entD*, *entE*, and *entF*) must direct the synthesis of enzymes which function in nonribosomal peptide biosynthesis. One of the genes would

---

Charles F. Earhart, Section of Molecular Genetics and Microbiology, The University of Texas at Austin, Austin, TX 78712–1095.

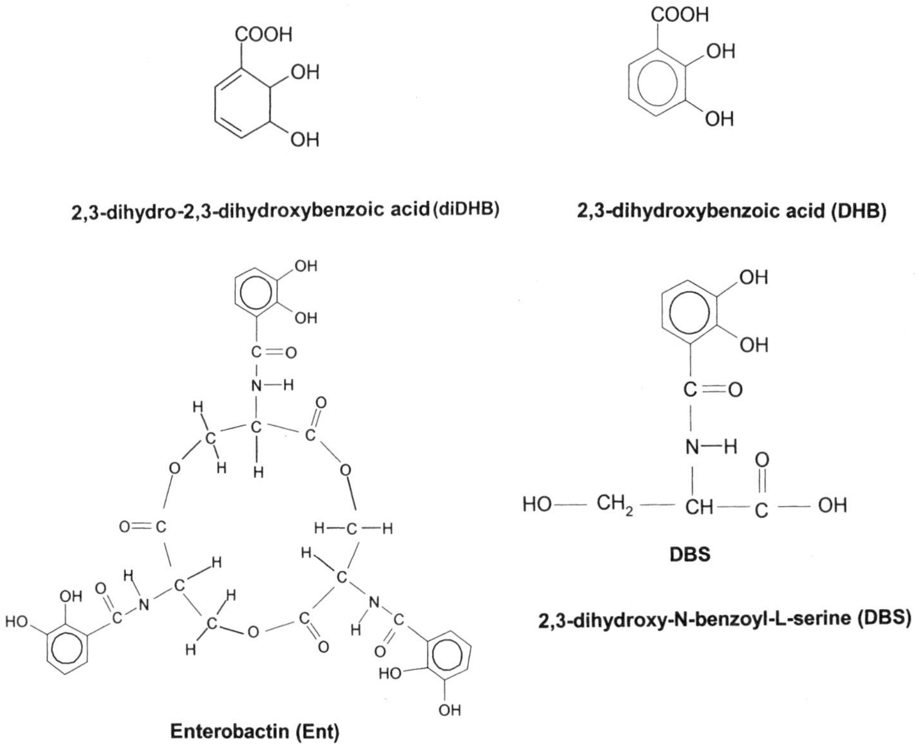

FIGURE 1 Structures of diDHB, DHB, DBS, and Ent.

clearly encode a bifunctional protein (*entB*), another gene (*wzz*) apparently has nothing to do with iron assimilation, and a gene for an efflux pump (*entS*) is present. In summary, many aspects of the Ent system that were laboriously determined would now be easily understood from sequence data alone. In many cases, of course, basic information leading to the classification of these families came from studies of the Ent system, as did much of the seminal biochemical and physiological data regarding iron uptake.

Work with the Ent system also led to the understanding that many minimal media have too little iron, with the consequence that bacteria growing in them are iron starved in their later growth stages. Also, it became standard procedure to screen for iron uptake systems by looking for new outer membrane polypeptides in the 80-kDa range in cells grown in iron-depleted media.

Because the Ent system played a basic role in current ideas concerning iron assimilation, many of the topics covered in this chapter, such as siderophore biosynthesis and transport, are covered in greater detail and broader context in other chapters of this book. Also, there are a number of recent reviews that include detailed information on the Ent system; these are listed at the end of the chapter. Therefore, in this chapter, the intent is to briefly review well-documented aspects of the Ent system and then to concentrate on some remaining major questions concerning Ent-related iron transport.

ENTEROBACTIN BIOSYNTHESIS

Ent is a cyclic trilactone composed of three 2,3-dihydroxybenzoylserine (DBS) units (Fig. 1). At neutral pH, when the ferric ion is ligated by the oxygens of the three diphenolic groups, the Fe^{3+}-Ent complex has a net charge of -3. At low pH values, Fe^{3+} chelation by Ent is via salicylate rather than catecholate complexing. That is, the *m*-hydroxyl oxygens become protonated and the carbonyl oxygens of the amide groups replace the three *m*-hydroxyl oxygens in the ligation. Under acidic conditions, the Fe^{3+}-Ent complex is therefore neutral. The ability of Ent to bind Fe^{3+} is unsurpassed among siderophores and other iron chelators, and it readily removes iron from proteins, insoluble iron complexes, and other siderophores.

The Ent gene cluster contains six genes (*entA* to *entF*) (Fig. 2) required for Ent biosynthesis. Before biosynthesis of Ent can occur, the Ent-specific precursor dihydroxybenzoic acid (DHB) must be produced. This is accomplished by the EntC, EntB and EntA proteins, which, in a three-step pathway, convert chorismate to DHB. This initial pathway has been known for over 30 years, and the individual reactions and enzymes have been studied in detail. Chorismate, the major branch point metabolite in aromatic biosynthesis, is an important connection linking Ent biosynthesis to overall cellular metabolism. Specifically, chorismate also serves as the substrate for the first enzyme specific for the synthesis of Phe and Tyr (chorismate mutase), Trp (anthranilate synthase), ubiquinone (chorismate lyase), and menaquinone (isochorismate synthase [MenK]). The first Ent-specific enzyme (EntC) is, not surprisingly, related to several of these enzymes and, like MenK, is an isochorismate synthase; *E. coli* therefore has two genes encoding enzymes that carry out the identical reaction. The enzymes are presumably required under different growth conditions; MenK is needed only under anaerobic conditions and is synthesized preferentially then, whereas EntC is most necessary in aerobic, low-iron environments, and its biosynthesis is regulated accordingly (see below). EntC is a 43-kDa enzyme that is active as a monomer. EntB is a bifunctional enzyme, participating both in DHB synthesis, as an isochorismatase yielding 2,3-dihydro-2,3-dihydroxybenzoate (diDHB) and pyruvate, and in the Ent synthase reaction, as an aryl carrier protein (EntG activity). It is isolated as a homotrimer of 32.6-kDa subunits. EntA (diDHB dehydrogenase) completes DHB synthesis by a NAD^+-dependent oxidation step. EntA, an octamer of 26,249-kDa subunits, is a member of the short alcohol-polyol-sugar dehydrogenase family.

The in vivo synthesis of Ent (3L-Ser + 3DHB + 6ATP → Ent + 6AMP + $6PP_i$) requires the products of four genes, *entB*, *entD*, *entE*, and *entF* (Fig. 3). Three amide bonds occur in Ent, and this suggested early on that the process would be similar to that for peptide antibiotic synthesis. However, details of the terminal reactions in Ent biosynthesis were not available until recently, when a decade-long spate of research concerning nonribosomal peptide synthesis culminated in a molecular understanding of the process, not only for Ent but also for many other siderophores and peptide antibiotics (see chapter 2).

EntD is a 4'-phosphopantetheine (P-pant) transferase; it is specific for and modifies both EntF and EntB posttranslationally by adding P-pant from coenzyme A to a conserved Ser residue, forming holo-EntF and holo-EntB. The presence of P-pant in these proteins enables them to function as acyl and aryl carriers, respectively. EntD (23.5 kDa) is thus not immediately involved in Ent synthesis; the only

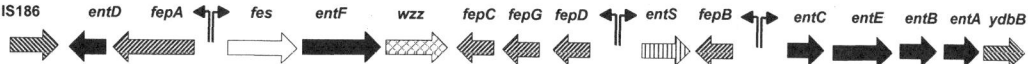

FIGURE 2 Enterobactin gene cluster (not to scale). Symbols: bent arrows, iron box with direction of transcription indicated; diagonally striped arrows, transport genes; solid arrows, biosynthetic genes; open arrow, gene for iron release; vertically striped arrow, gene for Ent excretion pump; cross-hatched arrow, gene determining the length of O-Ag. Modified from Earhart (1996).

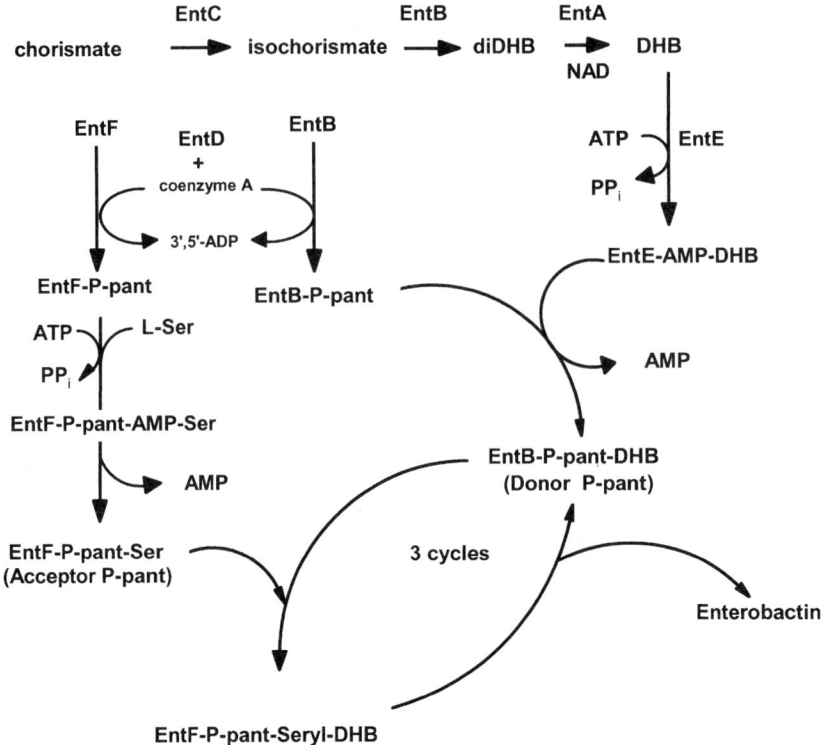

FIGURE 3 Synthesis of Ent from DHB and Ser. diDHB, 2,3-dihydro-2,3-dihydroxybenzoic acid; P-pant, 4′-phosphopantetheine.

proteins required for in vitro Ent synthesis (the Ent synthetase reaction) are EntE, holo-EntF, and holo-EntB. Procedures that readily identified EntE, EntF, and EntB failed to detect EntD. That EntD is a member of the recently discovered P-pant transferase family partially explains this, since EntD presumably is not needed in the same quantities as the Ent synthetase proteins.

The roles of EntB, EntE, and EntF are best understood in the context of domains known to be present in proteins concerned with nonribosomal peptide synthesis. These domains are designated as follows: peptidyl carrier protein (PCP), which contains the Ser-linked P-pant resulting from P-pant transferase activity; adenylation (A), which utilizes ATP to form acyladenylates and transfers the activated moiety to the appropriate PCP domain; and condensation (C), which catalyzes the synthesis of amide bonds. Thioesterase (TE) domains accomplish chain termination.

EntE is a 59.2-kDa monomer that has an A domain; specifically, it is a DHB-AMP ligase, activating DHB and then performing an intermolecular transfer of the DHB to the P-pant of EntB. In EntB, the P-pant tethers an aromatic carboxylic acid rather than an amino acid so that the domain is termed an aryl carrier protein domain (ArCP) rather than a PCP. The isochorismatase activity and the ArCP function of EntB are found in the amino and carboxyl termini of EntB, respectively. The two functions of EntB can be physically independent of one another; truncated EntB proteins lacking their carboxy one-third have isochorismatase activity, and carboxy-terminal fragments are functional as ArCP. This fact led to the erroneous idea from genetic experiments that what is now known as an ArCP was the product of a separate gene (*entG*). EntF (142 kDa, active as a monomer) has four domains, C, A, PCP, and TE. It is, in nonribosomal peptide biosynthesis terminology, a complete module. The A domain activates L-Ser and transfers it intramolecularly to the P-pant of the PCP domain. The EntF C domain then accepts DHB, by intermolecular transfer from the ArCP domain of EntB, and catalyzes amide bond formation with Ser, yielding DHB-Ser-S-EntF thioester. This DHB-Ser (DBS) moiety is then transferred to the TE domain, where it is bound to Ser in an oxoester linkage. An additional round of DBS synthesis places DBS in the just emptied EntF C domain. An ester bond is formed between this DBS and the DBS in the TE domain, yielding a DBS dimer in an oxoester linkage in the TE domain. An additional round of DBS synthesis, transfer to EntF, and ester bond formation produces a TE-bound linear DBS trimer; this trimer is the cyclized and released from EntF. The TE domain of EntF thus serves as a catalyst for elongation by forming ester bonds between adjacent DBS moieties and also functions to carry out intramolecular cyclization and release of DBS trimers.

Ent synthetase is not a stable complex, contrary to original expectations. Gel filtration, immunoprecipitation, and cross-linking experiments have all failed to detect associations among Ent synthetase components. Kinetic data from in vitro Ent synthetase experiments are consistent with the idea the three proteins interact transiently. Similarly, there is little evidence to support the notion that Ent synthetase is membrane associated. The Ent proteins can be isolated by normal procedures for soluble cytoplasmic proteins, no membrane is required for the Ent synthetase reaction, and several nonionic detergents that solubilize membrane have no effect on the reaction. That some EntB, EntE, and EntF proteins can be osmotically shocked out of cells was thought to provide indirect evidence for membrane association or compartmentalization of Ent synthetase. However, recent work has explained this selective release of certain cytoplasmic proteins purely on the basis of size; peptidoglycan is postulated to act as a molecular sieve, such that cytoplasmic proteins with native sizes of less than 100 kDa are released by osmotic shock whereas larger proteins or protein complexes are not. For Ent synthetase proteins, the extent of release is consistent with this explanation; the smaller proteins (EntB [97.8 kDa] and EntE [59.3 kDa]) are found in greater amounts in shockates than is EntF (142 kDa). This explanation for the release of certain cytoplasmic proteins by osmotic shock provides additional evidence against the idea the EntB, EntE, and EntF form a long-lived complex in vivo. If they did, they would not be present in shockates.

A minor fraction of EntB, EntE, and EntF can be found associated with membrane fractions isolated from iron-starved *E. coli*. This association was not adventitious, since the Ent synthetase proteins were not removed by additional membrane washes and were not present intracellularly at artificially high expression levels. Additionally, two fractionation procedures localized the protein to a membrane fraction whose density is intermediate between those of the inner and outer membranes. The significance of these observations is not clear, but at this time they provide the only evidence suggesting an association, however temporary or dynamic, between the membrane and Ent synthetase.

Some progress regarding the excretion of Ent has been made with the observation that cells deficient in *entS* (*ybdA*) release decreased amounts of Ent. Genomic comparisons indicate that EntS is a member of the major facilitator superfamily of export pumps; it is predicted to have 12 transmembrane segments and to use proton motive force as an energy source. Cells with mutations in *entS* grow in iron-poor media, but this is attributed to their excretion of Ent breakdown products (DBS monomers, dimers, and trimers), which can act as secondary siderophores (see below). Concerns regarding possible detrimental effects of being unable to excrete Ent from the cytoplasm are mitigated by evidence that Fes (ferrienterobactin esterase) apparently degrades intracellular Ent to DBS derivatives, which then leave cells by an unknown mechanism. Although this is a long-standing problem, studies of Ent excretion are still in the early stages; other components that may function with the EntS pump, such as membrane fusion proteins and outer membrane porins, have not been identified, and it is unknown whether other efflux pumps recognize Ent as a substrate.

SECONDARY SIDEROPHORES

It is likely that our understanding of iron uptake by products of Ent cluster genes has been limited by focusing on Ent, the defining and strongest siderophore of this system; in its basic form, one molecule of Ent is responsible for the import of only one ferric ion. The Ent of each Fe^{3+}-Ent complex that enters the cytoplasm is degraded by Fes to the linear forms $(DBS)_3$, $(DBS)_2$, and, ultimately, DBS. The necessity of one Ent molecule for every iron ion transported seems remarkably wasteful of carbon skeletons and energy, particularly when even the smallest estimate for the number of iron ions required per cell cycle is approximately 100,000. The Ent system may not be as profligate as first appears, however, since there is a variety of evidence that compounds related to Ent can act as secondary siderophores and function in routine iron uptake. The remainder of this section discusses the results of studies showing that Ent breakdown products as well as at least one Ent precursor (DHB) can serve as siderophores.

When the supernatants from stationary-phase wild-type *E. coli* cultures that have been grown in iron-deficient media are analyzed, the majority of the catecholate-containing compounds are $(DBS)_3$, $(DBS)_2$, DBS, and DHB; i.e., Ent is a relatively minor compound. In fact, the high concentration of degradation products complicated the initial determination of Ent structure. Several factors contribute to this low abundance of Ent: (i) Ent is inherently unstable, with a $t_{1/2}$ of approximately 30 min at 37°C in Tris buffer (pH 8), and (ii) Fes cleaves Ent and Fe^{3+}-Ent, and the resulting breakdown products are secreted into the medium. The relative amounts of the several catecholates in the supernatant therefore depend on the status of the *fes* gene and, because Fe^{3+}-Ent is stable, whether iron is absent from the medium or present in a relatively inaccessible form, e.g., bound to nonphysiological chelators such as 2,2'-dipyridyl, nitrilotriacetic acid, or ethylenediamine-N,N'-diacetic acid. Ent can remove iron from these chelators and therefore, as Fe^{3+}-Ent, avoid spontaneous hydrolysis. In addition, the relative distribution of linear DBS derivatives in the supernatant has been shown to vary depending on whether Fe^{3+}-Ent can enter the cells, even when the ratio of breakdown products to Ent is similar. Lastly, because Fes hydrolyzes Ent as well as Fe^{3+}-Ent, Fes may well play a protective role by degrading Ent before it can be secreted under conditions where Ent is being overproduced. There have been few systematic analyses of the profiles of catecholate molecules during growth under iron-starved conditions, and the limited results are curious. Ammerlaan performed a study of a transport-defective mutant grown in deferrated minimal medium supplemented with limiting iron (0.5 µM $FeCl_3$). Under these iron-poor conditions, little iron is available to stabilize Ent, and the Fe^{3+}-Ent complexes that do form cannot enter the cell and be degraded. The major species in the medium progressed from linear DBS-containing compounds and DHB to DHB and Ent as the culture reached stationary phase (8 h) until, at 24 h, DBS, $(DBS)_2$, and $(DBS)_3$ overwhelmingly predominated. In summary, supernatants of iron-starved cells contain many catecholate derivatives, whose relative abundance is subject to great variation with growth conditions and growth stage but in which Ent generally constitutes a minor fraction.

DBS, $(DBS)_2$, and $(DBS)_3$ can act to promote growth in low-iron media, although they have considerably lower affinities for Fe^{3+} than Ent does. Similarly, DHB is a siderophore for some organisms, such as *Brucella abortus*, and for *E. coli* it may provide a means of bringing iron into the periplasm, although subsequent transport into the cytoplasm seems to require that the iron bound by DHB be transferred to DBS-containing compounds.

Linear DBS derivatives evidently require, like Fe^{3+}-Ent, specific TonB-dependent receptor proteins. In addition to FepA (80 kDa), two other iron-regulated outer membrane proteins, Cir (74 kDa) and Fiu (83 kDa), can be used for uptake of DBS-Fe^{3+} and, presumably, the larger linear DBS-containing molecules as well. The structural genes for Cir and Fiu are not located in the Ent gene cluster. In *Salmonella*, Cir may be involved in the uptake of the catechol siderophore salmochelin (see chapter 14), but no other siderophores that use these receptors have been identified. Additional evidence that Cir and Fiu can transport catecholates comes from studies with catechol-substituted antibiotics.

Although the molecules and their outer membrane receptor proteins have been identified, many questions about secondary siderophores remain. For example, it is not known (i) how they are secreted or if EntS has a role in the process, (ii) if the DBS compounds invariably result from degradation of Ent or whether Ent synthetase sometimes releases intermediates, (iii) why DHB is so prevalent in the supernatants and the degree to which its synthesis is coupled to the Ent synthetase reaction, and (iv) how secondary siderophore-iron complexes traverse the cytoplasmic membrane. With respect to the last question, it is likely but not rigorously proven that the FepBCDG transporter is used.

Some insight into secondary siderophore physiology, particularly secretion, may be gleaned from data obtained with an unusual class of pseudorevertants, which grow in iron-deficient media despite lacking the ability to synthesize DHB or Ent (*entD entF entA*) or transport Ent (*fepA*). Pseudorevertants of these multiple mutants show no detectable OmpF and OmpC proteins, and the suppressor mutations mapped to the *ompB* (*envZ ompR*) region. Suppressor mutations could also occur in *ompF*, but the absence of other outer membrane proteins, including Cir, had no effect on iron uptake. Loss of both Cir and Fiu was not tested, however. Evidence that pseudorevertants excreted a molecule with siderophore activity was adduced from cross-feeding experiments. Synthesis of the postulated siderophore was iron regulated, it was able to remove iron from nitrilotriacetic acid and 2,2′-dipyridyl and dicitrate but not Fe^{3+}-Ent or ferrichrome, and energy requirements for its uptake mimicked those of the Ent system. One possibility for the siderophore is that although strains with *entA* mutations cannot make DHB, they can make its immediate precursor, diDHB. When this compound is grossly overproduced by using an *entA* strain harboring a multicopy plasmid bearing *entC* and *entB*, it is excreted. If this compound does function as a siderophore and is responsible for the ability of the pseudorevertants to grow in low-iron media, it is unclear why porin deficiencies are necessary for its release.

UPTAKE OF FERRIENTEROBACTIN

Ferrienterobactin enters the cell via an unusually complex transport system. This is necessitated by the fact that Fe^{3+}-Ent is slightly too large to utilize the general porins in the outer membrane; consequently, a dedicated outer membrane receptor protein and energy are required to transport Fe^{3+}-Ent into the periplasm. From the periplasm, an ABC transporter, specifically a member of the importer subfamily that uses a periplasmic binding protein, accomplishes passage through the cytoplasmic membrane. A review of FepA, the outer membrane receptor protein, is provided in chapter 4, while chapters 7 and 10 discuss the means by which Fe^{3+}-Ent crosses the outer membrane. Details of the Fe^{3+}-Ent binding protein (FepB) and how ABC transporters function in ferrisiderophore uptake are supplied in chapters 8 and 11.

Fe^{3+}-Ent binds with high affinity to outer membrane protein FepA. At 80 kDa, FepA is atypically large for a protein constituent of the outer membrane but is similar to most such proteins in that it probably exists as a trimer. It is exported by the Sec translocase, and the proprotein has a signal sequence of 22 amino acids. Colicins B and D also use FepA as a receptor, so that *fepA* mutations are readily obtained by selecting for resistance to one of these colicins.

The crystal structure of FepA was recently determined; two domains were found, a large (22-strand) β-barrel (amino acids 154 to 724) and an N-terminal globular domain (amino acids 1 to 153). The latter domain inserts into and completely occludes the barrel pore; this latter property has led it to be described as a plug or cork. The length of this plug is such that it traverses the height of the protein and membrane. That is, it has loops exposed on the outer surface as well as a sequence present in the periplasm. This N-terminal domain had not been predicted from topology studies and seems to be unique to outer membrane active transporters such as FepA. FepA proteins that lack the plug as a result of genetic manipulation behave as large, constantly open pores. Remarkably, this phenotype can be achieved by deletions that remove certain portions of the β-barrel. These latter deleted regions, which include periplasmic and outer loops as well as transmembrane segments, must play a role in maintaining the plug inside the core.

FepA alone can bind Fe^{3+}-Ent and colicins B and D, but for these ligands to penetrate the outer membrane, energy is required. This energy is provided by a complex of three proteins, TonB, ExbB, and ExbD, and mutations that confer colicin resistance are often found in *tonB*. The TonB ExbBD complex is not specific for FepA but, rather, recognizes and energizes six or seven other receptors as well. FepA is thus a member of the TonB-dependent outer membrane transporter family. In the case of FepA, the energy transduced by the complex acts to release Fe^{3+}-Ent from its FepA binding site and to modify the FepA plug conformation so as to allow passage of Fe^{3+}-Ent through the receptor pore and into the periplasm. The proton motive force provides the energy for this process; ExbB and ExbD are cytoplasmic membrane proteins with homologies to motor proteins (MotA and MotB) necessary for flagellar rotation, which also utilizes the proton gradient. ExbBD-mediated entry of protons into the cytoplasm presumably alters the conformation of TonB. Topological studies of TonB show that it has an N-terminal sequence that can anchor it in the cytoplasmic membrane followed by a long periplasmic domain capable of spanning the periplasm. The rate-limiting step in Fe^{3+}-Ent uptake is thought to be its passage through the outer membrane.

On entry into the periplasm, Fe^{3+}-Ent is bound by its specific binding protein, FepB. FepB, like all periplasmic binding proteins (PBPs), has a high affinity for its ligand and serves to concentrate and present Fe^{3+}-Ent to its specific cytoplasmic membrane permease. The requirement for FepB is absolute, and evidence that FepB binds Fe^{3+}-Ent was adduced both from studies in which FepB was localized to the outer membrane and from assays with purified FepB. The latter work also demonstrated that FepB has, for a PBP, an unusually high affinity for its ligand Fe^{3+}-Ent (K_D = 30 nM) and that FepB also recognizes desferri-Ent and the left-handed stereoisomer of Fe^{3+}-Ent (ferrienantio-Ent) but not related catecholate siderophores. Indirect genetic and physiological evidence makes it likely that FcpB is also required for iron transport by means of the Ent-related compounds DBS and DHB. Based on its amino acid sequence, FepB is classified in cluster 8 of the PBP family. The best-studied member of this family is FhuD, the binding protein for ferrichrome and several other hydroxamate siderophores, whose structure was recently determined. It can be anticipated that FepB will have a bilobed structure, as do all PBPs. However, the FhuD results suggest that FepB differs from most PBPs since its two domains are joined by a long α-helix rather than two or three β-strands and since the binding site does not undergo a large closing and opening on binding and releasing Fe^{3+}-Ent, respectively.

Transport of Fe^{3+}-Ent into the cytoplasm occurs by means of an ABC transporter. In addition to FepB, three proteins, FepD, FepG, and FepC, all located in the cytoplasmic membrane, are required. More than 1,000 ABC transporters are known; those necessary for siderophore, heme, and vitamin B_{12} uptake constitute a subfamily. The basic ABC transporter consists of four domains, two hydrophobic and two that bind ATP. For the Fe^{3+}-Ent importer, the two hydrophobic domains are distinct polypeptides (FepD and FepG) and the ATP binding domains consist of two FepB proteins. FepD and FepG are integral membrane proteins that are so hydrophobic that they fail to enter sodium dodecyl sulfate-polyacrylamide gels after normal solubilization protocols. In contrast, FepB is a peripheral membrane protein, located primarily in the cytoplasm, that is readily detectable by routine procedures for membrane proteins. Usually, all genes for ABC transporter components are in one operon. For the Fe^{3+}-Ent transporter, however, *fepB* is separated from the *fepDGC* operon by the presence of *entS*, which encodes the Ent exporter protein (Fig. 2).

Recent studies have determined the structure of the vitamin B_{12} ABC transporter (BtuCD) and suggested an uptake mechanism for these transporters. BtuCD is similar to the Fe^{3+}-Ent transporter in (i) being dependent on a PBP, (ii) belonging to the siderophore family of ABC transporters, and (iii) having hydrophobic domains with a total of 20 rather than the usual 10 to 14 transmembrane α-helices. It differs in that its two hydrophobic domains are present in one polypeptide (BtuC, a pseudoheterodimer).

The BtuCD structural studies, in conjunction with work on ABC transporters for histidine and maltose, have led to a plausible general mechanism for the functioning of these complexes. As applied to the Fe^{3+}-Ent transporter, the basic architecture and conformational changes would be as follows. The hydrophobic domains FepD and FepG would form a membrane-spanning translocation channel that could exist in open and closed forms. FepC proteins would be firmly attached to the cytoplasmic aspects of these hydrophobic domains. The two FepC proteins would form ATP binding motifs at their interface and each nucleotide binding site would require sequences on each polypeptide; the FepC domains could exist in two conformations, only one of which has complete ATP binding sites. Transport of Fe^{3+}-Ent would occur when FepB bearing Fe^{3+}-Ent binds to periplasmically exposed regions of FepD and FepG, possibly by charge-charge interactions. This association would have several conformational consequences; the FepG-FepD channel would open, in the process altering the FepC-FepC interface so as to form active ATP binding sites, and FepB would release Fe^{3+}-Ent. Fe^{3+}-Ent would unidirectionally traverse the channel, since its entry into the periplasm would be blocked by FepB, and be released into the cytoplasm. ATP binding and hydrolysis in the newly generated complete binding sites would complete the uptake cycle by inactivating these sites by altering the FepC-FepC interface, thereby also closing the FepD-FepG channel and releasing FepB.

REGULATION

E. coli proteins containing iron are required for many important functions, including DNA synthesis, respiration, amino acid biosynthesis, and tricarboxylic acid cycle functioning. Uptake of iron must be carefully controlled, however, since excessive levels of mobile intracellular iron can, particularly under aerobic conditions, lead to the production of harmful levels of the hydroxyl radical. The necessary iron homeostasis is accomplished primarily by the regulatory protein Fur. Chapter 13 in this book and the review by Andrews et al. include a broad overview as well as molecular details of global regulation by Fur.

Briefly, Fur is a 17-kDa protein that functions as a homodimer; each monomer can bind one Fe^{2+} ion, and the resulting Fe^{2+}-Fur complex is an active negative regulator of iron-responsive genes. Holorepressors bind 19-bp consensus DNA sites termed Fur or iron boxes; these operators are generally located within the pertinent promoter sites. Recent evidence indicates that these minimal iron boxes can bind two Fur dimers. The iron box is now viewed as consisting of three hexamers, each with the sequence GATAAT and with the third hexamer being in reverse order and separated from the first two by a single nucleotide. When two holorepressors bind, they are thought to occupy 13-bp binding sites (6-1-6) on opposite faces of the DNA and to overlap the middle hexamer in this binding. Some iron boxes have additional hexamers, and in these cases additional Fur molecules can be bound, increasing the prospect of more complex and modulated expression of these Fur-regulated genes.

Under conditions of iron deprivation, Fur acts so that proteins necessary for iron assimilation are expressed while the synthesis of iron storage proteins and nonessential iron-containing proteins is reduced. The induction process is straightforward; inadequate intracellular concentrations of Fe^{2+} prevent holorepressor formation; consequently, promoters with iron boxes will be available. The decreased synthesis of other proteins in the absence of Fe^{2+}-Fur is probably an indirect effect of negative regulation by the holorepressor; i.e., no case in which Fur acts as a transcriptional activator has been documented. For example, several iron storage and iron-containing proteins that appear to require Fe^{2+}-Fur for their synthesis in fact have their abundance negatively regulated by a 90-nucleotide RNA, RyhB. RyhB apparently functions as an antisense RNA that, with the assistance of the Hfq protein, hybridizes with the mRNA for these proteins, thereby preventing translation. The promoter region for RyhB contains an iron box. Therefore, the apparent positive regulation by Fur of certain proteins is actually a reflection of negative regulation by Fur of a translational inhibitor.

Fur regulates at least 50 *E. coli* genes including all Ent cluster genes and *fiu* and *cir*, the genes for the receptor proteins that recognize Ent-related siderophores. For the Ent cluster genes, there are three regulatory regions (Fig. 2), each with divergent promoters and Fur operators. There are two distinct promoters, each having its own iron box, for the *fepB* and *entCEBA* transcripts. In the other two regulatory regions, however, the promoters overlap and there may be just one centrally located iron box.

Regulation by Fur is complex, and control of Fur synthesis is also complicated. The average number of Fur molecules per exponentially growing *E. coli* cell is 5,000; this number doubles during the stationary phase and under conditions of redox stress. The *fur* gene is autoregulated, which may explain its abundance in the stationary phase, and, in the latter circumstance, its transcription is stimulated by superoxide-induced activation of SoxRS and by hydrogen peroxide oxidation and activation of OxyR. That is, Fur is a member of both the SoxRS and OxyR regulons; its increased abundance during oxidative stress presumably decreases the intracellular concentration of free iron and therefore the production of hydroxyl radicals. Even given the fact that Fur affects more than 50 *E. coli* K-12 genes and that the minimum member of Fur monomers bound to an iron box is four, Fur is remarkably prevalent for a regulatory protein. Perhaps, as suggested by De Lorenzo et al. in chapter 13, by binding free iron it also acts directly to block the generation of hydroxyl radicals. Whether or not Fur proteins exist in sufficient quantity to accomplish this sequestration function, it is likely that with adequate intracellular iron, conditions under which Fur concentrations are high will also be characterized by minimum synthesis of Ent cluster proteins.

Iron availability is dependent on many factors in addition to the absolute iron concentration in the environment. Well-known examples of this fact include the iron-deficient environment in eukaryotic hosts as a result of the presence of host iron binding proteins and the low solubility of ferric ion at neutral pH and above, such that many minimal media are iron poor. In addition to situations where iron is present but relatively unavailable, so that induction of specific transport systems is necessary for its assimilation, it is possible that there are growth conditions whereby uptake systems function relatively ineffectively and therefore need to be expressed to an unusually great extent. Possible examples of this have been observed in microarray data obtained with *Salmonella* growing in broth (Luria-Bertani [LB] broth plus glucose) or broth supplemented with either 0.6 or 1.5% agar. Fur-regulated genes examined, including almost all Ent cluster genes and *cir*, were more highly transcribed in cells growing on agar than in broth. (*Salmonella* has no *fiu* gene, and *entF* was not tested.) Typical data showed that, for example, *entABCE* transcription was 15- to 45-fold higher on 0.6% agar than in broth whereas *entD*, in another operon, had its transcription elevated 6- to 11-fold. Not only did the data show depression of Fur regulated genes, but also the transcriptional patterns noted were consistent with the operon organization. Iron-regulated genes were transcribed two-to fourfold more frequently in cells grown on 0.6% agar than in those grown on hard (1.5%) agar. The enhanced synthesis seen in cells grown on hard agar compared to that is broth-grown cells may result from slower diffusion of iron-containing complexes in solid media. The extreme induction that occurs on 0.6% agar medium is concomitant with the onset of cellular swarming activity; possibly, uptake of iron is difficult when cells have both aggregated and produced copious amounts of exopolysaccharide.

There are reports of two-component regulatory systems (TCSs) influencing the expression of Ent cluster genes. These findings are plausible, since TCSs are generally concerned with environmental sensing and some iron-containing proteins participate in functions that are affected by environmental conditions. In a recent systematic study, a strain in which *arcB*, which encodes a membrane-bound histidine kinase that responds to low oxygen by phosphorylating ArcA, was deleted had enhanced transcription of all *ent* genes, *fepA*, and *fepB*. Other TCSs also affected some Ent cluster genes; for example, a strain with defective porin regulation ($\Delta envZ\ \Delta ompR$) was upregulated for *fepA* and *entCEBA*, and a strain with a defective response to envelope stress ($\Delta cpxRA$) transcribed *entB* and *entA* excessively. Several other TCSs also had altered transcriptional regulation of one or two Ent cluster genes, but only with the ArcB strain were the results consistent with known operon structure and gene function. This survey was performed exclusively with cells growing logarithmically in LB broth at 37°C with aeration; it is possible that other growth conditions would have identified additional TCSs that influence iron assimilation.

The EnvZ OmpR TCS was initially characterized as a regulator of the porin genes, *ompF* and *ompC*, but is now recognized as a global regulator. In the analyses mentioned above, deletions of *envZ-ompR* altered the transcription of 125 genes. In iron studies, the initial suppressor mutations that gave rise to pseudo-revertants (see above) occurred in this TCS and certain *envZ* alleles resulted in reduced expression of many outer membrane and periplasmic proteins, including FepA, Cir, Fiu and FepB. Thus far, however, it appears that most effects of *envZ ompR* on iron assimilation are indirect, arising from alterations in the structure and transport properties of the outer membrane.

In addition to Fur, promoter strength variations, and possible involvement of TCSs, there is evidence for additional regulatory devices that affect the abundance of individual Ent-related proteins. In most cases, the mechanism(s) for this additional regulation is only speculative, as is evident in the following brief summaries.

The *fepA* and *entD* genes appear to be co-transcribed (Fig. 2). Synthesis of both proteins is iron regulated, and in the intercistronic space preceding *entD*, there is no Fur box and there are no properly positioned -35 and -10 promoter sequences. Despite being encoded by the same operon, FepA and EntD represent the two extremes in prevalence for iron-regulated genes. FepA, the rate-limiting protein in Fe^{3+}-Ent uptake, is highly expressed, and *fepA* codon usage is consistent with this. FepA was the first protein encoded by the Ent cluster to be observed. In contrast, EntD, with its posttranslational modification activity, is needed in only small amounts and was the last and most difficult of the Ent cluster products to be detected. The initiation codon of *entD* is UUG, and the gene is rich (ca. 8%) in rare codons; EntD was finally overproduced and isolated by the Walsh group, who optimized the initiation codon to AUG and similarly altered the codons for the next six amino acids. In addition to codon usage, repetitive extragenic palindrome (REP) sequences may help to explain the huge disparity in the numbers of FepA and EntD proteins made. There are two REP sequences in inverted orientation in the more than 140 bp separating *fepA* and *entD*. Each REP can form a stem-loop structure; two REP sequences in inverted orientation can hybridize into a larger and stronger stem-loop, wherein each REP sequence forms a side of the stem. One postulated function for REP sequences is stabilization of upstream mRNA, and unusual stability of the *fepA* mRNA has been noted. In addition, transcriptional analyses show that there is at least four times more *fepA* mRNA than *entD* mRNA. The data are consistent with the REP sequences playing a regulatory role by hindering degradation of the *fepA* portion of the transcript.

fes and *entF* appear to be coordinately transcribed, and the transcript, like that for *fepA entD*, has a long leader. The *fes-entF* intercistronic space (222 bp) is occupied almost entirely by a 216-bp ORF. The termination codon for this ORF (UGA) completely overlaps the *entF* initiation codon (AUG). The function of this ORF, like that of similarly placed ORFs upstream of *fepB* and *cir*, is unknown. A strong intrinsic transcriptional terminator site begins 15 bp downstream of the *entF* termination codon.

The next gene in the Ent cluster encodes the enigmatic protein FepE. FepE, a membrane protein originally described and named as being involved in Fe^{3+}-Ent transport, is now thought to be a member of the Wzz protein family. Wzz proteins determine the number of O-antigen (O-Ag) subunits present in lipopolysaccharide molecules. The *fepE* homologue wzz_{fepE} of *S. enterica* serovar Typhimurium is well studied and much like *fepE*; wzz_{fepE} and *fepE* have 74% amino acid identity. Like *fepE*, wzz_{fepE} is located in the Ent gene cluster, and both genes have a lower than average G+C content for an Ent cluster gene. Wzz_{fepE} regulates the extent of polymerization of O-Ag subunits so that very long ($\geq$100-subunit) O-Ags are produced, and it is likely that FepE does the same. FepE is poorly expressed even when present on a multicopy plasmid, and there is no evidence that its synthesis is iron regulated; sequence data support these findings since there is no obvious iron box or promoter in the 214 bp separating *entF* from *fepE*. It appears that *fepE*/wzz_{fepE} was adventitiously horizontally inserted into the Ent gene cluster. Plausible reasons can be proposed whereby it would or would not be advantageous to coregulate the lipopolysaccharide length with the induction of an iron transport system, but since *fepE* has not been shown to be iron regulated, it appears that *E. coli* has not chosen to use them.

The *fepD-entS* (*ybdA*) intergenic space is a second bidirectional promoter region in the Ent cluster (Fig. 2). The region is comparatively small, with 23 bp separating the transcriptional start sites and 103 bp between initiation codons. The *fepD* and *entS* promoters are unusual; the -35 sequences are poorly conserved, and, as is frequently the case in such situations, both -10 sequences are extended by an immediately upstream TGn sequence (where n is any deoxynucleotide). The two extended -10 sequences overlap extensively (8 of 10 bp). The Fur operator is large (25 bp), with four contiguous hexameric motifs, and offset, so that with two Fur dimers bound it would block only the *entS* -35 region. Binding of additional Fur dimers at this site with increasing repressor concentrations has been demonstrated in vitro, and the Fur footprints on DNA are correspondingly enlarged.

The *fepDCG* operon encodes the Fe^{3+}-Ent ABC transporter, in which two molecules of the ATPase FepC are expected for every FepD and FepG protein. *fepD* and *fepG* overlap at their stop/start codons, as do *fepG* and *fepC*. Nonetheless, translational coupling has been demonstrated only for FepD and FepG; inability to synthesize FepD prevents FepG but not FepC production. This is perhaps an indication of the differential expression that occurs from the transcript.

The third bidirectional promoter region, *fepB-entC*, differs from the other two in that the promoters do not overlap and each promoter has its own iron box. The iron box for *fepB* overlaps its -10 promoter determinant, while that for *entC* is atypically located in DNA that is transcribed early in the *entC* leader. At high Fur concentrations, however, the Fur footprint includes the ribosome binding site for the *entC* transcript. FepB has a distinctive leader: (i) it is long (213 nucleotides); (ii) it includes two box C elements, each having three copies of a monomer, in inverted orientation, permitting the formation of a strong stem-loop secondary structure; and (iii) it contains an ORF for a 60 or 68-amino-acid polypeptide whose termination codon occurs in the *fepB* ribosome binding site. Alterations in this leader, including many deletions and base pair changes designed to either interrupt translation of the ORF or eliminate possible large secondary structures, all had relatively minor effects on FepB expression. Unexpectedly, the largest effects on FepB synthesis yet found for a site in the 5′ untranslated region occur when an enhancer element (ε) centered 12 bp upstream of the ribosome binding site is eliminated. The translational efficiency of *fepB* mRNA is reduced fivefold.

PERSPECTIVE

Much remains to be learned about the functioning of the Ent system. Details concerning the regulation of many individual Ent cluster genes are unclear; for instance, the significance of the precisely placed ORFs upstream of *fepB*, *cir*, and *entF* and the roles of the palindromic REP and box C sequences remain to be determined. Also, it should be established if, on induction, Ent is the first siderophore excreted or if, perhaps, Ent precursors such as DHB and linear forms of DBS are released first. The excretion path for secondary siderophores is unknown, as are molecular details of Ent release. More generally, the way in which the Ent system is integrated into overall cell metabolism so that it can respond to changes in pH, oxygenation, and osmolarity remains in question.

Nonetheless, the research of greatest general interest will probably concern Ent system expression under natural conditions. Heretofore, almost all Ent research has been conducted with planktonically growing "rough" cells. Evidence that it is more difficult to obtain iron when growth is occurring on a solid surface, a common condition in *E. coli* ecology, and hints that exopolysaccharide (capsule or slime layer) restricts iron uptake suggest that Ent-mediated iron uptake is even more important for normal cell growth than was previously thought. In particular, iron assimilation in biofilms presents itself as a difficult but essential problem for study.

ACKNOWLEDGMENTS

I thank Elizabeth Earhart for drawing the figures, Sandra Miller for typing the manuscript, and Rasika Harshey for engaging in stimulating discussions and for providing data prior to publication.

SUGGESTED READING

Ammerlaan, M. C. 1994. *Studies on Enterobactin Synthetase*. Ph.D. dissertation. University of Texas, Austin.

Andrews, S. C., A. K. Robinson, and F. Rodriguez-Quinones. 2003. Bacterial iron homeostasis. *FEMS Microbiol. Rev.* **27:**215–237.

Christoffersen, C. A., T. J. Brickman, I. Hook-Barnard, and M. A. McIntosh. 2001. Regulatory architecture of the iron-regulated *fepD-ybdA* bidirectional promoter region in *Escherichia coli. J. Bacteriol.* **183:**2059–2070.

Earhart, C. F. 1996. Uptake and metabolism of iron and molybdenum, p. 1075–1090. *In* F. C. Neidhardt, R. Curtiss III, J. L. Ingraham, E. C. C. Lin, K. B. Low, B. Magasanik, W. S. Reznikoff, M. Riley, M. Schaechter, and H. E. Umbarger (ed.),

Escherichia coli and *Salmonella: Cellular and Molecular Biology*, 2nd ed. ASM Press, Washington, D.C.

Franke, D., G. A. Sprenger, and M. Muller. 2001. Synthesis of functionalized cyclohexadiene-*trans*-diols with recombinant cells of *Escherichia coli*. *Angew. Chem. Int. Ed.* **40:**555–557.

Furrer, J. L., D. N. Sanders, I. G. Hook-Barnard, and M. A. McIntosh. 2002. Export of the siderophore enterobactin in *Escherichia coli*: involvement of a 43 kDa membrane exporter. *Mol. Microbiol.* **44:**1225–1234.

Hancock, R. E. W., K. Hantke, and V. Braun. 1977. Iron transport in *Escherichia coli* K-12. 2,3-Dihydroxybenzoate-promoted iron uptake. *Arch. Microbiol.* **114:**231–239.

Hantash, F. M. 1998. *The Shock-Sensitive Proteins of the* Escherichia coli *Enterobactin System*. Ph.D. dissertation. University of Texas, Austin.

Hantash, F. M., and C. F. Earhart. 2000. Membrane association of the *Escherichia coli* enterobactin synthase proteins EntB/G, EntE, and EntF. *J. Bacteriol.* **182:**1768–1773.

Hantke, K. 1990. Dihydroxybenzoylserine—a siderophore for *E. coli*. *FEMS Microbiol. Lett.* **67:**5–8.

Köster, W. 2001. ABC transporter-mediated uptake of iron, siderophores, heme and vitamin B12. *Res. Microbiol.* **152:**291–301.

Lavrrar, J. L., and M. A. McIntosh. 2003. Architecture of a Fur binding site: a comparative analysis. *J. Bacteriol.* **185:**2194–2202.

Locher, K. P., A. T. Lee, and D. C. Rees. 2002. The *E. coli* BtuCD structure: a framework for ABC transporter architecture and mechanism. *Science* **296:**1091–1098.

Lundrigan, M., and C. F. Earhart 1981. Reduction in three iron-regulated outer membrane proteins and protein a by the *E. coli* K-12 *perA* mutation. *J. Bacteriol.* **146:**804–807.

Murray, G. L., S. R. Attridge, and R. Morona. 2003. Regulation of *Salmonella typhimurium* lipopolysaccharide O antigen chain length is required for virulence: identification of FepE as a second Wzz, *Mol. Microbiol.* **47:**1395–1406.

Nikaido, H., and E. Y. Rosenberg. 1990. Cir and Fiu proteins in the outer membrane of *Escherichia coli* catalyze transport of monomeric catechols: study with β-lactam antibiotics containing catechol and analogous groups. *J. Bacteriol.* **172:**1361–1367.

O'Brien, I. G., G. B. Cox, and F. Gibson. 1971. Enterochelin hydrolysis and iron metabolism in *Escherichia coli*. *Biochim. Biophys. Acta* **237:**537–549.

Oshima, T., H. Aba, Y. Masuda, S. Kanaya, M. Sugiura, B. L. Wanner, H. Mori, and T. Mizuno. 2002. Transcription analysis of all two-component regulatory system mutants of *Escherichia coli* K-12. *Mol. Microbiol.* **46:**281–291.

Pickett, C. L. 1983. *Iron Transport in* Escherichia coli: *a Characterization of Transport Mutants and a Genetic Description of the Enterochelin Gene Cluster*. Ph.D. dissertation. University of Texas, Austin.

Sprencel, C., Z. Cao, Q. Zengbiao, D. C. Scott, M. A. Montague, N. Ivanoff, J. Xu, K. M. Raymond, S. M. C. Newton, and P. E. Klebba. 2000. Binding of ferric enterobactin by the *Escherichia coli* periplasmic protein FepB. *J. Bacteriol.* **182:**5359–5364.

Stephens, D. L., M. D. Choe, and C. F. Earhart. 1995. *Escherichia coli* periplasmic protein FepB binds ferrienterobactin. *Microbiology* **141:**1647–1654.

Vazquez-Laslop, N., H. Lee, R. Hu, and A. A. Neyfakh. 2001. Molecular sieve mechanism of selective release of cytoplasmic proteins by osmotically shocked *Escherichia coli*. *J. Bacteriol.* **183:**2399–2404.

Walsh, C. T., J. Liu, F. Rusnak, and M. Sakaitani. 1990. Molecular studies on enzymes in chorismate metabolism and the enterobactin biosynthetic pathway. *Chem. Rev.* **90:**1105–1129.

Wang, Q., J. G. Frye, M. McClelland, and R. M. Harshey. 2004. Gene expression patterns during swarming in *Salmonella typhimurium*: genes specific to surface growth and putative new motility and pathogenicity genes. *Mol. Microbiol.* **52:**169–187.

Winkelmann, G., A. Cousier, W. Beck, and H. Jung. 1994. HPLC separation of enterobactin and linear 2,3-dihydroxybenzoylserine derivatives: a study on mutants of *Escherichia coli* defective in regulation (*fur*), esterase (*fes*) and transport (*fepA*). *BioMetals* **7:**149–154.

TRANSPORT BIOCHEMISTRY OF FepA

Phillip E. Klebba

10

> "It was the best of times, it was the worst of times . . ."
>
> Charles Dickens, *A Tale of Two Cities*.

In most ways the processes of prokaryotes underlie and explain the origins of biochemistry in the eukaryotic kingdom, and we doubly profit, in research and in life, from their ingenious inventions, mechanisms, and symbioses. But immense misery, debilitation, and mortality, caused by bacterial disease, casts another visage of the same phenomenon, the face of pathogenesis. The microbial world is a place beautiful beyond imagination, filled with intricately folded, delicately assembled polymeric entities that encode and catalyze the reactions of life, a place where large and small molecules randomly and instantaneously collide in solution or on membranous surfaces, allowing a fundamental property of biology to take hold: the propensity to bind. The complementary surfaces of solutes, substrates, ligands, proteins, and nucleic acids, perfected by evolution, align in brief, prolonged, or permanent associations that conjoin them so that biological chemistry may occur. These exacting affiliations involve every conceivable type of noncovalent interaction, often producing sophisticated binary, tertiary, or higher-order associations among certain participants but not among others that are slightly different in size or structure. The Darwinian side of this molecular dance is that the binding reactions may also preface a pirouette of death and destruction: bacteriophage recognize their hosts and lyse them; bacteriocins bind to surface receptors and kill their target cells; antibiotics penetrate the cell envelope and block basic biochemistry. Our perceptions of such events in bacterial membranes rapidly advanced in the last decade, from the inferences of molecular biology and biochemistry and the structures of crystallography. While each of these approaches enhanced our vision, each was also limited, resulting in an incomplete view of how ferric siderophore receptors transport their ligands. Research on FepA well illustrates the methodological and conceptual dichotomy: we are now well acquainted with its exquisite form, but this knowledge did not reveal the secrets of its transport biochemistry. The rudimentary understanding of its attributes that does exist only grudgingly emerged from a synthesis of data from several techniques and laboratories.

METAL TRANSPORTERS IN THE OUTER MEMBRANE

All gram-negative bacteria obtain iron by using high-affinity, energy-dependent, multiprotein, multifunctional, TonB-dependent receptor

---

Phillip E. Klebba, Department of Chemistry & Biochemistry, University of Oklahoma, 620 Parrington Oval, Norman, OK 73019, and Faculté de Medecine, Institut Necker Enfants Malades, 156 rue de Vaugirard, 75015 Paris, France.

Iron Transport in Bacteria, Edited by Jorge H. Crosa, Alexandra R. Mey, and Shelley M. Payne
© 2004 ASM Press, Washington, D.C.

complexes in their cell envelopes. Some of their components are unique to specific uptake systems (outer membrane [OM] receptors), and some are common to all the systems (TonB, ExbB, and ExbD). Transport of metals, including Fe^{3+} in the form of ferric siderophores and cobalt as vitamin B_{12}, requires the input of energy and the actions of TonB. Besides their activity as metal transporters, OM proteins like FepA, FhuA, and BtuB act as receptors for toxins and viruses. FepA is the cognate surface receptor for colicins B (ColB) and D (ColD); FhuA adsorbs ColM and phages T1, T5, and ϕ80. These OM proteins are not porins in the original sense of the word because they do not contain a nonspecific open channel; however, one of their structural elements is a porin-like, amphiphilic β-barrel through which iron ultimately passes into the cell (C-domain). Their other structural entity, a 150-residue globular N-terminal domain (N-domain), resides within and presumably regulates molecular traffic through the channel. Ligand binding activates such OM proteins to a transport-competent state, hence their designation as ligand-gated porins (LGPs).

The metals accumulated by LGPs across the OM in an energy-dependent manner against a concentration gradient are essential to the survival of bacteria. Iron acquisition is a focal point of bacterial pathogenesis. Not just microorganisms, but most organisms, require iron for a multiplicity of metabolic processes, including glycolysis, energy generation, DNA synthesis, and defense against toxic reactive oxygen species. Bacteria actively seek iron from animal hosts by secreting siderophores that release the metal from cellular fluids and tissues. Multicomponent transport systems either internalize the bacterial ferric siderophores, or strip the metal from eukaryotic iron proteins. Both gram-negative and gram-positive bacteria are susceptible to growth inhibition by iron deprivation. The passage of ligands through gated porins is thus a prototypic problem in membrane biology, relevant to such diverse questions as macromolecular recognition, signal and energy transduction, protein secretion, and membrane protein structure and function.

STRUCTURAL FEATURES OF METAL TRANSPORTERS

For a more complete description of the crystallographic data for FepA and its homologs and paralogs, see chapter 4. This chapter provides an overview of these structures as a basis for mechanistic discussions that follow. The crystals of FepA, FhuA, FecA, and BtuB all showed the hollow, 22-strand antiparallel β-barrel of the C-domain, which spans the outer membrane and contains large extracellular loops that function in ligand binding, and the structurally distinct globular N-domain, which folds into the barrel interior, blocking access to the periplasm. The N-domain contains α- and β-structure, as well as loops that rise to the top of the outer vestibule of the C-domain near the bound ligand. When ferrichrome binds to FhuA, when vitamin B_{12} binds to BtuB, and presumably (their crystal structures did not reveal this phenomenon) when ferric enterobactin (FeEnt) and ferric citrate (FeCit) bind to FepA and FecA, respectively, residues in these loops undergo minor changes that are propagated through the N-domain to residues at the periplasmic interface of the OM. The most obvious resulting N-domain rearrangements occur in the region containing the TonB box, so called because it was postulated to be involved in signal transduction to the accessory protein TonB. Thus, these structural changes are proposed to create a signaling pathway linking ligand recognition to mediation of transport by TonB.

N-Domain

The overall topologies of the FepA, FhuA, FecA, and BtuB N-domains are similar, despite folding differences that localize some analogous residues in different places (Color Plate 13). The N-domain is a four-strand β-sheet whose individual β-strands are connected by loops, turns, and short α-helices. It contains two loops at its exterior most surface: the first, NL1, projects up from the αβ-structure toward the opening of the pore vestibule in all four proteins. The second, NL2, is similar to NL1 in size and projection in FepA but smaller and less exposed in FhuA, FecA, and BtuB. Besides folding differently, the N-domain loops of the four receptors are distinctive in composition: those of FepA contain a preponderance of Arg residues at the top whereas FhuA, FecA, and BtuB do not; instead, they encode a few aromatic residues in the same relative positions. The different folding of the N-domains achieves similar but distinct three-dimensional forms: that of FepA is more elongated, while those of FhuA, FecA, and BtuB are compact and more spherical.

OM metal transport systems involve multiple protein components, requiring, in addition to the cell surface receptor proteins, the cell envelope proteins TonB, ExbB, and ExbD (see chapter 7). The functions of, and associations among, these proteins that ultimately facilitate active transport against a concentration gradient are incompletely defined, but they probably involve the N-terminal globular domains of the receptor proteins. Before crystallographic depictions became available, the TonB box in the N-domain of siderophore receptors was theorized to physically interact with TonB. However, significant gaps still exist in the explanation of how such a physical interaction may occur.

C-Domain

Like general and specific porins, LGPs contain an antiparallel, amphiphilic β-sheet that circumscribes an aqueous transmembrane channel. This revelation from crystallography was not surprising, and in fact the complete structures of OM β-barrels, including their short reverse turns on their periplasmic surface and large loops on their external surface, were predicted well in advance of their crystallographic solutions. The major unexpected finding of the X-ray results was the disposition of the N terminus within the membrane channel created by the C terminus. The transmembrane pore domains of OM metal transporters are formed by approximately 575 amino acids, and the loops of this β-barrel perform the crucial first steps in transport, identification, and selection of the proper metal complex. The receptor-ligand complexes observed by crystallography depict the end results of the binding process in vitro, when the receptor and ligand reach thermodynamic equilibrium, and these data do not provide much information about the initial recognition and discrimination of ligands. The FepA crystal structure did not illustrate the disposition of bound ligands, but it is known from site-directed mutagenesis and biochemistry that basic and aromatic amino acids in the surface loops adsorb ligands on the basis of ionic and hydrophobic bonds and ring stacking.

Inside the pores of OM metal transporters, between the barrel walls and the surface of the N-domains, ion pairs exist that may stabilize the N terminus within the channel. These bonds involve charged residues that are conserved within the sequences of LGPs: basic residues in the N-domain, and acidic and basic residues in the C-domain. These ion pairs are configured such that they may function in the transport mechanism, holding the N-domain in the C-domain at neutral pH but promoting either its expulsion from the pore or a conformational reorganization of its structure at low pH.

FIRST STAGE OF FeEnt UPTAKE: BINDING

FeEnt adsorbs to FepA in at least two steps, which led to the perception that two binding regions exist in the surface vestibule: an exterior site designated B1 and an interior site designated B2.

Analysis by Site-Directed Mutagenesis

Both FepA and FhuA (but not FecA) contain aromatic amino acids around the entrance to their pore vestibules. FepA contains seven such residues, all tyrosines; FhuA has six aromatic amino acids encircling the mouth of its channel. Although the crystal structures of FepA and FhuA did not attribute mechanistic importance to these residues, their prominence in the loops suggested a role in ligand recognition and binding. The individual replacement of 17 aromatic residues in the two binding regions of FepA by Ala (W101A, Y217A, Y260A, Y272A, Y285A, Y289A, W297A, Y309A, F329A, Y472A, Y478A, Y481A, Y488A, Y495A, Y540A, Y553A, and Y638A) revealed the importance of these amino acids in the binding of FeEnt.

SITE B1

Alanine replacements of aromatic residues in the extremities of five different loops that surround the mouth of the outer vestibule (L3, L5, L7, L8, and L10) affected the affinity of the FeEnt binding reaction. Single substitutions for certain targets had small effects (Y217A, Y285A, Y289A, W297A, Y309A, and Y553A), whereas others (Y481A and Y638A) increased the binding K_d as much as 20-fold. Some mutations produced equivalent reductions in both binding and transport, but others caused disproportionate deficiencies in the latter: F329A, Y638A, Y478A, and W101A increased the transport K_m 20-, 100-, 500-, and 1,000-fold, respectively. Other experiments identified K483 in the extremity of FepA L7 as a participant in ligand binding. Together, the data underscored the significance of the B1 region in initial binding and highlighted the contributions of side chain aromaticity and/or positive charge in the creation of affinity for FeEnt. The experiments also found single substitution mutations in the surface loops that affected FeEnt transport without having significant effects on its binding.

SITE B2

Mutagenesis experiments identified the central roles of Y260 and R316, deep within the vestibule, in the overall affinity for FeEnt and/or in the movement of FeEnt through the transmembrane channel. The mutation Y260A increased K_d of the FepA-FeEnt binding reaction 100-fold and the transport K_m 300-fold; R316A had minor effects on binding affinity but decreased the overall transport affinity 20-fold.

A Two-Site Binding Model

The mutagenesis results supported a hypothesis of two ligand binding regions in the FepA vestibule, populated by aromatic and basic residues that probably function through ionic and hydrophobic interactions with the negatively charged, aromatic siderophore. Many results converge on this two-site model, beginning with the identification of the sites by mutagenesis but also including measurements of fluorescence quenching engendered by ligand adsorption in vitro and in vivo, biphasic association and dissociation kinetics, loosely and tightly bound populations of FeEnt, and conformational motion in L7. The data describe or derive from structural changes that occur when FeEnt adsorbs to FepA and passes to the interior of the receptor. These findings portray the surface loops as dynamic entities, predisposed to an "open" conformation that captures metal complexes by the attractive forces of hydrophobic or ionic bonds and then closes around the molecule as binding equilibrium occurs (prior to transport), creating the closed state that was crystallographically observed.

ESR AND FLUORESCENCE SPECTROSCOPIC OBSERVATIONS IN VITRO

FepA interconverts between two distinct structural forms that were seen by electron spin resonance (ESR) spectroscopy. Fluorescence spectroscopy reiterated the existence of two receptor states during binding and facilitated the characterization of site-directed substitution mutants, establishing that alterations in the exterior B1 site impaired the initial binding stage while mutations in the interior B2 site affected the second binding phase.

FLUORESCENCE SPECTROSCOPIC OBSERVATIONS IN VIVO

ESR spectroscopy also observed conformational changes in FepA during ligand uptake, but its low sensitivity made the technique difficult to employ in vivo. Fluorescence spectroscopy solved this problem but not without procedural modifications. FepA residue E280C (in L3) was specifically modified by nitroxide compounds in vivo but was not labeled under the same conditions with Cys-specific fluorescent reagents. The analysis of nine other unpaired Cys substitution mutations at positions either on the cell surface or in the periplasm showed that fluorescein maleimide (FM) specifically labeled W101C, S271C, F329C, and S397C, which all reside in cell surface-exposed loops of FepA. Among these, cells expressing FepAS271C-FM and S397C-FM, at the extremities of L3 and L5, transported FeEnt normally.

LPS O-Antigen. Laboratory *E. coli* strains produce rough lipopolysaccharide (LPS), without an O-antigen. In such strains, fluors penetrated the OM bilayer and resisted washing procedures, precluding the specific labeling of target Cys residues. Conversely, pMF19, which carries a rhamnosyl transferase that confers production of the LPS O chain ($wbbL^+$) to rough mutants, allowed specific FM labeling of accessible Cys mutants. These findings permitted fluorescence measurements of ligand uptake in vivo.

Affinity of FepA and FhuA for Ligands. It was troubling that the affinities of purified FepA and FhuA for FeEnt and ferrichrome, respectively, were 100- to 1000-fold lower than in bacterial cells. In vitro, in detergents, the K_d was 0.01 to 0.1 μM, whereas radioisotopic methods showed that in vivo, the K_d was 0.1 to 0.2 nM. This lower affinity in vitro raised questions about crystallographic data from proteins that were purified in detergents, and the in vivo fluorescence system verified the subnanomolar K_d in live bacteria, confirming the distortions engendered by purification in the interactions of FepA with ligands. At the K_d measured in vivo, purified FepA is less than 1% saturated with FeEnt.

Kinetics of Adsorption and Desorption. Studies of FeEnt dissociation from FM-labeled FepA in vivo demonstrated that the kinetics of ligand release nearly perfectly mirror the biphasic adsorption reaction. Bound FeEnt dissociates from FepA in two stages, with nearly the same kinetic parameters that were found for its adsorption (for the B1 site, $k_1 \simeq k_2 = 0.03/s$; for the B2 site, $k_3 \simeq k_4 = 0.003/s$[Color Plate 14]). Radioisotopic methods confirmed this result, although with less precision.

LOOSELY AND TIGHTLY BOUND FeEnt

Radioisotopic studies of ^{59}FeEnt dissociation from FepA showed two subpopulations of bound ligands, one of which was exchangeable with (nonradioactive) FeEnt and one that was nonexchangeable. The former fraction of loosely bound ferric siderophore was associated with the B1 site, because the simultaneous mutational replacement of Y272 and F329 also released it. Thus, again, evidence supported the concept of binding sites that adsorb FeEnt with different affinities.

CONFORMATIONAL MOTION IN L7 DURING FeEnt BINDING

Experiments with a lysine-specific cleavable cross-linking reagent, sulfo-EGS, revealed conformational motion in the surface loops of FepA during adsorption of FeEnt. Sulfo-EGS reacted with two lysines, K332 and K483, and at least two other unidentified Lys residues in the surface loops. The reagent cross-linked K483 in FepA L7 to the major OM proteins OmpF, OmpC, and OmpA. FeEnt binding did not prevent the modification of K483 but blocked its cross-linking to OmpFC and OmpA. These data reiterated that the loops undergo motion in vivo of approximately 15 Å, from a ligand-free open state to a ligand-bound closed state. The coupling of FepA L7 to OmpF, OmpC, or OmpA was TonB independent but was inhibited by cyanide, suggesting a potential distinction between the two dependencies.

X-ray data later showed comparable motion in L7 and L8 of the closely related protein, FecA, during binding of its ligand, FeCit. As seen for FepA, the extent of L7 motion was about 15 Å. Together the studies documented a deviation in the FepA crystal structure from what occurs in vivo in the native environment of the gram-negative bacterial OM.

Selectivity of FepA for Ligands

Recent experiments investigated ligand selection by FepA. The receptor discriminates FeEnt, ferrichrome, Fe-agrobactin, Fe-corynebactin, and Fe-TRENCAM. In measurements of adsorption equilibria, the synthetic FeEnt analog Fe-TRENCAM, which differs from the native *E. coli* siderophore only in the composition of the backbone from which the catecholate chelation groups arise, bound to FepA with 100-fold-lower affinity (K_d = 20 nM). Likewise, only the pure catecholate siderophores Fe-TRENCAM and Fe-corynebactin inhibited ^{59}FeEnt adsorption, and only when provided at significantly higher concentrations than FeEnt itself. The lower affinity of FepA for Fe-TRENCAM in these assays showed the ability of the receptor to select FeEnt, even among its close structural homologs. FepA rejects heterologous iron complexes at the initial interaction with its surface loops, remaining unoccupied and available for productive reactions when it encounters a proper ligand. The inability of ferrichrome and Fe-agrobactin to even slightly impair FeEnt binding emphasized that heterologous metal complexes do not mistakenly adsorb to FepA, because their chemistry does not match the binding determinants in its loops.

Overview of FeEnt Binding to FepA

In lieu of crystallographic data for the determinants and process of ligand binding, a combination of other approaches yielded a consistent and somewhat surprising depiction of the molecular events that occur in the interaction between FeEnt and FepA. As expected from the chemistry of the metal complex, evidence exists that ionic and hydrophobic bonds within two ligand interaction sites, both containing aromatic and basic residues, provide the main components of affinity. The two sites are conformationally related: FeEnt adsorbs to predominantly aromatic residues in the extremities of the loops, initiating conformational changes that close the loops around the ligand and move it deeper into the vestibule, to the inner site, where it sits above the N-domain at binding equilibrium, ready for transport. The magnitude of the loop movements is about 15 Å; it is likely that multiple loops participate in the capture of the ligand in a manner akin to the capture of prey by the tentacles of a sea anemone.

SECOND STAGE OF FeEnt UPTAKE: INTERNALIZATION

The original crystal structures of FepA and FhuA posed the following questions: what types of changes occur in the N-domain to allow the passage of a molecule as large as a siderophore and what biochemical events trigger the structural metamorphosis? Other, more recent structures, i.e., those of FecA and BtuB, provided no substantially new insights into these key aspects of the LGP transport mechanism. However, the conservation of certain residues that occur in the barrel and N-terminal domains of FepA and FhuA, relative to the diversity seen in their surface loops, suggests that once bound, siderophores pass through LGP channels by a common mechanism. The FeEnt binding reaction is fully independent of TonB and energy: it occurs with identical kinetics and thermodynamics in *tonB* strains or energy-deficient or poisoned wild-type bacteria. In the transport stage, an unknown sequence of events that does require TonB and energy results in other conformational changes involving the N-domain that open a path in the OM bilayer through which FeEnt passes into the periplasm (Color Plate 14).

Interactions with TonB

Although not a siderophore receptor, BtuB is structurally similar and may serve as a model for the transport mechanism and TonB dependence of all LGPs. On the basis of genetic suppression and chemical cross-linking, the TonB box of BtuB was proposed to specifically interact with TonB at or near residue Q160. ESR studies of TonB box residue side chain mobility and susceptibility to relaxing agents further showed conformational differences in the presence and absence of bound vitamin B_{12}. These data led to the widely endorsed conclusion that vitamin B_{12} binding to BtuB stimulates a conformational change in its TonB box, resulting in signal transduction to TonB residue Q160 by direct physical contact, triggering TonB-mediated energy transfer from the inner membrane to the OM receptor protein, and promoting ligand internalization through the BtuB channel. Nevertheless, the LGP crystal structures, especially that of BtuB itself (Color Plate 15), are not fully consistent with this theory. In support of the theory, ligand binding to FhuA, FecA, and BtuB altered the position of their TonB boxes, in the last two proteins moving it from the barrel wall to the center of the channel. However, other results are less supportive. (i) Relative to the TonB box regions of FepA and FecA, which are similar, the shape of the BtuB TonB box is completely different, casting doubt on the notion that TonB uniformly recognizes and interacts with this sequence in all metal transporters; the significant variation that occurs among TonB box sequences reinforces this uncertainty. (ii) In BtuB, the TonB box resides inside the barrel and does not extrude on ligand binding. (iii) The conformational change that occurs within the BtuB TonB box when vitamin B_{12} binds is small and nondescript, moving its N-terminal extremity only a few angstroms. (iv) The TonB C terminus itself is too large to enter the barrel of BtuB and contact the TonB box of LGP. (v) It manifests no apparent clefts, grooves, or complementary surfaces to sustain the interaction. (vi) The position of TonB residue Q160, beneath the dimeric αβ structure of the TonB C terminus, may preclude direct physical contact with the BtuB TonB box within the β-barrel. The upshot is that either the theory of protein protein interactions between TonB and the TonB box of BtuB is somehow flawed or the crystallized forms of the LGPs are different from the functional proteins in vivo. Only additional data on iron transport mechanics will resolve these questions.

The functionality of the TonB box region of LGPs was potentially undermined by several studies. Receptor proteins lacking the complete N-domain, including the TonB box, showed TonB-dependent uptake of ferric siderophores, challenging the significance of the TonB box to iron transport. However, other experiments suggested that these data were compromised. This study showed that siderophore-receptor deficient *E. coli* strains may contain cryptic fragments of the N terminus of the missing receptor that somehow complement other mutant polypeptides of the same or different receptors expressed from plasmids. In other words, in the generation of *fhuA* and *fepA* hosts for plasmids, spontaneous chromosomal mutations were unknowingly isolated that retained an active N-terminal domain. This phenomenon was initially seen as the ability to confer colicin B sensitivity to the C-terminal β-barrel domain of FepA (Fepβ), presumably by inserting within it and facilitating quasi-normal, TonB-dependent uptake. A similar phenomenon, alpha-complementation, was demonstrated for the product of *E. coli lacZ*, β-galactosidase. It is worthwhile to try to explain the complementation phenomenon. One of the strains in question, *E. coli* KDF541 (spontaneous *fepA*, *fhuA*, and *cir*), contains a chromosomal fragment of *fepA* corresponding to and encoding its N-terminal 363 amino acids. This finding is not surprising, because the mutation in *fepA* of KDF541 was restricted to the *fepA* locus, not intruding, for example, into *fes*, a contiguous gene needed for the transport of FeEnt. The production of a 363-residue N-terminal fragment of FepA entails, in addition to the N-terminal 150 residues, 213 residues of the β-barrel (the first eight β-strands and four loops). How might complementation occur? One possibility is that the entire 360-residue polypeptide enters the plasmid-encoded empty Fepβ barrel, and another is that the N-terminal 150 amino acids enter and the remaining 210 residues hang in the periplasm. However, it is not obvious how such spontaneous assembly may confer functionality. Either alternative appears to preclude the postulated interactions with TonB, which are assumed to occur at the periplasmic rim of the barrel. In addition, if the N termini of FepA and FhuA specifically interact with TonB, then why does not this fragment, which is postulated to localize to the periplasm, bind to it and inhibit TonB, as the C terminus of TonB inhibits the activity of the receptor proteins? Previous work showed that 180 to 200 N-terminal amino acids are sufficient to localize FepA to the OM bilayer; therefore, it is likely that a 363-residue fragment of FepA will reside in the OM, not in the periplasm. Furthermore, quantitative analyses with ^{125}I-protein A showed no evidence of the 363-residue protein fragment in cells expressing Fepβ or Fepβ2, which both transported FeEnt. Quantitative analysis is important in the interpretation of the data, because, for example, 20% of wild-type binding and uptake activity was observed for Fhuβ, whereas under the growth conditions employed, the fragment was not detectable and its level therefore never exceeded 1%. Finally, the results with the isolated N-domain of FepA are relevant: it does not fold properly, and its affinity for FeEnt is 10,000-fold lower than is that of the intact, native receptor protein (K_d values of 5 μM and 0.2 nM, respectively). Therefore, not only does coexpression of the N- and C-domains of the receptor ameliorate the defects of the empty barrel, but also it corrects the anomalies of the globular domain. One clear conclusion from these findings is that the proposed complementation is not well understood; another is the need to eliminate the possibility of plasmid-derived artifacts, by generating precise site-directed deletions of the metal transporter structural genes and using these strains as hosts for plasmids carrying *fepA* or *fhuA* genes of interest.

Ligand Internalization

The actions of TonB on siderophore receptors presumably promote conformational changes in them that allow the internalization of metal complexes bound within their vestibules. Because the N-domain completely fills the C-domain, leaving no discernible opening, gap, or pore through which FeEnt might pass, structural changes must occur during transport. The very high affinity of FepA for its substrates should not be forgotten during consideration of its transport mechanism. FeEnt binds to FepA with a subnanomolar K_d that translates into a dissociation half-life of over 1 min. This calculation conflicts with the experimentally observed 20-s turnover number of FepA and argues for protein conformational change as an essential part of the transport mechanism that (i) undermines the affinity of the siderophore-receptor binding interaction, (ii) originates or unveils a channel to the periplasm, and (iii) facilitates internalization of the metal complex from the surface loops through the pore.

The driving force in siderophore internalization is another relevant consideration. Because LGPs apparently bind only a single molecule at any time and transport against a concentration gradient, their uptake thermodynamics differ from those of general or specific porins, which transport by mass action. The perception of FepA as a gated porin does not imply that siderophore receptors function by facilitated diffusion; siderophore concentrations in the natural environment are too low to drive diffusion-mediated accumulation at a rate sufficient to support growth: their entry into the cell occurs by energy-dependent active transport. So the following question arises: once bound in the surface loops of FepA, what force drives FeEnt inward, into the channel that presumably forms? The architecture of the LGP may provide an answer. The elaborate loops of the β-barrel have the potential to close the pore vestibule on the exterior, which conceptually solves several kinetic, thermodynamic, and physiological problems. First, if a channel to the periplasm arises as the ligand-receptor binding interaction collapses, then loop closure prevents escape of the solute to the exterior. Second, such a mechanism eliminates the need for a driving force of transport: the ferric siderophore enters the periplasm by random diffusion and accumulates within by adsorption to its binding protein (FepB). Finally, surface loop closure during the opening of a large channel precludes the nonspecific entry of other potentially toxic large molecules (e.g., detergents [see below]).

In theory, the complete expulsion of the N-domain from the β-barrel, as in the ball-and-chain mechanism, considerably simplifies the uptake process. First, the alternative idea, conformational changes in the N-domain while it remains resident within the β-barrel that creates a new pore, seems contrary to some of the dogmas of biology: simplicity, efficiency, and economy. A transmembrane channel already exists, so why not use it rather than create another? Next, movement of the globular domain into the periplasm will create inwardly directed fluidic forces on the loops that may both alter their conformation (abrogating binding affinity) and pull the metal complex into the periplasm. In this way, energy that may function to expel the globular domain from the barrel directly translates into the driving force for ligand internalization. Finally, the crystal structures of FepA, FhuA, FecA, and BtuB illustrate that the N termini of these proteins fold into analogous forms with an ostensibly appropriate shape for insertion into, and expulsion from, the channel domain (Color Plate 13). Given the variation in primary structure among these polypeptides, their transformation into, or participation in, four distinct channels, each with comparable permeability but different transport specificity, seems much less likely. Thus, a variety of arguments support the notion of mechanistically important conformational dynamics involving the N- and C-domains of LGPs. However, unlike the relatively well-characterized adsorption reaction, experimental data for the ensuing ferric siderophore transport reaction are currently sparse or lacking. Siderophore receptors present a methodological challenge: delineation of the changes that occur in receptor protein tertiary structure during ligand passage through their normally closed channels. Nevertheless, spectroscopic systems permit direct observations of ligand passage through the OM bilayer. Especially in light of its enhanced sensitivity and instrumentation, site-directed fluorescence labeling is a preferred method. This procedure observes the quenching of fluors attached to genetically engineered sites in the target protein. The quenching effects originate from changes in the environment of the fluor during ligand binding and/or transport. The fluorescence system effectively monitored FeEnt uptake by live bacteria and determined its dependence on ligand concentration, pH, carbon sources, and temperature and its susceptibility to metabolic poisons. As expected, FepA was sensitive to depletion of the proton motive force, but it was noteworthy that unlike for vitamin B_{12} transport through BtuB, inhibitors of electron transport and phosphorylation also blocked FeEnt uptake. These data, along with the availability of this novel spectroscopic system that measures only OM transport, suggest that reevaluation of siderophore receptor transport energetics is timely and appropriate.

ANOTHER VIEW: THE IMPORTANCE OF SELECTIVE PERMEABILITY

To briefly paraphase results from a decade of work by Nikaido and colleagues, the structure of the OM acts as a barrier to the nonspecific penetration of large or hydrophobic molecules, in part because of the hydrophilicity of LPS on its surface and in part because of the architecture of the proteins that reside in the OM. Without doubt, the presence of the N-domain of LGP within their β-barrel channels ensures one of the overriding principles of bacterial cell envelope physiology, that of selective permeability.

Structure of General Porins

As a result of work in the laboratories of Nikaido, Schulz, and Rosenbusch, the properties of general porins are well defined. They contain 16-strand, amphiphilic β-barrels that traverse the OM bilayer and a series of hydrophilic, antigenic loops that protrude to the cell surface. General porins are trimers, and in the context of this trimeric structure, their surface loops form a large vestibule, akin to a funnel, on the cell surface. The general porin vestibule collects solutes and deflects them to three underlying, open, nonspecific, hydrophilic channels (one per monomer) that traverse the OM bilayer to the periplasm. General porin channels contain an unusual loop within their interior, which was initially discovered by genetic studies; this loop is called the transverse, eyelet, or L3 loop. L3 crosses into the pore and forms a constriction zone, narrowing the effective diameter from about 20 Å at the exterior to 10 Å at the constriction. This distinctive structural feature defines the functionality of OmpF by creating the diffusional limit (600 Da) of its open channels, permitting the entry of nutritionally important solutes (amino acids, purine and pyrimidine bases, sugars, and certain vitamins) but excluding potentially lethal bile salts from the gut.

Structure of Specific Porins

Other porin structures reiterate the importance of a transverse loop to selective permeability. Specific porins, like LamB, are similar to general porins in many ways (primarily because they contain an open transmembrane channel), but they differ from general porins because they manifest affinity for a particular substrate. The structures of LamB (whose substrate is maltose or maltodextrins) and ScrY (whose substrate is sucrose) are known. LamB is structurally comparable to the general porins, with two exceptions. Its singularity arises from the facts that its β-barrel contains 18 strands rather than 16 and that some of its surface loops are so large that they almost close the channel vestibule at the cell surface. Among the surface loops, L9 is of particular interest because it folds over each LamB monomer to form an "umbrella" that physically shields the underlying pore. On the other hand, like a general porin, LamB contains an amphiphilic β-barrel, a series of hydrophilic antigenic loops that surround the maltoporin trimer, and a transverse loop (L3) that restricts its channel. Thus, the transporter achieves specificity for sugars while maintaining the permeability barrier of the OM to large hydrophobic molecules.

N-Domains of Metal Transporters

The unexpected disposition of the N-domain of FepA within its transmembrane channel may primarily represent another adaptation to maintain the OM permeability barrier. Whereas the L3 (transverse) loops of general and specific porins fold inward to narrow the interior diameter of their channels, the N-domains of the LGP completely close their pores. Although this modified architecture of siderophore receptors suggests their distinctiveness among OM proteins, the design exemplifies further evolution on the same theme: restriction of channel permeability. In LGP, a novel globular assembly, the N-domain, fully regulates solute passage through the pore. This is necessary because a channel like that of OmpF, but large enough to accommodate the size of a metal complex like FeEnt, will also compromise cell viability in the host by permitting the entry of bile salt detergents through the OM. Thus, although the function of the FepA N-domain was perceived as being adapted to the ligand such that it forms or participates in a FeEnt-specific channel, this remains unsubstantiated. It is certain, though, that the overall architecture of FepA and its structural relatives fundamentally maintains the overall OM permeability barrier. The idea that LGP transport occurs by the ball-and-chain mechanism, in which the N-domain has little affinity for the metal complex bound in the vestibule but acts nonspecifically to achieve its internalization, is consistent with this important theme. In this mechanism, the huge LGP channels always remain closed to the exterior: in the absence of ligand, the N-domain fills their pores, and during its internalization, the surface loops close as the N-domain exits, nonspecifically pulling the metal complex into the periplasm.

ACKNOWLEDGMENTS

I thank S. M. Newton for scientific collaborations and helpful discussions, and I thank his colleagues at the University of Oklahoma, Raj Annamalai, Jennifer Allen, Matt Bauler, Zhenghua Cao, Xunqing Jiang, Bo Jin, Wallace Kaserer, Li Ma, Marjorie Montague, Marvin Payne, Zengbiao Qi, Daniel C. Scott, Yi Shao,

Padma Thulasiraman, Ron Williams, and Paul Warfel, for their experimental contributions.

This work was supported by OCAST grant 00072, NSF grant MCB9709418, and NIH grants GM53836 and 1P20RR1182.

SUGGESTED READING

Annamalai, R., B. Jin, Z. Cao, S. M. C. Newton, and P. E. Klebba. 2004. Recognition of ferric catecholates by FepA. *J. Bacteriol.* **186:**3578–3589.

Braun, M., H. Killmann, and V. Braun. 1999. The beta-barrel domain of FhuAΔ 5–160 is sufficient for TonB-dependent FhuA activities of *Escherichia coli*. *Mol. Microbiol.* **33:**1037–1049.

Buchanan, S. K., B. S. Smith, L. Venkatramani, D. Xia, L. Esser, M. Palnitkar, R. Chakraborty, D. van der Helm, and J. Deisenhofer. 1999. Crystal structure of the outer membrane active transporter FepA from *Escherichia coli*. *Nat. Struct. Biol.* **6:**56–63.

Cao, Z., P. Warfel, S. M. Newton, and P. E. Klebba. 2003. Spectroscopic observations of ferric enterobactin transport. *J. Biol. Chem.* **278:**1022–1028.

Chimento, D. P., A. K. Mohanty, R. J. Kadner, and M. C. Wiener. 2003. Substrate-induced transmembrane signaling in the cobalamin transporter BtuB. *Nat. Struct. Biol.* **10:**394–401.

Jiang, X., M. A. Payne, Z. Cao, S. B. Foster, J. B. Feix, S. M. Newton, and P. E. Klebba. 1997. Ligand-specific opening of a gated-porin channel in the outer membrane of living bacteria. *Science* **276:**1261–1264.

Klebba, P. E. 2003. Three paradoxes of ferric enterobactin uptake. *Front. Biosci.* **8:**1422–1436.

Liu, J., J. M. Rutz, P. E. Klebba, and J. B. Feix. 1994. A site-directed spin-labeling study of ligand-induced conformational change in the ferric enterobactin receptor, FepA. *Biochemistry* **33:**13274–13283.

Murphy, C. K., and P. E. Klebba. 1989. Export of FepA::PhoA fusion proteins to the outer membrane of *Escherichia coli* K-12. *J. Bacteriol.* **171:**5894–5900.

Newton, S. M., J. D. Igo, D. C. Scott, and P. E. Klebba. 1999. Effect of loop deletions on the binding and transport of ferric enterobactin by FepA. *Mol. Microbiol.* **32:**1153–1165.

Nikaido, H., and M. Vaara. 1985. Molecular basis of bacterial outer membrane permeability. *Microbiol Rev.* **49:**1–32.

Payne, M. A., J. D. Igo, Z. Cao, S. B. Foster, S. M. Newton, and P. E. Klebba. 1997. Biphasic binding kinetics between FepA and its ligands. *J. Biol. Chem.* **272:**21950–21955.

Scott, D. C. 2001. Mechanism of ferric enterobactin transport through FepA: a bacterial venus flytrap. Ph.D. thesis. University of Oklahoma, Norman.

Scott, D. C., Z. Cao, Z. Qi, M. Bauler, J. D. Igo, S. M. Newton, and P. E. Klebba. 2001. Exchangeability of N termini in the ligand-gated porins of *Escherichia coli*. *J. Biol. Chem.* **276:**13025–13033.

Ullmann, A., D. Perrin, F. Jacob, and J. Monod. 1965. Identification, by in vitro complementation and purification, of a peptide fraction of *Escherichia coli* β-galactosidase. *J. Mol. Biol.* **12:**918–923. (In French.)

Usher, K. C., E. Ozkan, K. H. Gardner, and J. Deisenhofer. 2001. The plug domain of FepA, a TonB-dependent transport protein from *Escherichia coli*, binds its siderophore in the absence of the transmembrane barrel domain. *Proc. Natl. Acad. Sci. USA* **98:**10676–10681.

Vakharia, H. L., and K. Postle. 2002. FepA with globular domain deletions lacks activity. *J. Bacteriol.* **184:**5508–5512.

FERRICHROME- AND CITRATE-MEDIATED IRON TRANSPORT

Volkmar Braun, Michael Braun, and Helmut Killmann

11

Bacteria have to cope with the extreme insolubility of Fe^{3+} under aerobic conditions at neutral pH. In equilibrium with the Fe^{3+}-hydroxy polymer, the free Fe^{3+} concentration is less than 10^{-9} M. Fe^{3+} can be solubilized by siderophores, low-molecular-weight iron chelators synthesized and secreted by bacteria and fungi (see chapter 1). Bacteria use siderophores produced by other microorganisms, as well as their own siderophores.

Ferrichrome was the first siderophore to be characterized. It was identified as a brownish red compound containing iron in a study of cytochromes of the fungus *Ustilago sphaerogena* in 1952. This hydroxamate compound is a cyclic hexapeptide with three contiguous glycine residues and three N^5-acetyl-L-N^5-hydroxyornithine residues (Fig. 1). In the iron transport-active ferrichrome derivatives ferricrocin and ferrichrysin, one and two glycine residues, respectively, are replaced by serine residues.

Ferrichrome is used as an iron source by fungi and by gram-positive and gram-negative bacteria. The first bacterial studies were performed with *Salmonella enterica* serovar Typhimurium. Growth promotion and transport of [^{3}H]ferrichrome were determined with a mutant that was defective in biosynthesis of the siderophore enterobactin, rendering the cells dependent on an added siderophore. In addition, the antibiotic albomycin, which at that time was known to be structurally similar to ferrichrome even though the exact structure was not determined until 1982, was used to select mutants that no longer use ferrichrome as an iron source. Twelve mutant types were selected on the basis of growth on various hydroxamate siderophores. The mutations (*sidA* to *sidM*, except *sidJ*) were mapped by conjugation; eight were cotransducible with *panC*, which is located near the *fhuACDB* genes of *Escherichia coli*. The *sid* genes of *S. enterica* are now designated *fhuACDB*.

Ferrichrome transport was initially investigated to learn about its role in microbial iron metabolism. In 1975 it was shown that *tonA* (now *fhuA*) mutants resistant to phage T5 do not transport radioactive [^{55}Fe^{3+}]ferrichrome and that ferrichrome inhibits the killing of cells by colicin M. Previously, it had been shown that phage T5 and purified colicin M bind to isolated and purified FhuA (TonA) and that FhuA is located in the outer membrane. By this means, ferrichrome transport was linked to

---

Volkmar Braun, Michael Braun, and Helmut Killmann, Mikrobiologie/Membranphysiologie, Universität Tübingen, D-72076 Tübingen, Germany.

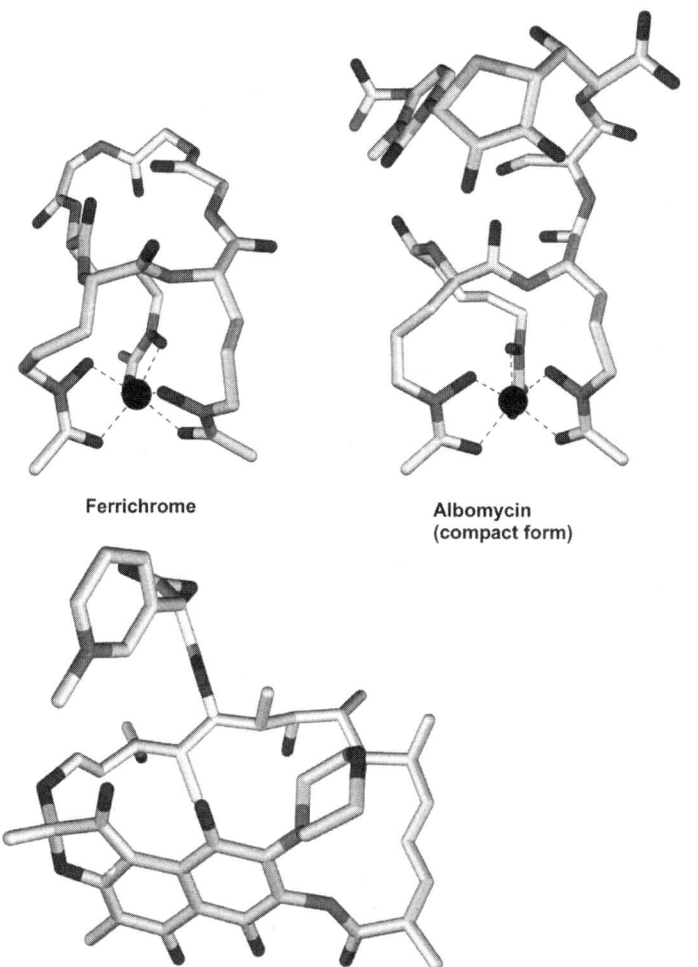

FIGURE 1 Structures of ferrichrome, albomycin, and rifamycin CGP 4832, as revealed by X-ray analysis of the FhuA cocrystals.

the *fhuA* gene, which was shown to encode the FhuA outer membrane protein. Ferrichrome also inhibited the binding of phage φ80 to cells. FhuA became a model protein for the elucidation of active, energy-consuming transport across the outer membrane of gram-negative bacteria, which was studied particularly in *E. coli* K-12.

The early phage geneticists noted a relationship between *tonA* and *tonB*. Mutations in either of the genes conferred resistance to phage T1, whereas *tonA* mutations alone (not *tonB* mutations) confer resistance to phage T5. It was then shown that in addition to a functional *tonB* gene, energy in the form of an energized cytoplasmic membrane is required for phage T1 and phage φ80 infections *tonB* mutants and energy-starved *tonB*$^+$ cells adsorb the phages reversibly but do not trigger DNA release from the phage heads and uptake of phage DNA into the cytoplasm. Energization of phage infection could proceed by the electron transport chain of the cytoplasmic membrane in ATPase-negative mutants or by ATP hydrolysis in heme-negative mutants. The energy inhibitors carbonyl cyanide-*m*-chlorophenylhydrazone (CCCP), 2,4-dinitrophenol (DNP), and azide inhibit the irreversible step in phage adsorption. The linkage between the energy requirement and TonB for phage infection led to the proposal that the role of TonB might involve coupling of the energized membrane state to phage infection. Since phage T1 host range mutants infect *tonB* mutants, DNA uptake across the cytoplasmic membrane does not require TonB. Ferrichrome inhibits the binding of phage T5 to *E. coli* cells, but a ferrichrome concentration 10-fold higher than that needed for complete inhibition of a *tonB* mutant is required to inhibit *tonB*$^+$ cells. Even a 1,000-fold-higher ferrichrome concentration inhibits phage T5 adsorption to the *tonB*$^+$ cells only by 40%. Since phage T5 infection does not require *tonB*, it was concluded that FhuA in energized TonB$^+$ cells assumes a conformation different from that in TonB$^-$ cells and that the altered FhuA conformation strongly reduces the interference of ferrichrome with phage T5 adsorption.

FERRICHROME IS TRANSPORTED BY THE MULTIFUNCTIONAL FhuA PROTEIN ACROSS THE OUTER MEMBRANE

The FhuA protein transports ferrichrome, the structurally related antibiotic albomycin, and the structurally unrelated antibiotic rifamycin CGP 4832 (a synthetic derivative of rifamycin) and serves as receptor for phages T1, T5, φ80, and UC-1 and for colicin M (a toxic protein) and microcin J25 (a toxic peptide). All functions require TonB activity except for infection by phage T5. The crystal structure of FhuA reveals a β-barrel composed of 22 antiparallel β-strands (residues 161 to 714 of the mature protein) and a globular domain (residues 1 to 160), termed the cork or plug, that inserts from the periplasmic side into the β-barrel and completely closes the β-barrel channel (see also chapter 4). Comparison of the crystal structures of FhuA loaded with ferrichrome and of unloaded FhuA (Fig. 2) reveals major long-range structural changes after ferrichrome binding. The switch helix (residues 21 to 32) is unwound, and residues E19, S20, and W22 are shifted 17 Å from their former α-carbon position. This movement exposes the TonB box of FhuA (residues 6 to 11) to the periplasm, where all of TonB except for the short N-terminal segment in the cytoplasmic membrane resides. The TonB box is not seen in the crystal structure and therefore is probably flexible. The mutations I9P (isoleucine replaced by proline) and V11D in the TonB box inactivate FhuA but do not affect the TonB-independent infection by phage T5. The mutation Q160K in TonB partially restores the transport of ferrichrome by FhuA(I9P), which suggests interaction of the Q160 region with the TonB box of FhuA.

The crystal structures of FhuA do not reveal movement of surface loops after loading with ferrichrome; this is in contrast to the strong movement of surface loops observed in FecA after binding of (Fe^{3+} citrate)$_2$ (see chapter 4). It is possible that crystal forces favor the open conformation. Evidence for loop movement was obtained with a fluorescence label linked to residue 336 of loop 4; upon binding of ferrichrome, fluorescence quenching is observed. Additional evidence for conformational changes in surface loops comes from binding of phages T1 and φ80 to loop 4, which occurs only irreversibly, accompanied by DNA release from the phage heads, in energized TonB$^+$ cells. This finding implies that loop 4 assumes a conformation that differs from the unenergized conformation, to which the phages bind reversibly without triggering a DNA release. It also means that TonB interacts with FhuA and triggers a conformational change in the absence of ferrichrome. The activity of FhuA involves strong long-range structural changes that take place throughout the molecule from the periplasm, where TonB interacts with FhuA, to the most prominent loop 4 at the cell surface and in the opposite direction from the ferrichrome binding site 20 Å above the outer lipid boundary of the outer membrane to the periplasm, where the switch helix unwinds and interaction with TonB is facilitated.

Transport across the outer membrane is energized by the proton motive force of the cytoplasmic membrane. It is thought that FhuA responds to the proton gradient across the

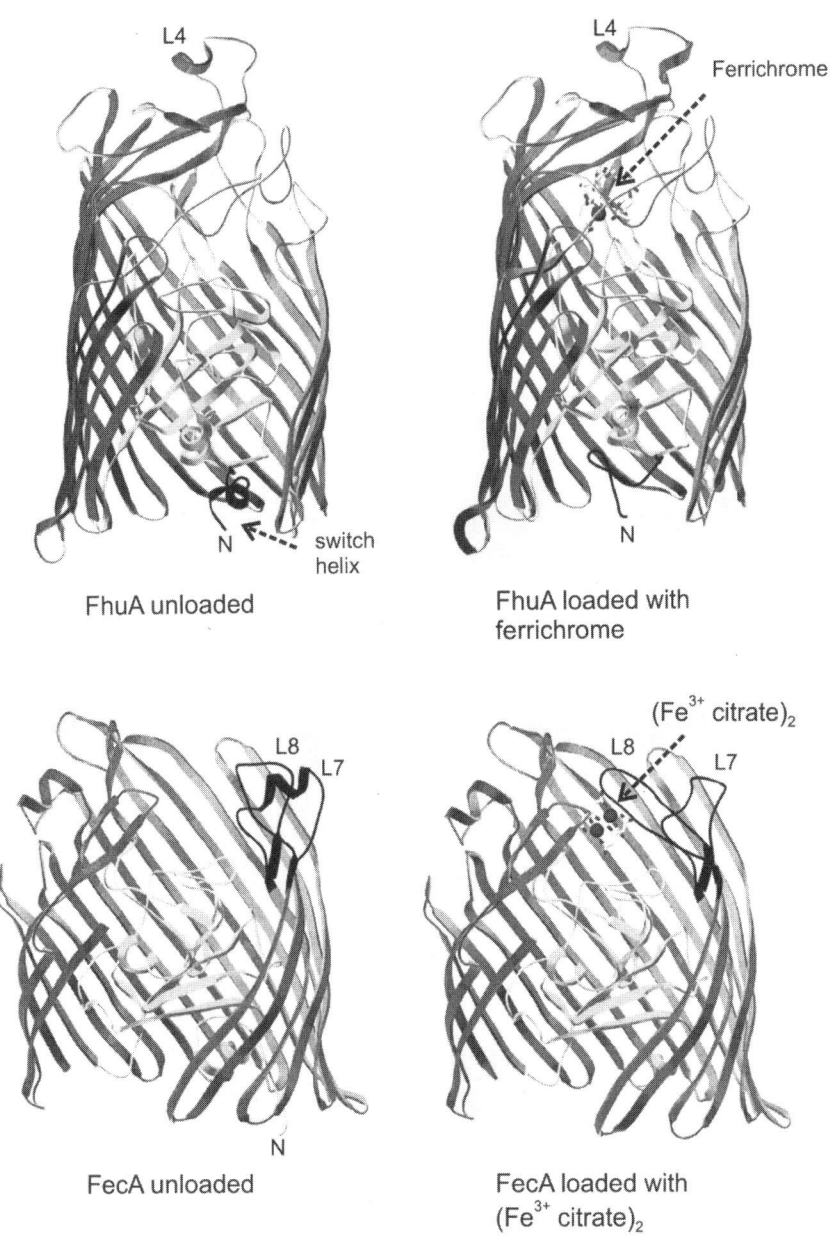

FIGURE 2 Crystal structures of the FhuA and FecA outer membrane proteins unloaded and loaded with their substrates. A portion of the β-strands (dark grey) was deleted to improve the view of the globular central domain (cork; light grey). FhuA1–18 and FecA1–79 are not seen in the crystal structure, which suggests a flexible structure. Structurally important regions and the amino termini (N) are indicated.

cytoplasmic membrane. Energy stored as conformational energy in TonB is transmitted to the outer membrane transport proteins. TonB has to trigger not only the release of ferrichrome from its binding site but also the movement of the cork so that a continuous channel is formed between the outer cavity and the periplasmic cavity for diffusion of ferrichrome, the antibiotics albomycin and rifamycin CGP 4832, and microcin J25 into the periplasm. If colicin M (molecular weight, 27,000) also crosses the outer membrane through the FhuA channel, the entire cork must be expelled into the periplasm. The same holds true if the DNA of phages T1, T5, ϕ80, and UC-1 passes through the FhuA β-barrel channel. Since a few colicin M molecules suffice to kill a cell and a single phage DNA molecule is sufficient to infect a cell, inactivation of FhuA by expelling the cork would probably not disrupt the outer membrane integrity. However, the cork is firmly incorporated in the β-barrel through seven salt bridges and approximately 60 hydrogen bonds. A rather large energy barrier would have to be overcome to release the cork from the β-barrel, but the energy barrier could be lowered if the hydrogen bonds between the amino acids are replaced stepwise by hydrogen bonds formed with water molecules. Even if this is a rather slow process, it may be sufficient for FhuA transport activity when one takes into account the low FhuA transport rate of less than 10 ferrichrome molecules per FhuA molecule per min. This is more than enough to satisfy the iron requirement of a cell ($\sim 10^5$ iron ions per 30-min generation time) since there are approximately 10^4 FhuA molecules per cell under iron-limiting conditions. Unless release of the cork is disproved experimentally, this possibility has to be considered.

Excision of the Cork Opens a Channel in FhuA

Removal of the cork should convert FhuA into an open channel through which substrates flow into the periplasm. This prediction was examined by excision of residues 5 to 160, which encompass the entire cork. A His-tagged derivative of FhuAΔ5–160, purified on Ni^{2+}-nitrilotriacetate (NTA) agarose, increases the conductance of artificial lipid bilayer membranes for KCl. The conductance increase does not occur in discrete steps, which shows that the channels formed are not stable. The conductance tracings indicate rapid and frequent opening and closing of channels, with a most frequent conductance step of approximately 0.5 nS in 1 M KCl. FhuAΔ5–160 exhibits the same properties when isolated from gels after solubilization in 2.7% sodium dodecyl sulfate (SDS) at 50°C and purification by SDS-polyacrylamide gel electrophoresis (PAGE). Cells containing FhuAΔ5–160 display an increased sensitivity to antibiotics for which the outer membrane forms a permeability barrier, such as novobiocin, erythromycin, rifamycin, and vancomycin. FhuAΔ5–160 confers to a *lamB* deletion strain lacking maltoporin an enhanced growth on maltotriose and growth on maltotetraose and maltopentaose, which diffuse through the outer membrane via FhuAΔ5–160.

The Activity of FhuAΔ5–160 Is Restored by Coexpression of Inactive FhuA and FhuA Fragments

FhuAΔ5–160 synthesized in a FhuA⁻ strain restores all FhuA activities except sensitivity to microcin J25. When expressed at a level similar to that of wild-type FhuA, FhuAΔ5–160 confers wild-type FhuA sensitivities to the phages and colicin M, and the ferrichrome transport rate is at least 30% of that of wild-type FhuA. Complementation of FhuAΔ5–160 by the plasmid-encoded FhuA1–160 fragment results in a high ferrichrome transport rate, which indicates that the N-terminal fragment reconstitutes wild-type FhuA. This finding opens the possibility that the FhuA⁻ strains used for measuring the FhuAΔ5–160 activities provide the cork fragment that is missing in FhuAΔ5–160 and complement FhuAΔ5–160 to wild-type FhuA. This possibility was apparently ruled out by using the *E. coli* H1857 *fhuA* mutant, which in a previous study was characterized as a complete *fhuA* deletion. The idea of complementation as the cause of the activity of a corkless outer membrane transporter in another system was put forward by Vakharia and Postle to explain their finding that FepAΔ17–150 does not confer sensitivity to colcin B in one *fepA* mutant, in contrast to the same FepAΔ17–150 protein conferring colicin B sensitivity to another *fepA* mutant. Intentional complementation of FhuAΔ5–160 by FhuA1–160 and the findings with FepA prompted us to reexamine the previously used *fhuABCD* deletion mutant H1857. The *fhuA* gene in the deletion mutant was shown not to be entirely deleted but to contain a frameshift mutation after residue 357 and thus to be able to synthesize a fragment containing the cork. A newly constructed complete *fhuA* deletion mutant exhibits no FhuA activity when it synthesizes plasmid-encoded FhuAΔ5–160.

Complementation of FhuAΔ5–160 by FhuA1–160 raises a few questions. At which stage (during ribosomal biosynthesis in the cytoplasm, translocation across the cytoplasmic membrane, or insertion into the outer membrane) is FhuA1–160 incorporated into FhuAΔ5–160? Is the FhuA fragment synthesized by *E. coli* H1857 inserted as such or only after cleavage to the size of the cork? This latter question is even more pertinent for the complementation of FhuAΔ5–160 by full-size but inactive mutant FhuA. In the latter case, the cork of FhuA might occasionally insert into cosynthesized FhuAΔ5–160 instead of into the β-barrel to which it is covalently linked. Since the link between the cork and the β-barrel is exposed to the periplasm, this kind of mechanism cannot be excluded, but it is sterically difficult to envisage. We prefer the idea that the region between the globular cork and the β-barrel represents a transition zone between two completely different structures. Since zones between such domains are frequently subject to proteolysis, we therefore suggest that FhuA and C-terminally truncated FhuA fragments are proteolytically cleaved in the periplasm, resulting in a free cork domain in the periplasm that can insert into FhuAΔ5–160.

Transport of Antibiotics by FhuA

Albomycin is a structural analogue of ferrichrome (Fig. 1). The Fe^{3+} binding center corresponds to ferrichrome and binds to the same site in FhuA as ferrichrome does. The antibiotically active thioribosyl moiety, which inhibits serine-tRNA synthetase, can assume two different conformations when bound to FhuA: a compact formation (Fig. 1) and an extended conformation. Fixation of the thioribosyl moiety to FhuA does not prevent albomycin transport, but the transport rate is lower than that of ferrichrome. Albomycin is transported across the cytoplasmic membrane by the FhuCDB proteins and serves as a suitable tool for the isolation of albomycin-resistant transport mutants. Since albomycin enters the cytoplasm by active transport and not by diffusion like most antibiotics, its MIC is very low, approximately 100-fold lower than that of ampicillin. In the cytoplasm, the thioribosyl moiety is cleaved from the iron carrier by peptidases, a necessary step for antibiotic action.

The rifamycin derivative CGP 4832 is transported by FhuA across the outer membrane but not by FhuCDB across the cytoplasmic membrane. Accumulation in the periplasm is sufficient to increase the cytoplasmic concentration, resulting in a 200-fold-lower inhibitory concentration than that of unmodified rifamycin, which is not actively transported across the outer membrane. The amazing recognition of CGP 4832 by FhuA is revealed by the crystal structure of FhuA loaded with CGP 4832. Although there is no structural resemblance between CGP 4832 and ferrichrome (Fig. 1), CGP 4832 is bound, with a single exception, by the same amino acid side chains that bind ferrichrome. Eight additional amino acid side chains also bind CGP 4832, but this extensive fixation does not prevent transport. CGP 4832 transport demonstrates that binding of compounds in the FhuA binding site for ferrichrome is sufficient for subsequent transport. The assumed structural transition of FhuA in response to the proton motive force of the cytoplasmic membrane changes the geometry of the amino acid side chains of the binding center such that ferrichrome, albomycin, and CGP 4832 are released. Subsequent diffusion through the open β-barrel channel does not seem to be dependent on specific interactions with amino acid side chains of the channel.

TRANSPORT OF FERRICHROME ACROSS THE CYTOPLASMIC MEMBRANE

Ferrichrome is transported across the cytoplasmic membrane by the periplasmic binding-protein-dependent transport system that represents a subfamily of the ABC transporters (see chapter 8). The bacterial ABC importers are composed of a periplasmic binding protein, one or two different (homodimer, heterodimer, or pseudoheterodimer) integral membrane proteins, and one or two ATPases that face the cytoplasm and supply the systems with energy. The almost identical design of all ABC transporters points to a common origin from bacteria to humans.

The FhuCDB proteins are encoded by the *fhuCDB* genes located downstream of the *fhuA* gene and transcribed in this order. The Fhu proteins transport not only ferrichrome across the cytoplasmic membrane but also a number of other ferric siderophores of the hydroxamate type, including ferric coprogen (transported across the outer membrane by FhuE) and ferric aerobactin (transported across the outer membrane by Iut). The *fhuC* and *fhuD* genes and the *fhuD* and *fhuB* genes are translationally coupled with overlapping stop-start codons, which guarantees the proper stoichiometry of the transport proteins.

Ferrichrome transport systems are widespread among gram-negative bacteria, e.g., *E. coli*, *Salmonella* spp., *Pantoea agglomerans*, and *Vibrio cholerae*, and gram-positive bacteria, e.g., *Bacillus subtilis* and *Staphylococcus aureus*. Because of the lack of a periplasm in gram-positive bacteria, the binding protein for ferrichrome transport is anchored to the cytoplasmic membrane by a lipid of the murein lipoprotein type. The FhuCDB system of *E. coli* K-12 is one of the best-characterized iron ABC transporters.

The Periplasmic Binding Protein FhuD

SUBSTRATE BINDING

The siderophore binding protein FhuD (266 amino acids aa, 29.7 kDa) is synthesized as a precursor with a signal sequence and is then processed and exported into the periplasmic space. Binding of iron (Fe^{3+}) hydroxamates to the mature FhuD protein has been studied by several types of experiments.

(i) Accumulation of [$^{55}Fe^{3+}$] ferrichrome in the periplasm of intact cells of a FhuD-overproducing strain, which, due to a mutation in the integral membrane protein FhuB, is unable to translocate the substrate into the cytoplasm, has been shown. A newly developed three-layer oil technique that separates cells from the assay medium according to their density greatly facilitates the determination of Fe^{3+}-siderophore transport into the periplasm. The routinely used washing of cells on membrane filters with 0.1 M LiCl removes ferrichrome from the periplasm but retains ferrichrome bound to FhuA.

(ii) Radiolabeled FhuD contained in the periplasmic fraction is protected against proteolytic degradation by proteinase K and trypsin only in the presence of ferric hydroxamates that are transported by FhuCDB.

(iii) A purified FhuD derivative lacking the signal sequence can reconstitute the ferrichrome transport of a *fhuD* mutant after transfer into the periplasm by osmotic shock treatment. It binds ferric hydroxamates in vitro, as shown by protease protection experiments. The dissociation constants, estimated from the concentration-dependent decrease in the intrinsic fluorescence intensity of His-tagged FhuD, are 0.3 μM for ferric coprogen, 0.4 μM for ferric aerobactin, 1 μM for ferrichrome, and 5.4 μM for albomycin. Ferrichrome A, ferrioxamine B, and ferrioxamine E, which are taken up poorly via the Fhu system, have dissociation constants of 79, 36, and 42 μM, respectively. The FhuD W68L mutant transports coprogen, but it transports ferrichrome only poorly; accordingly, the binding constants are 2.2 μM for coprogen and 156 μM for ferrichrome. The binding constants largely reflect the transport rates and indicate that the transport specificity for the various ferric hydroxamates is strongly determined by FhuD.

THE STRUCTURE OF FhuD REVEALS A NOVEL TYPE OF FOLD

The crystal structure of FhuD differs from the many known crystal structures of periplasmic binding proteins. The N- and C-terminal domains are each built of five β-sheets sandwiched between layers of α-helices. The two domains are connected via a single kinked α-helix that runs the remaining length of the protein. The domain interface in FhuD is predominantly hydrophobic, which suggests that binding and release of the siderophores are not accompanied by large-scale opening and closing of the binding site, and therefore differs from most binding proteins, which act as a Venus flytrap by closing after binding of the substrates and opening after release of the substrates. The ligand binding site lies in a shallow groove between the N- and C-terminal domains of the bilobate protein. Binding of gallichrome, a structural analogue of ferrichrome, to FhuD is mediated by hydrophilic and hydrophobic interactions and involves only three direct hydrogen bonds. The binding site is exposed at the surface of the protein.

Binding of gallichrome to FhuD differs from binding of ferrichrome to FhuA, in which 10 individual amino acid residues located deep inside the protein participate. Binding of ferric coprogen, ferric Desferal, and albomycin occurs via the iron coordination centers of the siderophores, with reorientation of amino acid side chains to accommodate the various structures. The antibiotically active thioribosyl moiety of albomycin is exposed to the water phase, is not seen in the crystal structure, and is apparently flexible. In FhuA, it is fixed by four residues inside the protein and assumes two different conformations.

The FhuB Integral Membrane Protein

FhuB was previously considered to be unique among the integral membrane proteins in that it is about twice the size (659 amino acids [aa] 70 kDa) of integral membrane proteins of other ABC transporters. Recently, FhuB-like integral membrane proteins have been identified in *V. cholerae*, *Rhizobium leguminosarum*, and *Rhodobacter capsulatus*. FhuB of *E. coli* consists of two homologous domains connected by a linker region of 21 aa residues. The linker can be cleaved, and the two halves assemble to form an active transporter. Both halves, FhuB[N] and FhuB[C], are essential for transport, no activity is observed with single domains.

Several areas of striking sequence similarity are found in the primary structures of hydrophobic components of ABC transporters. One of the conserved regions includes an invariant glycine residue at a distance of about 100 aa from the C terminus. The conserved Gly corresponds to a conserved Gly contained in the "E AA --- G --------- I - LP" motif, also designated the EAA motif; however, EAA reads GLA in FhuB[N] and GMA in FhuB[C]. The G226E mutation in FhuB[N] and the G559V and G559E mutations in FhuB[C] strongly reduce ferrichrome transport activity. In two other transport-inactive mutant proteins, FhuB (P60L) and FhuB (G426R), the mutated half, still covalently bound to the wild-type half of the protein, can be replaced by the separately synthesized wild-type half, resulting in a transport-competent protein. Apparently, the inactive portion of the protein is displaced by the active wild-type portion within the cytoplasmic membrane.

The topology of FhuB differs from the predicted transmembrane proteins of other ABC transporters in that each half consists of 10 membrane-spanning regions. The transmembrane topology model of FhuB is derived from genetically constructed FhuB–β-lactamase hybrid proteins in which N-proximal segments of FhuB are fused to the BlaM β-lactamase devoid of its signal sequence. Cells become resistant to ampicillin if the fusion site is located in a periplasmic site of FhuB. Multiple sequence alignment of 33 transmembrane ABC transporter proteins was also used to construct the model.

The 20-transmembrane-helix model of FhuB was recently confirmed by the crystal

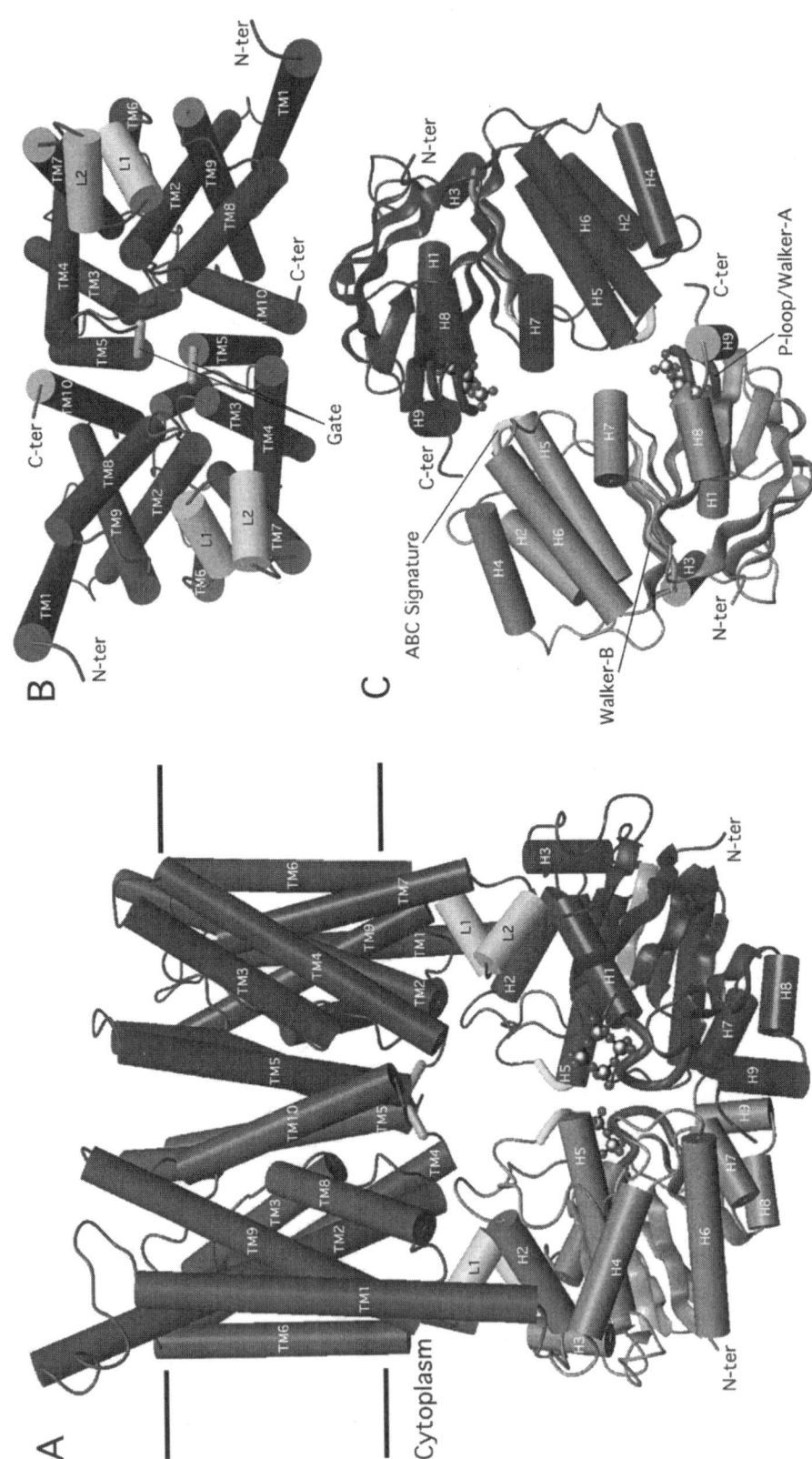

FIGURE 3 Structure of the vitamin B$_{12}$ transporter. The complete transporter is assembled from two membrane-spanning BtuC subunits and two ABC cassette BtuD subunits. Helices are drawn as cylinders and are labeled consecutively for each subunit, and strands are shown as ribbons. The amino and carboxy termini are indicated (N-ter and C-ter, respectively). (A) Side view of the full transporter. (B) View onto the cytoplasmic face of BtuC. (C) Bottom view onto the ABC cassette BtuD. Reprinted from Locher et al. (2002) with permission from the publisher.

structure of the BtuCD transporter for vitamin B$_{12}$ (Fig. 3). The two identical membrane-spanning subunits of BtuC are in close contact with each other and together provide 20 transmembrane helices grouped around a translocation pathway that is closed to the cytoplasm. Alignment of the sequences of FhuB[N] and FhuB[C] with that of BtuC, taking into account the proposed transmembrane model of FhuB, reveals a remarkable degree of correspondence (Fig. 4). Since the mechanism of vitamin B$_{12}$ transport is similar to that of the ferric siderophore transport, the BtuC and FhuB transmembrane arrangement is probably representative of the structure of the integral membrane proteins of ferric siderophores and heme transporters. They form a subfamily among the bacterial ABC importers.

The interface between the two BtuC subunits is formed by transmembrane helix 5 (TM5) and TM10. The four TM5/TM10 helices are assembled such that they form a cavity that spans two-thirds of the lipid bilayer, opens to the periplasmic space, and is large enough to accommodate vitamin B$_{12}$ (Fig. 3). The turn between TM4 and TM5 is proposed to form the gate at the center of the transporter. The large periplasmic loop between TM5 and TM6

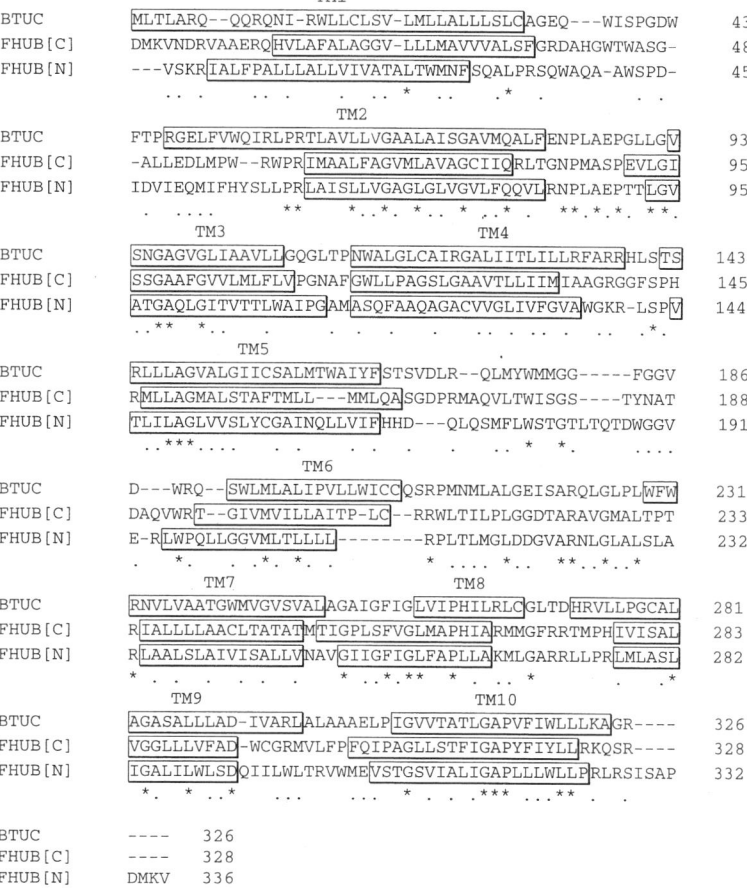

FIGURE 4 Sequence homology between the N-terminal half (FhuB[N]) and the C-terminal half (FhuB[C]) of FhuB and the BtuC protein. The transmembrane regions of BtuC (Fig. 3) are indicated and compared with the predicted transmembrane arrangement of FhuB.

is poorly resolved and is thought to serve as the binding site for the periplasmic binding protein. The cytoplasmic loop between TM6 and TM7 of each BtuC subunit folds into two short helical stretches (called L-loops) connected by a sharp bend. Both helices make extensive contacts with the ATPase BtuD. The L-loops represent the EAA-loop, which has been proposed to form the interface to the ATPases of the histidine and maltose ABC transporters.

The FhuC ATPase

The FhuC amino acid sequence (265 aa, 29 kDa) contains two highly conserved sequences typical of ATPases, the so-called Walker A and Walker B motifs. ATP hydrolysis with an isolated FhuC protein was demonstrated. Amino acid replacements K50E in the A motif and D172N result in FhuC derivatives that no longer transport ferrichrome and albomycin.

BtuD in a cocrystal with BtuC (Fig. 3) exhibits a polypeptide fold similar to that in the previously determined crystal structures of HisP and MalK. It can therefore be assumed that the FhuC protein has a similar fold and consists of a central six-strand β-sheet surrounded by nine α-helices and a peripheral, three-strand β-sheet. According to this model, the side chains making specific contacts to the L-loops of FhuB are located in helices h2 and h3, which surround the Q-loop of FhuC; the Q-loop is a conserved region among the ATPases. The proposed contacts between the ATPase monomers of HisP and MalK in the isolated ATPases differ from the functional relevant contacts in the BtuCD complex; this illustrates the limitation of conclusions drawn from crystal structure determinations.

INTERACTION OF THE COMPONENTS INVOLVED IN FERRICHROME UPTAKE

Interaction of the Binding Protein FhuD with the Outer Membrane Receptor FhuA

Solubilized FhuA is not retained by His-tagged FhuD bound to a Ni^{2+}-NTA-agarose column, and synthetic FhuD peptides do not bind to isolated FhuA; these results suggest that there is no physical interaction between FhuA and FhuD. Ferrichrome is translocated across the outer membrane independently of the following FhuCDB-mediated transport across the cytoplasmic membrane.

Interaction of the Binding Protein FhuD with the Integral Membrane Protein FhuB

Substrate-loaded FhuD delivers ferric hydroxamates to the FhuB transport protein in the cytoplasmic membrane. Several experimental approaches indicate a physical interaction of FhuD with FhuB: (i) in spheroplasts, FhuD protects radioactively labeled overproduced FhuB from being degraded by trypsin and proteinase K; (ii) purified His-tagged FhuD added to spheroplasts can be chemically cross-linked to overproduced radiolabeled FhuB, resulting in a product that corresponds in size to the FhuB-FhuD dimer in an SDS-polyacrylamide gel; and (iii) synthetic peptides of 10 and 20 aa, identical in sequence to FhuB regions, bind to purified FhuD in a modified peptide enzyme linked immunosorbent assay and BIAcore assay and inhibit in vivo ferrichrome transport. For the transport experiments, the peptides are brought into the periplasm by diffusion through channels in the outer membrane formed by the FhuA deletion mutant FhuAΔ322–355. One of the FhuB peptides, SMFLWSTGTL (Fig. 4), that interacts with FhuD covers the TM5/TM6 interhelix loop, which is located in the periplasm and is only partially ordered in the electron density map. It is highly likely that this peptide is contained in a periplasmic loop of FhuB that interacts with FhuD. Interestingly, the binding experiments reveal interaction of only the FhuB[N] but not the FhuB[C] TM5/TM6 interhelix loop with FhuD. Another FhuB peptide that strongly interacts with FhuD in the enzyme-linked immunosorbent assay and the BiaCore assay and that inhibits ferrichrome transport is located in TM6 and in the TM6/TM7 interhelix loop, which contacts the FhuC ATPase. The interacting region starts with the peptide

VILLAITPLC, contained in TM6, and ends with the peptide RWLTILPLGG, contained in the interhelix region (Fig. 4). Since a part of this region is exposed to the cytoplasm, it is difficult to envisage the binding of FhuD unless a major structural rearrangement in FhuB occurs after the interaction of FhuD with FhuB. This rearrangement would have to permit FhuD to insert to a certain extent into the FhuB channel so that it approaches the interaction site between FhuB and FhuC. In such a model, FhuD could trigger ATP hydrolysis by direct interaction with FhuC or through a short FhuB peptide that connects the FhuD binding site to the FhuC binding site on FhuB. However, it cannot be excluded that the binding assays gave a false-positive result, even though binding is highly specific. Random PCR mutagenesis resulted in three single FhuB mutants, two of which are located in this region [FhuB(T527P) and FhuB(L548F) of Fig. 4], with a ferrichrome transport rate of less than 50% that of the wild type.

A hypothetical model was proposed for vitamin B_{12} transport. In the model, substrate-loaded BtuF (FhuD) binds to BtuC (FhuB), which forms an open channel toward the periplasm and a closed channel toward the cytoplasm. On binding of ATP, the two BtuD (FhuC) ATPase molecules move toward each other, and, like a clothespin, the channel closes to the periplasm and opens to the cytoplasm. During this rearrangement, the substrate is released from the periplasmic binding protein and moves through the BtuC channel into the cytoplasm. Then the transporter returns to its resting state through the release of ADP and inorganic phosphate, dissociation of the binding protein, closing of the gate, and reorientation of the ABC cassettes to their original position.

Interaction of the Integral Membrane Protein FhuB with the ATPase FhuC

Interaction of FhuC with FhuB is demonstrated by dominant negative effects of plasmid-encoded FhuC derivatives that have single-amino-acid replacements in the putative ATP binding domains (K50E and D172N) that interfere with ferrichrome transport energized by chromosomally encoded wild-type FhuC. Furthermore, immunoelectron microscopy with anti-FhuC antibodies shows a FhuB-mediated association of FhuC with the cytoplasmic membrane.

Overexpression of plasmid-encoded FhuB derivatives mutated in the conserved glycine residues at position 226 and 559 in the L-loop involved in the interaction with FhuC inhibit the transport of ferrichrome of a $FhuB^+$ strain. Presumably, the FhuB derivatives compete with wild-type FhuB for interaction with FhuC.

CITRATE-MEDIATED Fe^{3+} TRANSPORT

The citrate-dependent iron transport system was first recognized in mutants of *E. coli* K-12 that could not synthesize enterobactin, then called enterochelin. Citrate stimulates growth during iron deficiency by replacing enterobactin as an iron chelator. Growth on 1 mM citrate induces the synthesis of a citrate-dependent iron transport system. A mutant that does not show a growth response to citrate does not transport iron supplied as ferric citrate. Appearance of an outer membrane protein (FecA) induced by ferric citrate and lack of this protein in transport-negative mutants paved the way for the molecular characterization of the Fec transport system. A citrate-dependent iron transport system was identified in *E. coli* K-12, *E. coli* B, *Shigella flexneri* 2a, and *Aerobacter aerogenes*. All *E.coli* and *Klebsiella pneumoniae* strains isolated from animals with bovine mastitis express FecA, as revealed by anti-FecA antibodies, and FecA synthesis is induced during growth in the presence of citrate. Homologs of the *E. coli* ferric citrate transport genes (*fec* genes) are found in a number of bacterial genomes, but the entire set of the seven *fec* genes is incomplete. Most, if not all, of the *fec* homologs probably encode transport proteins for iron and iron complexes other than ferric citrate. This also applies for the cases where iron limitation is overcome by addition of citrate, since iron might be solubilized by citrate but not transported as ferric citrate. Certain *E. coli* strains contain a plasmid-encoded citrate transport system that has no resemblance to the ferric citrate transport system.

The complex solution chemistry of iron citrate did not allow the chemical entity that is transported to be deciphered until the crystal structure of the FecA protein revealed (Fe^{3+}-citrate)$_2$ bound to FecA (Fig. 5). (Fe^{3+}-citrate)$_2$ binds to 10 residues of FecA located in a cavity that lies well above the outer boundary of the outer membrane lipid bilayer. Binding of (Fe^{3+}-citrate)$_2$ induces strong long-range structural transitions in FecA. Surface loops 7 and 8 are moved 11 and 15 Å, respectively, and cover the entry of the surface cavity (Fig. 2), preventing the escape of (Fe^{3+}-citrate)$_2$ back to the external milieu. In the FecA region exposed to the periplasm, a short helix is unwound; this might facilitate the binding of the TonB box of FecA to TonB by discrimination between citrate-loaded and unloaded FecA. Unlike most other outer membrane transport proteins, the TonB box DALTV is not located close to the N-terminal end of mature FecA but is found at residues 80 to 85. The point mutants DAPTV, DALTG, DALTR, DANTV, and GTNTV abolish FecA transport activity, which indicates the importance of the TonB box for interaction with TonB and for transport activity. This conclusion is supported by the in vivo formation of disulfide bridges between cysteine residues introduced into the TonB box and cysteine residues introduced into region 160 of TonB. Structural changes induced in FecA by TonB are proposed to exert two effects: (i) (Fe^{3+}-citrate)$_2$ is released from its binding sites, and (ii) the globular domain (residues 80 to 221) moves to open the channel of the β-barrel so that (Fe^{3+}-citrate)$_2$ can diffuse into the periplasm. Facilitated diffusion could occur along amino acid side chains that line the inside of the β-barrel. Once inside the periplasm, (Fe^{3+}-citrate)$_2$ binds to the FecB protein located in the periplasm. Isolated FecB is protected by ferric citrate and also by iron from degradation by proteinase K, which makes the ligand bound to FecB uncertain. It could mean that on FecB, (Fe^{3+}-citrate)$_2$ dissociates into Fe^{3+} and citrate and only iron is taken up into the cytoplasm; this is consistent with the observation that 10 times more radioactive $^{55}Fe^{3+}$ than [^{14}C]citrate is taken up into the cytoplasm. The amount of ^{14}C label associated with the cells is not influenced by growth of cells in the presence of citrate, but $^{55}Fe^{3+}$ uptake is induced by citrate.

The FecB protein delivers iron to the two very hydrophobic proteins FecB and FecC in the cytoplasmic membrane, which, together with the FecE protein, form an ABC transporter (Fig. 6). FecE contains the two Walker motifs typical for ATP binding proteins. It can be specifically photoaffinity labeled with 8-azido-[α-^{32}P] ATP, which supports the concept that FecE is an ATPase and ATP is the energy source for transport of iron across the cytoplasmic membrane.

INDUCTION OF THE FERRIC CITRATE TRANSPORT SYSTEM BY A TRANSMEMBRANE SIGNALING DEVICE

Induction by Ferric Citrate

The *fecABCDE* transport genes are transcribed in this order and form an operon with the promoter upstream of *fecA* (Fig. 6). Transcripts of the *fec* transport operon cover mainly *fecA* and *fecIR*; the level of *fecA* mRNA increases during iron deprivation in the presence of ferric cit-

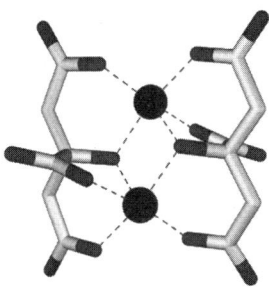

FIGURE 5 Crystal structure of dinuclear ferric citrate as it is bound to the FecA protein.

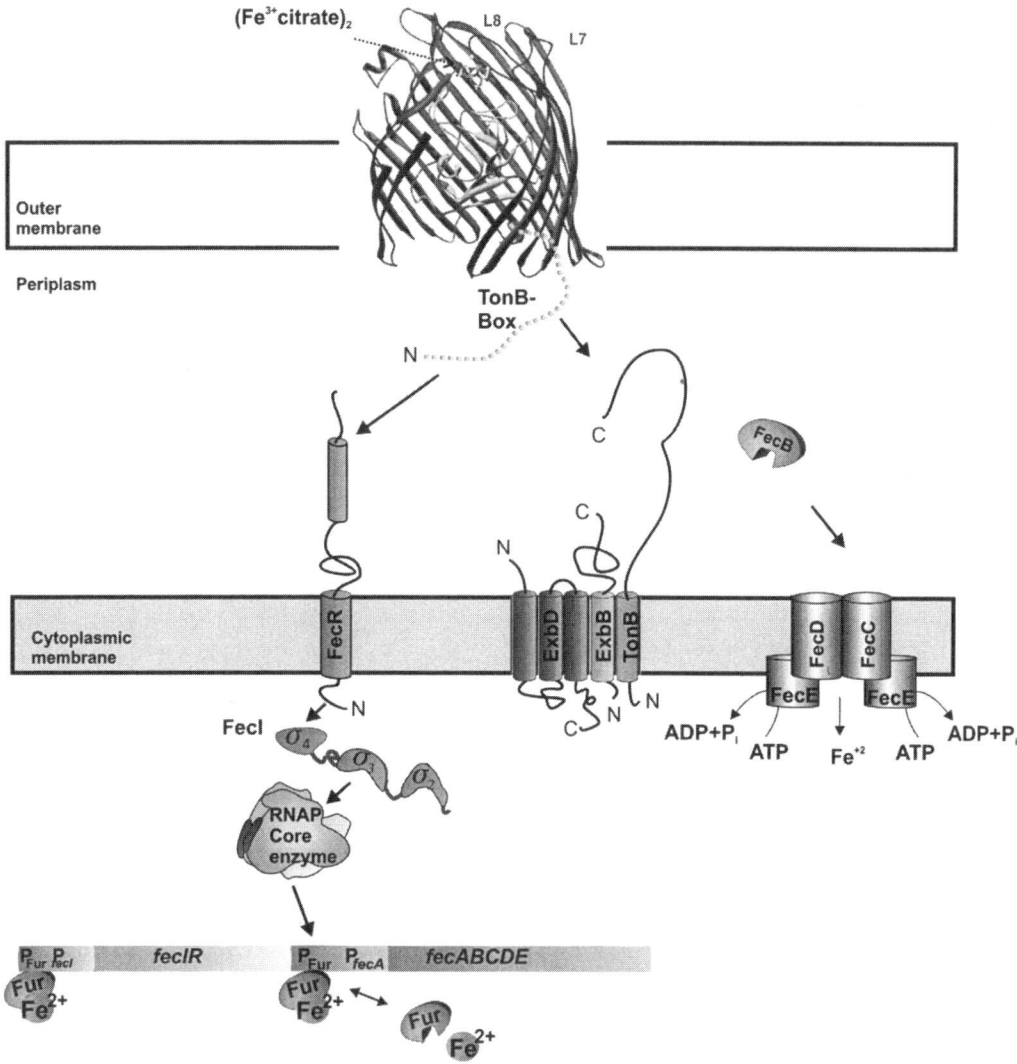

FIGURE 6 Subcellular location, transmembrane arrangement, and interaction of the FecIRABCDE proteins. $(Fe^{3+}\text{-citrate})_2$ binds to FecA in the outer membrane and elicits a signal that is transmitted by FecR across the cytoplasmic membrane into the cytoplasm, where it activates the FecI σ factor. FecA also transports ferric citrate into the periplasm. Transport and signal transduction require energy from the proton motive force of the cytoplasmic membrane, mediated by the TonB, ExbB, and ExbD proteins. The N-proximal segment of FecA interacts with TonB and with FecR. FecI binds to the β' subunit of the RNA polymerase, which binds to the promoter of the *fecABCDE* operon. The Fur protein loaded with Fe^{2+} represses the transcription of *fecIR* and *fecABCDE*.

rate, and the level of *fecIR* mRNA increases only during iron limitation. A hairpin structure downstream of *fecA* strongly reduces transcription of the downstream *fecBCDE* genes. The low transcription level of the *fecBCDE* genes requires reverse transcription-PCR to detect transcripts. In contrast, *fecA* is highly transcribed, which guarantees that uninduced cells will contain a minimal level of FecA able to respond to ferric citrate and initiate transcription of the *fec* transport genes. The *fecIR* genes form a transcriptional unit distinct from the downstream *fecABCDE* operon.

Of the Fec proteins, only the FecA outer membrane protein can be observed after separation of the membrane proteins by SDS-PAGE, and its synthesis depends on the presence of ferric citrate in the medium. Citrate is not sufficient for induction, as revealed by experiments in which the iron in the medium is complexed with deferri-ferrichrome. In a mutant devoid of ferrichrome transport to avoid repression of *fecABCDE* transcription by iron delivered by ferrichrome, citrate does not induce FecA synthesis or synthesis of β-galactosidase from a chromosomal *fecB-lacZ* fusion under the control of the *fecA* promoter. The Fec system is not induced under anaerobic conditions; the reason for this has not been explored.

Seven genes have been identified as essential for the control of *fec* transcription: *fecA*, *fecI*, *fecR*, *fur*, *tonB*, *exbB*, and *exbD* (Fig. 6). *fecA*, *tonB*, *exbB*, and *exbD* are also required for transport of ferric citrate across the outer membrane. However, transport across the outer membrane is not necessary for induction since a transport-negative *fecA* point mutant (*fecA38*) shows residual induction and another point mutant (*fecA4*) displays TonB-independent induction but TonB-dependent transport. Mutants with *fecBCDE* deletions are fully inducible, which indicates that ferric citrate does not have to be transported into the cytoplasm. This finding is consistent with transport of iron, but not citrate, into the cytoplasm and induction of the Fec system by ferric fluorocitrate and ferric phosphocitrate, both of which transport iron very poorly. Accumulation of citrate in the cytoplasm in an *icd* mutant deficient in isocitrate dehydrogenase activity does not induce the Fec system. On exposure of cells to high concentrations of ferric citrate, sufficient amounts of ferric citrate diffuse through the porins into the periplasm and are transported into the cytoplasm by constitutively expressed plasmid-encoded FecBCDE transport proteins without induction of the Fec system. Ferric citrate induces transcription of the *fecABCDE* transport genes without entering the periplasm or cytoplasm. Binding of $(Fe^{3+}\text{-citrate})_2$ to FecA is sufficient to elicit transcription initiation.

Purified FecI and RNA polymerase bind to a 75-bp DNA fragment containing the -10 and -35 regions upstream of *fecA*, as revealed by retardation of the electrophoretic mobility of the DNA fragment. Band shift of the *fecA* promoter fragment can also be demonstrated with cell lysates obtained from $fecI^+$ $fecR^+$ $fecA^+$ cells grown in low-iron medium in the presence of ferric citrate. In vitro *fec* mRNA synthesis requires FecI and RNA polymerase.

Reduction of transcription of a chromosomally encoded *fecB-lacZ* reporter gene by a *fecA* promoter DNA fragment provides a tool for the isolation of mutants in the promoter that no longer inhibit transcription. Such mutations are located around $+10$ of the *fecA* transcription start site, and site-directed mutants created by PCR support the unusual importance of the $+10$ region for transcription of *fecA*. The -10 region of the *fecA* promoter shows little sequence similarity to other -10 regions, in contrast to the -35 region, which displays high sequence similarity to extracytoplasmic function (ECF) σ factors and tolerates quite a number of mutations without affecting transcription.

The Signaling Pathway

The massive structural changes in FecA on binding of $(Fe^{3+}\text{-citrate})_2$ presumably generates a signal that is transmitted across the outer membrane. However, these structural changes are not sufficient to trigger the signal that initiates transcription. The Ton system, composed of TonB, ExbB, and ExbD, is required and couples the proton motive force of the cytoplasmic membrane to active outer membrane transport. Dissipation of the electrochemical potential of the cytoplasmic membrane by CCCP (0.01 mM) or DNP (1 mM) inhibits transcription, not as a result of a general energy shortage but specifically related to the induction of *fec* transport gene transcription. Transcription of the TonB-independent *fecA4* mutant is not affected by CCCP and DNP. It is assumed that additional structural changes occur through interactions with TonB or that the structural changes induced by $(Fe^{3+}\text{-citrate})_2$ binding are modified by the TonB system.

FecA contains an extra N-proximal portion (residues 1 to 79) that is not found in outer membrane transport proteins that only transport ferric siderophores and are not involved in transcription regulation (Fig. 6). Deletion of residues 14 to 68 abolishes induction but fully retains transport. Overproduction and secretion into the periplasm of a fragment (residues 1 to 100, including 33 residues of the signal sequence) inhibits induction but not transport. This fragment is located in the periplasm, which agrees with the crystal structure in which the joint between this fragment and the FecA globular portion is exposed to the periplasm. The fragment consisting of residues 1 to 79 is not observed in the crystal structure, presumably because it is flexible.

The signal initiated in FecA needs to reach the cytoplasm, where transcription of the *fec* genes takes place. The signal must cross the periplasmic space and the cytoplasmic membrane. It is transferred across the periplasmic space by the N terminal portion of FecA and the C-terminal portion of FecR, both of which are located in the periplasm. Residues 1 to 84 of FecR are located in the cytoplasm, residues 85 to 100 span the cytoplasmic membrane, and residues 101 to 317 are located in the periplasm (Fig. 6). The LexA-based bacterial two-hybrid system reveals specific in vivo interaction of $FecA_{1-79}$ with $FecR_{101-317}$. In vitro, $(His)_{10}$-FecR (FecR with 10 histidine residues fused to the N terminus) bound to a Ni^{2+}-agarose column retains FecA on the column and $(His)_{10}$-FecR and FecA coelute from the column. FecA with a deletion in the periplasmic N terminus is unable to bind to $(His)_{10}$-FecR, further demonstrating the involvement of the N terminal extension of FecA in FecR interaction.

The transmembrane segment of FecR is thought to transmit the signal across the cytoplasmic membrane. In the cytoplasm, the signal activates the FecI σ factor. $FecR_{1-85}$ induces FecI-mediated transcription of the *fec* transport genes independently of ferric citrate. Using the LexA two-hybrid system, it has been demonstrated in vivo that FecI binds to $FecR_{1-85}$, $FecR_{1-58}$, and $FecR_{9-85}$. Residues 9 to 58 seem to be sufficient for the binding of FecR to FecI. In vitro, $FecR-(His)_6$ (six histidine residues fused to the C terminus) bound to a Ni^{2+}-agarose column binds FecI. The specificity of the FecI interaction with $FecR_{1-85}$ has been demonstrated by using randomly generated single point mutants that no longer transcribe the *fecB-lacZ* reporter gene and fail to bind to FecI. All FecR mutants have the tryptophan residues that are highly conserved among FecR homologs replaced by other amino acids. The results demonstrate that region 4 of FecI specifically interacts with the N-proximal region of FecR and that this interaction is necessary for FecI to function as a σ-factor.

The complete signaling cascade from the cell surface into the cytoplasm consists of the structural change in FecA after binding of $(Fe^{3+}\text{-citrate})_2$, transfer of the transcription initiating signal to the periplasm with the help of TonB and the proton motive force of the cytoplasmic membrane, interaction of FecA with FecR in the periplasm, signal transfer across the cytoplasmic membrane by FecR, and interaction of FecR with FecI in the cytoplasm. There is no evidence for a chemical modification of FecI by FecR. It is unlikely that a fragment as small as $FecR_{9-58}$ has enzymatic activity.

The FecI σ Factor

The sequence of FecI is homologous to the sequence of ECF σ-factors, which regulate genes that determine cell envelope functions. For this reason, they are designated ECF σ factors. ECF σ factors belong to the class of σ^{70} factors composed of structurally and functionally conserved subregions. A FecI deletion analysis using the LexA two-hybrid system has revealed that regions 4.1 and 4.2 of FecI interact with FecR$_{1-85}$. Deletion of region 4 results in an inactive FecI that does not interact with FecR$_{1-85}$. Region 4 of FecI expressed from a multicopy plasmid competes with chromosomally encoded FecI for FecR interaction, thereby resulting in reduction of FecI activity. Random PCR mutagenesis of *fecI* covering regions 2.4 to 4 results in mutants with reduced interaction with FecR. All mutations are located in region 4, and all single mutants contain a leucine-to-proline substitution, which suggests that conformational changes reduce FecI activity.

FecI binds to the β′ subunit of RNA polymerase (Fig. 6), as revealed in vivo by using the LexA two-hybrid system and in vitro by binding of FecI to a His-tagged β′ fragment (β′$_{1-313}$) on Ni^{2+}-NTA-agarose. FecR$_{1-85}$ enhances the binding of FecI to β′; FecR of cells grown without ferric citrate diminishes the binding of FecI to β′. β′$_{1-313}$ competes with the binding of FecI to β′. A FecI point mutant, FecI(K155E), displays 9% of the transcription activity of wild-type FecI and binds β′ poorly.

FecI exclusively mediates the transcription of the *fec* transport genes. Of the seven *E. coli* σ factors, only the FecI RNA polymerase complex transcribes a *fecA* promoter DNA fragment. The amount of FecI estimated in this study amounts to less than 1% of the amount of σ^{70}. In contrast, SigX of *Bacillus subtilis*, which is 25% identical to FecI, fully complements an *E. coli* mutant devoid of FecIR. SigX is active in the absence of ferric citrate and does not require RsiX, which suppresses SigX activity.

A BLAST search reveals that FecI, together with other σ factors, forms a subgroup among the ECF σ factors with a score of more than 123 among gram-negative bacteria such as *Pseudomonas* species, *Bordetella* species, and *Xanthomonas campestris*. FecI species with a lower sequence similarity are found mainly in *Pseudomonas* species, *Agrobacterium tumefaciens*, *Bordetella bronchiseptica*, *Caulobacter crescentus*, streptomycetes, and mycobacteria.

REGULATION OF THE FERRIC CITRATE TRANSPORT SYSTEM BY IRON

Under iron-limiting growth conditions resulting in low intracellular iron levels, the Fur repressor protein is unloaded and transcription of *fecIR* occurs. *fecIR* transcription does not respond to ferric citrate. Iron limitation also plays a direct role in the induction of the transcription of the *fecABCDE* transport genes. However, iron limitation is not sufficient, and ferric citrate must be present in the medium to ensure transcription initiation. The Fec transport proteins are then synthesized, and iron is taken up until an intracellular surplus of iron not incorporated into heme and proteins containing heme or iron-sulfur centers is reached. The surplus of ferrous iron binds to Fur (see chapter 13), which represses transcription of the *fecIR* genes (Fig. 6). Fe^{2+}-Fur also represses the transcription of the *fecABCDE* genes; this guarantees a fast response to the sufficient iron load of the cells. If regulation occurred only through FecIR, induction would last until the FecIR proteins are diluted to a level at which they no longer can cause induction. DNase I footprint analysis with Mn^{2+}-Fur reveals that 30 nucleotides of the *fecIR* promoter on the coding strand and 38 nucleotides of the *fecA* promoter are protected from degradation. Mn^{2+} is used for the footprint analysis, because it is less prone than Fe^{2+} to oxidation. *fecIR* transcription is not autoregulated; this is atypical of ECF σ factors.

MECHANISM OF FERRIC CITRATE TRANSCRIPTION REGULATION

The activity of most ECF σ factors seems to be controlled by anti-σ factors. In the absence of the anti-σ factors, the σ factors initiate transcription without extracytoplasmic signals. There is no evidence that FecR acts as an anti-σ factor. In the absence of FecR, there is virtually no *fecABCDE* transcription. In contrast, cells containing the cytoplasmic portion of

FecR, FecR$_{1-85}$, and even fragments of FecR$_{1-85}$, have a high constitutive transcription of the *fecABCDE* genes. Cells containing longer FecR derivatives, extending from residues 1 to 273 (of a total of 317 residues), which do not interact with FecA, also transcribe *fecABCDE* constitutively, but at a level lower than that with the cytoplasmic FecR fragment. Although FecR is necessary for FecI activity, it cannot be ruled out that FecR acts as an anti-σ factor. If FecI is unstable, spontaneously denatures, precipitates, or is degraded by proteases, binding to FecR could maintain FecI in a stable conformation. When the signal from FecA occupied by (Fe^{3+}-citrate)$_2$ arrives through FecR, FecR undergoes a conformational change that might result in dissociation of FecI from FecR and immediate binding of FecI to the RNA polymerase core enzyme. In this model, FecR acts both as a chaperone for FecI and as an anti-σ factor, since FecI is kept in an active conformation or assumes an active conformation with FecR but cannot exert activity while bound to FecR.

SUMMARY AND PERSPECTIVES

Iron transport through the outer membrane as ferrichrome- and ferric citrate-siderophore complexes is catalyzed by the FhuA and FecA proteins and across the cytoplasmic membrane by the FhuCDB and FecBCDE proteins, respectively. The energy for transport across the outer membrane is provided by the proton motive force of the cytoplasmic membrane, and the energy for transport across the cytoplasmic membrane is provided by ATP hydrolysis. Energy transfer from the cytoplasmic membrane to the outer membrane is mediated by the TonB-ExbB-ExbD proteins. TonB binds to a conserved region of FhuA and FecA, termed the TonB box, which is exposed to the periplasm, as shown by the crystal structures of FhuA and FecA. The crystal structures of FhuA and FecA are very similar and consist of a β-barrel composed of 22 antiparallel β-strands and a globular domain, termed a cork or plug, that inserts from the periplasmic side into the β-barrel and completely closes the β-barrel channel. Binding of ferrichrome and ferric citrate induces large long-range structural changes in FhuA and FecA without opening a channel. It is thought that the channel is opened by energy input from the cytoplasmic membrane, accompanied by the release of ferrichrome and ferric citrate from their binding sites. On binding of ferric citrate to FecA, surface loops move a long distance to close the ferric citrate entrance, thereby preventing the escape of ferric citrate to the medium. On binding of ferrichrome to FhuA, no such surface loop movements are observed in two crystal structures of the protein. The structural basis for understanding the mode of action of the ABC transporters FhuCDB and FecBCDE is provided by the crystal structure of the BtuCD ABC transporter for vitamin B$_{12}$. The BtuC transmembrane protein consists of 20 α-helices and is representative of ferric siderophore transporters across the cytoplasmic membrane, as has been previously demonstrated for the FhuB protein. FecA not only serves as a transport protein but also is involved in a signaling cascade from the cell surface to the cytoplasm to initiate transcription of the *fecABCDE* transport genes. The signal is transmitted by FecA across the outer membrane and through FecR across the cytoplasmic membrane into the cytoplasm, where the FecI σ factor is activated.

Knowledge about the proteins that transport ferric siderophores, their interaction, and their subcellular location provides insights into the transport mechanism, but much work remains to be done before an understanding of the molecular mechanism of substrate translocation across the outer membrane and the cytoplasmic membrane is gained. Although crystal structures of the proteins in the entire pathway are now available (FhuA in the outer membrane, FhuD in the periplasm, and BtuCD in the cytoplasmic membrane), they provide only a framework for a detailed biochemical and genetic analysis. Crystal structures are static, yet transport is a dynamic process involving many rearrangements of amino acid side chains and entire polypeptide regions. The structural alterations during transport will be revealed only by detailed biochemical and biophysical studies. They will provide the ultimate insights into the functions of the transport proteins. How TonB measures and responds to the proton motive force of the cytoplasmic membrane, what "energized" TonB means, and how it elicits a structural change in the outer membrane transporters to open the channel and release the ferric siderophores from their binding sites are completely unknown. Interaction between the periplasmic binding proteins and the transporters in the cytoplasmic membrane and how these interactions trigger ATP hydrolysis remain to be uncovered. It is also not known what causes the periplasmic binding protein to transfer the substrate to the cytoplasmic membrane transporter. Furthermore, it is unclear whether the binding proteins alone determine transport specificity or whether the cytoplasmic membrane transporters also recognize the substrates. The various steps in opening and closing of the presumed transmembrane channel and their coupling to ATP hydrolysis await definitive answers, even though impressive progress has been made in investigating the maltose transport system. The extent to which the transport mechanism elucidated for one system applies for other systems, as exemplified by the FhuD protein that unlikely binds and releases the substrate according to the "Venus flytrap" mechanism and by the 20-transmembrane-helix arrangement of FhuB, which differs from the proposed 12-helix model of most other cytoplasmic membrane ABC transporters, also remain to be studied.

ACKNOWLEDGMENTS

We thank Karen A. Brune for critical reading of the manuscript and Susanne Mahren for designing Fig. 6.

This work was supported by the Deutsche Forschungsgemeinschaft (BR 330/19-1, BR 330/20-1, and the Forschergruppe "Bakterielle Zellhülle: Synthese, Funktion und Wirkort") and the Fonds der Chemischen Industrie.

SUGGESTED READING

Braun, V. 1999. Active transport of siderophore-mimicking antibacterials across the outer membrane. *Drug Resist. Updates* **2:**363–369.

Braun, V. 1995. Energy-coupled transport and signal transduction through the gram-negative outer membrane via TonB-ExbB-ExbD-dependent receptor proteins. *FEMS Microbiol. Rev.* **16:**295–307.

Braun, V. 1997. Surface signaling: novel transcription initiation mechanism starting from the cell surface. *Arch. Microbiol.* **237:**325–331.

Braun, V., K. Hantke, and W. Köster. 1998. Bacterial iron transport: mechanisms, genetics, and regulation, p. 67–145. *In* A. Sigel and H. Sigel (ed.), *Metal Ions in Biological Systems.* Marcel Dekker, Inc., New York, N.Y.

Braun, V., H. Killmann, E. Maier, R. Benz, and V. Braun. 2002. Diffusion through channel derivatives of the *Escherichia coli* FhuA transport protein. *Eur. J. Biochem.* **269:**4948–4959.

Braun, V., S. Patzer, and K. Hantke. 2002. TonB-dependent colicins and microcins: modular design and evolution. *Biochimie* **84:**365–380.

Braun, V., S. Mahren, and M. Ogierman. 2003. Regulation of the FecI-type ECF sigma factor by transmembrane signalling. *Curr. Opin. Microbiol.* **6:**173–180.

Braun, M., F. Endriβ, H. Killmann and V. Braun. 2003. In vivo reconstitution of the FhuA transport protein of *Escherichia coli* K-12. *J. Bacteriol.* **185:**5508–5518.

Davidson, A. L. 2002. Mechanism of coupling of transport to hydrolysis in bacterial ATP binding cassette transporters. *J. Bacteriol.* **184:**1225–1233.

Ferguson, A. D., V. Braun, H. P. Fiedler, J. W. Coulton, K. Diederichs, and W. Welte. 2000. Crystal structure of the antibiotic albomycin in complex with the outer membrane transporter FhuA. *Protein Sci.* **9:**956–963.

Ferguson, A. D., R. Chakraborty, B. S. Smith, L. Esser, D. van der Helm, and J. Deisenhofer. 2002. Structural basis of gating by the outer membrane transporter FecA. *Science* **295:**1715–1719.

Ferguson, A. D., E. Hofmann, J. W. Coulton, K. Diederichs, and W. Welte. 1998. Siderophore-mediated iron transport: crystal structure of FhuA with bound lipopolysaccharide. *Science* **282:**2215–2220.

Ferguson, A. D., J. Ködding, G. Walker, C. Bös, J. W. Coulton, K. Diederichs, V. Braun, and W. Welte. 2001. Active transport of an antibiotic rifamycin derivative by the outer-membrane protein FhuA. *Structure* **9:**707–716.

Härle, C., I. Kim, A. Angerer, and V. Braun. 1995. Signal transfer through three compartments: transcription initiation of the *Escherichia coli* ferric citrate transport system from the cell surface. *EMBO J.* **14:**1430–1438.

Killmann, H., G. Videnov, G. Jung, H. Schwarz, and V. Braun. 1995. Identification of receptor binding sites by competitive peptide mapping:

phages T1, T5, and φ 80 and colicin M bind to the gating loop of FhuA. *J. Bacteriol.* **177**:694–698.

Kim, I., A Stiefel, S. Plantör, A. Angerer, and V. Braun. 1997. Transcription induction of the ferric citrate transport genes via the N terminus of the FecA outer membrane protein, the Ton system and the electrochemical potential of the cytoplasmic membrane. *Mol. Microbiol.* **23**:333–344.

Köster, W. 2001. ABC transporter-mediated uptake of iron, siderophores, heme and vitamin B_{12}. *Res. Microbiol.* **152**:291–301.

Locher, K. P., A. T. Lee, and D. C. Rees. 2002. The *E. coli* BtuCD structure: a framework for ABC transporter architecture and mechanism. *Science* **296**:1091–1098.

Locher, K. P., B. Rees, R. Koebnik, A. Mitschler, L. Moulinier, J. P. Rosenbusch, and D. Moras. 1998. Transmembrane signaling across the ligand-gated FhuA receptor: crystal structures of free and ferrichrome-bound states reveal allosteric changes. *Cell* **95**:771–778.

Mademidis, A., H. Killmann, W. Kraas, I. Flechsler, G. Jung, and V. Braun. 1997. ATP-dependent ferric hydroxamate transport system in *Escherichia coli:* periplasmic FhuD interacts with a periplasmic and with a transmembrane/cytoplasmic region of the integral membrane protein FhuB, as revealed by competitive peptide mapping. *Mol. Microbiol.* **26**:1109–1123.

Moeck, G. S., and L. Letellier. 2001. Characterization of in vitro interactions between a truncated TonB protein from *Escherichia coli* and the outer membrane receptors FhuA and FepA. *J. Bacteriol.* **183**:2755–2764.

Ogierman, M., and V. Braun. 2003. In vivo cross-linking of the outer membrane ferric citrate transporter FecA and TonB, studies of the FecA TonB box. *J. Bacteriol.* **185**:1870–1885.

Ratledge, C., and L. G. Dover. 2000. Iron metabolism in pathogenic bacteria. *Rev. Microbiol.* **54**:881–941.

Schneider, R., and K. Hantke. 1993. Iron-hydroxamate uptake systems in *Bacillus subtilis*: identification of a lipoprotein as a part of a binding protein dependent transport system. *Mol. Microbiol.* **8**:111–121.

Vakhari, H. L., and K. Postle. 2002. FepA with globular deletions lacks activity. *J. Bacteriol.* **184**:5508–5512.

FERROUS IRON TRANSPORT

Klaus Hantke

12

Transport systems for both ferric and ferrous iron have been identified in bacteria (Fig. 1). Many bacteria produce siderophores to mobilize insoluble ferric iron from their surroundings. More than 500 different siderophores have been described, yet many are probably still unknown. Even in the well-studied species *Bacillus subtilis*, the siderophore corynebactin, first identified in *Corynebacterium glutamicum*, was detected only recently. Similarly, in *Salmonella enterica*, the major siderophores were found recently to be salmochelins and not enterochelin (enterobactin), as thought for more than 30 years. Many of these often brown to red or blue iron complexes have found wide scientific interest, as documented in this book. Ferrous iron uptake has not been studied equally well, although one has to assume that in evolution, ferrous iron uptake systems dominated during the oxygen-poor conditions characteristic of the planet during early life 2.7 billion years ago. These systems remain important even now, particularly in anaerobic environments, where ferrous iron is the main iron source. Under anaerobic conditions, ferrous iron is stable and is much more soluble than ferric iron, which allows the efficient transport of ferrous iron without its being complexed by ligands.

THE Feo TRANSPORT SYSTEM

One of the first ferrous iron transport systems identified in bacteria, Feo, was found in *Escherichia coli*. *feo* mutants are unable to take up iron in an [^{55}Fe]ferrous iron transport assay, although no growth defects have been observed in these mutants under aerobic or anaerobic conditions. However, under aerobic growth conditions, the *feo* mutant exhibits derepression of Fur-regulated ferric iron uptake systems. Loss of repression by the transcriptional regulator Fur (see chapter 13) indicates that the iron content of the *feo* mutant cells is lower than that of wild-type cells. This suggests that even under aerobic conditions, the Feo system is required to maintain normal cellular iron levels.

After the identification of the Feo transport system, it took another 6 years for the *feo* genes to be cloned and sequenced because the *feo* operon is deleterious for *E. coli* cells, even when cloned on a low-copy-number plasmid. Therefore, the *feo* operon was cloned from a *feoA*::Tn5 mutant. Analysis of the operon indicates that the transport system is encoded by three genes, *feoABC* (Fig. 2). *feoA* encodes a

---

Klaus Hantke, Mikrobiologie/Membranphysiologie, Universität Tübingen, Auf der Morgenstelle 28, D-72076 Tübingen, Germany.

Iron Transport in Bacteria, Edited by Jorge H. Crosa, Alexandra R. Mey, and Shelley M. Payne
© 2004 ASM Press, Washington, D.C.

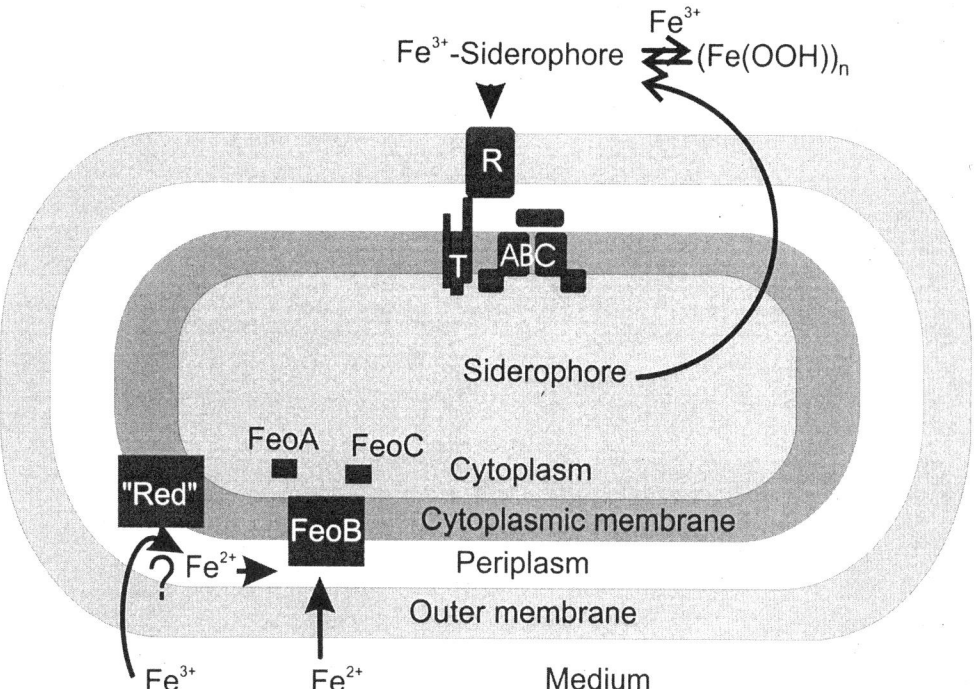

FIGURE 1 Schematic model of iron uptake in a gram-negative bacterium. Cells produce siderophores, which complex ferric iron. The Fe^{3+}-siderophore complexes are taken up via a receptor protein, R, in the outer membrane. The uptake is dependent on the TonB complex (T). A binding protein-dependent ABC transporter allows uptake through the cytoplasmic membrane. Fe^{2+} uptake through the cytoplasmic membrane may be accomplished by FeoB, possibly with the help of FeoA and FeoC, the latter being found only been found in *Enterobacteriaceae*. In oxygen-containing environments, Fe^{3+} may be converted to Fe^{2+} by generally uncharacterized reductive processes ("Red") at the cell surface or in the periplasm. In gram-positive bacteria, the picture is very similar, but the outer membrane receptor (R) and the TonB complex (T) are missing.

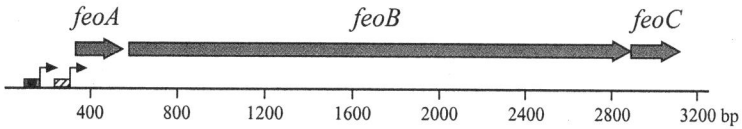

Fur binding site (not to scale), showing the direction of transcription

Fnr binding site (not to scale), showing the direction of transcription

FIGURE 2 Organization of the *feo* genes in *E. coli* K-12. Putative Fur- and Fnr-binding sites (not to scale) and the direction of transcription are shown.

small protein of 75 amino acids, and *feoB* encodes a membrane protein of 773 amino acids. *feoC* (annotated in the *E. coli* K-12 genome as *yhgG* = b3410) had been overlooked in the first analysis of the *feo* operon, but recent results indicate that FeoC (78 amino acids) is part of the Feo uptake system.

The major transport protein FeoB contains a typical nucleotide-binding motif in the first 160 N-terminal amino acid residues, and there are at least 7 carboxy-terminal α-helices that embed the protein in the cytoplasmic membrane. Although there is evidence that the transport of ferrous iron is energized by ATP hydrolysis, the nucleotide-binding motif in FeoB has more similarities to the GTP-binding sites found in Ras proteins and elongation factors than to the ATP-binding sites in bacterial ABC transporters. FeoB has four of the five GTPase signature motifs G1 to G5 (Fig. 3). The G5 motif, which is the least highly conserved sequence among the G proteins, has not been identified in FeoB. The high similarity of the FeoB N-terminal domain I (amino acids 1 to 160) to the GTP/GDP-binding site of eukaryotic G proteins, many of which couple membrane processes, suggested that the protein might be a GTPase. The N-terminal fragment of the protein, which contains domain I and the highly conserved G protein motifs, specifically bound GTP but not ATP, and it hydrolyzed GTP. The GTPase turnover rate is similar to that of Ras-like small, regulatory G proteins. In contrast to Ras-like G proteins, FeoB has a low affinity for GTP, similar to the affinities reported for members of the Era family of bacterial GTPases. Additional evidence for GTP binding has been obtained by mutating aspartate D123 in the GTP-binding site to asparagine. In other GTP-binding proteins, this type of mutation changes the ligand specificity from guanine nucleotides to xanthine nucleotides; the expected change in the binding specificity of FeoB was observed. In addition, the mutant FeoB D123N fails to complement an *E. coli feoB* mutant in iron transport, which indicates that GTP/GDP binding is necessary for iron uptake.

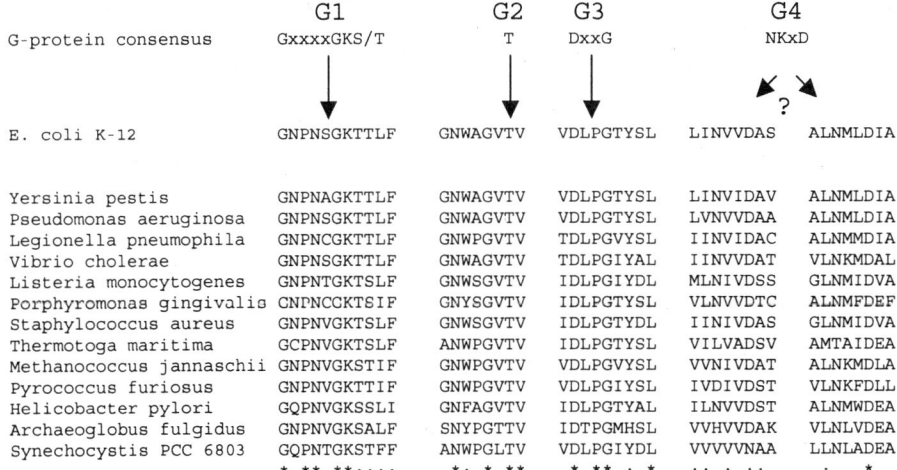

FIGURE 3 FeoB contains a G protein. The FeoB sequences found in >95 microorganisms, including pathogenic bacteria, cyanobacteria, and archaebacteria, contain highly conserved G-protein signature motifs within domain I of NFeoB (the first 160 amino acids). While the first three consensus motifs are unambiguous, two putative copies of the fourth G-protein consensus motif (NXXD) are observed (amino acids 91 to 94 and amino acids 120 to 123 of *E. coli*). Conserved amino acids are indicated by asterisks, highly similar ones are indicated by colons, and similar ones are indicated by dots.

Although it is clear that FeoB and the associated GTPase activity are required for ferrous iron transport, the mechanism of transport and precise function of FeoB are unknown. FeoB does not have homology to known metal ion transporters, and it is not known whether FeoB can bind or transport iron. G proteins can bind Mg^{2+}, and on cleavage of GTP and release of inorganic phosphate, the binding affinity of Mg^{2+} is reduced. It remains to be determined whether the magnesium-binding site in the G domain of FeoB binds Fe^{2+} and participates directly in transport or whether this domain is involved in transport only indirectly, if at all. The energy source for Fe^{2+} uptake also remains to be determined. In *Helicobacter pylori*, iron uptake by FeoB is inhibited when whole cells of *H. pylori* are treated with the protonophore carbonyl cyanide 4-(trifluoromethoxy) phenylhydrazone (FCCP), the ATPase inhibitor 1,3-dicyclohexylcarbodiimide (DCCD), or the ABC transporter inhibitor vanadate, suggesting that ATP is the energy source for transport. If *H. pylori* FeoB, like its *E. coli* counterpart, has only GTPase activity, the ATPase requirement may indicate that Feo activates a downstream ferrous transport ATPase. Alternatively, cleavage of GTP might drive the transport of ferrous iron by FeoB. It has been argued that the kinetic properties of FeoB in vitro make it unlikely that GTP hydrolysis is the energy source for iron transport, because this process is very slow and would not satisfy the iron needs of the cell; however, modulation by other factors, such as FeoA or FeoC, might change the properties of the GTPase domain of FeoB. Another possibility is that cleavage and/or binding of GTP/GDP Plays a regulatory role, linking ferrous iron uptake to the GTP/GDP status of the cell. The uptake of positively charged ion Fe^{2+} across the cytoplasmic membrane might be coupled to the proton gradient.

Regulation of *feo* Gene Expression

Ferrous iron uptake is repressible by iron and is derepressed in a *fur* mutant. The promoter region upstream of the *feo* genes (Fig. 2) has homology to the consensus Fur-binding site (see chapter 13), and Fur binding to this region of the DNA has been shown in vivo.

There is also a consensus Fnr-binding site in the *feo* promoter region. Fnr activates genes required for optimal growth under anaerobic conditions, and there is evidence for Fnr regulation of *feo*. A *feo-lacZ* fusion was expressed at lower levels in an *fnr* mutant than in the Frn$^+$ parent strain. Regulation of *feo* by Fnr may help couple ferrous iron uptake to growth in anaerobic environments where ferrous, rather than ferric, iron predominates.

Distribution of *feo* Genes in Bacteria

In 93 bacterial genomes screened in a BLAST search, more than 45 FeoB-like protein sequences were found. FeoB homologs are found in genomes of archaea and gram-positive and gram-negative bacteria (COGs, or phylogenetic classification of proteins encoded in complete genomes). Domains I (containing the G-protein motifs) and III (containing the hydrophobic transmembrane-spanning segments) are highly conserved among the analyzed FeoB proteins (Fig. 3). In 14 cases, a homolog of *feoA* is found closely associated with a *feoB* homolog. The *E. coli* FeoA protein has low similarity (30% identity) to the C-terminal domain of DtxR-like proteins (see chapter 22), but the possible significance of this homology is unknown.

In contrast to the wide distribution of *feoA* and *feoB*, *feoC* is found only in members of the *Enterobacteriaceae*. A *feoABC* operon has been identified in all members of the *Enterobacteriaceae* sequenced to date (except in the highly insect-adapted *Buchnera* species), including *Salmonella enterica* serovar Typhimurium and *Yersinia pestis*. The cysteine-rich motif of FeoC, EEPDGCLSGSCKSCPEG, might function as the first binding site for iron coming into the cell via FeoB.

Although Feo homologs can be found in many genomes, experimental data have confirmed the involvement of the Feo proteins in iron transport in only a few bacteria other than the *Enterobacteriaceae*. In *H. pylori*, FeoB appears to be the major high-affinity iron acquisition system. Growth of *H. pylori* under iron-limiting conditions is impaired in a *feoB* mutant and cannot be restored by ferrous iron salts. Growth of the *feoB* mutant can be restored by providing ferric iron sources, including iron citrate, and iron-loaded human transferrin and lactoferrin. Because both Fe^{3+} and Fe^{2+} serve equally well as iron sources for *H. pylori*, it has been suggested that the Feo system cooperates with ferric iron reductase activities of *H. pylori*. Fe^{3+} provided as a citrate salt is immediately reduced to Fe^{2+}, as evidenced by the formation of the magenta Fe^{2+}-ferrozine complex. It has been shown that FeoB-dependent ferrous iron transport in *Synechocystis* sp. strain PCC6803 is induced under iron-limiting growth conditions. In addition, the FeoB transport system is induced in mutants unable to transport ferric iron. However, attempts to isolate a double mutant deficient in ferric and ferrous iron uptake have failed, even though single mutants deficient in just one system are viable. *Legionella* spp., which were shown to be dependent on iron for growth when the culture conditions for these organisms were being established, have both FeoA and FeoB homologs. Growth of a *feoB* mutant *Legionella pneumophila* is reduced on iron-limited media, even though ferric iron uptake is not impaired, indicating a role of Feo in *Legionella* iron transport (see chapter 24).

FeoB and Pathogenicity

Since the supply of iron inside the host is a critical factor for many pathogenic bacteria, the importance of the Feo iron transport system for virulence has been studied in several pathogens. For *E. coli*, it has been shown that a *feoB* mutation reduces the ability of the strain to colonize the gut of mice. In *S. enterica* serovar Typhimurium, *feoB* mutants are outcompeted by the wild type during mixed colonization of the mouse intestine but the *feoB* mutation does not attenuate *S. enterica* serovar Typhimurium for oral or intraperitoneal infection of mice.

H. pylori, which is the cause of gastric mucosal inflammation, gastritis and ulceration, requires FeoB for growth under iron-limiting conditions. In a mouse model of *H. pylori* infection, *feoB* mutants were unable to colonize the mouse stomach following oral inoculation. This suggests that Feo-mediated ferrous iron uptake is critical for growth of *H. pylori* in vivo as well as in vitro. Similarly, growth of an *L. pneumophila feoB* mutant is reduced in vitro and in cultured cells. The *feoB* mutant is defective for growth in iron-depleted *Hartmanella vermiformis* amoebae, the likely host for *L. pneumophila* in nature, and in human U937 cell macrophages. In a mouse lung infection model, the wild-type strain outcompeted a *feoB* mutant. These results suggest that FeoB may play a role in intracellular survival and virulence of *L. pneumophila*.

OTHER Fe^{2+} TRANSPORT SYSTEMS

Fe^{2+} can also be transported by the CorA protein of *E. coli* and *S. enterica* serovar Typhimurium under certain conditions. CorA is a divalent cation transporter mainly for Mg^{2+}, but Co^{2+}, Mn^{2+}, Ni^{2+}, and Fe^{2+} can also be taken up. *S. enterica*, but not *E. coli* K-12, also has the Sit transport system, which can transport ferrous iron in addition to manganese (see chapter 14).

In *E. coli*, competition between zinc and iron uptake under metal-replete conditions occurs, indicating that a nonspecific metal transporter might exist in *E. coli*. This system is distinct from the CorA protein and may be analogous to the *FET4* gene product found in yeast. *FET4* encodes a low-affinity transporter for iron, copper, cobalt, manganese, and zinc, which allows metal acquisition under metal-replete conditions. This interesting observation indicates that eukaryotic cells do not always use highly specific uptake systems to satisfy their needs for divalent trace elements. Such a nonspecific uptake system in *E. coli* has not yet been identified.

An elaborate ferrous iron uptake system has been described in yeast. It consists of Fet3, the multicopper oxidase, which oxidizes extracellular ferrous iron, which is then transported into the cell through the permease Ftr1. Interestingly, a homolog of the Fet3 protein exists in several bacterial species. In *E. coli*, the homolog CueO protects cells from copper stress and also oxidizes ferrous iron. It is not known whether there is also a ferric iron transport system similar to that found in yeast.

CONCLUDING REMARKS

Interest in ferrous iron transport has increased during recent years. In oxygen-restricted environments, it is a major means of iron uptake. Ferrous iron uptake is also important in the presence of oxygen since many bacteria can reduce extracellular ferric iron. This aspect has not been dealt with in this chapter. The striking similarity of Feo to G proteins and the wide distribution of this protein make future research on the function of this protein in ferrous iron transport an exciting task.

ACKNOWLEDGMENTS

I thank Volkmar Braun (Tübingen) and Karen A. Brune (Konstanz) for critical reading of the manuscript.

My work was supported by the Deutsche Forschungsgemeinschaft and the Fonds der Chemischen Industrie.

SUGGESTED READING

Cowart, R. E. 2002. Reduction of iron by extracellular iron reductases: implications for microbial iron acquisition. *Arch. Biochem. Biophys.* **400**:273–281.

Hantke, K. 1987. Ferrous iron transport mutants in *Escherichia coli* K-12. *FEMS Microbiol. Lett.* **44**:53–57.

Hantke, K. 1997. Ferrous iron uptake by a magnesium transport system is toxic for *Escherichia coli* and *Salmonella typhimurium*. *J. Bacteriol.* **179**:6201–6204.

Hantke, K., G. Nicholson, W. Rabsch, and G. Winkelmann. 2003. Salmochelins, new siderophores of *Salmonella enterica* and uropathogenic *Escherichia coli* strains, are recognized by the outer membrane receptor IroN. *Proc. Natl. Acad. Sci. USA* **100**:3677–3682.

Kammler, M., C. Schön, and K. Hantke. 1993. Characterization of the ferrous iron uptake system of *Escherichia coli*. *J. Bacteriol.* **175**:6212–6219.

Katoh, H., N. Hagino, A. R. Grossman, and T. Ogawa. 2001. Genes essential to iron transport in the cyanobacterium *Synechocystis* sp. strain PCC 6803. *J. Bacteriol.* **183**:2779–2784.

Kim, C., W. W. Lorenz, J. T. Hoopes, and J. F. Dean. 2001. Oxidation of phenolate siderophores by the multicopper oxidase encoded by the *Escherichia coli yacK* gene. *J. Bacteriol.* **183**:4866–4875.

Marlovits, T. C., W. Haase, C. Herrmann, S. G. Aller, and V. M. Unger. 2002. The membrane protein FeoB contains an intramolecular G protein essential for Fe(II) uptake in bacteria. *Proc. Natl. Acad. Sci. USA* **99**:16243–16248.

May, J. J., T. M. Wendrich, and M. A. Marahiel. 2001. The *dhb* operon of *Bacillus subtilis* encodes the biosynthetic template for the catecholic siderophore 2,3-dihydroxybenzoate-glycine-threonine trimeric ester bacillibactin. *J. Biol. Chem.* **276**:7209–7217.

Outten, F. W., C. E. Outten, J. Hale, and T. V. O'Halloran. 2000. Transcriptional activation of an *Escherichia coli* copper efflux regulon by the chromosomal MerR homologue, cueR. *J. Biol. Chem.* **275**:31024–31029.

Parkhill, J., B. W. Wren, N. R. Thomson, R. W. Titball, M. T. Holden, M. B. Prentice, M. Sebaihia, K. D. James, C. Churcher, K. L. Mungall, S. Baker, D. Basham, S. D. Bentley, K. Brooks, A. M. Cerdeno-Tarraga, T. Chillingworth, A. Cronin, R. M. Davies, P. Davis, G. Dougan, T. Feltwell, N. Hamlin, S. Holroyd, K. Jagels, A. V. Karlyshev, S. Leather, S. Moule, P. C. Oyston, M. Quail, K. Rutherford, M. Simmonds, J. Skelton, K. Stevens, S. Whitehead, and B. G. Barrell. 2001. Genome sequence of *Yersinia pestis*, the causative agent of plague. *Nature* **413**:523–527.

Patzer, S. I., and K. Hantke. 1998. The ZnuABC high-affinity zinc uptake system and its regulator Zur in *Escherichia coli*. *Mol. Microbiol.* **28**:1199–1210.

Pohl, E., R. K. Holmes, and W. G. Hol. 1999. Crystal structure of a cobalt-activated diphtheria toxin repressor-DNA complex reveals a metal-binding SH3-like domain. *J. Mol. Biol.* **292**:653–667.

Roberts, S. A., A. Weichsel, G. Grass, K. Thakali, J. T. Hazzard, G. Tollin, C. Rensing, and W. R. Montfort. 2002. Crystal structure and electron transfer kinetics of CueO, a multicopper oxidase required for copper homeostasis in *Escherichia coli*. *Proc. Natl. Acad. Sci. USA* **99**:2766–2771.

Robey, M., and N. P. Cianciotto. 2002. *Legionella pneumophila feoAB* promotes ferrous iron uptake and intracellular infection. *Infect. Immun.* **70**:5659–5669.

Stojiljkovic, I., M. Cobeljic, and K. Hantke. 1993. *Escherichia coli* K-12 ferrous iron uptake mutants are impaired in their ability to colonize the mouse intestine. *FEMS Microbiol. Lett.* **108**:111–115.

Tatusov, R. L., D. A. Natale, I. V. Garkavtsev, T. A. Tatusova, U. T. Shankavaram, B. S. Rao, B. Kiryutin, M. Y. Galperin, N. D. Fedorova, and E. V. Koonin. 2001. The COG database: new developments in phylogenetic classification of proteins from complete genomes. *Nucleic Acids Res.* **29**:22–28.

Tsolis, R. M., A. J. Baumler, F. Heffron, and I. Stojiljkovic. 1996. Contribution of TonB- and

Feo-mediated iron uptake to growth of *Salmonella typhimurium* in the mouse. *Infect. Immun.* **64:** 4549–4556.

Velayudhan, J., N. J. Hughes, A. A. McColm, J. Bagshaw, C. L. Clayton, S. C. Andrews, and D. J. Kelly. 2000. Iron acquisition and virulence in *Helicobacter pylori:* a major role for FeoB, a high-affinity ferrous iron transporter. *Mol. Microbiol.* **37:** 274–286.

Waters, B. M., and D. J. Eide. 2002. Combinatorial control of yeast *FET4* gene expression by iron, zinc, and oxygen. *J. Biol. Chem.* **277:**33749–33757.

MODE OF BINDING OF THE Fur PROTEIN TO TARGET DNA: NEGATIVE REGULATION OF IRON-CONTROLLED GENE EXPRESSION

Víctor de Lorenzo, José Perez-Martín, Lucía Escolar, Graziano Pesole, and Giovanni Bertoni

13

The identification of the *fur* (for "ferric uptake regulation") gene in *Escherichia coli* in the early 1980s was a landmark in our understanding of how bacteria sense iron starvation as a major environmental signal for regulating not only iron transport and metabolism but also a large variety of other processes. The *fur* mutant strain displayed constitutive expression of all functions known at that time to be inhibited by iron, i.e., siderophore production and biosynthesis of some outer membrane proteins. The behavior of that mutant, like the one previously isolated but uncharacterized in *Salmonella enterica* serovar Typhimurium, suggested that a metallo-dependent repression was at the basis of the control exerted by iron on many, if not all, Fe-responsive genes. The *fur* gene of *E. coli* was subsequently mapped, cloned, and sequenced, its protein product was purified, and some basic aspects of the regulation mechanism were elucidated. Following the early studies of the *fur* gene and protein of *E. coli*, a few orthologous systems from other bacteria (notably those of *Pseudomonas aeruginosa*, *Bacillus subtilis*, and *Vibrio anguillarum*) underwent the same scrutiny. A low-resolution structure of the Fur protein of *Rhizobium leguminosarum* was published in 2002. This was followed by a high-resolution structure of the *P. aeruginosa* protein, a long awaited landmark in the field of gene regulation by iron. At the time of the writing of this review, Fur-DNA cocrystals were not available, thus leaving open some questions about the mode by which Fur recognizes target DNA in the genome of the corresponding bacteria (see below).

Even before any sequence or structural data were available, the genetic results supported a model for Fur-mediated metallo-regulation (Fig. 1) that has remained basically correct since then. The model is that Fe^{2+} binds directly to the Fur protein, which in turn, acquires a configuration capable of binding a specific DNA sequence. Since most known Fur binding sites overlap the -10 and -35 hexamers of cognate promoters, this scheme allows transcriptional control of a large number of genes through a classic repression mechanism (Fig. 1). This simple regulatory scheme has been highly conserved. With relatively minor variations, homologs of the Fur protein can be found in many gram-negative and gram-positive bacteria. Most of these homologues are able to complement an *E. coli fur* mutant, hence suggesting

---

Víctor de Lorenzo, José Perez-Martín, and Lucía Escolar, Centro Nacional de Biotecnología del CSIC, Campus Universidad Autónoma, Madrid 28043, Spain. *Graziano Pesole and Giovanni Bertoni*, Dipartimenti di Genetica e Fisiología, via Celoria 26, 20133 Milan, Italy.

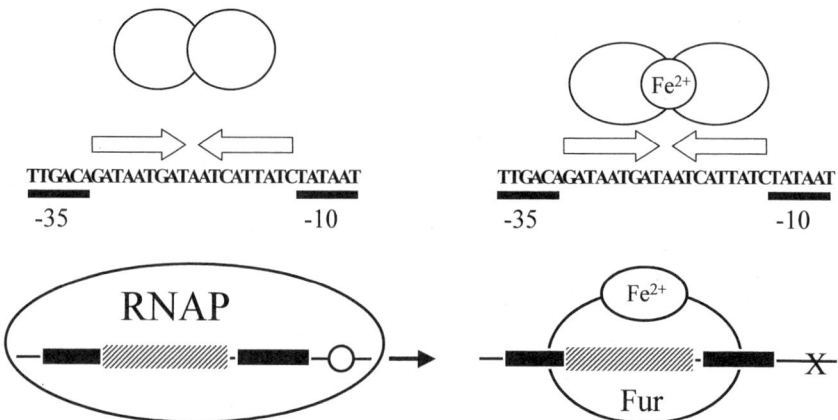

FIGURE 1 Standard model of Fur-mediated repression of metallo-regulated genes. Under iron-rich conditions, Fur binds the divalent ion, acquires a configuration able to bind target DNA sequences (generally known as Fur boxes or iron boxes [Fig. 4]), and inhibits transcription from virtually all the genes and operons repressed by the metal. In contrast, when iron is scarce, the equilibrium is displaced to release Fe^{2+}, the RNA polymerase accesses cognate promoters, and the genes for the biosynthesis of siderophore and other iron-related functions are expressed. In some cases (notably in *P. aeruginosa*), Fur controls the expression of at least one dedicated sigma factor (PvdS), which, in turn, causes a discrete set of genes to be expressed.

that the molecular mechanisms that control transcriptional regulation by iron are shared by many microorganisms.

Fur is not the only iron-responsive repressor known to exist in bacteria. High-G+C-content gram-positive microorganisms (typically *Corynebacterium* spp. but also *Mycobacterium* and *Streptomyces* spp.) contain a separate type of Fe-responsive repressor, whose prototype is DtxR. This protein was first discovered as the diphtheria toxin regulator in corynebacteria, but it controls not only expression of the toxin but also a whole collection of iron transport and metabolism genes. The DtxR orthologs IdeR (in mycobacteria) and SirR (in staphylococci) negatively regulate multiple genes in a fashion similar to that observed in Fur-regulated systems. Interestingly, DtxR displays little or no sequence similarity to Fur and binds a distinct target DNA operator, very different from that known as the Fur box or iron box (see below), but it acts in vivo and in vitro as an Fe^{2+}-dependent repressor. The crystal structure of *Corynebacterium diphtheriae* DtxR is known in detail. As discussed below, comparison of the three-dimensional organization of Fur and DtxR has revealed structural similarities and hints about how iron-binding repressors may exert their biological function. Both proteins control the expression of a large collection of genes that operate in iron-sequestering systems. However, besides siderophore production and intake, Fur and DtxR translate iron starvation into the production of proteases which degrade iron-containing proteins, uptake systems which acquire iron directly from heme groups, or whole systems for internalization of host iron-binding proteins.

This type of negative regulation might not be exclusive for the iron regulon. Another metal-dependent regulatory protein named Zur (for "zinc uptake regulator") is worth a comment in this respect. Similarly to iron, Zn is an element that, depending on its concentration, can be an essential micronutrient or can be highly toxic, and its intake is regulated by this Zur protein. The Zur protein is a Zn-responsive repressor that has considerable sequence similarity to Fur and controls the expression of genes putatively involved in Zn transport and metabolism. Not much is known about the repression mechanism, but some Zur-binding sequences have been described, and they appear similar to the Fur box. This may allow a degree of regulatory cross-regulation as well as the possibility for each repressor to recognize and regulate specifically its own set of genes.

GENETICS OF IRON REGULATION

That Fur is a general regulator in *E. coli* and many other bacteria is inferred by the large number of genes that are subject to transcriptional repression by iron in Fur$^+$ strains and that no longer respond to the metal ion in their Fur$^-$ counterparts. Many of the genes derepressed under iron starvation encode metal acquisition functions, thus ensuring that the high-affinity iron transport systems are induced. However, there is increasing evidence that many other genes with functions not directly related to iron (respiration, chemotaxis, the tricarboxylic acid cycle, glycolysis, methionine biosynthesis, phage-DNA packaging, DNA synthesis, purine metabolism, and redox stress resistance) are also repressed by Fur. Thus, Fur is an authentic global regulator and the genes under Fur control can be considered to form a superregulon.

In *E. coli*, the *fur* gene forms part of the *fldA-fur* bicistronic operon. This association is intriguing, since the *fldA* gene encodes a flavin-containing protein involved in redox reactions. The *fldA-fur* transcript is induced by the SoxRS system in response to superoxide-induced redox stress. In addition to the shared upstream promoter, the *fur* gene has an OxyR-responsive promoter, which is inducible by H_2O_2. This *fur* promoter is different from the probably much weaker one(s) first detected within a shorter region upstream of *fur* gene. This regulatory configuration with various binding sites through an extended promoter region (Fig. 2) reveals a link between redox-related stress and iron homeostasis. It may be sufficient to alter the function or the intracellular concentrations of Fur to make cells more sensitive to redox and oxidative damage and thus to disrupt the ability of a pathogenic microbe to survive the host oxidative defenses. In addition, the *fur* gene is autoregulated and perhaps weakly controlled as well by the cyclic AMP receptor protein (CRP), since one Fur box and one CAP box are located right upstream of the *fur* gene. In silico analysis of the genomic sequence of the *fldA-fur* chromosomal site also suggests the presence of a binding site(s) for MarA. This is a SoxS homologue that mediates resistance to multiple environmental hazards such as antibiotics, disinfectants, and oxidative-stress agents by modulating the expression of a large number of genes in *E. coli*. Experimental confirmation of these predictions is still needed.

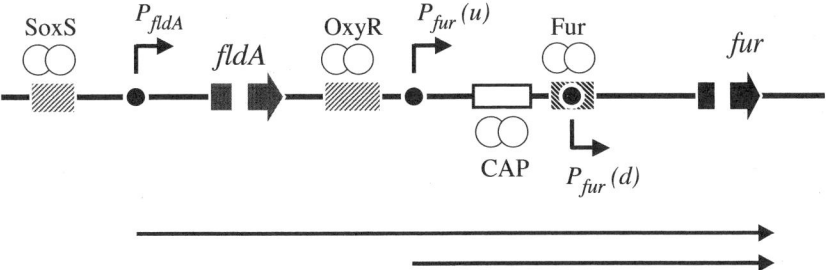

FIGURE 2 Organization of the *fldA-fur* promoter region in *E. coli*. The noncoding DNA upstream and downstream area of the nearby *fldA* and *fur* genes is enlarged. Sites for binding of each relevant regulatory protein (SoxS, OxyR, Fur, and the predicted catabolite activation protein [CAP]) are indicated (the size and stoichiometry are symbolic), as well as the transcription start site from each of the promoters. Note that *fur* can be expressed through a weak downstream promoter, $P_{fur}(d)$, modulated by CAP and Fur or by a strong OxyR-activated promoter, $P_{fur}(u)$, located further upstream. Finally, *fur* can be cotranscribed with *fldA* from a still further upstream SoxS-activated promoter.

Although this chapter deals primarily with negative regulation by iron, it should be noted that several *E. coli* genes (including *acnA*, *bfr*, *ftnA*, *fumA*, *fumB*, *sdhCDAB*, and *sodB*) and other bacteria are positively controlled by Fe-Fur. In general, Fur acts as a repressor and its positive effects in certain promoters are predominantly the result of indirect rather than direct effects. One exception is the ferritin (*pfr*) gene of *Helicobacter pylori* in which expression of *pfr* is repressed by the iron-free form of Fur through a direct interaction of the apo-Fur with the *pfr* promoter at a distinct (and noncanonical) Fur-binding site. Under this scheme, binding of Fe^{2+} to the apo-Fur removes repression and results in activation of *pfr*. Fur can also activate directly some promoters of *Neisseria meningitidis*. For many other genes, including the well-studied *E. coli sodB* gene, neither apo-Fur nor Fe-Fur binds directly to the promoter, making it unclear how Fur induces *sodB* at the transcriptional and posttranscriptional levels. The most significant recent discovery in this respect has been the observation that a small noncoding RNA (sRNA) named *rhyB* is under the negative control of Fe-Fur. Since *acnA*, *bfr*, *ftnA*, *sdh*, and *sodB* are derepressed in *rhyB* mutants, it is likely that such an sRNA has a capacity to block translation or to decrease transcript stability of these iron-induced genes. Thus, these genes are negatively regulated by an sRNA that is itself negatively regulated by Fur.

Since iron influences so many processes in the cell, it is tempting to consider Fur a global regulator which adjusts the entire metabolism in response to changes in environmental iron rather than a very specific transcription factor for a few iron transport gene promoters. Being a general regulator, however, suggests that binding sites are not limited to a very specific sequence but also to more relaxed and abundant DNA targets.

EVOLVING COMPLEXITY IN IRON REGULATION

Although Fur directly controls the expression of most of the iron regulon in *E. coli*, additional regulators modulate the expression of some genes. The first example of siderophore biosynthesis and iron transport genes being regulated by positive and negative factors was discovered in *V. anguillarum*. In this case, the transcriptional activator AngR is necessary for expression of the siderophore transport (*fat*) gene cluster at low iron level. A further level of complexity is reflected by the existence in some bacteria of dedicated sigma factors for expression of a subset of iron-responsive genes. For *E. coli*, this became evident through studies on the expression of the *fec* system of iron dicitrate transport. *fecIR* and *fecABCDE* are repressed by Fur-Fe^{2+} and induced by the *fec*-specific FecI-FecR system in response to ferric dicitrate. Within this transport system, FecI is a sigma factor of the class known as extracytoplasmic function or extracytoplasmic stress factors while FecR spans the cytopasmic membrane, with domains facing both the periplasm and the cytoplasm. According to the current model, ferric citrate binding by the outer membrane receptor FecA induces a conformational change that is transmitted to FecR. This releases FecI from its complex with FecR, and FecI then activates the transcription of the *fec* genes. The FecI complex with RNA polymerase appears to be absolutely specific for the *fec* operon. FecR acts somewhat differently from an anti-sigma factor, since it not only inhibits FecI but also converts the sigma factor into a configuration competent to activate transcription when needed.

Extracytoplasmic functions related to iron metabolism are found in many other bacteria. One well-studied example is the PvdS sigma factor protein of *P. aeruginosa*, which is required for the synthesis of the siderophore pyoverdine. Interestingly, the genomes of both *P. aeruginosa* and *P. putida* contain about one dozen *fecIR*-like gene clusters which are adjacent to known or putative siderophore receptors or other genes of unknown function. Their physiological roles are still to be discovered.

Finally, antisense RNAs have also evolved to exert a degree of negative control of some siderophore-related genes. *V. anguillarum* strain 775 uses two systems to down-regulate the expression of the iron-anguibactin uptake system determined by the *fat* operon. One is the well-characterized control by the Fur protein, and the other is mediated by an antisense RNA, called RNAα. This is a long antisense RNA molecule encoded within the iron transport gene, *fatB*. The concentration of RNAα increases during iron-rich conditions and binds the complementary *fatB* mRNA transcribed from the *fatDCBA* operon. Such an RNA-RNA pairing stimulates the degradation of the corresponding transcript, thus inhibiting the expression of the siderophore transport genes. Transcription of RNAα requires a functional Fur protein, but RNAα expression is independent of iron. The high levels of RNAα in the presence of iron are not due to increased transcription. On the contrary, iron somehow increases RNAα stability, and as a consequence, iron-anguibactin transport genes are repressed both transcriptionally by Fur and posttranscriptionally by RNAα; the process is reversed under iron starvation conditions.

Fur PROTEIN

The Fur protein of *E. coli* is a 17-kDa DNA-binding protein which interacts reversibly with target sequences of the genome in a fashion dependent on the presence of intracellular Fe^{2+}. Fur appears to be an abundant protein, with 5,000 copies per *E. coli* cell during logarithmic growth, rising to 10,000 copies in stationary phase; similar levels (2,500 to 7,500 molecules) of Fur were found in *Vibrio cholerae*. This relatively large number of proteins sharply contrasts with the generally low concentrations of other regulators in bacterial cells. It has been estimated that >90 genes in *E. coli* are controlled by Fur. From this, it follows that occupation of all potential target sites demands a much higher concentration of the factor than other, more specific regulators. The large number of molecules of Fur may also reflect the unusual fashion in which this protein binds DNA. Fur has the ability to polymerize along the DNA, and this may relate to the fact that a large number of genes are strongly controlled by the regulator and many others are weakly controlled. It is also tempting to envision additional functions of Fur, perhaps as a storage protein for binding free ferrous iron in the cell contributing to defense against redox damage (see above).

The Fur protein appears to be a dimer in solution, regardless of the presence or absence of Fe^{2+}. The protein thus has distinct sites for dimerization, DNA binding, and reversible interaction with divalent metals. The repressor binds one ferrous ion per subunit but can also reversibly bind related metals (typically Mn) in vitro, a property that may not be physiologically relevant, given the low concentration of these metals in vivo. Since Fe^{2+} is difficult to maintain in a reduced state under aerobic conditions, the fact that Mn^{2+} can replace Fe^{2+} in vitro has greatly facilitated studies of the repression mechanism in cell-free systems. Divalent-metal binding increases the affinity of Fur for its DNA operators ca. 1,000-fold (see below). It appears that various Fur proteins of different origins contain at least one structural, nonregulatory zinc ion per monomer as part of their chemical structure. The regulatory metal-binding site also responds to Zn. A possible role for zinc suggests that results of DNA-protein interaction experiments performed with the earlier preparations of purified Fur, which was used after extensive dialysis with EDTA, may need to be reevaluated. The presence and role of Zn in the Fur structure under natural conditions is still a debated issue, since Zn may not appear in Fur proteins from all bacteria.

Structure of Fur and DtxR

Attempts to determine the three-dimensional structure of the Fur protein date back to the mid-1980s, although they were flawed by the tendency of the protein to precipitate when concentrated in excess. Only recently have X-ray structures of Fur variants become available. At the time of writing this chapter, the Fur protein of *P. aeruginosa* (Fur_{PA}) complexed with Zn ions was the only three-dimensional structure that had been determined at a resolution of 1.8 Å. This structure confirms the requirement for one structural Zn atom surrounded by a robust coordination sphere clearly involved in maintaining the overall protein architecture. However, there is a second Zn atom per monomer, which most probably occupies the functional regulatory site. The differentiation of such types of Zn atoms was the result of combining crystallographic results with the spectroscopic data in solution. However, the most salient feature of the structure is the remarkable similarity of Fur to the known structure of the DtxR repressor of *C. diphtheriae*. The basic fold contains two domains: an N-terminal DNA-binding module of 83 amino acids and a C-terminal dimerization domain (residues 84 to 135). The DNA-binding domains of Fur_{PA} and DtxR share the same basic fold and can be superimposed in at least 60 C atoms of two helices and two strands included in the domain. However, the dimerization domains of Fur_{PA} and DtxR are quite different (Fig. 3). The Fur structure shows a novel dimerization module that is also involved in metal binding, leading to relative orientations between the DNA recognition helixes quite unlike those in DtxR. The overall architecture of Fur_{PA} and DtxR is thus similar in their DNA-binding elements, but the structural divergence of the dimerization domain leads to a DNA recognition process that is completely different in each repressor.

Since the Fur_{PA} structure is the only one available at the time of writing this chapter, it is not possible to generalize the domain organization and the mode of functioning found in this protein to every other Fur variant. However, it is plausible that the gross three-dimensional design of the repressor follows the generally good conservation of the primary and secondary structures of all the Fur proteins examined so far.

Fur BINDING SEQUENCE

The Fur target sites, like the Fur proteins, are highly conserved in many bacteria. The consensus target site, the so-called Fur box or iron box, consists of a 19-bp inverted repeat

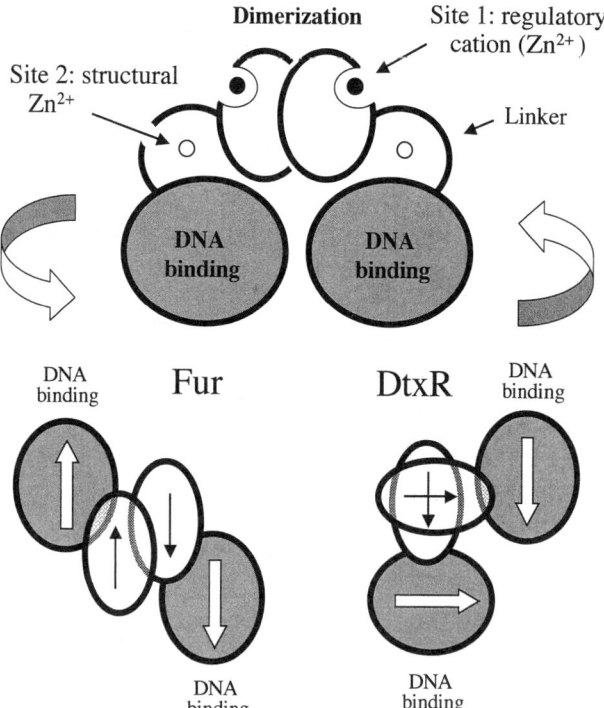

FIGURE 3 Domain structure of the Fur protein of *P. aeruginosa*. The sketch at the top is based on Pohl et al. (2003). The Fur_{PA} protein clearly has two domains: an N-terminal DNA-binding module (structurally similar to that of DtxR), and a distinct dimerization domain. The two modules are linked by a peptide, which is arranged around a permanently bound Zn atom. This important atom at metal-binding site 2 assists proper positioning between the DNA-binding and dimerization modules. The site for the regulatory divalent metal ion (presumably iron in vivo) is called site 1 and is occupied by Zn in the Fur_{PA} protein crystal. The ion at site 1 is coordinated by two residues following the loop that links the DNA-binding and dimerization domains. Metal binding could lead to a conformational change at this location, resulting in a motion changing the relative orientation of the DNA-binding domains of the dimer and thus increasing the affinity for the target DNA sequence. The drawings at the bottom symbolize the differences between the relative orientations of the otherwise very similar DNA-binding domains of Fur_{PA} and DtxR.

(5′GATAATGATAATCATTATC3′). The early definition of such a 19-bp consensus Fur box (Fig. 4) was the result of the DNase I footprint analyses of several Fur-binding sites. Placement of this 19-bp minimal Fur operator overlapping or downstream of a heterologous promoter ensures that its transcriptional activity is regulated by iron. This, together with the dimeric nature of the protein, suggested a mode of Fur-DNA interaction similar to that for classical bacterial repressors, in which a protein dimer recognizes a 19-bp palindromic DNA sequence. This notion was soon challenged by the results of hydroxyl radical footprinting of Fur within the promoter of the aerobactin siderophore operon of *E. coli*, which revealed that protein binding to its target site gives rise to a distinct pattern of two protected and four nonprotected base pairs which is repeated three times along the primary binding site. Furthermore, footprinting and methylation protection studies, as well as atomic-force and electron microscopy studies, indicated that multiple Fur molecules wrap around the double helix in a screw-like manner, extending into regions that do not appear to match the Fur box consensus. Further, the pattern of interactions between the Fur protein of *V. anguillarum* and the target *fatD* promoter challenged a simple repressor model of the Fur-DNA recognition mechanism. These observations triggered considerable efforts to understand the process and to propose alternative models.

As shown in Fig. 4, the same 19-bp sequence can be viewed as the combination of three adjacent repeats 5′NAT(A/T)AT3′, a view that is more consistent with the protection pattern revealed by hydroxyl radical footprinting on the aerobactin promoter. This view implies that the consensus sequence could be recognized by Fur as three repeats of 6 bp rather than as a 19-bp palindrome. While a minimum of three repeats is required to produce an effective Fur-binding site, their relative orientation and their number may not be so important. A key point is that only the sum of DNA-protein and protein-protein interactions contributed by at least three interaction sites may endow the complex with enough strength and specificity to be fully functional. More recent footprinting results with the unusually long promoter of the aerobactin operon still indicate that the hexamer NAT(A/T)AT is the basic unit of Fur recognition; up to 18 tandem hexanucleotide motifs were identified in the extended Fur-binding site of the aerobactin promoter. This interpretation of the Fur-binding site would appear to explain the ability of Fur binding to extend into adjacent sites lacking identifiable Fur boxes and to apparently polymerize along the DNA duplex. In the absence of stoichiometry data, it was hypothesized that each 6 bp could bind a Fur dimer, a notion that is now entirely ruled out by the structural data on the Fur$_{PA}$ protein. In any case, the view of Fur boxes as arrays of shorter sequence motifs would also explain why the target DNA frequently tolerates so many changes and why no individual bases essential for the interaction have been found in mutagenesis experiments of the consensus 19 bp. On the other hand, this type of interaction endows the protein with the ability to behave both as a very specific repressor and as a more general regulator.

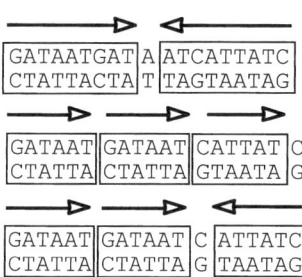

FIGURE 4 Interpretation of the canonical or consensus 19-bp Fur box as an array of hexameric sequences. The scheme shows alternative interpretations of the 19-bp consensus Fur-binding site (Fur box) in *E. coli*. These include a palindromic sequence composed of two 9-bp inverted repeats, an array of three directed 6-bp repeats, with the third repeat being imperfect, or an array of two directed repeats and one inverted repeat of the invariable sequence GATAAT.

The clarification of exactly how individual Fur dimers are arrayed along the double helix at Fur-binding sites and precisely how many nucleotide repeats are required to bind a single Fur dimer has been the subject of more recent studies, which have not been devoid of controversy. It was proposed that the Fur protein of *B. subtilis* (Fur$_{BS}$) binds the target 19-bp consensus sequence in a fashion very similar to that known to occur for DtxR (Fig. 5). Under this concept, the 19-bp consensus sequence was seen as the result of two overlapping 7–1–7 inverted repeats, which are recognized by two Fur dimers bound to opposite sides of the DNA helix. Such an opposite binding may not explain the ability of Fur to polymerize in one dimension through arrays of suboptimal sequences like those found in the aerobactin promoter of *E. coli*. However, this view is attractive (Fig. 5) and explains some caveats of the three-hexamer model. Nevertheless, it is clear that the modes of DNA binding of Fur and DtxR ought to be different, if only because the positions of the two DNA-binding domains with respect to each other in the functional dimer are completely different. A further reinterpretation of the Fur box is based on a variety of studies of the interaction between the repressor and the bidirectional *fepDGC-entS* region of *E. coli*, as well as on an analysis of its interaction with synthetic DNA sites. It is suggested that Fur$_{EC}$ binding sites are overlapping 13-bp 6–1–6 motifs (Fig. 4 and 5). This last arrangement would allow two Fur dimers to bind to each Fur box, on opposite faces of the double helix at sites displaced by approximately half a helical turn. This arrangement could explain the corkscrew manner in which Fur appears to wrap around the DNA duplex. Both the 7–1–7 and the 6–1–6 palindrome models are compatible with the recent estimation of the stoichiometry of the Fur$_{EC}$ complex with a canonical target site (Fig. 4): two Fur dimers/ 19-bp consensus sequence. However, the overlapping-palindrome models do not explain satisfactorily why the 13 15, or 19 bp proposed to be bound by Fur give extensive OH— radical or DNase I footprinting (minimum 30 ± 3 bp in most promoters) or how Fur can cooperatively bind suboptimal sequences adjacent to a consensus operator. All these caveats will doubtless be addressed in the near future.

CONSEQUENCES OF THE DISTINCT MODE OF INTERACTION OF Fur WITH ITS TARGET SEQUENCES

While some of the minor divergences between the different models discussed above could be attributed to peculiarities of the Fur proteins from different origins, the ultimate understanding of the Fur-DNA interactions requires accessing the cocrystal of Fur bound to its canonical 19-bp consensus target (Fig. 4). In the meantime, the mechanism of DNA recognition that emerges from the structure of the available Fur$_{PA}$ protein (see above) provides some interesting insights which may reconcile otherwise disparate features of the problem. The crystal structure of the Fur$_{PA}$ dimer clearly suggests not only structural similarities to DtxR but also similarities in the way in which the two proteins bind their operators (Fig. 3). For instance, when the predicted DNA-binding helices of the Fur$_{PA}$ protein dimer were docked to DNA in silico, they appeared to fit optimally into the major groove and thus cover a minimal sequence of about 17 bp. This leaves open some questions, particularly about the extension of the protection exerted by Fur on its target. With the model of Fig. 5, the binding of a single Fur$_{PA}$ dimer would not be expected

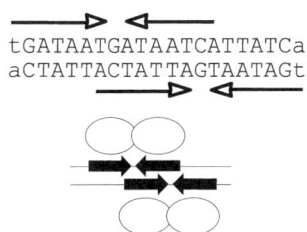

FIGURE 5 Interpretation of the consensus Fur-binding site as two overlapping palindromes that are recognized at opposite sides of the DNA helix. This model can accommodate a minimal recognition sequence of 21 bp [2 × (7–1–7)] or a shorter segment [2 × (6–1–6)]. The lower part of the figure indicates how the primary binding site would be assembled. Note that the central portion of the sequence is bound simultaneously at the two somewhat opposite sides of the helix.

to protect more than 20 bp in a DNase I footprinting assay. However, Fur has never been shown to protect less than 27 to 30 bp. This problem is solved if two Fur$_{PA}$ dimers were docked to DNA with a pseudo-twofold symmetry of the DNA sequence. If we consider the 19-bp consensus sequence (Fig. 4) to be the result of two overlapping Fur operators (as is the case with the DtxR-binding site), the prediction is that each Fur dimer could be bound to the major groove of the same sequence yet, approximately on opposite sides, in a fashion which avoids protein overlap. The best Fur$_{PA}$-DNA docking in silico was obtained with a spacing of 5 bp, so that the two dimers were located on opposite sides of the DNA helix. If this is the case, there would be a significant overlap of the DNA-binding domains of the two bound dimers, and this would suggest a conformational change in the DNA on Fur binding to accomodate such a tetramer in a short sequence. This change could dramatically stabilize the Fur-tetramer/DNA complex, thereby explaining why the typical Fur-binding sites include systematically a longer sequence. Like the DtxR regulator, two dimers of Fur$_{PA}$ may bind to a single operator that is no less than 27 bp in length. Further structural studies are needed to determine the fine architecture of the Fur-DNA complex.

The model that emerges from the three-dimensional assembly of Fur$_{PA}$ on DNA somehow integrates the proposals discussed above. On one hand, it supports the idea that a Fur tetramer binds two overlapping and opposite operators. However, it is also consistent with the suggestion that Fur recognizes arrays of three or more hexameric sites. In this case, the central hexamer region would be in contact with both dimers whereas the flanking regions would be recognized by only one dimer. Under this view, additional Fur dimers could bind when adjacent hexamers are present in extended operators (such as the 100 bp of the aerobactin promoter region of *E. coli* or the 63 bp of the *pvdS* promoter of *P. aeruginosa*). Since at least two dimers, and frequently more (see below), are needed to cause full repression of iron starvation-inducible genes, this model accounts for the need for a high concentration of Fur in cells, perhaps in combination with the suggested role of Fur in the tolerance of redox or oxidative stress (see above).

Some topical observations are still difficult to explain with this general view, in particular, the interaction of the Fur protein of *V. anguillarum* with its target site at the *fatD* promoter. In this case, some bases were hypersensitive to DNase I on Fur binding, suggesting that the bound sites underwent some type of Fur-induced bending. This is difficult to reconcile with the view of the interaction that emerges from studying the Fur$_{PA}$ structure. In fact, the corkscrew/overlapping-dimer structure would rather tend to straighten out otherwise bent DNA following Fur binding, an observation which has been repeatedly made with the Fur repressor of *E. coli*. These observations highlight the importance of determining the basis of the discrepancies between the interactions of Fur proteins of *E. coli*, *V. anguillarum*, and *P. aeruginosa* with virtually identical DNA target sites.

ORGANIZATION OF Fur TARGET SEQUENCES IN *E. COLI*

Regardless of the specific mechanism by which Fur binds its target DNA, it is possible to describe Fur sites in a genome as an array of the hexamer NAT(A/T)AT discussed above. This affords a new interpretation of many previously reported Fur sites. Figure 6 shows DNA segments of *E. coli* which have proven to be protected by the Fur protein, along with others simply known to be regulated by iron. Extended sites might tolerate a degree of divergence in the sequences involved, which could be compensated by the higher overall affinity. This view immediately suggests that depending on the number and the sequence matches of the hexamer arrays, different genes may have an entirely different response to iron conditions. Promoters bearing a minimal site close to the 19-bp consensus sequence (three hexameric repeats) would express cognate genes in a fashion subject to a simple on/off switch

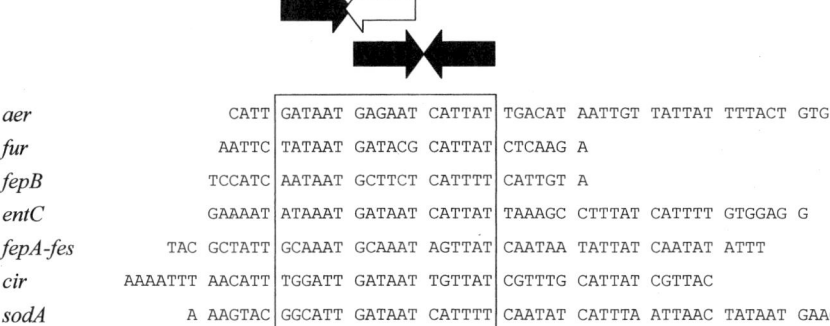

FIGURE 6 View of iron-responsive promoters of *E. coli* as arrays of hexameric sequences. The promoters shown here include those of the siderophore systems aerobactin (*aer*) and enterochelin (*fepB*, *fepA*, and *fes*), as well as those of the *fur* gene itself, the receptor for colicin I (*cir*), and the receptor for superoxide dismutase (*sodA*). The three repeats which comprise the consensus Fur box include the most highly conserved sequences. However, in addition, many other conserved bases [according to the 6-bp minimal motif NAT(A/T)AT discussed in the text] can be found within the protected sequences neighboring the core Fur box. It is likely that such adjacent sequences are not coincidental but are arrayed in a configuration of 6-bp repeats with a potential to interact specifically with the Fur protein as a whole. These additional contacts might strengthen the overall binding of the DNA segment to the regulator and explain why the protection is not limited to the consensus Fur box. It seems, therefore, that once a minimum of three repeats is assembled, a further increase in the number of repeats, in any orientation, may extended Fur-binding sites with higher affinities for the protein resulting from their increasingly cooperative occupation.

mechanism rather than to a gradual responsiveness to iron starvation. In contrast, in the cases where Fur binds to extended arrays of binding sites (which can be present in variable numbers and with different degrees of conservation), a wider window of responses can be generated. Thus, the extent of the repression of each iron-regulated promoter would vary as a function of the composition of the arrays, permitting a gradual response to iron deprivation. This means that depending on the specific intracellular concentration of the ion, as well as the redox or oxidative conditions of the cells (see above), some genes or gene groups would be inhibited while others would be derepressed. The general picture that emerges from all these considerations supports the function of Fur as a general regulator.

The genetic and biochemical technology used so far has identified only promoters subject to a relatively strong control by iron (i.e., displaying very strong Fur-DNA interactions). It is likely, however, that many other genes come under the influence of iron through the presence of less canonical, weaker binding sites spread over the corresponding promoter sequences. Thus, some genes may show a mild regulation or coregulation by iron while others are regulated by a strong repression/induction switch. The details of repression will become clearer as global responses to iron stress in different bacteria are examined by DNA chip technology and other genomic or proteomic approaches.

SIMILARITY OF Fur BOXES TO OTHER REGULATORY ELEMENTS?

The recent availability of bacterial genomes allows massive comparisons of DNA sequences that can be the target of regulatory proteins. In particular, some Fur-binding sites and some UP elements seem to occasionally coincide in the chromosome of *E. coli*. UP elements are AT-rich DNA sequences first described as upstream promoter segments that strongly stimulated the transcription of ribosomal promoters such as P*rrnB*. It was later discovered that such sites in many types of promoters interacted with the C-terminal domain of the α subunit of RNAP (αCTD) and facilitated the recruitment of the holoenzyme to the −35/−10 sequence. More recent research has defined a consensus sequence for full UP sites, which can be further refined as the addition of two (proximal and distal) subsites. When properly phased with the rest of the promoter elements, the perfect UP consensus sequence can increase transcription in vivo by a factor of 330. Comparison of the Fur box sequence with the UP consensus reveals little straightforward homology (Fig. 7). However, most of the bases in the Fur box are conserved with respect to the distal consensus subsite. When making genome searches, it should be noted that some moderately good αCTD binding sites do not match the consensus well but may have bases that are second-best choices rather than the consensus. In this respect, the comparison of Fig. 7 is somewhat conservative, since the consensus sequences do not take into consideration second-best choices within the αCTD-binding site. This raises the possibility that some UP elements might also fine-tune the corresponding promoters in response to iron. An in silico survey of Fur-UP overlaps revealed a collection of about 25 cases where such coincidence was evident. Fur-binding sites may perform as UP-like sites when placed at a suitable distance from the RNA polymerase binding site so that transcriptional inhibition occurs by blocking the access of the αCTD to such the site. In any case, the Fur-UP overlaps reveal an unexpected repertoire of channels for integration of iron starvation signals in given promoters. This highlights iron availability as one major environmental signal to which hundreds of genes adapt their expression.

ACKNOWLEDGMENTS

We thank J. Crosa, J. Helmann, and M. Vasil for making available results prior to publication.

This work was supported by EU contracts QLK3-CT-2002–01933, QLK3-CT-2002–01923, and INCO-CT-2002–1001; by grant BIO2001–2274 of the Spanish Comisión Interministerial de Ciencia y Tecnología (CICYT), and by the Strategic Research Groups Program of the Autonomous Community of Madrid.

SUGGESTED READING

Andrews, S. C., A. K. Robinson, and F. Rodriguez-Quiñones. 2003. Bacterial iron homeostasis. *FEMS Microbiol. Rev.* **27:**215–237.

Baichoo, N., and J. D. Helmann. 2002. Recognition of DNA by Fur: a reinterpretation of the Fur box consensus sequence. *J. Bacteriol.* **184:**5826–5832.

Baichoo, N., T. Wang, R. Ye, and J. D. Helmann. 2002. Global analysis of the *Bacillus subtilis* Fur regulon and the iron starvation stimulon. *Mol. Microbiol.* **45:**1613–1629.

Chai, S., T. J. Welch, and J. H. Crosa. 1998. Characterization of the interaction between Fur and the iron transport promoter of the virulence plasmid in *Vibrio anguillarum*. *J. Biol. Chem.* **273:**33841–33847.

Chen, Q., A. M. Wertheimer, M. E. Tolmasky, and J. H. Crosa. 1996. The AngR protein and the siderophore anguibactin positively regulate the expression of iron-transport genes in *Vibrio anguillarum*. *Mol Microbiol.* **22:**127–134.

Chen, Q., and J. H. Crosa. 1996. Antisense RNA, Fur, iron and the regulation of iron transport genes in *Vibrio anguillarum*. *J. Biol. Chem.* **271:**1885–1891.

Delany, I., G. Spohn, R. Rappuoli, and V. Scarlato. 2001. The Fur repressor controls transcription of iron-activated and -repressed genes in *Helicobacter pylori*. *Mol. Microbiol.* **42:**1297–1309.

de Lorenzo, V., F. Giovannini, M. Herrero, and J. B. Neilands. 1988. Metal ion regulation of gene expression: Fur repressor-operator interaction at the promoter region of the aerobactin system of pColV-K30. *J Mol. Biol.* **203:**875–884.

de Lorenzo, V., M. Herrero, F. Giovannini, and J. B. Neilands. 1988. Fur (ferric uptake regulation) protein and CAP (catabolite-activator protein) modulate transcription of *fur* gene in *Escherichia coli*. *Eur. J. Biochem.* **173:**537–546.

```
UP consensus   NNAAAWWTWTTTTNNNAAANNN
proximal                   AAAAAARNR
distal         NNAWWWWWTTTTTN
Fur box        GATAATGATAATCATTATC
```

FIGURE 7 Fur boxes and UP elements. The figure shows an alignment between the consensus UP site and the Fur box. W = A or T; R = A or G; N = any base. If the same UP-like site is occupied by Fur-Mn^{2+} (in vitro) or Fur-Fe^{2+} (in vivo), the RNAP may fail to bind the downstream −10 and −35 hexamers. In this way, Fur would hinder the UP-like element created by its own recognition sequence.

Dubrac, S., and D. Touati. 2002. Fur-mediated transcriptional and post-transcriptional regulation of FeSOD expression in *Escherichia coli*. *Microbiology* **148:**147–156.

Escolar, L., J. Perez-Martin, and V. de Lorenzo. 1999. Opening the iron box: transcriptional metalloregulation by the Fur protein. *J. Bacteriol.* **181:**6223–6229.

Escolar, L., J. Pérez-Martín, and V. de Lorenzo. 2000. Evidence of an unusually long operator for the Fur repressor in the aerobactin promoter of *Escherichia coli*. *J. Biol. Chem.* **275:**24709–24714.

Hantke, K. 1981. Regulation of ferric iron transport in *Escherichia coli* K12: isolation of a constitutive mutant. *Mol. Gen. Genet.* **182:**288–292.

Hantke, K. 2001. Iron and metal regulation in bacteria. *Curr. Opin. Microbiol.* **4:**172–177.

Lavrrar, J. L., C. A. Christoffersen, and M. A. McIntosh. 2002. Fur-DNA interactions at the bidirectional *fepDGC-entS* promoter region in *Escherichia coli*. *J. Mol. Biol.* **322:**983–995.

Lavrrar, J. L., and M. A. McIntosh. 2003. Architecture of a Fur binding site: a comparative analysis. *J. Bacteriol.* **185:**194–202.

Masse, E., and S. Gottesman. 2002. A small RNA regulates the expression of genes involved in iron metabolism in *Escherichia coli*. *Proc. Natl. Acad. Sci. USA* **99:**4620–4625.

Ochsner, U. A., P. J. Wilderman, A. I. Vasil, and M. L. Vasil. 2002. GeneChip expression analysis of the iron starvation response in *Pseudomonas aeruginosa*: identification of novel pyoverdine biosynthesis genes. *Mol. Microbiol.* **45:**1277–1287.

Outten, C. E., D. A. Tobin, J. E. Penner-Hahn, and T. V. O'Halloran. 2001. Characterization of the metals receptor sites in *Escherichia coli* Zur, an untrasensitive zinc(II) metalloregulatory protein. *Biochemistry* **40:**10417–10423.

Pohl, E., J. C. Haller, A. Mijovilovich, W. Meyer-Klaüke, E. Garman, and M. Vasil. 2003. Architecture of a protein central to iron homeostasis: crystal structure and spectroscopic analysis of the ferric uptake regulator. *Mol. Microbiol.* **47:**903–915.

Stork, M., M. Di Lorenzo, T. J. Welch, L. M. Crosa, and J. H. Crosa. 2002. Plasmid-mediated iron uptake and virulence in *Vibrio anguillarum*. *Plasmid* **48:**222–228.

Wee, S., J. B. Neilands, M. L. Bittner, B. C. Hemming, B. L. Haymore, and R. Seetharam. 1988. Expression, isolation and properties of Fur (ferric uptake regulation) protein of *Escherichia coli* K-12. *Biol. Metals* **1:**62–68.

Zheng, M., B. Doan, T. D. Schneider, and G. Storz. 1999. OxyR and SoxRS regulation of *fur*. *J. Bacteriol.* **181:**4639–4643.

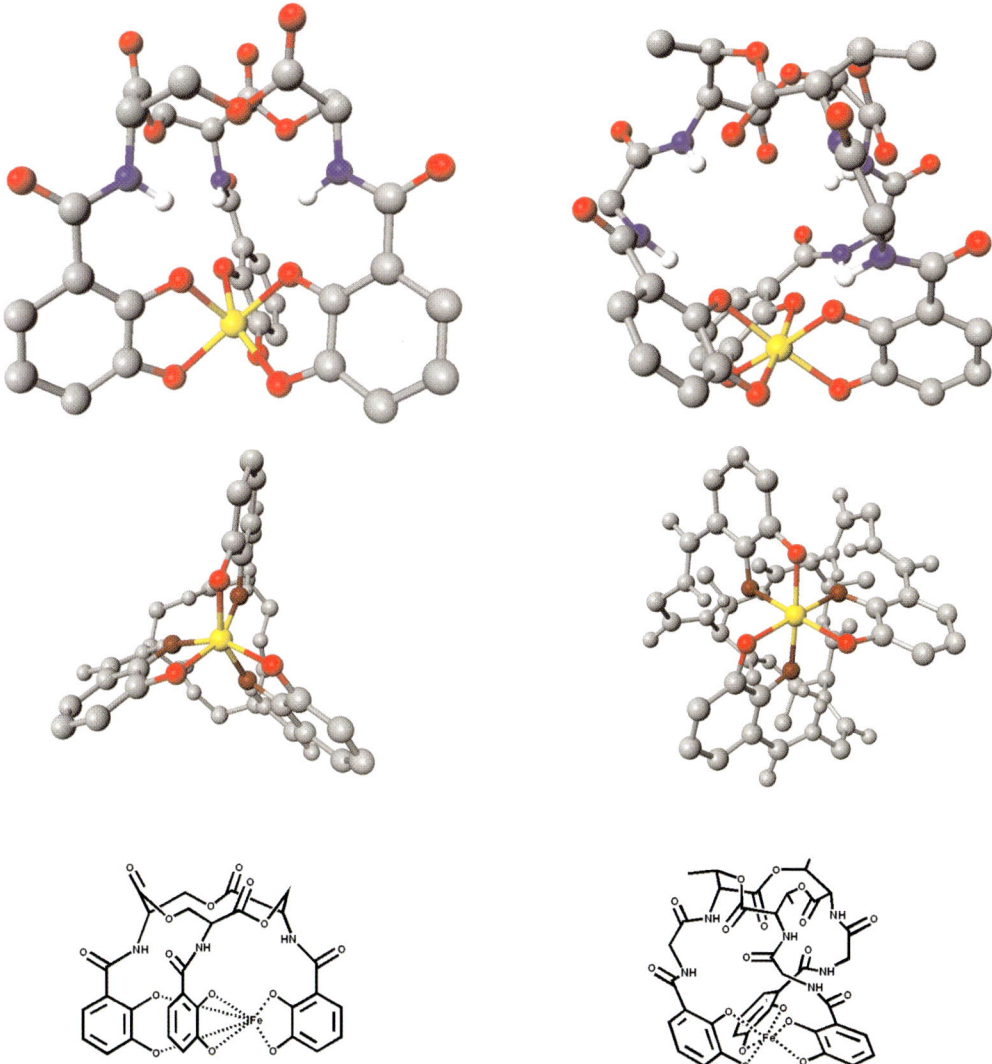

COLOR PLATE 1 (Chapter 1) (Top) Model of ferric enterobactin based on the crystal structure of vanadium(IV) enterobactin (left), and structure of ferric corynebactin as deduced from circular dichroism spectra and computer modeling (right). (Middle) Viewed down the pseudo-three-fold axis, the chirality of each metal center can be seen. Ferric enterobactin is Δ, while ferric corynebactin is Λ. Also apparent from this view is the difference in geometry around the iron center. While ferric corynebactin is nearly octahedral, ferric enterobactin is distorted toward a trigonal prismatic geometry. Note that only the ferric tris(catechol) center is colored. The three oxygen atoms on the front face of the octahedron are red; the three oxygen atoms on the back face are dark red. (Bottom) Under each structure is a schematic drawing that shows the molecular conformation.

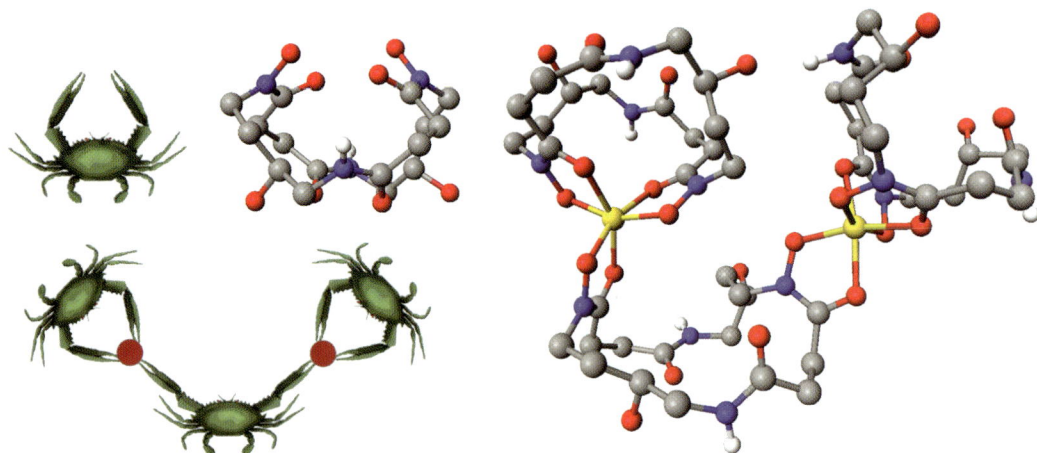

COLOR PLATE 2 (Chapter 1) Structural comparisons show the preorganization of alcaligin for metal binding. The structure of the free ligand (at left) is essentially the same as for the FeL^+ metal complex. The root mean square deviation in atom positions is 0.227Å. The structure of the dimeric Fe_2L_3 complex (at right) requires that the central ring twist from its ground state structure, which is preorganized to bind a common metal center, in order to bridge two metal ions. The cartoon illustrates this effect and is a reminder that the word "chelate" is derived from chelos, or crab's claw.

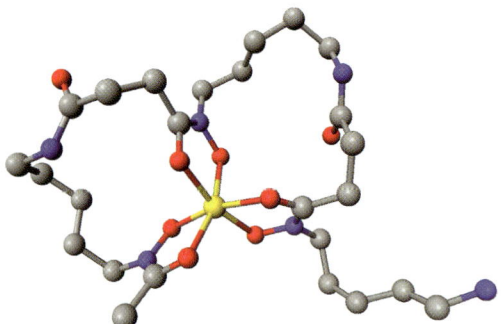

COLOR PLATE 3 (Chapter 1) Crystal structure of Fe(III) ferrioxamine B. The structure shown has a Δ chirality.

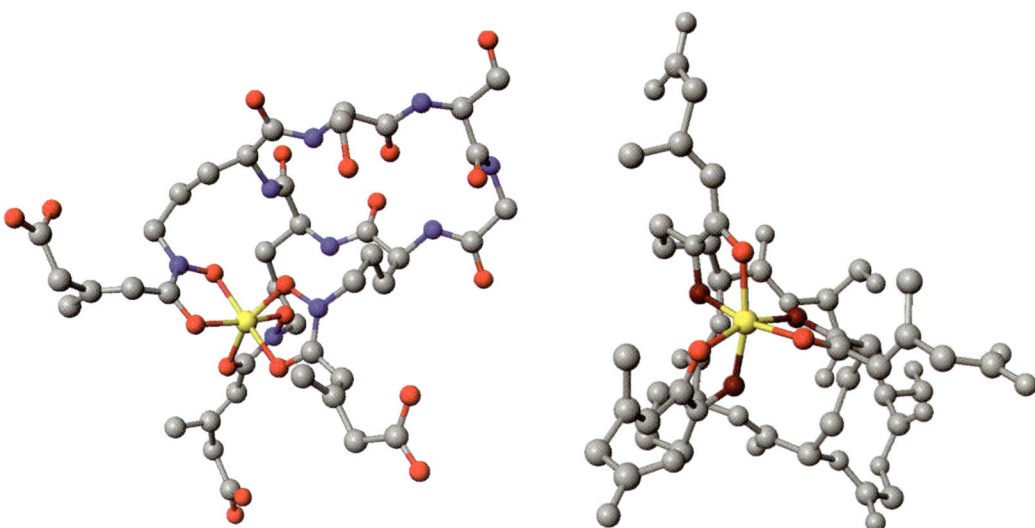

COLOR PLATE 4 (Chapter 1) (Left) Crystal structure of Fe(III) ferrichrome A. Due to disorder of two of the three methylglutaconic acid moieties and one seryl hydroxyl group, the original crystal structure was incomplete. These moieties were added into this Figure for clarity. (Right) A view down the pseudo-threefold axis showing the Δ chirality of the complex. Note that only the ferric tris(catechol) center is colored. The three oxygen atoms on the front face of the octahedron are red; the three oxygen atoms on the back face are dark red.

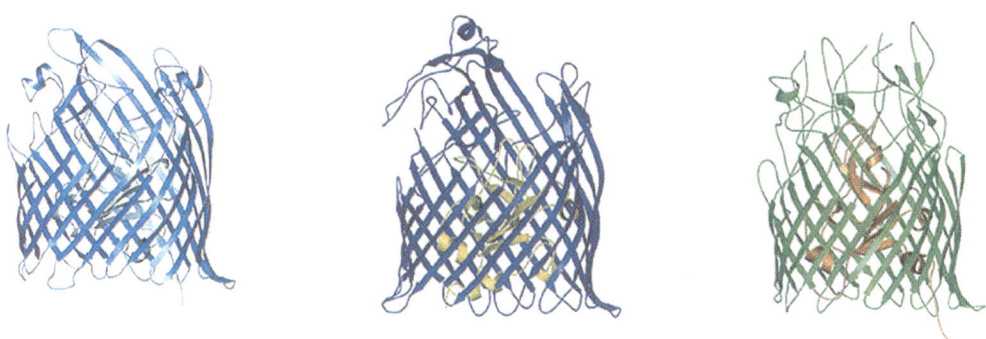

COLOR PLATE 5 (Chapter 4) Crystal structures of FecA (left), FhuA (middle), and FepA (right). The globular domain is drawn in a different color in each structure. Reprinted from Ferguson et al. (2002) with permission from the publisher.

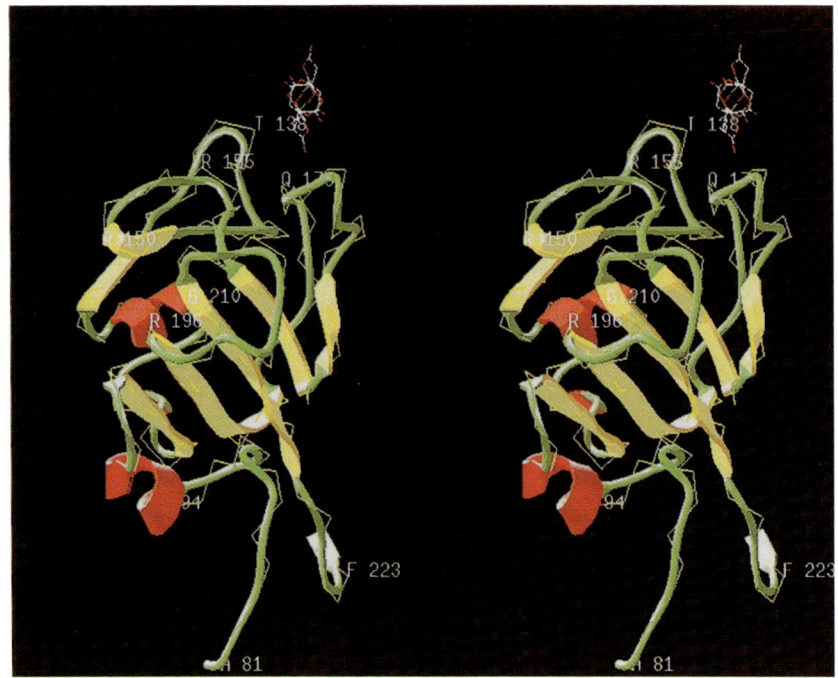

COLOR PLATE 6 (Chapter 4) Stereo view of the globular domain in FecA. The secondary structure is shown, with the helix in red, the β-strand in yellow, and the coil in green. Apices A (residue 138), B (residue 155), and C (residue 176) are indicated, as well as Arg150 and Arg196 involved in the lock region, the beginning and end of the β5-β6 loop (residues 196 and 210), the last residue of the domain (residue 223) and the first residue which could be located in the structure (residue 81). The diferric-dicitrate in the liganded structure is shown to visualize the possible channel for transport.

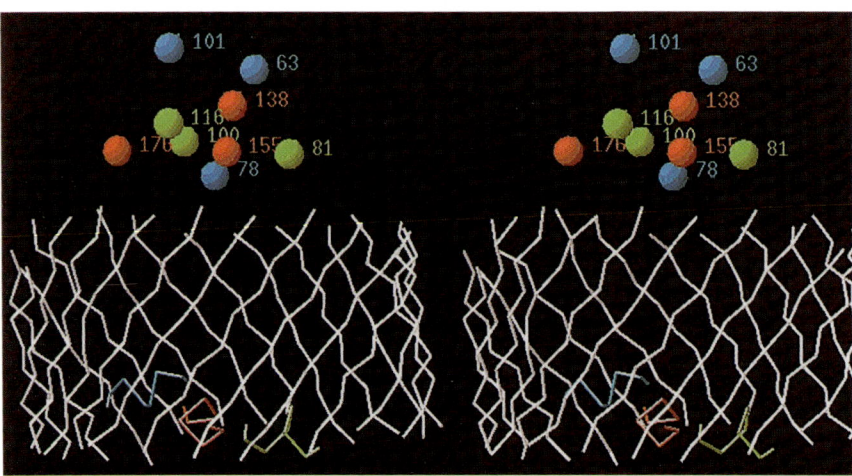

COLOR PLATE 7 (Chapter 4) Stereo view of the apices and switch helices in the three structures. The apices are indicated by residue number. The relative locations with respect to the barrel are shown, after a least-squares fit of the three structures, for FepA (blue [63, 78, 101]), FhuA (green [81, 100, 116]), and FecA (red [138, 155, 176]).

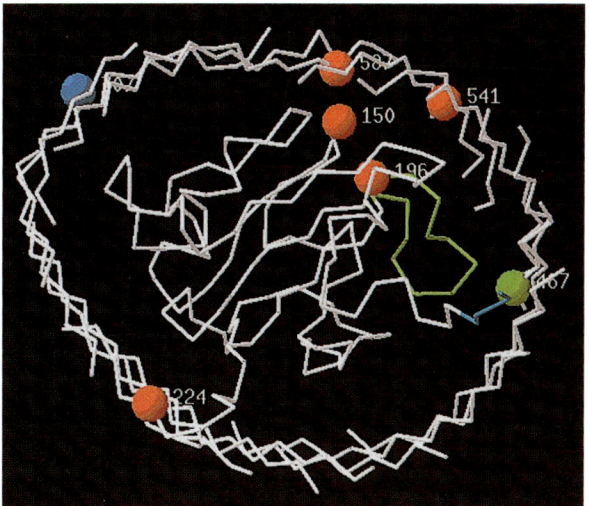

COLOR PLATE 8 (Chapter 4) Reference points in the structure of FecA. The view is from the bottom of the barrel. Arg150 and Arg196 of the globular domain and Glu541 and Glu587 on the barrel (red) constitute the lock region. The first residue of the barrel is also indicated. The TonB box and the possible second site of interaction with TonB are in blue, and the β5-β6 loop and strictly conserved Arg467 lining the putative channel are in green. These reference points occur in all three structures at identical locations.

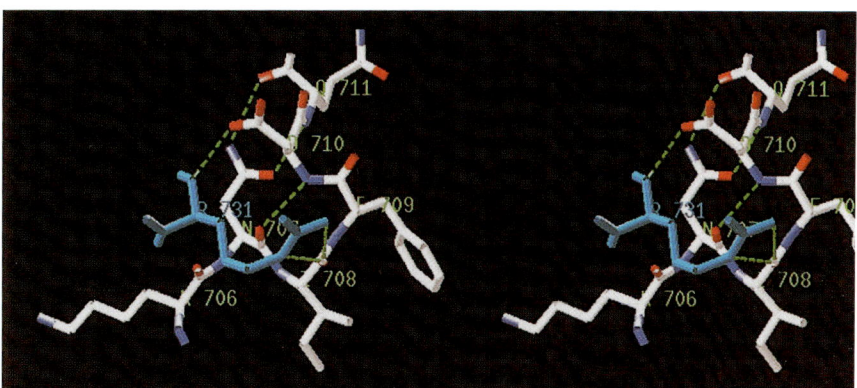

COLOR PLATE 9 (Chapter 4) Stereo view of the possible second site of interaction with TonB in FecA. The strictly conserved residues 707 to 710, at the beginning of loop 11, are in CPK colors, and the identical Arg731, at the end of loop 11, is in blue. This structural component superposes with the corresponding structures in FepA and FhuA.

A.

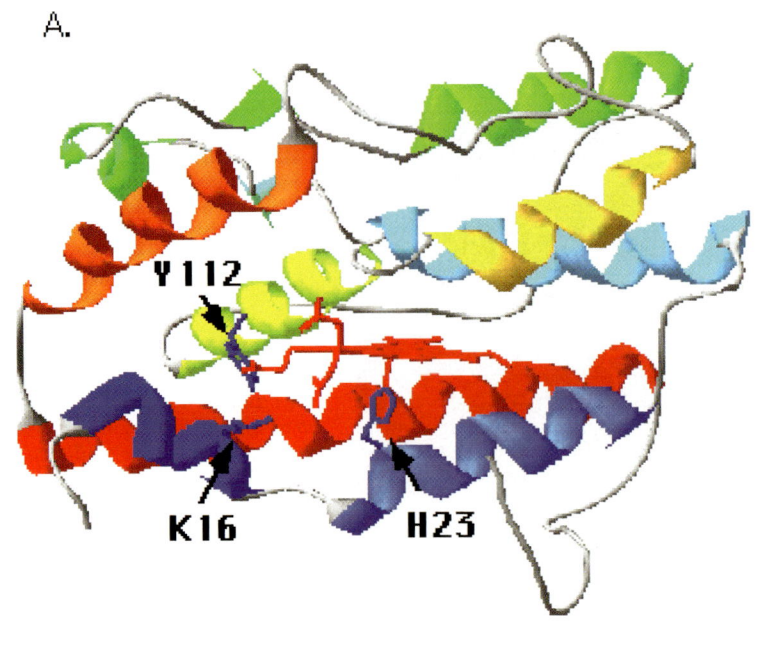

B.

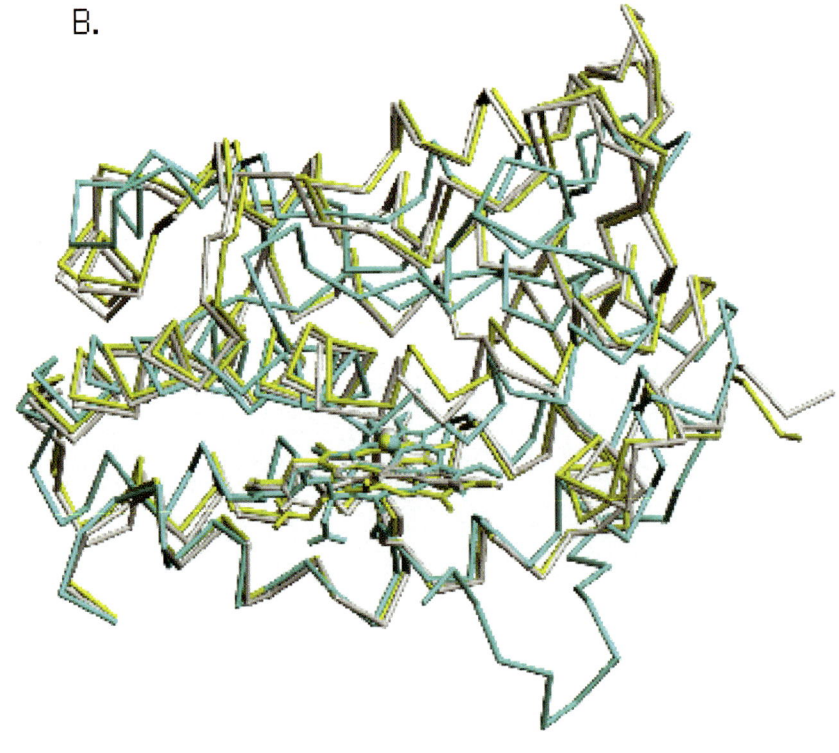

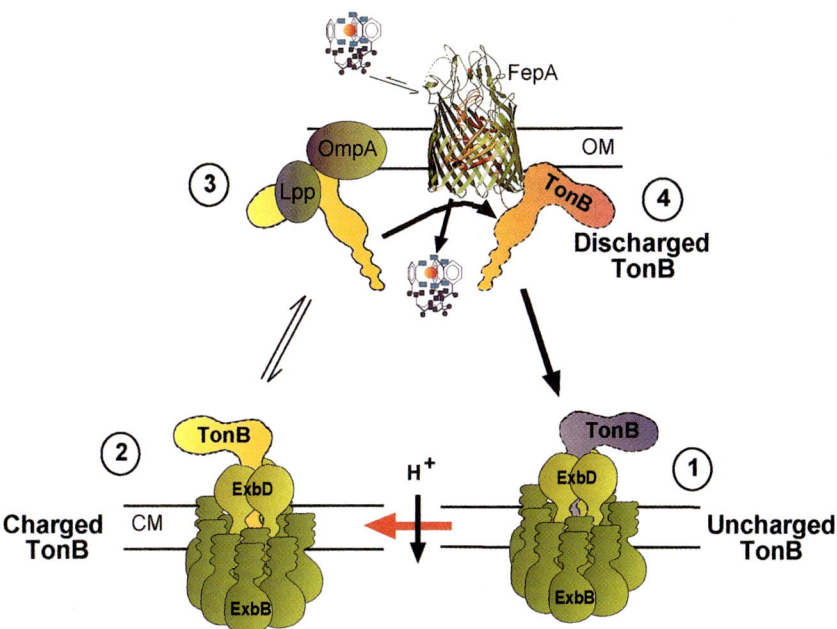

COLOR PLATE 11 (Chapter 7) Current model for TonB-dependent energy transduction. For a description, see the text. Reprinted from Postle and Kadner (2003) with permission of the publisher.

COLOR PLATE 10 (Chapter 6) (A) Crystal structure of HemO. The His23 ligand of the proximal helix (light and dark blue) coordinates the heme molecule (red) within the molecule. Residues Lys16 and Tyr112 coordinate the propionates of the heme molecule for regioselectivity by HemO. (B) Alignment of truncated human HO-1 (yellow) and *N. meningitidis* HemO (blue), demonstrating the high degree of structural similarities. Adapted from Schuller et al. (1999 and 2001).

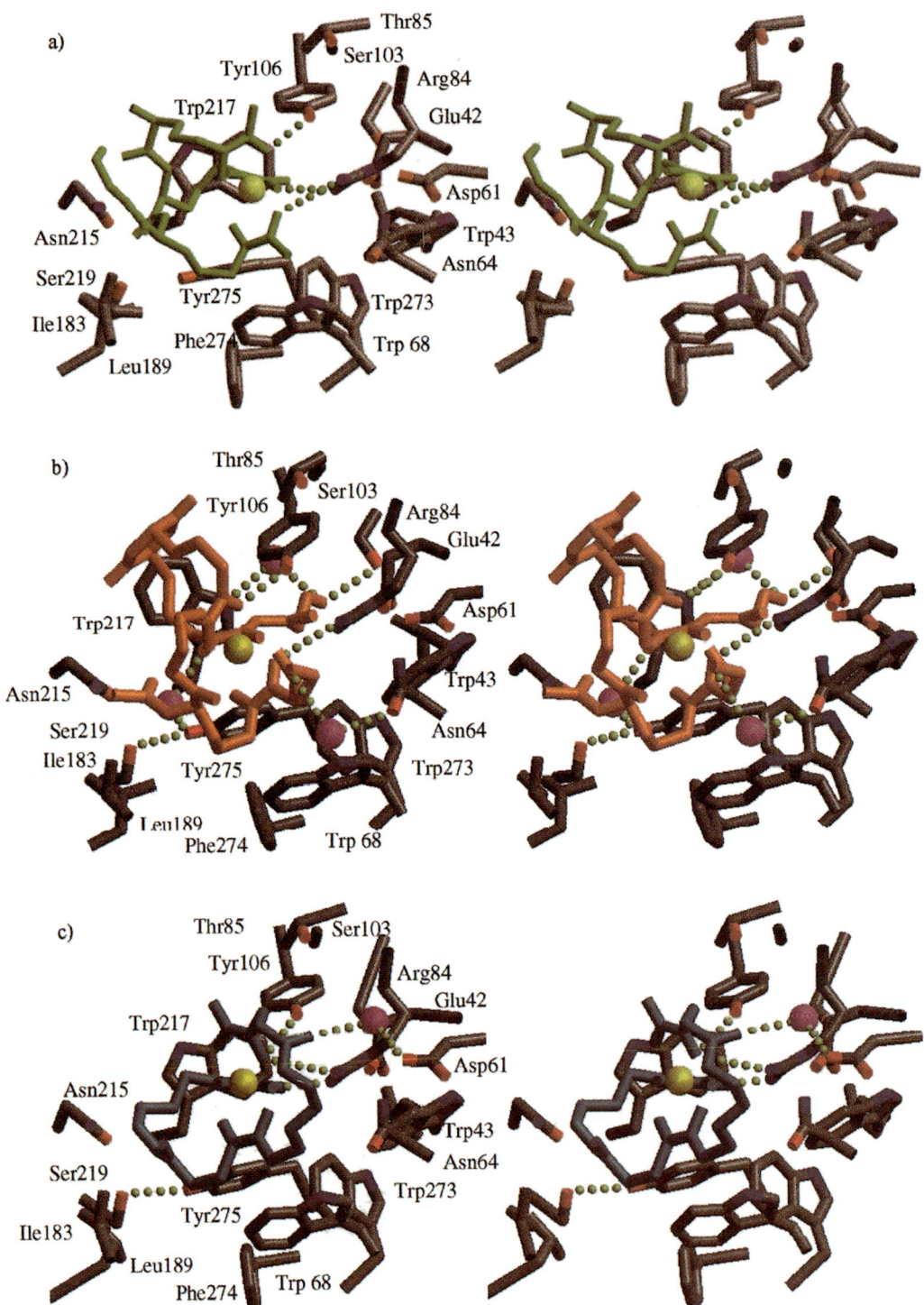

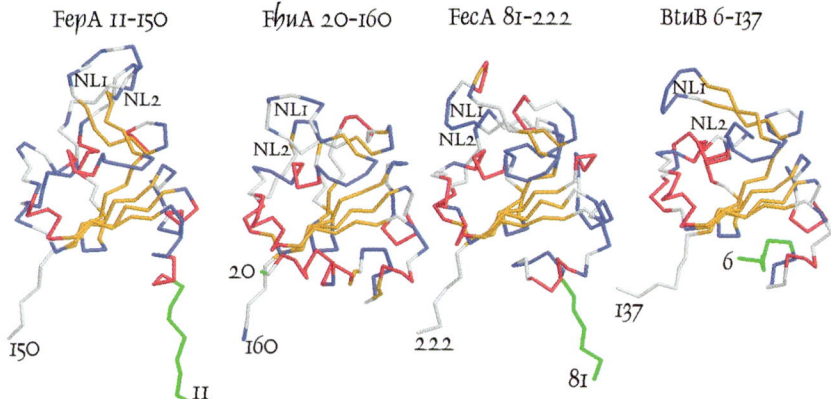

COLOR PLATE 13 (Chapter 10) The FepA, FhuA, FecA, and BtuB N termini. A four-strand β-sheet is the structural basis of all four N-domains, which each contain a loop (NL1) that projects into their surface vestibules. α-Helical regions are red, β-sheets are gold, turn domains are blue, and the TonB box regions of the receptor proteins are green. Note the similarity of the FepA and FecA TonB box sequences (residues 11 to 18 and 81 to 87, respectively), which adhere to the wall of their β-barrels in the ligand-free state, and the dissimilarity of the BtuB TonB box structure (crystallographic coordinates are from the Protein Data Bank [http://www.rcsb.org/pdb]).

COLOR PLATE 12 (Chapter 8) Stereo view of the binding sites in the albomycin (a), coprogen (b), and Desferal (c) complexes. The important hydrogen bonds involved in binding each of these hydroxamate-type sideophores are labeled. Albomycin is green, coprogen is orange, and Desferal is blue in each of their respective complexes. The proportion of albomycin visible is identical to that of ferrichrome. Reprinted from Clarke, Braun, et al. (2002) with permission from the publisher.

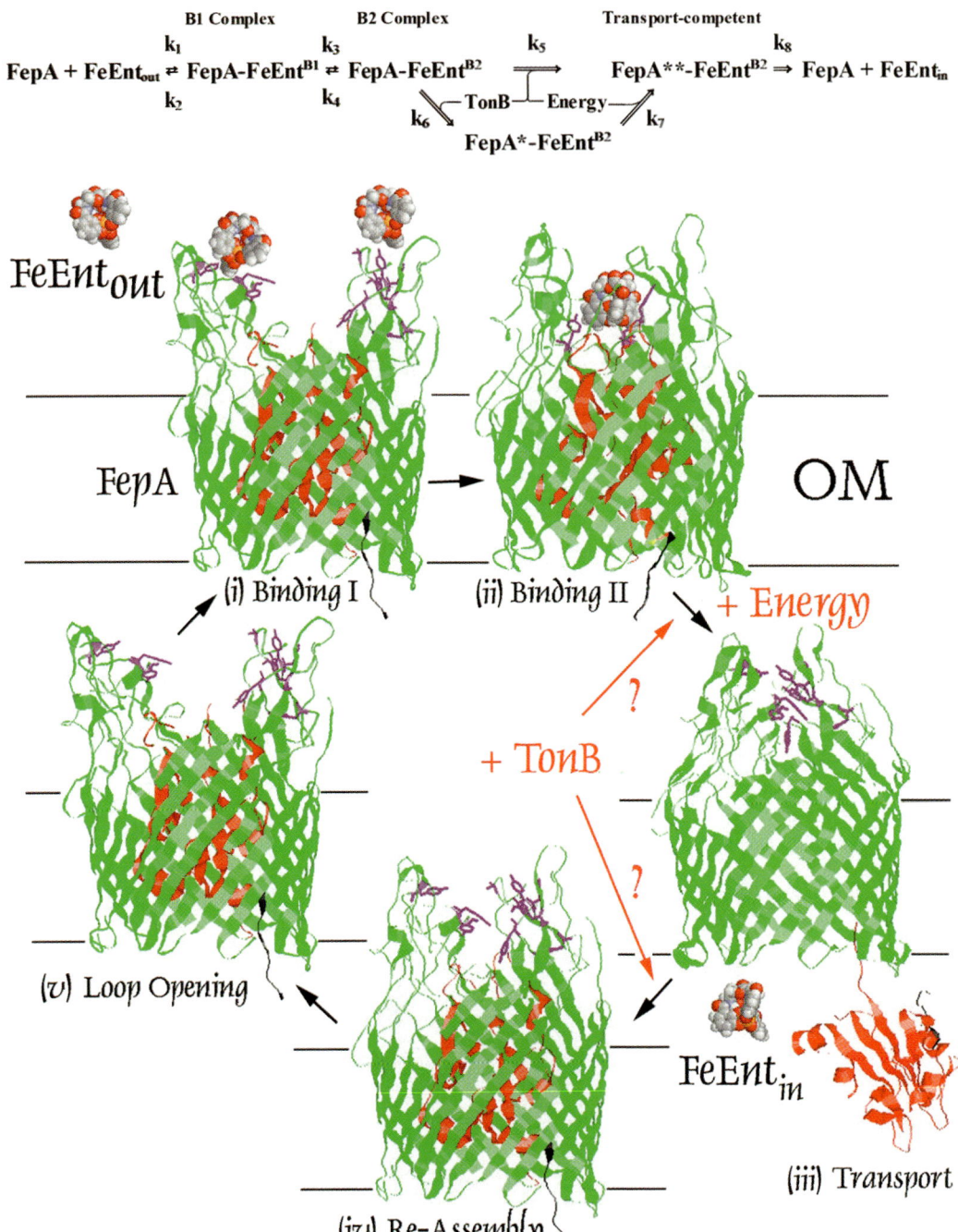

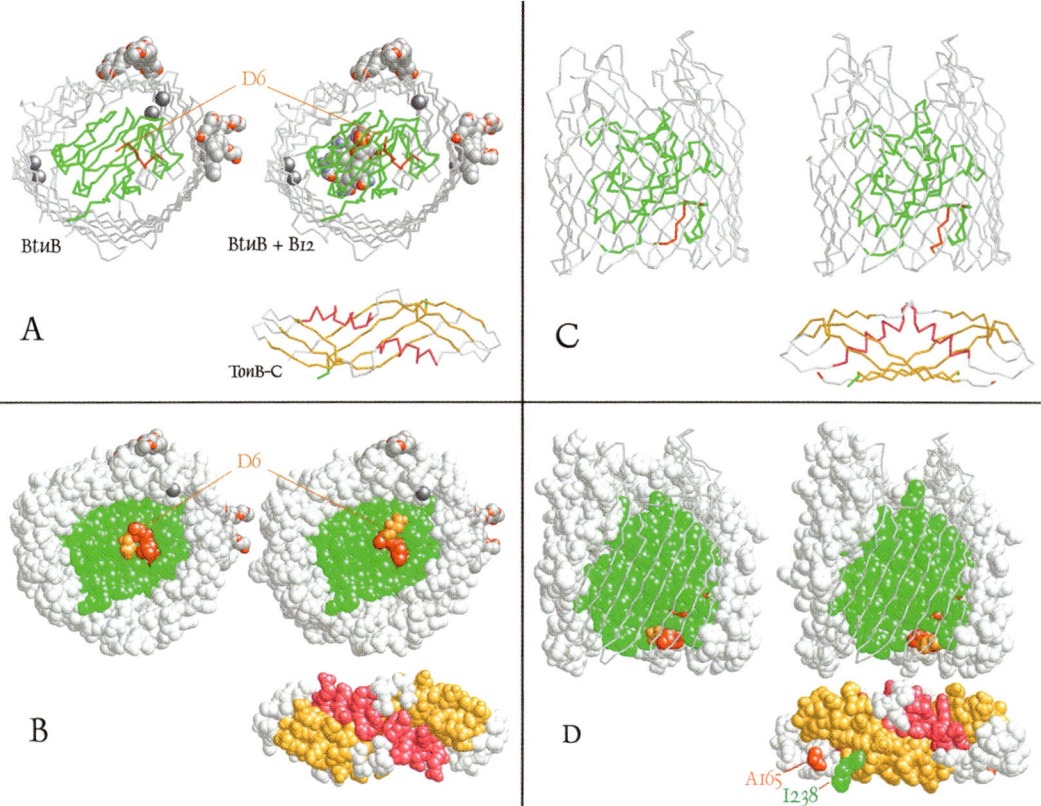

COLOR PLATE 15 (Chapter 10) Crystal structures of the TonB C terminus and BtuB in the absence and presence of B_{12}. For the TonB C terminus, helices are red and β-sheets are gold. A165, the N-terminal-most solved residue, 5 residues downstream from Q160, is red. I238, the penultimate amino acid, is green. For BtuB, the β-barrel is white, the N-domain is green, the TonB box is red, and amino acid D6, the first residue of the TonB box, is yellow. The figure illustrates the conformational motion that occurs in response to B12 binding: D6 flips over (A to D). The TonB box is within the barrel and does not extrude on ligand binding, and the TonB C-domain is too large to enter the barrel.

COLOR PLATE 14 (Chapter 10) Model of FeEnt transport through FepA. (Top) Formal representation of the FepA transport process. Constants k_1 to k_4 are experimentally defined: fluors attached to L3 reflect both the first and second binding stages ($k_1 = 0.02/s$; $k_3 = 0.005/s$;), which are both reversible ($k_2 = 0.03/s$; $k_4 = 0.003/s$ [see the text]). (Bottom) The FepA transport cycle is depicted as a series of conformational stages that result in binding and internalization of FeEnt. The representations of FepA originated from its crystal structure, but they are postulated forms that were not crystallographically demonstrated. By analogy to FhuA and FecA, FeEnt binding may relocate the TonB-box region of FepA away from the β-barrel wall. Such movement may signal receptor occupancy to TonB, but another view is that TonB box movement away from the barrel wall allows the N-domain to dislodge from the channel. Next, the ligand passes through the C-domain channel (Transport). Theory and experiments suggest, but so far do not explicitly prove, that input of energy is required at this stage. Similarly, TonB may or may not function during this phase of the transport reaction. A variety of findings raise the possibility that the N-domain exits the pore during ligand uptake, but this idea is not fully substantiated: structural changes that facilitate ligand transport may take place in the N-domain while it is resident in the channel. After transport, the receptor re-assembles, either by reinsertion of the N-domain into the β-barrel or by structural changes in situ within the pore, another potential phase for the input of energy and/or TonB. Lastly, the loops re-open to a state of maximum receptivity to ligands.

IRON TRANSPORT SYSTEMS IN PATHOGENIC BACTERIA

IV

PATHOGENIC *ESCHERICHIA COLI*, *SHIGELLA*, AND *SALMONELLA*

Shelley M. Payne and Alexandra R. Mey

14

The gram-negative enteric pathogens are a closely related group of bacteria. On the basis of genome analyses, *Escherichia coli* and *Shigella* can be considered a single species, and their classification as separate genera is largely historical. The *Salmonella* species are evolutionarily more distant but share many characteristics with the *E. coli* group. The enteric pathogens usually initiate infection following ingestion, causing diseases ranging from relatively mild enteritis to dysentery and septicemia. The extent of invasion and the nature of the disease depend on the specific set of virulence factors expressed by these pathogens. The pathogenic members of the enteric bacteria discussed in this chapter, and the diseases they cause, are summarized in Table 1.

The enteric pathogens can be found in a variety of niches within the host. Some strains are restricted to the intestinal tract, while others are invasive and cause systemic infections. There are examples of intracellular pathogens, as well as pathogens that colonize only the epithelial cell surfaces. Given their abilities to survive and replicate within a wide range of environments, it is not surprising that these pathogens express a diverse assortment of iron acquisition systems. Many of the iron transport systems originally described in *E. coli* K-12 (discussed in the earlier chapters of this book) are found also in the enteric pathogens. These include systems for enterobactin synthesis and transport and uptake systems for ferrichrome, ferric citrate, and ferrous iron. The major distinguishing feature of the pathogenic members of the group is the number and variety of additional iron transport systems present, many of which are encoded on plasmids or within pathogenicity islands. The major iron acquisition systems present in the enteric pathogens are listed in Table 2.

SIDEROPHORE-MEDIATED IRON TRANSPORT SYSTEMS

Enterobactin

Most of the enteric pathogens can use the catechol siderophore enterobactin for iron transport. The structure, function, and regulation of this system are essentially the same in the pathogenic isolates as in *E. coli* K-12 and are described in more detail in chapter 9. Enterobactin, a cyclic trimer of 2,3-dihydroxybenzoylserine (DHBS), is assembled via a nonribosomal peptide synthesis pathway by the products of the *entA(B/G)CDEF* genes. En-

---

Shelley M. Payne and Alexandra R. Mey, Department of Molecular Genetics and Microbiology, The University of Texas at Austin, Austin, TX 78712.

TABLE 1 The gram-negative enteric pathogens and the diseases they cause

Organism	Disease
Escherichia coli	
Intestinal isolates	
K-12	Nonpathogenic
Enterotoxigenic *E. coli* (ETEC)	Watery, cholera-like diarrhea
Enteropathogenic *E. coli* (EPEC)	Gastroenteritis
Enteroaggregative *E. coli* (EAggEC)	Persistent diarrhea
Enterohemorrhagic *E. coli* (EHEC)	Bloody diarrhea, hemorrhagic colitis, hemolytic-uremic syndrome
Enteroinvasive *E. coli* (EIEC)	Bacillary dysentery
Extraintestinal pathogenic *E. coli* (ExPEC)	
Neonatal meningitis *E. coli* (NMEC)	Septicemia, meningitis
Uropathogenic *E. coli* (UPEC)	Urinary tract infection
Salmonella enterica	
S. enterica serovar Typhi	Typhoid fever
S. enterica serovar Typhimurium	Gastroenteritis
S. enterica serovar Enteritidis	Gastroenteritis
Shigella	
S. flexneri	Bacillary dysentery
S. dysenteriae	Bacillary dysentery
S. sonnei	Bacillary dysentery
S. boydii	Bacillary dysentery

TABLE 2 Iron acquisition systems in the enteric pathogens

Strain[b]	Presence of iron acquisition system[a]:									
	Ent/Fep	Iuc/Iut	Ybt	Fhu	Iro	Sit	Fec	Feo	Cor	Shu (Chu)
E. coli										
Intestinal isolates										
K-12	+	−	−	+	−	−	+	+	+	−
EPEC E2348/69	+	−	−	+	−	−	−	+	+	+
EAggEC O42	+	−	+	+	−	+	+	+	+	+
EHEC O157:H7	+	−	+	+	−	−	−	+	+	+
Extraintestinal isolates										
UPEC CFT073	+	+	+	+	+	+	−	+	+	+
NMEC O18ac:H7:K1	+	−[c]	+	+	+	+	−	+	+	+
S. enterica serovar Typhimurium	+	+[d]	−	+[e]	+	+	−	+	+	−
S. bongori	+	−	−	+	−	+	−	+	+	−
S. flexneri	−	+	−	+	−	+	+[f]	+	+	−
S. dysenteriae type 1	+	−	−	+	+	+	+	+	+	+
S. sonnei	+	+	−	+	−	+	+	+	+	+[g]
S. boydii	−	+	−	+	−	+	+	+	+	−

[a] Ent/Fep, enterobactin synthesis and transport; Iuc/Iut, aerobactin synthesis and transport; Ybt, yersiniabactin synthesis and transport; Iro, salmochelin synthesis and transport; Sit, ferrous iron and manganese transport; Fec, ferridicitrate transport; Feo, ferrous iron transport; Cor, magnesium transport; Shu (Chu), heme transport.

[b] Strain abbreviations as in Table 1. Sequence data for EPEC E2348/69 and EAggEC O42 was obtained from the Wellcome Trust Sanger Institute (http://www.sanger.ac.uk/Projects/Escherichia_Shigella/); data for NMEC O18ac:H7:K1 were obtained from the University of Wisconsin *E. coli* Genome Project (http://www.genome.wisc.edu/sequencing/rs218.htm). Sequence data for *E. coli* K-12, EHEC O157:H7, and UPEC CFT073 are available through the National Institute for Biotechnology Information genome database (http://www.ncbi.nlm.nih.gov/PMGifs/Genomes/micr.html).

[c] Although this particular NMEC strain does not carry the aerobactin locus, these genes are found in many other systemically invasive *E. coli* K1 isolates.

[d] When present in *Salmonella* species, the aerobactin synthesis genes are typically plasmid encoded.

[e] *S. enterica* serovars Typhimurium and Enteritidis also encode a receptor, FoxA, for the ferrioxamine family of hydroxamates. This receptor is not found in any other enteric pathogens.

[f] The *fec* genes are present only in a subset of *S. flexneri* isolates (see the text).

[g] Although not all *S. sonnei* strains contain the *shu* locus, all strains tested used heme as an iron source, indicating that a different heme uptake system is present in at least a subset of *S. sonnei* strains.

terobactin is secreted into the environment, where it binds ferric iron with extremely high affinity. The iron-enterobactin complex binds to its receptor, FepA, on the surface of the bacterial cells and is transported across the outer membrane in a TonB-dependent manner (see chapters 7 and 9). FepB shuttles the siderophore through the periplasm to the cytoplasmic membrane permease, consisting of FepC, FepD, and FepG, which subsequently transports the ferrienterobactin into the cytoplasm. Once the siderophore is in the cytoplasm, iron is removed and enterobactin is hydrolyzed by an esterase, Fes. Expression of the biosynthesis and transport genes is tightly regulated by iron through Fur (see chapter 13). In the presence of iron, Fe-Fur binds to sites termed Fur boxes in the promoters of each of the *ent* operons and represses transcription. Thus, the enterobactin biosynthesis and transport system is maximally expressed under low-iron conditions.

Among the enteric pathogens, the *ent* and *fep* genes have been sequenced and the synthesis and transport of enterobactin have been characterized in *Salmonella enterica* and *Shigella flexneri*. The enterobactin locus is present also in every pathogenic *E. coli* strain sequenced to date (Table 2). The location and organization of the *ent* and *fep* genes within these genomes are the same as in *E. coli* K-12 (Fig. 1A) and exhibit very high sequence conservation with respect to the *E. coli* K-12 genes (Fig. 1B). Although the genes are present in all *Shigella* species, some strains of *Shigella* fail to synthesize or transport enterobactin. In *S. flexneri*, for example, mutations or small deletions are found in each of the *ent-fep* operons, effectively eliminating synthesis and transport of the siderophore.

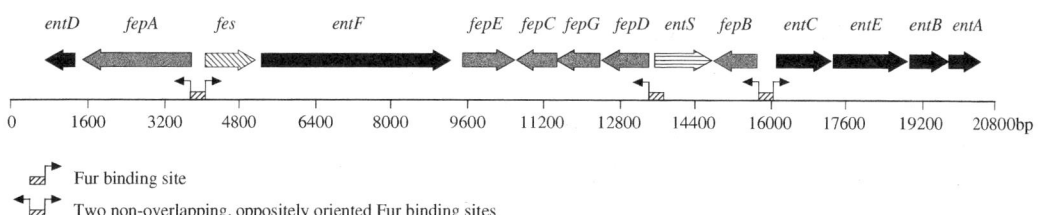

A Organization of the enterobactin locus showing genes: entD, fepA, fes, entF, fepE, fepC, fepG, fepD, entS, fepB, entC, entE, entB, entA across 0–20800 bp, with Fur binding sites indicated.

Fur binding site

Two non-overlapping, oppositely oriented Fur binding sites

Enterobactin synthesis (black arrows)
- *entA* (2,3-dihydro-2,3-dihydroxybenzoate dehydrogenase)
- *entB* (isochorismatase)
- *entC* (isochorismate synthase)
- *entD*
- *entE* } (enterobactin synthase)
- *entF*

Enterobactin transport (grey arrows)
- *fepA* (TonB-dependent outer membrane receptor)
- *fepB* (periplasmic binding protein)
- *fepC*
- *fepD* } (inner membrane permease complex)
- *fepG*

Other (hatched arrows)
- *fes* (enterobactin esterase)
- *entS* (enterobactin secretion)

B

Strain	FepA	EntA
E. coli O157:H7	99(99.5)	99(100)
E. coli CFT073	99(99.5)	98(99)
S. flexneri 2a	99(99.5)	97(98)
S. enterica serovar Typhi	82(89)	88(92)
S. enterica serovar Typhimurium	82(89)	89(93)

FIGURE 1 (A) Organization of the enterobactin locus in *E. coli*, *Shigella*, and *Salmonella*. Genes in this cluster are represented by thick arrows. The arrows indicate the direction of transcription and are shaded according to function. The small hatched boxes with right-angle arrows show the location and direction of the Fur binding sequences (not to scale). (B) FepA and EntA amino acid homologies between *E. coli* K-12 and selected *E. coli*, *Shigella*, and *Salmonella* isolates. The numbers indicate percent identity (percent similarity) to the corresponding *E. coli* K-12 protein.

In addition to enterobactin, the precursor molecule 2,3-dihydroxybenzoic acid and enterobactin breakdown products such as the linear dimer and trimer of DHBS can provide iron to the enterics. These are transported through the outer membrane receptors Cir and Fiu.

Salmochelins

The salmochelins are a newly described family of catechol siderophores produced by several members of the *Enterobacteriaceae*. These catechol siderophores consist of two or three (Fig. 2A) DHBS moieties linked by glucose residues. Glucose forms the bridge between the serine hydroxyl group of one DHBS molecule and the carboxylate of another. The proteins involved in the synthesis and transport of salmochelin are encoded by the *iro* locus, a Fur-regulated locus consisting of *iroB*, *iroC*, *iroD*, *iroE*, and *iroN* (Fig. 2B). IroB has homology to glycosyl transferases and is probably involved in attaching the glucose residue to DHBS. IroC is a member of the ABC transporter family but has higher homology to eukaryotic multidrug resistance export proteins than to bacterial iron importers. IroD has sequence similarity to Fes,

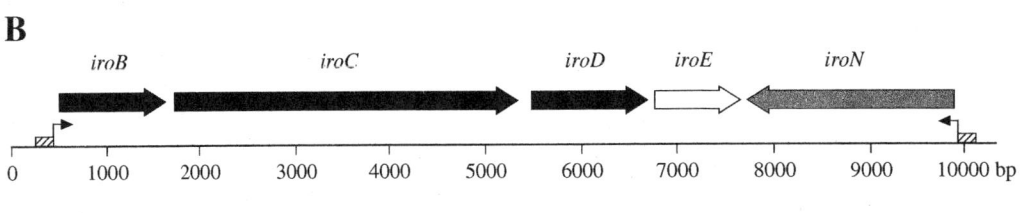

Fur binding site

Salmochelin synthesis (black arrows)
  iroB (glycosyl transferase)
  iroC (ATP-binding cassette exporter)
  iroD (Fes-like protein)

Salmochelin transport (grey arrow)
  iroN (TonB-dependent outer membrane receptor)

Unknown (white arrow)
  iroE (putative periplasmic hydrolase)

FIGURE 2 (A) Structure of the catechol siderophore salmochelin. The molecule containing three DHBS moieties is shown, but two subunits linked by a single glucose also occur. (B) Organization of the *iro* locus for salmochelin production and transport. The thick arrows indicate the direction of transcription and are shaded according to their roles in biosynthesis or transport. The small hatched boxes with right-angle arrows show the location and direction of the Fur binding sequences (not to scale).

the enterobactin esterase, but appears to be involved in siderophore biosynthesis rather than in the breakdown of the siderophore after transport into the cell. IroE has homology to periplasmic hydrolases. The precise roles of each of the gene products remain to be determined, but mutations in *iroBC* or in *iroD* eliminate the synthesis of salmochelins, indicating a role in biosynthesis. IroN, a 78-kDa outer membrane protein, has homology to TonB-dependent outer membrane siderophore receptors, including FepA, and is responsible for the transport of salmochelins into the bacterial cell. IroN can transport other catechol siderophores and has some functions in common with FepA. Both can transport enterobactin, but they differ in their ability to transport siderophores produced by other organisms. IroN mediates the uptake of corynebactin, a siderophore made by corynebacteria (see chapter 22), while FepA transports one of the myxobacterial siderophores, myxochelin C.

The *iro* genes were first described in *S. enterica* and are present in almost all *Salmonella* strains with the exception of *Salmonella bongori*. Among the *E. coli-Shigella* group, the *iro* locus is present in *Shigella dysenteriae* type 1 and some of the pathogenic *E. coli* strains, including the uropathogenic *E. coli* (UPEC) strains, but is absent from the genomes of the other *Shigella* species and *E. coli* K-12. In UPEC strain 536, the *iro* genes are found within a pathogenicity island, PAI III. This island, located at the *thrW* tRNA gene, also encodes S-fimbriae.

Aerobactin

While the majority of enteric pathogens make and use catechol siderophores, most *S. flexneri* and *Shigella boydii* strains do not. These catechol-negative strains are not lacking in siderophore-mediated iron transport, however, but instead produce a hydroxamate siderophore, aerobactin. Aerobactin was first detected in *Aerobacter* (*Klebsiella*) *aerogenes* and in extraintestinal strains of *E. coli* and was subsequently identified in a variety of enteric bacteria. This siderophore is commonly produced by clinical isolates of *E. coli*, as well as by some clinical isolates of *Salmonella*. Among the *Shigella* species, aerobactin is synthesized by most *S. flexneri* and *S. boydii* strains but is only rarely found in cultures of *S. dysenteriae* or *Shigella sonnei*. Some *Shigella* isolates synthesize both enterobactin and aerobactin.

The synthesis of aerobactin from lysine and citrate requires the products of four genes, *iucABCD* (Fig. 3A), and is described in detail in chapter 2 (see Fig. 10 in chapter 2). Following aerobactin synthesis and secretion into the extracellular milieu, the iron-aerobactin complex binds to a specific outer membrane receptor, IutA. IutA has homology to the TonB-dependent outer membrane receptors, and transport of aerobactin across the outer membrane requires TonB. Aerobactin is bound in the periplasm and transported across the cytoplasmic membrane by the generic hydroxamate transport system consisting of FhuB (periplasmic binding protein) and FhuCD (inner membrane permease). While the aerobactin synthesis genes *iucABCD* and the outer membrane receptor gene *iutA* form a single operon, the *fhuBCD* genes form a distinct genetic cluster and are not linked to siderophore biosynthesis genes. Expression of the aerobactin genes, like the enterobactin genes, is negatively regulated by the Fur protein as described in chapter 13, and Fur binding sites are found in the promoter of the aerobactin operon.

Limited DNA sequence analysis of aerobactin genes from different species indicates that the DNA sequences are conserved. The enzymes responsible for aerobactin biosynthesis in the different species are almost identical, with >90% amino acid identity in many cases (Fig. 3B). The receptors, although still highly similar, show greater variation than do the biosynthesis proteins. This likely reflects the surface exposure of the receptor proteins and their potential recognition by the host immune system. Greater variation of these surface antigens would help the pathogen escape antibodies produced during previous exposure to other IutA$^+$ enteric pathogens.

Although the aerobactin genes are highly conserved, the location of the operon within the genomes varies widely (Fig. 3B), and the

A

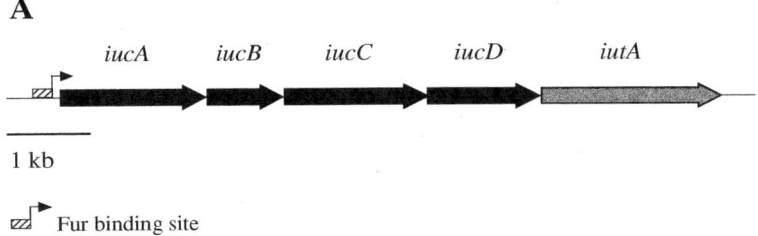

1 kb

▱▸ Fur binding site

Aerobactin synthesis (black arrows)

  iucA (aerobactin synthase)
  iucB (N^6-hydroxylserine: acetyl coenzyme A N^6-transacetylase)
  iucC (aerobactin synthase)
  iucD (lysine: N^6-hydroxylase)

Aerobactin transport (grey arrow)

  iutA (TonB-dependent outer membrane receptor)

B

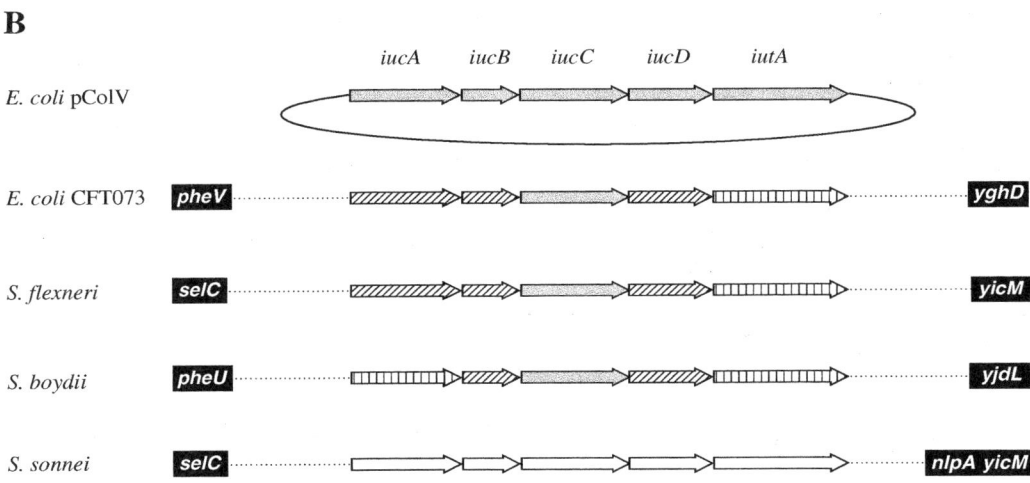

Identity with pColV-encoded protein:

  ▤ >95%
  ▨ 90-95%
  ▥ 85-89%
  ☐ No sequence data available

FIGURE 3 (A) Organization of the aerobactin locus in the enteric pathogens. The thick arrows indicate the direction of transcription of each of the open reading frames. The shading of the arrows reflects the function of these genes in aerobactin synthesis or uptake. The small hatched box with a right-angle arrow shows the location and direction of the Fur binding sequence (not to scale). (B) Comparison of the location of the aerobactin locus in selected *E. coli* isolates and *Shigella* species. The black boxes represent sequences also found in the *E. coli* K-12 genome. These sequences are the likely sites of insertion of the aerobactin island in each strain. The dotted lines delineate DNA of various lengths between the aerobactin locus and the ends of the island. The thick arrows indicate the direction of transcription of the aerobactin genes and are shaded according to the degree of homology of each gene product to the equivalent *E. coli* pColV-encoded protein.

DNA sequences immediately flanking the genes are distinctly different. The first aerobactin genes characterized in *E. coli* were found on a plasmid, pColV, and additional aerobactin-encoding plasmids have been found in *Salmonella* and other enteric genera. In enteroinvasive *E. coli* (EIEC) and in extraintestinal *E. coli* (ExPEC) K1 and UPEC, as well as in *S. flexneri*, *S. boydii*, and *S. sonnei*, the genes are found on the chromosome; however, their precise location has been determined only in a few of these pathogens (Fig 3B).

The pColV and *A. aerogenes* aerobactin genes are flanked by copies of the insertion sequence IS*1*, and copies of IS*1* and other insertion sequences are found upstream and downstream of some of the chromosomal aerobactin genes in *E. coli* and *Shigella*. The distances between the genes and the insertion sequences, and the DNA sequences between the aerobactin genes and IS*1* elements are different, indicating that the genes are unlikely to have been spread as a composite transposon. The association with IS*1* and other IS elements may simply reflect the large number of insertion sequences found in many enteric pathogens. In most cases, the aerobactin genes appear to be part of a larger pathogenicity island or island-like region (Fig. 3B). In *S. flexneri*, the genes are located immediately downstream of the tRNA gene *selC*, a common site for pathogenicity islands in enteric pathogens. The *selC* locus also appears to be the site of insertion of the aerobactin genes in *S. sonnei* strains that are aerobactin positive. In *S. boydii*, however, the *selC* region lacks an island and is essentially identical to the corresponding region in *E. coli* K-12. Instead, the *S. boydii* aerobactin genes map to the *pheU* tRNA locus. In both *S. flexneri* and *S. boydii*, the presence of the aerobactin island is associated with deletions of sequences found in *E. coli* K-12 downstream of the tRNA gene.

Yersiniabactin

Yersiniabactin, produced by many highly virulent *Yersinia* isolates (see chapter 15), is also synthesized by some *Salmonella* and *E. coli* strains but not by *Shigella* species. In the enteric pathogens, yersiniabactin synthesis and transport proteins are encoded within a region called the high-pathogenicity island (HPI), based on its homology to the *Y. pestis* high-pathogenicity island. The yersiniabactin system is present in isolates of *Salmonella enterica* subspecies III and VI but is absent in subspecies I, which includes the pathogenic serovars Typhi and Typhimurium. Among *E. coli* pathotypes, yersiniabactin is produced by most strains of enteroaggregative *E. coli* (EAggEC), and by ExPEC strains such as UPEC and the highly invasive K1 isolates that cause septicemia and meningitis. The HPI and associated yersiniabactin production are less commonly associated with EIEC, enterotoxigenic *E. coli* (ETEC) or enteropathogenic *E. coli* (EPEC) strains and are very rarely found in enterohemorrhagic *E. coli* (EHEC) strains.

Xenosiderophore Transport Systems

Most enteric bacteria not only synthesize receptors for their own, endogenous siderophores but also produce receptors that recognize xenosiderophores, i.e., siderophores produced by other microorganisms. Notably, the receptor and cytoplasmic permease for the fungal hydroxamate siderophore ferrichrome are widely distributed among enteric bacteria, even though this siderophore is not made by these organisms. For bacteria that are found in the environment as well as within the mammalian host, it may be advantageous to be able to use a siderophore present in the environment. For organisms that are rarely found outside the host, it is more difficult to conceive a possible advantage of having a transport system for a soil fungal siderophore such as ferrichrome.

The ferrichrome receptor FhuA and the FhuBCD transporters are described in chapters 4 and 11. These proteins are functional in *E. coli*, *Shigella*, and *Salmonella* and allow these bacteria to use ferrichrome as a sole source of iron. The gene encoding the receptor is unlinked to the genes encoding the periplasmic binding protein (FhuB) and the cytoplasmic membrane permease (FhuC and FhuD). The FhuA receptor binds some colicins, microcins, and bacteriophages, but with respect to siderophore recognition, it appears to be specific for ferrichrome. In contrast, the periplasmic and cytoplasmic membrane proteins function in the transport of all the hydroxamate siderophores, including aerobactin, used by these bacteria. The Fhu proteins, like the other enteric iron transport proteins, are synthesized in response to iron limitation, and the genes are negatively regulated by Fur.

S. enterica serovars Typhimurium and Enteritidis are able to use another class of hydroxamates, ferrioxamines, via the FoxA receptor. This receptor is usually absent from other *Salmonella* strains, and uptake of the ferrioxamines has been used to distinguish serovars Typhimurium and Enteritidis from the other *Salmonella* serovars. Neither *E. coli* nor *Shigella* species encode the ferrioxamine receptor FoxA. In *Salmonella*, FoxA mediates the uptake of ferrioxamines B, E, and G, but, surprisingly, only ferrioxamine G transport requires TonB function. All three compounds require FhuB, FhuC, and FhuD for transport through the inner membrane.

Two additional outer membrane proteins with homology to siderophore receptors, IreA, and Iha, are present in some pathogenic *E. coli* strains. IreA was first identified in an *E. coli* blood isolate and was subsequently found in several other pathogenic *E. coli* strains but not in normal fecal isolates. This outer membrane protein has homology to *E. coli* Cir and to *Vibrio cholerae* IrgA, both of which are TonB-dependent receptors for the siderophore enterobactin and/or its breakdown products. A Fur box is present in the putative promoter of *ireA*, and its expression is iron and pH regulated. Relative to expression in broth culture, expression of *ireA* is increased in human blood, urine, and ascites. Although it is likely that this protein functions in iron acquisition, the ligand for this receptor has not been identified. Like IreA, Iha has sequence similarity to the *V. cholerae* enterobactin receptor IrgA. Iha is found in EHEC O157:H7 and other pathogenic *E. coli* strains but has not yet been shown to function in iron acquisition. This protein has been proposed to play a role in adherence of EHEC isolates to host epithelial cells.

NONSIDEROPHORE IRON TRANSPORT SYSTEMS

Heme and Hemoglobin

The most abundant potential iron source for bacteria within vertebrate hosts is heme. The majority of the 3 to 4 g of iron within an adult human is in the form of heme, largely as hemoglobin but also in other heme proteins. Although heme is not taken up by *E. coli* K-12, many pathogenic *E. coli* and *Shigella* strains have the ability to transport heme and use it as a sole source of iron. Interestingly, heme transport appears to be completely absent in *Salmonella* species.

The best-characterized loci encoding heme transport systems in the enterics are the *shu* and *chu* loci found in *S. dysenteriae* type 1 and *E. coli* O157:H7, respectively. The *shu* and *chu* heme transport loci are essentially identical at the DNA level and will be referred to here as the *shu* locus. The *shu* heme transport locus is found in several intestinal and extraintestinal *E. coli* isolates; however, it is not uniformly present in pathogenic *E. coli* and *Shigella* strains (Table 2). Many of the pathogens that do not possess the *shu* locus, such as *S. sonnei*, are nevertheless capable of using heme as an iron source, indicating that there is at least one additional heme transport system in this group of bacteria.

The *shu* island in *S. dysenteriae* is an approximately 10-kbp region flanked by *yhiD* and *yhiF*, genes of unknown function that are located adjacent to each other near min 78 of the *E. coli* chromosome (Fig. 4). The *shu* island is located at precisely the same position in *E. coli* O157:H7. The *shu* locus consists of eight genes, all of which are predicted to be under the control of iron and Fur, although this has been demonstrated only for the *shuA* gene encoding the *S. dysenteriae* heme receptor ShuA. ShuA is an outer membrane protein that binds heme and requires TonB for heme transport. The sequences of *shuT*, *shuU*, and *shuV* suggest that they encode a periplasmic protein-dependent ABC transporter system that transports

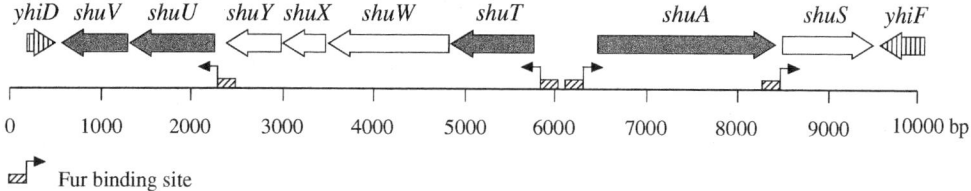

Heme transport (grey arrows)
  shuA (TonB-dependent outer membrane receptor)
  shuT (periplasmic binding protein)
  shuU ⎱ (inner membrane permease complex)
  shuV ⎰

Unknown (white arrows)
  shuW (non-functional in *S. dysenteriae* Type 1 due to a premature stop codon)
  shuX
  shuY
  shuS

ORFs flanking the *shu* locus, present also in *E. coli* K-12 (hatched arrows)
  yhiD
  yhiF

FIGURE 4 Organization of the *S. dysenteriae shu* heme transport locus. The thick arrows show the direction of transcription of the *shu* genes. The shading of the arrows reflects what is known about the role of each *shu* gene in heme uptake. The hatched boxes with right-angle arrows show the location and direction of the Fur binding sequences (not to scale).

heme across the cytoplasmic membrane. The functions of *shuS*, *shuW*, *shuX*, and *shuY* are unknown, yet homologs of these genes are found in the heme transport loci of other enteric bacteria. These genes appear to encode soluble proteins. These proteins may increase the efficiency of utilization of heme as an iron source and may be involved in heme storage or transfer to other proteins. ShuS is a heme binding protein that can rapidly transfer heme to heme oxygenases such as the *C. diphtheriae* HmuO protein; however, ShuS does not itself have heme oxygenase activity. No heme oxygenase protein has yet been identified in enteric bacteria.

A puzzling aspect of these heme transport genes is that only *shuA* is required to confer upon an *E. coli* K-12 strain the ability to grow with heme as the iron source. A possible explanation for this is that although *E. coli* K-12 does not use heme as an iron source, many closely related *E. coli* strains do, and *E. coli* K-12 may contain remnants of a heme transport system.

The *shuA* gene does not confer efficient heme uptake when transferred into *Salmonella* in the absence of the other genes, supporting the conclusion that *Salmonella* strains lack any heme transport functions.

While *Shigella* species do not use hemoglobin efficiently, many *E. coli* strains can grow with hemoglobin as a source of heme and iron in vitro. No hemoglobin binding receptor on the surface of heme-utilizing *E. coli* cells has been identified, but *E. coli* strains produce extracellular proteases that may degrade hemoglobin and facilitate the release of heme for subsequent uptake by the bacterium. Hbp, a plasmid-encoded protease produced by a clinical isolate of *E. coli*, degrades hemoglobin and binds the released heme. This protein is a member of the immunoglobulin A1 protease family and appears to function as a hemophore (see chapter 3), as well as a protease, for delivery of heme to *E. coli*.

Some of the pathogenic *E. coli* strains produce hemolysins that may increase the availability of heme to invading pathogens. In one strain of *E. coli*, expression of the plasmid-encoded hemolysin is iron and Fur regulated, with maximal expression observed under low-iron conditions. The coordinate regulation of hemolysin production and heme uptake suggests that the hemolysin and heme transport systems may function in concert in vivo to provide iron to the pathogen.

Ferric Dicitrate

The ferric dicitrate transport system, encoded by the *fec* genes, provides a source of iron for selected strains of *E. coli* (Table 2), but this iron transport system is not found in *Salmonella*. Interestingly, *Salmonella*, but not *E. coli*, transports citrate for use as a carbon source, and this transport system is distinct from the Fec system used to transport Fe dicitrate. Fe dicitrate is transported across the outer membrane via the TonB-dependent receptor FecA and into the cytosol via the FecBCDE periplasmic binding protein-dependent ABC transporter system (see chapter 11).

Each of the *Shigella* species contains *fec* genes, but the genes are not uniformly present. The *fec* locus is typically present in isolates of *S. flexneri* 1a and *S. dysenteriae* type 1 but is usually absent from *S. flexneri* 2a strains. Nevertheless, the *Shigella fec* genes have been best characterized in an *S. flexneri* 2a isolate. In this strain, the *fec* genes are associated with a multiple-drug resistance element within a larger, pathogenicity island-like locus. The island, which also includes an integrase gene and a number of phage genes, is located at the *serX* tRNA gene. In several other Fec$^+$ *S. flexneri* strains that have been tested, the *fec* genes are linked to the antibiotic resistance element, suggesting that when present, the *fec* genes have been acquired as part of this pathogenicity island.

α-Keto- and α-Hydroxyacids

α-Ketoacids, which are primary metabolites of prokaryotic and eukaryotic cells, can form ferric complexes and supply iron for bacterial growth. These compounds are less efficient than siderophores in delivering iron but are able to stimulate the growth of certain *E. coli* and *Salmonella* strains in low-iron media. Use of α-ketoacids has not been reported for *Shigella*. The use of these compounds as iron sources is TonB dependent, suggestive of transport through an outer membrane receptor; however, a specific receptor has not yet been identified. *S. enterica* serovar Typhimurium excretes α-ketoacids, and the amount secreted increases significantly during growth under iron-limited conditions. Additionally, a *fur* mutant hyperexcretes α-ketoacids, even when grown in the presence of iron. Thus, this system resembles siderophore-mediated iron acquisition, both in its TonB dependence and in its regulation of expression by iron and Fur.

Ferrous Iron

Ferrous iron, found primarily under anaerobic conditions or at acidic pH, is much more soluble than ferric iron and therefore does not require a chelator for solubilization and transport into the bacterial cell. Transport of ferrous iron appears to be important in several environments, including areas within the host. Ferrous iron transport systems are energy dependent and are predicted to be associated primarily with the cytoplasmic membrane. It is not clear how ferrous iron is transported across the outer membrane. There is no evidence for the involvement of TonB or any specific outer membrane receptor in this process, and the ferrous iron may diffuse through general porins. It is also not known whether ferric iron in the periplasm can be reduced to the ferrous state for subsequent transport across the inner membrane via ferrous iron transporters.

The major ferrous iron transporter in the *Enterobacteriaceae* is the Feo system, which is described in detail in chapter 12. The Feo system is encoded by the *feoABC* genes, and these genes are found in *E. coli*, *Shigella*, and *Salmonella* species, as well as in many other species. Mutations in either *feoA* or *feoB* reduce the growth under low-iron conditions of strains lacking siderophore-mediated iron transport.

FeoB is a large, GTP binding cytoplasmic membrane protein that is required for efficient transport of ferrous iron. The roles of FeoA and FeoC have yet to be determined. Feo resembles other enteric iron transport systems in its regulation by Fur. In addition to a Fur binding site, the *feo* promoter contains a binding site for Fnr, a transcriptional activator of genes required for anaerobic metabolism. Loss of Fnr leads to reduced expression of *feo*, indicating that the *feo* operon responds to oxygen as well as iron levels. This regulation may reflect the predominance of ferrous, rather than ferric, iron under anaerobic conditions.

Another potential ferrous iron transporter is the Sit system, which may mediate both manganese and ferrous iron transport. This system has been most extensively studied in S. *enterica* serovar Typhimurium and is also found in *Shigella* and in several pathogenic E. *coli* isolates (Table 2; Fig. 5B). The S. *enterica* serovar Typhimurium *sitABCD* genes are on the SPI1 pathogenicity island and were originally identified as a putative iron transport system. The Sit system consists of a periplasmic binding protein (SitA), an ATP binding protein (SitB), and the inner membrane permease (SitC and SitD) (Fig. 5A). The S. *enterica* serovar Typhimurium Sit system was shown to transport both manganese and ferrous iron in vitro. Because the concentration of iron required for transport via Sit is significantly greater than physiological concentrations, the *Salmonella* Sit system is proposed to function in manganese rather than ferrous iron transport in vivo. The expression of *sitA* is repressed in the presence of either iron or manganese. The iron repression is mediated by Fur. The S. *enterica* serovar Typhimurium *sitA* promoter also has a putative binding site for MntR, a manganese-binding transcriptional regulator, and thus the manganese repression of *sitA* has been proposed to be mediated by MntR. S. *flexneri* SitA is homologous to S. *enterica* serovar Typhimurium SitA (Fig. 5B). It is also encoded in a four-member operon and is negatively regulated by either iron or manganese. The *Shigella* Sit system may play a greater role in iron transport than does the *Salmonella* Sit system; in the absence of the aerobactin and Feo systems, the S. *flexneri sit* genes allowed the acquisition of iron by *Shigella* both in vitro and in vivo.

Ferrous iron can be transported also through the magnesium transporter CorA, a putative integral cytoplasmic membrane protein with a large periplasmic domain. In both E. *coli* and *Salmonella*, an increase in the transport of ferrous iron is seen in cells grown under low-magnesium conditions. Maximal ferrous iron uptake under these conditions requires CorA but not the Feo system; however, other ferrous iron transport systems probably contribute to uptake under low-magnesium conditions as well. *corA* is found in all *Shigella* species tested, and the sequences of the E. *coli*, S. *enterica* serovar Typhimurium, and S. *flexneri* 2a CorA indicate a high degree of conservation.

ROLE OF THE IRON TRANSPORT SYSTEMS IN THE BIOLOGY OF THE ENTERIC PATHOGENS

Iron is not freely available in the mammalian host, and it is further withheld in response to infection. There is epidemiological evidence that increasing the amount of iron in humans increases the risk and severity of infection with bacterial pathogens, including *Salmonella*. This supports the concept that competition between the host and pathogen for iron is a critical component of the host-pathogen interaction. Members of the E. *coli*-*Shigella*-*Salmonella* group colonize a variety of habitats within the host, and these varied environments are reflected in the number and diversity of iron transport systems present in these organisms (Table 2).

A number of studies have addressed the potential roles of the various iron transport systems in the pathogenesis of the enteric pathogens. Isolates of these pathogens are consistently found to produce siderophores, which have high affinities for ferric iron and may be important in removing iron from many host iron-protein complexes. While all the clinical isolates produce one or more siderophores, there is not a strict correlation between the par-

A

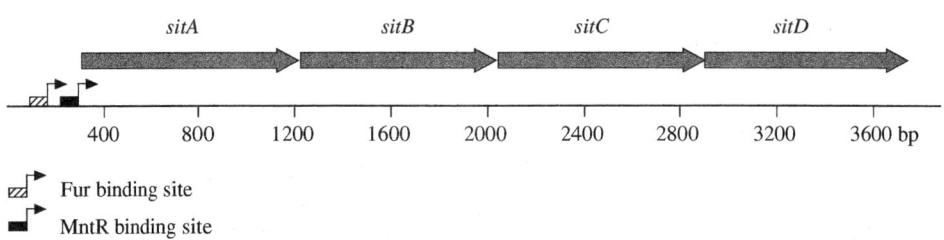

◪▸ Fur binding site
◾▸ MntR binding site

Ferrous iron/manganese transport
  sitA (periplasmic binding protein)
  sitB ⎫
  sitC ⎬ (inner membrane permease complex)
  sitD ⎭

B

Protein	Identity (%)	Similarity (%)
S. enterica serovar Typhimurium SitA		
S. enterica serovar Typhi SitA	100	100
S. flexneri 2a SitA	74	85
E. coli CFT073 SitA	66	76
S. enterica serovar Typhimurium SitB		
S. enterica serovar Typhi SitB	98	99
S. flexneri 2a SitB	74	83
E. coli CFT073 SitB	74	83
S. enterica serovar Typhimurium SitC		
S. enterica serovar Typhi SitC	99	99.5
S. flexneri 2a SitC	84	91
E. coli CFT073 SitC	84	90
S. enterica serovar Typhimurium SitD		
S. enterica serovar Typhi SitD	99	99.5
S. flexneri 2a SitD	66	83
E. coli CFT073 SitD	65	83

FIGURE 5 (A) Organization of the *sit* operon for ferrous iron and manganese uptake. The thick arrows show the direction of transcription of the four *sit* genes. The small boxes with right-angle arrows show the location and direction of the Fur and MntR binding sites (not to scale). (B) Homology of Sit proteins from selected *Salmonella*, *E. coli*, and *Shigella* isolates to the *S. enterica* serovar Typhimurium Sit system. The *S. flexneri* and *E. coli* CFT073 *sitA* genes are essentially identical at the nucleotide level; however, an apparent frameshift mutation in the CFT073 *sitA* gene creates a divergent carboxy terminus in CFT073 SitA.

ticular siderophore produced and the type of infection or the habitat of the strain. Strains causing extraintestinal infections are more likely to produce aerobactin, either alone or along with enterobactin. This is not absolute, however, since strains that produce only enterobactin have also been isolated from a variety of extraintestinal sites. Although enterobactin has a higher affinity for iron than does aerobactin in vitro, the production of aerobactin has several potential advantages over the production of enterobactin in vivo. Enterobactin has been shown to bind to serum albumin, and antibodies against enterobactin have been found in humans. Recently, it has been shown that the human neutrophil lipocalin NGAL is a catechol binding compound and can effectively bind enterobactin. The function of NGAL may be to sequester iron in neutrophil granules. Enterobactin bound to particular host proteins might not be available for iron transport by a pathogen, and free enterobactin has a short half-life. In addition, the hydroxamates such as aerobactin have a higher affinity for iron at low pH than do enterobactin and other catechols. Thus, it is likely that depending on the niche in which a bacterium is found, a specific siderophore may provide an advantage in iron acquisition and survival.

Nonsiderophore iron transport systems are also important in certain environments. Feo is widely distributed in enterics, and many of the pathogenic strains encode systems for heme transport and Sit systems. In particular, the presence of Sit may correlate with intracellular survival or growth. Sit is present in *Salmonella*, *Shigella*, EIEC, and UPEC isolates but is only rarely found in strains that are strictly extracellular.

Because many enteric pathogens encode multiple iron uptake systems, it has been difficult to assess the precise contribution of any single iron transport system to pathogenesis. One general approach to dissecting the role of iron transport systems is to construct mutants defective in one or more systems and compare these to the wild-type strain in animal models. These studies have provided valuable insights into iron transport by pathogens; however, the results must be interpreted with caution, given the limitations of the animal models. The following sections summarize the current state of knowledge with respect to the role of iron transport systems in the pathogenesis of *Salmonella*, *Shigella*, and *E. coli*.

Salmonella

Within the *Salmonella* group, there are pathogens of humans as well as other animals. Depending on the strain and its sites of replication within the host, *Salmonella* species produce a variety of diseases, ranging from mild enteritis associated with colonization of the epithelial surface by *S. enterica* serovar Enteritidis to systemic typhoid fever caused by *S. enterica* serovar Typhi (Table 1).

In the host, *Salmonella* may be found in a number of distinct environments: in the gastrointestinal tract, in the bloodstream, or within cells, predominantly within monocytes. The organisms are acquired by ingestion and are relatively invasive. The invasion may be limited to a localized area of the intestine and the underlying lymphoid tissue, or the bacteria may be carried to the liver, spleen, and other organs via infected macrophages. When growing intracellularly within the macrophage, salmonellae remain within a phagocytic vacuole that is modified by the bacterium. Thus, the intracellular environment experienced by *Salmonella* is different from that encountered by *Shigella* species and EIEC strains, which grow free within the cytosol of the host epithelial cells.

Most of the studies of iron transport and pathogenesis in *Salmonella* have been done with *S. enterica* serovars Typhimurium and Typhi. *S. enterica* serovar Typhi, which causes typhoid fever, is normally a strictly human pathogen, but virulence studies can be done with the mouse model. *S. enterica* serovar Typhimurium, which causes gastroenteritis in humans, is a natural pathogen of certain mouse strains, producing a typhoid-like disease. Thus, *S. enterica* serovar Typhimurium in the mouse can be used as a model of human typhoid fever but is less useful for understanding gastroenteritis.

In using the mouse model for *Salmonella* pathogenesis, it is important to consider the *Nramp1* (natural resistance-associated macrophage protein 1) genotype of the mouse strain. *Nramp1*, originally named *Ity*, was first identified as a locus associated with resistance or susceptibility of mice to systemic infection with *S. enterica* serovar Typhimurium. It was later shown that *Nramp1* encodes an integral membrane protein in phagocytes that is recruited to the phagosome membrane following phagocytosis. This protein is a divalent cation transporter that transfers iron from the phagosome to the cytoplasm. Lack of this protein may increase the iron available to *Salmonella* multiplying in the vesicle, allowing increased growth of the bacteria in this environment. Because the mouse *Nramp1* genotype affects the availability of iron to *Salmonella* within the phagosome, *Nramp1* status may influence the outcome of experiments to test the role of iron transport systems in *Salmonella* pathogenesis. Differences at the *Nramp1* locus, as well as other genotypic differences between the mouse strains used in virulence studies, could explain some of the apparent discrepancies in the literature about the importance of particular iron transport systems in *Salmonella* pathogenesis.

Salmonella isolates have been surveyed for the presence of a number of transport systems that may function in iron uptake (Table 2). There are conflicting reports of the relative importance of enterobactin in the virulence of *S. enterica* serovar Typhimurium in the mouse model. It was originally reported that enterobactin biosynthesis mutants were attenuated in CFW mice following intraperitoneal inoculation and that coadministration of purified enterobactin restored virulence. A subsequent study that used the same, as well as additional, *ent* mutants did not find differences in virulence by the intravenous and intraperitoneal routes of infection. However, because the two studies were not carried out with identical mouse strains, it cannot be ruled out that genotypic differences may have influenced the requirement for enterobactin in this model. Mutations in *aroA*, a gene required for enterobactin synthesis, reduce the virulence of *S. enterica* serovars Typhimurium and Typhi in the mouse. The *aroA* mutation also eliminates several biosynthetic pathways, including aromatic amino acid biosynthesis, and the loss of these capabilities may be partially or fully responsible for the reduction in virulence. The *aroA* and *ent* mutations employed in these studies also abolish the synthesis of salmochelin, making it difficult to determine the contributions of enterobactin independent of salmochelin. Although a requirement for the catechol siderophores has not been clearly established in vivo, there is evidence that the enterobactin transport system is expressed in *S. enterica* serovar Typhi during growth in the human host. Serum from patients with typhoid fever contains antibodies against the *S. enterica* serovar Typhi FepA homolog, suggesting that catechol uptake may be important during infection.

The role of other siderophores in diseases caused by *Salmonella* has not been systematically researched. Aerobactin is not commonly produced by *Salmonella*, but strains that do produce this siderophore are often associated with systemic infections. Thus, aerobactin may play a role in the pathogenesis of some *Salmonella* isolates, but this has not been shown experimentally.

A role for the ferrioxamine receptor FoxA in pathogenesis has been shown for *S. enterica* serovar Typhimurium. A *foxA* mutant showed reduced colonization of rabbit ileal loops and was attenuated in mice when administered by the intragastric or intravenous route. Ferrioxamine, which is not produced by *Salmonella*, would not be expected to be present in the host, suggesting that there may be another ligand for FoxA in vivo.

While no single siderophore has been shown to be essential for *Salmonella* virulence, TonB-dependent transport is required. *S. enterica* serovar Typhimurium *tonB* mutants grew poorly in serum and were attenuated for virulence by the intragastric route. The mutant was not less virulent by the intraperitoneal route, nor was it defective for growth in cultured human cells. The *tonB* mutant of *S. enterica* serovar Typhi exhibited a greater virulence defect than did the *tonB* mutant of *S. enterica* serovar Typhimurium. The *S. enterica* serovar Typhi *tonB* mutant was attenuated in mice injected intraperitoneally and was defective for growth in HeLa cells and human monocytic cells. The defect in virulence of *S. enterica tonB* mutants may be explained in part by loss of FoxA function in the absence of TonB. However, not all transport through FoxA is TonB dependent; for example, ferrioxamines B and E can be transported via FoxA in the absence of TonB, albeit at lower levels.

The TonB-independent ferrous iron and manganese transport systems Feo and Sit are required for full virulence of *S. enterica* serovar Typhimurium in susceptible mouse strains. A *feoB* mutant was shown to be outcompeted by the wild-type strain in a mixed infection of the mouse intestine. When administered intravenously, a *feoB* mutant was less virulent and a *feo sit* double mutant was avirulent, indicating that the ferrous iron transport systems are important also during the systemic stages of *S. enterica* serovar Typhimurium infection. Thus, ferrous iron may be a significant source of iron for *Salmonella* growing in vivo.

Shigella

The *Shigella* species are the most homogenous group among the enteric pathogens with respect to pathogenesis. These organisms cause dysentery, an inflammatory disease of the colon, following ingestion of even small numbers of the bacteria (Table 1). They gain access to host tissue though the M cells of the gut and use plasmid-encoded invasins to enter the adjacent epithelial cells. *Shigella* species lyse the phagocytic vacuole and multiply within the cytoplasm; they must therefore be adapted to growth in the distinct environment of the cytoplasm of the eukaryotic cell. The bacteria spread from cell to cell by polymerizing host cell actin, which propels the bacteria directly into the adjacent cell, where they continue to replicate in the cytosol. *Shigella* typically produces localized lesions and only rarely spreads to deeper tissues or produces septicemia.

Humans are the primary host for *Shigella*, and there is no animal model that mimics the natural disease. The guinea pig Serény test, in which the bacteria are inoculated onto the conjunctival sac, where they invade the epithelial cells and provoke inflammation, exhibits the best correlation with ability to cause disease in humans, but it cannot reproduce gut colonization and growth in intestinal epithelial cells. Shigellae are able to replicate within tissue culture cells; however, the environment of human epithelial cells in artificial medium also fails to accurately represent the intestinal environment. In particular, the iron status of the cultured cells may be different from that of in vivo epithelial cells. Nevertheless, the use of cultured cells allows measurements of invasion, intracellular multiplication, and cell-to-cell spread (plaque formation). Despite the limitations of currently used model systems, it has been instructive to look at the effects of mutations in the iron transport systems on the behavior of *Shigella* in these systems.

Because all four species of *Shigella* (*S. dysenteriae*, *S. flexneri*, *S. sonnei*, and *S. boydii*) cause such similar diseases and occupy the same niche within the host, it was somewhat surprising to find that there were significant differences in the repertoire of iron transport systems found in each of these species (Table 2). The only two iron transport systems that are common to all members of this group are Feo and Sit. Feo is found in all the enterics, but Sit has a more restricted distribution. The correlation between the presence of the *sit* genes, found in *Salmonella*, *Shigella*, EIEC, and UPEC, and growth in the intracellular environment of the host suggests that Sit may play a role in adaptation of these pathogens to life inside the host cell.

S. flexneri, the most extensively characterized of the *Shigella* species, synthesizes aerobactin as its only siderophore. Aerobactin biosynthesis and transport mutants appear unaffected in their invasion and growth within cultured cells compared with the wild-type strain. In models that require some growth in extracellular compartments, including invasion of the chicken embryo from the allantoic fluid, the Sereny test, or infection of ligated rabbit ileal loops, the mutant required a longer time or a larger inoculum to produce the same effect as the wild-type strain. This suggests that siderophore production may be an advantage for growth in extracellular compartments of tissues, but siderophores are not required for *Shigella* to grow in the host cell cytosol.

The lack of a siderophore requirement for growth of *S. flexneri* in the intracellular environment was confirmed by measuring the growth and intercellular spread (plaque formation in monolayers) of a *tonB* mutant of *S. flexneri*. The *tonB* mutant, like the aerobactin mutant, was unaffected in invasion or plaque formation, suggesting that TonB-independent iron acquisition is important for growth of *S. flexneri* in the intracellular environment.

Analysis of additional iron transport mutants of *S. flexneri* indicates that no single system is required for growth within the host cells. In *S. flexneri*, loss of the aerobactin, Sit, and Feo systems resulted in loss of ability to form plaques in epithelial cell monolayers. Loss of two of the three systems resulted in a reduction in plaque size, but single mutants formed wild-type plaques. The ability of systems transporting either ferrous (Feo) or ferric (aerobactin) iron to allow intracellular growth indicates that both forms of iron must be present within the cultured cells. Similar results were obtained when the mutants were assessed in the guinea pig Sereny model; eliminating a single iron transport system had little effect, but mutations in multiple iron transport systems resulted in reduced virulence.

The intracellular expression of *S. flexneri* iron transport systems has also been measured. The aerobactin system was not induced above basal levels when the bacteria were growing in the cytosol of eukaryotic cells; however, the *sit* genes and *fhuA*, encoding the ferrichrome receptor, were upregulated in this intracellular environment. These genes may be responding to a lower level of iron starvation or may be differentially expressed in response to specific iron sources in the host cell. Whether FhuA plays a role in growth in the intracellular environment is not known. Its known ligand, ferrichrome, would not be expected to be present in the host cells, but it is possible that FhuA has an additional ligand or other function.

Similar analyses of mutants of *S. dysenteriae* type 1 indicate that there are differences among the *Shigella* species with respect to the iron transport systems required for intracellular growth. In contrast to the *S. flexneri tonB* mutant, a *tonB* mutant of *S. dysenteriae* failed to grow in the cytosol of cultured cells or to spread to adjacent cells, although it invaded normally. This mutant also gave a negative result in the Sereny test. The specific TonB-dependent transport systems that are required for *S. dysenteriae* intracellular growth have not been identified. *S. dysenteriae* produces enterobactin, but it does not require synthesis and transport of the siderophore for intracellular growth. Heme is a possible source of intracellular iron, but mutations in heme transport also have no effect on invasion or intracellular replication. The *iro* locus is present in *S. dysenteriae* type 1 strains but is not found in other *Shigella* species. Thus, although the Iro systems may contribute to *S. dysenteriae* iron acquisition in the host, it cannot be essential for the production of dysentery by other *Shigella* species.

E. coli

Pathogenic *E. coli* strains are associated with a variety of different diseases and cause both intestinal and extraintestinal infections (Table 1). Because *E. coli* strains cause such a wide array of diseases, it is not surprising that there is considerable variety in the number and types of iron transport systems in these bacteria (Table 2). Even among strains that colonize the same site and cause the same disease in the host, there are few absolute correlations between the presence of a particular system and the ability to cause a particular disease. The presence of the Ent/Fep, Fhu, and Feo systems is characteristic of all *E. coli* strains, and one or more of these may be important for some stage of colonization and production of disease. These are rarely the only systems found, suggesting that additional systems may provide some growth advantage in a specific niche in the host. For diseases for which there is an animal model, the relative importance of each system in pathogenesis can be assessed; however, good animal models are not available for some of the *E. coli* diseases, and for these, the importance of the iron transport systems is largely based on epidemiological data. Therefore, it is not known whether the higher proportion of some factors in a pathotype is due to association of the factor with disease or whether it reflects the more clonal nature of some pathotypes.

INTESTINAL *E. COLI* ISOLATES

E. coli strains causing intestinal infections include ETEC, EPEC, EAggEC, cell-detaching *E. coli* (CDEC), EHEC, and EIEC (Table 1). ETEC strains colonize the small intestine and secrete a cholera-like toxin or a heat-stable toxin, resulting in watery diarrhea. EPEC strains interact with the epithelium of the small intestine and produce attaching and effacing lesions. EAggEC, CDEC, and EHEC strains are also relatively noninvasive and produce disease from their site of interaction with the epithelial surface. The EHEC strains tend to cause more severe syndromes such as hemorrhagic colitis and hemolytic-uremic syndrome due to the secretion of Shiga toxins. The EIEC strains cause a form of dysentery that is clinically indistinguishable from the disease caused by the *Shigella* species. These strains carry a virulence plasmid that is essentially the same as the large virulence plasmid found in the *Shigella* species, and while the EIEC strains differ from the *Shigella* strains biochemically and serologically, the virulence factors and mechanism of pathogenesis are largely the same.

Most of the known iron uptake systems are represented among the strictly intestinal pathogenic *E. coli* strains. Genotypic and phenotypic analyses of intestinal *E. coli* strains indicate that the pathogens are somewhat more likely to produce aerobactin than are the nonpathogenic fecal isolates and that aerobactin production is more common among resident than transient *E. coli* organisms in the intestine. The ferrous iron transport system Feo has been shown to promote the colonization of the mouse small intestine, indicating that ferrous iron may be important for growth in this environment. The *Yersinia* HPI that encodes yersiniabactin-mediated iron acquisition has been identified in many pathogenic *E. coli* strains. In a study of diarrhea patients in China, *E. coli* strains harboring the *Yersinia* HPI were among the pathogens most frequently isolated, suggesting that the *Yersinia* HPI may be a common feature of pathogenic intestinal *E. coli* strains. The *Yersinia* HPI is present in EAggEC but is less commonly associated with ETEC, EPEC, and EIEC strains and has been found in only a few EHEC pathotypes. The ability to use heme as an iron source is a feature of the EHEC strains. Considering the hemorrhagic nature of the more serious forms of disease caused by this pathogen, it is likely that heme transport provides a growth advantage to these bacteria. However, the lack of a small-animal model that mimics the human diseases has precluded the ability to test this directly. Heme transport systems have been identified in a number of other noncommensal intestinal *E. coli* strains. Some of these also produce hemolysins that are capable of lysing erythrocytes and other host cells, releasing hemoglobin and other intracellular iron sources. Among the intestinal *E. coli* strains, the Sit system is found primarily in EIEC isolates, and this may correlate with the intracellular life-style of this pathogen. The Sit system has been identified in only one *E. coli* isolate predicted to be strictly extracellular (EAggEC O42) (Table 2), and the role of the Sit system in this strain has not been investigated.

EXTRAINTESTINAL *E. COLI* ISOLATES

The ExPEC strains include the disseminating *E. coli* strains, which are associated with septicemia and meningitis, and the UPEC strains, which cause urinary tract infections. The possible roles of iron and iron transport systems in the pathogenesis of extraintestinally disseminating strains of *E. coli* have been examined in several model systems. Numerous studies have shown that administration of iron or heme with the bacterial inoculum promotes tissue invasion and systemic spread of ExPEC strains in mouse models. As the iron transport systems were elucidated, the contributions of these systems were assessed. The initial discovery of aerobactin synthesis in an invasive strain of *E. coli* was followed by the demonstration that introduction of the aerobactin operon into a strain lacking siderophore synthesis was sufficient for the production of septicemia by *E. coli* in a mouse model. Loss of the aerobactin system was associated with loss of virulence. The importance of aerobactin in systemic disease is supported by epidemiological evidence that shows a correlation between production of aerobactin and isolation of the strain from an extraintestinal site. Similarly, a comparison of ExPEC strains carrying the *Yersinia* HPI island with their isogenic HPI$^-$ mutants in subcutaneously inoculated mice showed that the mutants were much less virulent. Systemic spread and virulence were restored by reintroducing the wild-type gene. The yersiniabactin system, encoded within the *Yersinia* HPI, is frequently found in ExPEC strains, suggesting that it may play some role in spread from the intestine. Heme transport systems are a common feature of ExPEC strains. This is perhaps not surprising, considering the tissue damage associated with colonization of extraintestinal sites.

As the complete DNA sequences become available for pathogenic *E. coli* strains, a more comprehensive assessment of the iron transport systems contributing to virulence in a single strain will be possible. The sequence of UPEC strain CFT073 has been determined, and the iron transport systems present can be largely identified from the sequence (Table 2). In this UPEC strain, genes for the production and transport of both enterobactin and aerobactin are present, as is the Shu (Chu) heme transport system. In a mouse model of ascending urinary tract infection, loss of any one of these systems did not severely impact the ability of the strain to infect the kidneys but did reduce its ability to compete with the wild-type strain in a mixed infection. Loss of both the enterobactin and aerobactin transport systems did result in reduced infectivity, indicating that the production of at least one of the siderophores is required. A *tonB* mutant, unable to use any of these systems, was highly attenuated in this model and failed to compete with the wild type for colonization or to infect the kidneys. Interestingly, this UPEC strain also carries the Sit system. Although UPEC strains are not obligate intracellular pathogens, recent data suggest that these pathogens do invade uroepithelial cells, possibly as a mechanism for evading the host immune response. The Sit system may be important for iron acquisition during this intracellular phase of the UPEC life cycle. A complete picture of the contribution of each of the iron transport systems to urinary tract infections should emerge from studies of the remaining iron transport genes and a refinement of the animal model.

ANIMAL-PATHOGENIC *E. COLI*

Some *E. coli* strains are nonhuman pathogens that cause intestinal or systemic infections in a specific animal host. These strains can be studied in their natural or related hosts and can be infected by the normal route of infection. While the results of these studies cannot be extrapolated to the human pathogens, they do offer important insights into host-pathogen interactions. *E. coli* strains isolated from the blood of bacteremic neonatal calves are typically aerobactin positive. Germfree lambs can be infected with bovine-pathogenic strains, and the effects of aerobactin uptake has been assessed in this model. In a mixed infection with an IutA$^+$ strain and the isogenic IutA$^-$ strain, the aerobactin transporter outcompeted the IutA$^-$ strain in the lamb intestine. Lower numbers of the *iutA* mutant were found in mesenteric lymph nodes, and the mutant was not found in deeper tissues and organs. Further, aerobactin was detected in the intestine, liver, spleen, kidneys, urine, cerebrospinal fluid, and bile, indicating that it was produced by the pathogenic *E. coli* in vivo.

Colibacillosis in chickens is caused by avian-pathogenic *E. coli*. This disease begins as a respiratory tract infection and progresses to a systemic infection. The avian-specific strains that cause colibacillosis have been analyzed for the presence and expression of iron transport systems. Aerobactin production is common among these strains, and aerobactin genes are expressed during experimental infection. Expression of the aerobactin and the *iro* system genes was detected in the avian host by selective transcript capture and enrichment of pathogen-specific sequences expressed in vivo. Immunization of chickens with the aerobactin receptor protected against a lethal challenge, supporting the model that the aerobactin genes are expressed and play a role in systemic avian infections with *E. coli*.

SUMMARY

While the precise roles of each iron transport system in transmission, colonization, survival, and spread of enteric pathogens in the host cannot be unambiguously defined, several features of iron transport in the enteric pathogens are clear. One is the relative abundance of iron transport systems in the pathogens. No enteric pathogens with only a single iron transport system have been identified, and most of the pathogens have a minimum of five distinct iron acquisition systems. Among this group of pathogens, all appear to make at least one siderophore. Heme transport is common among the *E. coli-Shigella* group, and Sit is a characteristic of pathogens able to grow within host cells. As more of the iron transport systems are characterized in the enteric pathogens, their functions in the interaction between the pathogen and its host should become clearer.

ACKNOWLEDGMENTS

Studies reported in this chapter represent contributions over the years by a number of members of our laboratory, including Nicola Davies, Kathleen Lawlor, Melody Mills, Erin R. Murphy, Georgiana Purdy, Stephanie Reeves, Laura Runyen-Janecky, Michael Schmitt, Alfredo Torres, and Elizabeth Wyckoff.

These studies were supported by grant AI16935 from the National Institutes of Health.

SUGGESTED READING

Baumler, A. J., R. M. Tsolis, A. W. van der Velden, I. Stojiljkovic, S. Anic, and F. Heffron. 1996. Identification of a new iron regulated locus of *Salmonella typhi*. *Gene* **183**:207–213.

Boyer, E., I. Bergevin, D. Malo, P. Gros, and M. F. Cellier. 2002. Acquisition of Mn(II) in addition to Fe(II) is required for full virulence of *Salmonella enterica* serovar Typhimurium. *Infect. Immun.* **70**:6032–6042.

Clermont, O., S. Bonacorsi, and E. Bingen. 2001. The *Yersinia* high-pathogenicity island is highly predominant in virulence-associated phylogenetic groups of *Escherichia coli*. *FEMS Microbiol. Lett.* **196**:153–157.

de Lorenzo, V., A. Bindereif, B. H. Paw, and J. B. Neilands. 1986. Aerobactin biosynthesis and transport genes of plasmid ColV-K30 in *Escherichia coli* K-12. *J. Bacteriol.* **165**:570–578.

Der Vartanian, M., B. Jaffeux, M. Contrepois, M. Chavarot, J. P. Girardeau, Y. Bertin, and C. Martin. 1992. Role of aerobactin in systemic spread of an opportunistic strain of *Escherichia coli* from the intestinal tract of gnotobiotic lambs. *Infect. Immun.* **60**:2800–2807.

Dozois, C. M., F. Daigle, and R. Curtiss III. 2003. Identification of pathogen-specific and conserved genes expressed in vivo by an avian pathogenic *Escherichia coli* strain. *Proc. Natl. Acad. Sci. USA* **100**:247–252.

Goetz, D. H., M. A. Holmes, N. Borregaard, M. E. Bluhm, K. N. Raymond, and R. K. Strong. 2002. The neutrophil lipocalin NGAL is a bacteriostatic agent that interferes with siderophore-mediated iron acquisition. *Mol. Cell* **10**:1033–1043.

Hantke, K. 1997. Ferrous iron uptake by a magnesium transport system is toxic for *Escherichia coli* and *Salmonella typhimurium*. *J. Bacteriol.* **179**:6201–6204.

Hantke, K., G. Nicholson, W. Rabsch, and G. Winkelmann. 2003. Salmochelins, siderophores of *Salmonella enterica* and uropathogenic *Escherichia coli* strains, are recognized by the outer membrane receptor IroN. *Proc. Natl. Acad. Sci. USA* **100**:3677–3682.

Kingsley, R., W. Rabsch, P. Stephens, M. Roberts, R. Reissbrodt, and P. H. Williams. 1995. Iron supplying systems of *Salmonella* in diagnostics, epidemiology and infection. FEMS Immunol. Med. Microbiol. **11**:257–264.

Lam-Yuk-Tseung, S., and P. Gros. 2003. Genetic control of susceptibility to bacterial infections in mouse models. *Cell. Microbiol.* **5**:299–313.

Luck, S. N., S. A. Turner, K. Rajakumar, H. Sakellaris, and B. Adler. 2001. Ferric dicitrate transport system (Fec) of *Shigella flexneri* 2a YSH6000 is encoded on a novel pathogenicity island carrying multiple antibiotic resistance genes. *Infect. Immun.* **69:**6012–6021.

Reissbrodt, R., R. Kingsley, W. Rabsch, W. Beer, M. Roberts, and P. H. Williams. 1997. Iron-regulated excretion of α-keto acids by *Salmonella typhimurium*. *J. Bacteriol.* **179:**4538–4544.

Runyen-Janecky, L. J., S. A. Reeves, E. G. Gonzales, and S. M. Payne. 2003. Contribution of the *Shigella flexneri* Sit, Iuc, and Feo iron acquisition systems to iron acquisition in vitro and in cultured cells. *Infect. Immun.* **71:**1919–1928.

Torres, A. G., P. Redford, R. A. Welch, and S. M. Payne. 2001. TonB-dependent systems of uropathogenic *Escherichia coli*: aerobactin and heme transport and TonB are required for virulence in the mouse. *Infect. Immun.* **69:**6179–6185.

Tsolis, R. M., A. J. Baumler, F. Heffron, and I. Stojiljkovic. 1996. Contribution of TonB- and Feo-mediated iron uptake to growth of *Salmonella typhimurium* in the mouse. *Infect. Immun.* **64:**4549–4556.

Vokes, S. A., S. A. Reeves, A. G. Torres, and S. M. Payne. 1999. The aerobactin iron transport genes in *Shigella flexneri* are present within a pathogenicity island. *Mol. Microbiol.* **33:**63–73.

Williams, P. H. 1979. Novel iron uptake system specified by ColV plasmids: an important component in the virulence of invasive strains of *Escherichia coli*. *Infect. Immun.* **26:**925–932.

Wyckoff, E. E., D. Duncan, A. G. Torres, M. Mills, K. Maase, and S. M. Payne. 1998. Structure of the *Shigella dysenteriae* haem transport locus and its phylogenetic distribution in enteric bacteria. *Mol. Microbiol.* **28:**1139–1152.

Zhou, D., W.-D. Hardt, and J. E. Galan. 1999. *Salmonella typhimurium* encodes a putative iron transport system within the centrisome 63 pathogenicity island. *Infect. Immun.* **67:**1974–1981.

YERSINIA

Robert D. Perry

15

The genus *Yersinia* is part of the *Enterobacteriaceae* family and consists of environmental species (*Yersinia aldovae*, *Y. bercovieri*, *Y. frederiksenii*, *Y. intermedia*, *Y. kristensenii*, *Y. mollaretii*, and *Y. rohdei*), a primary fish pathogen (*Y. ruckeri*), two enteropathogenic species (*Y. enterocolitica* and *Y. pseudotuberculosis*), and the plague bacillus (*Y. pestis*). Yersiniosis (caused by the enteropathogens) and plague are zoonotic diseases.

Y. enterocolitica is found in the gastrointestinal tract of asymptomatic animals, with pigs as a common reservoir of serotypes pathogenic for humans (O:3, O:9, and O:5,27). The disease is acquired primarily by consumption of contaminated food or water. The ability to grow at refrigeration temperatures has resulted in food-borne epidemics and infection from transfusion of contaminated blood. Except for sporadic epidemics, disease from *Y. enterocolitica* is infrequent. The most common disease is a mild, self-limiting enteritis with symptoms of an invasive inflammatory diarrhea. In children younger than 5 years, symptoms include fever, diarrhea, and abdominal pain. In adults and children older than 5 years, fever with abdominal pain often mimics appendicitis. A reactive polyarthritis sequela can occur. Bacteria multiply in the intestinal lumen and invade gut-associated lymphoid tissues, where they cause ulcers in the terminal ileum and abscesses on Peyer's patches. Mesenteric lymphadenopathy is a common finding. Disseminated yersiniosis (sepsis with infection of the liver and spleen) is rare but can occur in elderly patients and those with predisposing conditions.

Yersiniosis from *Y. pseudotuberculosis* is contracted by contact with infected animals or their excrement or by eating contaminated food. The incidence of disease is greatest between November and February—the same seasonality observed in animals. *Y. pseudotuberculosis* causes a relatively mild, self-limiting lymphadenitis that also mimics appendicitis. Although a septicemia that is often fatal can occur from inflammation of the mesenteric lymph nodes, the majority of such patients have predisposing conditions: liver disease, hemachromatosis, or diabetes.

Plague, caused by *Y. pestis*, is infamous for three pandemics, but it primarily affects rodents and is transmitted from animal to animal by fleas. After the flea has ingested an infected blood meal, the organism grows in the flea midgut (stomach), eventually colonizing and

---

Robert D. Perry, Department of Microbiology, Immunology, and Molecular Genetics, MS415 Medical Center, University of Kentucky, Lexington, KY 40536–0298.

Iron Transport in Bacteria, Edited by Jorge H. Crosa, Alexandra R. Mey, and Shelley M. Payne
© 2004 ASM Press, Washington, D.C.

blocking a valve, the proventriculus, between the stomach and esophagus. This blockage is critical to the transmission of plague. The blocked flea feeds more aggressively but cannot pump blood into the blocked stomach. These feeding attempts dislodge bacteria colonizing the proventriculus, and contaminated blood flows back into the site of the fleabite. From there, Y. pestis cells spread through the lymphatic system to the regional lymph node of the animal, where they multiply to high levels. This causes the characteristic swollen lymph nodes (buboes) of bubonic plague. Organisms then spread via the blood to the liver and spleen, where they again proliferate to high levels before initiating a septicemia. Infection of a flea via a contaminated blood meal completes the life cycle of this obligate parasite. Y. pestis is never free-living but is entirely dependent on its flea vector and mammalian hosts.

Humans are an accidental dead-end host for bubonic plague (infection from a flea bite), which has a 2- to 8-day incubation period before the onset of symptoms: fever, chills, headache, and weakness. An enlarged bubo signals a potential bacteremia, and death can occur 2 to 4 days later. In septicemic plague, no bubo is observed and fever is the primary symptom of a high-level bacteremia. Infection of the lungs (secondary pneumonic plague) can lead to aerosol spread of the disease in humans but generally not in animals. This route of transmission is highly contagious, although the exposure range is considered to be limited (gener-

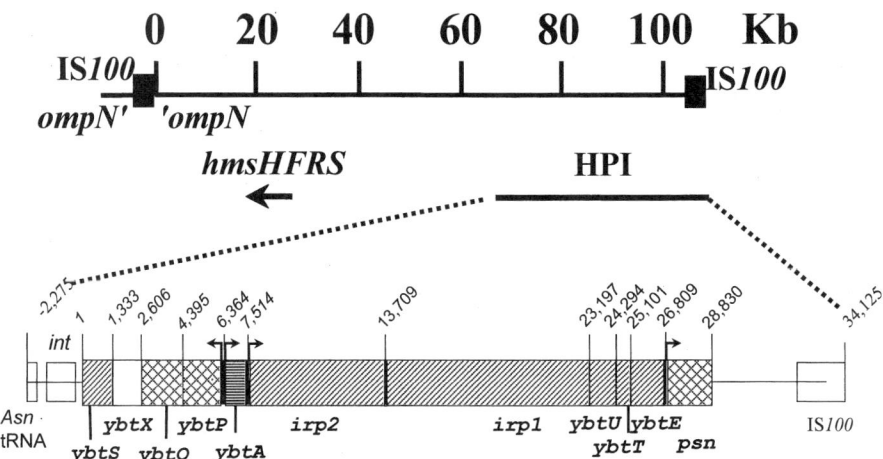

FIGURE 1 Genetic organization of the *pgm* locus and the HPI pathogenicity island of *Y.pestis*. IS 100 elements (black boxes) define the ends of the *pgm* locus. Numbering of the *ybt* locus within the HPI corresponds to nucleotide numbers for the start codons of *ybtA*, *irp2*, *irp1*, *ybtU*, *ybtT*, *ybtE*, and *psn* and for the stop codons of *ybtS* (start of *ybt* locus), *ybtX*, *ybtQ*, *ybtP*, and *psn* (end of the *ybt* locus). The *Yersinia* HPIs are extremely similar in the core region (tRNA$_{asn}$ to *psn*), with two differences in the *Y. enterocolitica* HPI: an ERIC (enterobacterial repetitive intergenic consensus) sequence in the *ybtA* promoter region and an inactive putative integrase gene. Genes of the *ybt* locus encoding transport proteins (cross-hatched boxes), biosynthetic enzymes (boxes with diagonal lines), the regulatory protein YbtA (box with horizontal lines), and *ybtX* (unknown function [open box]) are shown with arrows indicating promoter regions and the direction of transcription. Each promoter has proven Fur and YbtA binding sites (not shown). Molecular masses of Ybt proteins are indicated in kilodaltons. The unprocessed and processed masses of Psn are shown. High-molecular-weight proteins 1 and 2 (HMWP1 and HMWP2) are encoded by *irp1* and *irp2*, respectively.

ally less than 2 m). The onset of primary pneumonic plague occurs 1 to 6 days after infection by aerosolized respiratory droplets. Symptoms include fever, chest pain, lymphadenopathy, and coughing up of sputum with large numbers of *Y. pestis* cells, polymorphonuclear leukocytes, and blood. Although the organisms can be found in the bloodstream and internal organs, death occurs from pneumonia. Pneumonic plague is fatal unless the appropriate antibiotics are given within 20 h of the onset of symptoms. Bubonic plague has a ~50% mortality rate without treatment and a ~14% rate with proper therapy. Septicemic plague has a higher fatality rate than bubonic plague, possibly because the symptoms are less diagnostic and the correct antibiotic treatment is therefore delayed.

The virulence properties of the human pathogenic *Yersinia* species include iron acquisition systems. In fact, *Y. pestis* was one of the early pathogens in which the connection between iron metabolism and disease outcome was established. In 1956, Jackson and Burrows observed that "nonpigmented" (Pgm$^-$) mutants of *Y. pestis* were avirulent in mice unless injected along with iron or hemin. These Pgm$^-$ isolates, which failed to bind sufficient hemin to form greenish-brown colonies at 26°C, probably arose from deletion of a 102-kb chromosomal region now termed the *pgm* locus (Fig. 1). This region contains the hemin storage (*hms*) locus, which is required for the characteristic hemin adsorption. The *hms* locus is required for blockage of the flea proventricular valve but is not involved in iron or heme acquisition by the bacterium. Instead, the loss of mammalian virulence is due to deletion of the yersiniabactin (Ybt) siderophore-dependent iron transport system encoded within a high-pathogenicity island (HPI) located in the *pgm* locus. The Ybt system plays an important role in the pathogenesis of *Y. pestis* and *Y. enterocolitica*. Iron and hemoprotein transport systems in these two organisms have been extensively studied, and several systems in addition to Ybt are involved in pathogenesis. Since *Y. pseudotuberculosis* and *Y. pestis* are closely related, many systems encoded by *Y. pestis* are also present in *Y. pseudotuberculosis* but have not been extensively examined. The genomes of *Y. pestis* KIM (biotype mediaevalis) and CO92 (biotype orientalis) have been sequenced, and analysis has revealed 12 potential iron or hemoprotein transport systems.

SIDEROPHORE-DEPENDENT IRON TRANSPORT SYSTEMS

The *Y. pestis* genome encodes three potential siderophore biosynthesis systems, two of which are nonribosomal peptide synthetase (NRPS) systems. Only one of these has been demonstrated to produce a siderophore, yersiniabactin (Ybt). Genes *iucA* through *iucD*, which are required for non-NRPS-mediated synthesis of the hydroxamate siderophore aerobactin, are also present in the *Y. pestis* genome.

The Ybt Iron Transport System

The Ybt systems of *Y. pestis* and *Y. enterocolitica* are nearly identical and have been extensively characterized. Although not as well studied, the *Y. pseudotuberculosis* Ybt system is probably the same as the Ybt system of the other two species. Ybt production uses a mixed NRPS-polyketide (PK) synthesis mechanism that assembles salicylate, three cysteines, a malonyl linker group, and three methyl groups into a siderophore composed of salicylate, one thiazolidine, and two thiazoline rings (Fig. 2). Ybt is similar in structure to pyochelin of *Pseudomonas aeruginosa* and anguibactin of *Vibrio anguillarum*. YbtD (a putative phosphopantetheinyl transferase encoded outside the HPI) activates HMWP1 and HMWP2, presumably by trans-

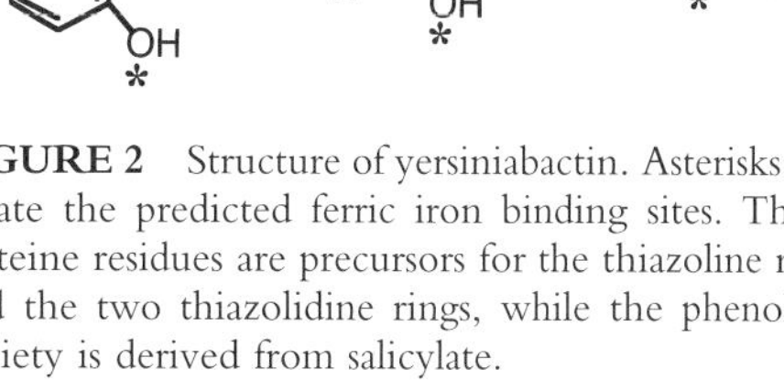

FIGURE 2 Structure of yersiniabactin. Asterisks indicate the predicted ferric iron binding sites. Three cysteine residues are precursors for the thiazoline ring and the two thiazolidine rings, while the phenolate moiety is derived from salicylate.

ferring the 4′-phosphopantethein (P-pant) moiety of coenzyme A to specific sites on these NRPS and PK enzymes. The P-pant moieties serve as tethers for the amino acid or acyl groups that will be assembled into the siderophore. YbtS is probably required for the synthesis of salicylate from chorismate. YbtE adenylates salicylate and transfers the activated compound to HMWP2, which possesses cyclization and condensation domains required for incorporation of salicylate and two cysteine molecules. This partial structure is passed on to HWMP1, where NRPS and PK/fatty acid synthase domains add the malonyl group linker and the final thiazoline ring. The middle thiazoline ring is reduced by YbtU to form a thiazolidine ring. A terminal thioesterase domain of HMWP1 probably releases the completed siderophore from the enzyme complex. An external thioesterase, YbtT, is required for the in vivo synthesis of Ybt and may serve to remove aberrant or mischarged structures from the enzyme complex. Enzymatic assembly of the Ybt siderophore is described in detail in chapter 2.

The mechanism by which Ybt is secreted from bacteria is unknown (Fig. 3). Although YbtX has some similarities to EntS, which is required for the export of enterobactin (see chapter 9), a strain with a mutation in *ybtX* still secreted Ybt and had no observed in vitro defect. Components of the uptake system, YbtP, YbtQ, and Psn, are not required for export of the siderophore.

The formation constant of Ybt with ferric iron is 4×10^{36}, which is higher than for a number of other siderophores; Ybt does not bind ferrous iron. In vitro studies indicate that Ybt is able to remove iron from transferrin and lactoferrin; strains with a defective Ybt system are unable to use these compounds as sources of inorganic iron. Once the Ybt-Fe complex is formed, the outer membrane (OM) receptor

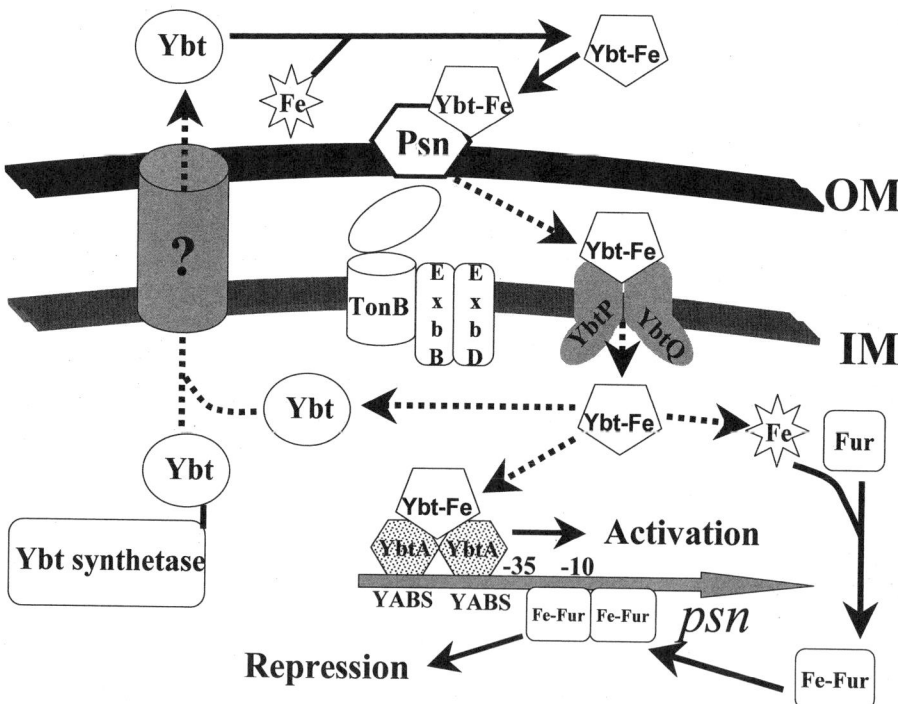

FIGURE 3 Model of the Ybt system. Dashed arrows designate predicted mechanisms, steps, or substrate transported that have not been experimentally demonstrated. It is uncertain whether a PBP is required for uptake, and none is shown in this model.

Psn (termed FyuA in *Y. enterocolitica*) and the TonB system are required for translocation across the OM. This receptor and TonB are also required for transport of the bacteriocin pesticin across the OM. The *tonB* genes of *Y. enterocolitica*, and now *Y. pestis*, have been characterized and encode typical enteric TonB proteins. In *Y. pestis*, *exbB* and *exbD* appear to form an operon and are not adjacent to *tonB*. Passage across the inner membrane (IM) involves an unusual ABC transport system (Fig. 3) composed of two fused-function proteins, YbtP and YbtQ, that have both IM permease and ATP hydrolase domains. Such fused-function ABC transporters are generally found in export systems. Both YbtP and YbtQ are required for iron uptake via the Ybt system. A periplasmic binding protein (PBP) may or may not be required for uptake. The HPIs encoding the Ybt system do not contain a gene likely to encode such a protein; however, it is possible that a Ybt-specific PBP may be encoded elsewhere in the genome. The mechanism for the removal of iron from the siderophore is undetermined; reduction to ferrous iron or degradation of Ybt are the two most likely methods. Although it has been clearly demonstrated that Ybt is involved in the accumulation of iron, translocation of the siderophore into the bacterial cell has not been experimentally determined.

In vitro, mutations in the Ybt system reduce the ability of *Y. pestis* strains to accumulate iron and to grow at 37°C on Fe-chelated agar. In strains unable to synthesize the siderophore, addition of purified Ybt overcomes these defects. Strains with mutations in genes encoding transport functions still produce Ybt and display more severe iron uptake defects than do strains unable to synthesize Ybt, probably because Ybt chelates residual iron in the medium, making it unavailable to other *Y. pestis* iron transport systems. In vivo, the *Y. pestis* Ybt system is essential during the early stages of infection (spread via the lymphatic system and growth in the lymph nodes) in mice injected subcutaneously to mimic bubonic plague. It is notable that the Ybt system is completely dispensable after this stage of the disease. Ybt$^-$ mutants are fully virulent if injected intravenously (Table 1).

Y. enterocolitica and *Y. pseudotuberculosis* mutants have been constructed that are unable to synthesize Ybt but respond to exogenously added siderophore. Ybt transport mutants of both species are unable to utilize Ybt. In vivo, Ybt$^-$ strains of both species show a loss of virulence when administered by the subcutaneous and intravenous infection routes (Table 1).

In *Y. pestis* KIM10+ and CO92, and *Y. enterocolitica* 8081 (ftp://ftp.sanger.ac.uk/pub/pathogens/ye.), the organization of the *ybt* gene locus is essentially identical. The loci differ in the distance between the coding sequence for the Ybt receptor (*psn* in *Y. pestis* and *fyuA* in *Y. enterocolitica*) and YbtD. In *Y. pestis* KIM10+, *psn* and *ybtD* are separated by approximately 170.5 kb, while the distance between these two genes is over 337 and 17.7 kb, respectively, in *Y. pestis* CO92 and *Y. enterocolitica* 8081. Nearly all of the Ybt system (Fig. 1) is encoded by four operons within the HPI: *psn*, *irp2-irp1-ybtU-ybtT-ybtE*, *ybtA*, and *ybtP-ybtQ-ybtX-ybtS*. Again, it appears that similar mechanisms are used by all three human-pathogenic species of *Yersinia* to regulate the expression of the Ybt genes. In *Y. pestis*, the Fur-Fe complex represses each of the operons (see chapter 13). In addition to this typical form of negative regulation for iron transport systems, YbtA and Ybt stimulate transcription from the *psn*, *irp2*, and *ybtP* promoters while repressing transcription from the *ybtA* promoter. In contrast, transcription from *Y. pestis* *ybtD*, encoded outside the HPI and *pgm* locus, is not repressed by iron or activated by YbtA. These results suggest that *ybtD* may have been recently recruited into the Ybt system.

YbtA is an AraC-family transcriptional regulator that probably binds to 18-nucleotide repeated sequences within each of the four *ybt* promoters. Altering one repeat sequence of the putative YbtA binding site in the *psn* promoter significantly reduced its transcriptional activity under Fe-deficient conditions. My working hypothesis is that YbtA binds the Ybt-Fe com-

TABLE 1 Effects of iron and hemoprotein transport systems on the pathogenesis of *Yersinia*

Strain	Relevant traits[a]	LD$_{50}$ by indicated infection route[b]	
		s.c. injection	i.v. injection
Y. pestis			
KIM5–2053.11+ (pCD1::*yopJ*::Mu*d*I1734)	Ybt$^+$Yfe$^+$Yfu$^+$Hmu$^+$Has$^+$ (YopJ$^-$Psa$^-$)	120	NT
KIM5-2046.41 (pCD1::*yopJ*::Mu*d*I1734)	Ybt$^-$ ($\Delta irp2$) Yfe$^+$Yfu$^+$Hmu$^+$Has$^+$ (YopJ$^-$Psa$^-$)	>1.3 × 10^7	NT
KIM5 (pCD1Ap)+	Ybt$^+$Yfe$^+$Yfu$^+$Hmu$^+$Has$^+$ (YopJ$^+$Psa$^+$)	<8	NT
KIM5–2045.7 (pCD1::*yopJ*::Mu*d*I1734)	Ybt$^-$ (Δpsn) Yfe$^+$Yfu$^+$Hmu$^+$Has$^+$ (YopJ$^-$Psa$^-$)	>5.4 × 10^6	NT
KIM5–3173 (pCD1::*yopJ*::Mu*d*I1734)	Ybt$^-$ (Δpgm) Yfe$^+$Yfu$^+$Hmu$^+$Has$^+$ (YopJ$^-$Psa$^-$)	NT	<12
KIM5–2031.12 (pCD1::*yopJ*::Mu*d*I1734)	Ybt$^-$ (Δpgm) Yfe$^-$($\Delta yfeAB$) Yfu$^+$ Hmu$^+$Has$^+$ (YopJ$^-$Psa$^-$)	NT	>1.7 × 10^7
KIM5–2031.11+ (pCD1::*yopJ*::Mu*d*I1734)	Ybt$^+$Yfe$^-$($\Delta yfeAB$) Yfu$^+$Hmu$^+$ Has$^+$(YopJ$^-$ Psa$^-$)	8.7 × 10^3	NT
KIM5–2031.12+ (pCD1Ap)	Ybt$^+$ Yfe$^-$ ($\Delta yfeAB$) Yfu$^+$Hmu$^+$ Has$^+$ (YopJ$^+$ Psa$^+$)	74.3	NT
KIM5–2082.11+ (pCD1Ap)	Ybt$^+$Yfe$^-$($\Delta yfeAB$) Yfu$^-$Hmu$^+$ Has$^+$ (YopJ$^+$ Psa$^+$)	<82	NT
KIM5–2044.21+ (pCD1::*yopJ*::Mu*d*I1734)	Ybt$^+$Yfe$^+$Yfu$^+$Hmu$^-$ ($\Delta hmuP'$-V) Has$^+$(YopJ$^-$ Psa$^-$)	42	NT
KIM5–2081.1+ (pCD1Ap)	Ybt$^+$Yfe$^+$Yfu$^+$Hmu$^-$ ($\Delta hmuP'$-V) Has$^-$($\Delta hasR$-E) (YopJ$^+$ Psa$^+$)	<4.2	NT
Y. pseudotuberculosis			
IP 2790Sm	Ybt$^+$	3.6 × 10^6	6.9
IP 2790 H$^-$	Ybt$^-$ (*irp2*::pJMA13)	>2.1 × 10^9	1.9 × 10^3
PB1 Pst^s	Ybt$^+$	1.6 × 10^5	20
PB1 Pst^r	Ybt$^-$ (*psn*)	>10^7	10^3
Y. enterocolitica			
WA-314	Ybt$^+$Yfu$^+$	NT	5 × 10^2
WA *fyuA*(pYV08)	Ybt$^-$ (*fyuA*) Yfu$^+$	NT	>5 × 10^6
WA *yfuB*(pYV08)	Ybt$^+$Yfu$^-$ (*yfuB*::*kan*)	NT	7 × 10^2
WA Pst^s	Ybt$^+$	2.3 × 10^3	10^2
WA Pst^r	Ybt$^-$ (*psn*)	1.1 × 10^5	2.3 × 10^3

[a] In addition to differences in iron transport systems, some *Y. pestis* strains had mutations in *yopJ* and *psa* encoding YopJ and an adhesin, respectively. Mutations in these two genes decreased virulence >15-fold compared to the wild-type strain. *Y. pseudotuberculosis* and *Y. enterocolitica* Pst^r are resistant to pesticin due to a mutation in the Ybt OM receptor.

[b] LD$_{50}$, 50% lethal dose; s.c., subcutaneous; i.v., intravenous; NT, not tested.

plex to activate transcription of the regulated genes (Fig. 3). If a siderophore-Fe complex is required for maximal expression, this would prevent activation by Ybt before it was secreted from the bacterial cell. Addition of exogenous Ybt to an *irp2*::*kan* mutant, KIM6–2046.1, carrying a transcriptional *ybtP*::*lacZ* fusion (on pEUYbtP) stimulated β-galactosidase activity 4-fold within 10 min and 7.5-fold (maximum activation) within 30 min, compared to a culture with no Ybt added. The amount of exogenous Ybt that can stimulate transcription from the *ybtP* promoter 2-fold in 20 min is 500-fold lower than that required for growth stimulation of KIM6–2046.1. These results indicate that Ybt is an extremely effective signal molecule for activation of Ybt biosynthesis and transport. My laboratory has also used the *ybtP*::*lacZ* reporter to assess the effects of 13 different *ybt* mutations on the expression of *ybt* genes. Based on β-galactosidase activity from the *ybtP::lacZ* reporter, these strains were divided into five different regulatory classes (Table 2). Class I has normal regulation and contains the Ybt$^+$ strain KIM6+ as well as the Δ*ybtX* mutant (which displays no in vitro defect in the Ybt system). Mutants lacking YbtA comprise the class II group and show almost no transcriptional activity from the *ybtP::lacZ* reporter. Class III includes all the mutants defective in Ybt transport and exhibits β-galactosidase activity that is moderately higher than that in wild-type cells (Table 2). This may be due to increased iron deprivation as a result of Ybt binding the trace levels of iron in the medium and thereby preventing its acquisition by other *Y. pestis* transport systems. Class IV includes a group of mutants that produce no detectable Ybt in a bioassay for growth stimulation. Strains with these mutations (Table 2) exhibit ~18-fold lower transcriptional activity than does Ybt$^+$ KIM6+, presumably because they do not produce the siderophore. Class V also contains mutants that do not produce detectable levels of Ybt by bioassay; however, these strains have β-galactosidase activity that is ~25-fold-higher than that of class IV mutants, ~1.4-fold higher than that of wild-type cells, and ~71% of the activity displayed by class II mutants (Table 2).

The results with class II and IV mutants conform to our regulatory model. For transport mutants (class III), minute quantities of Ybt may nonspecifically enter the cell to activate the Ybt system. In both *Y. pestis* and *Y. enterocolitica*, mutations in the Ybt IM permease genes (*ybtPQ* in *Y. pestis*, and *irp6* and *irp7* in *Y. enterocolitica*) upregulated the expression of YbtA-dependent reporters. A *Y. enterocolitica irp8* (*ybtX* in *Y. pestis*) mutant was similar to wild-type cells, while an *irp6,7,8* triple mutant also displayed upregulated expression. This suggests that Irp8/YbtX does not serve as an alternative Ybt-Fe transporter that is sufficient for signal transduction but not for utilization of the siderophore. Class V strains may either produce minute amounts of Ybt sufficient to activate transcription (Δ*ybtT* and *irp1-2086* mutations) or produce an aberrant molecule that retains signaling activity (i.e., *ybtS::kan* mutation). The lack of YbtT may lead to the accumulation

TABLE 2 Expression of a *ybtP* promoter gene fusion in *Y. pestis* derivatives[a] grown in Fe-deficient medium

Class	Relevant characteristics	% of activity in wild-type strain[b]
I	Ybt$^+$	100
	Δ*ybtX* (Ybt$^+$)	104
II	Δ*pgm* (Ybt$^-$)	2
	ybtA::kan (Ybt$^-$)	1
III	Δ*psn* (Ybt transport mutant)	181
	Δ*ybtP* (Ybt transport mutant)	188
	tonB::kan (Ybt transport mutant)	162
IV	*irp1::kan* (Ybt biosynthesis mutant)	5
	irp2::kan (Ybt biosynthesis mutant)	5
	Δ*ybtE* (Ybt biosynthesis mutant)	5
	Δ*ybtU* (Ybt biosynthesis mutant)	7
	Δ*ybtD* (Ybt biosynthesis mutant)	6
V	*ybtS::kan* (Ybt biosynthesis mutant)	126
	Δ*ybtT* (Ybt biosynthesis mutant)	131
	irpl-2086 (HMWP1 with defective thioesterase domain; Ybt biosynthesis mutant)	154

[a] All strains carried the reporter plasmid pEUYbtP (*ybtP::lacZ*).
[b] Averages of Miller units from four or more samples compared to wild-type Ybt$^+$.

of mischarged products on the Ybt synthetase complex, with just enough authentic Ybt produced to stimulate transcription. The terminal thioesterase domain of HMWP1 (which is inactivated by the *irp1-2086* mutation) is thought to release the completed siderophore from the enzyme complex. However, chemical cleavage of the ester bond may produce enough secreted Ybt to activate transcription without stimulating bacterial growth. Finally, in the absence of YbtS, an alternate phenolate ring structure may be inefficiently substituted for salicylate, which could yield Ybt-like molecules that retain signaling activity.

Although both *Y. enterocolitica* and *Y. pseudotuberculosis* regulate the expression of the *ybt/irp* genes in a similar manner to *Y. pestis* (activation of *ybt* gene transcription by exogenous Ybt), there are significant differences. A mutation in the Ybt OM receptor causes a loss of siderophore production in *Y. enterocolitica* and *Y. pseudotuberculosis* but not in *Y. pestis*. In *Y. enterocolitica*, an *fyuA* mutant does not express green fluorescent protein from an *fyuA*::*gfp* reporter. This is similar to the *Pseudomonas* pyochelin system, where interaction of the OM receptor with the siderophore transmits a signal for the activation of pyochelin gene expression (see chapter 19). However, a *Y. pestis* Ybt OM receptor mutant overproduces Ybt. This discrepancy may be explained by a difference in the permeability of the OMs of these bacterial species. The OMs of *Y. enterocolitica* and *Y. pseudotuberculosis* are relatively impermeable to small hydrophobic molecules compared to the OM of *Y. pestis*. Although *Y. enterocolitica* has an ERIC (enterobacterial repetitive intergenic consensus) sequence inserted upstream of the *ybtA* promoter, which may affect the expression of this gene, it seems likely that functional levels of the YbtA regulator are produced. It is unlikely that siderophore-dependent activation of *ybt/irp* genes would occur in the absence of YbtA. Finally, studies with *Y. enterocolitica*, using a *fyuA* translational fusion to a reporter gene, found that cells growing in the peritoneal cavity of mice exhibited strong expression while cells growing in the spleen had moderate expression. Cells recovered after growth in the liver and intestinal lumen showed the lowest level of expression. Compared to expression under Fe-deficient conditions in vitro, *Y. enterocolitica* cells had threefold-higher expression after growth in the peritoneal cavity.

The FhuBCD Siderophore-Dependent Iron Transport Systems

The ABC transport system FhuBCD is required for the use of a number of siderophores including aerobactin, ferrichrome, and ferrioxamine B. *Y. enterocolitica* uses both ferrioxamine B and ferrioxamine E and expresses an OM receptor, FoxA, for these siderophores (Table 2). Thus, treatment of iron overload with desferrioxamine increases the risk of septicemia, since *Y. enterocolitica* can use this iron chelator. *Y. pestis* cells do not encode a FoxA receptor, and desferrioxamine B inhibits the growth of *Y. pestis* cells. There are conflicting reports on ferrioxamine use by *Y. pseudotuberculosis*, but *Y. frederiksenii*, *Y. intermedia*, and *Y. kristensenii* all utilize ferrioxamine. Both *Y. pestis* and *Y. enterocolitica* use exogenously supplied aerobactin and ferrichrome as iron sources. The *Y. enterocolitica* ferrichrome OM receptor, termed FcuA, is 25% identical and 32% similar to FhuA, the *Escherichia coli* ferrichrome receptor. The *Y. enterocolitica* FcuA receptor recognizes ferrichrome, ferricrocin, ferrichrysin, ferrirubin, and ferrirhodin siderophores. A homologue to *fcuA* is present in the *Y. pestis* genome. The genetic organization of the *Y. enterocolitica* and *Y. pestis fcuA* and *fhuCDB* operons differs from that in *E. coli*, in that *fcuA* is not located upstream of *fhuCDB* in the yersiniae (see chapter 10). The predicted amino acid sequences of FhuB and FhuC in *Y. enterocolitica* 8081 are 86 and 95% identical, respectively, to their counterparts in *Y. pestis*.

Y. pestis also encodes an *iucABCD iutA* operon (Table 2) similar to that in *E. coli* (see chapter 13), but with one difference. Although the IutA OM receptor for aerobactin is functional in *Y. pestis*, a frameshift mutation is present in *iucA* that is probably responsible for the inability to produce aerobactin. This mutation may also affect the expression of the downstream genes *iucB* to *iucD*, since the cloned *Y. pestis* locus did not complement an *E. coli* strain with an insertion in *iucB*. A survey of 50 *Y. enterocolitica* strains failed to find any that produced aerobactin. The aerobactin biosynthetic operon and *iutA* gene in *Y. pseudotuberculosis* IP32953 (http://bbrp.llnl.gov/bbrp/html/microbe.html) seem intact, while they are absent from the *Y. enterocolitica* 8081 genome. In *Y. pestis*, mutations in either *tonB* or the *fhuCDB* locus prevent the use of ferrichrome and aerobactin as iron sources. While *Y. pseudotuberculosis* is able to use aerobactin as an iron source, there are conflicting reports about the ability of *Y. enterocolitica* to use this siderophore. The ability to utilize exogenous siderophores would probably be beneficial for *Y. enterocolitica* in environments with mixed bacterial populations. In contrast, the presence of such systems in *Y. pestis* may merely indicate the organism's *Enterobacteriaceae* heritage, since this organism has a relatively isolated life cycle—growing in the flea midgut and in rodent lymph and blood systems as well as internal organs (all normally sterile environments).

The Enterobactin-Dependent Iron Transport System

Y. enterocolitica utilizes exogenous enterobactin as an iron source and contains genes orthologous to *E. coli fepDGCB*, encoding the PBP and cytoplasmic membrane permease, and *fes*, which encodes enterobactin esterase (Table 3) (see chapter 8). The *Y. enterocolitica fepD*, *fepG*, and *fes* genes complement *E. coli* strains with corresponding mutations. Mutations in *fepD* or *fes* prevent *Y. enterocolitica* from using exogenous enterobactin. Although the *Y. enterocolitica* locus has a gene order similar to that in *E. coli*, it is missing *orf43* and the enterobactin-biosynthetic genes. With *Y. enterocolitica fepDGB* as probes, Southern blot hybridization indicated that these sequences were present in biotypes IA, Ib, and II but not in biotype IV of *Y. enterocolitica*. This method also failed to detect similar genes in *Y. pestis* and *Y. pseudotuberculosis*. A search of the *Y. pestis* CO92 and KIM10+ genomes did not locate genes homologous to *fepD* or *fes*; however, an isolated *fepB* homologue gene was found. Thus use of exogenous enterobactin seems to be restricted to *Y. enterocolitica*. A gene homologous to the OM receptor for enterobactin, *fepA*, is located downstream of *fes* in *Y. enterocolitica*. However, *Y. enterocolitica fepA'* has a frameshift mutation that yields a truncated 40-kDa product. It has been suggested that CccA, a receptor for catecholate and dihydroxybenzoylserine (DHBS), or other iron-repressible OM proteins may substitute for the truncated FepA OM receptor, which is probably nonfunctional. A study with antibodies against *E. coli* FepA suggests that some *Y. enterocolitica* strains do have a full-length *fepA* gene product.

The Ysu and Ynp Systems

The components of the Ysu and Ynp *Y. pestis* putative siderophore biosynthesis systems show the highest similarities to other siderophore biosynthetic enzymes; however, neither system has been demonstrated to produce a siderophore or be involved in iron acquisition. Genes encoding proteins that may be involved in transport and other functions are present in both loci. Compared to *Y. pestis* KIM10+, the *ynp* locus has been rearranged in *Y. pestis* CO92, with *y3404* (*YPO1011* in CO92) and *y3406* (*YPO1012*) being over 285 kb away from the remainder of the *ynp* genes. BLAST searches with the predicted amino acid sequences of three *ynp* and four *ysu* genes indicate that these systems are probably absent from *Y. enterocolitica* 8081.

YsuE and YsuG are most closely related to the *Bordetella* alcaligin biosynthetic enzyme AlcC, while YsuH and YsuI (Fig. 4) show similarities to AlcB and AlcA, respectively. YsuJ possesses a decarboxylase domain found in a wide variety of proteins including some involved in siderophore biosynthesis. Although

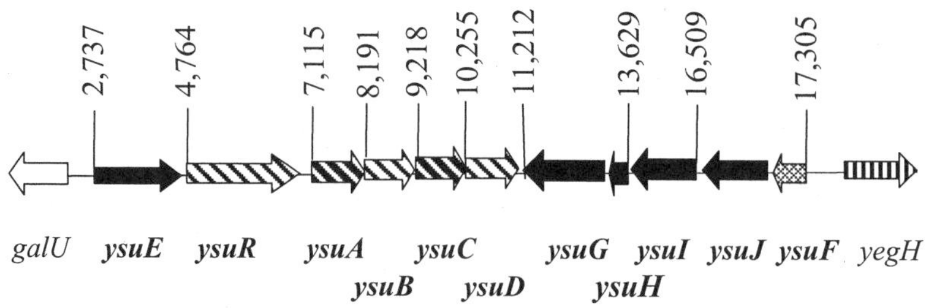

	YsuE	YsuR	YsuA	YsuB	YsuC	YsuD	YsuG	YshH	YsuI	YsuJ	YsuF	YegH
kDa	69.5	83.9 (79.2)	40.6 (34.9)	36.4	35.9	31.5	71.9	22.5	50.1	57.7	30.3	60.3
Putative Function	SBE	OM receptor	PBP	IMP	IMP	ATPase	SBE	SBE	SBE	SBE	iron reductase	exporter

FIGURE 4 Genetic organization of the *ysu* locus of *Y. pestis*. Numbering indicates base pairs of a cloned DNA fragment. Arrows indicate the ORFs and direction of transcription. Putative biosynthetic enzymes are indicated by solid arrows, and putative transport functions are indicated by diagonally striped arrows. YsuF (cross-hatched arrow) shows similarity to FhuF, an Fe^{3+}/siderophore reductase in *E. coli*. YegH contains domains present in export proteins and is a homologue of *E. coli* YegH. *ysuE* is intact in *Y. pestis* CO92; in *Y. pestis* KIM10+, *ysuE1* and *ysuE2* are probably pseudogenes resulting from a frameshift mutation. Deduced protein molecular masses are given in kilodaltons; YsuR and YsuA have predicted signal sequences, and the numbers in parentheses are sizes after processing. SBE, siderophore-biosynthetic enzyme; IMP, inner membrane permease; ATPase, ATP hydrolase.

ysuE is intact in *Y. pestis* CO92, it has undergone a frameshift mutation in strain KIM10+, yielding two open reading frames (ORFs) encoding proteins of 318 and 287 amino acids. Because of this *ysuE* pseudogene *Y. pestis* KIM10+ probably does not produce a Ysu-associated siderophore; however, the transport system encoded within the *ysu* locus may still be functional. YsuR is a putative OM receptor, while *ysuA* through *ysuD* encode a PBP, two permeases, and an ATP hydrolase, respectively. Finally, YsuF is related to FhuF, an *E. coli* reductase that has been implicated in removal of iron from hydroxamate siderophores. If YsuF possesses this activity, it may be responsible for removing iron from a number of siderophores (possibly Ybt).

We have tentatively designated the second potential siderophore-dependent system Ynp (for "*Yersinia* non-ribosomal peptide synthesis"). The *ynp* locus spans ~30 kb and includes genes encoding putative NRPS enzymes, an OM receptor, an ABC transport system, and a protein similar to YbtX (Table 3). There are seven ORFs whose potential products have similarities to yersiniabactin, pyochelin, and/or anguibactin biosynthetic enzymes; five ORFs encode NRPS domains, one possesses a thioesterase domain, and one resembles the putative reductases NrpU of *Proteus mirabilis* and YbtU. However, several ORFs may be disrupted or degenerate. In *Y. pestis* KIM10+, one of the putative NRPS proteins (Y3416) has an internal 6-amino-acid deletion and a 50-amino-acid region that is not 100% conserved compared to its counterpart in CO92 (YPO0776) (Table 3). Y3411 is a small ORF that could have resulted from a frameshift mutation. There is also an IS*100* insertion between two of the ORFs (Y3406

TABLE 3 Components of the Fhu, enterobactin, and Ynp systems

System	Protein	Selected characteristics[a]
Fhu[c]	IucA	66.8 kDa (*Y. pseudotuberculosis* only); aerobactin biosynthesis
	IucB	36.8 kDa; aerobactin biosynthesis
	IucC	66.5 kDa; aerobactin biosynthesis
	IucD	50.3 kDa; aerobactin biosynthesis
	IutA	80.3/77.6 kDa; aerobactin OM receptor
	FcuA	84.0/78.2 kDa (*Y. pestis*); 81.8/78.0 kDa (*Y. enterocolitica*); hydroxamate OM receptor
	FoxA[b]	78.6/75.8 kDa; ferrioxamine OM receptor
	FhuB	70.3 kDa; IM permease
	FhuC	28.7 kDa; ATP hydrolase
	FhuD	39.7/38.0 kDa; PBP
Enterobactin[b]	FepA'	40-kDa truncated product; defective enterobactin OM receptor in strain WA
	FepB	Enterobactin binding protein
	FepD	37.3 kDa; IM permease
	FepG	35.8 kDa; IM permease
	FepC	29.6 kDa; ATP hydrolase
	Fes	39.8 kDa; enterobactin esterase
	CccA	65kDa; OM receptor for catecholate and DHBS
Ynp[c]	Y3404	79.0/76.5 kDa; OM receptor
	Y3406	124.8 kDa; NRPS domains
	Y3410	239.4 kDa; NRPS domains
	Y3411	12.5 kDa; similarity to HMWP2 and AngR
	Y3412	62.4 kDa; NRPS domains
	Y3416	215.6 kDa; NRPS domains
	Y3418	36.1 kDa; dehydrogenase/reductase domain
	Y3419	41.6 kDa; similar to NrpU and YbtU
	Y3420	28.9 kDa; thioesterase domain
	Y3421	49.3 kDa; similar to YbtX
	Y3422	65.7 kDa; IM permease/ATP hydrolase
	Y3423	66.1 kDa; IM permease/ATP hydrolase

[a] Molecular masses of proteins with signal peptides are shown as unprocessed/processed.
[b] FoxA and the enterobactin system are present in *Y. enterocolitica* but not in *Y. pestis*.
[c] Proteins in the Fhu and Ynp system use ORF designations and molecular masses from the *Y. pestis* KIM10+ genome sequence

and Y3410) that may abrogate synthesis of a product. The *ynp* locus also contains genes encoding a putative OM receptor with a TonB box, two ABC transporters with permease and ATPase domains, as well as protein with similarity to YbtX (Table 3). Like the organization of the Ybt system, the gene for the putative OM receptor is separated from genes for the ABC transporters by NRPS genes, there is no ORF corresponding to a PBP, and the ABC transporters are followed by a YbtX-like ORF.

P-pant transferases initiate the assembly of siderophores, polyketides, and fatty acids by modifying acyl, aryl, or peptidyl carrier protein domains. The Ynp system does not encode this essential activity. A BLAST search of the *Y. pestis* genome identified only two potential P-pant transferases: YbtD (see above) and a homologue of the *E. coli* acyl carrier protein, which is essential for fatty acid synthesis. If the Ynp system synthesizes a product, one of these P-pant transferases must catalytically activate the biosynthetic enzymes. However, the transport and iron reductase functions encoded by these two loci may be functional even if siderophores are not produced.

SIDEROPHORE-INDEPENDENT IRON TRANSPORT SYSTEMS

In addition to the systems described above, the *Y. pestis* genome encodes a number of ABC transport systems with significant similarities to iron or siderophore transport systems. Since none of these systems is associated with genes encoding biosynthetic enzymes for siderophores and since the ligands for most of these systems have not been experimentally determined, they are included in this section on siderophore-independent iron transport systems.

The Yfe Iron and Manganese ABC Transporter

The *Y. pestis yfe* locus (Fig. 5) was identified by growth restoration of an enterobactin-negative *E. coli* strain under Fe-chelated conditions. Based on sequence similarities, the YfeA through YfeD ABC transporter is a member of the Mn and Zn transport cluster (TC 3.A.1.15). Other members of this cluster include *Synechocystis* MntCAB, *Treponema pallidum* TroABCD, and *Streptococcus pneumoniae* PsaBCAD and AdcCBA. Some of the members of this cluster transport iron or other cations. The Yfe system transports both iron and manganese and consists of a PBP (YfeA), a heterodimeric permease (YfeC and YfeD), and an ATP hydrolase (YfeB) (Fig. 6). YfeE is a putative IM protein of unknown function. No OM receptor or porin has been identified for this system. Insertions or deletions in the *yfeA* to *yfeD* operon and in *yfeE* abrogate growth restoration in *E. coli*. Mutations in Y1892, a predicted cytoplasmic protein encoded between *yfeA* to *yfeD* and *yfeE* (Fig. 5), do not affect the Yfe system. In *Y. pestis yfe* mutants, growth and iron transport defects are observed only in Ybt$^-$ strains. *Y. pestis* strains with a $\Delta yfeA$, $\Delta yfeAB$, or $\Delta yfeBCDE$ mutation show a 50% reduction in their ability to grow across Fe-chelator gradient plates compared to a Yfe$^+$ strain. A defect in *yfeE* is less severe than are other mutations in the Yfe system. A $\Delta yfeE$ strain takes 24 h longer than the Yfe$^+$ strain to achieve full growth on gradient plates containing ovotransferrin. The growth defects observed on gradient plates are due to iron chelation; supplementation with iron but not with Mn or Zn restored growth to Yfe$^-$ mutants.

Energy-dependent iron uptake by a strain with a $\Delta yfeAB$ mutation was reduced ~60% after 40 min of incubation compared to the Ybt$^-$ Yfe$^+$ parental strain. Mn and Zn competed with iron for uptake through the Yfe system. Transport studies with Mn revealed that the Yfe system accumulated this ion and that iron but not Zn competed for Mn uptake. Neither TonB nor HasB (a TonB-like protein; see the discussion of the Has system, below) is required for iron uptake via the Yfe system.

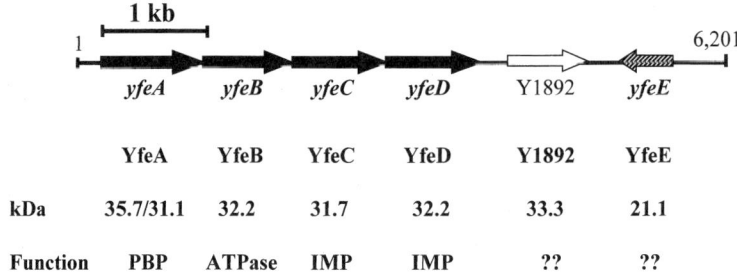

FIGURE 5 Genetic organization of the *yfe* locus of *Y. pestis*. Numbering in base pairs corresponds to the size of a cloned DNA fragment. Arrows indicate the ORFs and direction of transcription. Deduced protein molecular masses are given in kilodaltons. YfeA has a predicted signal sequence, and sizes indicated are for the unprocessed/processed forms. IMP, inner membrane permease; ATPase, ATP hydrolase.

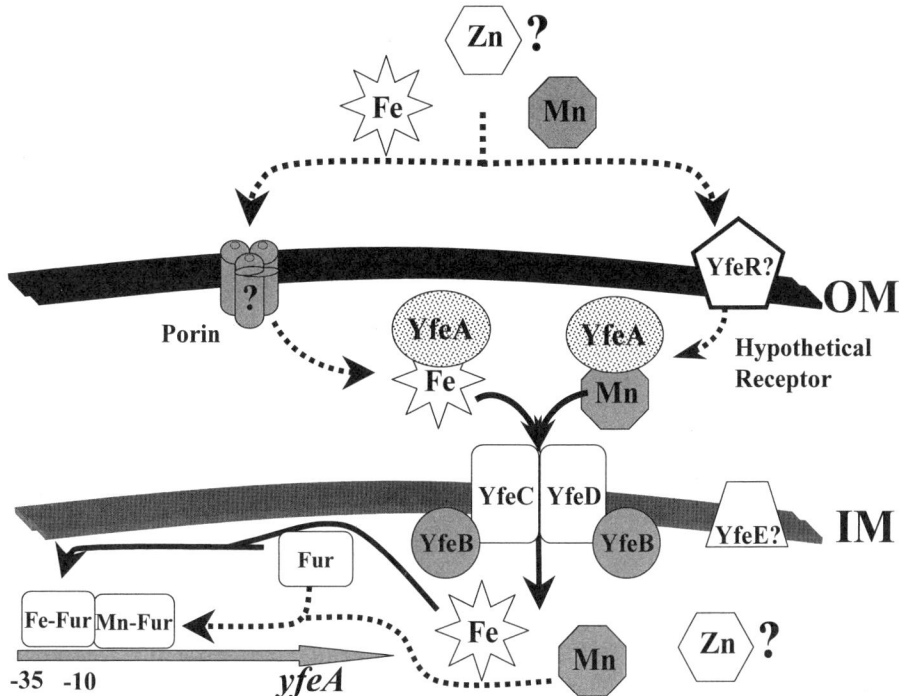

FIGURE 6 Model of the Yfe iron and manganese transport and regulatory system of *Y. pestis*. No OM component for the Yfe transport system has been identified. The YfeR receptor and porin are alternative speculative channels through the OM. The question mark indicates the uncertainty about Zn uptake by the Yfe system. Dashed arrows identify mechanisms or steps that have not been determined experimentally.

Compared to a Ybt$^+$ Yfe$^+$ fully virulent strain, a Ybt$^+$ Δ*yfeAB* strain of *Y. pestis* exhibited a ~70-fold loss of virulence in mice infected subcutaneously (Table 1). In infection by an intravenous route that bypasses the early lymphatic stages of bubonic plague, a Ybt$^-$ mutant is fully virulent but a Ybt$^-$ Yfe$^-$ mutant is completely avirulent (Table 1). This suggests that the Yfe transport system is more effective than the Ybt siderophore-dependent system in the later stages of bubonic plague. Thus, one explanation for multiple iron transport systems in pathogens is that some transport systems function more efficiently than others in specific organ systems or environments.

yfeA to *yfeD* and *yfeE* are two separate operons with no closely linked OM receptor or porin gene (Fig. 5). The *yfeA* promoter region has a strong consensus Fur binding site (FBS), while a weaker potential FBS is present within the *yfeE* promoter region. Studies using both transcriptional and translational reporter gene fusions have demonstrated that the *yfeA* promoter is Fur regulated and repressed by iron and Mn. Growth with excess iron caused a fivefold repression, while excess Mn resulted in a ca. twofold repression. The *yfeA* promoter activity was unaffected by the addition of Zn to the growth medium. Repression of the *yfeA* promoter by Mn required Fur. However, this does not appear to be a general characteristic of Fur-regulated promoters, since several other Fur- and Fe-repressible *Y. pestis* and *E. coli* promoters did not display Mn repression. We speculated that YfeE might modulate the expression of the YfeABCD system. However, expression from the *yfeA* promoter in a Δ*yfeE* strain is essentially identical to that in a YfeE$^+$ strain. Thus, YfeE does not regulate the expression of the Yfe system, and the function of this protein remains enigmatic. Finally, recent studies with a *yfeE::lacZ* fusion have demonstrated that expression from this promoter is not affected by the Fe, Mn, or Zn status of *Y. pestis* cells. Using primers specific for *Y. pestis yfeA*, an appropriately sized product was detected by PCR in *Y. enterocolitica* and *Y. pseudotuberculosis*.

The *yfeA* to *yfeE* genes are present in the *Y. enterocolitica* 8081 genome and are highly conserved. However, it has not been experimentally demonstrated whether this system functions in *Y. pseudotuberculosis* or *Y. enterocolitica*.

The Yfu Iron Transport System

The *Yersinia* Yfu system belongs to the FeT cluster (TC 3.A.1.10) of ABC transporters, which includes *Serratia marcescens* Sfu and *Neisseria* Fbp. Yfu was characterized first in *Y. enterocolitica* and subsequently in *Y. pestis* as a cloned locus that enhanced the growth of an *E. coli* enterobactin-biosynthetic mutant under Fe-deficient conditions. DNA hybridization analysis suggests that *Y. pseudotuberculosis* may encode a similar system. The *Y. enterocolitica yfu* locus encodes four proteins: YfuA to YfuD. YfuA, YfuB, and YfuC are a PBP, IM permease, and ATP hydrolase, respectively (Table 4), with >75% identity to the *Y. pestis* YfuABC proteins. YfuD, present in *Y. enterocolitica* but not in *Y. pestis*, has significant similarity to the *E. coli* hypothetical protein YahN and relatively low similarities to several transporter proteins or membrane channels and contains a LysE-type translocator domain. One member of the LysE family effluxes lysine. YfuD is not essential but does enhance the function of the *Y. enterocolitica* YfuABC system in *E. coli*. In *Y. enterocolitica*, the Ybt system masks Fe-deficient growth defects due to mutation of the Yfu system. Recent studies using Fe-chelator gradient plates have demonstrated growth defects in *Y. pestis yfu* mutants but only when the Ybt and Yfe systems are mutated.

In gram-negative bacteria, iron ABC transporters such as Yfu are often associated with OM receptors that bind transferrin or lactoferrin. These OM receptors show a high degree of similarity among the different bacteria and between transferrin and lactoferrin receptors. However, searches of the *Y. pestis* KIM10+ and CO92 genomes failed to identify any putative OM receptors for these host iron binding proteins. Whatever the ligand is for the *Y. enterocolitica* Yfu system, its translocation through the OM appears to be TonB independent.

TABLE 4 Components of selected siderophore-independent iron transport systems

Iron transport system	Protein	Mass (kDa)[a]	Selected characteristics
Yfu	YfuA	36.3/33.7	PBP
	YfuB	56.5	IM permease
	YfuC	37.6	ATP hydrolase
	YfuD[b]	22.7	LysE-type transporter domain
Yiu	YiuA	42.0/38.2	PBP
	YiuB	38.2	IM permease
	YiuC	28.6	ATP hydrolase
	YiuR	73.8/70.8	TonB-dependent OM receptor
Fit	FitA	35.5/30.8	PBP
	FitB	34.1	IM permease
	FitC	34.0	IM permease
	FitD	28.3	ATP hydrolase
Feo	FeoA	8.8	Undetermined
	FeoB	83.9	IM permease
	FeoC	9.2	Undetermined

[a] Molecular masses predicted from *Y. pestis* KIM10+ genome sequence; unprocessed/processed molecular mass given.
[b] YfuD is encoded in *Y. enterocolitica* but not in *Y. pestis*. The molecular masses of the *Y. pestis* and *Y. enterocolitica* YfuA to YfuC proteins are nearly identical.

No virulence defects have been detected in *Y. enterocolitica* or *Y. pestis yfu* mutants (Table 1). In *Y. enterocolitica*, any defect may be masked by the Ybt system. Similarly, a *Y. pestis* Ybt$^+$ Yfu$^-$ mutant is fully virulent. In addition, a *Y. pestis* Ybt$^+$ Yfe$^-$ Yfu$^-$ mutant is as pathogenic as its Ybt$^+$ Yfe$^-$ Yfu$^+$ parent.

Although there is a small gap between *yfuC* and *yfuD* in *Y. enterocolitica*, *yfuA* to *yfuD* appear to form an operon with a putative FBS in the promoter region upstream of *yfuA*. In *Y. pestis*, *yfuA* to *yfuC* probably constitutes an operon. Fusion of the *yfuA* promoter region with *lacZ* resulted in Fur-dependent and Fe-repressible expression of β-galactosidase activity. Thus, the Yfu system is Fur regulated in *Y. pestis* and probably also in *Y. enterocolitica*.

The Yiu Iron Transport System

Sequence analysis of the *Y. pestis* CO92 and KIM10+ genomes identified the *yiu* locus that contains four ORFs. The *Y. enterocolitica* 8081 genome possesses three of these genes but lacks *yiuA*. YiuA, YiuB, and YiuC are putative PBP, IM, permease, and ATP hydrolase proteins, respectively, for an ABC transporter. YiuR is a putative TonB-dependent OM receptor with sequence similarity to IrgA, an enterobactin receptor in *Vibrio cholerae*. YiuA to YiuC have similarities to *Corynebacterium diphtheriae* Irp6A to Irp6C proteins, which are required for corynebactin-dependent iron uptake. YiuC is related to FepC, the ATP hydrolase required for enterobactin uptake. The cloned *Y. pestis* locus enhances Fe-deficient growth of an *E. coli* enterobactin-deficient mutant, and YiuR is not required for this growth stimulation. Studies with Fe-chelator gradient plates have demonstrated that a Δ*yiuABCR* mutation affects the growth of *Y. pestis*, but only in strains defective in the Ybt, Yfe, and Yfu systems. Any one of these three iron transport systems masks growth defects due to the Yiu system. Thus, there appears to be a hierarchy of *Y. pestis* iron transport systems, with Ybt being the most effective and Yiu being the least effective of the inorganic iron transport systems tested. A mutant defective in *yiuR* also exhibits an Fe-deficient growth defect in a Ybt$^-$ Yfe$^-$ Yfu$^-$ background. However, this strain grows better than a *Y. pestis* Ybt$^-$ Yfe$^-$ Yfu$^-$ YiuABCR$^-$ strain. These data suggest that the Yiu ABC transporter can function independently of the YiuR receptor.

The *yiuABC* operon lies upstream of *yiuR*; both the *yiuABC* and *yiuR* promoter regions contain putative FBSs. Fur and iron regulation of the *yiuABC* promoter has been experimentally demonstrated.

The Putative Fit ABC Transport System

Sequence analysis of the *Y. pestis* CO92 and KIM10+ genomes identified a locus with four ORFs that we have tentatively designated *fit*. FitA and FitD are putative PBP and ATP hydrolase ABC transport proteins, respectively, while FitB and FitC resemble IM permeases (Table 4). The deduced protein sequences are identical in CO92 and KIM10+, except for differences in the assigned start sites, and are highly conserved in *Y. enterocolitica* 8081. Sequence similarities for all four proteins are to putative or proven iron transport systems, with the highest similarities to siderophore-type transporters. However, no experimental analysis of this system is currently available. Its inclusion as an iron transport system in *Yersinia* spp. is based entirely on sequence analysis.

The Putative Fiu ABC Transport System

In *Y. pestis* KIM10+, *y2837*, *y2839*, and *y2842* encode another potential iron ABC transporter whose genetic locus is tentatively designated the *fiu* locus. After disregarding differences in assigned start sites, the deduced amino acid sequences of Y4043 through Y4046 from KIM10+ are identical to those in *Y. pestis* CO92. However, these genes are absent from the *Y. enterocolitica* 8081 genome. The PBP component, Y2842, shows similarity to a number of hypothetical PBPs, some annotated as siderophore or iron transporters. For experimentally proven iron transporters, *Y. pestis* YiuA (see above) and *C. diphtheriae* Irp6A (uptake of the siderophore corynebactin) show the highest similarities to Y2842. Y2839, a putative IM permease, possesses a FecCD transport subfamily domain. Of transporter components with a demonstrated function, *Y. pestis* YiuB (see above) and *C. diphtheriae* Irp6B have the greatest similarities to Y2839. Y2837, the ATP hydrolase component, has significant similarity to *E. coli* FhuC. While these amino acid sequence similarities suggest that this potential ABC transporter is involved in iron acquisition, there is no supporting experimental evidence.

The Feo Ferrous Iron Transport System

The ferrous iron transport (Feo) system of *E. coli* has been characterized by Klaus Hantke's research group (see chapter 12). We used the *E. coli* FeoB amino acid sequence to search the *Y. pestis* genomes for homologous proteins. The *Y. pestis* FeoB protein is 74% identical and 85% similar to that of *E. coli* K-12 FeoB. We constructed an in-frame deletion in *feoB*. The *Y. pestis feoB* mutant with a functional Ybt system did not exhibit growth defects under any of the tested conditions. However, a Ybt$^-$ FeoB$^-$ strain demonstrated iron-deficient growth defects compared to its Ybt$^-$ FeoB$^+$ parent. The *feoB* mutant was defective in growth at 30 and 37°C across gradients plates, containing nitrilotriacetic acid, that were incubated under reduced-oxygen conditions. In addition, the *feoB* mutant was unable to grow as well as its Ybt$^-$ FeoB$^+$ parental strain in static cultures containing the iron chelator EDDA. Thus, the Feo ferrous iron transport system is functional in *Y. pestis*.

The genetic organization of this system in *Y. pestis* KIM10+ and CO92 and *Y. enterocolitica* 8081 is the same as that of *E. coli*, with three ORFs (*feoABC*) comprising a putative operon. FeoA and FeoC are predicted to be small proteins (Table 4), but they are conserved in enteric organisms that contain *feoB*. However, the role(s) of FeoA and/or FeoC in iron transport has not been determined, nor has the promoter region(s) in the *feo* locus been analyzed.

HEME TRANSPORT SYSTEMS

The Hmu/Hem Heme Uptake System

The Hem system of *Y. enterocolitica* was the first ABC transporter involved in heme uptake and hemoprotein utilization to be characterized (Table 5). It is a member of a growing family

TABLE 5 Components of the Hmu/Hem and Has systems

Heme transport system	Protein	Mass (kDa)[a]	Selected characteristics
Hmu (Hem)	HmuP' (HemP)	4.5 (6.5)	Unknown; not essential
	HmuR (HemR)	74.2/71.2 (75.1/72.1)	OM receptor
	HmuS (HemS)	39.1	Heme binding?
	HmuT (HemT)	29.6/26.8	PBP
	HmuU (HemU)	35.5	IM permease
	HmuV (HemV)	29.6	ATP hydrolase
Has putative hemophore system	HasR	93.5	OM receptor for hemophore
	HasA	22.2	Secreted hemophore
	HasD	65.4	ATP hydrolase for HasA secretion
	HasE	49.0	IM permease for HasA secretion
	HasB	29.4	TonB-like protein
	HasF (TolC)	50.6/48.3	OM export component for hemophore

[a] Unprocessed/processed molecular mass. Molecular masses are for *Y. pestis* proteins, except for HemP and HemR (in parentheses). *Y. enterocolitica* HemS, HemT, HemU, and HemV are very similar in molecular mass to Hmu proteins.

of similar ABC transporters found in enteric bacteria. The Hmu heme transporter of *Y. pestis* is nearly identical to the *Y. enterocolitica* Hem system. HmuP′ is 40 amino acids smaller than HemP; however, this difference is insignificant since HemP is not an essential component of this ABC transporter. A TonB-dependent OM receptor (HmuR/HemR) binds hemin and a wide variety of hemoproteins, including hemoglobin, myoglobin, hemoglobin-haptoglobin, heme-hemopexin, and heme-albumin. A *Y. pestis* HmuR$^-$ mutant is unable to use hemin or any of these hemoproteins as iron sources. The mechanism for receptor recognition of these proteins is unknown but may involve binding to the heme moiety, even though most of the porphyrin structure is not surface exposed when complexed with hemoglobin. Two histidine residues in HemR are important in the utilization of hemoproteins. Alteration of the HemR H461 residue eliminates use of hemoproteins and severely reduces the accumulation of free heme. Similarly, a mutation at H128 abolishes the ability to use hemoproteins as iron sources. Both histidine residues are surface accessible and are likely to be involved in heme binding. It is presumed, but not definitely proven, that only the heme moiety is translocated into the periplasm, where it is bound by the PBP HmuT/HemT. The heme is then transported across the IM by the permease-ATP hydrolase complex of HmuU/HemU and HmuV/HemV (Fig. 7). [^{14}C]hemin uptake experiments and growth studies of an *E. coli* porphyrin biosynthetic mutant expressing the Hmu system suggest that the intact heme moiety is transported into the bacterial cytoplasm. In *Y. enterocolitica*, a *hemT* mutant could still use heme as a source of iron, while the growth of *hemU* and *hemV* mutants on heme was severely reduced. The ability of Δ*hmuR*, Δ*hmuP′RSTUV*, or *hmu*

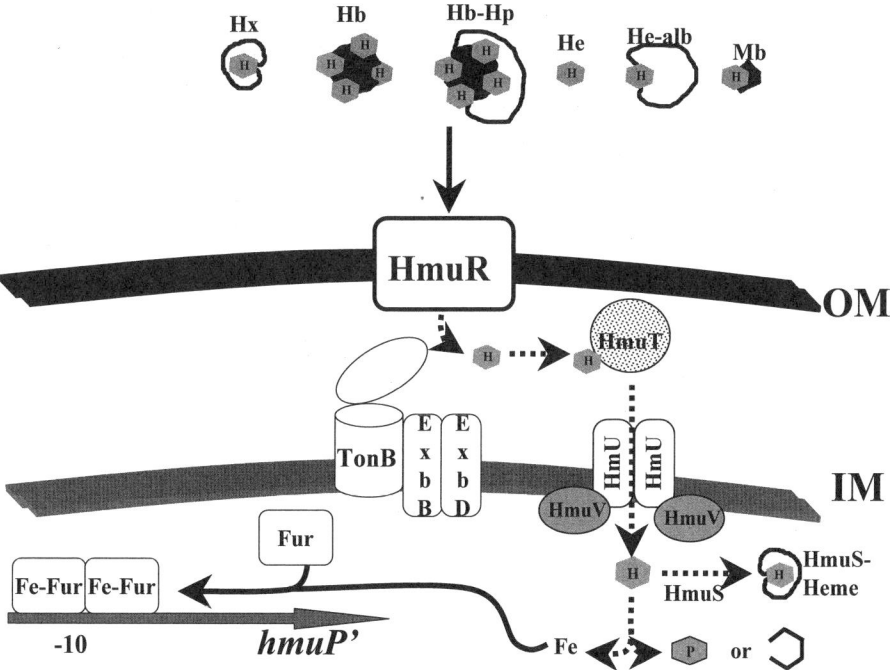

FIGURE 7 Model of the Hmu/Hem hemoprotein transport and regulatory system. Only Hmu designations are used for clarity. Dashed arrows designate predicted functions or mechanisms that have not been fully demonstrated experimentally. Abbreviations: Hx, heme-hemopexin; Hb, hemoglobin; Hb-Hp, hemoglobin-haptoglobin; He, heme; He-alb, heme-albumin; Mb, myoglobin.

T::kan mutants to grow in medium containing EDDA and ≥2.5 μM hemoglobin initially led us to propose that a secondary low-affinity hemoglobin transport system existed in *Y. pestis*. However, we recently determined that the Ybt siderophore system was responsible for the observed growth enhancement in the Δ*hmuP'RSTUV* strain. Thus, it may be necessary to reevaluate the growth responses of the various *hmu* and *hem* mutations in a Ybt⁻ background. A function for the enigmatic HmuS/HemS cytoplasmic protein has recently been proposed based on work with the related ShuS protein of *Shigella dysenteriae* (see chapter 6). HmuS/HemS probably serves as a storage complex for excess heme by binding cytoplasmic heme and possibly forming a multimeric complex reminiscent of ferritin, thereby preventing heme toxicity. It is not clear whether the yersiniae extract inorganic iron from heme. Rather, acquired heme might be used directly in respiratory-chain proteins and other heme-requiring enzymes, thus reducing the need for inorganic iron in heme biosynthesis. Three ORFs upstream of *hmuP'* (Fig. 8) are conserved in other enteric heme ABC transporters. Y0547 is a 437-amino-acid, 49.2-kDa protein containing a domain found in oxygen-independent coproporphyrinogen III oxidases. However, a strain in which *y0547* had been deleted still grew in a medium containing hemoglobin and EDDA. Recently, a gene, *hutZ*, was identified in *V. cholerae* as part of a heme utilization operon and was implicated in the ability to use heme efficiently as an iron source, although it lacks detectable heme oxygenase activity (see chapter 16). However, searches of the *Y. pestis* genome sequences failed to identify ORFs with significant similarities to the genes encoding HutZ, eukaryotic heme oxygenase, or HmuO (heme oxygenase of *C. diphtheriae*). These data support the hypothesis that *Yersinia* species may not remove inorganic iron from acquired heme.

A *hemR* mutation in *Y. enterocolitica* did not affect virulence in mice infected by the intravenous route. Similarly, in a model of bubonic plague, *Y. pestis* Δ*hmuP'RSTUV* cells grown at 26°C in the presence of hemin (to simulate growth in the flea gut) and injected subcutaneously into mice retained full virulence (Table 1). Although not required for bubonic plague in a laboratory mouse model, the Hmu hemoprotein transport system is important for growth in J774 macrophage-like cells. After infection of a J774 monolayer, wild-type *Y. pestis* cells grown at 37°C under Fe-deficient conditions undergo an initial decrease in cell number followed by a growth phase and eventual destruction of the J774 monolayer. The Ybt and Yfe inorganic iron transport systems are not required for the growth of *Y. pestis* in J774 cells under these conditions. However, the Hmu⁻ mutants fail to fully recover from the initial decline in cell numbers. Whether this intracellular growth requirement for the Hmu system

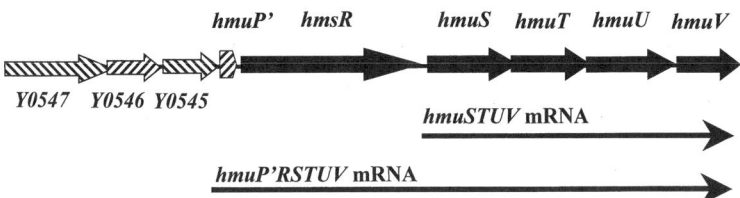

FIGURE 8 Genetic organization of the ~6.9-kb *hmu* locus of *Y. pestis*. Sizes of the Hmu proteins are shown in Table 5. Predicted proteins Y0547 (437 residues; 49.2 kDa), Y0546 (181 residues; 20.2 kDa), and Y0545 (216 residues; 23.2 kDa) correspond to W, X, and Y proteins in other enteric hemoprotein ABC transport systems. The genetic organization of *Y. enterocolitica hemPRSTUV* is identical. However, analysis of the *hemS* promoter region and sequencing of the region upstream of *hemP* has not been performed.

occurs in vivo or in nonphagocytic cells remains to be determined.

The genetic organization of the *hmu* locus and the three upstream ORFs (*Y. pestis* KIM10+ Y0545 to Y0547) are shown in Fig. 8; this region is highly conserved in *Y. pestis* CO92 and *Y. enterocolitica* 8081. A duplication of 15 bp near the predicted start site of HemV suggests that the protein may be slightly larger than HmuV. Also, *hemR* is predicted to encode 11 additional amino acids compared to the predicted coding sequence for *hmuR*. A second locus encoding sequences similar to HemR (52% identity and 67% similarity), HemS (53% identity and 69% similarity), HemT (60% identity and 75% similarity), and part of HemP is present in *Y. enterocolitica* 8081 but not in the *Y. pestis* genomes. The *hemR* and *hemS*-like sequences apparently do not compensate for *hemR* and *hemS* mutations; thus, they are unlikely to function in heme uptake and utilization. However, it is possible that the *hem T*-related sequence is responsible for the ability of a *hemT* mutant to use hemin as an iron source.

Promoter regions upstream of *hmuP'* and *hmuS* have been identified, and primer extension analysis has identified potential transcriptional start sites. The *hmuP'* promoter probably initiates transcription of the entire *hmu* operon, while transcription from the *hmuS* promoter yields a *hmsSTUV* mRNA. Putative FBSs are located in both promoters. Transcriptional reporter gene fusions indicate that the *hmuP'* but not the *hmuS* promoter is regulated by Fur and repressed by excess iron. The *hmuS* promoter has low-level constitutive activity. Western blot analysis confirms that HmuR is strongly repressed by iron while HmuS displays only modest iron repression, possibly due to iron-regulated expression from the *hmuP'* promoter.

Studies with a *Y. enterocolitica hemR* translational fusion to *gfp* found that cells growing in the liver and intestinal lumen had the lowest level of expression while cells isolated from the spleen expressed moderate levels of the HemR-green fluorescent protein fusion. Bacterial cells recovered from the peritoneal cavity of infected mice exhibited the highest level of expression. Similar expression patterns were observed for a *yfuA* reporter construct (see the discussion of Ybt above). These results could indicate the existence of tissue-specific signals that regulate the in vivo expression of both systems. Alternatively, the expression patterns may simply reflect the relative levels of bacterial iron stress in these diverse organ systems.

The Has Hemophore System

The Has hemophore system, which was initially characterized in *S. marcescens*, secretes a small protein (HasA) via a type I secretion system (HasD, HasE, and HasF/TolC). HasA functions as a hemophore to bind heme and hemoglobin, and this complex is bound by an OM receptor (HasR) (see chapter 3 for a detailed analysis of this system). *Y. pestis* encodes a Has hemophore system (Table 5) similar to the *S. marcescens* Has system. $HasA_{Yp}$ is secreted from *E. coli* by a hybrid type I secretion system consisting of HasDE from *Serratia* and TolC from *E. coli*. However, $HasA_{Yp}$ secretion could not be detected in a variety of *Y. pestis* strains or in *E. coli* carrying the cloned *Y. pestis hasRA-DEB* locus. The deduced amino acid sequences of $HasD_{Yp}$, $HasE_{Yp}$, and $HasF_{Yp}$ do not contain any obvious defects to explain this result. Furthermore, $HasA_{Yp}$ and $HasR_{Yp}$ are synthesized in an in vitro transcription-translation reaction. $HasA_{Yp}$ did bind heme and hemoglobin but was not recognized by the $HasR_{Sm}$ receptor. Perhaps not unexpectedly, the growth of a *Y. pestis hmu has* mutant in iron-deficient medium supplemented with hemoglobin was not significantly different from that of its *hmu has*$^+$ parent. Finally, a *Y. pestis hmu has* mutant retained full virulence in mice infected subcutaneously (Table 1). The overall evidence suggests that the *Y. pestis* Has system is not functional, at least under the experimental conditions employed.

HasB is a TonB-like protein, and in *S. marcescens* either TonB or $HasB_{Sm}$ allows heme uptake through the $HasR_{Sm}$ TonB-dependent receptor. However, $HasB_{Sm}$ cannot substitute for TonB in the acquisition of inorganic iron. Our studies indicate that $HasB_{Yp}$ also cannot substitute for TonB in either the Ybt or Hmu transport system in *Y. pestis*.

The *Y. pestis hasRADEB* locus contains an FBS in the promoter region and is regulated by Fur and iron. The *Y. pestis hasF* (*tolC*) gene is located 281 kb from the *hasRADEB* locus and does not possess a readily identifiable FBS in the promoter region. Sequences similar to *hasRADEB* are also present in the *Y. enterocolitica* 8081 genome and are organized as in *Y. pestis* KIM10+ and CO92, with one exception. *Y. enterocolitica* 8081 has four tandem copies of *hasA*-related sequences, all more closely related to *Y. pestis* than to *S. marcescens hasA*. These *hasA*-like genes have diverged from each other, with the "*hasA1*" and "*hasA4*" predicted amino acid sequences being similar to each other (42% identity and 52% similarity) and with "*hasA3*" being related to "*hasA2*" (42% identity and 51% similarity).

IRON REGULATION AND STORAGE

The *Y. pestis* Fur protein (16.7 kDa) controls the expression of the Ybt, Yfe, Yfu, Yiu, Has, and Hmu systems. Iron repression and Fur regulation of the Feo, Fit, Fhu, Ynp, and Ysu systems has not been examined. The Ybt and Hem systems of *Y. enterocolitica* are also regulated by Fur. *Y. pestis* and *Y. enterocolitica* Fur proteins are highly similar to *E. coli* and other enteric Fur proteins. The one unique aspect of Fur regulation in *Y. pestis* is the Fur-dependent Mn repression of the *yfeA* to *yfeD* promoter region. It remains to be determined whether this regulation is due to characteristics of the *Y. pestis* Fur protein, the *yfeA* to *yfeD* promoter, or both. Fur regulation is described in detail in chapter 13.

A *Y. pestis fur* mutant is unable to grow under Fe-surplus conditions unless the Ybt system has been deleted (Δpgm); even then, these cells grow poorly under these conditions. Presumably, constitutive overexpression of the Ybt system leads to the accumulation of iron to toxic levels. Two-dimensional gel electrophoresis of cell extracts from *Y. pestis* Fur$^+$ and Fur$^-$ strains identified a number of Fe-repressible polypeptides that were regulated by Fur. However, at least nine Fe-repressible polypeptides were regulated independently of Fur. Searches of the *Y. pestis* genome have failed to identify any alternative Fe-responsible regulators other than PmrA and PmrB, which induce the expression of their target genes under high-iron growth conditions in *Salmonella*. The PmrAB system might be responsible for regulating the expression of two Fe-inducible polypeptides that are not affected by mutations in *fur*. Alternatively, these proteins could be induced by regulators responsive to the oxidative stress that would be caused by the accumulation of excess iron in *fur* mutants. Four polypeptides were induced in a Fur-dependent manner under surplus iron conditions. It has recently been shown that in *E. coli*, Fur-Fe represses the transcription of *rhyB*, which encodes an antisense mRNA that prevents the synthesis of several proteins. Thus, these proteins are normally highly expressed only during growth with excess iron. *Y. pestis* contains a *rhyB* gene, but its expression or role in iron regulation has not been determined.

The *Y. pestis* genome contains genes for bacterioferritin (*bfrA*), bacterioferritin-associated ferridoxin (*bfd*), and ferritin (*ftnA*), encoding proteins of 18.2, 7.2, and 19.2 kDa, respectively. A *Y. pestis* KIM6+ high-molecular-mass complex that bound inorganic iron was resolved to a 19-kDa cytoplasmic polypeptide that could be BfrA, FtnA, both, or neither. Although the Hms phenotype of *Y. pestis* was earlier proposed as a hemin storage function, it does not serve this purpose.

SUMMARY

Y. pestis, *Y. enterocolitica*, and probably *Y. pseudotuberculosis* encode a multitude of proven or putative iron transport systems. For the entero-pathogenic yersiniae, the systems that allow the utilization of exogenous siderophores may enable these bacteria to compete more effectively with other microbes in the intestinal lumen and during residence outside hosts. The rationale for multiple systems in *Y. pestis*, with its relatively isolated life cycle in the flea gut and normally sterile tissues of mammals, is less clear. Some of the *Y. pestis* iron transport systems are required in different tissues. The Ybt system is absolutely essential during the lympathic stage of bubonic plague, while the Yfe transporter is important during the later stages of disease. However, none of the other inorganic or hemoprotein transport systems appear to play a role in disease initiated by subcutaneous injection. Perhaps one or more of these systems has some relevance for pneumonic plague or is important for disease in other animals. Finally, some of these systems (e.g., the aerobactin-Fhu system) may simply reflect the *Enterobacteriaceae* heritage of *Y. pestis*.

ACKNOWLEDGMENTS

This chapter is a compilation of research results from numerous research programs that could not be fully acknowledged due to a limit on the number of literature citations. The laboratories of Volkmar Braun, Robert Brubaker, Elisabeth Carniel, Klaus Hantke, Jürgen Heesemann, Roy Robins-Browne, Pamela Sokol, Igor Stojiljkovic, and Christopher Walsh have all made significant contributions to our understanding of iron metabolism in *Yersinia*. In my laboratory, Jennifer Abney, Bill Baker, Vincent Bertolino, Heather Jones, Idefonso Mier, Jr., Mike Nagiec, and Paul Schuetze, as well as Scott Bearden, Alexander Bobrov, Jacqueline Fetherston, Valérie Geoffroy, Shimei Gong, Olga Kirillina, James Lillard, Jr., Michael Pendrak, Teanna Staggs, and Jan Thompson, have all contributed published or unpublished observations described in this chapter. The sequence data for *Y. enterocolitica* 8081 was produced by the *Yersinia enterocolitica* Sequencing Group at the Sanger Institute and can be obtained from ftp://ftp.sanger.ac.uk/pub/pathogens/ye. The sequence data for *Y. pseudotuberculosis* IP32953 was generated at Lawrence Livermore National Laboratories and the Pasteur Institute; contigs of this genome sequence can be accessed at http://bbrp.llnl.gov/bbrp/html/microbe.html.

Research in my laboratory has been supported by Public Health Service grants AI25098, AI33481, and AI42738.

SUGGESTED READING

Bäumler, A., R. Koebnik, I. Stojiljkovic, J. Heesemann, V. Braun, and K. Hantke. 1993. Survey on newly characterized iron uptake systems of *Yersinia enterocolitica*. *Zentbl. Bakteriol.* **278:** 416–424.

Bearden, S. W., and R. D. Perry. 1999. The Yfe system of *Yersinia pestis* transports iron and manganese and is required for full virulence of plague. *Mol. Microbiol.* **32:**403–414.

Bobrov, A. G., V. A. Geoffroy, and R. D. Perry. 2002. Yersiniabactin production requires the thioesterase domain of HMWP2 and YbtD, a putative phosphopantetheinylate transferase. *Infect. Immun.* **70:**4204–4214.

Brem, D., C. Pelludat, A. Rakin, C. A. Jacobi, and J. Heesemann. 2001. Functional analysis of yersiniabactin transport genes of *Yersinia enterocolitica*. *Microbiology* **147:**1115–1127.

Carniel, E. 2001. The *Yersinia* high-pathogenicity island: an iron-uptake island. *Microbes Infect.* **3:** 561–569.

Carniel, E., A. Guiyoule, I. Guilvout, and O. Mercereau-Puijalon. 1992. Molecular cloning, iron-regulation and mutagenesis of the *irp2* gene encoding HMWP2, a protein specific for the highly pathogenic *Yersinia*. *Mol. Microbiol.* **6:**379–388.

Fetherston, J. D., V. J. Bertolino, and R. D. Perry. 1999. YbtP and YbtQ: two ABC transporters required for iron uptake in *Yersinia pestis*. *Mol. Microbiol.* **32:**289–299.

Gong, S., S. W. Bearden, V. A. Geoffroy, J. D. Fetherston, and R. D. Perry. 2001. Characterization of the *Yersinia pestis* Yfu ABC iron transport system. *Infect. Immun.* **67:**2829–2837.

Jacobi, C. A., S. Gregor, A. Rakin, and J. Heesemann. 2001. Expression analysis of the yersiniabactin receptor gene *fyuA* and the heme receptor *hemR* of *Yersinia enterocolitica* in vitro and in vivo using the reporter genes for green fluorescent protein and luciferase. *Infect. Immun.* **69:**7772–7782.

Koebnik, R., K. Hantke, and V. Braun. 1993. The TonB-dependent ferrichrome receptor FcuA of *Yersinia enterocolitica*: evidence against a strict coevolution of receptor structure and substrate specificity. *Mol. Microbiol.* **7:**383–393.

Perry, R. D., J. Abney, I. Mier, Jr., Y. Lee, S. W. Bearden, and J. D. Fetherston. 2003. Regulation of the *Yersinia pestis* Yfe and Ybt iron transport systems, p. 275–283. *In* M. Skurnik, K. Granfors, and J. A. Bengoechea (ed.), *The Genus* Yersinia: *Entering the Functional Genomic Era*. Kluwer Academic/Plenum Publishers, New York, N.Y.

Perry, R. D., S. W. Bearden, and J. D. Fetherston. 2001. Iron and heme acquisition and storage systems of *Yersinia pestis*. *Recent Res. Dev. Microbiol.* **5:**13–27.

Rakin, A., E. Saken, D. Harmsen, and J. Heesemann. 1994. The pesticin receptor of *Yersinia enterocolitica*: a novel virulence factor with dual function. *Mol. Microbiol.* **13:**253–263.

Rossi, M.-S., J. D. Fetherston, S. Létoffé, E. Carniel, R. D. Perry, and J.-M. Ghigo. 2001. Identification and characterization of the hemophore-dependent heme acquisition system of *Yersinia pestis*. *Infect. Immun.* **69:**6707–6717.

Saken, E., A. Rakin, and J. Heesemann. 2000. Molecular characterization of a novel siderophore-independent iron transport system in *Yersinia*. *Int. J. Med. Microbiol.* **290:**51–60.

Schubert, S., D. Fischer, and J. Heesemann. 1999. Ferric enterochelin transport in *Yersinia enterocolitica*: molecular and evolutionary aspects. *J. Bacteriol.* **181:**6387–6395.

Stojiljkovic, I., and D. Perkins-Balding. 2002. Processing of heme and heme-containing proteins by bacteria. *DNA Cell Biol.* **21:**281–295.

Straley, S. C., and M. N. Starnbach. 2000. *Yersinia*: strategies that thwart immune defenses, p. 71–92. *In* M. W. Cunningham and R. S. Fujinami (ed.), *Effects of Microbes on the Immune System*. Lippincott Williams & Wilkins, Philadelphia, Pa.

Stuart, S. J., J. K. Prpic, and R. M. Robins-Browne. 1986. Production of aerobactin by some species of the genus *Yersinia*. *J. Bacteriol.* **166:**1131–1133.

Thompson, J. M., H. A. Jones, and R. D. Perry. 1999. Molecular characterization of the hemin uptake locus (*hmu*) from *Yersinia pestis* and analysis of *hmu* mutants for hemin and hemoprotein utilization. *Infect. Immun.* **67:**3879–3892.

VIBRIO

Manuela Di Lorenzo, Michiel Stork, Alejandro F. Alice, Claudia S. López, and Jorge H. Crosa

16

Gram-negative marine bacteria that belong to the genus *Vibrio* are commonly found as etiological agents of disease in humans and animals. The most important members of this genus are *Vibrio cholerae* and the marine pathogens *V. vulnificus*, *V. anguillarum*, *V. harveyii*, and *V. parahaemolyticus*. *V. vulnificus* causes hemorrhagic septicemias with high fatality rates in humans as well as marine animals, while *V. anguillarum* is responsible for the hemorrhagic septicemic disease vibriosis in salmonid fish. *V. parahaemolyticus* is normally found in the marine and estuarine environment but can cause gastroenteritis, wound infection and septicemia in humans (Table 1). The vibrios differ from many bacterial species in having two chromosomes. These bacteria have multiple iron transport systems, and genes for these systems are located on both of the replicons.

The ability of vibrios and other pathogenic bacteria to cause disease depends on many factors that work in concert to establish the pathogen in the vertebrate host. One of these is its ability to compete with the host organism for iron. Iron is an essential metal for nearly all living systems; however, in biological fluids it exists only as a complex with iron-binding proteins, making it essentially unavailable. Therefore, to establish an infection, invasive microorganisms must depend on their ability to use the iron found in the host, which is complexed to high-affinity iron-binding proteins such as transferrin or lactoferrin or is part of heme in the red cells. As must be clear by now to the reader of this book, to accomplish this endeavor, microorganisms possess multiple mechanisms for iron acquisition. These include receptors in the outer membrane that allow them to acquire the iron directly from the complexes with the high-affinity iron-binding eukaryotic proteins or, as in the case of heme, to directly transport it inside the cell. Another common strategy among microorganisms is the synthesis of siderophores. These compounds are low-molecular-weight, high-affinity iron-binding molecules that can scavenge ferric iron away from the high-affinity iron-binding proteins in the host, such as transferrin and lactoferrin, and present it as a ferri-siderophore complex to the cognate outer membrane protein receptors and subsidiary iron transport proteins.

Although in some of the *Vibrio* species a siderophore or heme-mediated mechanism of iron uptake could play an important role in virulence, additional virulence factors such as

---

Manuela Di Lorenzo, Michiel Stork, Alejandro F. Alice, Claudia S. López, and Jorge H. Crosa, Department of Molecular Microbiology L220, Oregon Health and Science University, 3181 SW Sam Jackson Park Road, Portland, OR 97239.

TABLE 1 Pathogenic *Vibrio* species

Microorganism	Disease	Host
V. alginolyticus	Cellulitis and acute otitis media or externa. Life-threatening bacteremia in immunocompromised individuals.	Humans
V. anguillarum	Terminal hemorrhagic septicemia (vibriosis).	Fish
V. cholerae	Cholera. Watery diarrhea, loss of circulation and blood volume. Metabolic acidosis, potassium depletion, and ultimately vascular collapse and death.	Humans
V. fluvialis	Watery diarrhea.	Humans
V. harveyi	Vibriosis.	Fish and oysters
V. parahaemolyticus	Diarrhea, gastroenteritis abdominal cramps, nausea, vomiting, headache, fever, and chills may be associated with infections caused by this organism.	Humans and oysters
V. vulnificus	Mild gastroenteritis, wound infection, and septicemia.	Humans and eels

toxins, pili and flagella, hemolysins and cytolysins, lipopolysaccharide, and capsules are also required for the development of the disease.

In this chapter we describe siderophore and heme-mediated iron uptake systems in the vibrios and discuss their roles as components of the virulence of these bacteria.

SIDEROPHORE-MEDIATED IRON UPTAKE

Production of siderophores is one mechanism that enables bacteria to acquire iron within the vertebrate host or other iron-limited environment. Siderophores are low-molecular-weight compounds that coordinate ferric ions with high affinity and specificity, with their association constants ranging from 10^{22} to 10^{50} M. During an infection, some siderophores scavenge the required iron from the high-affinity iron-binding proteins of the host.

Siderophores are synthesized from carboxylic acids and amino acids via a nonribosomal peptide synthetase (NRPS) mechanism (see chapter 2). These enzymes catalyze the formation of a wide variety of peptides such as siderophores and antibiotics in the absence of an RNA template. They are multimodular enzymes that work as an enzymatic assembly line in which the order of the modules determine the order of the amino acids in the peptide. The finished siderophore molecule is secreted into the extracellular milieu, where it forms a tight complex with ferric iron. The ferri-siderophore is then recognized by a specific receptor in the outer membrane and internalized to the periplasm. The energy for this transport is provided by the proton motive force generated in the inner membrane and transduced by the TonB-ExbBD complex to the outer membrane receptor (see chapter 7). Once in the periplasm, the ferri-siderophore is bound by the periplasmic binding protein, which shuttles the ferri-siderophore to the permease proteins located in the inner membrane. The ferri-siderophore is transported across the inner membrane by using the energy provided by ABC transporters. In the cytoplasm, the ferric iron is released from the siderophore and reduced to ferrous iron.

The genes for siderophore biosynthesis and transport are expressed only under iron-limiting conditions. As observed with other gram-negative bacteria, expression of the iron uptake systems in vibrios is regulated by the Fur protein. In the presence of iron, Fur binds to the Fur box in the promoter regions of iron-responsive genes and represses transcription (see chapter 13). When the iron concentration becomes limiting, iron is released from Fur. The apo-Fur no longer binds to the promoters, and the repression is relieved.

In this section, we discuss the anguibactin system of *V. anguillarum*, the vibriobactin system of *V. cholerae*, the vulnibactin system of *V. vulnificus*, and the vibrioferrin system of *V. parahaemolyticus* (Fig. 1).

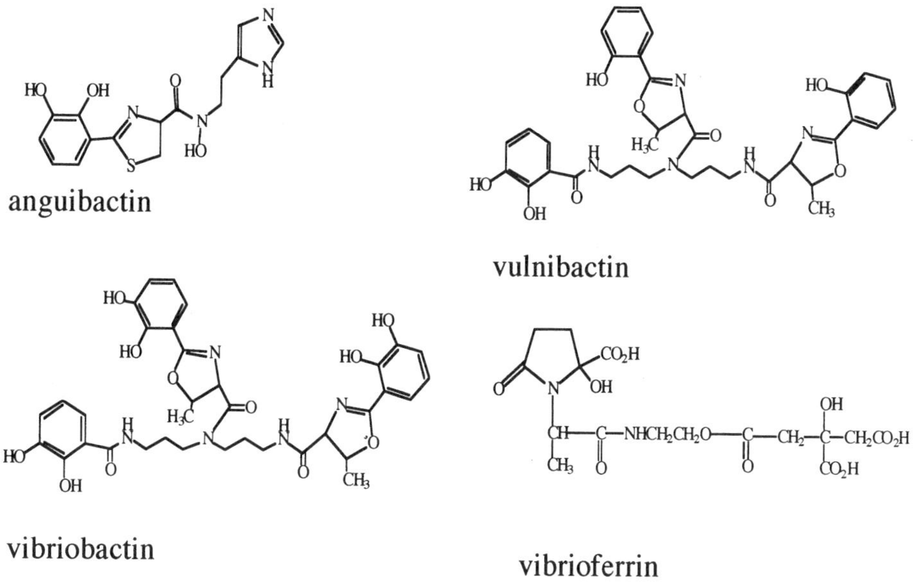

FIGURE 1 Structures of siderophores produced by *Vibrio* species.

V. anguillarum

A key feature necessary for many serotype O1 *V. anguillarum* strains to survive and cause an infection in the vertebrate host is the possession of the ~65-kb virulence plasmid. The plasmid provides the bacterium with genes for the synthesis and transport of the siderophore anguibactin. In contrast, serotype O2 *V. anguillarum* strains rely on a chromosomally encoded siderophore system that produces a siderophore distinct from the anguibactin produced by the O1 strains.

Anguibactin is synthesized from 2,3-dihydroxybenzoic acid (DHBA), L-cysteine, and N-hydroxyhistamine. The ω-N-hydroxy-ω-N-[[2'-(2",3"-dihydroxyphenyl) thiazolin-4'-yl]carboxy] histamine anguibactin molecule contains both the catechol DHBA and the N-hydroxyl group of hydroxyhistamine. As a consequence, anguibactin can be classified as a catechol as well as a hydroxamate siderophore.

BIOSYNTHESIS

The current pathway for anguibactin biosynthesis was proposed based on the structure of the siderophore as well as on the genetic, biochemical, and in silico studies (Fig. 2). In *V. anguillarum* strain 775, the genes for these proteins are carried on the pJM1 plasmid. Most of the enzymes involved in anguibactin biosynthesis have domains with similarity to those for NRPS.

The anguibactin precursors DHBA and L-cysteine are activated by the adenylation (A) domains of the AngE protein and the AngR protein, respectively. Activated DHBA is tethered on the phosphopantetheinylate arm (pPant) of the aryl carrier protein domain of the AngB/G protein, while activated cysteine is tethered to the pPant of the peptidyl carrier protein (PCP) domain of the AngM protein. The AngR protein also posses a PCP domain on which cysteine could be tethered, but this PCP domain may not be functional because in this domain an essential serine is replaced by an alanine. The condensation (C) domain of AngM catalyzes the formation of the peptide bond between DHBA and cysteine. In this or later steps of anguibactin biosynthesis, cysteine is cyclized to a thiazoline ring by one of the two cyclization (Cy) domains of AngN. Another Cy domain is found in AngR, but also this domain is probably not functional since an

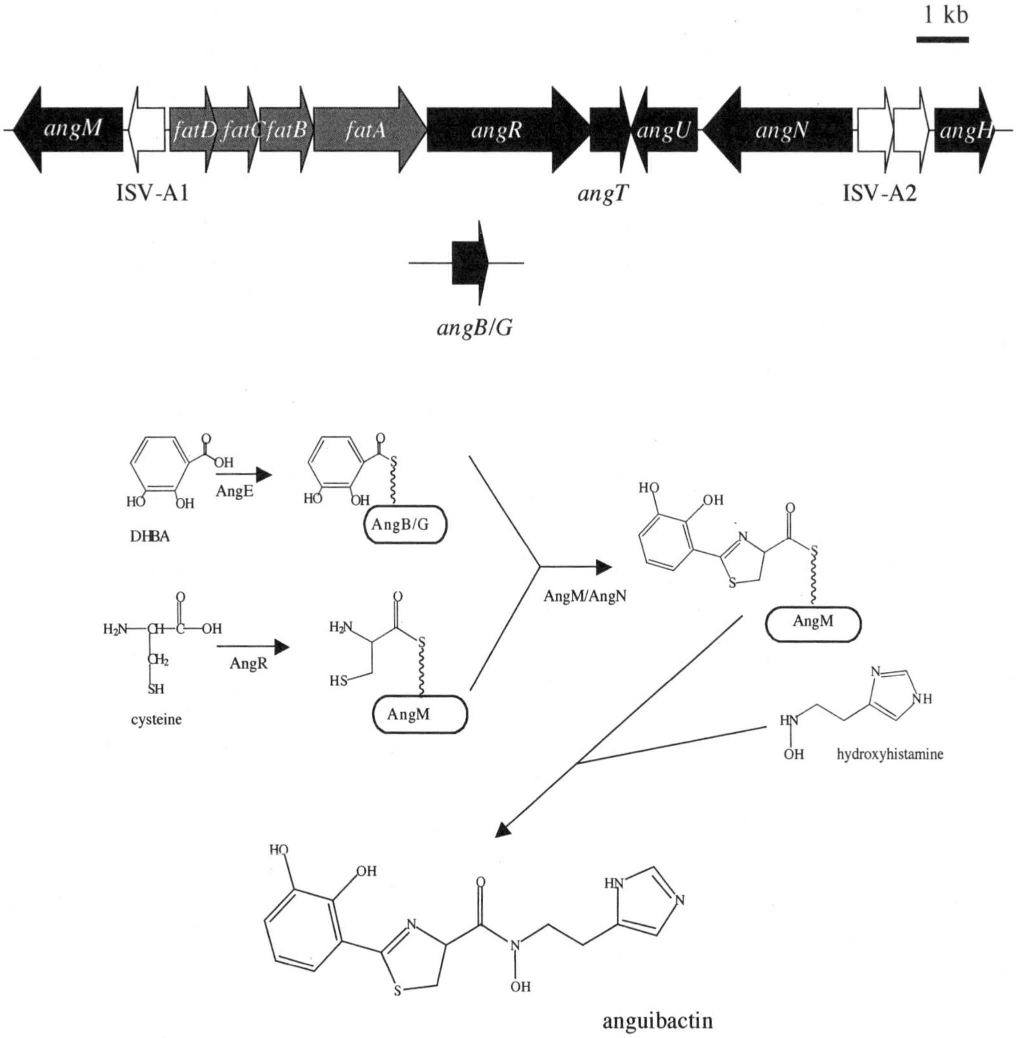

FIGURE 2 Genetic organization of *V. anguillarum* anguibactin biosynthetic genes and proposed biosynthetic pathway.

essential aspartic acid is replaced by asparagine. In other systems, the full-length peptide is released from the PCP domain of the last module by a thioesterase (TE) domain. In anguibactin biosynthesis, a TE domain was found in the AngT protein. However, mutation of the *angT* gene results only in a decrease of anguibactin production, suggesting that this thioesterase is not essential for release of the final product. In our model, the dihydroxyphenylthiazoline dipeptide is released from the PCP domain of AngM by nucleophilic attack of N-hydroxyhistamine, resulting in the free anguibactin molecule. N-Hydroxyhistamine is obtained by modification of histidine catalyzed by two tailoring enzymes, the monooxygenase AngU and the histidine decarboxylase AngH.

TRANSPORT

Ferric anguibactin is transported by the FatABCD proteins. The genes encoding these proteins are located on the virulence plasmid and are transcribed as a single polycistronic mRNA with the biosynthesis genes *angR* and *angT* (ITB operon). The ferric anguibactin receptor FatA is an 86-kDa protein that is essential for anguibactin transport. The FatA amino acid sequence is similar to other outer membrane receptors involved in iron transport, and a TonB box was identified at its amino-terminal end. This TonB box is one of the interaction sites between the outer membrane receptor and the energy transducer TonB. The chromosomes of *V. anguillarum* harbor two *tonB* genes: *tonB1* and *tonB2*. A mutant with a mutation in *tonB1* is able to transport anguibactin, indicating that either both TonB systems can mediate the uptake of anguibactin or only TonB2 is required. Once in the periplasm, the ferric anguibactin binds the periplasmic binding protein FatB, a 35-kDa lipoprotein that is anchored in the inner membrane. FatB shuttles ferric anguibactin to the inner membrane permease FatCD, which transports the ferric anguibactin complex to the cytoplasm. The ATPase necessary for anguibactin transport in *V. anguillarum* has not yet been identified.

REGULATION

In *V. anguillarum* the siderophore-mediated iron uptake system is negatively regulated by the chromosomally encoded Fur protein. It has been shown that on binding at the Fur box of the ITB operon promoter, Fur bends the DNA, blocking RNA polymerase binding. The positive regulators, AngR protein and products encoded in the *trans*-acting factor (TAF) region, act at the same promoter. AngR is a bifunctional protein that participates in both the regulation and biosynthesis of anguibactin. The TAF products are encoded in a region of the virulence plasmid noncontiguous to the ITB operon. In addition to the positive regulators AngR and TAF, the ITB operon promoter is positively regulated by the siderophore anguibactin itself. Furthermore, within the ITB operon, the ratio of *fat* and *ang* gene expression is ~50:1. The reduced relative expression of the *ang* genes within this operon is probably due to regulation by an antisense RNA, RNAβ.

V. cholerae

The causative agent of cholera, *V. cholerae*, has multiple iron transport systems. Most *V. cholerae* strains produce and use the catechol siderophore vibriobactin (Fig. 1). Vibriobactin contains three 2,3-dihydroxybenzoyl residues, two of which are linked to the norspermidine (NS) backbone via cyclized L-threonine moieties, whereas the third is directly linked to the NS. With its three catechol residues, vibriobactin is expected to form a 1:1 complex with ferric iron.

The vibriobactin genes are located on chromosome I in two separate gene clusters. Each cluster contains both biosynthetic genes and genes for vibriobactin transport. In addition, one of the clusters includes the genes for the synthesis of DHBA from chorismate. Mutants defective in either synthesis or uptake of vibriobactin have been assayed for virulence in an infant-mouse model of cholera. These mutants showed only a modest reduction in growth in the host and in virulence, suggesting that one or more of the other iron transport systems play the major role in iron acquisition in vivo. Vibriobactin production is highest among environmental isolates of *V. cholerae*, as might be expected if the siderophore is more important for growth and survival in the environment than in the vertebrate host.

BIOSYNTHESIS

From genetic and biochemical studies, it has been determined that the biosynthesis of vibriobactin occurs via an NRPS mechanism, albeit a nonstandard one (Fig. 3; see also chapter 2). Norspermidine, the backbone of the vibriobactin molecule, has no free carboxylate moiety for thioester attachment to the PCP domain of an NRPS. As a consequence, intermediates to the finished product need to be free, soluble molecules instead of pPant-tethered thioesters of typical NRPSs.

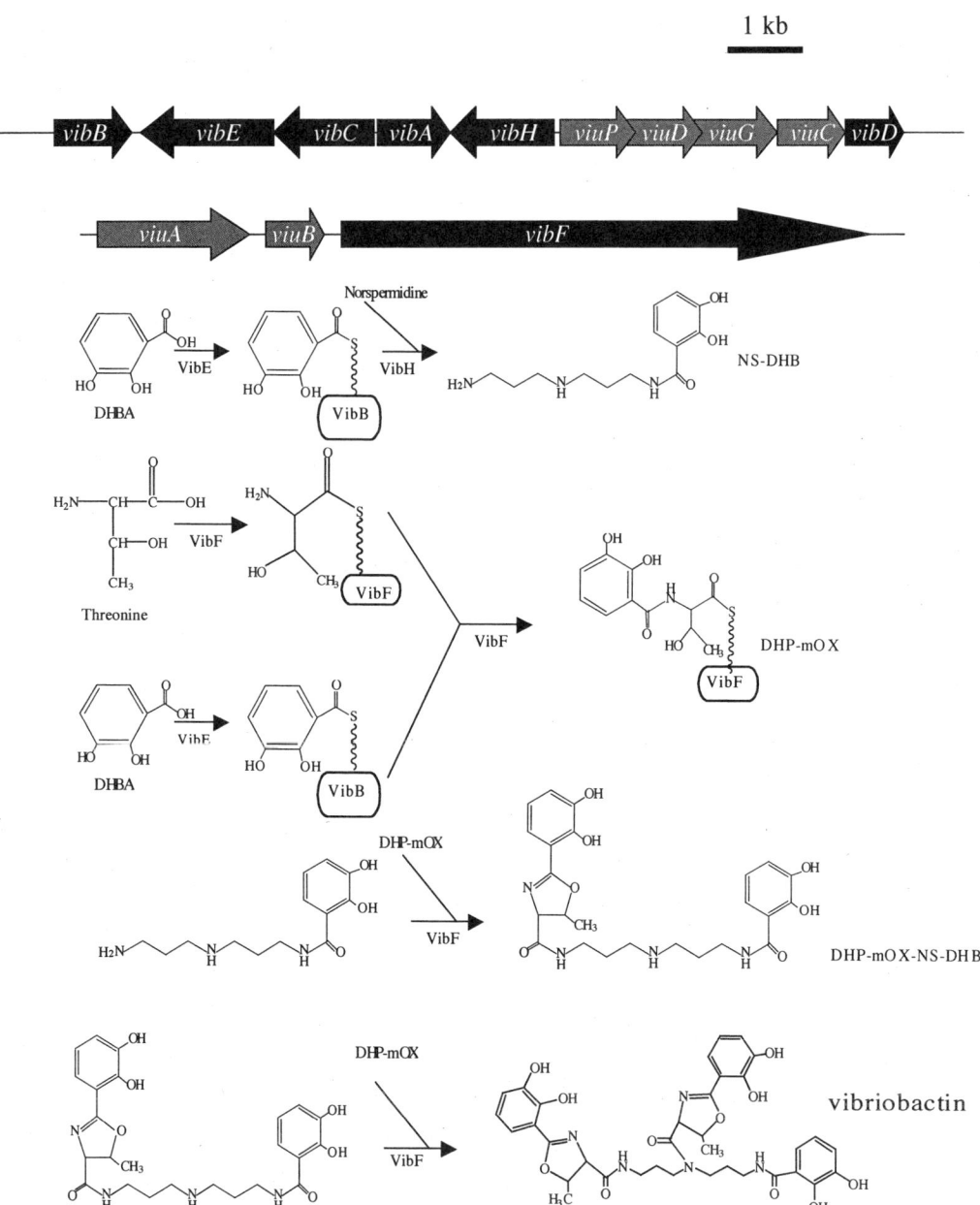

FIGURE 3 Genetic organization of *V. cholerae* vibriobactin biosynthetic genes and biosynthetic pathway.

VibD is required for vibriobactin biosynthesis and is thought to attach the pPant moiety to the ArCP and the PCP domains of VibB and VibF, respectively. The A domain of VibE activates each DHBA and tethers it to the ArCP domain of VibB. One of the three DHBA residues loaded on VibB is transferred to one of the primary amines of the norspermidine backbone in a reaction catalyzed by the condensation domain of VibH. The other two DHBA residues are linked to L-threonine by the VibF protein. The 270-kDa VibF NRPS with its predicted six domains, Cy-Cy-A-C-PCP-C, is the core of the vibriobactin assembly line. The A domain of VibF activates L-threonine and transfers it to the PCP domain of the same protein. The Cy domains of VibF catalyze the condensation of DHBA with the amino group of the L-threonine and the heterocyclization of threonine to an oxazoline ring to yield dihydroxyphenyl-methyloxazolinyl (DHP-mOX) loaded on the PCP domain of VibF. In the next step, the C2 domain of VibF transfers the heterocyclic acyl DHP-mOX group to the primary amine of NS-DHB, generating a free intermediate (DHP-mOX–NS-DHB). A second DHP-mOX group is then transferred by one or both C domains of VibF to the secondary amine of DHP-mOX–NS-DHB to yield vibriobactin.

TRANSPORT

The vibriobactin receptor ViuA is encoded in the same genetic cluster as the vibriobactin utilization protein ViuB. ViuB complements an *Escherichia coli fes* mutant, which removes iron from the ferri-enterobactin complex in the cytoplasm. From these data, it is thought that ViuB functions to remove iron from ferri-vibriobactin and enterobactin complexes. Additional proteins involved in the transport of vibriobactin are encoded in the other vibriobactin gene cluster. This region contains four open reading frames (ORFs), *viuPDGC*. ViuP has amino acidic sequence homology to the periplasmic binding protein of *E. coli* FepB. ViuP is a lipoprotein, like the periplasmic binding protein FatB in *V. anguillarum*. ViuD and ViuG are probably the cytoplasmic membrane permease proteins; as has been reported for other inner membrane protein pairs, ViuD and ViuG have homology to each other. The fourth ORF in the cluster, *viuC*, encodes a protein with homology to ATPase proteins of ABC transporters. Both Walker motifs A and B can be found in ViuC, indicating that it could be the ATPase component of the inner membrane complex. Expression of these four genes in *E. coli* complemented mutations in the *fepBDGC* genes. A mutation in *viuA*, the vibriobactin outer membrane receptor gene, resulted in a strain completely defective in vibriobactin transport. In contrast, mutants with mutations in *viuP* and in *viuG* still could transport vibriobactin, albeit to a lower extent. The ability of *viu* mutants to use vibriobactin suggested the presence of an additional system for the transport of catechol siderophores in *V. cholerae*. This system, identified as *vctPDGC*, is linked to the enterobactin receptor gene *vctA* (see below). Both VctPDGC and ViuPDGC systems can transport either vibriobactin or enterobactin; thus, the ligand specificity of these inner membrane transporters is lower than that of the outer membrane receptors.

REGULATION

The vibriobactin outer membrane receptor is iron regulated, and regulation of *viuA* expression by iron is dependent on the *fur* gene. A sequence with homology to the *E. coli* Fur box is located upstream of the other transport genes, indicating a likely role of Fur in their regulation. The *vibF* gene is also regulated by Fur, but there is also a twofold reduction in expression in an *irgB* mutant.

V. vulnificus

V. vulnificus is an opportunistic marine pathogen that can cause a fatal septicemic disease in humans and eels. Primary septicemia is often associated with diseases which result in iron overload, such as cirrhosis, hemochromatosis, and thalassemia. In virulence studies, iron-overloaded mice are used to mimic these conditions. Iron uptake is important in causing an infection, since a mutation in *venB*, encoding an isochorismate lyase, resulted in an increased 50% Lethal dose (LD_{50}). *V. vulnificus* tests positive in both the Arnow test (catecholates) and the Csáky test (hydroxamates). Vulnibactin contains only catechol residues, indicating that strains of *V. vulnificus* must have two siderophores, one a catechol and the other a hydroxamate compound. The hydroxamate siderophore has not yet been characterized.

The catechol siderophore vulnibactin (*N*-[3-(2,3-dihydroxybenzamido)propyl]-1,3-bis[2-(2-hydroxyphenyl)-*trans*-5-methyl-2-oxazoline-4-carboxamido]propane) consists of one residue of DHBA, two residues of salicylic acid, and two residues of L-threonine on an NS backbone (Fig. 1). This siderophore is very similar to vibriobactin, the siderophore of *V. cholerae*; the differences are that vulnibactin has two salicylic acid residues and one molecule of DHBA whereas vibriobactin has three DHBA residues.

BIOSYNTHESIS

So far, only one gene, *venB*, in the biosynthesis of the *V. vulnificus* catechol siderophore has been described. The VenB protein is an isochorismate lyase that is involved in the synthesis of DHBA, one of the precursors of vulnibactin, and in tethering of the activated DHBA. From the genome sequence, genes encoding putative NRPSs can be identified in a single large cluster on chromosome II (Fig. 4). The domains found are similar to those of *V. cholerae* but with some differences. For example, VibH (VV20844) and VibE (VV20836) homologs are present whereas the domains found in VibF of *V. cholerae* are encoded in two adjacent ORFs (VV20830 and VV20831). The main difference in the structures of these two siderophores is that the two DHBA residues that are linked to NS through threonine in vibriobactin are replace by salicylate moieties in vulnibactin. Genes for the synthesis and assembly of salicylate are therefore expected to be found in *V. vulnificus*, while these genes are not present in *V. cholerae*. Within the biosynthesis vulnibactin gene cluster, there is an ORF (VV20839) with homology to *pchB*, which is involved in the biosynthesis of salicylate from isochorismate in *Pseudomonas aeruginosa*. Downstream of the *pchB* homologue are ORFs with an aryl carrier protein domain (VV20840) and an A domain (VV20841), which could function in the activation of salicylate. Thus, the pathway for vulnibactin biosynthesis should be similar to that of vibriobactin, with additional domains for the assembly of salicylate in the siderophore molecule. Figure 4 shows a hypothetical model of this pathway based on our in silico findings.

TRANSPORT

Ferric vulnibactin is transported to the periplasm through the receptor VuuA. The VuuA amino acid sequence reveals a TonB box. There are three TonB systems in *V. vulnificus*. It is not clear if one, two, or all three TonBs can provide the energy to the vulnibactin receptor. The system that transports vulnibactin across the inner membrane has not been identified, but it is likely to be similar to that used by other bacterial siderophores. On chromosome II of *V. vulnificus*, noncontiguous to the vulnibactin biosynthesis cluster VV21011 to VV21016, there are ORFs with homology to ferric hydroxamate transport proteins. Perhaps vulnibactin shares this transport system to enter the cytoplasm with the hydroxamate siderophores.

REGULATION

The genes studied in the vulnibactin system, *vuuA* and *venB*, are regulated by Fur in a manner similar to that described for other gram-negative bacteria. No other positive or negative regulators for iron uptake in *V. vulnificus* have been identified so far.

V. parahaemolyticus

V. parahaemolyticus is normally found in the marine and estuarine environments. In humans it can cause gastroenteritis, wound infection, and septicemia; iron uptake may be important for this bacterium to proliferate in the host. *V.*

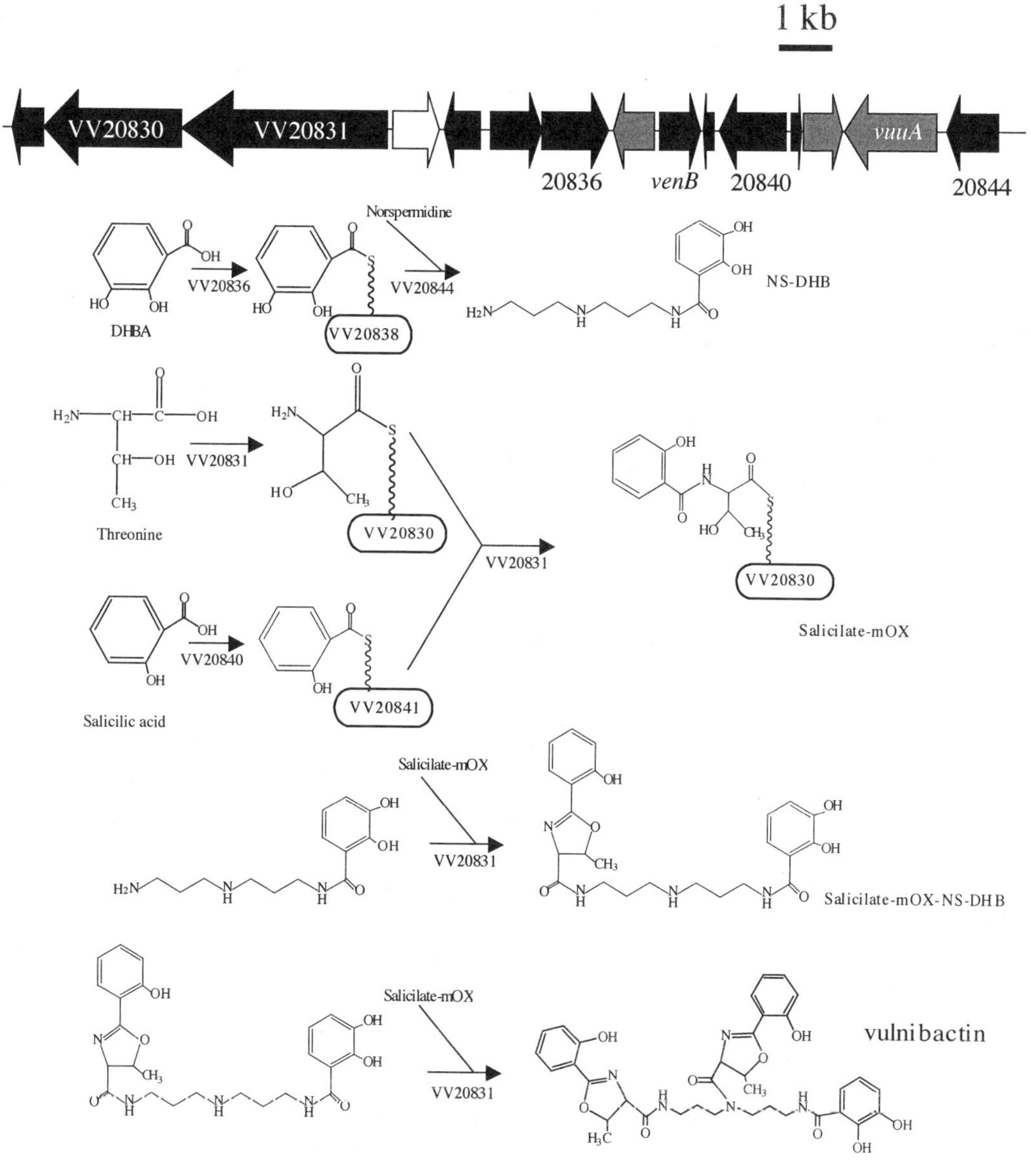

FIGURE 4 Genetic organization of *V. vulnificus* vulnibactin biosynthetic genes and proposed biosynthetic pathway.

parahaemolyticus produces the siderophore vibrioferrin, which is neither a hydroxamate nor a catechol siderophore and thus represents a new class of siderophores. Vibrioferrin is an acidic, hydrophobic molecule that is synthesized from alanine, citric acid, ethanolamine, and 2-ketoglutaric acid. Vibrioferrin is able to scavenge the iron from 30% saturated human transferrin but not from human lactoferrin even when it is fully iron saturated. A spontaneous vibrioferrin mutant is not able to grow in the presence of human transferrins as the sole iron source. The mechanism of synthesis of vibrioferrin remains to be elucidated.

TRANSPORT

The 78-kDa outer membrane receptor PvuA has been identified and shown to bind radiolabeled vibrioferrin. Moreover, a mutant with a mutation in *pvuA* is no longer able to grow in the presence of the iron chelator ethylenediamine-di-(*o*-hydroxyphenyl acetic acid) (EDDA). The polyclonal antiserum against PvuA was able to detect a 78-kDa outer membrane protein of *V. alginolyticus*. This is not surprising since it has been reported that some *V. alginolyticus* strains are able to produce vibrioferrin.

REGULATION

By growing *V. parahaemolyticus* on agar plates containing manganese, it was possible to isolate *fur* mutants, since high levels of manganese are toxic for any cell containing a functional Fur protein. Studies with these *fur* mutants revealed that the *pvuA* gene and vibrioferrin production are regulated by Fur.

HEME-MEDIATED IRON UPTAKE SYSTEMS

Heme release occurs when cells are damaged and the intracellular material is liberated. In serum, heme is bound to hemoglobin, hemopexin, albumin, and lipoproteins. Hemoglobin, which is the source of most circulating heme, is tightly bound to haptoglobin, a heterotetrameric glycoprotein. In this way, the amount of free heme, a potential source of iron for pathogens, in the extracellular environment is limited.

Microorganisms have developed diverse strategies to obtain heme. The first heme receptor identified in the family *Vibrionaceae* was *V. cholerae* HutA. Subsequently, two additional heme receptors HutR and HasR were identified. The transcription of *hutA* and *hutR* genes is induced under iron-limiting conditions. Transport by these receptors is dependent on TonB; however, the TonB box region of HasR does not have similarity to the TonB boxes of any other heme receptors in the genus *Vibrio*. *V. cholerae* has two distinct TonB proteins, named TonB1 and TonB2; while HutA and HutR interact with both proteins, HasR apparently uses only TonB2.

HutA and HutR carry the FRAP motif, which is well conserved among heme receptors, although it is also present in many non-heme receptors, indicating that it may be important for receptor function independent of ligand specificity. These proteins also contain the NPNL consensus sequence, located C-terminal to the FRAP motif, and the conserved histidine residue in the FRAP/NPNL region, which could be involved in heme binding. *V. cholerae* HasR is similar to the HasR hemophore receptors from *P. aeruginosa* and *Serratia marcescens*. It has 50% identity to the FRAP/NPNL consensus motifs, although it does not have the conserved histidine, suggesting that it uses a different mechanism for heme binding or transport.

The *hutA hutR hasR* triple mutant was completely defective in heme transport. Interestingly, this strain did not show a reduced ability to compete with its vibriobactin-defective parental strain in an infant-mouse model, indicating that additional sources of iron are available in the mammalian host.

Other genes involved in heme transport in *V. cholerae* are *hutBCD*, which encode a periplasmic binding protein (HutB) and cytoplasmic membrane permeases (HutC and HutD). These genes are located downstream of, and in an operon with, the *tonB1*, *exbB1*, and *exbD1* genes. All these genes are expressed from a single Fur-regulated promoter upstream of *tonB1*. Although the HutBCD proteins, together with the TonB1 system and HutA, permit *E. coli* to transport heme, they are not essential for heme utilization in *V. cholerae*, suggesting that an additional mechanism for the transport of heme across the inner membrane may be present.

Three additional genes, *hutWXZ*, form an operon that is divergently transcribed from the *tonB1* operon. The expression of these genes is regulated by iron in a Fur-dependent manner. One of these genes, *hutZ*, is required for efficient utilization of heme as an iron source. HutZ is a heme-binding protein that may act in heme trafficking and/or heme storage. The role of this protein in heme utilization is under investigation.

In *V. vulnificus*, the gene encoding the heme receptor HupA was cloned and sequenced. Transcription of this gene is negatively regulated by iron and is constitutive in a *fur* mutant. The *hupA* promoter contains a 19-bp dyad-symmetric sequence homologous to the *E. coli* Fur-binding site. Upstream of this gene, in reverse orientation, is a gene (*hupR*) encoding a protein with homology to the LysR family of bacterial transcriptional activator proteins. HupR may be a positive regulator of *hupA* transcription under low-iron conditions in the presence of heme. Strains of *V. vulnificus* that can infect eels also use heme and hemoglobin as iron sources.

The outer membrane protein HuvA of *V. anguillarum* is essential for heme uptake. A *huvA* mutant showed a decrease in virulence in comparison with the wild-type strain in experimental infections, but only when the fish were overloaded with heme. The wild-type and HuvA$^-$ strains had similar lethality in normal fish. This would suggest that heme transport does not play a major role in vivo but that increased heme availability could enhance infection.

Using different *V. cholerae* genes involved in heme utilization as probes against chromosomal DNA from several pathogenic vibrios, it was found that the heme iron utilization systems of *V. parahaemolyticus*, *V. alginolyticus*, and *V. fluvialis* are similar at the DNA level to that of *V. cholerae*. Underscoring the findings at the DNA levels, it was also shown that some of the heme utilization proteins of *V. cholerae*, *V. parahaemolyticus*, and *V. alginolyticus* are functionally interchangeable.

OTHER TRANSPORTERS IN THE VIBRIOS

In addition to vibriobactin, *V. cholerae* can utilize different siderophores synthesized by other microorganisms, including enterobactin synthesized by several members of the *Enterobacteriaceae*, ferrichrome produced by the fungus *Ustilago sphaerogena*, and schizokinen produced by *Bacillus megaterium* and *Anabaena variabilis*. The outer membrane receptor for ferrichrome uptake has been identified, and the predicted FhuA protein has homologies to different siderophore receptors from diverse bacteria, especially FhuA from *E. coli*. However, it presents differences in a region known to form the "gating loop," and it also may contain an adjacent loop that would be an alternate binding site for ferrichrome or albomycin. Genetically, the *fhuA* gene is located in an operon (*fhu*) with *fhuC*, *fhuB*, and *fhuD*, whose transcription is iron regulated by the Fur protein.

The outer membrane receptors for enterobactin in *V. cholerae* are IrgA and VctA. Neither the *irgA* nor the *vctA* single mutants are defective in enterobactin utilization; however, the *irgA vctA* double mutant is unable to utilize enterobactin as iron source. The IrgA receptor has similarity to other TonB-dependent receptors, and its expression is negatively regulated by Fur and positively regulated by IrgB, the product of a divergently transcribed upstream gene, *irgB*. Previously it was described that an *irgA*::Tn*phoA* mutant had a decreased competitive index and a 100-fold-increased LD_{50} in an infant-mouse model relative to its parental strain (O395). However, that mutation could not be complemented by the cloned *irgA* genes, and a newly constructed *irgA* knockout mutant strain did not show reduced LD_{50} or reduced competition in the infant-mouse model, demonstrating that this IrgA is not a virulence factor in *V. cholerae*.

The second enterobactin receptor, VctA, has similarity to the enterobactin receptor from *Neisseria gonorrhoeae*, FetA, and to heme receptors from diverse microorganisms. Upstream of *vctA* and divergently transcribed is an ORF (VCA0231), which encodes a protein with similarity to the AraC/XylS family of transcriptional regulators with unknown function. Upstream of that gene are four genes (*vctPDGC*) encoding periplasmic binding protein-dependent ABC transport system.

V. anguillarum can utilize ferric iron as a complex with either ferrichrome or enterobactin; however, it can synthesize neither of these two siderophores. *V. vulnificus* can utilize desferrioxamine B (a siderophore from *Streptomyces*) as an iron source. Recently, it was determined that an outer membrane protein of 78 kDa is induced in *V. vulnificus* in the presence of this siderophore, and N-terminal sequencing of that protein proved that it was neither HupA nor VuuA. However, the involvement of this protein, or HupA, in *V. vulnificus* pathogenicity was not tested.

ENERGY-TRANSDUCING COMPLEXES

Ferric iron, in complex with either siderophores or heme, is actively transported across the outer membrane through the substrate-specific outer membrane receptor. Energy required for this process is provided by the proton motive force generated in the cytoplasmic membrane, which is in turn transduced by the TonB-ExbBD proteins to the outer membrane receptors (see chapter 7). The TonB-ExbBD proteins form a complex in the inner membrane, and TonB interacts directly with ligand-bound receptors. TonB is thought to induce a conformational change in these receptors that promotes internalization of the bound substrate. A characteristic of *Vibrio* species is the presence of more than one set of *tonB-exbB-exbD* genes. These fall into two distinct groups, TonB1 and TonB2, based on genetic organization and the similarity of their amino acid sequences. In Fig. 5 the genetic arrangement of the different TonB systems in *V. cholerae*, *V. anguillarum*, and *V. vulnificus* is shown.

The two TonB systems have overlapping functions; however, in some instances only one of the systems can facilitate the uptake of a specific ligand. In *V. cholerae*, either of the two TonB systems can function in the uptake of heme, vibriobactin, and ferrichrome, while the TonB1 system is specifically required for the uptake of the siderophore schizokinen and the TonB2 system is essential for enterobactin uptake. None of these ligands was transported in a strain defective for both TonB systems, and the double *tonB* mutant shows a significantly decreased ability to colonize the intestine of suckling mice. This shows that TonB-dependent iron transport systems are required for growth in the host; it remains to be determined but which systems are relevant in vivo.

HEMOLYSINS: ARE THEY THERE TO OBTAIN IRON?

Some of the vibrios produce hemolysins, and the iron released by lysed red blood cells represents a significant potential source of iron. Hemolysins might also be important for virulence in other ways. For example, *V. parahaemolyticus* produces various hemolysins. One of these is the thermostable direct hemolysin (TDH), which is responsible for the hemolytic activity observed on Wagatsuma agar (Kanagawa phenomenon [KP]) and is the major virulence factor in food-borne *V. parahaemolyticus*-induced gastroenteritis. Because KP activity is observed in strains isolated from patients but not in most environmental strains, KP has been used as a virulence marker for this bacterium. Certain KP-negative strains are still capable of causing gastroenteritis, and these produce a TDH-related hemolysin. Both type of hemolysins are pore-forming toxins. Furthermore, a thermolabile hemolysin has been found in this species, which also displays phospholipase activity.

In addition to the TDH, another thermostable hemolysin (δ-VPH) has recently been found in several *V. parahaemolyticus* strains and a related protein (Vc-δTH) was also identified in *V. cholerae* serotype O1. A *V. cholerae* Vc-δTH mutant was analyzed in an infant-mouse cholera model, and a role for this hemolysin in *V. cholerae* O1 pathogenesis was not observed. The predominant hemolysin synthesized in *V. cholerae* O1 and non-O1 strains is HlyA, which can permeabilize different eukaryotic cells and synthetic lipid vesicles by forming a pentameric transmembrane diffusion channel. The hemolysin may be involved in the pathogenesis of *V. cholerae*, since the purified protein induces secretion of fluid in a rabbit ligated ileal loop model. In El Tor strains, synthesis of the hemolysin is regulated by iron via Fur. Thus, this protein might increase the availability of heme,

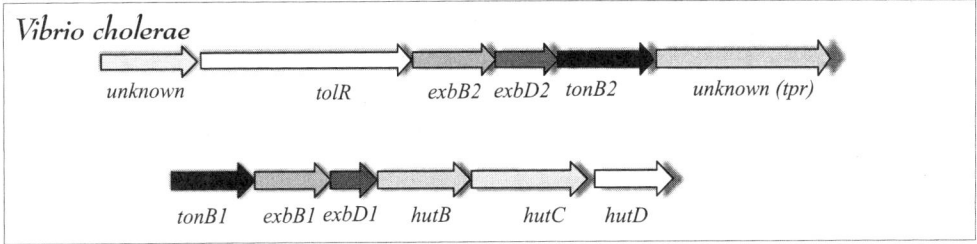

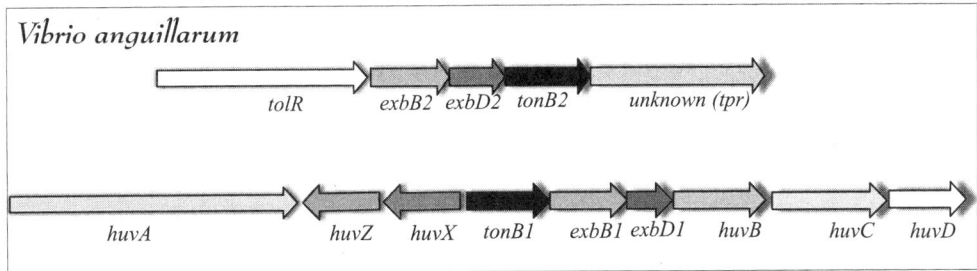

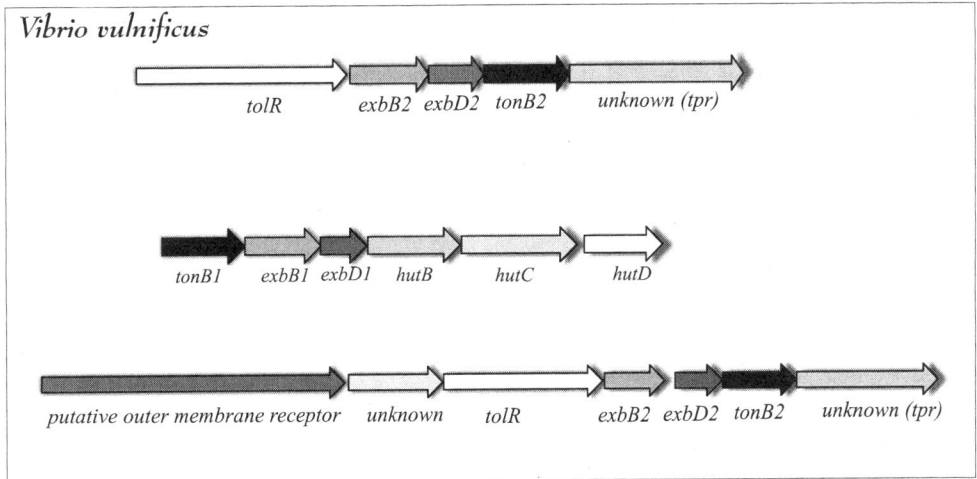

FIGURE 5 Genetic organization of the different TonB system genes in *V. cholerae*, *V. anguillarum*, and *V. vulnificus*. For *V. cholerae*, the *exbB2*, *exbD2*, and *tonB2* genes are on chromosome 1 while the *tonb1*, *exbB1*, and *exbD1* genes are on chromosome 2. The *hutB*, *hutC*, and *hutD* genes encode periplasmic and inner membrane proteins involved in heme transport. For *V. anguillarum*; *huvB*, *huvC*, and *huvD* encode periplasmic and inner membrane proteins involved in heme transport; *huvZ*, and *huvX* encode proteins involved in heme transport; and *huvA* encodes the outer membrane heme receptor. For *V. vulnificus*, the *exbB2* (VV20362), *exbD2* (VV20361), *tonB2* (VV20360) and *tonB1* (VV21614), *exbB1* (VV21613), *exbD1* (VV21612), *hutB* (VV21611), *hutC* (VV21610), and *hutD* (VV21609) genes are placed in chromosome 2. Another *tonB2* (VV10847) gene [together with the *exbB2* (VV10485) and *exbD2* (VV10846) genes] is placed in chromosome 1. These genes are in a putative operon with a gene that would encode an outer membrane receptor (VV10842). The numbers of the genes are from genomic sequence of the *V. vulnificus* CMCP6 strain (GenBank accession numbers, AE016796 and AE016795).

hemoglobin, or other intracellular iron sources under iron-limiting conditions.

In *V. vulnificus*, there is a cytolytic hemolysin encoded by the *vvhA* gene. Its role as a potential virulent factor has been described by using the purified protein. However, it appeared that the hemolysin is less important in vivo than would be predicted, since there is no correlation between hemolysin production and virulence. Likewise, it has been confirmed that the *vvhA* gene is expressed under certain conditions such as stationary phase or in the presence of glucose. It was also demonstrated that cyclic AMP receptor protein activates the expression of the *V. vulnificus* hemolysin gene by binding directly to the *vvhBA* promoter and that the hemolytic activity can be increased in a dose-dependent manner by addition of cyclic AMP. However, a definitive role for VvhA in *V. vulnificus* pathogenesis remains to be demonstrated.

ACKNOWLEDGMENT

We thank Elizabeth Wyckoff for reading and editing of the manuscript.

SUGGESTED READING

Actis, L. A., W. Fish, J. H. Crosa, K. Kellerman, S. R. Ellenberger, F. M. Hauser, and J. Sanders-Loehr. 1986. Characterization of anguibactin, a novel siderophore from *Vibrio anguillarum* 775(pJM1). *J. Bacteriol.* **167**:57–65.

Aso, H., S. Miyoshi, H. Nakao, K. Okamoto, and S. Yamamoto. 2002. Induction of an outer membrane protein of 78 kDa in *Vibrio vulnificus* cultured in the presence of desferrioxamine B under iron-limiting conditions. *FEMS Microbiol. Lett.* **212**:65–70.

Bang, Y. B., S. E. Lee, J. H. Rhee, and S. H. Choi. 1999. Evidence that expression of the *Vibrio vulnificus* hemolysin gene is dependent on cyclic AMP and cyclic AMP receptor protein. *J. Bacteriol.* **181**:7639–7642.

Braun, V. 1995. Energy-coupled transport and signal transduction through the gram-negative outer membrane via TonB-ExbB-ExbD-dependent receptor proteins. *FEMS Microbiol. Rev.* **16**:295–307.

Crosa, J. H. 1989. Genetics and molecular biology of siderophore-mediated iron transport in bacteria. *Microbiol. Rev.* **53**:517–530.

Crosa, J. H. 1980. A plasmid associated with virulence in the marine fish pathogen *Vibrio anguillarum* specifies an iron-sequestering system. *Nature* **284**:566–568.

Crosa, J. H., and C. T. Walsh. 2002. Genetics and assembly line enzymology of siderophore biosynthesis in bacteria. *Microbiol. Mol. Biol. Rev.* **66**:223–249.

Fouz, B., R. Mazoy, M. L. Lemos, M. J. del Olmo, and C. Amaro. 1996. Utilization of hemin and hemoglobin by *Vibrio vulnificus* biotype 2. *Appl. Environ. Microbiol.* **62**:2806–2810.

Funahashi, T., K. Moriya, S. Uemura, S. Miyoshi, S. Shinoda, S. Narimatsu, and S. Yamamoto. 2002. Identification and characterization of *pvuA*, a gene encoding the ferric vibrioferrin receptor protein in *Vibrio parahaemolyticus*. *J. Bacteriol.* **184**:936–46.

Genco, C. A., and D. W. Dixon. 2001. Emerging strategies in microbial haem capture. *Mol. Microbiol.* **39**:1–11.

Goldberg, M. B., S. A. Boyko, and S. B. Calderwood. 1990. Transcriptional regulation by iron of a *Vibrio cholerae* virulence gene and homology of the gene to the *Escherichia coli fur* system. *J. Bacteriol.* **172**:6863–6870.

Griffiths, G. L., S. P. Sigel, S. M. Payne, and J. B. Neilands. 1984. Vibriobactin, a siderophore from *Vibrio cholerae*. *J. Biol. Chem.* **259**:383–385.

Keating, T. A., C. G. Marshall, and C. T. Walsh. 2000. Reconstitution and characterization of the *Vibrio cholerae* vibriobactin synthetase from VibB, VibE, VibF and VibH. *Biochemistry* **39**:15522–15530.

Köster, W. L., L. A. Actis, L. S. Waldbeser, M. E. Tolmasky, and J. H. Crosa. 1991. Molecular characterization of the iron transport system mediated by the pJM1 plasmid in *Vibrio anguillarum* 775. *J. Biol. Chem.* **266**:23829–23833.

Linkous, D. A., and J. D. Oliver. 1999. Pathogenesis of *Vibrio vulnificus*. *FEMS Microbiol. Lett.* **174**:207–214.

Mazoy, R., C. R. Osorio, A. E. Toranzo, and M. L. Lemos. 2003. Isolation of mutants of *Vibrio anguillarum* defective in haeme utilisation and cloning of *huvA*, a gene coding for an outer membrane protein involved in the use of haeme as iron source. *Arch. Microbiol.* **179**:329–338.

Mey, A. R., and S. M. Payne. 2001. Haem utilization in *Vibrio cholerae* involves multiple TonB-dependent haem receptors. *Mol. Microbiol.* **42**:835–849.

Mey, A. R., E. E. Wyckoff, A. G. Oglesby, E. Rab, R. K. Taylor, and S. M. Payne. 2002. Identification of the *Vibrio cholerae* enterobactin receptors VctA and IrgA: IrgA is not required for virulence. *Infect. Immun.* **70**:3419–3426.

Occhino, D. A., E. E. Wyckoff, D. P. Henderson, T. J. Wrona, and S. M. Payne. 1998. *Vibrio chol*-

erae iron transport: haem transport genes are linked to one of two sets of *tonB, exbB, exbD* genes. *Mol. Microbiol.* **29:**1493–1507.

Okujo, N., M. Saito, S. Yamamoto, T. Yoshida, S. Miyoshi, and S. Shinoda. 1994. Structure of vulnibactin, a new polyamine-containing siderophore from *Vibrio vulnificus*. *Biometals* **7:**109–116.

O'Malley, S. M., S. L. Mouton, D. A. Occhino, M. T. Deanda, J. R. Rashidi, K. L. Fuson, C. E. Rashidi, M. Y. Mora, S. M. Payne, and D. P. Henderson. 1999. Comparison of the heme iron utilization systems of pathogenic vibrios. *J. Bacteriol.* **181:**3594–3598.

Seliger, S. S., A. R. Mey, A. M. Valle, and S. M. Payne. 2001. The two TonB systems of *Vibrio cholerae*: redundant and specific functions. *Mol. Microbiol.* **39:**801–812.

Stork, M., M. Di Lorenzo, T. J. Welch, L. M. Crosa and J. H. Crosa. 2002. Plasmid-mediated iron uptake and virulence in *Vibrio anguillarum*. *Plasmid* **48:**222–228.

Taniguchi, H., S. Kubomura, H. Hirano, K. Mizue, M. Ogawa, and Y. Mizuguchi. 1990. Cloning and characterization of a gene encoding a new thermostable hemolysin from *Vibrio parahaemolyticus*. *FEMS Microbiol. Lett.* **55:**339–345.

Wright, A. C., L. M. Simpson, and J. D. Oliver. 1981. Role of iron in the pathogenesis of *Vibrio vulnificus* infections. *Infect. Immun.* **34:**503–507.

Wyckoff, E. E., A. M. Valle, S. L. Smith, and S. M. Payne. 1999. A multifunctional ATP-binding cassette transporter system from *Vibrio cholerae* transports vibriobactin and enterobactin. *J. Bacteriol.* **181:**7588–7596.

NEISSERIA

Cynthia Nau Cornelissen and P. Frederick Sparling

17

The genus *Neisseria* consists of two human pathogens, *Neisseria gonorrhoeae* (gonococcus) and *Neisseria meningitidis* (meningococcus), as well as several commensal species, including *Neisseria lactamica*. Because of their high levels of DNA sequence conservation, the two pathogenic species are considered members of the same genomospecies. In spite of this sequence conservation, the two pathogens typically cause dramatically different diseases, ranging from localized infection of the genital tract (gonococcus) to explosive septicemia with or without meningitis and death (meningococcus).

N. *gonorrhoeae* is the etiologic agent of the sexually transmitted disease gonorrhea, which is characterized by a relatively low-grade, albeit purulent infection of the urethra in males and of the endocervix in females. Complications include epididymitis in males and salpingitis and pelvic inflammatory disease in females. Gonococcal conjunctivitis occurs in adults as a result of autoinoculation and in neonates as a result of passage through the vagina of an infected mother. Bacteremic gonococcal disease, or disseminated gonococcal infection, occurs in about 0.5 to 1.0% of patients with untreated mucosal infection at present and is often accompanied by arthritic and/or dermatologic manifestations. By contrast, *N. meningitidis* is an etiological agent of bacterial meningitis but is also considered a member of the normal flora of the human nasopharynx. Dissemination from the site of colonization can result in fulminant meningococcemia, a rapidly fatal manifestation, and/or in meningococcal meningitis, resulting from transmigration of *N. meningitidis* across the blood-brain barrier. While *N. meningitidis* strains remain susceptible to antimicrobial therapies, appropriate treatment often is not effective, as evidenced by fatality rates of about 10% in most industrialized countries.

Several virulence factors have been identified that could contribute to this discrepancy in disease potential. *N. meningitidis* expresses a polysaccharide capsule, the genes for synthesis of which are not present in the genome of *N. gonorrhoeae*. Likewise, genes encoding a putative RTX toxin and glutathione peroxidase are limited to *N. meningitidis* and could contribute to the virulence of this pathogen, although the mechanisms by which they do so currently are underfined. Other genetic differences between gonococci and meningococci that might help

Cynthia Nau Cornelissen, Department of Microbiology and Immunology, Medical College of Virginia Campus, Virginia Commonwealth University, Richmond, VA 23284. *P. Frederick Sparling*, Department of Medicine, University of North Carolina at Chapel Hill, Chapel Hill, NC 27599–7031.

explain the observed differences in their clinical behaviors include an additional porin (PorA) in meningococci. A recent study identified eight small genetic islands in *N. meningitidis* that are absent from the genome of *N. gonorrhoeae*; these islands include genes that encode proteins with similarity to a siderophore receptor and members of a leukotoxin secretion system. Insertional mutagenesis of the island that contains the putative siderophore receptor gene resulted in a mutant that could not cause bacteremia in an infant-rat model of meningococcal infection, which strongly suggests that these small genetic islands contribute to the virulence of meningococci. These results are consistent with many earlier reports that iron is crucial for meningococcal disease and suggest that the different pathogenic potentials of gonococci and meningococci could be due, in part, to distinct iron acquisition strategies.

Despite the discrepant disease manifestations, the life-styles of the pathogenic *Neisseria* species are remarkably similar. Both pathogens begin the infectious process at the mucosal epithelium; the target of colonization by *N. gonorrhoeae* is the genital mucosa, while the target of *N. meningitidis* infection is the nasopharynx. Following adherence to nonciliated epithelial cells, both microorganisms are capable of growth within and transcytosis through the epithelial cell layer, from which they gain access to the subepithelial tissues. In addition to epithelial cells, the pathogenic *Neisseria* species are capable of entering other cell types, including professional phagocytes such as neutrophils and monocytes. Although the pathogenic *Neisseria* species are considered classical extracellular pathogens, intraepithelial cell growth was documented recently, and it is possible that invasion of and growth within host cells plays an important role in pathogenesis.

A wide variety of virulence factors have been described and characterized in the pathogenic *Neisseria* species. The meningococcus is surrounded by a polysaccharide capsule, which aids in serum resistance. The presence of this structure in the meningococcus and its absence in the gonococcus could explain certain major differences in disease potential, especially meningitis. In normal hosts, meningitis typically is caused by encapsulated bacteria due to the antiphagocytic properties of capsules and the key role of polymorphonuclear leukocytes in defense of the cerebrospinal fluid. The presence of capsules also could help explain the differences in modes of transmission, because of the effects of capsules in preventing desiccation: the meningococcus is transmitted from person to person via respiratory droplets, whereas the gonococcus requires intimate sexual contact for transmission. There is also evidence that the meningococcal capsule interferes with invasion of epithelial cells but confers a selective advantage in the bloodstream. Thus, modulating the expression of the capsule would be advantageous to *N. meningitidis* during the course of infection. Indeed, capsule expression may be down-regulated on contact with epithelial cells and then subsequently up-regulated following transcytosis through the epithelial cell layer. Although gonococci do not express a polysaccharide capsule, they decorate lipooligosaccharide (LOS) with sialic acid, which contributes to serum resistance; in this sense, sialylation may be considered an alternative to capsule expression in the gonococcus. However, as with capsule, sialylation of LOS interferes with invasion. Gonococci are capable of rapid variation of LOS structures that are essential for sialylation, resulting in alternate states that either enhance invasion (lack of sialic acid) or enhance serum resistance (presence of sialic acid).

Multiple surface adhesins have been characterized as contributors to the infectious process in the pathogenic *Neisseria* species. These adhesins include a type IV pilus, a putative tip-associated adhesin, and a family of invasins called Opa proteins. The pilus, composed of the subunit protein PilE, and the tip-associated adhesin, PilC, together mediate binding to epithelial cells, endothelial cells, and professional phagocytes. The Opa proteins, which facilitate binding to and invasion of various cell types by virtue of interaction with host cell receptors in the carcinoembryonic-antigen-cellular adhesion molecule (CEACAM) family or in the heparin sulfate proteoglycan group, are encoded by a family of genes. The tissue tropism of the cell is at least in part determined by which *opa* gene is expressed at any particular time. The mechanism that modulates expression of the *opa* genes is known as slipped-strand mispairing, and genes that are subject to this form of regulation are referred to as contingency loci. During DNA replication, the length of a repeated tract of nucleotides can be increased or decreased, which gives rise to changes in the reading frame of the structural gene encoding an Opa protein. At any one time, the neisseriae express between zero and three (or more) Opa proteins simultaneously, the remainder of the genes being out of frame and therefore not translated into protein. Like the Opa proteins, the type IV pilus component PilE and the tip-associated adhesin PilC are subject to high-frequency variation, PilE variation is mediated, in part, by slipped-strand mispairing, resulting in variable on-off expression (phase variation), and additionally via a RecA-dependent gene conversion event that results in recombination between one of several silent (PilS) loci and the only pilus expression locus, PilE. The PilC protein, like the Opa proteins, is subject to variation by slipped-strand mispairing, but in this case, it occurs due to the presence of a homopolymeric repeat in the promoter region. Changes in the number of nucleotides in this tract result in variation in promoter strength and therefore in modulation of expression. Thus, the pathogenic *Neisseria* species possess multiple adherence-related virulence factors, whose variable, and perhaps coordinated, expression leads to changes in tissue tropism and the capacity to invade numerous cell types.

Two other neisserial virulence factors impact invasion and intracellular survival. The product of the *porB* gene is classified as a typical gram-negative outer membrane porin but with other, somewhat unexpected properties. This protein has been reported to translocate from the bacterial outer membrane into the host cell membrane, where it forms ion-gated channels. In various host cell types, these channels allow transient Ca^{2+} influx from the extracellular milieu, whose effects range from impact on invasion, to exocytosis of endosomes, to modulating apoptosis. The other neisserial virulence factor thought to be important in intracellular growth is the immunoglobulin A1 (IgA1) protease. As its name implies, the original function ascribed to this protein was cleavage of IgA1, found on mucosal surfaces. However, like the porin protein, this virulence factor has recently been assigned new functions in pathogenesis. Expression of IgA1 protease results in diminished levels of intracellular Lamp1, a lysosomal marker, and subsequently leads to increased intracellular survival by the members of the pathogenic *Neisseria* species. Recent evidence suggests that there is a delicate interplay between PorB and IgA1 protease, whereby PorB causes transient Ca^{2+} fluxes, resulting in perturbed endosome trafficking and exocytosis of Lamp1 to the cell surface, where IgA1 protease cleaves it, potentiating intracellular survival.

In conjunction with these classical virulence factors, the pathogenic *Neisseria* species express a wide array of iron acquisition systems. Although the *Neisseria* species are not known to synthesize or secrete detectable siderophores, they are capable of iron acquisition from host transferrin, lactoferrin, and hemoglobin without the synthesis of a siderophore intermediate. Iron acquisition from these host proteins occurs via a process that is dependent on the expression of specific outer membrane receptors. In addition to these sources of iron, the pathogenic *Neisseria* species are capable of using heme, siderophores made by other microorganisms (xenosiderophores), and a variety of keto acid-iron complexes as sole iron sources. The contribution of iron to the pathogenesis of these pathogens was recognized in early studies, in which it was reported that administration of iron dextran resulted in dramatically enhanced infection rates in a mouse model of meningococcal infection. More recently, a gonococcal mutant defective in iron acquisition from transferrin or lactoferrin was demonstrated to be incapable of eliciting signs and symptoms of urethritis in human male volunteers. This result is in marked contrast to the results of studies in which the contributions of some of the other classical virulence factors have been tested in the same human challenge model. A mutant lacking the IgA1 protease was indistinguishable from the wild-type strain in its capacity to cause urethritis, and a PilE mutant and a RecA mutant, independently tested in the human challenge model, were both capable of colonization and of causing mild urethritis in human males. While variants selected for by human "passage" were invariably Opa$^+$, no particular Opa protein was predominant, suggesting that while Opa expression was important, the exact identity of the protein was not. Together, these results highlight the importance of iron acquisition in the pathogenesis of the neisseriae and support efforts at further characterization of the systems that enable the procurement of this necessary nutrient.

NONSIDEROPHORE IRON TRANSPORT SYSTEMS: UTILIZATION OF TRANSFERRIN AS AN IRON SOURCE

All isolates of *N. gonorrhoeae* tested to date are capable of transferrin-iron utilization. The process of transferrin-iron acquisition is energy dependent and requires direct contact between the iron-binding protein and the bacterial cell surface. This latter piece of evidence formed the basis for the conclusion that transferrin-iron acquisition in the pathogenic neisseriae is a receptor-mediated process that does not involve the synthesis of a siderophore intermediate. Human transferrin is specifically recognized by the gonococcus, since no other mammalian forms of transferrin are bound or utilized as an iron source. The exquisite species specificity of the receptor undoubtedly contributes to the restricted host range of this pathogen.

The Transferrin-Binding Proteins, TbpA and TbpB

Two iron-repressed, transferrin-binding proteins (Tbps) were isolated and characterized from the gonococcus by transferrin-affinity chromatography. These proteins, originally called Tbp1 and Tbp2, were approximately 100 and 85 kDa in molecular mass, respectively. Using a combination of techniques, including cloning by repair and screening expression libraries with Tbp1-specific antisera, the gene that encoded the 100-kDa protein was identified, sequenced, and ultimately named *tbpA*. The product of the *tbpA* gene, subsequently renamed TbpA, was identified as a member of the TonB-dependent family of outer membrane transporters by virtue of its similarity with the ferric enterobactin and vitamin B$_{12}$ transporters of *Escherichia coli*. Mutants incapable of TbpA expression were unable to grow on transferrin as the sole iron source, consistent with the proposal that TbpA serves as the portal of entry for iron through the gram-negative outer membrane (Fig. 1).

The *tbpB* gene was identified in *N. gonorrhoeae* based primarily on its proximity to the *tbpA* gene. *tbpB* is located immediately upstream of *tbpA*, with the genes being separated by an intergenic region of 86 bp (Fig. 2). The predicted protein contains a putative signal II cleavage site, suggesting that TbpB is a lipoprotein. This was confirmed for gonococcal TbpB by intrinsic labeling with [^{14}C] palmitate. TbpB is proposed to be surface exposed and loosely tethered to the outer leaflet of the outer membrane via its lipid tails (Fig. 1); however, it is possible that there are as yet unidentified transmembrane domains that allow this protein to span the outer membrane. Sequencing *tbpB* revealed no strong homologues, leaving the membrane topology and precise structure of the TbpB protein currently unresolved. An important attribute of TbpB not shared by TbpA is the ability to discriminate between apo- and holo-transferrin. Thus, this bacterial lipoprotein, like its eukaryotic counterpart, preferentially binds the ferrated form of transferrin, suggesting that it recognizes the "closed" structure of transferrin, a conformation assumed when iron is bound. This specificity for ferrated transferrin could explain the observation that mutants lacking TbpB are capable of transferrin-iron acquisition, albeit at a lower rate. Gonococcal mutants lacking TbpB were capa-

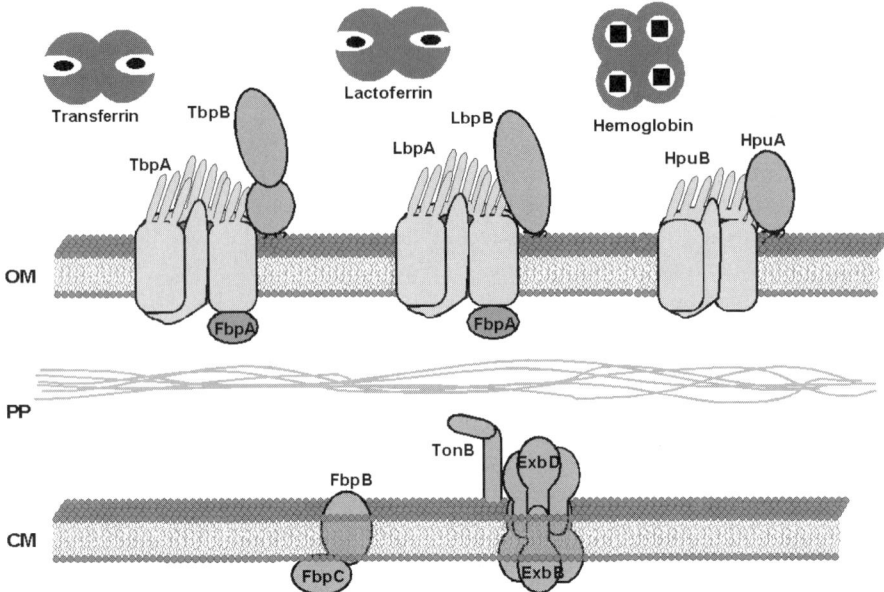

FIGURE 1 Model of iron acquisition systems expressed by the *Neisseria* species. TbpA and TbpB are outer membrane (OM) proteins that bind to transferrin and subsequently facilitate iron removal and internalization. LbpA and LbpB are similar proteins that function as the lactoferrin receptor. HpuA and HpuB are homologous to the Tbp and Lbp receptor proteins and facilitate the utilization of hemoglobin as a heme and iron source. Another hemoglobin receptor, HmbR (not shown), is expressed by *N. meningitidis* and does not conform to this two-protein receptor paradigm, as discussed in the text. FbpA is a periplasmic (PP), protein and FbpB and FbpC are shown as integral cytoplasmic membrane (CM) proteins. This group of proteins facilitates the periplasmic and cytoplasmic transport of iron, donated by transferrin or lactoferrin. The TonB, ExbB, and ExbD energy transduction system is shown within or tethered to the cytoplasmic membrane. This group of proteins is required for energization of all three iron acquisition systems shown. See the text for details.

ble of 50% of wild-type levels of iron uptake from transferrin. Thus, while TbpB is not absolutely required for transferrin-iron uptake, its presence makes the process more efficient.

Other Proteins Required for Transferrin-Iron Internalization

Following initial contact of transferrin with the bacterial outer membrane, the participation of several periplasmic and cytoplasmic proteins is required to accomplish transferrin-iron internalization. TbpA is a member of the TonB-dependent family of transporters, and TonB is required for the internalization event (see chapter 7). Gonococcal *tonB*, *exbB*, and *exbD* homologues have been identified, cloned, and mutagenized. Mutagenesis of each of these genes resulted in strains that were incapable of iron acquisition from transferrin. These mutants remained ligand-binding competent, but growth on transferrin-iron and internalization of iron from transferrin was completely inhibited. In addition, these mutants were dramatically hindered in transferrin release, suggesting that TonB-derived energy played a crucial role in ligand release from the receptor. A mutant in the TonB box of TbpA was similarly impaired in ligand release and transferrin-iron internalization. Transferrin affinity purification of gonococcal TbpA resulted in the copurification of TonB, consistent with a stable interaction between these proteins; however, the TonB-box mutant of TbpA was incapable of establishing or facilitating this interaction.

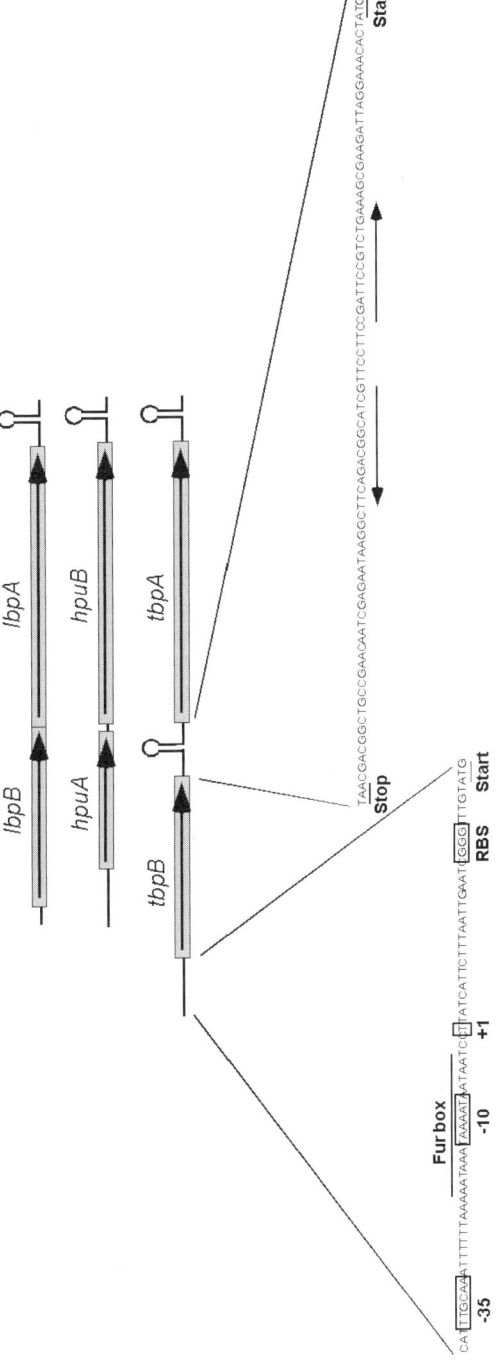

FIGURE 2 Genetic arrangement of *tbp* genes. By comparison with the *lbp* genes, encoding the lactoferrin receptor, and with the *hpu* genes, encoding the hemoglobin receptor, only the *tbp* genes are separated by an intergenic region. This region consists of 86 bp and includes a region of dyad symmetry (arrows) capable of forming a stem-loop structure in mRNA. The promoter region that precedes *tbpB* is also shown; it includes canonical promoter, overlapped by a putative Fur binding site (Fur box). The transcriptional start site, ribosome start site (RBS), and start codon are also shown.

In addition to energy transduction via TonB, ExbB, and ExhD, transferrin-iron acquisition in the neisseriae requires the ferric-binding protein, FbpA. The *fbp* (ferric-binding protein) gene cluster, located at a locus distant from the *tbp* gene cluster in the pathogenic *Neisseria* species, encodes proteins that have similarity to the periplasmic binding protein-dependent cytoplasmic permease systems such those for maltose uptake and histidine import. FbpA has been demonstrated to receive ^{55}Fe delivered to the cell via transferrin, lactoferrin, or citrate, implicating this protein in periplasmic transport of ferric iron. Interestingly, the FbpA protein, which has been cystallized, bears a striking resemblance to a single lobe of transferrin, leading to speculation that this protein is the "primordial" transferrin, existing prior to the putative gene duplication event from which the current bilobed transferrin arose. FbpA binds ferric iron with an affinity near that of transferrin, and the iron is coordinated with similar residues and an anion. These observations make it feasible for FbpA to serve as the periplasmic transporter of ferric iron after it has been stripped from transferrin and passed into the periplasm through the pore in TbpA (Fig. 1).

The genes located downstream of *fbpA* (*fbpB* and *fbpC*) encode proteins that resemble cytoplasmic permeases and ATPases, respectively. The FbpB gene product (Fig. 1) resembles the permeases MalF and HisQ, while the FbpC gene product (Fig. 1), like its homologues, contains a highly conserved Walker box, consistent with ATP binding by this component. It is proposed that the FbpC protein provides the energy, in the form of ATP hydrolysis, to drive uptake through the FbpB protein, which serves as the cytoplasmic membrane permease. These steps are directly analogous to those known to occur in the process of ferric siderophore and vitamin B_{12} uptake in gram-negative bacteria (see chapters 8 and 11).

Mechanism of Iron Uptake from Transferrin

Based on computer predictions, observed sequence diversity, and similarity to other TonB-dependent receptors, a two-dimensional topology model for gonococcal TbpA was developed. This model, like the crystal structures of FepA and FhuA, depicts 22 transmembrane β-strands connected by 11 surface-exposed loops. Putative loops 4 and 5 are critical for ligand-binding functions, since their deletion resulted in TbpA mutants lacking the ability to bind transferrin or utilize it as an iron source. Interestingly, loop 5 is one of the most variable in terms of sequence identity and length, leading to the hypothesis that vulnerable binding epitopes are camouflaged with and surrounded by antigenic diversity. Loop 5 also retains transferrin-binding domains, since this loop, when reconstituted ex vivo with a fusion partner, was capable of specific, dose-dependent transferrin binding.

Because no TbpB homologues have been identified for which the crystal structure is known, no topology models have been developed for this protein. The amino-terminal half of TbpB is required to form a stable transferrin-binding domain, which withstands heating and treatment with sodium dodecyl sulfate. A lower-affinity binding domain was identified in the less diverse carboxy-terminal half of meningococcal TbpB and may be characteristic of TbpBs proteins in general. The observations that meningococcal TbpB contains two transferrin-binding domains and internal sequence redundancy led to the concept that, like the ligand it binds, TbpB is a bilobed protein (Fig. 1).

Several observations suggest that TbpA and TbpB function together to bind transferrin and relieve it of its iron. Both proteins bind transferrin on the cell surface, but they do so differently, with different sensitivities to ionic conditions and different specificities but with similarly high affinities. The protease accessibility of TbpB is dramatically influenced by the presence and TonB-dependent energization of TbpA, indicating that the two proteins are at least in close proximity to one another in the outer membrane. Transferrin affinity isolation experiments yield TbpB only in the presence of TbpA (under high-ionic-strength conditions), consistent with the hypothesis that these proteins form a complex. Likewise, the in vitro characteristics of purified meningococcal Tbp proteins suggest that these proteins form a stable complex. Thus, the available evidence suggests that TbpA and TbpB, while capable of independent, distinct interactions with ligand, operate as a unit in the outer membrane to facilitate ligand binding and possibly iron extraction.

A working model of the transferrin-iron acquisition process is as follows. The ferrated form of human transferrin interacts with TbpA and TbpB, which perhaps form a complex in the gonococcal outer membrane. Using TonB-derived energy, the conformations of TbpA and TbpB change, facilitating iron release from transferrin. The process of transferrin binding and/or transferrin-iron removal might be made more efficient by the presence and participation of TbpB. The precise steps in the iron-stripping event, wherein transferrin is deferrated and the ferric iron atom is removed at the cell surface, remain a mystery. Some aspect of the stripping or ligand release event probably depends upon TonB-derived energy. The ferric iron atom, once removed, is thought to traverse the outer membrane via the pore within TbpA and subsequently to be passed off to FbpA, which waits in the periplasm below. Ferric-FbpA probably diffuses across the periplasm and then interacts with the FbpB and FbpC proteins, which together facilitate the passage of iron into the cytoplasm.

Genetics and Regulation of the Transferrin-Iron Acquisition System in *N. gonorrhoeae*

The neisserial *tbp* gene locus is the only one described to date in which the *tbp* genes are separated by any significant distance. The gonococcal *tbpB* gene is separated from *tbpA* by 86 bp, although the two are cotranscribed in a single bicistronic operon. Within the intergenic region is a region of potential secondary structure consisting of 35 nucleotides (Fig. 2), with a calculated free energy of dissociation of -23.6 kcal. The presence of a secondary structure in the bicistronic operon or in a *tbpB*-specific monocistronic mRNA could be important for maintaining the optimum stoichiometry of *tbp*-specific mRNA species and could translate to effects on the stoichiometry of the functional Tbp complex. Given that this genetic arrangement of the *tbp* genes is unique to the pathogenic *Neisseria* species, it is plausible that the intergenic region and its impact on Tbp synthesis could be particularly important to the biology of these organisms.

The mechanism by which the Tbp proteins are expressed as a function of iron stress has been studied by using both *N. meningitidis* and *N. gonorrhoeae*. The operons in both species are preceded by putative promoters containing sequences with significant similarity to Fur-binding sites. The Fur (ferric uptake regulator) protein is a global regulator that represses gene transcription in the presence of large internal pools of ferrous iron (see chapter 13). The neisserial *fur* genes have been characterized, and a point mutation in gonococcal *fur* resulted in decreased iron-dependent repression of the Tbp proteins. While the Fur-binding site has not been genetically characterized in the *tbp* locus, it is clear that Fur mediates repression of these genes. To date, iron regulation is the only well-characterized mechanism that influences expression of the transferrin receptor components. Decreased external pH appears to negatively impact the expression of the Tbp proteins; however, the mechanism by which it does so is unclear. Addition of ligand has no effect on Tbp expression, unlike the situation with some siderophore transporters, whose expression is increased in the presence of substrate (see chapter 20). Unlike many other genes in the chromosomes of the neisseriae, there is no evidence that the *tbp* genes are "contingency loci."

UTILIZATION OF LACTOFERRIN AS AN IRON SOURCE

The Lactoferrin LbpBA Receptor Is Similar to the Transferrin Receptor

Gonococci and meningococci express a receptor for binding lactoferrin that is, in many respects, very similar to the transferrin receptor.

The major difference from the transferrin receptor, which is uniformly present in gonococci and meningococci, is that only about one-half of gonococci are able to use lactoferrin as a source of iron, although all meningococci can use this iron source. Inability to use lactoferrin as an iron source is due to loss of expression of the receptor. Loss by some naturally occurring gonococcal isolates of the receptor is surprising, since lactoferrin is abundant on mucosal sites and might be presumed to be the first and most important source of essential iron for mucosal pathogens.

The lactoferrin receptor is composed of a 98-kDa integral membrane protein, LbpA, and an 80-kDa putative lipoprotein, LbpB. The predicted amino acid sequence of LbpA is 46% identical to that of TbpA in gonococcal strain FA19, and is 32% identical to that of FA19 TbpB. LbpB proteins differ from TbpB proteins in the presence of one or two strikingly acidic domains in the C-terminal half of LbpB and in a 21-amino-acid C-terminal extension that is absent in TbpB. LbpB proteins in meningococci and gonococci are 81% identical, but the acidic domains are only 51% identical. The structure and function of the lactoferrin receptor have not been studied as extensively as in the transferrin receptor. Only human lactoferrin binds to the receptor.

It is known that LbpA is essential for the function of the lactoferrin receptor, since no in vitro or naturally occurring isolate that does not express LbpA is able to bind lactoferrin or to grow on lactoferrin as a sole iron source. The ability to specifically bind lactoferrin, and not transferrin, must be due to variation in one or more of the exposed loops of LbpA, but little work has been done to elucidate the nature of the lactoferrin-binding domains in the receptor. Receptor binding is complex, with a high-affinity (K_d, 5 nM) and a low-affinity (K_d, 45 to 550 nM) site; LbpA is essential to high-affinity binding. There is a typical TonB box in LbpA, and the function of the receptor in iron uptake from lactoferrin is dependent on TonB. Presumably, LbpA serves as a gated porin for the entry of iron released from lactoferrin at the cell surface. Since iron entry is dependent on the periplasmic transporter FbpA, just as for iron released from transferrin, the pathways for iron entry into the cytoplasm appear to be identical for transferrin and lactoferrin (Fig. 1). The mechanism of release of iron from the ligand after lactoferrin is bound to the receptor is unknown; it presumably does not involve local changes in pH, since lactoferrin holds onto iron above pH 4.5. There is no evidence that lactoferrin is cleaved during binding to the receptor, and one is left with the possibility that unspecified conformational changes attendant to binding are important to iron release from lactoferrin.

LbpB differs from TbpB primarily in the runs of acidic residues in LbpB, which are rich in aspartic and glutamic acids. At first glance, these acidic domains might be expected to help bind lactoferrin, which differs from transferrin in its very basic N terminus. In fact, the evidence for a binding role of LbpB is equivocal at best. Strains expressing only LbpA grow on lactoferrin normally, although they bind about 40% as much lactoferrin and take up about 40% as much iron from lactoferrin as isolates that express both LbpB and LbpA. This may not be due to loss of LbpB function, however, since strains that lack LbpB (either by deletion of *lbpB* or by insertion of a nonpolar *aphA-3* cassette in *lbpB*) express only about 40% as much LbpA in semiquantitative Western blots. Thus, the defect in iron uptake and binding in strains lacking LbpB could be due to reduced expression of LbpA. Reduction in LbpA expression in strains with a deletion in *lbpB* is not due to decreased transcription of *lbpA*, based on levels of PhoA expressed in *lbpA phoA* transcriptional fusions. Reasons for the reduced synthesis of LbpA in strains lacking LbpB are not known. There is no evidence that LbpB specifically affects the binding of ferrated lactoferrin, unlike TbpB, which recognizes ferrated transferrin preferentially. Thus, despite extensive effort, it is not clear that LbpB plays any role in binding and iron uptake. LbpB can be purified from lactoferrin affinity columns in the absence of LbpA, which is the only direct evidence for a functional role for LbpB in the receptor.

Genetics and Regulation of the Lactoferrin Receptor

The *lbpA* and *lbpB* genes are arranged in tandem, in the order *lbpB lbpA*, and are presumed to be an operon since insertion of the omega element into *lbpB* interrupts the expression of both *lbpB* and *lbpA*. A single promoter with a typical Fur box precedes *lbpB*, and, as expected, expression of the lactoferrin receptor is regulated by iron. Unlike the transferrin operon, there is no intervening sequence between *lbpB* and *lbpA* (Fig. 2).

The inability of some gonococci to utilize lactoferrin is due to an inability to express a receptor for binding lactoferrin, a consequence of a variety of mutations in the genes for the receptor (Fig. 3). Among 40 tested recent clinical gonococcal isolates that grew on lactoferrin, all expressed LbpA whereas only 12 also expressed LbpB. Among 47 isolates that did not grow on lactoferrin, none expressed LbpA although 5 expressed LbpB. This confirmed the essential role of LbpA and the apparent lack of role for LbpB, or at least the relative unimportance of LbpB to receptor function. The basis for loss of LbpB included a variety of point mutations in *lbpB* and also apparent frameshift mutations due to slipped-strand mispairing in a polycytidine tract in *lbpB*: each of eight tested isolates expressing LbpB had 7, 10, or 13 C's in *lbpB*, whereas many LbpB-nonexpressing isolates had 8 or 9 C's in this tract.

The most common basis for loss of expression of the entire receptor (loss of both LbpB and LbpA) was a deletion of 2.7 kb that included all of *lbpB* and most of the 5' end of *lbpA*, resulting in isolates that expressed neither LbpB nor LbpA (Fig. 3). Half of the gonococci that did not express a functional lactoferrin receptor had the same *lbpBA* deletion. DNA sequencing showed that the deletion was identical in each of 14 tested strains that made neither LbpB nor LbpA. The junctions of the deletions were identical, as were the upstream and downstream sequences for over 200 bp up-

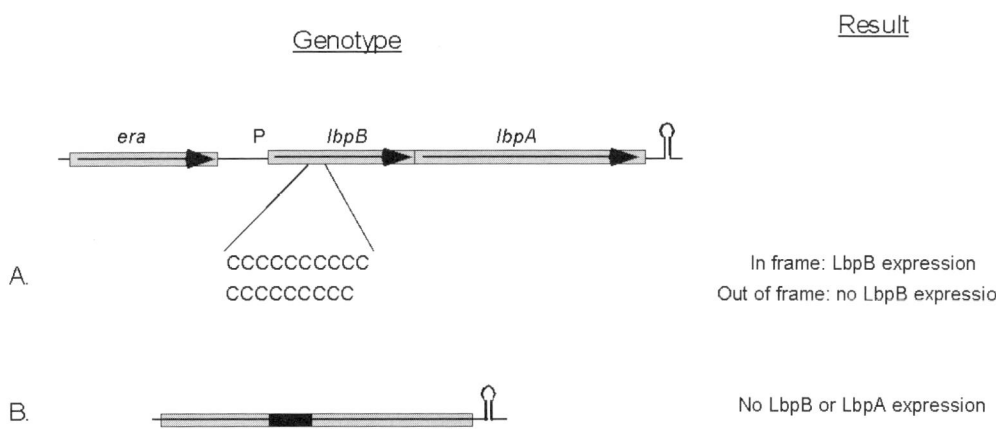

FIGURE 3 Naturally occurring mutations in *lbpB*. Two major classes of *lbpB* mutants have been characterized that result in no expression of the LbpB protein. (A) A polypyrimidine tract consisting of 10 C residues results in expression of the LbpB protein, while a decrease in the number of C residues to 8 or 9 shifts the *lbpB* gene out of frame and prevents LbpB expression. Expression is found with C strings of 7 or 13 residues as well. (B) A deletion of 2.7 kb with a concomitant insertion of 41 residues of a repeated element (black box) results in a gene fusion between the upstream *era* gene, which encodes a GTP-binding protein, and the *lbpA* gene. This fusion event results in no expression of either LbpB or LbpA. Not shown are various point mutations in *lbpB* or *lbpA*, which also result in loss of expression in some isolates. P indicates the approximate position of the iron-regulated promoter that drives the expression of *lbpBA*.

stream and over 900 bp downstream. Each deletion contained an identical 41-bbp insert at the junction point. Although the 41-bp element is a member of the family of Correia repeats found widely in the *Neisseria* genomes, this particular sequence appeared to be rare outside of strains with the *lbpBA* deletion. Isolates carrying this deletion had many different porin antigenic variants (both P.1A and P.1B serovars) and at least four distinct nutritional auxotypes. These data suggest possible transfer in nature of a single deletion between genetically diverse strains. Horizontal spread by in vivo transformation has been invoked to account for variations in other gonococcal genes, and gonococci are highly transformable. The data do not discriminate definitively between the horizontal-spread model and the possibility that an ancestral *lbpBA* deletion might have survived for a very long time, sufficient for accumulation of mutations in biosynthetic and porin genes. The prevalence of an identical deletion in geographically and genetically diverse isolates suggests that there is a selective advantage for isolates carrying a deletion of the lactoferrin receptor genes.

Function of the Lactoferrin Receptor in Humans

A strain FA1090 derivative that expresses neither the transferrin nor the lactoferrin receptor was unable to establish urethral infection in male volunteers, in contrast to the parental strain FA1090, which makes the transferrin receptor but fails to make the lactoferrin receptor due to the prototypical deletion of the *lbp* genes. Another derivative of FA1090, which expresses a fully functional lactoferrin receptor but not a transferrin receptor, was able to infect volunteers and had an apparent 50% infective dose similar to that of FA1090. This proved that there is sufficient lactoferrin on the uninflamed male urethral mucosa to support growth and that the lactoferrin receptor is functional in vivo. To determine whether there might be a selective advantage in vivo for the strain that made both the transferrin and lactoferrin receptors, male volunteers were challenged with an approximate 50:50 mixture of isogenic strains that expressed either both receptors or only the transferrin receptor, at a dose near the 50% infective dose. By the time the experiment was terminated, 100% of colonies from each of the five volunteers who were infected expressed both receptors. In three volunteers, the recovered organisms expressed both receptors from the first positive culture, but in two volunteers there was a gradual increase in the proportion of recovered bacteria that expressed both receptors. This strongly suggested that infection was initiated not by a single organism but by both input strains and that there was a selective advantage in expressing both receptors.

The question then remains: if there is a selective advantage for gonococci that express both receptors, at least in the male urethra, why do not all gonococci express both receptors? Is there a functional disadvantage to expressing the lactoferrin receptor in some circumstances? There might be immunological or other pressures against the lactoferrin receptor, either in more prolonged infections or in other ecological sites such as the female genital tract. Since the proportion of isolates that make the lactoferrin receptor are identical between men and women there is little evidence of a relative disadvantage to expression of the receptor in women. LbpB is predicted to be strongly antigenic based on the long runs of acidic residues in the protein. If the lactoferrin receptor is not as crucial to the biology of infection as the transferrin receptor, as appears to be the case, immune responses might result in preferential selection for isolates that do not make the lactoferrin receptor proteins. There could also be other pressures against the receptor, such as generation of oxygen-derived toxic hydroxyl radicals due to binding lactoferrin close to the bacterial cell surface. This is all speculation, however, and there is no answer to the question at the moment. It also is equally unclear why the lactoferrin receptor is universally expressed by all meningococci; this presumably reflects differences in the ecological sites and the relative importance of lactoferrin in the nasopharynx.

UTILIZATION OF HEMOGLOBIN-BOUND HEME AS AN IRON SOURCE

Two Receptors for Hemoglobin: HpuAB and HmbR

All meningococci and gonococci can utilize either free heme or heme bound to hemoglobin as an iron source. Other heme proteins such as the cytochromes are not useful sources of heme iron. Normally, there is little free heme in vivo, since heme is complexed with either hemoglobin or various enzymes, or with the serum carrier proteins hemopexin or albumin. Moreover, although free heme can be used as an iron source, it is toxic, and only narrow ranges of heme concentration support growth of meningococci or gonococci in vitro. Free-heme uptake in vitro requires energy but neither TonB nor FbpA, and there is no documented receptor for free heme. Since gonococci and meningococci cannot use heme complexed to either hemopexin or albumin, it is doubtful that free heme plays a role in the normal physiology of iron uptake for these organisms. The important source of heme for these bacteria is hemoglobin.

Ability to use hemoglobin is due to expression of receptors for hemoglobin, and in the neisseria, there are two distinct receptors that differ in structure and specificity. All gonococci express a receptor designated HpuAB, which is built on the same general model as the transferrin and lactoferrin receptors, and many meningococci express a very similar receptor (Fig. 1). Meningococci also frequently express another, quite different hemoglobin receptor designated HmbR, composed of a single 89-kDa protein. In gonococci, *hmbR* is present only as an unexpressed pseudogene. Many meningococci produce both HpuAB and HmbR. Hemoglobin normally is intracellular, and when released from dying red blood cells it is bound in vivo to the carrier protein haptoglobin. The HpuAB receptor binds hemoglobin with or without haptoglobin, as well as free haptoglobin. In contrast, HmbR is specific for hemoglobin, and neither haptoglobin nor hemoglobin-haptoglobin complexes are bound. The HpuB and HmbR receptors share certain general properties, including dependence on TonB for function. Both evidently function well, since the rates and final extent of growth in vitro on hemoglobin as a sole iron source with either of the hemoglobin receptors are equal to those obtained with transferrin or lactoferrin as sole iron sources.

HpuA/HpuB Receptor

HpuA is a lipoprotein of 36 kDa, and HpuB is an 85-kDa integral outer membrane protein with typical TonB box domains (Fig. 1). (The terminology of the lipoproteins is unfortunate, with the designation A for Hpu and B for Tbp and Lbp.) Mature HpuA has relatively little sequence identity to the lipoproteins LbpB and TbpB (22 and 30%, respectively) and is only about one-half the size of the other two iron receptor lipoproteins. Likewise, mature HpuB is only 36% identical to LbpA and 33% identical to TbpA, and much of the identity is in the TonB box domains. An important contrast to the transferrin and lactoferrin receptors is that both HpuA and HpuB are essential for the uptake of heme from bound hemoglobin; in the other receptors, the lipoprotein either plays a helping role (TbpB) or appears to be without function (LbpB). One explanation for the essential role of both proteins in the HpuAB receptor is that there seems to be a single binding site for hemoglobin on the receptor, made up of HpuA and HpuB together, in contrast to demonstrably independent binding of transferrin to each of TbpA and TbpB. Another difference is that HpuAB binds hemoglobin from many different species equally well whereas the transferrin and lactoferrin receptors are absolutely specific for the human ligand. The binding affinity of HpuAB for hemoglobin is about 150 μM, which is about 10 to 20 times lower than that of the transferrin receptor for its ligand. Heme competes only slightly with hemoglobin for binding to the receptor, suggesting that the receptor recognizes globin rather than heme. It is not known how heme is released from hemoglobin, but it is clear that heme traverses the outer membrane, periplasm, and cytoplasmic membrane intact, since binding hemoglobin to HpuAB provides essential porphyrins to gonococci that cannot synthesize their own porphyrins. TonB-dependent heme transport must involve transit through a pore in HpuB, and there presumably are a periplasmic carrier and cytoplasmic transporters, but their identities are unknown.

Genetics and Regulation of HpuAB

Expression of HpuAB, like that of other iron ligand receptors, is regulated by iron through the transcriptional repressor Fur. In addition, HpuAB (and HmbR) undergo high-frequency phase variation due to slipped-strand mispairing that affects the length of a homopolymeric tract, as with *opa*, *pilC*, and other genes as described above. In *hpuAB*, there is a G tract within *hpuA*, and when it numbers 10, the *hpuAB*, genes are expressed. When there are one or two more or fewer G's, the genes are out of translational frame and neither protein is expressed. Isolates that have been passed in vitro several times rapidly switch to a predominantly receptor "off" phase and do not use hemoglobin or bind hemoglobin. The frequency of phase variation varies in different strains between 10^{-3} and 10^{-6}, in part due to variations in mismatch repair systems that affect the rate of slipped-strand mispairing.

Recently, mutants of *hpuB* were isolated that restored hemoglobin-specific iron uptake in a gonococcal strain with a deletion in *hpuA*, which otherwise renders the strain unable to grow on hemoglobin. These mutations occurred in several sites in *hpuB*, probably within the putative transmembrane barrel. Although they did not increase the binding of hemoglobin, which remained less than that of the wild type, growth on hemoglobin was normal. Perhaps these mutations change the conformation of HpuB to one that would normally result only from interaction with HpuA, either enhancing the release of heme from hemoglobin and/or enhancing the entry of heme through the gated pore within HpuB. Curiously, uptake of heme from bound hemoglobin was totally abolished in the mutants when serum albumin was added; normally, the HpuAB receptor is not affected by addition of albumin, which binds free heme. Introduction of wild-type *hpuA* restored resistance to serum albumin. Thus, HpuA apparently protects heme from binding by serum proteins after it is released from hemoglobin, in addition to assisting in the formation of a binding site for hemoglobin. Gain-of-function mutants such as these undoubtedly will prove highly useful to our understanding of structure-function relationships once a crystal structure for HpuAB is available.

Function of HpuAB in Humans

The function of the HpuAB receptor in humans is unclear. Human challenge experiments have all been done with HpuAB "off" variants, and there is no evidence in experiments with over 30 tested male volunteers for selection in vivo of receptor "on" switch variants. However, there may be a functional role for the receptor in women. Isolates were collected from women in a clinic, and expression of HpuAB in vitro after two or at most three passes in the laboratory was correlated with the stage of the menstrual cycle in the women at the time of culture. Although the majority of isolates were in the receptor "off" phase, were significantly more were in the "on" phase during the first 2 weeks of the cycle than in the last half of the cycle. The inference was that the availability of hemoglobin from menses provided a selective advantage to isolates that expressed HpuAB.

SIDEROPHORE-MEDIATED IRON UPTAKE

Neisseria Species Do Not Produce Siderophores

Many bacteria produce siderophores, as discussed elsewhere in this book. Evidence is strong that neither gonococci nor meningococci make siderophores, based on our inability to detect them by a variety of standard assays and, more compellingly, by their inability to grow on media that contain ferric iron but lack simple iron solubilizers such as citrate. The addition of hemoglobin, transferrin, or lactoferrin to such media supports excellent growth, and thus gonococci do not produce extracellular compounds that solubilize ferric iron and make it available to the organism, the basic definition of a siderophore. Moreover, genomic sequences show no evidence of functional siderophore biosynthesis genes. Early studies showing that gonococci and meningococci produce hydroxamate siderophores were flawed by contamination with small amounts of hydroxamates in crude growth media.

FetA Is a Functional Siderophore Receptor

Meningococci and gonococci express a 76-kDa Fur-regulated outer membrane protein that formerly was designated FrpB but now has been designated FetA in recognition of its function as a phenolate siderophore receptor. The gonococcal version of the protein lacks an exposed loop that is present in meningococcal FetA, and there are other variations in surface-exposed loops, but the protein otherwise is well conserved. Early studies failed to discern a function for this protein, which was expected to be a receptor based on unambiguous homology to the family of TonB-dependent receptors. However, a small domain of the protein is similar to the binding domain of the *E. coli* enterobactin receptor, and a monoclonal antibody specific for the enterobactin receptor cross-reacts with the neisserial protein. On this basis, despite a lack of overall similarity to the enterobactin receptor FepA, consideration was given to the possibility that "FrpB" might be a siderophore receptor. Subsequently, specific binding of enterobactin to gonococci was demonstrated, and enterobactin supported growth. Immediately downstream of *fetA* is *fetB*, which encodes a protein in the family of periplasmic transporters, and also *fetCD*, which encode proteins related to ATP-dependent cytoplasmic membrane transporters. All are linked in a single operon. Polar insertion mutations in *fetA* abolish enterobactin binding and enterobactin-dependent growth, as well as expression of the downstream genes. Insertional mutagenesis of *fetB* does not affect enterobactin binding but abolishes growth. Mutation of *fetCD* also prevents growth when enterobactin is the sole iron source. The binding affinity of FetA for enterobactin is relatively low (K_d, about 5 μM), which is about a 1,000-fold lower than that of the transferrin receptor for transferrin and 30-fold lower than that of HpuAB for hemoglobin. This suggests that another phenolate siderophore produced by bacteria in the gastrointestinal tract or elsewhere might be the normal ligand for FetA, although there is no direct evidence for this at present. It also is possible that FetA has other, undetected functions in pathogenesis. There is no doubt, however, that FetA is a functional phenolate siderophore receptor.

Expression of *fetA* is regulated not only by Fur but also by high-frequency slipped-strand mispairing (a familiar theme) in a run of cytidines within the *fetA* promoter, which affects promoter strength and the amount of FetA produced. Thus, as in the case of both HpuAB and LbpB, FetA expression has two components: iron- and Fur-dependent repression of transcription and phase variation due to variable lengths of hompolymeric tracts. When the homopolymeric tracts occur in a structural gene (*hpuA* or *lbpB*), phase variation results in either strong expression or no expression due to translational frameshifting. When the tract occurs in a promoter (*fetA*), phase variation affects the amount of protein expressed. Earlier estimations that FetA was produced in many, but not all, meningococci and gonococci were flawed by a failure to account for phase variation, and it is likely that all pathogenic *Neisseria* species produce FetA. It does not seem to be important in experimental infection of male volunteers; since input (challenge) strains are always in the "low-expression" phase and remain so in the strains reisolated from the volunteers. If there were an important function in the male urethra, one would expect selection for the "high-expression" FetA phase in vivo.

OTHER TonB-DEPENDENT IRON UPTAKE SYSTEMS

Inspection of neisserial genomic sequences shows, in addition to the known receptors, several open reading frames that encode putative proteins with sequence similarity to members of the TonB-dependent family. Three of these have been studied: one was not expressed, but the other two were expressed in the outer membranes of either occasional gonococci or the majority of meningococci and gonococci. Insertional mutagenesis failed to reveal any defects in iron uptake. This does not mean that they do not have functions, merely that none was detected. Indeed, iron-dependent intraepithelial cell growth is impeded by mutation of *tonB* but not by mutation in the transferrin, lactoferrin, or hemoglobin receptors and not by mutations in the heme biosynthetic pathway. This indicates that TonB is crucial to iron scavenging inside cells and strongly suggests that one or more of the known or predicted outer membrane proteins discovered by genomic sequencing function inside epithelial cells to scavenge iron from an unknown source. This also points out clearly that although we have learned much, a lot remains to be discovered.

REDUNDANCY IN IRON UPTAKE SYSTEMS IN THE PATHOGENIC *NEISSERIA* SPECIES: IMPLICATIONS AND FUTURE DIRECTIONS

One of the curiosities in most bacteria is the apparent redundancy in their systems for accomplishing crucial steps in infection. For example, there are multiple adherence ligands including Pili, Opa, LOS, and probably others, as well as multiple host receptors for binding these adherence ligands. Resistance to complement is achieved by capsules or sialylation of LOS, by domains on PorB that bind complement regulatory proteins such as C4b and factor H, and by immunogens such as reduction-modifiable protein (RMP), which induce the formation of non-complement-fixing antibodies that protect other surface antigens from attack by antibody and complement. Redundancy in these functions argues that they are highly important to the organism, either in escaping certain host defenses or in enabling infection at different niches or at different times in vivo. Iron-scavenging systems are no exception. There are at least five independent receptor-ligand systems, one for transferrin, one for lactoferrin, two for hemoglobin, and one for phenolate siderophores produced by other bacteria. There apparently is a sixth for iron scavenging inside epithelial cells. Why are there so many receptors? Why do four of them (LbpB, HpuAB, HmbR, and FetA) undergo phase variation in addition to Fur-dependent regulation? Can they function simultaneously, or do they operate independently in vivo?

It is difficult to answer these questions by performing experiments with animal models, because no model closely approximates human biology. Neither the transferrin nor the HpuAB receptor is important to the pathogenesis of infection in the mouse female genital tract. Much more relevant are observations of natural human infection, which suggest at least three things: the transferrin receptor plays an important role, because it is expressed on 100% of gonococci and meningococci and, unlike the other iron receptors, does not undergo phase variation; the lactoferrin receptor is less important overall, because 50% of isolates that compete successfully in nature do not make the receptor; and the HpuAB receptor is important in women, because it is in the "on" phase more often early in the menstrual cycle, when its ligand is most abundant.

Since menses also provides serum transferrin to the genital mucosa, the latter observation argues that gonococci use both transferrin and hemoglobin simultaneously and have at least a slight growth advantage when both receptors are functionally engaged. The same general conclusion, i.e., that expression of both the transferrin and lactoferrin receptors enables a strain to outcompete its isogenic sibling that expresses only the transferrin receptor, is supported by evidence from experiments with male volunteers. This should not be surprising, since one presumes that nature is conservative and there would not be redundancy in expression of iron receptors unless there were a selective advantage for doing so. Use of a *fetA*-PhoA reporter proved that growth on transferrin only partially represses *fetA* expression, and thus the phenolate receptor is available even when growth rates are maximal on another iron source. Other experiments showed that growth on transferrin had minimal effects on expression of the lactoferrin receptor. Thus, provision of iron on natural ligands such as transferrin does not saturate Fur, and other receptors can be expressed at the same time.

It remains to be shown whether expression of particular iron receptors is crucial at different anatomical and physiological sites in the host, although the available data suggest that there is a unique receptor for intracellular growth. It will be interesting to discern the functional roles played by the receptors, keeping in mind that receptors in other bacteria sometimes have more than one function. Up to this point, attention on the iron receptors in the neisseriae has understandably focused exclusively on iron transport, but there could be functions in adherence or invasion or other properties in addition to iron uptake. It also will be interesting to try to better understand structure-function relationships in the various receptors, which should be assisted by determinations of three-dimensional structures based on X-ray crystallography. Finally, there are obvious opportunities for development of vaccines against the iron receptors. The transferrin receptor is being pursued as a potential vaccine for meningococcal and gonococcal diseases, and others such as FetA or HpuAB may have applied uses of this sort as well.

ACKNOWLEDGMENTS

Our unpublished experiments were supported by grants AI31496 and AI26837 to P.F.S. and AI147141 and AI39523 to C.N.C. from the National Institutes of Health.

SUGGESTED READING

Adhikari, P., S. A. Berish, A. J. Nowalk, K. L. Veraldi, S. A. Morse, and T. A. Mietzner. 1996. The *fbpABC* locus of *Neisseria gonorrhoeae* functions in the periplasm-to-cytosol transport of iron. *J. Bacteriol.* **178:**2145–2149.

Anderson, J. E., M. M. Hobbs, G. D. Biswas, and P. F. Sparling. 2003. Opposing selective forces for expression of the gonococcal lactoferrin receptor. *Mol. Microbiol.* **48:**1325–1337.

Bayliss, C. D., D. Field, and E. R. Moxon. 2001. The simple contingency loci of *Haemophilus influenzae* and *Neisseria meningitidis*. *J. Clin. Investig.* **107:**657–662.

Boulton, I. C., M. K. Yost, J. E. Anderson, and C. N. Cornelissen. 2000. Identification of discrete domains within gonococcal transferrin-binding protein A that are necessary for ligand binding and iron uptake functions. *Infect. Immun.* **68:**6988–6996.

Carson, S. D., B. Stone, M. Beucher, J. Fu, and P. F. Sparling. 2000. Phase variation of the gonococcal siderophore receptor FetA. *Mol. Microbiol.* **36:**585–593.

Carson, S. D. B., P. E. Klebba, S. M. C. Newton, and P. F. Sparling. 1999. Ferric enterobactin binding and utilization by *Neisseria gonorrhoeae*. *J. Bacteriol.* **181:**2895–2901.

Chen, C. J., D. Mclean, C. E. Thomas, J. E. Anderson, and P. F. Sparling. 2002. Point mutations in HpuB enable gonococcal HpuA deletion mutants to grow on hemoglobin. *J. Bacteriol.* **184:**420–426.

Cohen, M. S., and J. G. Cannon. 1999. Human experimentation with *Neisseria gonorrhoeae:* progress and goals. *J. Infect. Dis.* **179:**S375–S379.

Cornelissen, C. N., J. E. Anderson, and P. F. Sparling. 1997. Energy-dependent changes in the gonococcal transferrin receptor. *Mol. Microbiol.* **26:**25–35.

Cornelissen, C. N., M. Kelley, M. M. Hobbs, J. E. Anderson, J. G. Cannon, M. S. Cohen, and P. F. Sparling. 1998. The transferrin receptor expressed by gonococcal strain FA1090 is required for the experimental infection of human male volunteers. *Mol. Microbiol.* **27:**611–616.

Dehio, C., S. D. Gray-Owen, and T. F. Meyer. 1998. The role of neisserial Opa proteins in interactions with host cells. *Trends Microbiol.* **6:**489–495.

Jerse, A. E., E. T. Crow, A. N. Bordner, I. Rahman, C. N. Cornelissen, T. R. Moench, and K. Mehrazar. 2002. Growth of *Neisseria gonorrhoeae* in the female mouse genital tract does not require the gonococcal transferrin or hemoglobin receptors and may be enhanced by commensal lactobacilli. *Infect. Immun.* **70:**2549–2558.

Kenney, C. D., and C. N. Cornelissen. 2002. Demonstration and characterization of a specific interaction between gonococcal transferrin binding protein A and TonB. *J. Bacteriol.* **184:**6138–6145.

Klee, S. R., X. Nassif, B. Kusecek, P. Merker, J.-L. Beretti, M. Achtman, and C. R. Tinsley. 2000. Molecular and biological analysis of eight genetic islands that distinguish *Neisseria meningitidis* from the closely related pathogen *Neisseria gonorrhoeae. Infect. Immun.* **68:**2082–2095.

Larson, J. A., D. L. Higashi, I. Stojiljkovic, and M. So. 2002. Replication of *Neisseria meningitidis* within epithelial cells requires TonB-dependent acquitision of host cell iron. *Infect. Immun.* **70:**1461–1467.

Lewis, L. A., E. Gray, Y.-P. Wang, B. A. Roe, and D. W. Dyer. 1997. Molecular characterization of *hpuAB*, the haemoglobin-haptoglobin-utilization operon of *Neisseria meningitidis. Mol. Microbiol.* **23:**737–749.

Masri, H. P., and C. N. Cornelissen. 2002. Specific ligand binding attributable to individual epitopes of gonococcal transferrin binding protein A. *Infect. Immun.* **70:**732–740.

Merz, A. J., and M. So. 2000. Interactions of pathogenic neisseriae with epithelial cell membranes. *Annu. Rev. Cell Dev. Biol.* **16:**423–457.

Nassif, X., C. Pujol, P. Morand, and E. Eugene. 1999. Interactions of pathogenic *Neisseria* with host cells. Is it possible to assemble the puzzle? *Mol. Microbiol.* **32:**1124–1132.

Ram, S., M. Cullinane, A. M. Blom, S. Gulati, D. P. McQuillen, B. G. Monks, C. O'Connell, R. Broden, C. Elkins, M. D. Pangburn, B. Dahlback, and P. A. Rice. 2001. Binding of C4b-binding protein to porin: a molecular mechanism of serum resistance of *Neisseria gonorrhoeae. J. Exp. Med.* **193:**281–295.

Ronpirin, C., A. E. Jerse, and C. N. Cornelissen. 2001. The gonococcal genes encoding transferrin binding proteins (Tbp) A and B are arranged in a bicistronic operon but are subject to differential expression. *Infect. Immun.* **69:**6336–6347.

Schryvers, A. B., and I. Stojiljkovic. 1999. Iron acquisition systems in the pathogenic *Neisseria. Mol. Microbiol.* **32:**1117–1123.

Turner, P. C., C. E. Thomas, C. Elkins, S. Clary, and P. F. Sparling. 1998. *Neisseria gonorrhoeae* heme biosynthetic mutants utilize heme and hemoglobin as heme sources but fail to grow within epithelial cells. *Infect. Immun.* **66:**5215–5223.

HAEMOPHILUS

Daniel J. Morton and Terrence L. Stull

18

The genus *Haemophilus* consists of several species of small pleomorphic gram-negative organisms, isolated as pathogens or commensals of a single host species. The haemophili are aerobic or facultatively anaerobic and have a growth requirement for one or both of two growth factors in blood, the X factor (protoporphyrin IX [PPIX] or heme) and the V factor (NAD or NADP). (Note that heme is correctly defined as ferrous PPIX while hemin is ferric PPIX; however, for the purposes of this chapter, "heme" is used as a general term and does not indicate a particular iron valence state.) The type species, *Haemophilus influenzae*, is the most significant human pathogen in the genus, causing meningitis, otitis media, and epiglottitis in children and pneumonia in adults with preexisting conditions such as chronic obstructive pulmonary disease or AIDS. Meningitis caused by *H. influenzae* has been typically associated with type b capsular strains and has been largely eradicated in the developed world following the introduction of vaccines based on the type b capsule. *H. influenzae* requires both the X and V factors for aerobic growth in vitro. Most strains reportedly do not require heme for anaerobic growth, although there is an apparent requirement for an iron source under these growth conditions. A second species requiring both the X and V factors, *H. haemolyticus*, is not a significant pathogen. *H. ducreyi*, the etiologic agent of the sexually transmitted disease chancroid, requires only the X factor, as do *H. aphrophilus* (X factor required for primary isolation only), an infrequent cause of endocarditis and brain abscesses, and the canine commensal *H. haemoglobinophilus*. Other species require only the V factor. These include infrequent causes of human disease such as *H. parainfluenzae*, *H. parahaemolyticus*, and *H. paraphrophilus*, as well as species isolated from other mammalian and avian hosts, including the pathogens *H. parasuis*, *H. somnus*, and *H. paragallinarum*.

HEME AND IRON REQUIREMENT AND ACQUISITION

Haemophili are each isolated from only one host species. Thus, the iron and heme acquisition systems of haemophili are adapted to the host species and may be important pathogenic factors. Studies of iron acquisition in the haemophili have been performed primarily

---

Daniel J. Morton, Department of Pediatrics, University of Oklahoma Health Sciences Center, Oklahoma City, OK 73104. *Terrence L. Stull*, Department of Pediatrics and Department of Microbiology/Immunology, University of Oklahoma Health Sciences Center, Oklahoma City, OK 73104.

Iron Transport in Bacteria, Edited by Jorge H. Crosa, Alexandra R. Mey, and Shelley M. Payne
© 2004 ASM Press, Washington, D.C.

with *H. influenzae*; therefore, this chapter focuses on *H. influenzae*, with reference to other species as appropriate.

A heme source appears to be essential only for aerobic growth of *H. influenzae*. Typically, the heme requirement for unrestricted growth of *H. influenzae* for aerobic growth on complex media is 2 to 10 μg of heme/ml. Anaerobic growth on heme-free media is the norm, although one strain reportedly failed to thrive following repeated anaerobic subculture on a heme-depleted medium. The majority of studies regarding the heme dependence of *H. influenzae* under anaerobic growth conditions have been performed with complex media, raising the possibility that residual heme is sufficient to support growth. We have successfully subcultured four strains of *H. influenzae* anaerobically for up to 20 generations on a chemically defined medium without heme, with no effect on their ability to grow.

The heme requirement of *H. influenzae* results from the lack of all enzymes in the biosynthetic pathway of PPIX. Most strains possess ferrochelatase, the enzyme that catalyzes the insertion of iron into PPIX to form heme as well as the reverse reaction, and will grow if provided with PPIX and an iron source. However, certain strains have an absolute growth requirement for an exogenous supply of heme in vitro and may lack ferrochelatase. About 4 to 8% of strains fail to grow when supplied with PPIX plus an iron source; however, lack of a ferrochelatase gene or ferrochelatase activity has not been directly demonstrated. Failure of one strain to grow on PPIX plus iron was overcome by supplying the periplasmic iron transport system (*fbpABC* [see below]) in *trans*, indicating that the strain possessed ferrochelatase but was compromised in elemental iron acquisition. In general, the *H. influenzae* in vitro requirement for a porphyrin source can be satisfied by a heme source or PPIX plus a utilizable iron source, except in some strains possibly lacking ferrochelatase.

Certain strains of *H. influenzae* exhibit regulated or selectable expression of the ability to utilize PPIX plus iron for aerobic growth in vitro. Strains that initially failed to grow on PPIX and iron grew after 24 to 48 h; single-colony isolates from these turbid cultures grew well on PPIX plus iron. An analogous situation may occur in other haemophili. *H. aphrophilus* exhibits a heme requirement for initial isolation despite possessing the enzymes of the porphyrin biosynthesis pathway. The mechanism underlying this initial heme requirement and subsequent adaptation is unknown. *H. ducreyi* may require as much as 50 μg of heme per ml for growth. The basis for the requirement for high levels of heme has not been established but may involve the lack of a functional high-affinity heme acquisition system (see below).

Heme in the human host is intracellular in the form of hemoglobin and heme-containing enzymes such as catalase. Free hemoglobin, released by the lysis of erythrocytes, is rapidly bound by the serum protein haptoglobin ($K_d \sim 10^{-15}$), and the hemoglobin-haptoglobin complex is cleared from the circulation by the reticuloendothelial cells of the liver, bone marrow, and spleen. Free heme, derived principally from the degradation of methemoglobin, is bound by the serum proteins hemopexin ($K_d \sim 10^{-13}$) and albumin ($K_d \sim 10^{-8}$) and cleared from the circulation. *H. influenzae* utilizes hemoglobin as well as complexes of hemoglobin-haptoglobin, heme-albumin, and heme-hemopexin as heme sources for in vitro growth. Other heme-containing proteins are potential sources of heme in vivo; however, in vitro studies characterizing the utilization of heme-containing enzymes have been limited. For example, one isolate failed to grow significantly with catalase as the sole heme source, although in a second study catalase was used more efficiently than hemoglobin by two strains. Lactoperoxidase and cytochrome *c* failed to support growth, whereas myoglobin is as efficient as hemoglobin in supporting growth. The relevance of either catalase or myoglobin as a potential heme source in vivo is unclear.

Free iron, like heme, is not readily available in vivo; its availability is limited by binding to transport or storage proteins. Iron transport mechanisms are required for normal mammalian physiology, since the soluble ferrous form is oxidized to the ferric state that forms insoluble polymers in aqueous solution at neutral pH. Most iron in the body is intracellular as heme or ferritin. Iron in extracellular fluids and on mucosal surfaces is complexed by the iron-binding proteins transferrin and lactoferrin ($K_d \sim 10^{-22}$). Transferrin is bound by *H. influenzae* at the cell surface and functions as an iron source. Only a small number of strains are reported to bind and utilize lactoferrin as an iron source in vitro.

H. influenzae does not produce siderophores. The only *Haemophilus* species reported to produce these low-molecular-weight iron chelators are the heme-independent species *H. parainfluenzae*, *H. paraphrophilus*, and the procine pathogen *H. parasuis*.

A more detailed discussion of mechanisms mediating the utilization of the potential in vivo heme and/or iron sources by *H. influenzae* follows. Table 1 summarizes our understanding of the various proteins and/or genes that have been shown to be or are potentially involved in the acquisition of heme and/or iron, and Table 2 describes the results of studies of the roles of these proteins in pathogenicity.

HEME SOURCES

Unbound Heme

Several proteins have been implicated in the acquisition of heme in vitro. The protein HxuC, encoded by the first gene in the *hxuCBA* gene cluster, is essential for the utilization of heme when it is present at low levels (10 μg/ml) in vitro. HxuC is an approximately 80-kDa outer membrane protein initially identified as part of an operon encoding a heme-hemopexin utilization system; it is highly conserved at the amino acid level among *H. influenzae* strains (Table 1). HxuC possesses sequences characteristic of a TonB-dependent protein. In addition to their defect in growth on low levels of heme, mutants with mutations in *hxuC* are unable to utilize the heme-hemopexin complex, the heme-albumin complex, and hemoglobin in a background where other hemoglobin acquisition pathways are mutated.

These growth defects do not result from downstream effects on *hxuB* and *hxuA* since, in all cases except the heme-hemopexin complex, a *hxuBA* deletion mutant grows as well as the wild-type strain with each heme source. The effect of the *hxuC* mutation on the range of potential heme sources suggests that it is an essential component in the translocation of heme across the outer membrane when the heme concentration is low and that it functions in concert with specific receptors when heme is protein bound. Acquisition of heme at high concentrations in vitro is independent of HxuC. The presence in HxuC of sequences characteristic of TonB-dependent proteins indicates energy-dependent translocation of a substrate across the membrane and supports a proposed role in heme transport. We have recently shown that a type b strain harboring an *hxuC* insertion mutation is unaffected in the ability to establish infection or in the duration of infection in the infant-rat model but exhibits a significantly lower level of bacteremia than the wild-type strain does.

A second outer membrane protein implicated in heme acquisition is lipoprotein e(P4), the product of the *hel* gene. The potential role of this protein in heme acquisition was initially identified by transformation of an *H. influenzae* genomic plasmid library into an *Escherichia coli hemA* mutant background. The *hemA* mutation renders *E. coli* unable to synthesize heme unless the organism is supplied with exogenous aminolevulenic acid. In addition, the *E. coli* K-12 outer membrane is impermeable to heme or PPIX. Lipoprotein e(P4) was identified as complementing the *E. coli hemA* mutant for growth on heme. An *H. influenzae hel* mutant was unable to utilize heme, PPIX, or hemoglobin under aerobic growth conditions. Under anaerobic growth conditions, the mutants grew as well as the wild-type strain on each of these heme sources, probably since heme is unnecessary for anaerobic growth. Lipoprotein e(P4) was later shown to be required for the utilization of NAD and NADP, and the growth deficiencies of *hel* mutants described above were attributed to failure to utilize NAD

TABLE 1 Heme and iron acquisition related proteins of *H. influenzae*

Gene	Protein	Calculated molecular mass, kDa[a] (no. of strains)	Conservation (% identity)[b] (no. of strains)	Function	Identification method[c]	Localization[d] (TonB-dependent protein[e])
hxuA	HxuA	99–102 (7)	79.7–86.6 (6)	Heme-hemopexin binding protein	Genetic, biochemical	Secreted and OM (N)
hxuB	HxuB	63 (6)	94.9–98.1 (5)	Activator/secretor of HxuA	Genetic, biochemical	OM (N)
hxuC	HxuC	79–81 (7)	89.8–94.8 (6)	Transport of heme across the OM	Genetic	OM (Y)
hel	e(P4)	30–31 (2)	97.1 (1)	Transport of heme across the OM	Genetic	OM (N)
hbpA	HbpA	61 (2)	98.9 (1)	Transport of heme across the PS or CM?	Genetic, biochemical	PS or IM
tdhA (HI0113)	TdhA	85 (Hi)[f] (1) 84 (Hd)[f] (1)	92.8 (Hi) (1) 57.5 (Hd)	Potential OM heme-binding protein	Genetic, biochemical	OM (Y)
hgp	Hgp	113–124 (12)	44.8–91.9 (11)	Hemoglobin, hemoglobin/haptoglobin-binding protein	Genetic, biochemical	OM (Y)
HI1217	Hup	103 (2)	89.8 (1)	Hemoglobin-binding protein/heme utilization protein	Genetic, biochemical	OM (Y)
fbpA	FbpA	36 (2)	98.2 (1)	Periplasmic iron-binding protein	Genetic	PS
fbpB	FbpB	57 (4)	98.4–98.8 (3)	Inner membrane permease (transport of iron)	Genetic	IM
fbpC	FbpC	40 (4)	98 (3)	Nucleotide-binding protein; provides energy to FbpB for transport of iron across CM	Genetic	IM
tbpA	TbpA	103 (5)	92.9–94.9 (5)	Transferrin-binding protein (complex with TbpB)	Genetic, biochemical	OM (Y)
tbpB	TbpB	69–73 (10)	66.4–75.4 (9)	Transferrin-binding protein (complex with TbpA)	Genetic, biochemical	OM (N)
tonB	TonB	28–29 (2)	93.3 (1)	Energy transducer; transport of bound iron across the OM (complex with ExbB, ExbD)	Genetic	IM

(Continued)

rather than failure to use heme. However, this fails to explain the ability of a recombinant *E. coli* strain harboring *hel* to utilize heme. Lipoprotein e(P4) was subsequently identified as a phosphomonoesterase, a member of an enzyme superfamily called the DDDD phosphohydrolases (named for two pairs of essential aspartic acid residues in separate domains of the protein and separated by a variable spacer region). Site-directed mutagenesis demonstrated that all four aspartate residues (D64, D66, D181, and D185) are essential for phosphomonoesterase activity. Plasmids carrying e(P4) with mutations in these sites were assessed for the ability to complement the *hemA* mutation in *E. coli*; the D64 and D66 mutations failed to complement the mutant *E. coli*, while the D181 and D185 mutants supported growth. In addition, a putative heme-binding site at amino acids 45 to 50 (KVAFDH) was identified based

TABLE 1 *Continued*

Gene	Protein	Calculated molecular mass, kDa[a] (no. of strains)	Conservation (% identity)[b] (no. of strains)	Function	Identification method[c]	Localization[d] (TonB-dependent protein[e])
exbB	ExbB	17 (3)	96–100 (2)	See TonB	No data for *H. influenzae*	IM
exbD	ExbD	16 (3)	98.6–99.3 (2)	See TonB	No data for *H. influenzae*	IM
yfeA	YfeA	32 (2)	100 (2)	Part of potential inner membrane transport system for iron	Up-regulated under iron restriction	IM

[a] Molecular masses of gene products were calculated prior to any modification of the encoded product. The numbers in the table are ranges for all of the available fully sequenced genes. Numbers in parentheses indicate the number of complete gene sequences available. Sequences used were from Rd KW20 plus those listed below in footnote *b* for each protein.

[b] Values for percent identity were obtained by using the sequence alignment application, AlignX, of the vector NTI Suite v.7. Ranges for percent identity represent values obtained from pairwise alignments between the protein derived from *H. influenzae* Rd KW20 (GenBank accession no. L42023) and each additional sequence available for a given protein. A pairwise alignment was performed for each protein pair, using the applications default settings for the slow (more accurate) method. Numbers in parentheses represent the number of sequences used in pairwise alignments with the Rd KW20 protein. Strains used for comparison of each protein (GenBank accession no. or other source of sequence) are as follows. For HxuA: *H. influenzae* type b (Hib) DL42 (U08348), nontypeable *H. influenzae* (NTHi) N182 (U08349), Hib HI689 (AF536755), Hib E1a (AF545479), NTHi HI1388 (AF536754), NTHi 86–028NP (http://www.microbial-pathogenesis.org), and NTHi HI3224A (http://www.micro-gen.ouhsc.edu). For HxuB: Hib DL42 (U09840), Hib HI689 (AF536755), Hib E1a (AF545479), NTHi HI1388 (AF536754), and NTHi 86–028NP (http://www.microbial-pathogenesis.org). For HxuC: Hib DL42 (U09840), Hib HI689 (AF536755), Hib E1a (AF545479), NTHi HI1388 (AF536754), NTHi 86–028NP (http://www.microbial-pathogenesis.org), and NTHi HI3224A (http://www.micro-gen.ouhsc.edu). For e(P4): NTHi 86–028NP (http://www.microbial-pathogenesis.org). For HbpA: Hib DL42 (M84028). For TdhA: *H. ducreyi* (Hd) strain 35000 (AF052977) and NTHi 86–028NP (http://www.microbial-pathogenesis.org) alignment against 84% of protein sequence available. For Hgp: comparisons were made between HgpB from Rd KW20 and HgpC (HI0712) of Rd KW20, HgpD (HI0635) of Rd KW20, HgpE (HI1566) of Rd KW20, HgpA from Hib HI689 (U51922), HgpB from Hib HI689 (AF022910), HgpC from Hib HI689 (AF094574), HgpB from Hib E1a (AF259266), HgpA from NTHi N182 (AF221059), HgpB from NTHi N182 (AF221060), HgpC from NTHi N182 (AF221059), and HgpA from NTHi TN106 (U43198). For HI1217: NTHi HI3224A (http://www.micro-gen.ouhsc.edu). For FbpA: NTHi TN106 (S72674). For FbpB and FbpC: NTHi TN106 (S72674), NTHi HI3224A (http://www.micro-gen.ouhsc.edu), and NTHi 86–028NP (http://www.microbial-pathogenesis.org). For TbpA: Hib DL63 (U10882), Hib Eagan (U15051), Hib MinnA (U15052), NTHi PAK 12085 (U15053), and NTHi SB33 (U15058). For TbpB: Hib DL63 (U10882), Hib Eagan (U15051), Hib MinnA (U15052), NTHi PAK 12085 (U15053), NTHi SB12 (U15054), NTHi SB29 (U15055), NTHi SB30 (U15056), NTHi SB32 (U15057), and NTHi 86–028NP (http://www.microbial-pathogenesis.org). For TonB: NTHi TN106 (U04996). For ExbB: NTHi TN106 (U04996) and NTHi HI3224A (http://www.micro-gen.ouhsc.edu). For ExbD: NTHi TN106 (U04996) and NTHi 86–028NP (http://www.microbial-pathogenesis.org). For YfeA: NTHi HI3224A (http://www.micro-gen.ouhsc.edu) and NTHi 86–028NP (http://www.microbial-pathogenesis.org).

[c] Functions were defined by construction and analysis of mutants (genetic) or by binding to heme or heme- or iron-containing proteins (biochemical).

[d] Localization: OM, outer membrane; PS, periplasmic space; IM, inner membrane.

[e] Putative outer membrane proteins were examined for sequences characteristic of TonB-dependent proteins; Y, protein contains sequences characteristic of a TonB-dependent protein; N, protein does not contain sequences characteristic of a TonB-dependent protein.

[f] Hi, *H. influenzae*; Hd, *H. ducreyi*.

on homology to other heme-binding proteins. An F48C mutation in the heme-binding site was unable to support heme transport in the *E. coli hemA* mutant and also lacked phosphomonoesterase activity. Insertion of the F48C mutation within the heme-binding site of e(P4) into *H. influenzae* results in a mutant unable to grow aerobically on heme but retaining the ability to grow anaerobically. The precise role of lipoprotein e(P4) in heme acquisition by *H. influenzae* remains to be defined.

A heme-binding lipoprotein, designated HbpA, was initially identified by screening recombinant *E. coli* harboring a type b *H. influenzae* genomic library on heme-containing media and further analyzing *E. coli* strains exhibiting increased pigmentation due to heme binding. An approximately 51-kDa protein, identified in whole-cell lysates of the recombinant *E. coli* strain, was subsequently shown to bind to heme-agarose. The protein is antigenically conserved among heme-dependent

TABLE 2 Roles of heme and iron acquisition proteins in vivo

Gene[a]	Protein[a]	Immunogenicity	Protection in animal models	Pathogenicity of mutants
hxuA	HxuA	Antibodies present in convalescent-phase sera from Hib[b] meningitis patients, but not in acute-phase sera; antibodies in sera from rats immunized with Hib	ND[c]	Mutant is virulent in the infant rat
hxuB	HxuB	ND	ND	Mutant is virulent in the infant rat
hxuC	HxuC	ND	ND	Mutant shows reduced virulence in the infant rat
hel	e(P4)	Antibodies in sera from rabbits immunized with purified e(P4)	Anti-e (P4) antibodies do not protect the infant rat	ND
hgp	Hgp	Antibodies in sera from healthy adults; antibodies in sera from rats immunized with rHgpA	ND	Complete Hgp mutant shows attenuation in 30-day-old rats and the chinchilla model of otitis media (see the text)
HI1217	Hup	ND	ND	Mutant is virulent in the infant rat
tbpA	TbpA	Antibodies in sera from healthy adults; antibodies in sera from rabbits and guinea pigs immunized with rTbpA	Passive transfer of anti-rTbpA not protective in the infant rat	ND
tbpB	TbpB	Antibodies in sera from healthy adults; antibodies in convalescent-phase sera from patients with meningococcal meningitis; antibodies in rabbits and guinea pigs immunized with rTbpB	Passive transfer of anti-rTbp antibodies protects infant rats; immunization with rTbpB enhances clearance from the rat lung	ND
tonB	TonB	ND	ND	Mutant is avirulent in the infant rat

[a] Genes and proteins are as in Table 1.
[b] Hib, *H. influenzae* serotype b.
[c] ND, not determined.

Haemophilus species (excluding *H. haemoglobinophilus*) but not heme-independent species. Sequence analysis demonstrated that HbpA would have a predicted molecular mass of ~61 kDa, significantly higher than the observed mass on sodium dodecyl sulfate-polyacrylamide gel electrophoresis. This discrepancy is presumably due to lipid acylation altering the mobility of the protein. The protein shows significant homology to periplasmic peptide-binding proteins DppA and OppA of *E. coli* and *Salmonella enterica* serovar Typhimurium, respectively. The cellular location of HbpA has not been established, although several observations suggest that it is located in the periplasm and is probably anchored to the inner membrane. This evidence includes the failure of monoclonal antibodies raised against HbpA to react with whole cells of *H. influenzae*, the failure to detect HbpA in cell-free supernatant following osmotic shock, and the homology between HbpA and DppA. It has been reported, with no supporting sequence data, that adjacent to the HbpA gene in *H. influenzae* are four open reading frames (ORFs) encoding proteins of high homology (60 to 74% at the amino acid level) to DppB to DppD, additional proteins encoded by the *E. coli* dipeptide permease operon. The available genomic sequences from *H. influenzae* strains Rd KW20 (GenBank accession no. L42023), nontypeable 86-028NP (http://www.microbial-pathogenesis.org), and nontypeable HI3224A (http://www.micro-gen.ouhsc.edu) do not confirm this report. Immediately downstream of HbpA in both the Rd KW20 and 86–028NP sequences is a protein showing homology (21% identity for the Rd KW20 protein) to HugZ, a component of the *Plesiomonas shigelloides* heme utilization system. The role of HugZ has not been established, although it may be involved in protection against heme toxicity. There has been no further molecular analysis or assessment of mutants with mutations in HbpA. HbpA may be involved in the transport of heme across the inner membrane, although its role has yet to be established definitively.

A putative TonB-dependent outer membrane heme receptor from *H. ducreyi* was cloned by successfully complementing an *E. coli hemA aroA tonB* mutant expressing *H. ducreyi* TonB for growth on heme. The cloned gene was designated *tdhA* (for "TonB-dependent heme receptor") and encodes a protein of approximately 81 kDa. TdhA is homologous to the product of the ORF from the *H. influenzae* Rd KW20 genome sequence designated HI0113. Expression of TdhA in *H. ducreyi* occurred only under conditions of heme-restricted growth. Although it was possible to purify recombinant TdhA from *E. coli* by using heme-agarose affinity, attempts to use this approach to purify the native protein in *H. ducreyi* failed. Importantly, a *tdhA* mutant of *H. ducreyi* was unaltered in its ability to utilize heme, heme-albumin, or hemoglobin. We have similarly been unable to detect a heme-related phenotype in HI0113 mutants of three *H. influenzae* strains. A *tonB* mutant of *H. ducreyi* was also unaffected in its ability to utilize heme sources, except that it could not grow on hemoglobin. *H. ducreyi* requires concentrations of heme in the range of 50 µg/ml to grow. Although *tdhA* may bind heme in *H. ducreyi*, these data indicate that subsequent steps in a high-affinity heme acquisition pathway may be lacking and would be consistent with the high heme requirement for *H. ducreyi* growth.

Heme-Hemopexin Complexes

Utilization of the human heme-hemopexin complex is mediated by the products of the *hxuCBA* gene cluster. HxuA is a 100-kDa secreted protein that binds the heme-hemopexin complex; however, in certain strains, HxuA is associated predominantly with the outer membrane. Whether HxuA is functional in the outer membrane as a heme-hemopexin receptor or whether the presence in the outer membrane represents a transitory step in the secretion is not known. HxuB (an approximately 60-kDa outer membrane protein) is proposed to be involved in the secretion and/or activation of HxuA; HxuB exhibits homology to members of a family of transporters that function to secrete an effector across the outer membrane, including ShlB, the secretor/activator of the ShlA hemolysin of *Serratia marcescens*. After the binding of heme-hemopexin by secreted HxuA, the HxuA-heme-hemopexin complex is presumably bound by the bacterial cell for heme internalization. The outer membrane receptor for the HxuA-heme-hemopexin complex has not been identified, although the possibility remains that the receptor is HxuC. Mutations in *hxuA* and/or *hxuB* abolish the ability of the mutant strain to utilize heme-hemopexin. As mentioned above, mutations in *hxuC* also abolish the ability to utilize heme-hemopexin, presumably as a function of its conjectured role in a heme internalization pathway. An *hxuBA* mutant is not affected in its ability to cause bacteremia in the infant-rat model of invasive disease.

Three additional proteins were isolated by affinity purification with hemopexin from a type b strain of *H. influenzae*. These three proteins had approximate molecular masses of 57, 38, and 29 kDa. The 57- and 29-kDa proteins bound hemopexin after Western blotting. N-terminal amino acid sequencing of the 38-kDa protein identified it as the porin P2. No N-terminal amino acid sequence could be obtained from the 29-kDa protein, suggesting that the N terminus of this protein is blocked and leading other authors to speculate that the 29-kDa protein may be lipoprotein *e*(P4). No additional sequence data or molecular analyses are available for this reported complex of proteins, and their identity or potential role in heme acquisition is yet to be determined.

Heme-Albumin Complexes

HxuC is essential for the utilization of heme-albumin complexes. Four strains were unable to utilize heme-albumin complexes when an insertion mutation was present in *hxuC*. However, all four strains retained the ability to utilize the complex when *hxuC* was intact and *hxuA* and *hxuB* were completely deleted. These data indicate that the acquisition of heme from heme-albumin is distinct from acquisition from heme-hemopexin. Preliminary data from our laboratory indicate that heme-albumin complexes are bound at the *H. influenzae* cell surface and that binding is not affected by mutations in any of the genes of the *hxuCBA* gene cluster.

Hemoglobin

Hemoglobin is bound at the cell surface of *H. influenzae* in a heme-repressible manner. A family of homologous outer membrane proteins mediates hemoglobin utilization. The first identified member of this family was purified from *H. influenzae* type b strain HI689 by using a hemoglobin affinity purification procedure. An N-terminal amino acid sequence was successfully obtained from the purified protein; a degenerate oligonucleotide designed from the amino acid sequence was used to probe an *H. influenzae* HI689 plasmid library, and a clone was isolated. Sequencing of the isolated clone revealed a gene (*hgpA*) encoding a 120-kDa protein. The gene contained a series of 33 CCAA tetranucleotide repeating units immediately following the sequence encoding the leader peptide. These CCAA repeats were later shown to mediate variable expression of the protein through slip-strand mispairing (see below). Insertional mutation of *hgpA* had no effect on binding of hemoglobin by the mutant strain or on growth of the mutant strain with hemoglobin as the sole heme source. Subsequent experiments revealed that strain HI689 contained two additional genes, designated *hgpB* and *hgpC*, encoding proteins homologous to HgpA and similarly containing CCAA repeats. However, insertional mutation of all three *hgp* genes in the same strain had no effect on either hemoglobin binding or hemoglobin utilization. It was hypothesized that the insertion mutations may have left partially functional truncated proteins and/or fusion proteins; consequently, a second group of mutations were constructed to completely replace each gene with an antibiotic resistance cassette. Growth studies using the strain with complete deletions of all three *hgp* genes showed reduced ability to utilize low levels of hemoglobin, although the complete deletion mutant retained the ability to bind hemoglobin. The Hgp proteins were thus established to play a role in the acquisition of heme from hemoglobin, although it is clear from growth and binding studies that additional moieties provide alternative pathways.

HgpA is the prototype of a family of proteins that are widely distributed across the species and present in variable numbers in different strains. Each of the *hgp* genes identified to date contains a characteristic length of CCAA repeats. The complete genomic sequence of *H. influenzae* Rd KW20 revealed four genes containing CCAA repeats, and the genome of the nontypeable isolate 86–028NP (http://www.microbial-pathogenesis.org) reveals three homologous genes. We have partially or fully sequenced each of the *hgp* genes of an additional type b strain (E1a) containing two *hgp* genes and a nontypeable strain (HI1388) with three *hgp* genes. Another laboratory has sequenced three *hgp* genes of the nontypeable isolate N182 (GenBank accession no. AF221059 and AF221060). A survey of strains representing the genetic breadth of the species, using a $(CCAA)_6$ oligonucleotide as a probe in Southern analyses, revealed the presence of one to five regions of CCAA repeats in each strain. Although the presence of CCAA hybridization does not necessarily correlate with the presence of an *hgp* gene, Southern hybridization analyses, in conjunction with sequence data from several strains, demonstrate uniform presence but extensive variation in the *hgp* genes across the species *H. influenzae*. Table 1 shows that overall conservation of the Hgp proteins ranges from approximately 45 to 92%. However, there are regions of extremely high identity within the proteins, e.g., the region immediately downstream of the QPTN repeat region encoded by the CCAA nucleotide repeats (Fig. 1). The reason for the apparent redundancy in Hgp expression has not been experimentally established. We speculate that alteration in the expression of these genes may provide a mechanism for evasion of the host immune response. Alternatively, Hgp proteins may have variable affinities for different heme sources (see "Hemoglobin-haptoglobin complexes," below)

The CCAA repeat regions of the *hgp* genes mediate variable expression of the proteins. Removal or addition of one CCAA unit alters the reading frame of the protein, leading to introduction of multiple stop codons downstream of the CCAA repeats. Figure 2 shows the nucleotide and amino acid sequences of the N-terminal region of the *hgpB* gene from type b strain E1a. In frame 1, the gene encodes a full-length protein of approximately 113 kDa (~111 kDa following cleavage of the leader peptide). Deletion of one or two CCAA repeats results in the introduction of stop codons (Fig. 2). Using an *hgpA-lacZ* fusion, we demonstrated phase-variable expression of *lacZ* associated with alterations in the length of the CCAA repeat. Alterations from blue colonies expressing *lacZ* to white nonexpressing colonies occurred at a frequency of approximately 1%, while the frequency of shift from lack of expression to expression was approximately 0.6%. These findings are consistent with a slip-strand mechanism, since an in-frame gene may switch to either of two out-of-frame sequences while an out-of-frame gene may switch to one in-frame sequence or one out-of-frame sequence. The frequency of alteration in *lacZ* expression was unrelated to the growth conditions of the organisms, indicating that variation is not inducible by the presence of a particular heme source. The frequency of phase variation of a given Hgp may reflect the length of the CCAA repeat, which varies between 13 and 37 repeats in the sequenced genes. The effect of the repeat length on the rate of variation in Hgp expression has not been experimentally established for the *hgp* genes. However, the frequency of variation of the *mod* gene of *H. influenzae* strain Rd KW20, which contains an AGTC repeat region within the structural gene, increased linearly as the number of repeats increased from 17 to 38.

Since deletion of the entire complement of Hgp proteins of a strain does not abolish hemoglobin utilization, additional systems must be functional in the acquisition of heme from hemoglobin. Affinity purification using hemoglobin as the primary ligand yielded an additional protein of approximately 105 kDa, whose N-terminal amino acid sequence exhibited 100% identity to the sequence encoded by the ORF designated HI1217 from the strain Rd KW20 genomic sequence. The protein encoded by the HI1217 ORF is homologous to other heme/iron acquisition proteins and possesses sequences characteristic of TonB-dependent proteins; it was designated Hup (for "heme utilization protein"). A *hup* deletion mutant in the strain HI689 derivative lacking the Hgp proteins was successfully constructed. The *hup/hgp* deletion strain showed a reproducible decrease in the ability to utilize hemoglobin compared to the *hgp* deletion strain. However, the ability to utilize hemoglobin was not abolished. Thus, yet other outer membrane proteins may be involved in hemoglobin bind-

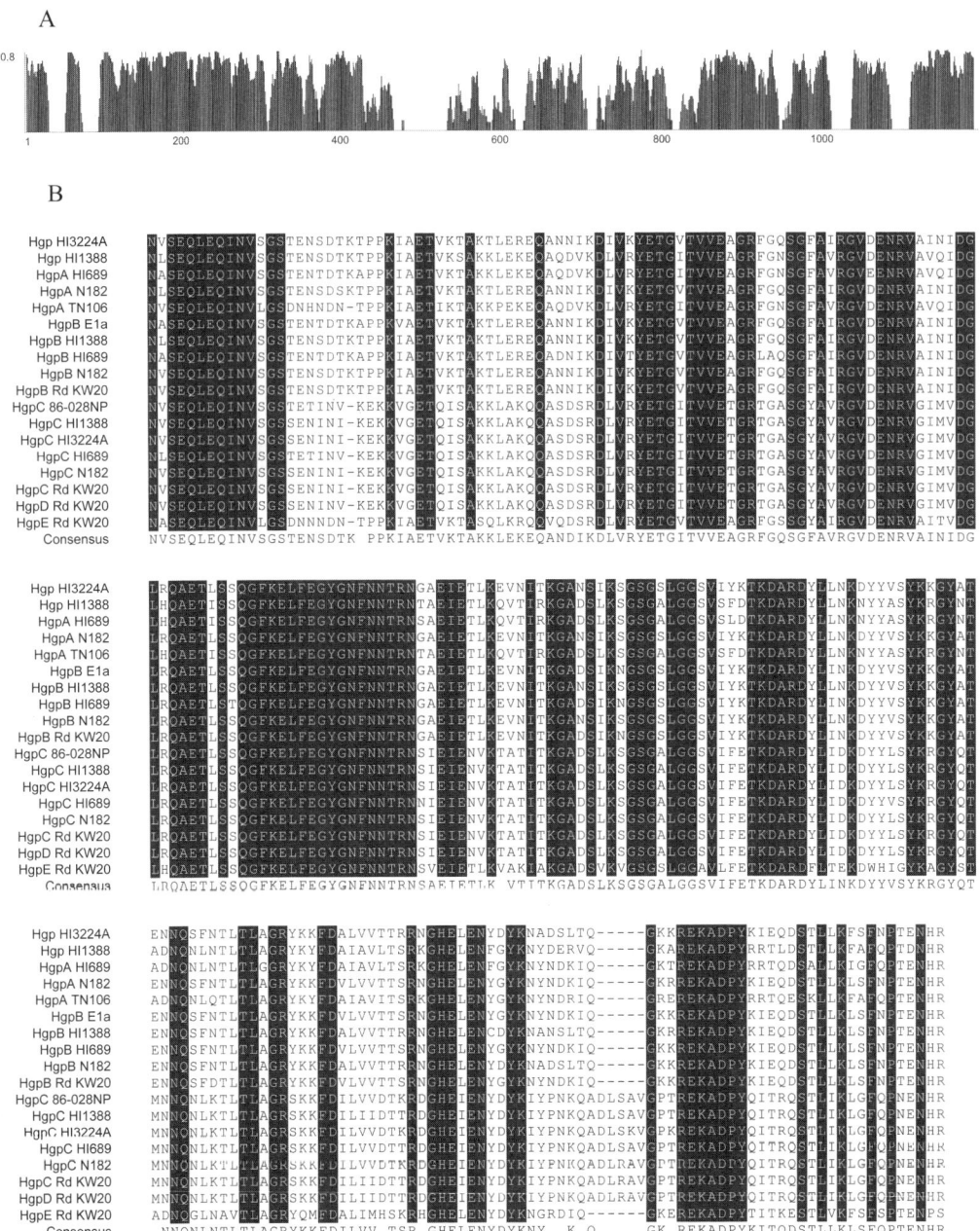

FIGURE 1 Alignment of the hemoglobin/hemoglobin-haptoglobin binding proteins (Hgp) of *H. influenzae*. (A) Similarity plot derived from a comparison of all available full-length Hgp sequences. Sequences used were from strains Rd KW20 (HgpB, HgpC, HgpE, and HgpD), HI689 (HgpA, HgpB, and HgpC), N182 (HgpA, HgpB, and HgpC), E1a (HgpB), and TN106 (HgpA) (see the Table 1 footnotes for appropriate GenBank accession numbers). The similarity plot was generated using the sequence alignment application, AlignX, of the Vector NTI suite v.7 (Informax, Bethesda, Md). (B) Alignment of the region directly downstream of the QPTN amino acid repeats encoded by the CCAA nucleotide repeats (approximately amino acids 100 to 400) for all available sequences. In addition to the complete Hgp sequences, partial Hgp sequences were available from the following strains: HI3224A, one unspecified Hgp and HgpC (http://www.micro-gen.ouhsc.edu); HI1388, one unspecified Hgp, HgpB, and HgpC; and 86–028NP, HgpC (http://www.microbial-pathogenesis.org). The similarity plot was generated using the sequence alignment application, AlignX, of the Vector NTI suite v.7.

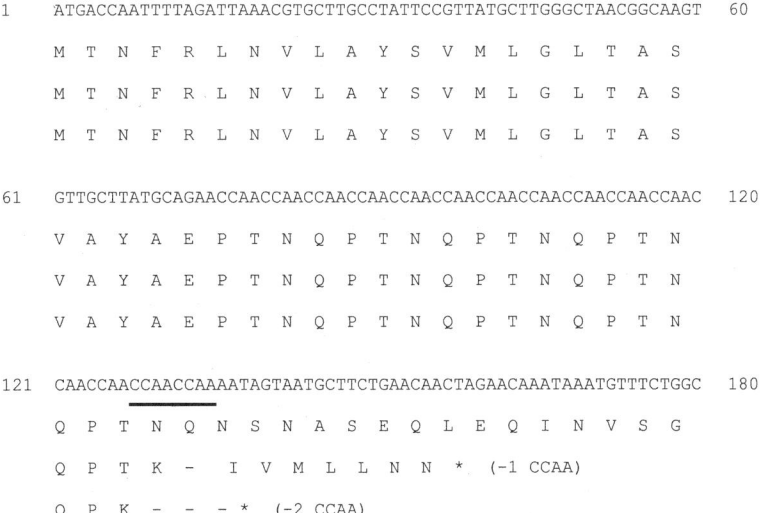

FIGURE 2 Nucleotide sequence of the portion of the *hgpB* gene from *H. influenzae* type b strain E1a encoding the N-terminal region of the protein. The nucleotide sequence as shown contains 15 CCAA repeats; one CCAA repeat has been added to the sequence as it was originally cloned to bring the gene into frame. The translation shown in frame 1 continues to produce a full-length protein of ~113 kDa. The introduction of stop codons following the removal of one or two CCAA repeats is shown. The underlined CCAA repeats are those that have been removed.

ing and utilization. The *hup* deletion mutant also exhibited a reduced ability to utilize low levels of heme and heme-hemopexin and thus may play a role in internalization of heme. In addition, neither the *hup* single deletion mutant nor the *hup/hgp* quadruple mutant demonstrated any reduction in virulence compared to the wild-type strain in 5-day-old infant rats. The role of Hup in the utilization of hemoglobin and/or other heme sources remains to be fully elucidated.

HxuC is essential for utilization of hemoglobin by *H. influenzae* in a background where the entire *hgp* complement of a strain is mutated. We have recently confirmed this observation by using complete *hgp* deletion backgrounds in three additional *H. influenzae* strains (HI689, E1a, and HI1388). However, mutation of *hxuC* in a strain retaining its *hgp* genes had no detectable effect on hemoglobin utilization. These data suggest the existence of two distinct pathways for hemoglobin utilization, one mediated by the Hgp proteins and a second mediated by HxuC and as yet unidentified proteins, possibly including Hup. It is also conceivable that HxuC itself may function as a hemoglobin receptor.

A hemoglobin-binding outer membrane protein, designated HgbA, has also been identified in *H. ducreyi*. This protein has a mass of approximately 108 kDa and exhibits the sequences characteristic of TonB-dependent receptors. An isogenic *hgpA* mutant is unable to grow on hemoglobin as the sole heme source and is also attenuated in both an animal model and a human model of *H. ducreyi* infection.

Hemoglobin-Haptoglobin Complexes

Utilization of the hemoglobin-haptoglobin complex by *H. influenzae* is mediated by the Hgp proteins. Growth studies with hemoglobin-haptoglobin as the sole heme source demonstrate that the HI689 complete *hgp* deletion mutant failed to grow on this heme source. These data were confirmed by experiments with complete deletion mutants in three additional strains (E1a, HI1388, and 86-028NP). Further studies of HI689 and its *hgp* deletion derivatives show that any one of the Hgp proteins is sufficient to support growth on hemoglobin-haptoglobin. Reported growth studies have been performed using commercially available haptoglobin purified from pooled human serum. Human haptoglobin is characterized by molecular heterogeneity and occurs as three major phenotypes: Hp1-1, Hp2-2, and Hp2-1. The polymorphism in haptoglobin arises from variant α chains ($α_1$ and $α_2$). The β subunit is invariant in all phenotypes. This variation in the α subunit gives rise to large differences in the molecular mass of the different haptoglobin phenotypes. Hp1-1 is an $α_1β$ dimer with a molecular mass of 86 kDa; Hp2-2 is a polymer of a basic $α_2β$ subunit and is the largest haptoglobin species, varying in size from 170 to 900 kDa; and Hp2-1 consists of two $α_1β$ units flanking a variable-length $α_2β$ polymer, yielding a species varying from 86 to 300 kDa. It is possible that the Hgp proteins have variable affinities for each of the haptoglobin phenotypes, to more efficiently utilize the haptoglobin expressed by a specific host.

The wild-type HI689 strain and its complete *hgp* deletion derivative have been compared in the infant-rat model of invasive infection. In 5-day-old infant rats, there is no difference in the ability to infect, the duration of infection, or the level of bacteremia between the wild type and the complete mutant strain. However, 30-day-old rats challenged with the wild-type strain became infected at a higher rate than those challenged with the mutant strain. Further, the *hgp* deletion mutant strain was cleared from the blood of 30-day-old rats more rapidly than was the wild-type strain, and the level of bacteremia was significantly lower in the mutant-challenged group. Plasma haptoglobin levels are at a nadir in 5-day-old infant rats. This finding may explain the lack of detectable difference in virulence between the wild-type and mutant strains in this age group. Haptoglobin levels in 30-day-old rats are similar to those in adults, and the haptoglobin may successfully sequester free hemoglobin from the mutant strain. Haptoglobin binding and sequestration of hemoglobin to protect 30-day-old rats from the mutant strain are consistent with the data derived from in vitro growth studies and with the observation that injection of human hemoglobin into these rats spares the clearance of the *hgp* mutant strain.

In addition, a complete *hgp* deletion mutant of a nontypeable *H. influenzae* strain exhibited reduced virulence compared to the wild-type strain in a chinchilla model of otitis media. Although the microenvironment of the middle ear during otitis media has not been fully characterized, there is evidence for the presence of both hemoglobin and haptoglobin in middle-ear secretions from human patients with otitis media and animals with experimentally induced otitis media. The hemoglobin-haptoglobin complex potentially represents one of the heme sources available to *H. influenzae* during otitis media and the factors involved in its acquisition may be important virulence determinants.

Summary of Heme Acquisition by *H. influenzae*

Figures 3 to 5 show a schematic of a proposed model of heme acquisition by *H. influenzae*. In this model, TonB-dependent outer membrane proteins involved in the acquisition of free heme (HxuC and the TdhA homolog HI0113), as well as heme from heme-hemopexin (HxuC), heme-albumin (HxuC), hemoglobin (HxuC, the Hgp proteins, and HI1217), and hemoglobin-haptoglobin (Hgps proteins) have been identified. Additionally, a secreted hemophore (HxuA) mediates heme-hemopexin utilization and may be involved in the direct binding of heme-hemopexin at the bacterial cell surface. The outer membrane lipoprotein *e*(P4) has also been implicated in the acquisition of heme, PPIX, and heme from hemoglobin, although its role has not been fully established. The periplasmic heme transport system has not been identified, although HbpA may be one component. Available genomic sequences do not contain any obvious candidates for a heme-specific ATP-binding cassette

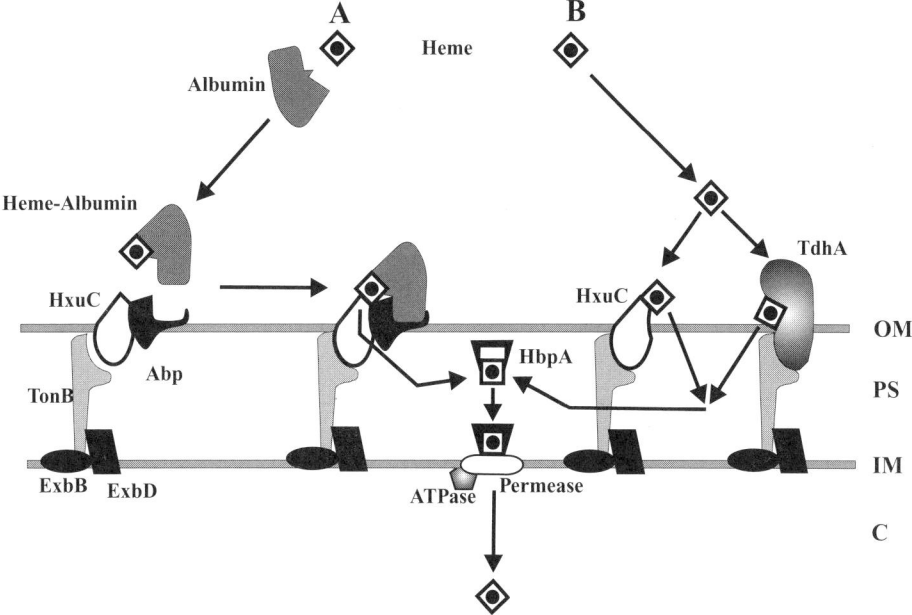

FIGURE 3 Schematic of proposed pathways for the acquisition of heme from the heme-albumin complex (A) and the acquisition of free heme (B) by *H. influenzae*. Heme derived from hemoglobin or heme-containing enzymes is bound by albumin at two binding domains (in this schematic, a single heme molecule is shown bound to albumin). In the acquisition of heme from heme-albumin (pathway A), the heme-albumin complex is bound by a proposed heme-albumin binding outer membrane protein (Apb). Abp is proposed to interact with HxuC in the TonB-dependent transport of heme across the outer membrane to the periplasmic transport system. Free heme is bound by either HxuC or TdhA, which are responsible for the TonB-dependent transport of heme across the outer membrane (pathway B). HxuC and TdhA may act independently or together. HxuC is essential for the acquisition of low levels of heme in vitro. The role of TdhA is proposed based on homology, but no heme-related phenotype has been demonstrated in a TdhA-lacking mutant. Internalization of heme involves a putative periplasmic system. In this model, heme binds to a pleriplasmic binding protein (possibly HbpA) that interacts with a permease to transport heme into the cytoplasm through a process requiring an ATPase. OM, outer membrane; PS, periplasmic space; IM, inner membrane; C, cytoplasm.

(ABC) transporter(s), an area warranting further investigation.

In summary, *H. influenzae* has developed extreme redundancy in its ability to acquire heme from many human protein sources.

IRON SOURCES

Free Iron

Certain strains of *H. influenzae* are unable to grow on media containing iron and PPIX. A genetic locus involved in the acquisition of iron by *H. influenzae* was identified by construction of a genomic library from a strain able to grow on iron plus PPIX. The library was transformed into a strain unable to grow on iron plus PPIX, and transformants growing on iron plus PPIX were selected. This strategy identified an operon, designated *hitABC*, which has subsequently been identified as the inner membrane permease transporting iron from the periplasm to the cytosol. The operon is highly homologous to the ferric-binding protein operon of *Neisseria gonorrhoeae* and is referred to here as *fbpABC*, in line with the accepted nomenclature for that organism. The *fbpABC* operon encodes an ABC transporter system for the

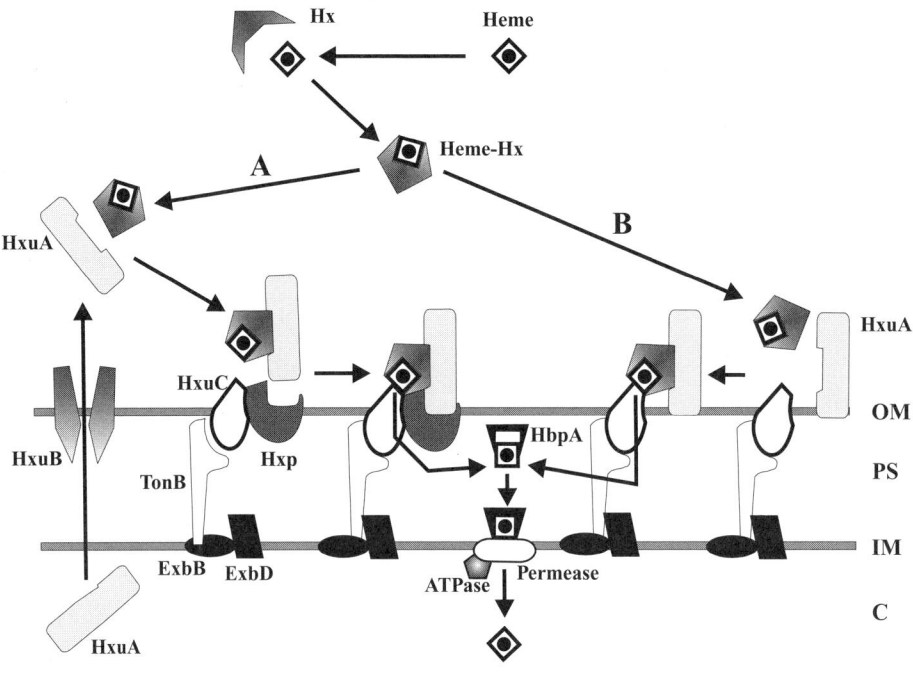

FIGURE 4 Schematic of proposed pathways for the acquisition of heme from the heme-hemopexin (Heme-Hx) complex by *H. influenzae*. Heme derived from hemoglobin or heme-containing enzymes is complexed by hemopexin (Hx) in a 1:1 molar ratio. The heme-hemopexin complex-binding protein, HxuA, is secreted from the bacterial cell by using the secretor protein, HxuB. Two alternative pathways of heme-hemopexin binding to the bacterial outer membrane are shown. In the first pathway (pathway A), secreted HxuA binds to the heme-hemopexin complex and the resulting HxuA-heme-hemopexin complex binds to an outer membrane binding protein. This model proposes the existence of an outer membrane HxuA binding protein (Hxp) interacting with HxuC to deliver heme to the periplasmic transport system in a TonB-dependent manner. Alternatively, HxuC may provide both HxuA binding and heme transport functions. In the second pathway (pathway B), the heme-hemopexin complex binds to membrane-anchored HxuA; this is followed by TonB-dependent delivery of heme to the periplasmic transport system through HxuC. Internalization of heme involves a putative periplasmic system. In this model, heme binds to a pleriplasmic binding protein (possibly HbpA) that interacts with a permease to transport heme into the cytoplasm through a process requiring an ATPase. OM, outer membrane; PS, periplasmic space; IM, inner membrane; C, cytoplasm.

transport of free iron across the inner membrane into the cytoplasm. The periplasmic iron-binding protein, FbpA, is thought to function by binding iron at the inner surface of the outer membrane and transferring the iron across the periplasmic space to the inner membrane permease, FbpB. FbpC, a putative nucleotide-binding protein, may interact with FbpB to provide energy for the transport of iron across the inner membrane via the binding and hydrolysis of a nucleoside triphosphate. A mutant with a mutation in FbpA is unable to use transferrin as an iron source, although it retains the ability to bind transferrin, and it also has a reduced ability to utilize ferric citrate as an iron source. Similarly, an *fbpC* mutant was unable to utilize transferrin-bound iron, ferric ammonium citrate, or ferric nitrate; however, it did retain the ability to use heme. Whether FbpA or FbpC mutants are attenuated in the acquisition of heme from any other heme source has not been reported. In view of the

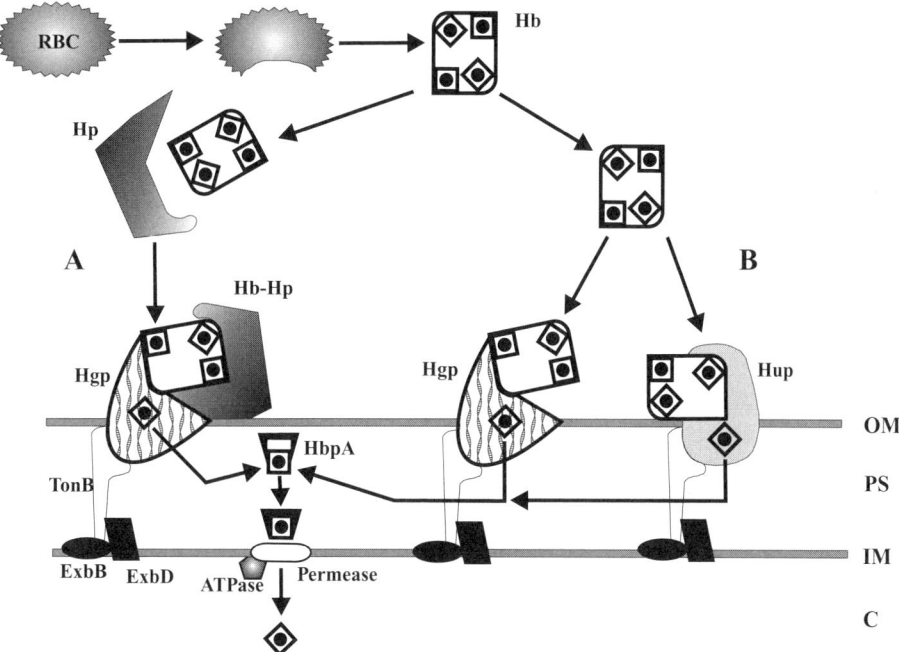

FIGURE 5 Schematic of proposed pathways for the acquisition of heme from hemoglobin (Hb) and the hemoglobin-haptoglobin complex (Hb-Hp) by *H. influenzae*. In pathway A, hemoglobin released by the lysis of red blood cells (RBC) is bound by the serum protein haptoglobin (Hp) in a 1:1 molar ratio. In localized areas of RBC lysis, sufficient hemoglobin may be present to saturate available haptoglobin; under such circumstances, uncomplexed hemoglobin may represent a significant heme source. The hemoglobin-haptoglobin complex is bound by one of the hemoglobin-haptoglobin-binding proteins (Hgp), this is followed by the TonB-dependent transport of heme across the outer membrane to the periplasmic transport system. In pathway B, hemoglobin may be bound by one of the Hgp proteins or Hup followed by the TonB-dependent transport of heme across the outer membrane to the periplasmic transport system. Since a mutant strain lacking expression of its Hgp proteins and Hup retains the ability to utilize hemoglobin, additional pathways for utilization of heme from hemoglobin exist. Internalization of heme involves a putative periplasmic system. In this model, heme binds to a pleriplasmic binding protein (possibly HbpA) that interacts with a permease to transport heme into the cytoplasm through a process requiring an ATPase. OM, outer membrane; PS, periplasmic space; IM, inner membrane; C, cytoplasm.

ability of the mutant strain to grow normally on heme, it seems likely that the *fbpABC* operon represents a periplasmic acquisition system for elemental iron and plays no role in heme acquisition.

Two periplasmic proteins, FbpA and YfeA, from *H. influenzae* strains grown in an iron-restricted medium were identified by sodium dodecyl sulfate-polyacrylamide gel electrophoresis. FbpA, an ~40-kDa protein, was present at high levels in five type b strains and six nontypeable strains, and YfeA was present at lower levels in most of the strains. YfeA, a 31-kDa protein, was found to be 100% homologous to *Yersinia pestis* YfeA, a component of a proposed periplasmic iron and/or manganese transport system (see chapter 15). The available *H. influenzae* genomic sequences, Rd KW20, 86–028NP (http://www.microbial-pathogenesis.org), and HI 3224A (http://www.micro-gen.ouhsc.edu), contain sequences encoding proteins homologous to the YfeABCD proteins of *Y. pestis*. While the *yfe* operon of *Y. pestis* supplied in trans to an *E. coli* siderophore-deficient mutant renders the host strain capable of utilizing iron chelated by EDDA, no data characterize the potential role of the Yfe homologs in *H. influenzae*.

Transferrin

It has been known since the mid-1980s that *H. influenzae* utilizes iron complexed to transferrin. The utilization of transferrin-bound iron was subsequently characterized as siderophore independent, contact dependent, and mediated by specific binding at the cell surface via a receptor complex consisting of two proteins, TbpA and TbpB, encoded by the *tbpBA* operon. TbpA is a protein of ~100 kDa with significant homology to TonB-dependent iron acquisition proteins. TbpB is a lipoprotein with variable molecular mass, has no membrane-spanning regions, and is thought to be anchored to the outer membrane via the amino-terminal lipid. Mutation of either TbpA or TbpB alone significantly reduces the binding of transferrin, while mutation of both together completely abolishes binding. While mutation of TbpA renders the mutant strain unable to utilize transferrin, mutation of TbpB significantly reduces the ability to obtain iron from transferrin but does not abolish it. Antibodies against both TbpA and TbpB of *H. influenzae* can be detected in sera from healthy adults. Passive transfer of antibodies against rTbpB protects infant rats against intraperitoneal challenge with *H. influenzae* type b, although anti-rTbpA antibodies are not protective. Immunization with an rTbpB enhances the clearance of nontypeable *H. influenzae* from the rat lung. There are no reported data on the fate of transferrin-binding protein mutants in animal models of infection.

Transferrin-binding proteins have been detected widely across the species *H. influenzae* and in other species of the genus. Of 83 *H. influenzae* strains 79 expressed proteins of between ~60 and ~90 kDa, reacting to anti-rTbpB antisera, while 32 of 32 strains expressed a ~100-kDa protein reactive to anti-rTbpA antisera; strains examined in these analyses included examples of all serotypes and nontypeable strains. Sequence of the *tbp* genes from several *H. influenzae* strains are available and demonstrate that TbpA is highly conserved at the amino acid level while TbpB is less highly conserved (Table 1).

The presence of transferrin-binding proteins has also been demonstrated in *H. influenzae* biogroup aegyptius (the cause of Brazilian purpuric fever), as well as *H. parasuis*, *H. somnus*, *H. paragallinarum*, and *H. avium*. Each of these species, including *H. influenzae*, is highly specific in utilizing transferrin derived from their host species. Additionally, binding of transferrin was reported in certain biotype III strains of *H. parainfluenzae* and a single isolate of *H. parahaemolyticus*, although utilization of transferrin was not investigated.

Lactoferrin

There is no convincing molecular evidence that *H. influenzae* expresses a receptor or specific uptake mechanism for human lactoferrin. There are conflicting reports in the literature about the ability of *H. influenzae* to utilize human lactoferrin as an iron source. Iron-repressible binding of human lactoferrin to nontypeable *H. influenzae* has been reported; however, the moieties mediating binding were not identified. In addition, there are no sequences in the available *H. influenzae* genomes encoding a putative lactoferrin-binding protein. At this time, there are no convincing data that lactoferrin represents a significant source of iron for *H. influenzae*.

Summary of Iron Acquisition by *H. influenzae*

Our current understanding of elemental iron acquisition by *H. influenzae* under iron-limiting conditions suggests that transport of transferrin-derived iron across the outer membrane of *H. influenzae* is mediated by TbpA and TbpB. Iron liberated from transferrin is subsequently transported across the cytoplasmic membrane by the products of the *fbpABC* operon. Little is known about the transport of elemental iron from other sources across the outer membrane, although subsequent steps may involve the *fbpABC* operon, since *fbpA* and *fbpC* mutants are altered in utilization of iron compounds. Utilization of other iron sources potentially available to *H. influenzae* in vitro, such as ferritin, has not been well investigated.

ACCESSORY PROTEINS

The transport of iron and heme bound to chelators or proteins across the gram-negative outer membrane is dependent on the TonB, ExbB, and ExbD proteins. These proteins form a cytoplasmic membrane complex that transduces energy across the periplasmic space to the protein receptors in the outer membrane. In *H. influenzae*, a *tonB* mutant was unable to grow on heme at 8.5 µg/ml, hemoglobin, hemoglobin-haptoglobin, heme-hemopexin, or transferrin. However, the mutant could grow on heme at 50 µg/ml. These data demonstrate that the high-affinity heme and iron acquisition pathways of *H. influenzae* are TonB-independent consistent with the sequences characteristic of TonB-dependent proteins present in many of the outer membrane proteins discussed above. There is also clearly a TonB-independent low-affinity heme acquisition, since a TonB mutant is able to grow at high heme levels. This uptake may involve passive transfer through a porin channel; it has been suggested that porin P2, with an apparent exclusion size of 1.4 kDa, forms pores that are large enough to allow the passage of heme. Interestingly, P2 has been isolated in an affinity isolation procedure with heme-hemopexin as the primary ligand. However, there is no other direct evidence that P2 mediates heme acquisition. Similarly, an *H. ducreyi tonB* mutant was unable to grow on hemoglobin but retained the ability to grow on the high levels of heme required for growth of wild-type strains, indicating that *H. ducreyi* may lack an intact high-affinity acquisition mechanism for heme.

REGULATION OF IRON AND HEME ACQUISITION MECHANISMS

Many of the proteins and/or binding phenotypes discussed above, such as the Tbp proteins and the Hgp proteins, are up-regulated under conditions of iron or heme limitation. Iron limitation of *H. influenzae* is generally achieved by addition of an iron chelator to the growth medium in the presence of PPIX as a porphyrin source. However, such a strategy also starves the organism for heme since elemental iron is not available for insertion into the porphyrin. We have shown that transferrin binding by *H. influenzae* is repressible by heme, rather than elemental iron, in the medium. Transferrin bound to organisms grown in a heme-restricted medium supplemented with various iron compounds at concentrations up to 10 mM but did not bind when the medium was supplemented with hemoglobin or heme. Additionally, binding to organisms grown in normal human cerebrospinal fluid, heat-inactivated normal human serum, or heat-inactivated human breast milk supplemented with 10 mM ferric nitrate, but not to organisms grown in the same fluids supplemented with heme, was demonstrated. We speculated that transferrin binding by *H. influenzae* might be regulated by a mechanism similar to Fur-mediated regulation in other bacteria but that the cofactor may be heme rather than elemental iron. Other authors reported that elemental iron, as well as heme, hemoglobin, and PPIX, repressed the expression of transferrin binding, as well as the binding of hemopexin and hemoglobin. However, these studies with elemental iron were done using a chemically defined medium containing PPIX, effectively ensuring that the organisms were not heme restricted when supplemented with iron. Thus, conclusions could not be drawn about iron regulation in the absence of heme, and further investigation is necessary to clarify this issue.

A *fur* gene has been identified in *H. influenzae*. Repeated attempts to construct an *H. influenzae fur* deletion or insertion mutant in our laboratory have failed. However, we have constructed a point mutation in the highly conserved HHDH motif (residues 86 to 90 in *E. coli* Fur, residues 88 to 92 in *H. influenzae* Fur), which may comprise part of the iron/cofactor-binding site of Fur, in three different *H. influenzae* strains. Preliminary analysis of these strains indicates a role for Fur in the regulation of both Tbp and Hgp expression. Fur-binding sites have been proposed to occur upstream of several of the genes described in this chapter, for example, *tbp*, but no experimental evidence has been presented on their validity.

Although there is increasing evidence for positive regulation of many bacterial iron and heme acquisition systems, there have been no reported studies of this issue in the haemophili. This is an area in need of research. An examination of the available genomic sequences reveals several homologues of the LysR family of positive transcriptional regulators that may provide an exciting avenue of research.

There are sparse data beyond phenotypic observations of the regulation of heme and iron acquisition pathways in the haemophili, and this area would benefit from additional investigation.

OTHER ISSUES

It is clear that a primary step in the acquisition of heme in the host would be release of the heme source from the cells where the vast majority of heme is sequestered. *H. ducreyi* expresses a heme-repressible hemolytic activity encoded by the *hhdB* and *hhdA* genes. The *H. ducreyi* hemolysin has subsequently been shown to lyse a wide range of cells in addition to erythrocytes. Expression of the *H. ducreyi* hemolysin is not necessary for pustule formation in human volunteers. Comparison of the HhdB and HhdA sequences to the available *H. influenzae* genomic sequences reveals a number of proteins with significant identity to the *H. ducreyi* proteins, which may represent candidate hemolytic factors.

IMPACT OF THE AVAILABLE GENOMIC SEQUENCES

There are currently three *H. influenzae* genomic sequences fully or partially available, i.e., the type d derivative strain Rd KW20 (GenBank accession no. L42023), nontypeable strain 86–028NP (http://www.microbial-pathogenesis.org), and nontypeable strain HI3224A (http://www.micro-gen. ouhsc.edu). Analysis of these genomes reveals additional genes encoding putative heme and/or iron acquisition proteins. For example, the product of the Rd KW20 ORF designated HI0973 shows homology to TbpB from many bacterial species, although there does not appear to be a HI0973 homologue in either the 86–028NP or the HI3224A genomic sequence. A second area of interest is the region from ORF HI1466 through ORF HI1472 in Rd KW20, which includes a putative TonB-dependent receptor and homologues of an ABC iron transport system. These observations indicate that despite the identification of numerous proteins involved in heme and/or iron acquisition by *H. influenzae*, additional pathways remain to be identified. The redundancy in the heme/iron acquisition pathways of *H. influenzae* has made conventional approaches of investigation, such as generation of random mutagenesis libraries, problematic. The availability of the genomic sequences and the consequent ability to identify specific genes of interest potentially allows a systematic investigation of the relevant phenotypes. Thus, a thorough mining of the genomic sequences may provide valuable insights into the heme or iron acquisition systems of *H. influenzae*.

CONCLUSION

H. influenzae has a specific growth requirement for a porphyrin source under aerobic growth conditions and for an iron source under anaerobic conditions. Because *H. influenzae* is highly adapted to its human host, it has developed highly specific acquisition mechanisms for these essential nutrients. Investigations of the *H. influenzae* heme and/or iron acquisition mechanisms reveal them to be highly redundant and at the same time exquisitely complex and refined. While specific pathways, such as the heme-hemopexin utilization system encoded by the *hxuCBA* operon, have been elucidated, only a rudimentary overall understanding of these complex systems is available. Phenotypic and genotypic analyses of the heme utilization pathways are complicated by the apparent redundancy in these systems. In addition, there is a paucity of information about the regulation of expression of these pathways, and although components of these systems are expressed in vivo, there is limited information about the importance of each of the pathways in colonization and infection. No investigations of the potential for variable expression of the heme and iron acquisition pathways of *H. influenzae* in the various possible niches of the organism have been reported. It would be of interest, for example, to compare the heme and/or iron acquisition pathway(s) predominantly used during nasopharyngeal colonization with those used during invasive disease such as bacteremia or localized infections such as otitis media or pneumonia. It is clear that this field is open to extensive and potentially exciting investigations.

SUGGESTED READING

Al-Tawfiq, J. A., K. R. Fortney, B. P. Katz, A. F. Hood, C. Elkins, and S. M. Spinola. 2000. An isogenic hemoglobin receptor-deficient mutant of *Haemophilus ducreyi* is attenuated in the human model of experimental infection. *J. Infect. Dis.* **181:** 1049–1054.

Cope, L. D., R. P. Love, S. E. Guinn, A. Gilep, S. Usanov, R. W. Estabrook, Z. Hrkal, and E. J. Hansen. 2001. Involvement of HxuC outer membrane protein in utilization of hemoglobin by *Haemophilus influenzae*. *Infect. Immun.* **69:** 2353–2363.

Cope, L. D., S. E. Thomas, Z. Hrkal, and E. J. Hansen. 1998. Binding of heme-hemopexin complexes by soluble HxuA protein allows utilization of this complexed heme by *Haemophilus influenzae*. *Infect. Immun.* **66:**4511–4516.

Cope, L. D., R. Yogev, U. Muller-Eberhard, and E. J. Hansen. 1995. A gene cluster involved in the utilization of both free heme and heme:hemopexin by *Haemophilus influenzae* type B. *J. Bacteriol.* **177:** 2655–2653.

Elkins, C., P. A. Totten, B. Olsen, and C. E. Thomas. 1998. Role of the *Haemophilus influenzae* Ton system in the internalization of heme from hemoglobin. *Infect. Immun.* **63:**2194–2220.

Evans, R. W., J. B. Crawley, C. L. Joannou, and N. D. Sharma. 1999. Iron proteins, p. 27–86. *In* J. J. Bullen and E. Griffiths (ed.), *Iron and Infection: Molecular, Physiological and Clinical Aspects*. John Wiley & Sons, Inc., New York, N.Y.

Gilder, H., and S. Granick. 1948. Studies on the *Hemophilus* group of organisms. Quantitative aspects of growth on various porphin compounds. *J. Gen. Physiol.* **31:**103–117.

Gray-Owen, S. D., and A. B. Schryvers. 1996. Bacterial transferrin and lactoferrin receptors. *Trends Microbiol.* **4:**185–191.

Hanson, M. S., C. Slaughter, and E. J. Hansen. 1992. The *hbpA* gene of *Haemophilus influenzae* type b encodes a heme-binding lipoprotein conserved among heme-dependent *Haemophilus* species. *Infect. Immun.* **60:**2257–2266.

Harkness, R. E., P. Chong, and M. H. Klein. 1992. Identification of two iron-repressed periplasmic proteins in *Haemophilus influenzae*. *J. Bacteriol.* **174:**2425–2430.

Hasan, A. A., J. Holland, A. Smith, and P. Williams. 1997. Elemental iron does repress transferrin, haemopexin and haemoglobin receptor expression in *Haemophilus influenzae*. *FEMS Microbiol. Lett.* **150:** 19–26.

Kirby, S. D., S. D. Gray-Owen, and A. B. Schryvers. 1997. Characterization of a ferric-binding protein mutant in *Haemophilus influenzae*. *Mol. Microbiol* **25:**979–987.

Loeb, M. R. 1995. Ferrochelatase activity and protoporphyrin IX utilization in *Haemophilus influenzae*. *J. Bacteriol.* **177:**3613–3615.

Morton, D. J., J. M. Musser, and T. L. Stull. 1993. Expression of the *Haemophilus influenzae* transferrin receptor is repressible by hemin but not elemental iron alone. *Infect. Immun.* **61:**4033–4037.

Morton, D. J., P. W. Whitby, H. Jin, Z. Ren, and T. L. Stull. 1999. Effect of multiple mutations in the hemoglobin- and hemoglobin-haptoglobin-binding proteins, HgpA, HgpB, and HgpC of *Haemophilus influenzae* type b. *Infect. Immun.* **67:** 2729–2739.

Morton, D. J., L. O. Bakaletz, J. A. Jurcisek, T. M. VanWagoner, T. W. Seale, P. W. Whitby, and T. L. Stull. 2004. Reduced severity of middle ear infection caused by nontypeable *Haemophilus influenzae* lacking the hemoglobin/hemoglobin-haptoglobin binding proteins (Hgp) in a chinchilla model of otitis media. *Microb. Pathog.* **36:**25–33.

Palmer, K. L., A. C. Thornton, K. R. Fortney, A. F. Hood, R. S. Munson, Jr., and S. M. Spinola. 1998. Evaluation of an isogenic hemolysin-deficient mutant in the human model of *Haemophilus ducreyi* infection. *J. Infect. Dis.* **178:**191–199.

Reilly, T. J., B. A. Green, G. W. Zlotnick, and A. L. Smith. 2001. Contribution of the DDDD motif of *H. influenzae e* (P4) to phosphomonoesterase activity and heme transport. *FEBS Lett.* **494:** 19–23.

Ren, Z., H. Jin, P. W. Whitby, D. J. Morton, and T. L. Stull. 1999. Role of CCAA nucleotide repeats in regulation of hemoglobin and hemoglobin-haptoglobin binding protein genes of *Haemophilus influenzae*. *J. Bacteriol.* **181:**5865–5870.

Sanders, J. D., L. D. Cope, and E. J. Hansen. 1994. Identification of a locus involved in the utilization of iron by *Haemophilus influenzae*. *Infect. Immun.* **62:**4515–4525.

Stull, T. L. 1987. Protein sources of heme for *Haemophilus influenzae*. *Infect. Immun.* **55:**148–153.

Thomas, C. E., B. Olsen, and C. Elkins. 1998. Cloning and characterization of *tdhA*, a locus encoding a TonB-dependent heme receptor from *Haemophilus ducreyi*. *Infect. Immun.* **66:**4254–4262.

Vogel, L., F. Geluk, H. Jansen, J. Dankert, and L. van Alphen. 1997. Human lactoferrin receptor activity in non-encapsulated *Haemophilus influenzae*. *FEMS Microbiol. Lett.* **156:**165–170.

Whitby, P. W., K. E. Sim, D. J. Morton, J. A. Patel, and T. L. Stull. 1997. Transcription of genes encoding iron and heme acquisition proteins of *Haemophilus influenzae* during acute otitis media. *Infect. Immun.* **65:**4696–4700.

PSEUDOMONAS

Keith Poole

19

Pseudomonas spp., which are gram-negative bacilli that typically inhabit terrestrial and aquatic environs, have gained notoriety as important human, animal, and plant pathogens, as well as agents of biocontrol. Among the pseudomonads, *Pseudomonas aeruginosa* is clearly the most significant as regards infectious disease in humans, generally causing opportunistic infections in individuals with compromised host defenses. While in many instances the infection is attributable to general immunosuppression, as in AIDS patients and neutropenic patients undergoing chemotherapy, *P. aeruginosa* is most commonly associated with, e.g., bacteremic infections in severe burn victims, chronic pulmonary infections in patients with cystic fibrosis (CF), hospital-acquired (nosocomial) pneumonias, and acute ulcerative keratitis, where in each case there is a failing of localized host defenses. The organism is also an important cause of nosocomial urinary tract and wound infections and is frequently associated with less severe, noninvasive diseases such as otitis externa and folliculitis. *P. aeruginosa* exhibits innate resistance to multiple antimicrobials, a property that has long complicated antipseudomonal chemotherapy and probably explains the observation that prior antibiotic usage is a predetermining factor for acquisition of nosocomial *P. aeruginosa* infections. The closely related phytopathogen, *Burkholderia* (formerly *Pseudomonas*) *cepacia*, is similarly resistant to many antimicrobials and also causes opportunistic and potentially debilitating infections in the lungs of CF patients as well as in individuals with chronic granulomatous disease.

While the pathogenesis of *B. cepacia* is poorly understood, several virulence factors have been described and are well characterized in *P. aeruginosa*, including a number of extracellular enzymes and proteases that appear to promote infection by compromising host physical barriers as well as host innate and acquired immune responses. Several of these, including the ADP-ribosylating exotoxin A and the proteases elastase and alkaline protease, are maximally expressed under conditions of iron limitation that predominate in the host, particularly during infection. While the iron control of elastase production appears to occur posttranscriptionally, both exotoxin A and alkaline protease expression is modulated by iron at the level of transcription in parallel with systems for iron acquisition and under the control of a common transcriptional regulator (see below). As with many other pathogenic organisms, iron acquisition by *P. aeruginosa* involves the

---

Keith Poole, Department of Microbiology and Immunology, Queen's University, Kingston, Ontario, Canada K7L 3N6.

Iron Transport in Bacteria, Edited by Jorge H. Crosa, Alexandra R. Mey, and Shelley M. Payne
© 2004 ASM Press, Washington, D.C.

synthesis and uptake of siderophore molecules and is a key determinant of the virulence of this organism. Indeed, siderophore and siderophore receptor production by clinical isolates of *P. aeruginosa* (and *B. cepacia*) during infection is well documented, and defects in siderophore biosynthesis and/or uptake correlate with loss of virulence in animal models of infection.

SIDEROPHORE-MEDIATED IRON TRANSPORT

P. aeruginosa, which synthesizes two known siderophores, pyoverdine (Fig. 1) and pyochelin (Fig. 2), is quite striking in the range of xenosiderophores (siderophores made by other bacteria) and chelators that it can use for iron acquisition. These include the bacterial siderophores aerobactin, enterobactin, and its precursor 2,3-dihydroxybenzoic acid and breakdown product *N*-(2,3-dihydroxybenzoyl)-L-serine, pyoverdine from other pseudomonads, and cepabactin; the fungal siderophores deferrioxamines, deferrichrysin, deferrirubin, and coprogen; synthetic chelators such as nitrilotriacetic acid; and naturally occurring chelators including citrate and *myo*-inositol hexakisphosphate (Fig. 3). *P. aeruginosa* also synthesizes salicylic acid, which binds and promotes iron uptake, although as a precursor of pyochelin it may not be intended as a siderophore as such.

Endogenous Siderophores

PYOVERDINE

Pyoverdine is characterized by a conserved dihydroxyquinoline chromophore to which is attached a small dicarboxylic acid (or its mo-

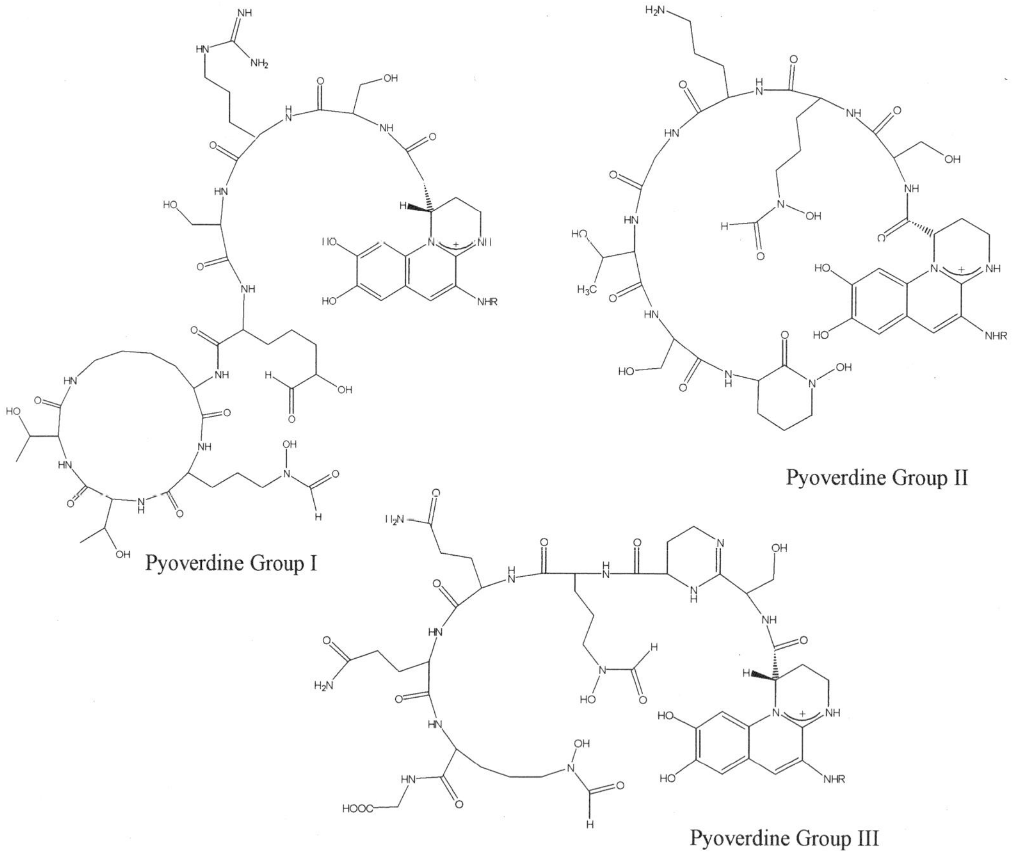

FIGURE 1 Pyoverdines of *P. aeruginosa*.

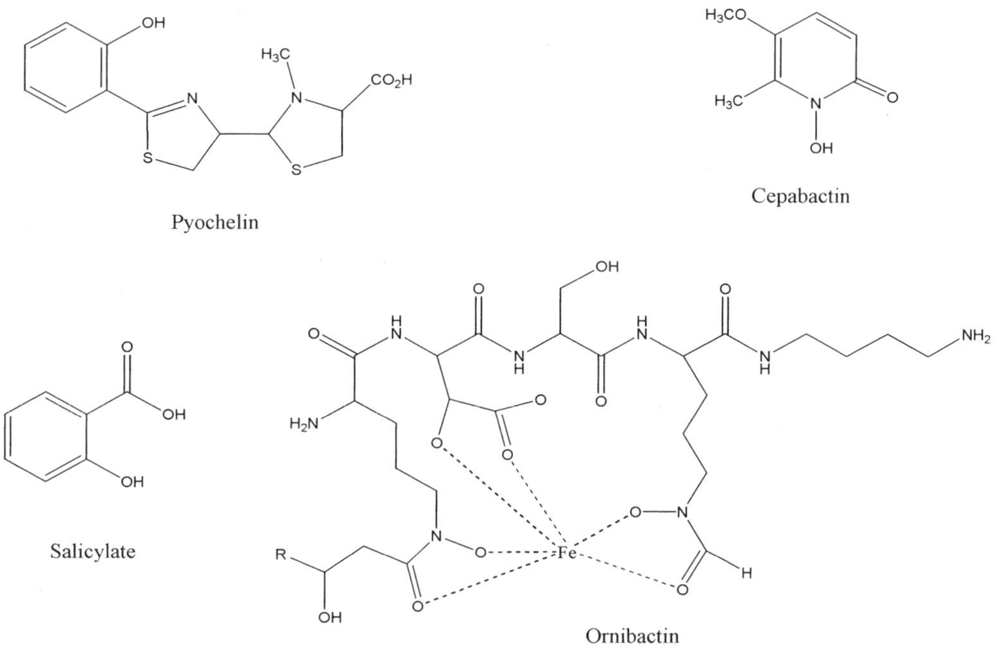

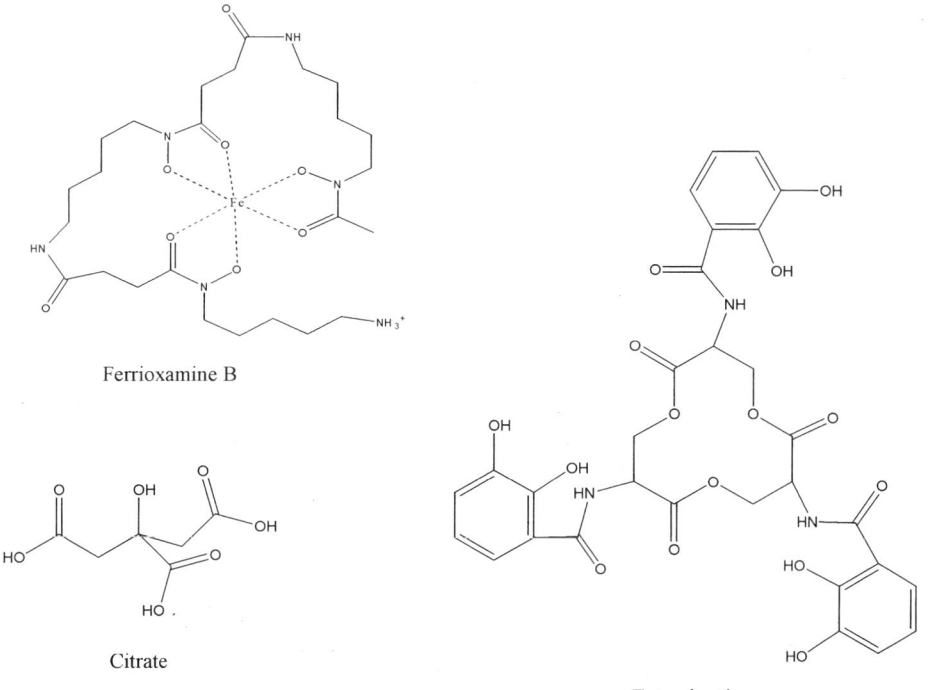

FIGURE 2 Endogenous siderophores of *Pseudomonas* and *Burkholderia* spp.

FIGURE 3 Heterologous bacterial siderophores used by *P. aeruginosa*.

noamide) and a peptide chain of variable length and composition (Fig. 1). Pyoverdine (also called pseudobactin when found in rhizosphere pseudomonads such as *Pseudomonas putida* and *Pseudomonas fluorescens* [see chapter 29]) is responsible for the characteristic fluorescence of *P. aeruginosa* and, indeed, all fluorescent pseudomonads. The peptide portion of individual pyoverdines can vary in length from 6 to 12 amino acids and may contain unusual amino acids (e.g., cyclo-δN-hydroxyornithine, formyl-δN-hydroxyornithine, and diaminobutyric acid) as well as D- and L-isomers of the more common amino acids. This variability in sequence and composition appears to explain the noted specificity of pyoverdine utilization in *P. aeruginosa* and other fluorescent pseudomonads, where, for example, a given strain of *Pseudomonas* often uses its own pyoverdine but not that of other *Pseudomonas* strains, and, indeed, suggests that the peptide moiety is involved in receptor recognition and binding. Three structurally distinct pyoverdines, called types I, II, and III, are known in *P. aeruginosa* (Fig. 1), although a fourth structure has recently been reported, differing from type III pyoverdine only by the absence of L-Gln. Pyoverdine binds Fe^{3+} with high affinity (10^{32} M^{-1}) in a 1:1 stoichiometry. Effective at acquiring iron from transferrin and lactoferrin,

TABLE 1 Genetics of pyoverdine biosynthesis

Gene	*Pseudomonas* genome designation[a]	Function or activity
pvdA	PA2386	L-Ornithine-N^5-oxygenase
pvdD	PA2399	Nonribosomal peptide synthetase
pvdE	PA2397	ABC transporter (pyoverdine export?)
pvdF	PA2396	Transformylase
pvdG	PA2425	Thioesterase
pvdL	PA2424	Nonribosomal peptide synthetase (chromophore biosynthesis?)
pvdI	PA2402	Nonribosomal peptide synthetase
pvdJ[b]	PA2400/PA2401[b]	Nonribosomal peptide synthetase
fpvA	PA2398	Ferric pyoverdin receptor
pvdS	PA2426	ECF sigma factor
pvcA	PA2254	? (chromophore biosynthesis?)
pvcB	PA2255	Oxygenase (chromophore biosynthesis?)
pvcC	PA2256	Hydroxylase (chromophore biosynthesis?)
pvcD	PA2257	Cytochrome (chromophore biosynthesis?)
ptxR	PA2258	LysR family transcriptional regulator
pvdQ	PA2385	Probable acylase
pvdP	PA2392	?
pvdM	PA2393	Probable dipeptidase
pvdN	PA2394	Probable aminotransferase
pvdO	PA2395	Probable Fe(III) reductase
	PA2412	Probable thioesterase
pvdH	PA2413	Probable aminotransferase

[a] As annotated by the *Pseudomonas aeruginosa* Community Annotation Project (see http://www.pseudomonas.com).
[b] Originally annotated as two genes (*pvdJ* and *pvdK*), resequencing of this region confirmed a single open reading frame designated *pvdJ*.

the siderophore is produced by *P. aeruginosa* growing in vivo, where it plays a vital role in growth and pathogenesis. Indeed, mutants with deficiencies in pyoverdine biosynthesis or transport are effectively avirulent in animal models of infection.

Biosynthesis. Several genes for the biosynthesis and transport of pyoverdine have been identified to date, generally clustered within a region of the *P. aeruginosa* chromosome referred to as the *pvd* locus (Table 1; Fig. 4). An operon implicated in synthesis of the chromophore, *pvcABCD*, does, however, map elsewhere. Mutations in cytochrome and heme biosynthetic genes also negatively impact pyoverdine biosynthesis, although the nature of the link to pyoverdine biosynthesis remains uncertain. While the *pvd* locus appears to be involved in the nonribosomal synthesis of the peptide moiety of pyoverdine, there are indications that genes for chromophore biosynthesis (*pvdL*; Table 1) occur here as well. This raises questions regarding the role of the *pvcABCD* operon in chromophore biosynthesis, especially given the lack of conservation of the *pvc* genes in other fluorescent (i.e., pyoverdine-producing) pseudomonads. Not surprisingly, genome analysis has identified many of the *pvd* genes in other fluorescent pseudomonads, including *Pseudomonas putida*, *Pseudomonas fluorescens*, and *Pseudomonas syringae*, where synthesis of the various pyoverdine peptide backbones by nonribosomal peptide synthetases appears to be the norm (see chapter 29).

Regulation. Pyoverdine production in *P. aeruginosa* is dependent on the product of a gene, *pvdS*, encoding a sigma factor essential for the expression of several *pvd* genes, as well as the *pvcABCD* genes. Related sigma factors also play a role in pyoverdine biosynthesis in other fluorescent pseudomonads (see chapter 29). PvdS is a member of a family of extracytoplasmic function (ECF) sigma factors that typically are regulated in response to environmental cues. The *pvd* pyoverdine biosynthetic genes carry a consensus iron starvation (IS) box (TAAAT-N_{16}-CGT) that is required for PvdS-mediated RNA polymerase binding to, and thus transcription from, *pvd* promoters (Fig. 4 and 5). PvdS control of *pvc* expression is indirect and is mediated by PtxR, a positive regulator of the LysR family that is required for *pvc* gene expression and whose expression is itself controlled by PvdS. This ECF sigma factor is also required for production of the extracellular virulence factors exotoxin A and endoprotease PrpL, as well as other iron-regulated nonsiderophore genes such as alkaline protease, which requires PvdS for full expression. The rationale for coregulation of these virulence factors with pyoverdine is for the most part unclear. Alkaline protease is, however, able to release iron from transferrin and thereby stimulate the growth of *P. aeruginosa*

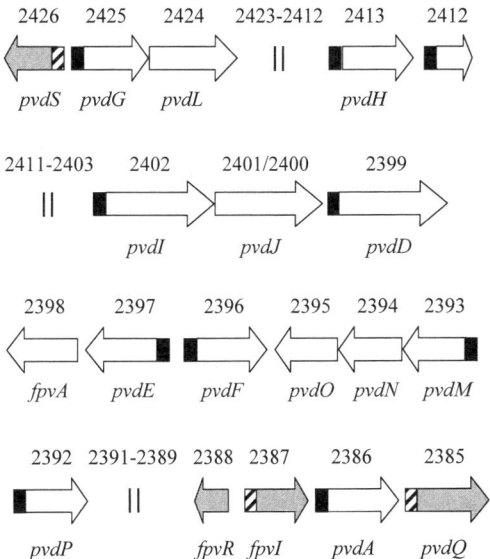

FIGURE 4 Genetic organization of known pyoverdine biosynthetic genes (see Table 1 for their description) within the *pvd* locus. Gene designations are given below the genes, and their corresponding PA designations (*Pseudomonas* Genome Project; http://www.pseudomonas.com) are shown above. Genes whose expression is positively impacted by PvdS are shown in white. Where present, iron starvation (IS, solid squares) and Fur (hatched squares) boxes are shown upstream of the biosynthetic genes. Intervening DNA is indicated by double vertical lines, while the open reading frames present within this intervening region are identified by their PA designations.

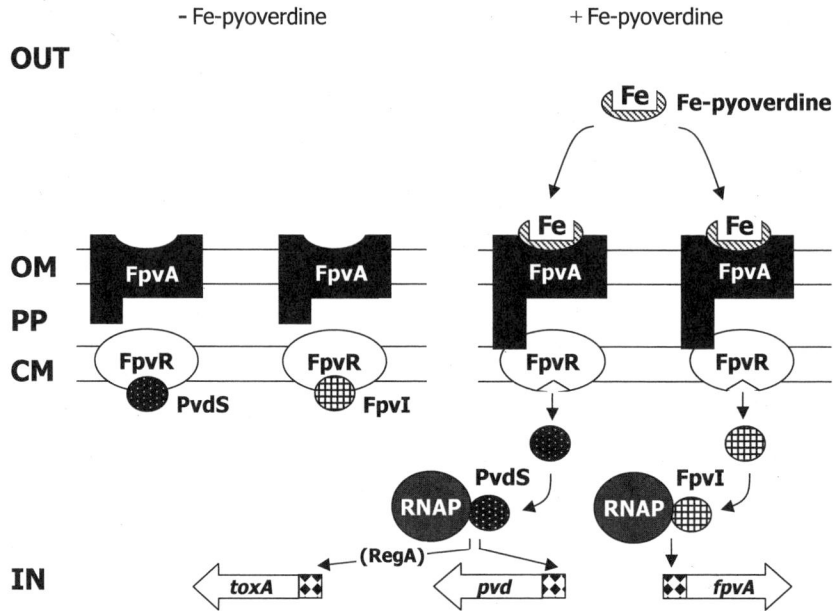

FIGURE 5 Schematic showing pyoverdine-dependent expression of its cognate biosynthetic (*pvd*) and receptor (*fpvA*) genes in *P. aeruginosa*. In the absence of pyoverdine (left), the ECF sigma factors necessary for *fpvA* (FpvI) and *pvd* (PvdS) expression are sequestered by the anti-sigma factor FpvR and are unavailable to stimulate *pvd* and *fpvA* gene expression. In the presence of pyoverdine (right), the iron complex of this siderophore interacts with its outer membrane receptor, FpvA, causing a conformational change that results in the release of the sigma factors by FpvR. These are then free to direct RNA polymerase (RNAP) to the *pvd* and *fpvA* genes and other pyoverdine-regulated genes. Note that the known PvdS/pyoverdine involvement in *toxA* (exotoxin A) expression is apparently mediated by the RegA transcriptional activator. OM, outer membrane; PP, periplasm; CM, cytoplasmic membrane.

cultures where transferrin is the sole source of iron, perhaps by making it available for binding and uptake by pyoverdine. PrpL is also able to degrade lactoferrin, although the contribution to iron acquisition and growth as a result of this remains to be studied. Significantly, expression of *pvdS* has been documented in vivo (e.g., during infection of the lungs of CF patients), and the sigma factor has been implicated as a virulence factor in an endocarditis model of infection.

Pyoverdine production in *P. aeruginosa* increases in response to iron limitation as a result of increased expression of the corresponding biosynthetic genes. The iron control of gene expression is mediated by the product of a gene, *fur*, which, like its well-characterized *Escherichia coli* counterpart (see chapter 13) is a repressor that employs Fe(II) as a corepressor and binds to a consensus operator sequence (Fur box) in the promoters of iron-regulated genes. The crystal structure of $Fur_{P.a.}$ shows the protein to be dimeric and composed of an N-terminal DNA-binding domain possessing a characteristic winged-helix motif and a C-terminal dimerization domain. Two metal-binding sites occur in the protein, one wholly within the dimerization domain and likely to be the regulatory Fe(II) binding site and a second connecting the dimerization and DNA-binding domains and appearing to be specific for Zn(II) and of structural importance. Recent data suggest, too, that $Fur_{P.a.}$ binds as two dimers to a single operator.

While *P. aeruginosa* Fur (Fur$_{P.a.}$) is >70% similar to *E. coli* Fur (Fur$_{E.c.}$) and their respective Fur boxes are very similar, if not identical, the former is distinguishable from the latter (and indeed from the Fur proteins of many other bacteria) by the absence of two cysteine-rich motifs conserved within the C terminus of most Fur repressors. Further, Fur$_{P.a.}$, unlike Fur$_{E.c.}$ is required for viability and is not autoregulated. These distinguishing characteristics of *P. aeruginosa* Fur appear to be common to pseudomonads. Their Fur proteins also lack the cysteine-rich motifs, and the Fur homologue of *P. putida*, like that of *P. aeruginosa*, is required for viability. The significance of these differences between *Pseudomonas* and other Fur proteins remains speculative but suggests that Fur plays a role in *P. aeruginosa* beyond simply controlling the expression of genes involved in iron acquisition. Indeed, a recent study highlighted the iron regulation (by Fur) of a number of genes in *P. aeruginosa*, including a variety of regulatory genes, genes encoding metabolic or detoxifying enzymes, and several genes/operons associated with iron acquisition, including those involved in the production and utilization of siderophores.

Intriguingly, the *pvd* biosynthesis and transport genes lack Fur boxes, although a Fur box is present upstream of *pvdS* and iron regulation of *pvdS* and Fur binding to its promoter have been confirmed. Iron regulation of the *pvd* genes thus occurs indirectly, as a result of Fur control of *pvdS* expression. Similarly, many other iron-regulated genes in *P. aeruginosa* lack Fur boxes and are iron regulated owing to a dependence on PvdS for expression. Cell density has also been implicated in the control of pyoverdine production, with defects in quorum sensing correlating with a reduction in pyoverdine levels and pyoverdine biosynthetic gene expression expression, possibly because PvdS is itself positively regulated by quorum sensing.

Transport. The ferric pyoverdine receptor of *P. aeruginosa* is a ca. 90-kDa outer membrane protein encoded by the *fpvA* gene and inducible under conditions of iron limitation. FpvA is exported to the outer membrane by the twin arginine translocation (TAT) system and displays sequences typical of receptors whose activities are dependent on the energy-coupling TonB protein (see chapter 7). A *tonB* gene that is required for siderophore-mediated iron transport (*tonB1*) has been found in *P. aeruginosa*. Two additional *tonB* genes (*tonB2* [PA0197] and PA0695 [http://www.pseudomonas.com]) have been identified in this organism, although neither has been shown to play a significant role in iron uptake. Unlike *tonB1*, these *tonB* genes are linked to homologues of the *exbB* (*exbB* and PA0693, respectively) and *exbD* (*exbD* and PA0694, respectively) genes whose products function in, e.g., *E. coli* as part of a complex with TonB in promoting energy-dependent siderophore-mediated iron uptake across the outer membrane (see chapter 7). Analysis of mutants has confirmed that *exbB* to *exbD* are not required for iron (or heme) acquisition, raising the possibility that TonB1 works with the *exb* homologues (i.e., PA0693 and PA0694) downstream of PA0695 or that it functions independently of such auxiliary proteins. In *E. coli*, ExbB-ExbD can be partially replaced by the products of the *tolQR* genes, and while homologues of these *tol* genes occur in *P. aeruginosa*, they appear to be essential, precluding a ready assessment of their contribution to TonB1-dependent iron uptake. Suggestions that TonB1 might function independently of ExbBD homologues are, perhaps, supported by the observation of a novel N-terminal extension on the protein that is absent in all other TonB homologues so far described. Still, while this extension might replace the function provided by these auxiliary proteins, the observed dependence of TonB1 on the *E. coli* ExbBD proteins for activity in this heterologous host suggests that TonB1 does, in fact, function with comparable accessory proteins in its native host. The functional significance, then, of this extension is unclear, although it does appear, at least, to be required for TonB1 stability, much as ExbBD contributes to TonB stability in *E. coli*. A number of C-terminal residues that are highly conserved among TonB proteins are essential for TonB1 function in pyoverdine-mediated iron uptake, perhaps playing a role in TonB1 interaction with the FpvA receptor.

Although it is generally accepted that siderophore receptors bind specifically to the ferrated siderophore, FpvA is known to bind iron-free pyoverdine with an affinity comparable to that of ferric pyoverdine. In fact, FpvA receptors on the surface of iron-limited *P. aeruginosa* are probably saturated with iron-free siderophore, even though only iron-bound pyoverdine actually enters cells. One suggestion is that iron-bound pyoverdine displaces iron-free pyoverdine on FpvA; such displacement has indeed been demonstrated both in vitro and in vivo. After ferric pyoverdine uptake and dissociation inside cells, pyoverdine is then recycled back into the extracellular medium, where it reloads onto FpvA. This may explain earlier observations that *P. aeruginosa* accumulates less pyoverdine than Fe^{3+} during transport assays.

Recently, the receptors for the type II and III pyoverdines (Fig. 1) have been identified (FpvAII and FpvAIII, respectively) and their genes have been cloned and sequenced. Like FpvA (also known as FpvAI), the FpvAII and FpvAIII receptors possess an N-terminal extension (see below), although they show limited sequence similarity and, unlike FpvAI, lack a TAT export sequence. Significantly, FpvAII and FpvAIII lack a number of aromatic residues implicated in type I pyoverdine binding by FpvAI and possibly forming a pyoverdine-binding pocket. While it is unclear if this difference relates to the presence of a cyclic substructure in type I pyoverdine that is absent in type II and III pyoverdines (Fig. 1), it is intriguing that aromatic residues play an important role in the binding of ferrichrome by the *E. coli* FhuA ferrichrome receptor and that ferrichrome, too, has a cyclic structure. Moreover, FpvAI shows substantial sequence similarity to FhuA, and it is, in fact, possible to derive a three-dimensional model of FpvAI by threading its sequence throught the available FhuA crystal structure.

Receptor-Mediated Gene Expression. FpvA possesses additional sequences at the N terminus that are absent from many ferric siderophore receptors but present in the well-characterized ferric dicitrate receptor, FecA, of *E. coli* and the ferric pyoverdine (pseudobactin) BN7/8 receptor, PupB, of *P. putida* WCS358. In these last two receptors, the N termini play a role in signal transduction, with siderophore binding to the cognate receptor ultimately inducing expression of the corresponding receptor gene. This occurs in, e.g., *E. coli* as a result of stimulation (by the FecA N terminus) of a two-component system that includes a cytoplasmic membrane-spanning "sensor," FecR, and a cytoplasmic ECF sigma factor, FecI, the latter of which promotes *fecA* gene expression (see chapter 11). In a similar fashion, FpvA mediates the well-known pyoverdine stimulation of FpvA receptor production in a process that also involves FpvIR, homologues of FecIR. With FpvI as the ECF sigma factor and FpvR functioning as an apparent anti-sigma factor (Table 2), FpvI stimulation of *fpvA* gene expression ultimately occurs in response to a signal(s) generated by FpvA on ferric pyoverdine binding (Fig. 5). FpvR as a strict anti-sigma factor would, however, contrast with FecR, which appears to be required for FecI activity.

FpvA also plays a role in pyoverdine biosynthesis, confirmed by observations of reduced pyoverdine production in FpvA-deficient strains, as a result of reduced expression of *pvd* genes. Similarly, defects in pyoverdine biosynthesis also negatively impact *pvd* gene expression, arguing for a positive role for pyoverdine in *pvd* gene expression (and, thus, pyoverdine biosynthesis). Indeed, pyoverdine stimulation of *pvd* gene expression has been noted, depending on the presence of FpvA and involving the aforementioned FpvR anti-sigma factor and PvdS. As with siderophore-dependent control of *fpvA* expression, then, siderophore interaction with FpvA initiates a signal transduction cascade involving the N terminus of FpvA,

TABLE 2 Known and putative iron receptors and their regulators in *Pseudomonas*[a]

Receptor[b]	Regulator(s)				
	FecI/PupI-like ECF sigma factor	FecR/PupR-like sensor/anti-sigma factor	Two-component sensor kinase	Two-component response regulator	Other
P. aeruginosa PAO					
FpvA	FpvI (Pig32)[c]	FpvR[d]			
FptA					PchR[e]
PfeA			PfeS	PfeR	
FecA (PA3901)[f]	PA3899[g,h]	PA3900[h]			
PA0151[g]					
PA0192					
FiuA	FiuI[h]	FiuR[h] (Pig17)[c]			
PirA			PirS (Pig 19)[c]	PirR (Pig 19)[c]	
PA1302[i]	PA1300[h]	PA1301[h]			
PA1322[h] (pig31; also PfuA)[c]					
PA1365	PA1363	PA1364			
UfrA (PA1910)	PA1912	PA1911			
PA1922					
PA2466	PA2468[h]	PA2467[h]			
PA2911					
PA3268					
HasR	PA3410[h] (Pig25)[c]	PA3409[h]			
PA4156					
PA4168					
PA4514[h] (pig12-ORF1; also PiuA)[c]					
PA4675					
PA4897[h]	PA4896[h]	PA4895[h]			
B. cepacia					
OrbA					

[a] Where genes have not been shown to function as receptors or regulators, the designation of putative receptors and regulators is based on similarities to known siderophore or heme receptors and regulatory proteins as annotated by the *Pseudomonas aeruginosa* Community Annotation Project (see http://www.pseudomonas.com), and their proposed involvement in regulating the indicated putative receptors is based solely on their proximity (i.e., immediately adjacent) to the putative receptor genes.

[b] FpvA, ferric pyoverdine receptor; FptA, ferric pyochelin receptor; PfeA, ferric enterobactin receptor; FecA, ferric dicitrate receptor; FiuA, ferrioxamine B receptor; PirA, putative ferric enterobactin receptor; UfrA, unknown ferric siderophore receptor; HasR, heme uptake receptor; OrbA, ornibactin receptor.

[c] First identified by Ochsner and Vasil, with their designation given in parentheses.

[d] Also involved in control of pyoverdine biosynthesis through its action on PvdS, an ECF sigma factor required for expression of pyoverdine biosynthetic genes.

[e] An AraC family regulator also involved in pyochelin production through its action on the *pch* genes of pyochelin biosynthesis.

[f] Role in ferric dicitrate uptake yet to be confirmed.

[g] Where receptor or regulator activity has yet to be confirmed experimentally, the gene designation provided by the *Pseudomonas* Genome Project (http://www.pseudomonas.com) is reported.

[h] Confirmed as iron regulated, i.e., expression increases under iron limitation.

[i] Probable heme utilization protein or receptor.

which triggers the release of, in this case, PvdS by FpvR, with the sigma factor then free to activate *pvd* gene expression (Fig. 5). Given the role of PvdS in exotoxin A expression, the FpvA-FpvR-PvdS cascade also mediates pyoverdine stimulation of toxin production (Fig. 5). With FpvR functioning as an apparent anti-sigma factor for both PvdS and FpvI, it is unclear whether the two sigma factors occupy common or different sites on FpvR.

Although *fpvA* expression is iron regulated, the gene, like the *pvd* genes, lacks a binding site for the Fur repressor responsible for iron regulation of gene expression in *P. aeruginosa*. As with the PvdS sigma factor required for *pvd* gene expression, however, the *fpvI* gene whose product promotes *fpvA* expression is iron regulated and contains a Fur-binding site in its promoter. Thus, iron regulation of *fpvA* occurs indirectly as a result of Fur control of FpvI.

PvdS, like its FecI homologue, belongs to a subgroup of ECF sigma factors that are iron regulated (by Fur) and implicated in iron acquisition processes and, thus, are dubbed iron starvation sigma factors. These appear to be widespread among the fluorescent pseudomonads (see Chapter 29); indeed, several examples are readily identifiable in the *P. aeruginosa* genome sequence (Table 2). Interestingly, these often occur in tandem with FecR/FpvR homologues and adjacent to putative siderophore receptor genes (Table 2). It is likely, therefore, that these regulatory gene pairs control expression of the receptor genes in response to hitherto unidentified, probably heterologous siderophores, whose binding (complexed with iron) to the cognate receptors ultimately triggers expression of the corresponding receptor genes (see below).

Uptake of Heterologous Pyoverdines. Although there is strain specificity of pyoverdine recognition and uptake, many fluorescent pseudomonads, including *P. aeruginosa*, are able to utilize some pyoverdines produced by other fluorescent pseudomonads. Different strains may produce structurally identical pyoverdines or pyoverdines with conserved features that are probably important for receptor recognition. Utilization of heterologous pyoverdines also occurs when strains are able to express individual receptors specific for heterologous pyoverdines. Only a limited number of structurally unique pyoverdines have been found to date in *P. aeruginosa* (Fig. 1), and any strain producing a particular pyoverdine type will obviously be able to utilize the same pyoverdine type produced by any other strain. Examples of *Pseudomonas* spp. utilizing structurally different heterologous pyoverdines are also known, although these pyoverdines share a conserved feature within the peptide moiety. Strains producing pyoverdines with a C-terminal cyclopeptide substructure, for example, tend to be able to utilize one another's pyoverdines despite the presence of differences in the amino acid sequence of the peptide moiety. Presumably, this cyclopeptide structure is important for recognition by a given strain's ferric pyoverdine receptor, and, as such, the receptor is able to recognize not only its own siderophore but also heterologous siderophores with the same feature. Thus, the FpvA ferric pyoverdine receptor of *P. aeruginosa* PAO1 accommodates iron complexed to the endogenous pyoverdine and the structurally different pyoverdine of *P. fluorescens* ATCC 13525, both of which demonstrate C-terminal cyclopeptidic moieties.

PYOCHELIN

Pyochelin is a condensation product of salicylic acid and two cysteinyl residues (Fig. 2; enzymology is described in chapter 2) that binds Fe(III) in a 2:1 stoichiometry with a rather low affinity in aqueous media (5×10^5 M^{-1}). Still, the siderophore promotes iron uptake in *P. aeruginosa*, is able to acquire iron from transferrin, and is implicated as a virulence determinant, although the last of these may relate to its ability to catalyze the formation of tissue-damaging free radicals.

Biosynthesis and Transport. The genes for pyochelin biosynthesis have been identified and occur in two separate operons,

TABLE 3 Genetics of pyochelin biosynthesis

Gene	*Pseudomonas* genome designation[a]	Function or activity
pchA	PA4231	Isochorismate synthase (salicylate biosynthesis)
pchB	PA4230	Isochorismate pyruvate lyase (salicylate biosynthesis)
pchC	PA4229	Thioesterase
pchD	PA4228	Adenylate-forming enzyme (activates salicylate)
pchR	PA4227	AraC family transcriptional regulator
pchE	PA4226	Dihydroaeruginoate synthase (nonribosomal peptide synthetase)
pchF	PA4225	Pyochelin synthetase (nonribosomal peptide synthetase)
pchG	PA4224	NADPH-dependent thiazoline reductase
pchH	PA4223	ABC transporter (pyochelin export)
pchI	PA4222	ABC transporter (pyochelin export)
fptA	PA4221	Ferric pyochelin receptor

[a] As annotated by the *Pseudomonas aeruginosa* Community Annotation Project (see http://www.pseudomonas.com).

pchDCBA, involved in the synthesis of the pyochelin precursor salicylic acid, and *pchEFGHI*, although *pchHI* may have an export rather than a synthesis function (Table 3). The *fptA* gene encoding the 75-kDa ferric pyochelin receptor occurs immediately downstream of *pchI*. Expression of *fptA* has been demonstrated in vivo, and the gene is inducible in vitro by human respiratory mucus, consistent with a role in in vivo iron acquisition, particularly in lung infections.

Regulation. Production of pyochelin and the ferric pyochelin receptor is inducible under conditions of iron limitation, as a result of Fur regulation of the *fptA* receptor gene and the *pchDCBA* and *pchEFGHI* biosynthetic operons. Pyochelin and FptA production also require the product of the *pchR* gene, encoding an iron-regulated (by Fur) AraC family regulator necessary for expression of the *fptA* and *pch* genes. Pyochelin also plays a positive role in both *fptA* and *pch* gene expression, with pyochelin-deficient strains exhibiting a marked deficiency in expression of these genes that is reversed on pyochelin supplementation, in a process mediated by PchR. In fact, PchR actually represses *fptA* expression in the absence of pyochelin while promoting its expression in the presence of this siderophore (Fig. 6). Thus, this regulator mediates the positive effect of pyochelin (or, more likely, ferric pyochelin) on *fptA* and *pch* gene expression and thus the production of pyochelin and its receptor (Fig. 6). PchR activity may be modulated by the FptA receptor, inasmuch as mutants lacking FptA show reduced *fptA* gene expression. Still, the resultant defect in ferric pyochelin uptake in such mutants would also negatively impact PchR stimulation of *fptA* expression should the iron-siderophore complex act directly on PchR. In the absence of genes encoding an obvious cytoplasmic membrane permease for ferric pyochelin, however (see below), the likelihood that the complex actually reaches the cytoplasm to do so is questionable. In any case, it is clear that siderophore-dependent expression of cognate transport genes is a common theme in *P. aeruginosa*, involving both endogenous (see also the discussion of pyoverdine, above) and heterologous (see below) siderophores. Presumably, detection of the ferrated siderophore (by the cognate receptor?) provides confirmation that a given siderophore is successful in acquiring iron in a particular environment and, thus, assurance of a return on further investment in this siderophore and its receptor. Given the multiplicity of endogenous and heterologous siderophores at the organism's disposal, this would allow it to up regulate the most appropriate system (i.e., the one most successful in obtaining iron) for iron acquisition at a given time and place. This is simi-

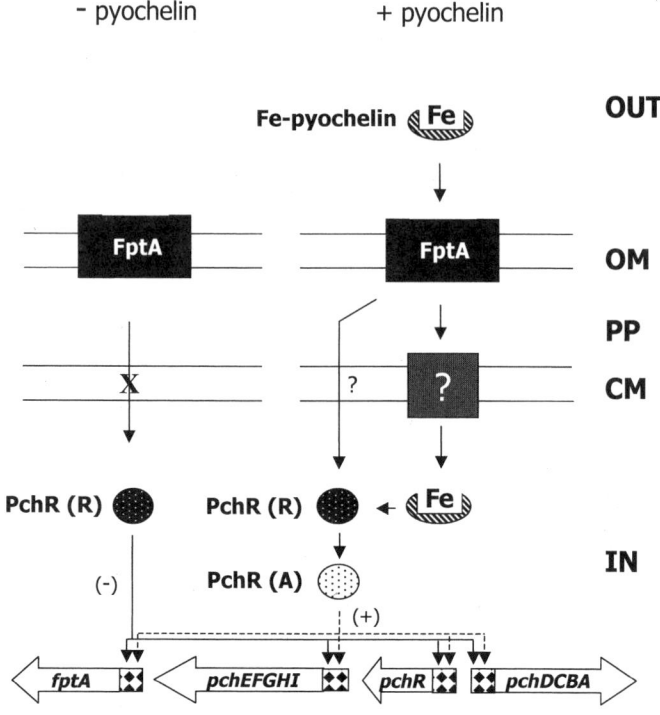

FIGURE 6 Schematic showing pyochelin-dependent expression of its cognate biosynthetic (*pch*) and receptor (*fptA*) genes in *P. aeruginosa*. In the absence of pyochelin (left), the regulatory protein, PchR, exists in a repressor form (R) and blocks the expression of *fptA* and, possibly, the pyochelin biosynthetic genes. In the presence of pyochelin (right) or, more probably, ferric pyochelin, PchR is converted to a transcriptional activator (A) of the receptor and biosynthetic genes. It is unclear, however, whether ferric pyochelin is a direct effector of PchR activator activity, following its transport across the cytoplasmic membrane via a hitherto unknown carrier, or whether it simply interacts with its receptor, FptA, on the cell surface and this is signaled to PchR via an as yet undiscovered signal transduction pathway. OM, outer membrane; PP, periplasm; CM, cytoplasmic membrane.

lar to the regulation of siderophore transport described for *Bordetella* (see chapter 20).

Xenosiderophore Transport

P. aeruginosa is able to utilize a variety of xenosiderophores, not only heterologous pyoverdines (see above) but also siderophores of other bacterial genera and fungi. The use of multiple siderophores by *P. aeruginosa* is consistent with the presence of multiple receptor homologues in the genome sequence (Table 2). Use of xenosiderophores appears, in many instances, to require induction of outer membrane receptors specific for the xenosiderophores. Siderophore induction of receptor genes is, in fact, seen in a number of bacteria (see chapter 20), and in *P. aeruginosa* it often involves regulatory protein pairs of the FecIR-FpvIR sort (see above and Fig. 5) or the sensor kinase/response regulator sort (see below and Fig. 7). Strikingly, homologues of such regulator genes are often linked to putative siderophore receptor genes in the *P. aeruginosa* genome (Table 2). Its is tempting to speculate, then, that the use of heterologous siderophores is common in *P. aeruginosa* because of the presence of an array of inducible receptors that respond specifically to these heterologous siderophores.

ENTEROBACTIN

A major catecholate siderophore of *E. coli* and other members of the *Enterobacteriaceae*, enterobactin (Fig. 3) (see chapter 9) promotes iron uptake into *P. aeruginosa* following its induction of PfeA, an outer membrane homologue of the *E. coli* FepA ferric enterobactin receptor (see chapter 10). Expression of *pfeA* requires the products of the linked *pfeRS* operon, which encodes a response regulator (PfeR) and histidine kinase (PfeS), components of the superfamily of environmentally responsive two-component regulatory protein pairs. Members of this superfamily utilize a phosphorelay mechanism to regulate gene expression, with an environmental stimulus triggering autophosphorylation of the histidine kinase and subsequent phosphoryl group transfer to the response regulator, which then activates target gene expression. In response to enterobactin, then, PfeRS function to activate *pfeA* expression via phosphotransfer form PfeS to PfeR (Fig. 7), with enterobactin ultimately promoting PfeR binding to the *pfeA* promoter region. Expression of both *pfeA* and *pfeRS* is iron regulated, mediated by the Fur repressor, whose binding to the *pfeA* and *pfeR* promoters has been demonstrated. As such, enterobactin induction of *pfeA* occurs only under conditions of iron limitation. Enterobactin inducibility of *pfeA* is independent of the PfeA receptor, distinguishing this system from the receptor-dependent ECF sigma factor-controlled systems described above.

PfeA-deficient mutants still demonstrate enterobactin induction of *pfeA*, arguing for a second system for the transport of ferric enterobactin across the outer membrane in *P. aeruginosa*, since the siderophore must reach the cytoplasmic membrane PfeS component to initiate the phosphorely mechanism (Fig. 7). Consistent with this, PfeA-deficient mutants display growth, albeit reduced, in an enterobactin-supplemented iron-restricted minimal medium. A possible candidate for this second uptake system is the product of the *pirA* gene (PA0931; http://www.pseudomonas.com) (Table 2), which displays substantial homology to PfeA. This gene is also linked to an operon encoding kinase/response regulator homologues of the superfamily of two-component regulators (PirRS; Table 2) and thus is also probably siderophore regulated. While it is unlikely that two systems would be present and specifically responsive to enterobactin in *P. aeruginosa*, it is possible that PirA promotes modest uptake of ferric enterobactin, sufficient to "activate" PfeRS and, ultimately *pfeA* gene expression (Fig. 7), while another, perhaps structurally related siderophore is its natural ligand.

Genes encoding homologues of the periplasmic (*fepB*, PA4159) and cytoplasmic membrane (*fepC, fepD,* and *fepG*; PA4158, PA4169, and PA4161, respectively) components of ferric enterobactin uptake in *E. coli* are readily identifiable in the *P. aeruginosa* genome se-

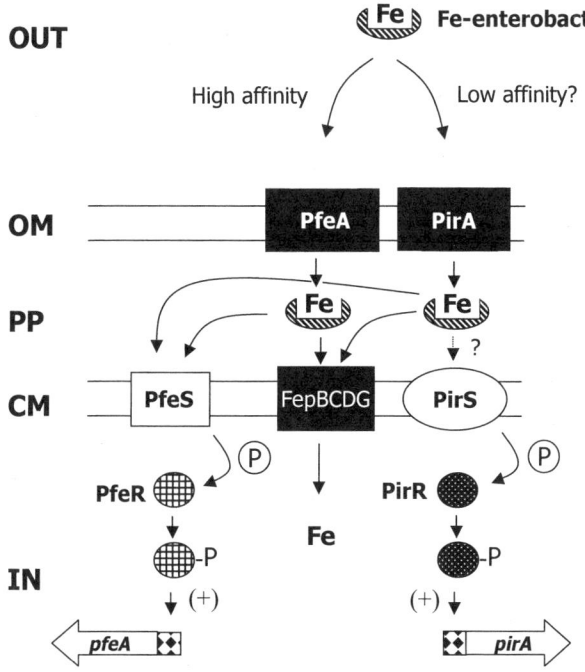

FIGURE 7 Schematic showing enterobactin-dependent expression of its cognate receptor gene *pfeA* in *P. aeruginosa*. Ferric enterobactin that is deposited in the periplasm by the PfeA receptor interacts with the PfeS sensor kinase, which ultimately phosphorylates the response regulator, PfeR, thereby activating it to drive the expression of the *pfeA* gene. This process is independent of any transport of the ferric siderophore complex across the cytoplasmic membrane by permease components (FepBCDG). A second receptor able to accommodate ferric enterobactin, PirA, has been proposed to explain ferric enterobactin uptake and stimulation of PfeSR in the absence of PfeA. Apparently associated with its own phosphorelay two-component regulatory system, PirSR, it is likely that PirA also transports another, as yet unknown, siderophore as its primary substrate and that this provides for up regulation, via PirSR, of the *pirA* gene. OM, outer membrane; PP, periplasm; CM, cytoplasmic membrane.

quence. Interestingly, *fepC* was identified as an iron-regulated gene in a recent DNA microarray study of iron starvation-inducible genes in *P. aeruginosa*, although the other *fep* gene homologues were not. It is likely that these genes will, like the *pfeA* receptor gene, be responsive to enterobactin and thus will not be inducible under iron limitation unless the siderophore is present. Why *fepC* was expressed under iron limitation in the absence of enterobactin is unclear, although it is noteworthy that the gene is transcribed divergently from the *fepBDG* homologues, perhaps because it plays an additional role in the cell unrelated to enterobactin uptake.

CITRATE

The ability of citrate (Fig. 3) to mediate iron uptake in *P. aeruginosa* is well established, and, as in *E. coli* (see chapter 11), citrate-mediated iron uptake in *P. aeruginosa* requires prior exposure to citrate. While the existence of a ferric citrate receptor has not yet been confirmed, a *fecA* homologue (PA3901) is present in the *P. aeruginosa* genome, adjacent to the *fecIR/fpvIR* homologues (PA3899 to PA3900) (Table 2) that probably mediate the citrate inducibility of *fecA*.

FERRIOXAMINE B

Ferrioxamine B (Fig. 3) promotes iron uptake in *P. aeruginosa* as a result of its induction of a high-molecular-weight iron-regulated outer membrane protein that probably serves as its receptor. The ferrioxamine B receptor gene, *fiuA* (PA0470) (Table 2), has recently been identified and is under the control of the FpvIR-like products of the adjacent *fiuIR* genes (Table 2).

Uptake Post-Outer Membrane

Siderophores and their cognate receptors function to deliver the ferric siderophore complex to the periplasm, where additional transport components are responsible for delivering the iron to the cytoplasm. In *E. coli*, there are siderophore-specific periplasmic and cytoplasmic membrane components of the ABC permease family for the transport of ferric dicitrate (FecBCDE) (see chapter 11) and ferric enterobactin (FepBCDG) (see chapter 9) and a less specific transport system that accommodates hydroxamate siderophores (FhuBCD) that are internalized by a variety of outer membrane receptors. Little is known, however, about Fe(III)-siderophore transport post-outer membrane in *P. aeruginosa*, although *fepBCDG* homologues have been identified in the *P. aeruginosa* genome sequence (see above). Strikingly, no obvious periplasmic or cytoplasmic membrane transporter homologues are identifiable in the ferric pyochelin biosynthesis and uptake locus, in the vicinity of the genes implicated in ferric dicitrate uptake, or adjacent to any of the receptor homologues shown in Table 2. Although candidate iron- and PvdS-regulated ABC transporter homologues are present in the *pvd* locus (e.g., PA2408 to PA2409), these genes occur in an apparent operon with genes known not to be involved in ferric pyoverdine uptake (PA2408 to PA2409).

The obvious similarities to pyoverdine biosynthesis and uptake in fluorescent rhizosphere organisms, along with observations that cloned receptor genes for heterologous pyoverdines are alone sufficient to enable these organisms to utilize these heterologous siderophores, suggest that periplasmic and cytoplasmic membrane permeases specific for a particular pyoverdine may not be required in *Pseudomonas*. While a broadly specific Fhu-like system might be responsible for accommodating any pyoverdine delivered to the periplasm, the lack of obvious periplasmic and cytoplasmic membrane transporter candidates for most siderophores in *P. aeruginosa* suggests that iron may be removed from siderophores (with the possible exception of enterobactin) in the periplasm and that an iron-specific, rather than siderophore-specific, transporter delivers the metal to the cytoplasm (Fig. 8). Indeed, early studies of ferric pyoverdine uptake indicated that iron and pyoverdine dissociated in the periplasm, although the mechanism was not identified. Release of iron from ferrated citrate, pyochelin, and pyoverdine via a reductive mechanism has been de-

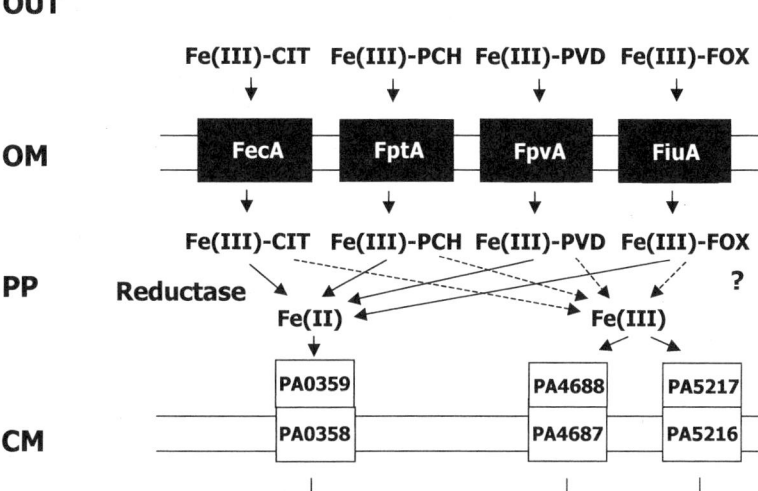

FIGURE 8 Schematic showing siderophore-mediated iron uptake in *P. aeruginosa* with a proposed common transporter for Fe operating at the cytoplasmic membrane. Siderophore-specific receptors are responsible for transport of the various iron-siderophore complexes across the outer membrane into the periplasm, where Fe is released as Fe^{2+} (via a reductive mechanism) or Fe^{3+} by an as yet unknown mechanism. Transporters specific for Fe^{2+} (PA0369/PA0358) or Fe^{3+} (PA5216/PA5217; PA4687/PA4688) are then responsible for transport across the inner membrane. CIT, citrate; PCH, pyochelin; PVD, pyoverdine; FOX, ferrioxamine B. The PA designations are those reported by the *Pseudomonas* Genome Project (http://www.pseudomonas.com).

tected in *P. aeruginosa*, although the reductase activities were for the most part cytoplasmic. Should iron be freed from siderophores in the periplasm, candidate iron-specific transporters include a homologue (PA4359/PA4358) of the *E. coli* FeoAB Fe(II) transporter (see chapter 12) and homologues of the *Serratia marcescens* SfuABC and *Haemophilus influenzae* HitABC Fe^{3+} (see chapter 18) uptake systems (PA4687/PA4688 and PA5216/PA5217, respectively, in the *P. aeruginosa* genome). Interestingly, the PA4359 to PA4358 and PA5216 to PA5217 genes are iron regulated, although clearly any involvement of an Fe^{2+} transporter in siderophore-mediated iron uptake would require an as yet undiscovered reductive release mechanism in the periplasm.

Siderophore-Mediated Iron Uptake in *B. cepacia*

Four siderophores are reportedly produced by *B. cepacia*, including pyochelin, cepabactin (a cyclic hydroxamate), salicylic acid (formerly azurechelin), and the ornibactins (Fig. 2). The organism is also able to use the *Burkholderia pseudomallei* siderophore, malleobactin, to promote iron uptake. Pyochelin production by *B. cepacia* has been correlated with severe pulmonary disease in CF patients although salicylic acid and ornibactins predominate in lung isolates from these patients, suggesting that they are the major contributors to iron acquisition and thus growth in vivo.

Ornibactins are linear hydroxamate-hydroxycarboxylate siderophores possessing a conserved tetrapeptide to which is attached one of three possible acyl groups (Fig. 2). Two genes involved in ornibactin biosynthesis have been identified, encoding homologues of pyoverdine biosynthetic genes; this probably reflects the common presence of amino acids in both siderophore structures. The gene for the ferric ornibactin receptor, *orbA*, has also been cloned (Table 2); consistent with its homology to the family of TonB-dependent receptors, *tonB*, *exbB*, and *exbD* genes have been identified in *B. cepacia* and their involvement in iron acquisition has been confirmed. Receptor mutants and mutants deficient in ornibactin biosynthesis exhibit reduced virulence in animal models of infection, highlighting the importance of this iron uptake system for in vivo growth and pathogenesis. Interestingly, CepR, a quorum-sensing regulator that positively influences the expression of its cognate homoserine lactone and an extracellular protease, negatively impacts ornibactin production; i.e., mutants show increased ornibactin levels and expression of at least one ornibactin biosynthetic gene. This contrasts with *P. aeruginosa*, where the LasIR quorum-sensing system positively impacts pyoverdine production.

HEME ACQUISITION

P. aeruginosa, like many bacteria, is able to acquire iron from heme and heme-containing proteins such as hemoglobin via the expression of cell surface receptors. The heme receptors, like siderophore receptors, are TonB dependent (see chapter 3). Two heme uptake systems have been detected in *P. aeruginosa*, encoded by the *phu* and *has* loci, and expression of each is iron regulated via Fur. The *phu* genes include *phuR*, encoding an outer membrane heme receptor (Table 2), and *phuSTUVW*, encoding a transporter of the ABC family. Deletion of *phuR* or *phuSTUV* greatly the reduces growth of *P. aeruginosa* with hemin or hemoglobin as the sole iron source, although residual growth is clearly apparent and is probably attributable to the products of the *hasRA* operon. The *hasA* gene encodes an extracellular heme-binding protein (hemophore; see chapter 3) that also facilitates the utilization of hemoglobin iron, while the *hasR* product is a receptor for heme (Table 2). HasA homologues are present in *S. marcescens* and *P. fluorescens*, where their export requires the products of a three-gene operon, *hasDEF*, encoding a type I secretion apparatus of the ABC family. Homologues of *hasDEF* occur as a putative operon in *P. aeruginosa*, immediately downstream of *hasR*, although their role in HasA export has not been addressed. Interestingly, recent proteome studies have shown that PhuR production is positively regulated by quorum sensing, as is production of "functional" HasA, apparently because the protease(s) responsible for HasA activation is itself controlled by quorum sensing. Thus, quorum-sensing mutants are unable to utilize hemoglobin as an iron source.

While mutants lacking both the *has*- and *phu*-encoded heme acquisition systems are virtually unable to utilize hemin or hemoglobin, suggesting that only these two systems function in heme acquisition in *P. aeruginosa*, a possible third heme receptor gene, PA1302, has been identified in the *P. aeruginosa* genome (Table 2). Intriguingly, this gene is linked to genes encoding FecR-/FpvR-like sensor (PA1301) and FecI/FpvI-like ECF sigma factor (PA1302) components (Table 2), which are strongly iron regulated and, perhaps, involved in PA1302 expression. Similarly, the *hasR* heme receptor gene occurs adjacent to sensor (PA3409) and ECF sigma factor (PA3410) genes (Table 2), which are also iron regulated. Still, given the direct Fur regulation of *hasR* expression (see above), any iron regulation of this receptor does not occur via iron/Fur control of PA3409 to PA3410. The role of these genes in *hasR* expression if any remains to be elucidated.

CONCLUDING REMARKS

The ability of *P. aeruginosa* to utilize multiple siderophores in its quest for iron clearly provides this organism with an important competitive advantage in nature but probably necessitates the use of multiple siderophore-specific receptors and a common intracellular transporter for iron. This provides for needed flexibility in terms of siderophore utilization without devoting substantial resources (and genes) to the transport of what may be infrequently encountered ligands. While much has been learned about *P. aeruginosa* iron acquisition, the details of intracellular iron transport, including the identity of the cytoplasmic membrane and periplasmic transporter(s), as well as the site and mechanism of iron-siderophore dissociation, remain to be elucidated. Not unexpectedly, expression of individual receptors is regulated and under the control, in most instances, of a family of iron starvation ECF sigma factors that appear to be particularly common in *P. aeruginosa*. These function to provide, ultimately, for siderophore-dependent expression of the various ferric siderophore receptor genes present in this organism, ensuring that of the repertoire of receptors available, only the most effective at a given time and place is expressed. This provides for iron acquisition in a variety of environments and in the presence of a variety of microbial competitors while preventing possible excess iron accumulation as a result of multiple iron-siderophore transporters operating simultaneously.

The connection between pyoverdine biosynthesis and uptake and expression of extracellular virulence factors such as exotoxin A, alkaline protease, and the PrpL endoprotease is intriguing and needs further investigation. These apparently disparate activities may, for example, play a common role in pyoverdine-mediated iron acquisition, with these virulence factors promoting in vivo iron release through damage or destruction of host cells and/or proteins that sequester iron. Alternatively, their activities on host cells and/or proteins could provide component parts (e.g., amino acids) of the siderophore to be used in its biosynthesis. Incorporation of growth medium-provided amino acids into pyoverdine has, for example, been demonstrated, and growth medium effects on pyoverdine production are known. This suggests that the nutritional status of the environment, as influenced, e.g., by the action of these toxins and enzymes, can positively impact pyoverdine production by *P. aeruginosa*. The enzymology of pyoverdine biosynthesis, especially biosynthesis of the chromophore, needs to be resolved, along with the function of the many iron- and PvdS-regulated genes of the *pvd* locus and the possible contribution of the PvdS-regulated virulence factors to this process. The significance of apopyovedine binding to the FpvA receptor with respect to ferric pyoverdine binding and uptake also remains to be elucidated, as does the site and mechanism of iron-pyoverdine dissociation inside the cell and the mechanism of siderophore recycling to the cell surface.

Finally, while the study of iron acquisition in *Pseudomonas* has, for the most part, focused on planktonic organisms, biofilms may well be the predominant mode of growth of this organism in nature. Thus, more effort needs to be devoted to understanding the details of siderophore biosynthesis and transport in biofilms, as well as their regulation and contribution to the growth of biofilm organisms. While iron-regulated outer membrane proteins have been identified in in vitro-grown biofilms, suggesting that ferric siderophore receptors are expressed, the significance of these for biofilm growth and survival is as yet unclear. A recent study of the negative impact of lactoferrin (and other iron chelators) on *P. aeruginosa* biofilm formation does, however, highlight the probable significance of iron availability for biofilm formation.

SUGGESTED READING

Arevalo-Ferro, C., M. Hentzer, G. Reil, A. Gorg, S. Kjelleberg, M. Givskov, K. Riedel, and L. Eberl. 2003. Identification of quorum-sensing regulated proteins in the opportunistic pathogen *Pseudomonas aeruginosa* by proteomics. *Environ. Microbiol.* **5:**1350–1369.

Beare, P. A., R. J. For, L. W. Martin, and I. L. Lamont. 2003. Siderophore-mediated cell signalling in *Pseudomonas aeruginosa*: divergent pathways regulate virulence factor production and siderophore receptor synthesis. *Mol. Microbiol.* **47:**195–207.

Crosa, J. H., and C. T. Walsh. 2002. Genetics and assembly line enzymology of siderophore biosynthesis in bacteria. *Microbiol. Mol. Biol. Rev.* **66:** 223–249.

Darling, P., M. Chan, A. D. Cox, and P. A. Sokol. 1998. Siderophore production by cystic fibrosis isolates of *Burkholderia cepacia*. *Infect. Immun.* **66:**874–877.

De Chial, M., B. Ghysels, S. A. Beatson, V. Geoffroy, J. M. Meyer, T. Pattery, C. Baysse, P. Chablain, Y. N. Parsons, C. Winstanley, S. J. Cordwell, and P. Cornelis. 2003. Identification of type II and type III pyoverdine receptors from *Pseudomonas aeruginosa*. *Microbiology* **149:**821–831.

DeWitte, J. J., C. D. Cox, G. T. Rasmussen, and B. E. Britigan. 2001. Assessment of structural features of the *Pseudomonas* siderophore pyochelin required for its ability to promote oxidant-mediated endothelial cell injury. *Arch. Biochem. Biophys.* **393:** 236–244.

Dorrestein, P. C., K. Poole, and T. P. Begley. 2003. Formation of the chromophore of the pyoverdine siderophores by an oxidative cascade. *Org. Lett.* **5:**2215–2217.

Gaille, C., C. Reimmann, and D. Haas. 2003. Isochorismate synthase (PchA), the first and rate-limiting enzyme in salicylate biosynthesis of *Pseudomonas aeruginosa*. *J. Biol. Chem.* **278:**16893–16898.

Lamont, I. L., P. A. Beare, U. Ochsner, A. I. Vasil, and M. L. Vasil. 2002. Siderophore-mediated signaling regulates virulence factor production in *Pseudomonas aeruginosa*. *Proc. Natl. Acad. Sci. USA* **99:**7072–7077.

Meyer, J. M. 2000. Pyoverdines: pigments, siderophores and potential taxonomic markers of fluorescent *Pseudomonas* species. *Arch. Microbiol.* **174:** 135–142.

Meyer, J. M., A. Stintzi, and K. Poole. 1999. The ferripyoverdine receptor FpvA of *Pseudomonas aeruginosa* PAO1 recognizes the ferripyoverdines of *Pseudomonas aeruginosa* PAO1 and *Pseudomonas fluorescens* ATCC 13525. *FEMS Microbiol. Lett.* **170:** 145–150.

Mossialos, D., U. Ochsner, C. Baysse, P. Chablain, J.-P. Pirnay, N. Koedam, H. Budzikiewicz, D. U. Fernàndez, M. Schäfer, J. Ravel, and P. Cornelis. 2002. Identification of new, conserved, non-ribosomal peptide synthetases from fluorescent pseudomonads involved in the biosynthesis of the siderophore pyoverdine. *Mol. Microbiol.* **45:**1673–1685.

Ochsner, U. A., Z. Johnson, and M. L. Vasil. 2000. Genetics and regulation of two distinct haem-uptake systems, *phu* and *has*, in *Pseudomonas aeruginosa*. *Microbiology* **146:**185–198.

Ochsner, U. A., P. J. Wilderman, A. I. Vasil, and M. L. Vasil. 2002. GeneChip(R) expression analysis of the iron starvation response in *Pseudomonas aeruginosa*: identification of novel pyoverdine biosynthesis genes. *Mol. Microbiol.* **45:**1277–1287.

Ongena, M., P. Jacques, E. de Pauw, and P. Thonart. 2002. Synthesis of peptide-modified pyoverdins by a fluorescent *Pseudomonas* strain grown in isoleucine-supplemented medium. *Lett. Pept. Sci.* **8:** 21–27.

Pohl, E., J. C. Haller, A. Mijovilovich, W. Meyer-Klaucke, E. German, and M. L. Vasil. 2003. Architecture of a protein central to iron homeostasis: crystal structure and spectroscopic analysis of the ferric uptake regulator. *Mol. Microbiol.* **47:**903–915.

Poole, K., and G. A. McKay. 2003. Iron acquistion and its control in *Pseudomonas aeruginosa*: many roads lead to Rome. *Front. Biosci.* **8:**d661–d686.

Ravel, J., and P. Cornelis. 2003. Genomics of pyoverdine-mediated iron uptake in pseudomonads. *Trends Microbiol.* **11:**195–200.

Redly, G. A., and K. Poole. 2003. Pyoverdine-mediated regulation of FpvA synthesis in *Pseudomonas aeruginosa*: involvement of a probable extracytoplasmic function sigma factor, FpvI. *J. Bacteriol.* **185:** 1261–1265.

Royt, P. W. 1990. Pyoverdine-mediated iron transport. Fate of iron and ligand in *Pseudomonas aeruginosa*. *Biol. Metals* **3:**28–33.

Schalk, I. J., M. A. Abdallah, and F. Pattus. 2002. A new mechanism for membrane iron transport in *Pseudomonas aeruginosa*. *Biochem. Soc. Trans.* **30:** 702–705.

Singh, P. K., M. R. Parsek, E. P. Greenberg, and M. J. Welsh. 2002. A component of innate immunity prevents bacterial biofilm development. *Nature* **417:**552–555.

Sokol, P. A., P. Darling, S. Lewenza, C. R. Corbett, and C. D. Kooi. 2000. Identification of a siderophore receptor required for ferric ornibactin uptake in *Burkholderia cepacia*. *Infect. Immun.* **68:** 6554–6560.

Takase, H., H. Nitanai, K. Hoshino, and T. Otani. 2000. Impact of siderophore production on *Pseudomonas aeruginosa* infections in immunosuppressed mice. *Infect. Immun.* **68:**1834–1839.

Vasil, M. L., and U. A. Ochsner. 1999. The response of *Pseudomonas aeruginosa* to iron: genetics, biochemistry and virulence. *Mol. Microbiol.* **34:** 399–413.

Visca, P., L. Leoni, M. J. Wilson, and I. L. Lamont. 2002. Iron transport and regulation, cell signalling and genomics: lessons from *Escherichia coli* and *Pseudomonas*. *Mol. Microbiol.* **45:**1177–1190.

Zhao, Q., and K. Poole. 2002. Mutational analysis of the TonB1 energy coupler of *Pseudomonas aeruginosa*. *J. Bacteriol.* **184:**1503–1513.

BORDETELLA

Timothy J. Brickman, Carin K. Vanderpool, and Sandra K. Armstrong

20

INTRODUCTION

Bordetella pertussis, *Bordetella bronchiseptica*, and *Bordetella parapertussis* are mammalian respiratory pathogens that are highly genetically related gram-negative ß-proteobacteria of the family *Alcaligenaceae*. *B. pertussis*, the agent of whooping cough (pertussis), is an obligate human pathogen with no known environmental or nonhuman reservoirs. *B. parapertussis* causes a pertussis-like syndrome in humans and respiratory infections in sheep, and *B. bronchiseptica* infects nonhuman mammals, causing canine kennel cough, atrophic rhinitis in swine, and snuffles in rabbits. Adherence and host colonization by these species is thought to be mediated primarily by cell surface molecules including filamentous hemagglutinin, pertactin, and fimbriae. Bordetellae multiply on the host respiratory mucosa and produce pertussis toxin (*B. pertussis*), tracheal cytotoxin, dermonecrotic toxin, and adenylate cyclase, which damage host cells or interfere with host immune function. The genes encoding most of the known virulence factors are coordinately regulated by the BvgAS two-component signal transduction pathway in response to uncharacterized environmental signals. Transcription of the Bvg-regulated virulence genes has not been reported to be affected by iron starvation stress.

B. pertussis, *B. parapertussis*, and *B. bronchiseptica* are widely studied pathogens, and their genomic DNA sequences have recently been determined by investigators at The Wellcome Trust Sanger Institute. Analysis of this new genomic information, in parallel with ongoing experimental research, will yield insights into the mechanisms of *Bordetella* iron retrieval in the host. The aim of this chapter is to summarize the current knowledge of iron acquisition in *B. pertussis*, and although it is not possible to cite all of the works by investigators in the field, their research contributions have been critical to the development of this knowledge base.

In Vivo Growth of *B. pertussis*

Growth of *B. pertussis* in humans is restricted to the mucosal epithelial surface, and the bacteria exhibit a remarkable tropism for the cilia of ciliated respiratory epithelial cells. Although the bacteria are eventually cleared, the characteristic coughing paroxysms can continue for weeks, presumably due, in part, to direct damage to the respiratory epithelium. *B. pertussis* cells produce certain cytopathic toxins that

---

Timothy J. Brickman, Carin K. Vanderpool, and Sandra K. Armstrong, Department of Microbiology, University of Minnesota Medical School, Minneapolis, MN 55455–0312.

Iron Transport in Bacteria, Edited by Jorge H. Crosa, Alexandra R. Mey, and Shelley M. Payne
© 2004 ASM Press, Washington, D.C.

cause host tissue damage, potentially leading to the release of iron sources. Tracheal cytotoxin is a disaccharide tetrapeptide that is released from the bacteria during peptidoglycan synthesis. This cytotoxin causes an increase in host nitric oxide production, and as a result, epithelial cellular tight junctions are damaged, leading to extrusion of the ciliated cells and denudation of the epithelium. The *Bordetella* adenylate cyclase/hemolysin is a bifunctional virulence factor that, on internalization by a eukaryotic cell, is activated by calmodulin and catalyzes the hyperproduction of cyclic AMP, leading to the disruption of cellular regulatory functions. Since the adenylate cyclase has minimal in vitro hemolytic activity, it is thought that its function as a hemolysin may not play a major role in *B. pertussis* virulence in the host. Another toxin produced by *B. pertussis* and related subspecies that causes considerable cytopathology is the heat-labile dermonecrotic toxin. This toxin causes alterations in cellular morphology and differentiation by catalyzing the deamidation or polyamination of eukaryotic Rho regulatory GTPases, resulting in their constitutive activation. The contribution of dermonecrotic toxin to *B. pertussis* pathogenesis in a human host is not known. However, when injected subcutaneously into mice, this toxin causes pronounced necrotic lesions, and its production by *B. bronchiseptica* is associated with the characteristic nasal turbinate atrophy in swine with atrophic rhinitis.

B. pertussis Iron Acquisition

On colonization of the respiratory epithelium by *B. pertussis* and during the course of infection, changes in the host environment occur and the iron sources available to the bacterium may fluctuate. It could be envisioned that, since the organisms have ready access to mucosal lactoferrin, *B. pertussis* cells would exploit that iron source by using their native siderophore. Furthermore, siderophores produced by respiratory commensals or temporary colonizers may provide other iron retrieval options to *B. pertussis* if the pathogen can transport these compounds. During infection, *Bordetella* toxins may alter the integrity of the epithelium, allowing serum components such as complement, and possibly lymphocytes and erythrocytes, to breach the mucosa. This serum and the host intracellular molecules released on cell lysis may yield additional iron sources such as hemoglobin, haptoglobin, hemopexin, and transferrin that could be used by *B. pertussis*.

In 1987, Redhead and colleagues reported that *B. pertussis* could use ovotransferrin as the sole iron source and that the cells could tightly bind ovotransferrin, transferrin, and lactoferrin. In a separate study by Camille Locht and colleagues, two candidate transferrin- and lactoferrin-binding proteins were identified in fractions derived from iron-starved *B. pertussis* and *B. bronchiseptica* cells; however, further characterization of those proteins has not been reported. Recent analysis of the *B. pertussis* genomic sequence failed to identify genes encoding obvious homologs of known microbial lactoferrin- and transferrin-binding receptor proteins. *B. pertussis* can acquire iron from transferrin, even when the transferrin is sequestered in a dialysis bag, suggesting that iron acquisition from these host proteins may occur via a siderophore. Siderophore activity was detected in iron-restricted cultures of *B. pertussis*, *B. parapertussis*, and *B. bronchiseptica* by using the universal chrome azurol S (CAS) assay. The siderophore activity was determined to be of the hydroxamate chemical class and was produced by virulent as well as avirulent *B. pertussis* strains.

Subsequent research determined that *B. pertussis*, *B. parapertussis*, and *B. bronchiseptica* all produce the macrocyclic dihydroxamate siderophore alcaligin (Fig. 1). *B. pertussis* also utilizes a variety of other iron sources including heme compounds and siderophores, such as enterobactin, produced by other microbial species (xenosiderophores). Of the known iron sources that *B. pertussis* can use, the genetic systems for utilization of only three sources (heme, alcaligin, and enterobactin) have been characterized. Several Fur-regulated *B. pertussis*

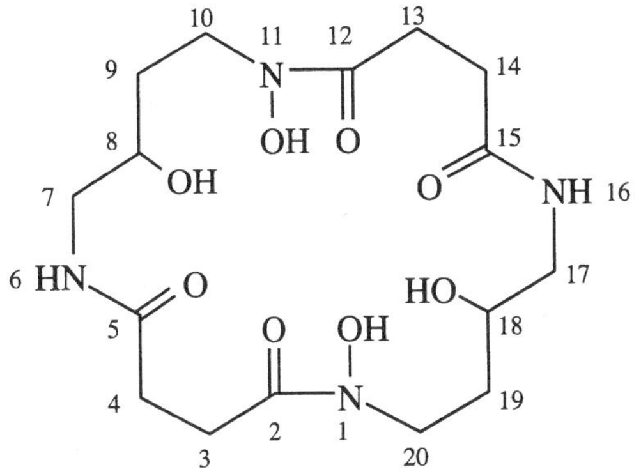

FIGURE 1 Molecular structure of alcaligin.

genes encoding putative siderophore receptors have been identified; however, the specific iron sources recognized by those receptors remain unknown. In silico analysis of the *B. pertussis* genome sequence has revealed a number of genes or gene clusters encoding predicted ferric siderophore transport systems. Although *B. pertussis* has an apparent multiplicity of genetic systems for the utilization of different iron sources, it not clear whether the organism exhibits preferential utilization of a given iron source at a specific tissue site or phase of the infection. Moreover, the iron compounds that are available to the organism in the host respiratory tract are not known. Our present understanding of the regulation of the three known *B. pertussis* iron retrieval systems indicates that the organism can detect specific iron sources in its immediate environment and, in response, can individually up regulate the genetic systems encoding functions for utilization of those iron sources. The fact that *B. pertussis* has evolved the ability to sense and respond to distinct iron compounds implies that this obligate pathogen encounters different iron sources during its life in a host.

ALCALIGIN SIDEROPHORE UTILIZATION

Using the CAS siderophore detection assay as well as the Csaky assay for hydroxamates, several groups have confirmed that virulent strains of *B. pertussis* and *B. bronchiseptica* both produced hydroxamate-class siderophores in response to iron starvation. Crystalline deferrisiderophores were purified from cultures of wild-type *B. pertussis* and *B. bronchiseptica* strains based on the benzyl alcohol-ether extraction procedure. Biological activity of the purified *Bordetella* siderophores was demonstrated by growth stimulation bioassays and [^{55}Fe]ferric siderophore uptake assays that indicated the involvement of a high-affinity transport system. Nuclear magnetic resonance spectroscopy and mass spectrometric analyses determined that the siderophores of *B. pertussis* and *B. bronchiseptica* were identical to the previously described siderophore alcaligin produced by the taxonomically related bacterial species *Alcaligenes* (now *Achromobacter*) *xylosoxidans* subsp. *xylosoxidans*. Alcaligin is produced at levels approximating 100 to 150 μM in iron-starved *B. pertussis* and *B. bronchiseptica* liquid cultures.

Alcaligin

Alcaligin, 1,8(S),11,18(S)-tetrahydroxy-1,6,11,16-tetraazacycloeicosane-2,5,12,15-tetrone, is a 20-member macrocyclic dihydroxamate siderophore with the molecular formula $C_{16}H_{28}N_4O_8$ (molecular weight 404) (Fig. 1). It is a cyclic dimer of two repeating units consisting of succinic acid and 1-amino-4-(N-hydroxylamino)-2(S)-butanol that exists primarily in a twofold-symmetric conformation in aqueous solution. Alcaligin is structurally similar to the cyclic dihydroxamate siderophores bisucaberin (22-member ring), produced by the marine bacterium *Pseudoalteromonas haloplanktis*, and putrebactin (20-member ring) produced by *Shewanella putrefaciens*, except that bisucaberin incorporates two residues of N-hydroxycadaverine and putrebactin has two residues of N-hydroxyputrescine instead of the hydroxyl-substituted N-hydroxyputrescine residues of alcaligin. As a tetradentate siderophore, alcaligin binds ferric iron at a 3:2 molar ratio (Fe_2Alc_3) at pH 6.0, whereas a less stable monomeric species (FeAlc) is predominant at pH 2.0.

The Fe$_2$Alc$_3$ complex has a stability constant estimated at 10^{37} M^{-1} at pH 6.0 based on EDTA displacement studies. Crystallographic analysis of ferric alcaligin elucidated the structural basis for its high stability: the free alcaligin ring is structurally preorganized for ferric iron binding. The Fe$_2$Alc$_3$ complex assumes a monobridged, U-shaped supramolecular structure composed of two terminal ligand fragments and one bridging fragment; thus, each iron atom is coordinated in a pseudooctahedral environment involving three hydroxamates. The alcaligin ring of the bridging ligand in the Fe$_2$Alc$_3$ complex is significantly twisted from its C_2 molecular symmetry, but the spatial positions of the ring atoms of the two terminal ligands are remarkably similar to those of the free alcaligin ligand. This structural preorganization precludes the need for marked conformational changes to bind iron and provides a 100-fold increase in metal complex stability over that of ferric complexes formed by the more flexible linear dihydroxamate siderophore rhodotorulic acid.

Studies that address the natural process of iron release from ferric alcaligin complexes have focused on a possible reductive mechanism. Siderophore denticity and preorganization afforded by cyclization of ferric hydroxamate siderophore complexes suggest that an increase in ligand denticity has a greater effect on redox potential than does the closure (cyclization) of the ligand backbone of the same denticity. The redox potential of ferric alcaligin might allow reductive iron removal under physiological conditions. Analysis of the kinetics of iron release from alcaligin indicated that preorganization of the iron-binding ligands of alcaligin, compared to the lack of structural preorganization of another tetradentate siderophore rhodotorulic acid, strongly influenced the ligand dissociation paths for their Fe$_2$L$_3$ ferric complexes.

Bordetella Alcaligin Biosynthesis Genes

In the development of screening methods for the identification of *Bordetella* siderophore mutants in our laboratory, a cross-species experimental approach was used. This approach was developed on the basis of the genetic relatedness of *B. pertussis* and *B. bronchiseptica* and the identity of alcaligin produced by both species and was necessitated by the inability to culture *B. pertussis* on CAS siderophore indicator agar. Siderophore-deficient mutants were identified by mass screening of *B. bronchiseptica* mini-Tn5 *lacZ1* random transposon insertion mutants on CAS agar. The *B. pertussis* homologs of the mutated siderophore genes were identified by DNA hybridization and mutant complementation experiments, and defined *B. pertussis* mutants were then constructed and phenotypically characterized.

One siderophore-deficient *B. bronchiseptica* mutant phenotype was complemented by a *B. pertussis* DNA region containing an open reading frame (ORF) whose product has amino acid sequence similarity to a family of pyridoxal phosphate-dependent amino acid decarboxylases. In biochemical studies, the mutant was shown to be defective in a constitutive ornithine decarboxylase. A *B. pertussis* ornithine decarboxylase gene (*odc*) mutant was also constructed and confirmed to be alcaligin deficient. The product of the reaction catalyzed by ornithine decarboxylase is putrescine, which was consistent with the structure-based prediction of putrescine as a likely precursor of alcaligin.

A gene cluster of dedicated alcaligin biosynthesis genes was identified through studies of another group of siderophore mutants (Fig. 2). Genetic complementation analyses confirmed the existence of three cotranscribed genes, *alcA*, *alcB*, and *alcC*, required for alcaligin production. The transcriptional start site of the *B. pertussis alc* genetic operon was localized to a promoter overlapping a Fur repressor-binding site. The steady-state level of *alc*-specific transcripts was elevated 18-fold under iron starvation compared with iron-replete growth conditions, demonstrating iron regulation at the transcriptional level. AlcA, AlcB, and AlcC have significant amino acid sequence similarities to enzymes involved in aerobactin siderophore biosynthesis in the family Enterobacteriaceae (Table 1). Based on these similarities, we postu-

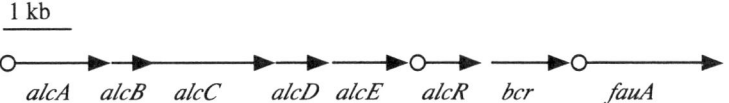

FIGURE 2 Genetic organization of the *Bordetella* alcaligin siderophore system. Arrows indicate the transcriptional orientations and spatial limits of genes, and open circles represent known Fur-regulated promoter-operator regions. Known AlcR-responsive control regions reside upstream of *alcA* and *fauA*. *bcr* has been renamed *alcS*.

late that AlcA is an oxygenase that catalyzes the hydroxylation of the alcaligin precursor putrescine produced from ornithine by the constitutive ornithine decarboxylase and that AlcB functions in an acylation step involving succinate. AlcC would function in one of the subsequent reactions, yielding alcaligin. The Locht research group determined the nucleotide sequence of the DNA region downstream from *alcC* and identified two additional ORFs presumed to be involved in alcaligin production, designated *alcD* and *alcE*. The

TABLE 1 *Bordetella* alcaligin system proteins and homologs

Bordetella alcaligin system genes			Homologs			
Protein	Function	GenBank accession no.	Protein	Function	GenBank accession no.	Identity
AlcA	Known alcaligin biosynthesis protein	Q44740	IucD	Aerobactin biosynthesis protein of *E. coli*; lysine-N6-hydroxylase/L-ornithine N5-oxygenase family	P11295	29% in 428 aa[a]
AlcB	Known alcaligin biosynthesis protein	AAB40619	IucB	Aerobactin biosynthesis protein of *E. coli*	Q47317	24% in 168 aa
AlcC	Known alcaligin biosynthesis protein	AAB40620	IucC	Aerobactin biosynthesis protein of *E. coli*	Q47318	29% in 595 aa
AlcD	Hypothetical alcaligin biosynthesis protein	CAA03891	BAB83801	Hypothetical protein of *V. parahaemolyticus*	AB066099	23% in 223 aa
AlcE	Known alcaligin biosynthesis protein	CAA03892	RSc2224	Hypothetical protein of *Ralstonia solanacearum*; dioxygenase (α subunit) oxidoreductase family	NP_52035	60% in 370 aa
AlcR	Known AraC/XylS family transcriptional regulator	AAC38169, CAA03893, AAL17615	PchR	Transcriptional regulator of pyochelin system genes of *P. aeruginosa*; AraC/XylS family	P40883	27% in 242 aa
Bcr (Aks)	Known alcaligin permease	CAA03894	Bcr	Sulfonamide and bicyclomycin resistance protein of *E. coli*; major facilitator family of membrane efflux pumps	P28246	29% in 391 aa
FauA	Known ferric alcaligin receptor protein	AAD26430	FpvA	Ferric pyoverdine receptor protein of *P. aeruginosa*; TonB-dependent receptor family	A40601	37% in 704 aa

[a] aa, amino acids.

function of AlcD is unknown, although it has similarities to several hypothetical proteins including a predicted *Vibrio parahaemolyticus* iron-sulfur cluster protein that maps adjacent to a *V. parahaemolyticus iutA* ferric aerobactin receptor protein gene homolog. AlcE is similar to the ring-hydroxylating dioxygenase family of iron-sulfur cluster proteins; therefore, we hypothesize that AlcE may be inovlved in catalyzing the two hydroxyl additions at alcaligin ring positions 8 and 18.

The AlcR-Positive Regulator and the *alcABCDER* Operon

In an experimental attempt to directly select for ferric alcaligin transport mutants, a pool of *B. bronchiseptica* random transposon insertion mutants was treated with the iron- and redox-activated quinone antibiotic streptonigrin on iron-restricted medium containing alcaligin as the sole iron source. A streptonigrin-resistant mutant was obtained that was defective not only in alcaligin utilization but also in alcaligin production as well as in the production of AlcC and the ferric alcaligin outer membrane receptor protein, FauA. Complementation analysis identified a gene encoding an AraC/XylS-family transcriptional regulator, AlcR, that mapped downstream from the alcaligin biosynthesis genes. AlcR was independently identified by the Locht group using a screening method for identification of *Bordetella* Fur-regulated promoters based on the Fur titration assay in *Escherichia coli*. AlcR is most similar to the known AraC-like regulators PchR of *Pseudomonas aeruginosa* and YbtA of *Yersinia pestis*, which are involved in regulation of the pyochelin and yersiniabactin iron transport systems, respectively. AlcR is also highly similar to the recently identified *Bordetella* BfeR regulator of enterobactin utilization genes. The pleiotropic phenotype of the *alcR* mutant was consistent with a role of AlcR as a positive regulator of alcaligin biosynthesis and transport genes. Promoter mapping and RNA hybridization analyses using *alc* region-specific probes with defined *alcA* promoter deletion mutants determined that *alcR* is transcribed primarily from the *alcABCDE* promoter and is also transcribed from a weaker secondary Fur-regulated promoter located immediately upstream of its coding sequence.

The AlcS Permease

A permease of the major facilitator superfamily class of proton motive force-dependent membrane efflux pumps, AlcS (previously known as OrfX or Bcr), is encoded by a monocistronic operon located between *alcR* and *fauA*. Expression of *alcS* is independent of the *alcABCDER* control region and is not repressible by iron. In phenotypic analyses, *alcS* null mutants were defective in (i) growth in iron-depleted medium; (ii) growth stimulation by iron salts, alcaligin, and hemin; (iii) alcaligin production; and (iv) transcriptional activation of *alcABCDER*. Interestingly, studies of an alcaligin-deficient Δ*alcA* Δ*alcS* double-mutant strain revealed that the nonpolar Δ*alcA* mutation suppressed the growth defects of *alcS* strains. This suppression was reversed by complementation of the Δ*alcA* mutation, confirming that the growth-defective phenotype of *alcS* mutants is associated with alcaligin production. Research on the *E. coli* enterobactin siderophore system by Mark McIntosh and colleagues demonstrated that the EntS membrane efflux pump is required for the efficient secretion of enterobactin. Our observations with the *Bordetella* mutants suggest that in the absence of a functional AlcS permease normally required for efficient export of newly synthesized alcaligin, alcaligin production is toxic to *Bordetella* cells.

The FauA Ferric Alcaligin Receptor

The *B. pertussis* outer membrane ferric alcaligin receptor was identified in our laboratory by using siderophore-deficient *P. aeruginosa* strain IA1 as a heterologous host system to screen for *B. pertussis* alcaligin utilization genes. An ORF downstream from *alcR* conferred to strain IA 1 the ability to utilize ferric alcaligin as a sole source of iron. Nucleotide sequencing identified a TonB-dependent receptor family homolog that was shown to be the ferric alcaligin receptor gene designated *fauA* (for "ferric alcaligin uptake"). In regulation studies, *fauA*, which has a functional Fur repressor-binding site, was shown to be Fur and iron repressible and to be dependent on the AlcR regulator for maximal expression. These analyses of FauA regulation explained our previous observations of the ferric alcaligin utilization defect in *alcR* mutants.

The *Bordetella* alcaligin system gene cluster spans an approximately 11-kb genomic region and includes the *alcABCDER* operon and the *alcS* and *fauA* genes (Fig. 2). The *alcA* and *fauA* genes are thought to delimit the alcaligin system gene cluster, since flanking ORFs are not predicted to be involved in iron transport. At this time, no other ferric alcaligin transport functions have been identified in *Bordetella* species, but candidate cytoplasmic membrane transporters have been identified in the *B. pertussis* nucleotide sequence database.

Transcriptional Repression of Alcaligin System Genes

By analogy to other known siderophore systems, it was hypothesized that regulation of *Bordetella* siderophore production involved a Fur homolog. To identify the *fur* locus, we isolated deregulated *B. bronchiseptica* iron transport mutants by selection for spontaneous manganese resistance, which had been shown to yield *fur* mutants in other bacterial species. The *B. pertussis fur* gene was cloned by complementation of the deregulated mutant phenotype, and its identity was confirmed by nucleotide sequencing. In concurrent studies, the Beall research group also identified and characterized the *B. pertussis fur* gene.

In gel mobility shift experiments performed in our laboratory, purified *B. pertussis* Fur protein bound probes representing the *alcA* upstream DNA region in a divalent-metal-dependent and sequence-specific manner. The *alcA* operator region exhibited multiple high-affinity Fur-binding sites; at least three electrophoretic species representing Fur-DNA complexes were resolved. Mathematical modeling of the Fur-*alcA* DNA gel shift data determined that the multiple Fur-binding events display a significant degree of positive cooperativity. The Fur protein also exhibited specific interactions with *alcR*-upstream DNA region probes spanning the *alcR* promoter identified by primer extension, which is adjacent to a single Fur-binding site. Fur affinity for the *alcA* and *alcR* operators was nearly identical based on K_d estimates derived from Bjerrum plots of equilibrium binding data. A third Fur-binding region was identified upstream of the *fauA* receptor gene; thus, Fur- and iron-mediated negative control of transcription is exerted at three known promoter-operator regions within the alcaligin system gene cluster: (i) upstream of the *alcABCDER* operon, (ii) upstream of *alcR*, and (iii) upstream of *fauA*.

Transcriptional Activation of Alcaligin System Genes

We had originally chosen mini-Tn5 *lacZ1* for mutagenesis of *B. bronchiseptica* not only because it is useful for insertional inactivation of genes but also because it generates operon fusions for transcriptional regulation studies. However, during initial characterization of alcaligin-deficient *B. bronchiseptica alc*::mini-Tn5 *lacZ1* insertion mutants, iron-regulated *lacZ* reporter gene expression was not observed. This was inconsistent with the experimental results with the parent strain demonstrating that alcaligin production and *alc* transcription were strongly iron repressible. Subsequent studies in our laboratory revealed that *alcR* was part of the Fur-regulated *alcABCDER* operon. AlcR, an AraC-like regulator that might require a small-molecule inducer, is necessary for maximal expression of alcaligin biosynthesis and transport activities. Supplementation of the alcaligin-deficient *B. bronchiseptica alc*::mini-Tn5 *lacZ1* cultures with purified alcaligin resulted in strong activation of *alc* transcription. Compensation for the polarity of the insertions on the cotranscribed *alcR* gene by supplying *alcR*$^+$ plasmids resulted in maximal alcaligin-responsive transcriptional activity. Transfer of an *alcR* null mutation to the *alc*::mini-Tn5 *lacZ1* strains eliminated transcriptional responsiveness to alcaligin, confirming that maximal *alc* operon transcriptional activity during iron starvation is alcaligin and AlcR dependent. Alcaligin does not increase *alc* transcription under iron-replete conditions, indicating that loss of Fur repression is a prerequisite for alcaligin inducer responsiveness and AlcR activation of transcription. A truncated *alcR* mutant allele was constructed that encodes only the AlcR inducer-binding and multimerization domain (amino acids 1 to 223) but lacks the C-terminal DNA-binding domain. This AlcR(1,223) derivative is defective for activation of transcription, and coexpression of *alcR*(1,223) with wild-type *alcR* in *trans*-complementation analyses strongly inhibits AlcR- and alcaligin-mediated activation of *alcA*. The *trans*-dominance of the truncated *alcR* allele provides genetic evidence for AlcR multimerization, consistent with the prediction that AlcR functions as a homodimer similar to other AraC family regulators. The *alcABCDER* operon and *fauA* promoters both require AlcR for maximal transcriptional activity. No other AlcR-regulated genes have been identified.

Alcaligin Sensing

Transcriptional activation of *Bordetella* genes by AlcR is remarkably sensitive to the presence of purified deferri-alcaligin inducer. In inducer titration experiments, AlcR showed a threshold concentration of 10 ng of alcaligin/ml for activation of *alc* operon transcription in iron-starved cultures. This alcaligin concentration is more than 1,000-fold lower than the concentration required for measurable growth stimulation in bioassays. Interestingly, *fauA* and *tonB* null mutants that cannot transport ferric alcaligin for growth are equally capable of responding transcriptionally to the same low levels of alcaligin inducer. These results indicate that high-affinity receptor-mediated uptake of ferric alcaligin via FauA or another TonB-dependent outer membrane receptor is not required for induction of *alc* transcription. Similarly, transport-proficient cells that are treated with the proton ionophore carbonyl cyanide *m*-chlorophenyl hydrazone to inhibit active transport are also fully transcriptionally responsive to alcaligin inducer. Since AlcR is an AraC-type regulator, probably requiring direct interaction with the alcaligin inducer, the siderophore must gain access to the cell cytoplasm for interaction with AlcR. The molecular mass of deferri-alcaligin is only 404 Da, consistent with the possibility that it could cross the outer membrane via passive diffusion through porins; however, the mechanism of inducer transport across the cytoplasmic membrane remains to be determined.

In the course of characterizing the inducer requirements of AlcR, we also observed that *B. bronchiseptica* AlcR was partially constitutive for activation of *alc* operon transcription when overproduced from multicopy plasmids, whereas *B. pertussis* AlcR retained strict alcaligin inducer dependence when overproduced. Functional analysis of *B. pertussis*-*B. bronchiseptica* AlcR chimeras engineered by domain swapping and reciprocal site-specific mutagenesis defined the molecular basis of this multicopy suppression phenotype as a single amino acid difference at position 103 in the N-terminal inducer-binding and multimerization domain of AlcR. Site-directed mutagenesis based on AlcR structural predictions resulted in the production of a mutant AlcR protein with a threonine substitution at this position that was fully constitutive for activation of *alcA*::mini-Tn*5 lacZ1* transcription, even at single-copy gene dosage; the *alcR*(Con) allele was phenotypically dominant when coexpressed with the wild-type *alcR*. These results defined amino acid 103 as a critical AlcR structural determinant of alcaligin-dependent transcriptional activation of the alcaligin system genes.

XENOSIDEROPHORE UTILIZATION

Many bacteria have functional transport systems for siderophores produced by other microbial species. Piracy and utilization of xenosiderophores expands the microbial iron retrieval repertoire and increases the chances of successful growth in a given environment by providing bacteria with a strong competitive edge. For example, it has long been known that *E. coli* K-12 strains not only are capable of producing and using the enterobactin and aerobactin siderophores but also can utilize the xenosiderophores ferrichrome, rhodotorulic acid, coprogen, and ferrioxamine B for iron acquisition. Similarly, the rhizosphere inhabitant *Pseudomonas putida* produces its own pyoverdine siderophore and can utilize enterobactin, aerobactin, and other pyoverdine structural variants. The importance of siderophore piracy to strict pathogens living within a host is not obvious. Xenosiderophore utilization by pathogens may allow them to profit by using the iron chelators produced by coinfecting or commensal microbes. However, direct in vivo evidence for such profiteering is lacking.

Although *B. pertussis* and *B. bronchiseptica* have been reported to use a variety of xenosiderophores, the only genetic system with a characterized specificity is that for the utilization of enterobactin and the enterobactin breakdown product 2,3-dihydroxybenzoylserine. *B. pertussis* and *B. bronchiseptica* can also use ferrichrome and desferroxamine B to obtain iron, and *B. bronchiseptica* has the capacity to utilize coprogen, schizokinen, ferricrocin, vicibactin, ferrichrysin, ferrirubin, protochelin, aerobactin, and several pyoverdins. Two putative siderophore receptor genes, *bfrZ* and *bfrA*, are present in *B. bronchiseptica* and are absent from the *B. pertussis* genome, suggesting that *B. bronchiseptica* has a broader siderophore utilization capacity than *B. pertussis*. Further analysis of the *B. pertussis* genomic sequence reveals the existence of eight candidate siderophore receptor genes for utilization of unknown ligands, four of which are *bfrB*, *bfrC*, *bfrD*, and *bfrE*. Interestingly, expression of the *B. pertussis bfrD* gene, located directly adjacent to the *bfrE* putative siderophore receptor gene, has been reported to be activated by the BvgAS signal transduction system involved in virulence gene regulation.

Enterobactin

B. pertussis and *B. bronchiseptica* use ferric enterobactin, a potent catechol-type siderophore that is a cyclic trimer of 2,3-dihydroxybenzoylserine. Enterobactin is synthesized by members of the *Enterobacteriaceae* and in certain *Streptomyces* species isolated from soil. There are bacterial pathogens in addition to *B. pertussis* and *B. bronchiseptica* that use enterobactin as a xenosiderophore but do not inhabit the intestinal tract, soil, or water and in fact are obligate pathogens that are generally restricted to the host mucosa. For example, *Neisseria meningitidis* and *Neisseria gonorrhoeae* do not synthesize siderophores but possess transport systems for utilization of ferric enterobactin (see chapter 17). The fact that these obligate mucosal pathogens have no known external environmental reservoirs but possess functional enterobactin transport systems implies that the ability to use this high-affinity iron chelator may confer a growth advantage in the host. Although it is clear that *N. gonorrhoeae* may encounter enterobactin-producing members of the fecal *Enterobacteriaceae* in its genital tract habitat, the ability of obligate human respiratory pathogens such as *N. meningitidis* and *B. pertussis* to use enterobactin suggests that this siderophore can also be found on the human respiratory mucosa. Enterobactin has an extremely high ferric ion stability constant of 10^{49} M^{-1}; in comparison, the *Bordetella* alcaligin siderophore has a ferric ion stability constant of 10^{37} M^{-1}. In complex microbial communities where enterobactin is present, lower-affinity siderophores such as alcaligin would be stripped of their iron by enterobactin and thus would be ineffective in retrieving iron for the producing organism. A microbe with a functional enterobactin transport system would then be able to access and utilize this chelated iron to satisfy its nutritional requirements. The ability of a microbe to use the highest-affinity siderophore in its environment confers one of the best competitive growth advantages, since a key, limiting nutrient is involved.

Bordetella Enterobactin Utilization Genes

The *B. pertussis* and *B. bronchiseptica* BfeA outer membrane receptor is required for utilization of ferric enterobactin and 2,3-dihydroxybenzoylserine. This receptor has significant amino

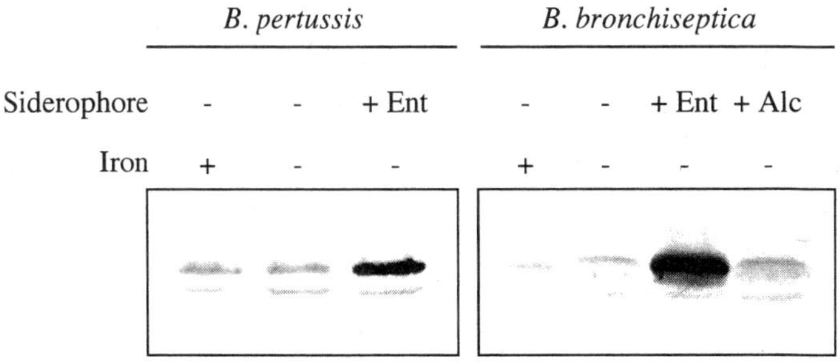

FIGURE 3 Enterobactin-inducible BfeA receptor protein production. Production of the ~80-kDa BfeA ferric enterobactin receptor protein was monitored by immunoblot analysis of *B. pertussis* and *B. bronchiseptica* total proteins by using a cross-reactive *E. coli* FepA-specific antiserum. Lanes: + Iron, iron-replete culture; − Iron, iron-depleted culture; − Iron +Ent, iron-depleted culture supplemented with enterobactin; − Iron +Alc, iron-depleted culture supplemented with alcaligin.

acid sequence identity to the ferric enterobactin receptors of *E. coli* (FepA) and *P. aeruginosa* (PfeA) and is also ~52% identical to the *Salmonella enterica* IroN outer membrane receptor for ferric catechols. The BfeA proteins of *B. pertussis* and *B. bronchiseptica* react in immunoblot analysis with antiserum raised to the *E. coli* ferric enterobactin receptor protein (Fig. 3). In both *Bordetella* species, BfeA was not produced in iron-replete cultures or, to any significant degree, in iron-depleted cultures, unless they were supplemented with purified enterobactin. This result is consistent with the reported observation by Phillip Klebba and colleagues of enhanced ferric enterobactin binding to iron-starved *B. bronchiseptica* cells that were precultured with the siderophore. We also observed that a *B. bronchiseptica fur* mutant produced BfeA at low levels regardless of the iron concentration in the medium but significantly increased BfeA expression in the presence of enterobactin. Together, these experimental results indicate that production of the BfeA ferric enterobactin receptor in *Bordetella* spp. is Fur repressible and enterobactin responsive.

Our analysis of the *B. pertussis* genomic database identified two ORFs, flanking the *bfeA* ferric enterobactin receptor gene, that are predicted to encode related ferric enterobactin utilization functions (Fig. 4; Table 2). Downstream of *bfeA* is an ORF (*bfeB*) in the same transcriptional orientation as *bfeA*, encoding a deduced protein with 32% identity to the product of the *S. enterica iroE* gene located in the *iroBCDEN* catechol siderophore utilization gene cluster. Similarly positioned downstream

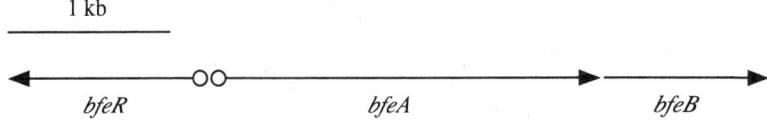

FIGURE 4 Genetic organization of the *Bordetella* enterobactin utilization system. Arrows indicate the spatial limits and transcriptional orientations of the *bfe* genes, and open circles designate the positions of predicted Fur-binding sites upstream from *bfeR* and *bfeA*.

TABLE 2 *Bordetella* enterobactin utilization system proteins and homologs

Bordetella enterobactin system genes			Homologs			
Protein	Function	GenBank accession no.	Protein	Function	GenBank accession no.	Identity
BfeR	Regulator of enterobactin transport and utilization; AraC/XylS family protein		AlcR	Regulator of alcaligin siderophore system genes of *B. pertussis* and *B. bronchiseptica*; AraC/XylS family protein	AAC38169	34% in 243 aa[a]
BfeA	Ferric enterobactin receptor protein; TonB-dependent receptor family	AAA98536	PfeA	Ferric enterobactin receptor protein of *P. aeruginosa*; TonB-dependent receptor family	Q05098	52% in 740 aa
BfeB	Hypothetical ferric enterobactin esterase/hydrolase		IroE	Predicted esterase/hydrolase of *Salmonella enterica* serovar Choleraesuis; esterase/hydrolase of αβ superfamily	AAK33132	32% in 243 aa

[a] aa, amino acids.

from the *P. aeruginosa pfeA* enterobactin receptor gene is an ORF (genome locus tag PA2689) potentially encoding a protein that is also highly similar to both IroE and the *B. pertussis* BfeB protein. BfeB also shows significant similarity to the Fes protein of *Erwinia chrysanthemi* and other members of the αβ-esterase/hydrolase fold subfamily of the serine active-site esterase/lipase superfamily. In *E. coli*, the Fes enzyme is required for the release of iron from ferric enterobactin after its transport to the cytoplasm. Other bacterial species capable of acquiring iron via enterobactin would be predicted to require a similar esterase or other activity for the release of iron from the ferric siderophore. The BfeB protein has the signature motif GXSXGG, which is common to members of the serine active-site esterase family (InterPro accession no. IPR000379). This motif lies within a 64-amino-acid segment of BfeB that is highly conserved among IroE, Fes, and other esterases. The BfeB protein also has a predicted secretion signal, suggesting that it may be localized to the cytoplasmic membrane or the periplasm. If BfeB is periplasmic, it is possible that it not only functions as the periplasmic binding protein component of the transport apparatus but also provides the esterase activity required for removal of iron from enterobactin.

Regulation of Enterobactin Utilization Genes

Divergently transcribed from *bfeA* is the *bfeR* gene, which is predicted to encode an AraC/XylS-like regulator that has significant identity to *Bordetella* AlcR, involved in alcaligin gene regulation, with the greatest similarity residing in the putative C-terminal DNA-binding domain. Since BfeA production in iron-starved cells was stimulated by the presence of enterobactin, BfeR was postulated to mediate enterobactin-responsive transcriptional activation of *bfeA* in a manner analogous to activation of *fauA* transcription by alcaligin and AlcR. Experiments in our laboratory with *B. pertussis* and *B. bronchiseptica bfeR* deletion mutants indicate that BfeR is required for efficient utilization of ferric enterobactin. Transcriptional analyses using *bfeA-lacZ* gene fusions demonstrated that upregulated *bfeA* expression in iron-starved *Bordetella* strains required both *bfeR* and enterobactin.

The ORFs flanking the three *bfe* genes are not predicted to be involved in iron metabolism. Therefore, any other genes potentially encoding ferric enterobactin utilization functions, such as cytoplasmic membrane permease components, are located outside the *bfe* gene cluster. The close spacing between the *bfeA* and *bfeB* genes and the lack of predicted promoter sequences upstream of *bfeB* suggest that these genes are translationally coupled. The *bfeR-bfeAB* intergenic region contains three putative Fur-binding sites, and a 0.4-kb DNA region containing these sites binds Fur strongly in Fur titration assays in *E. coli*. In *B. pertussis* and *B. bronchiseptica*, BfeR activates *bfeAB* transcription at this intergenic region by a mechanism requiring the enterobactin siderophore as the inducer.

Of the known ferric enterobactin utilization systems, the only other system reported to be positively transcriptionally regulated in response to enterobactin is that of *P. aeruginosa*. However, the mechanism of transcriptional regulation of ferric enterobactin transport genes in *P. aeruginosa* involves the PfeR-PfeS two-component signal transduction system rather than an AraC/XylS family regulator. The *Bordetella* BfeA and BfeB proteins and the homologous *P. aeruginosa* PfeA and PA2689 proteins have significant similarity to the *S. enterica* IroN and IroE proteins, respectively. Similar to *Bordetella* BfeA, IroN functions in *Salmonella* strains to transport ferric 2,3-dihydroxybenzoylserine complexes as well as ferric enterobactin. Additionally, in *S. enterica*, IroN serves as a receptor for corynebactin, the *Corynebacterium glutamicum* siderophore that is structurally similar to enterobactin and identical to bacillibactin produced by *Bacillus subtilis*. It is conceivable that the *B. pertussis* BfeA receptor has a similarly broad catechol specificity, allowing the organism to potentially utilize siderophores produced by coryneform commensals or by soil organisms that may transiently colonize the host, as well as enterobactin that may be available on the human respiratory mucosa.

HEME UTILIZATION

B. pertussis is considered a fastidious microbe, owing to its growth inhibition by compounds such as colloidal sulfur, peroxides, and certain fatty acids: compounds that contaminate glassware, are components of many culture media, and are present in agar. Early studies determined that supplementation of agar media with whole blood or erythrocytes could overcome growth inhibition of *B. pertussis*; additives such as serum, albumin, starch, and ionic exchange resins could also be substituted with moderate success. These additives have been presumed to bind or otherwise inactivate inhibitors in the agar medium, although it is not clear whether addition of blood or serum also provides other nutrients to *B. pertussis*. Today, many researchers continue to use Bordet's and Gengou's original 1906 formulation of potato extract agar that is supplemented with 20 to 25% sheep blood. Prior to 1939, the only liquid medium that would support the growth of *B. pertussis* contained blood or other substances derived from animal tissues, and investigators reported that addition of iron or hemin to defined liquid media resulted in enhanced growth.

More recently, the Dyer and Beall groups observed that hemin and hemoglobin served as sources of nutritional iron for *B. pertussis*. Nicholson and Beall showed that utilization of hemin and hemoglobin was TonB dependent in both *B. pertussis* and *B. bronchiseptica*, indicating that a specific high-affinity system was required for heme transport across the outer membrane. *B. pertussis* and *B. bronchiseptica* can use a variety of heme iron sources including hemin, hemoglobin-haptoglobin, heme-bovine serum albumin, and human hemoglobin, as well as hemoglobins from other animal sources including turkey, rabbit, bovine, and pig.

Bordetella Heme Utilization Genes

In *B. pertussis* and *B. bronchiseptica*, the BhuR outer membrane heme receptor is required for utilization of all of the known heme iron compounds. The apparent broad substrate specificity of BhuR is similar to that of some other TonB-dependent heme receptors, including

FIGURE 5 Genetic organization of the *Bordetella* heme iron utilization system. Arrows indicate the transcriptional orientations and spatial limits of genes, and open circles designate the positions of predicted promoter regions. A Fur-binding site resides upstream of *hurI*, and a predicted ECF σ factor-dependent promoter is upstream of *bhuR*.

Yersinia enterocolitica. HemR and *Y. pestis* HmuR, which mediate the utilization of hemin and hemoglobin, as well as myoglobin, catalase, and hemopexin. The genetic system encoding heme utilization functions in *B. pertussis* and *B. bronchiseptica* includes the *hurIR bhuRSTUV* genes (Fig. 5; Table 3). An orthologous gene cluster, *rhuIR bhuRSTUV*, is required for heme utilization in the more distantly related avian pathogen *Bordetella avium*.

The *hurIR bhuRSTUV* gene clusters of *B. pertussis* and *B. bronchiseptica* are virtually identical at the nucleotide sequence level. Furthermore, the *B. pertussis hurI* and *bhuRSTUV* genes can complement heme utilization defects in *B. bronchiseptica hurI* and *bhuR* mutants, respectively, indicating that the systems are functionally interchangeable. HurI and HurR are most similar to members of the iron starvation subfamily of extracytoplasmic function (ECF) regulators, including FecI and FecR of the *E. coli* ferric citrate uptake system and PupI and PupR of the *P. putida* WCS358 ferric pseudobactin uptake system. FecI and PupI are ECF σ factors whose activity is modulated by a physical interaction with the cytoplasmic membrane-bound FecR and PupR proteins, respectively. FecI and PupI are able to initiate the transcription of genes encoding iron transport machinery only under conditions of iron limitation in the presence of the specific iron compound. HurI contains a motif characteristic of ECF σ factors, while HurR contains the three highly conserved tryptophan residues shared by FecR and its homologs and has a predicted membrane-spanning region, consistent with its predicted cytoplasmic membrane localization.

TABLE 3 *Bordetella* heme iron utilization system proteins and homologs

Bordetella heme iron system genes			Homologs			
Protein	Function	GenBank accession no.	Protein	Function	GenBank accession no.	Identity
HurI	Known regulator; predicted ECF σ factor	AAM34715	FecI	ECF σ factor of *E. coli*	JV0111	46% in 153 aa[a]
HurR	Predicted sensor/regulator		FecR	Sensor/regulator of *E. coli*	B37804	29% in 332 aa
BhuR	Known TonB-dependent heme receptor	AAK38153	PhuR	TonB-dependent heme receptor of *P. aeruginosa*	AAC13289	28% in 787 aa
BhuS	Predicted heme utilization protein		HemS	Heme utilization protein of *Y. enterocolitica*	S54436	37% in 332 aa
BhuT	Predicted periplasmic heme-binding protein		HemT	Predicted periplasmic heme-binding protein of *Y. enterocolitica*	S54437	37% in 248 aa
BhuU	Predicted heme permease		HemU	Predicted heme permease of *Y. enterocolitica*	CAA54863	49% in 254 aa
BhuV	Predicted ATPase		HemV	Predicted ATPase of *Y. enterocolitica*	CAA54864	35% in 250 aa

[a] aa, amino acids.

The Bhu proteins are most similar to components of heme utilization systems in other gram-negative pathogens including *P. aeruginosa*, *Y. pestis* and *Y. enterocolitica*, *Shigella dysenteriae*, and *Vibrio cholerae*. BhuR is an outer membrane heme receptor with characteristics that are conserved among members of the microbial heme receptor family, including a TonB box C motif and the FRAP/NPNL amino acid sequence motif. BhuR contains an N-terminal "extension" which is unusual among heme receptors but is commonly found in receptors whose expression is regulated by ECF σ factors, such as the *E. coli* ferric citrate receptor FecA and the *P. putida* WCS358 ferric pseudobactin receptor PupB. This N-terminal region is required for the signaling process that leads to transcriptional initiation by the ECF σ factor. BhuS is predicted to be a cytoplasmic protein, and while its homologs have been termed heme-degrading factors in some systems, their role in heme degradation has not been demonstrated experimentally. Production of the *Y. enterocolitica* HemS protein in *E. coli* protected cells against heme toxicity, and the ShuS protein from *S. dysenteriae* binds heme, leading to the hypothesis that the role of these proteins in heme utilization may be to bind and sequester or store cytoplasmic heme. Other putative heme-degrading enzymes have not yet been identified in *Bordetella* species. The predicted BhuT, BhuU, and BhuV proteins show significant similarity to the periplasmic heme-binding proteins, cytoplasmic membrane permeases, and ATP-binding proteins, respectively, of many other bacterial iron assimilation systems.

Regulation of *Bordetella bhu* Genes

The *bhuRSTUV* genes appear to constitute an operon, since there are no obvious promoters or terminators within the cluster, and a deletion-insertion mutation in *bhuR* cannot be fully complemented by *bhuR* in *trans*, consistent with polarity on downstream *bhuSTUV* genes. The *hurR-bhuR* intergenic region is 210 nucleotides in length, and the *hurI* and *hurR* ORFs overlap by 22 nucleotides, suggesting that they also are in an operon and that translation of HurI and HurR is coupled. A predicted σ^{70}-like promoter with a potential Fur-binding site overlapping the -35 region was identified upstream of *hurI*, and promoter elements resembling those of other ECF σ factor-dependent promoters are found upstream of *bhuR*. Expression of heme iron utilization systems in gram-negative species is generally controlled by Fur repression. However, Fur titration assays demonstrated that the predicted *Bordetella bhuR* promoter region had little or no functional Fur-binding activity while the *hurI* promoter region had a strong Fur-binding site. The organization and the location of predicted Fur-binding sites and promoter elements within the *hurIR bhuRSTUV* gene cluster indicates that the *bhu* genes are iron repressed through the action of Fur at an operator site upstream of the *hurI* positive regulator gene.

The first experimental evidence supporting positive regulation of the *bhu* system was the observation that BhuR production by iron-starved bacteria was increased in response to hemin (Fig.

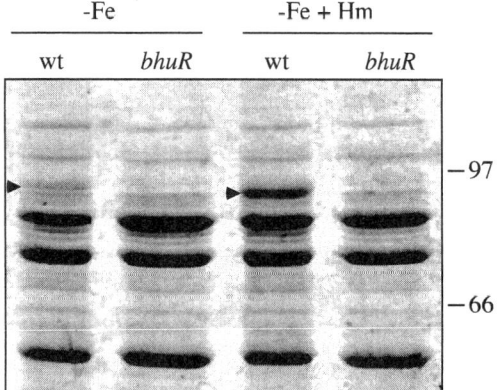

FIGURE 6 Heme-inducible BhuR receptor protein production. Sodium dodecyl sulfate-polyacrylamide gel electrophoresis of membrane fractions of a wild-type (wt) *B. bronchiseptica* strain and an isogenic *bhuR* null mutant cultured under iron-depleted conditions ($-$Fe) and iron-depleted conditions with hemin supplementation (5 μM) ($-$Fe $+$Hm) is shown. Positions of molecular mass markers (in kilodaltons) and BhuR (arrowheads) are indicated. Adapted from Vanderpool and Armstrong (2001) with permission.

6). Heme-inducible BhuR production was abrogated in a *B. bronchiseptica hurI* mutant strain; *hurI* supplied in *trans* restored heme-inducible BhuR production and also resulted in overproduction of BhuR, consistent with a *hurI* gene dosage effect. *B. pertussis* and *B. bronchiseptica hurI* mutant strains are significantly less proficient in utilization of hemin and hemoglobin than are their wild-type parental strains. However, the mutants exhibit low heme uptake levels, indicating that while HurI is required for maximal heme utilization, it is not absolutely required for heme iron acquisition.

Analyses of *bhuR-lacZ* transcriptional fusions in wild-type *B. bronchiseptica* cells demonstrated that *bhuR* transcription is activated in response to hemin and hemoglobin only under iron starvation conditions. Heme-responsive *bhuR* transcription was *hurI* dependent, in agreement with experiments demonstrating heme-responsive and *hurI*-dependent BhuR protein production. Overexpression of *hurI* in cells carrying a *bhuR-lacZ* fusion plasmid suppressed the inducer requirement, resulting in heme-independent *bhuR* expression. Promoter mapping and transcriptional analyses demonstrated that transcription from the *hurI* promoter was iron regulated and heme independent, while optimal expression from the *bhuR* promoter was *hurI* dependent and required both iron starvation and the heme inducer. Mutagenesis studies also indicated that iron-regulated transcription through the *hurR-bhuR* intergenic region contributes to *bhuR* expression in the absence of the heme inducer.

Consistent with the hypothesized role for BhuR in signal transduction, *B. bronchiseptica bhuR* mutant cells exhibited a loss of heme-responsive *bhuR* transcription. Heme-inducible *bhuR* transcription was restored to this mutant by *bhuR* alone in *trans*, even though the heme iron utilization defect was not rescued, suggesting that the processes of signaling and transport are separable. Similarly, studies in Terry Connell's laboratory of the *B. avium* heme utilization system have shown that the ECF σ factor RhuI activates the transcription of *bhuR* in response to hemin and that induction requires the BhuR heme receptor.

Experimental results and analogies to other iron acquisition systems indicate that the *Bordetella* heme utilization system is a hybrid system combining the heme transport machinery found in gram-negative pathogens with regulators characteristic of the iron starvation subfamily of ECF regulators. Our current understanding of these systems suggests the following model for *Bordetella bhu* gene regulation. Under iron-limiting conditions, Fur derepression at the *hurI* promoter allows transcription of *hurIR* and presumably low levels of readthrough transcription of *bhuRSTUV* so that some BhuR is present at the cell surface to sense heme. HurI remains inactive until BhuR senses heme in the environment and sends a signal through its N terminus to HurR and then to HurI, allowing a productive association between HurI and RNA polymerase and initiation of transcription at the *bhuR* promoter.

Increased production of the heme utilization machinery encoded by *bhuRSTUV* allows efficient uptake and utilization of heme iron sources. The advantage to *Bordetella* cells of maintaining the requirement for positive regulation of the heme transport system while most other pathogens have systems responsive to the iron starvation signal alone is not clear. However, in the context of *B. pertussis* pathogenesis, it seems plausible that heme may not be a readily available iron source during colonization of the intact mucosa of the nasopharynx and hence that full expression of heme transport machinery would be wasteful. *B. pertussis* may rely more heavily on heme iron during later stages of infection, when damage to the epithelium has occurred, liberating heme compounds from dead host cells. Coordinated regulation of expression of iron acquisition systems according to the availability of particular iron sources may be an important mechanism allowing efficient adaptation to a changing host environment.

FERRIMONE-INDUCIBLE IRON ACQUISITION IN VIVO

B. pertussis and *B. bronchiseptica* possess multiple mechanisms for the assimilation of iron: one process involving the action of a native siderophore, other mechanisms for utilizing xenosiderophores, and a siderophore-independent process to obtain heme iron. It is unknown which iron sources are utilized in vivo or whether certain iron sources are preferentially used at various stages of infection in the host. However, it is known that the TonB-ExbBD system is absolutely required for transport of ferric alcaligin, heme compounds, and xenosiderophores in *Bordetella* spp. A *B. pertussis* *tonB exbB* mutant was found to be markedly defective in colonizing mice, underscoring the importance of high-affinity iron transport in pathogenesis. One study using *Bordetella alcR* regulator mutants failed to demonstrate an effect of *alcR* mutation on mouse virulence, but since *alcR* mutants retain some low-level capacity to produce and transport alcaligin, this result did not eliminate the possibility that alcaligin utilization contributes to *Bordetella* virulence. Importantly, a study of neonatal swine infected with a *B. bronchiseptica alcA* siderophore biosynthesis mutant demonstrated significantly reduced colonization and corresponding nasal pathology, indicating that the alcaligin siderophore is required for maximal virulence in this animal model.

When nutritional iron is limiting, *B. pertussis* and *B. bronchiseptica* can produce alcaligin along with its requisite transport and utilization functions and can also produce proteins required for the utilization of host heme iron compounds and certain xenosiderophores, including enterobactin. Each of these distinct iron-scavenging systems is controlled by a different Fur- and iron-regulated positive transcriptional regulator protein that can sense and respond to the presence of the cognate iron source, ultimately resulting in elevated expression of the genes involved in its assimilation. Characterization of iron source sensing and transcriptional responsiveness in the alcaligin siderophore system has revealed that activation of transcription by AlcR can occur at extremely low concentrations of alcaligin, indicating that *Bordetella* species have evolved an extraordinary capacity to sense and respond to the presence of alcaligin in their environment. In this role as an inducer, alcaligin functions as an iron-related hormone, a signaling molecule that we have termed a ferrimone. Similarly, heme iron and enterobactin induce the expression of their respective utilization systems in *B. pertussis* and *B. bronchiseptica*. Using ferrimone-dependent positive control circuits, *Bordetella* cells coordinate environmental and intracellular signals to increase the expression of the cognate iron transport system genes under conditions in which the presence of a particular iron source is perceived. We postulate that the ability of *Bordetella* species to selectively activate the expression of the different iron systems contributes to its capacity to effectively adapt and multiply in the host environment during the course of an infection.

A critical question that has emerged from induction studies of the endogenous alcaligin siderophore relates to the biological role of iron source perception and signal transduction. One purpose of ferrimone-mediated induction may be to elevate the expression of genes involved in alcaligin production and transport, because the siderophore is perceived as an effective iron delivery mechanism under those particular environmental conditions. Conversely, the inducer-activator relationship could serve as a mechanism informing the cell that the siderophore is ineffective at retrieving iron at the current level of production and that siderophore production and transporter levels must therefore be elevated. Iron-free alcaligin is a potent inducer of alcaligin gene transcription, whereas ferric alcaligin does not induce; instead, its delivery eventually results in repression of transcription. Furthermore, induction is independent of the FauA ferric alcaligin receptor and the TonB system. Therefore, these observations appear to be consistent with the second model. Simply stated, induction may serve to increase siderophore production and transport when the bacterium perceives that the chelator has returned "empty-handed."

Further research is needed to characterize the key elements of ferrimone-inducible regulation of all of the *B. pertussis* iron transport systems and to determine their importance to growth in the human host. It will be of interest to define spatiotemporal aspects of tissue expression of the iron retrieval systems in a natural infection and to ascertain whether the systems play distinct, additive roles in pathogenesis even though they ultimately serve the same purpose of providing iron to the bacterium. Iron source-responsive positive regulation could be particularly important in situations when bordetellae can effectively access only a subset of potential iron sources or when the bacteria are confronted with a mixture of iron sources, some of which may be more effectively utilized than others in a particular microenvironment.

ACKNOWLEDGMENTS

Studies in our laboratory were supported by grant AI-30188 from the National Institute of Allergy and Infectious Diseases.

We thank the researchers at the Sanger Centre for generously providing access to *Bordetella* nucleotide sequences prior to annotation. We apologize to the researchers whose work was not cited due to space constraints.

SUGGESTED READING

Beall, B., and G. N. Sanden. 1995. A *Bordetella pertussis fepA* homologue required for utilization of exogenous ferric enterobactin. *Microbiology* **141**:3193–3205.

Beaumont, F. C., H. Y. Kang, T. J. Brickman, and S. K. Armstrong. 1998. Identification and characterization of *alcR*, a gene encoding an AraC-like regulator of alcaligin siderophore biosynthesis and transport in *Bordetella pertussis* and *Bordetella bronchiseptica*. *J. Bacteriol.* **180**:862–870.

Braun, V. 1997. Surface signaling: novel transcription initiation mechanism starting from the cell surface. *Arch. Microbiol.* **167**:325–331.

Brickman, T. J., and S. K. Armstrong. 2002. *Bordetella* interspecies allelic variation in AlcR inducer requirements: identification of a critical determinant of AlcR inducer responsiveness and construction of an *alcR*(Con) mutant allele. *J. Bacteriol.* **184**:1530–1539.

Brickman, T. J., H. Y. Kang, and S. K. Armstrong. 2001. Transcriptional activation of *Bordetella* alcaligin siderophore genes requires the AlcR regulator with alcaligin as inducer. *J. Bacteriol.* **183**:483–489.

Brickman, T. J., J.-G. Hansel, M. J. Miller, and S. K. Armstrong. 1996. Purification, spectroscopic analysis, and biological activity of the macrocyclic dihydroxamate siderophore alcaligin produced by *Bordetella pertussis* and *Bordetella bronchiseptica*. *BioMetals* **9**:191–203.

Giardina, P. C., L. A. Foster, S. I. Toth, B. A. Roe, and D. W. Dyer. 1997. Analysis of the *alcABC* operon encoding alcaligin biosynthesis enzymes in *Bordetella bronchiseptica*. *Gene* **194**:19–24.

Gorringe, A. R., G. Woods, and A. Robinson. 1990. Growth and siderophore production by *Bordetella pertussis* under iron-restricted conditions. *FEMS Microbiol. Lett.* **66**:101–106.

Hou, Z., C. J. Sunderland, T. Nishio, and K. N. Raymond. 1996. Preorganization of ferric alcaligin, Fe_2L_3. The first structure of a ferric dihydroxamate siderophore. *J. Am. Chem. Soc.* **118**:5148–5149.

Kang, H. Y., and S. K. Armstrong. 1998. Transcriptional analysis of the *Bordetella* alcaligin siderophore biosynthesis operon. *J. Bacteriol.* **180**:855–861.

Kirby A. E, N. D. King, and T. D. Connell. 2004. RhuR, an extracytoplasmic function sigma factor activator, is essential for heme-dependent expression of the outer membrane heme and hemoprotein receptor of *Bordetella avium*. *Infect. Immun.* **72**:896–907.

Locht, C., R. Antoine, and F. Jacob-Dubuisson. 2001. *Bordetella pertussis*, molecular pathogenesis under multiple aspects. *Curr. Opin. Microbiol.* **4**:82–89.

Moore, C. H., L. A. Foster, J. G. Gerbig, D. W. Dyer, and B. W. Gibson. 1995. Identification of alcaligin as the siderophore produced by *Bordetella pertussis* and *Bordetella bronchiseptica*. *J. Bacteriol.* **177**:1116–1118.

Murphy, E. R., R. E. Sacco, A. Dickenson, D. J., Metzger, Y. Hu, P. E. Orndorff, and T. D. Connell. 2002. BhuR, a virulence-associated outer membrane protein of *Bordetella avium*, is required for the acquisition of iron from heme and hemoproteins. *Infect. Immun.* **70**:5390–5403.

Nicholson, M. L., and B. Beall. 1999. Disruption of *tonB* in *Bordetella bronchiseptica* and *Bordetella pertussis* prevents utilization of ferric siderophores, haemin and haemoglobin as iron sources. *Microbiology* **145**:2453–2461.

Nishio, T., N. Tanaka, J. Hiratake, Y. Katsube, Y. Ishida, and J. Oda. 1988. Isolation and structure of the novel dihydroxamate siderophore alcaligin. *J. Am. Chem. Soc.* **110:**8833–8734.

Pradel, E., and C. Locht. 2001. Expression of the putative siderophore receptor gene *bfrZ* is controlled by the extracytoplasmic-function sigma factor BupI in *Bordetella bronchiseptica*. *J. Bacteriol.* **183:**2910–2917.

Pradel, E., N. Guiso, and C. Locht. 1998. Identification of AlcR, an AraC-type regulator of alcaligin siderophore synthesis in *Bordetella bronchiseptica* and *Bordetella pertussis*. *J. Bacteriol.* **180:**871–880.

Redhead, K., T. Hill, and H. Chart. 1987. Interaction of lactoferrin ahd transferrins with the outer membrane of *Bordetella pertussis*. *J. Gen. Microbiol.* **133:**891–898.

Register K. B., T. F. Ducey, S. L. Brockmeier, and D. W. Dyer. 2001. Reduced virulence of a *Bordetella bronchiseptica* siderophore mutant in neonatal swine. *Infect. Immun.* **69:**2137–2143.

Vanderpool, C. K., and S. K. Armstrong 2001. The *Bordetella bhu* locus is required for heme iron utilization. *J. Bacteriol.* **183:**4278–4287.

PORPHYROMONAS GINGIVALIS

Caroline Attardo Genco, Waltena Simpson, and Teresa Olczak

21

Periodontal diseases, the most common of the destructive oral inflammatory diseases, comprise a group of pathological states of the gingiva and the supporting structures of the periodontium. They are described as the bacterially initiated conversion of a healthy gingival region to one characterized by inflammation (gingivitis) and destruction of the supporting structures of the teeth (periodontitis). *Porphyromonas gingivalis*, the etiological agent of periodontal disease, is an anaerobe that requires iron in the form of heme for growth. This gram-negative microorganism flourishes under the anaerobic conditions found in the periodontal pocket. The oral cavity is a unique anatomical site in the body. It is composed of multiple epithelial and mucosal surfaces, as well as calcified hard tissues. These tissues are constantly bathed by saliva, a broth of glandular secretions in which variable levels of microorganisms and food particles are suspended. The gingival sulcus, a 0.5-mm-deep crevice (in healthy individuals) between the tooth and the gum, is the portal of entry and niche for a unique bacterial ecosystem composed of highly adapted microorganisms, most notably *P. gingivalis*.

P. gingivalis appears to colonize primarily the subgingival sites of the oral cavity but has also been recovered from the tongue, tonsils, saliva, and supragingival dental plaque samples. *P. gingivalis* constitutes less than 5% of the cultivable subgingival flora of individuals with gingivitis, but its numbers can increase dramatically in advanced periodontal disease. Since *P. gingivalis* is recovered almost exclusively from disease sites, its primary source of infection remains unknown. The precise mechanism by which *P. gingivalis* enters the human oral cavity and becomes established within the oral mucosa is not well defined. Evidence indicates that *P. gingivalis* is largely an opportunistic pathogen. Although studies support the transmission of *P. gingivalis* by intimate contact, they do not indicate that transmission results in disease. In addition, data do not exclude the possibility that pathogenic strains have virulence factors that are not present in strains colonizing healthy individuals. Transient changes in the host immune response and the indigenous microflora may be important cues to *P. gingivalis* to produce a set of factors that are required for pathogenesis. Moreover, the ability of *P. gingivalis* to colonize periodontal sites appears to correlate

Caroline Attardo Genco, Department of Medicine, Boston University School of Medicine, Section of Infectious Diseases, Boston, MA 02118. *Waltena Simpson,* Department of Biological Sciences, South Carolina State University, Orangeburg, SC 29117. *Teresa Olczak,* Institute of Biochemistry and Molecular Biology, Wrocław University, Przybyszewskiego 63/77, 51–148 Wrocław, Poland.

with the ability of the organism to obtain nutrients for growth and with the anaerobic microenvironment within this site.

P. gingivalis is essentially absent during periodontal health, but during disease progression to periodontitis it can account for a very significant percentage of the pathogenic microbiota. Eradication of *P. gingivalis* from the subgingival microbial population correlates with resolution of the disease. In addition, the serum immunoglobulin G antibody response to this organism is elevated in individuals with a history of periodontal diseases compared to that in control individuals. *P. gingivalis* has also been implicated as a risk factor for other conditions including cardiovascular disease and preterm delivery of low-birth-weight infants.

The availability of iron in gingival crevicular fluid (GCF) may be significant for *P. gingivalis* growth and virulence. *P. gingivalis* is capable of utilizing hemin, hemoglobin, myoglobin, transferrin, and inorganic iron as sole iron sources. In addition, hemoglobin bound to haptoglobin, and hemin complexed to hemopexin can be used as heme sources, indicating that the *P. gingivalis* must have a means of removing the hemin from these host iron-binding proteins. The absence or presence of heme has been demonstrated to influence the expression of a number of virulence factors in *P. gingivalis*. The effect of virulence factor expression may serve to enhance the pathogenicity of *P. gingivalis*. Among the *P. gingivalis* virulence factors influenced by heme levels are iron-heme-specific outer membrane proteins. Due to their importance, the specific mechanisms utilized by *P. gingivalis* for the accumulation of heme represent an area of intense investigation. Lactoferrin, transferrin, and hemoglobin are known constituents of GCF and probably support the growth of *P. gingivalis* in vivo. *P. gingivalis* utilizes hemoglobin, which serves as a source of heme, quite efficiently compared to its use of other iron sources. Although hemoglobin is immediately bound by haptoglobin in the host environment, the proteolytic capability of *P. gingivalis* may contribute to the degradation of this host hemoglobin-sequestering protein and liberate hemoglobin for binding and heme uptake into the cell. Thus, it is likely that the major source of iron for growth of *P. gingivalis* in GCF is hemoglobin.

IRON REQUIREMENTS OF *P. GINGIVALIS*

Similar to most pathogens, *P. gingivalis* requires iron for growth. Hemin is capable of satisfying the entire iron requirement of *P. gingivalis* and is an essential nutrient for the growth of the bacterium, playing a role mostly as the prosthetic group of a *b*-type cytochrome. Hemin consists of the tetrapyrrole molecule protoporphyrin IX (PPIX), into the center of which iron (Fe^{3+}) is ligated. The characteristic black pigmentation exhibited by *P. gingivalis* colonies grown on blood agar plates is attributed to the accumulation of hemin on the bacterial cell surface. This ability of *P. gingivalis* to store hemin appears to provide a nutritional advantage in the iron-limited environment of a healthy periodontal pocket. In addition to hemin, *P. gingivalis* is capable of utilizing hemoglobin, haptoglobin, myoglobin, hemopexin, and transferrin, as well as inorganic iron sources such as ferric and ferrous chloride. While the need for heme by *P. gingivalis* has been well documented, the precise mechanisms used by the organism to obtain this crucial nutrient are still poorly understood. Studies of this bacterium have revealed that it does not produce siderophores. However, the recent identification and characterization of several putative outer membrane heme receptors have provided significant insight into how this organism acquires heme. In addition to outer membrane heme receptors, *P. gingivalis* produces proteinases known collectively as gingipains, which, in addition to their ability to degrade extracellular matrix proteins, have recently been implicated in heme acquisition. Here, we discuss the roles of specific outer membrane proteins and gingipains in heme accumulation by *P. gingivalis*.

HEMIN BINDING AND UTILIZATION IN *P. GINGIVALIS*

Heme is a hydrophobic molecule, and the Fe-O-Fe bond, which can form, may result in dimerization at physiological pH. Formally, the term "heme" refers to reduced, ferrous [Fe^{2+}] iron pp IX whereas the term "hemin" refers to the oxidized, ferric [Fe^{3+}] form of the molecule. In an aqueous solution in the absence of proteins or reducing agents, iron protoporphyrin is found in its oxidized form [hemin]. Although the formal definitions distinguish "heme" and "hemin," the term "heme" is widely used to indicate iron protoporphyrin IX in either oxidation state.) The size of these heme dimers limits the ability of heme to readily traverse the bacterial outer membrane through porin channels without the involvement of some type of chaperone. In gram-negative bacteria, the outer membrane forms a permeability barrier for substrates of >600 Da. Due to these constraints, porin-mediated heme entry into the bacterial periplasm is highly unlikely.

Other than its accepted role as the prosthetic group of a b-type cytochrome, the biological functions of accumulated FePPIX of *P. gingivalis* are obscure since, paradoxically, FePPIX stored either in or on the bacterial cell surface poses a major problem because of its involvement in oxygen radical-mediated damage to DNA, proteins, and lipids. It has been speculated that cell surface FePPIX binding results from the close packing of monomeric iron porphyrin through hydrophobic interactions, but the exact mechanism of this binding, the chemical nature of the porphyrin-containing pigment, and the oxidation state of the iron are not known. Smalley et al. reevaluated this problem by using ^{57}Fe Mössbauer spectroscopy and demonstrated that FePPIX is bound to the bacterial cell in the mu-oxo dimeric form, [Fe^{3+}PPIX]$_2$O, and that the black pigment is also composed of this material. *P. gingivalis* cells were shown to generate mu-oxo dimers from oxyhemoglobin that was first converted into methemoglobin and then degraded. After degradation, a mixture of monomeric and mu-oxo dimeric FePPIX was generated and utilized in vitro by the bacterium. Formation of [Fe^{3+}PPIX]$_2$O via reaction of Fe^{2+}PPIX with oxygen and its toxic derivatives would serve as an oxidative buffer and permit *P. gingivalis* and other black-pigmented anaerobes to maintain a local anaerobic environment. Tying up free oxygen species with iron PPIX would also reduce and limit Fe^{2+}PPIX-mediated oxygen radical cell damage. Smalley et al. also speculated that the formation of a cell surface mu-oxo dimer layer may function as a protective barrier against assault by reactive oxidants generated by neutrophils. Further investigations of this hypothesis have revealed that *P. gingivalis* cells carrying a surface layer of mu-oxo heme were less susceptible to peroxidation by hydrogen peroxide. Thus, the binding of mu-oxo heme by *P. gingivalis* may specifically aid survival during neutrophil attack through inactivation of hydrogen peroxide. Selective interference with these mechanisms would offer the possibility of attenuating the pathogenicity of *P. gingivalis* and other iron PPIX-binding pathogens whose virulence is regulated by reactive oxidants.

P. gingivalis does not produce siderophores. Rather, this organism acquires iron and heme via outer membrane receptors and hemophore-like proteins, extracellular heme/hemoglobin/hemopexin-binding proteins that capture heme and shuttle it to a specific outer membrane receptor (see chapter 5), for iron and heme acquisition. The best-characterized mechanism by which bacteria acquire heme involves direct binding of heme or hemoproteins to specific outer membrane receptors. These heme acquisition systems require periplasmic protein-dependent transport machinery for the passage of heme through the cytoplasmic membrane. The periplasmic transport systems belong to the family of ABC (ATP-binding cassette) transporters, which utilize a periplasmic substrate-binding protein, one or two hydrophobic integral membrane-spanning proteins, and one or two hydrophilic proteins with ATPase activity. Following the binding of heme or host-sequestering proteins to specific receptors, heme is removed from the bacterial receptor and transported into the cell by an energy-requiring process. Energy for the transport of iron or heme across the outer membrane into the periplasmic space in most gram-negative organisms is provided by the TonB protein in association with the ExbB and ExbD proteins (see chapter 7). Receptors that require energy supplied via the TonB system are termed TonB-dependent receptors and have amino acid homology in several regions termed TonB boxes. Studies in our laboratory have demonstrated that both [^{14}C]hemin and [^{59}Fe]hemin are accumulated by *P. gingivalis*. This indicates that both the iron and the porphyrin ring (i.e., the entire hemin moiety) are taken into the cell. Our studies also demonstrated that hemin is transported into *P. gingivalis* by an energy-requiring process. The energy dependence of hemin transport in *P. gingivalis* suggests that a TonB analog anchored in the cytoplasmic membrane may function in the transport of hemin. A putative *tonB* gene has been identified in the *P. gingivalis* W83 genome (Institute for Genomic Research database), and thus *P. gingivalis* most probably utilizes a TonB-dependent mechanism for the transport of heme. However, to date, detailed structure-function studies of the *P. gingivalis* TonB system have not been conducted.

Bramanti and Holt provided some of the first data indicating that the expression of at least 10 surface proteins with molecular masses ranging from 26 to 83 kDa were observed in *P. gingivalis* strain W50 grown under hemin-depleted conditions. Growth of *P. gingivalis* strain W50 in the presence of normal to excess hemin levels (i.e., hemin replete) resulted in the down regulation of these proteins. In contrast, under heme-limiting conditions (i.e., hemin depleted), the expression of at least 12 outer membrane proteins was reported. Two of these "hemin-regulated outer membrane proteins," at approximately 26 kDa (OMP26) and 83 kDa (OMP83), were recognized by sodium dodecyl sulfate-polyacrylanide gel electriphoresis as major hemin-regulated *P. gingivalis* W50 outer membrane proteins. The investigators postulated that the hemin repression observed required the presence of PPIX. Growth of the hemin-depleted cultures in normal or hemin-replete medium resulted in the rapid repression of OMP26. Radioiodination studies revealed that OMP26 was no longer accessible to iodine labeling after less than 1 min of culture shift from hemin-depleted to hemin-replete conditions. However, OMP26 was not repressed in the presence of lactoferrin, transferrin, inorganic iron, zinc PPIX, or PPIX. Although OMP26 may function to bind hemin under hemin-depleted conditions and transport it across the outer membrane into the cytoplasmic region, a mutant with a mutation in the gene encoding OMP26 has not been generated and the precise role of OMP26 in hemin binding and utilization in *P. gingivalis* has not been defined.

Recently, the gene for a 35-kDa protein (HBP35), which appears to encode a hemin-binding protein, was identified in *P. gingivalis* strain 381. The HBP35 protein was also capable of binding selected protoporphyrins containing hemin. The HBP35 protein contains a typical heme regulatory motif that functions as a direct hemin-binding site. Insertional inactivation of the gene produced a beige mutant that exhibited little coaggregation. The HBP35-deficient mutant also exhibited decreased virulence. This suggests that the hemin-binding capability of this protein may be important for the expression of virulence in *P. gingivalis*.

In addition to hemin-binding proteins, a novel receptor involved in the acquisition of both hemin and heme from hemoglobin in *P. gingivalis*, HmuR, has been identified. The role of the *hmuR* gene in heme accumulation has been extensively studied through mutant construction and biochemical analysis of the protein. Several other putative outer membrane proteins, which may contribute to hemin acquisition in *P. gingivalis*, have been also recently described (Table 1). These include the Tla, Tlr, HemR, and IhtB proteins. Although experimental analysis indicates the involvement of these genes in hemin accumulation, with the exception of *tla* they have not been analyzed by the production of isogenic mutants.

THE Tla RECEPTOR

Aduse-Opoku et al. initially reported on the identification of the *tla* (for "TonB-linked adhesion") gene from *P. gingivalis* strain W50.

TABLE 1 Heme-binding proteins identified in *P. gingivalis*

Gene/protein	Description	Protein size (kDa)	*P. gingivalis* strain	Function determined through mutagenesis?
OMP26	Hemin binding	26	W50	No
OMP83	Hemin binding	83	W50	No
tla/Tla	TonB linked	118.7	W50	Yes
hemR/HemR	Hemin regulated	48	53977	No
hmuR/HmuR	Hemin utilization	73	A7436	Yes
ihtB/IhtB	Iron heme	32.4	W50	No
tlr/Tlr	TonB linked	~71[a]	W50	No
hbp35/HBP35	Coaggregation factor, hemin-binding protein	35	381	Yes
HbR	Hemoglobin receptor	19	33277	Yes

[a] Size of protein calculated by multiplying the number of amino acids (708) by the standard weight of an amino acid (i.e., 100 Da).

The C-terminal region of the Tla protein exhibits 98% identity to the arginine-specific protease Rgp. The N terminus of the Tla protein has regional similarity to TonB-dependent receptors, which are frequently involved in the translocation of hemin, iron, colicins, or vitamin B_{12} into the periplasm in other bacteria. The *tla* gene has been characterized through mutational analysis. An isogenic *tla* mutant was unable to grow in media containing low concentrations of hemin (<2.5 mg/liter), and hemin-depleted cells of this mutant failed to respond to hemin in an agar diffusion plate assay, suggesting a role for Tla in acquisition and utilization of low levels of environmental hemin. Interestingly, the isogenic *tla* mutant produced significantly less arginine-specific protease activity than did compared to the parental strain.

THE Tlr RECEPTOR

In an approach to the identification of a heme transport system of *P. gingivalis*, Slakeski et al. screened a *P. gingivalis* genomic library and identified the *tla* gene. On closer inspection, however, these investigators determined that the previously reported uncleotide sequence of the *tla* gene consisted of two separate sequences that had been combined, presumably through a cloning error or a genetic rearrangement. These investigators revealed that the C terminus of their strain W50-derived sequence, designated *tlr* (for "TonB-linked receptor"), possessed no similarity to the Rgp proteinase. In addition, the full sequence of the *tlr* gene is included in the Institute for Genomic Research *P. gingivalis* W83 database, whereas the sequence of *tla* is not. The *tlr* gene is located immediately downstream of four open reading frames (*htrABCD*) that encode a putative ATP-binding cassette system with sequence similarity to heme transport systems of other bacteria.

THE HemR RECEPTOR

Karunakarun et al. have described an additional gene, *hemR*, which was proposed to encode an outer membrane protein in *P. gingivalis* strain 53977. The N terminus of the *hemR* gene exhibited homology to genes involved in iron acquisition from several gram-negative microorganisms. However, the C terminus of the *hemR* gene exhibited homology to the *prtT* gene of *P. gingivalis*. PrtT encodes an arginine-specific protease whose nucleotide sequence is distirct from those of the major arginine-specific proteases possessed by *P. gingivalis*, HRgpA and RgpB. Since an isogenic *hemR* mutant could not be created and characterized, a role for the *hemR* gene in hemin accumulation in *P. gingivalis* cannot be established.

THE IhtB PROTEIN

Dashper et al. have recently reported on the characterization of the IhtB (for "iron heme transport") protein. This lipoprotein was localized to the cell surface of P. gingivalis strain W50 by Western blot analysis of an outer membrane protein preparation and by immunocytochemical staining of whole cells using IhtB peptide-specific antiserum. IhtB released from the cell surface was shown to bind hemin in a hemin-agarose assay. The growth of heme-limited (but not heme-replete) P. gingivalis cells was inhibited by preincubation with IhtB peptide-specific antiserum. The ihtB gene was located between an open reading frame encoding a putative TonB-linked outer membrane receptor (IhtA) and three open reading frames (ihtABCDE) that have sequence similarity to ABC operons in other bacteria. Analysis of the deduced amino acid sequence of IhtB showed significant similarity to the Salmonella enterica serovar Typhimurium protein CbiK, a cobalt chelatase that is structurally related to the ATP-dependent family of ferrochelatases. Molecular modeling indicated that the IhtB amino acid sequence could be threaded onto the CbiK fold, with the IhtB structural model containing the active-site residues critical for chelatase activity. These results suggest that IhtB may function to remove iron from heme prior to uptake by P. gingivalis. Before this can be definitively concluded, however, an ihtB mutant must be created and characterized.

HEMOGLOBIN BINDING AND UTILIZATION IN P. GINGIVALIS

The binding and accumulation of hemoglobin by P. gingivalis is currently the focus of investigations being conducted in several independent laboratories. Amano et al. found the binding of human hemoglobin by P. gingivalis strain 381 to occur rapidly, reversibly, and specifically, with an apparent K_d of 10^{-6} M. Hemoglobin binding was inhibited by unlabeled human hemoglobin but not by hemin and PPIX. The binding was only partially inhibited by human serum albumin, transferrin, lactoferrin, catalase, and cytochrome c, suggesting that binding may not occur through the heme moiety. The binding of hemoglobin increased considerably when the bacterium was grown under hemin-limited conditions, suggesting that the expression of hemoglobin-binding proteins may be negatively regulated by hemin.

Investigations of the binding of hemoglobin to the P. gingivalis envelope over a wide range of pH values (from 4.5 to 9.0) have also been performed. The binding activity in low-pH buffers was much higher than at high pH; the optimum pH values for hemoglobin binding were found to be 4.5 and 5.0. Since the hemoglobin bound to the envelope was found to dissociate at pH 8.5 and 9.0, binding was determined to be reversible. In a separate study, hemoglobin binding by P. gingivalis was demonstrated to decrease when the growth temperature was increased from 37 to 39°C. These findings are of considerable interest, since both temperature and pH increase in the periodontal pocket as it transforms from health to disease. Thus, the decreased ability of P. gingivalis to bind hemoglobin under these conditions may function to limit the potential accumulation of toxic heme molecules.

THE HmuR RECEPTOR

Recent investigations in our laboratory have resulted in the identification of a novel receptor, HmuR, which is involved in the uptake of free hemin and heme from hemoglobin in P. gingivalis. The hmuR (for "hemin utilization receptor") gene has homology to genes whose products are TonB-dependent outer membrane receptors involved in iron acquisition. Two regions of the translated open reading frame of HmuR (amino acids 33 to 39 and 135 to 170) exhibited extensive sequence homology to TonB boxes I and IV, respectively, with the most pronounced homology at the region corresponding to TonB box IV. The amino-terminal region of the hmuR gene exhibited 100% identity to that of the previously reported hemR gene. Interestingly, the carboxy-terminal regions of hemR and hmuR were strikingly dissimilar. Although previous studies have determined that hemR is present in strains 53977, W50, and 381, we were unable to PCR amplify the hemR gene from P. gingivalis strain A7436. This suggests that in this strain, heme transport can occur independently of HemR.

Southern blot analysis revealed that one copy of the *hmuR* gene is present in the *P. gingivalis* A7436 genome, as well as in *P. gingivalis* strains 381, 53977, and W50, with the N-terminal region of the gene product being highly conserved. An additional analysis using a C-terminal *hmuR* probe revealed that the C terminus of the *hmuR* product in various *P. gingivalis* strains demonstrated a similar banding pattern. The *hmuR* gene is contranscribed with an 0.429-kb open reading frame (*hmuY*) located immediately upstream of it. The transcription of the *hmuY-hmuR* operon has been demonstrated to be repressed by hemin.

Growth studies demonstrated that an isogenic *hmuR* mutant exhibits a decreased ability to grow with either bovine hemin or human hemoglobin as the sole iron source. The decreased growth of the mutant strain did not appear to result from a generalized growth defect since this strain grew as well as the parental strain when supplemented with ferric chloride as the sole iron source. This finding indicates that HmuR is specific for the uptake of heme-containing compounds such as hemin and hemoglobin but that the uptake of inorganic iron (i.e., ferric chloride) is mediated by another mechanism. Previously, we have demonstrated that hemoglobin can compete for the binding and accumulation of hemin in *P. gingivalis*, further suggesting that hemin and hemin and hemoglobin transport can occur via a common pathway. The decreased ability of the *hmuR* mutant strain to grow with hemin and hemoglobin as the sole source of iron correlated well with the phenotype of the organism on blood agar plates. Characterization of the isogenic *hmuR* mutant revealed that it bound significantly less hemoglobin than the parental strain did; however, binding was not completely diminished. The fact that the *hmuR* mutant did not exhibit a total lack of hemoglobin binding appears to be due to the presence of other proteins involved in hemoglobin binding in this strain.

The HmuR protein was further examined by analyzing purified recombinant HmuR protein and *Escherichia coli* cultures producing HmuR. Recombinant HmuR directly binds hemin ($K_d = 2.4 \times 10^{-5}$ M) and hemoglobin but not human serum albumin, hemopexin, haptoglobin, or transferrin. *E. coli* strains in which HmuR was overexpressed were able to bind hemoglobin. In addition, *E. coli* strains expressing membrane-associated recombinant HmuR were capable of binding porphyrins and metalloporphyrins such as hemin, zinc PPIX, and copper PPIX. This suggests that the active site of HmuR may have a histidine, which binds to the metal present in the porphyrin ring.

Amino acid comparions of the conserved motifs of several different hemoglobin or heme receptors and the *P. gingivalis* HmuR protein revealed that HmuR contains highly conserved domains, including the invariant histidine residues (His95 and His434) and the FRAP (in HmuR YRAP) and NPNL (in HmuR NPDL) amino acid boxes, which may be involved in heme and hemoglobin binding. To define the molecular interaction between HmuR and heme or hemoglobin, we constructed a series of mutants by site-directed mutagenesis. Four *hmuR* mutants were constructed with specific mutations in the highly conserved putative hemin-binding domains in HmuR. These included mutant I with the replacement of His95 by alanine, mutant II with the replacement of His434 by alanine, mutant III with the replacement of FRAP420–423 by FAAA, and mutant IV with the replacement of NPDL442–445 by NAAA. Hemin- and hemoglobin-binding experiments demonstrated that the HmuR NPDL domain is required for hemoglobin binding. During growth in a minimal medium with serum as the sole iron source, all four *hmuR* mutants showed reduced growth rates, indicating that mutations at conserved sites may interfere with the ability of HmuR to scavenge iron from serum proteins.

ROLE OF GINGIPAINS IN HEME ACQUISITION

Recent investigations have indicated a significant role for the *P. gingivalis* gingipain proteases in the acquisition of heme from hemoglobin. The gingipains of *P. gingivalis* specifically cleave substrates behind either arginine or lysine residues. Arginine-specific gingipains are encoded by two related genes, *rgpA* and *rgpB*, while the lysine-specific gingipain is encoded by a single genetic locus, *kgp*. Comparison of the primary structure of gingipains with protein sequences available in the databases has not revealed any significant similarity to known proteins, including other proteolytic enzymes. This indicates that the gingipains are unique proteases; for this reason, they have been assigned to a separate family (family C25) of cysteine proteinases. In addition to gingipains, *P. gingivalis* produces a number of exopeptidases belonging to the cysteine peptidase, serine peptidase, and metallopeptidase families, but they have not been well characterized.

The translated part of the *rgpA* gene encodes an N-terminal prosequence, a catalytic domain, and a large C-terminal domain (Fig. 1). The C terminus itself is composed of four domains (HA1 to HA4), which were originally named hemagglutinin domains because at least one was suspected to participate in hemagglutination. The initial translation product is apparently subjected to posttranslational processing, leading to the formation of several different molecular forms of the enzyme. The simplest form corresponds to the catalytic domain alone and may be generated by an interrupted transcription process. The most common form of the *rgpA* gene product, however, is the noncovalent but very stable complex of the N-terminal catalytic domain with a hemagglutinin/adhesion domain(s) derived from the C terminus of the initial translation product. This form is designated HRgpA, referring to the high molecular mass of the complex. Amino acid sequence analysis of its individual components indicates that proteolytic processing of the precursor protein occurs at four arginine sites and one lysine site. HRgpA can be found both membrane bound and as a soluble enzyme.

In contrast to HRgpA, RgpB is expressed as a preproenzyme lacking the majority of the hemagglutinin domains but otherwise closely related to the HRgpA product, with the N-terminal two-thirds of the primary structure of RgpB being nearly identical to HRgpA. The *rgpB* gene is expressed as a precursor that requires posttranslational modification by proteolytic cleavage of the prosequence. The *rgpA* and *rgpB* genes are conserved among laboratory strains and clinical isolates of *P. gingivalis*, since

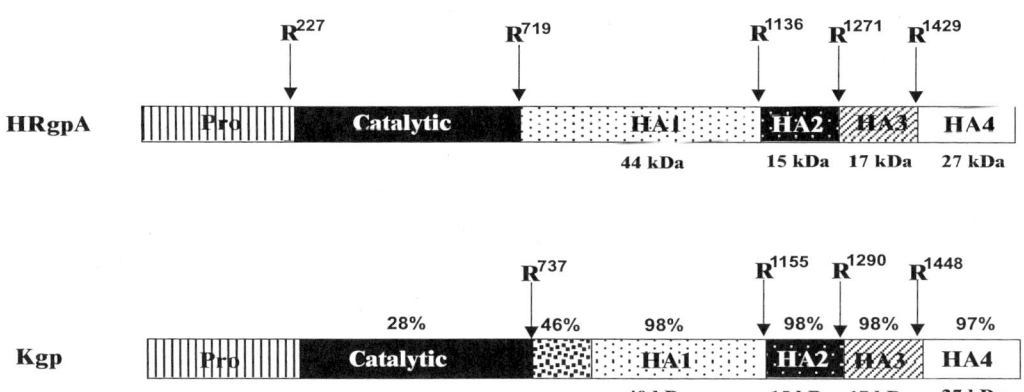

FIGURE 1 Structures of *P. gingivalis* gingipains HRgpA and Kgp. The sizes of the hemagglutinin domains (HA1 to HA4) are indicated. Amino acids involved in posttranslational processing are indicated by arrows. Areas of similar shading indicate sequences identical in both proteins. Percentages represent the degree of identity between the two compared regions of HRgpA and Kgp.

gene restriction analysis reveals only minor polymorphism. The similarities between the nucleotide and amino acid sequences of *rgpA* and *rgpB* suggest that the two genes may have been generated through the duplication of an ancestor *rgp* gene, insertion of the hemagglutinin domain region into one copy of the two resulting *rgp* genes (or deletion of the region from one *rgp*), and homologous recombination between the proteinase domains of the two *rgp* genes. The recent demonstration of nonreciprocal recombination in *P. gingivalis* supports this hypothesis.

Similarly to *rgpA*, the *kgp* gene encodes a polypeptide consisting of a prosequence, a catalytic domain, and the C-terminal extension harboring the hemagglutinin/adhesion domain(s) (Fig. 1). As is the case for the HRgpA initial translation product, the nascent Kgp polypeptide is apparently processed at multiple arginine sites and at a single lysine site, leading to the formation of the nonconvalent complex of the catalytic and hemagglutinin domains. Similar to HRgpA and RgpB, Kgp may be associated with the cell membrane or excreted as a soluble protein.

P. gingivalis gingipains are capable of degrading immunoglobulins, bactericidal proteins, and a number of extracellular matrix proteins; these include fibrinogen, kininogen, fibronectin, and collagen. HRgpA is also a very efficient enzyme in the generation, through direct cleavage of C5, of a potent chemotactic factor, C5a, that probably contributes to the significant leukocyte infiltration at *P. gingivalis*-induced periodontitis lesions. This gingipain also degrades C3, thereby eliminating the creation of C3-derived opsonins. Rgp proteins and Kgp rapidly cleave and inactivate interleukin-6 (IL-6), with Kgp being most potent in this regard. Studies have also shown that soluble gingipains are capable of converting IL-8 to a more potent species, while membrane-associated gingipains rapidly degrade IL-8.

While numerous studies have demonstrated the capability of gingipains to degrade a variety of human proteins, recent investigations have determined a role for gingipains in heme accumulation. The gingipains appear to be involved primarily in the acquisition of hemoglobin, in contrast to other iron/heme sources successfully utilized by *P. gingivalis*. Studies by Nakayama et al. have revealed that the hemagglutinin domain of HRgpA contains a 19-kDa protein, HbR, which binds hemoglobin. The HbR protein is encoded by an internal region (HGP15 domain [HA2 region in Fig. 1]) of HRgpA. The HbR protein is also present in Kgp, as well as in an additional lysine-specific proteinase, PrtP, and a hemagglutinin, HagA. The HA2 domain was shown to bind hemoglobin ($K_d \sim 10^{-9}$ M) as well as heme ($K_d \sim 10^{-8}$ M) very efficiently. Overexpression of the HGP15 domain of HRgpA in *E. coli* permitted this organism to bind hemoglobin in a pH-dependent manner, as is typical in *P. gingivalis*. Mutation of the HGP15 region resulted in the production of nonpigmented *P. gingivalis* cells and a decreased ability of these cells to absorb hemoglobin. Studies by these investigators have also shown that human lactoferrin is capable of removing HbR from *P. gingivalis* cells.

While some proteins involved with heme capture bind directly to the iron center, the HA2 domain of *P. gingivalis* recognizes heme by a mechanism that is porphyrin mediated. Studies have demonstrated that in vitro porphyrin binding to HA2 is iron independent. Porphyrins that differed from PPIX in only the vinyl group of the tetrapyrrole ring showed comparable effects in competing with hemoglobin for HA2. For porphyrins which differed from PPIX at both propionic acid side chains, the modification was detrimental. Correlation of porphyrin competition and growth recovery suggests that the HA2 domain may act as a high-affinity hemophore at the cell surface to capture porphyrin from hemoglobin. We have also conducted studies to define which domains of Kgp are required for hemoglobin and hemin utilization in *P. gingivalis* by the construction of C-terminally truncated Kgp mutants. *P. gingivalis* truncated mutants missing the complete hemagglutinin domains bound 50% less hemoglobin and hemin than did to the wild-type strain. Furthermore, the hemagglutinin domains are required for efficient catalytic activity of Kgp.

Additional studies have shown that the lysine-specific gingipain, Kgp, degrades hemoglobin and therefore may act as a "hemoglobinase." Moreover, an additional study in our laboratory has demonstrated that Kgp is capable of cleaving hemopexin, haptoglobin, and transferrin. Interestingly, *P. gingivalis* Kgp hemagglutinin domain truncation and deletion mutants exhibit a delayed degradation of hemoglobin, hemopexin, haptoglobin, and transferrin. We have also shown that growth of *P. gingivalis* in a minimal medium with normal human serum as a source of heme correlated not only with the ability of the organism to cleave these iron sources but also with an increase in Kgp enzymatic activity. These findings, coupled with those from other laboratories, seem to suggest that soluble Kgp may function as a heme scavenger or hemophore-like protein. As is typical of hemophores, the gingipains of *P. gingivalis* are secreted proteins. Also, the amino acid sequence of Kgp has no similarity to TonB-dependent outer membrane proteins. The hemophore HasA of *S. marcescens* interacts with an outer membrane receptor, HasR. Since Kgp may function as a hemophore in *P. gingivalis*, it may also be capable of interacting with an outer membrane receptor. Interestingly, enzyme-linked immunosorbent assays in our laboratory detected an interaction between HmuR and purified Kgp and HRgpA. Recombinant HmuR appeared to bind to Kgp more readily, since higher concentrations of HRgpA were needed to observe an interaction. Since the reactivity between HmuR and RgpB was not detected, it is hypothesized that the interaction between HmuR and Kgp or HRgpA actually occurs at the hemagglutinin domains of the gingipains.

The precise role of the Kgp-HmuR complex in heme acquisition was further elucidated through the construction and analysis of a *kgp* isogenic mutant and a *kgp hmuR* double mutant. Examination of the *kgp* isogenic mutant and the *kgp hmuR* double mutant revealed that the removal of these proteins, either singly or in combination, diminished the ability of the organism to bind hemoglobin. Interestingly, the *kgp hmuR* double mutant was still capable of binding hemoglobin, most probably due to the existence of additional hemoglobin-binding proteins. Also, Shi et al. showed that *P. gingivalis* mutants deficient in the *rgpA* and *kgp* genes were able to bind hemoglobin, probably due to the presence of other proteins with hemoglobin-binding activity. It has been reported that *P. gingivalis kgp* mutants display a nonpigmented colony phenotype and that this phenotype correlates with a decreased ability to bind hemoglobin. In agreement with these studies, the *kgp* isogenic mutant and the *kgp hmuR* double mutant constructed in *P. gingivalis* strain A7436 were observed to be nonpigmented when grown on blood agar plates. In contrast, the *hmuR* isogenic mutant was darkly pigmented on blood agar plates, most probably due to the absence of HmuR and an accumulation of hemoglobin and hemin by Kgp and HRgpA on the bacterial cell surface. Comparable levels of arginine-specific proteinase activities were detected in whole-cell fractions of both mutant strains and the parental strain, indicating that mutation of *kgp* did not affect the production of HRgpA and RgpB.

Physiological studies demonstrated that the *kgp* isogenic mutant exhibited decreased growth with hemin as the sole iron source. This suggests that Kgp may play a more crucial role in the utilization of hemin than was previously suspected. Lewis et al. have reported that Kgp binds hemin with high affinity ($K_d = 16$ nM). The tight binding and the ability of Kgp to interact with HmuR may indicate that Kgp serves to bind hemin and deliver it to HmuR. The *kgp hmuR* double mutant was diminished in its ability to grow with either hemin or hemoglobin as the sole iron sources. This indicates that both HmuR and Kgp are required for the optimal uptake of hemin and hemoglobin in *P. gingivalis*. This finding also suggests that while other proteins may be involved in the binding of hemoglobin, they are not sufficient to permit the uptake of these iron sources.

The hemagglutinin and adhesion domains of RgpA and Kgp have significant amino acid sequence homology to the hemagglutinin domain of HagA, a protein containing four contiguous direct repeats, each of which contains a functional hemagglutinin domain. *P. gingivalis* possesses four additional hemagglutinin genes (*hagB*, *hagC*, *hagD*, and *hagE*), and the ability to agglutinate erythrocytes is a feature that distinguishes this organism from other black-pigmented anaerobes. Therefore, it is likely that in addition to gingipains, hemagglutinins may aid in iron-heme utilization in *P. gingivalis*. Although the functions of multiple hemagglutinins and their contributions to *P. gingivalis* virulence are not well understood, they are postulated to promote adherence and colonization of tissue surfaces and, in conjunction with the proteases, promote attachment to and detachment from target human cells.

MODEL OF HEME TRANSPORT IN *P. GINGIVALIS*

Based on studies conducted in other laboratories as well as in our laboratory, we have derived a model to represent how HmuR and Kgp might be involved in the acquisition of heme and hemoglobin (Fig. 2). Red blood cells are degraded by a *P. gingivalis* hemolysin, resulting in the liberation of free hemoglobin and heme. Hemoglobin or heme can be bound directly by HmuR embedded within the *P. gingivalis* outer membrane. A direct, physical interaction of HmuR with a periplasmic energy-transducing protein, TonB, working in conjunction with the ExbB and ExbD proteins, may facilitate the internalization of heme into the periplasm. A recent search of the *P. gingivalis* W83 database at The Institute for Genomic Research allowed us to identify components of a putative hemin transport operon with homology to the *Yersinia enterocolitica* hemin transport system proteins. The *Y. enterocolitica* hemin utilization protein HemS, hemin permease (HemU), and ATP-binding hydrophilic protein (HemV) have 43, 44, and 53% homology respectively, to specific contigs in the *P. gingivalis* database. Thus, heme may be further transported into the cell by binding to an ATP-dependent permease (Fig. 2A). Once heme is in the cytoplasm, the removal of iron may occur via the action of a heme-degrading protein. Alternatively, hemoglobin can be bound by soluble Kgp, acting as a hemophore (Fig. 2B). The hemoglobin-Kgp complex can then interact with HmuR, which could facilitate the transport of heme into the cell, as previously described. A third mechanism of heme transport would involve the degradation of hemoglobin by soluble or membrane-bound Kgp. The liberated heme could then directly interact with HmuR, the outer membrane heme receptor (Fig. 2C), and be internalized.

REGULATION OF HEME UTILIZATION IN *P. GINGIVALIS*

We previously determined that expression of the *P. gingivalis hmuR* gene increased with passage into iron-depleted media and that transcript levels decreased when the bacteria were grown in iron-rich media. In contrast, *kgp* and *rgpA* gene expression was not tightly regulated by iron but, more probably, was regulated by cell density. We also detected an *hmuR* transcript in *P. gingivalis* cultures grown with hemin. The observation that transcription was still detected when bacteria were grown in the presence of heme suggests that transcription could be under dual control. This could involve the activation of transcription in the presence of heme and repression in the presence of iron. Furthermore, the promoter-operator region of the *hmuR* gene contains a putative Fur-binding sequence. In many microorganisms, the regulation of iron uptake genes and virulence factors is under the control of the ferric uptake regulator (Fur) protein, a transcriptional repressor of iron-regulated promoters by virtue of its Fe^{2+}-dependent DNA-binding activity (see chapter 13).

Homologs for the *fur* gene have been described for many gram-negative pathogens, including *Neisseria*, *Yersinia*, *Salmonella*, *Shigella*, *Vibrio*, *Pseudomonas*, *Campylobacter*, *Helicobacter pylori*, and *Haemophilus*. Most of these homologs are capable of complementing an *E. coli fur* mutant, suggesting that the molecular mechanisms that control transcriptional regulation by iron are shared by many microorga-

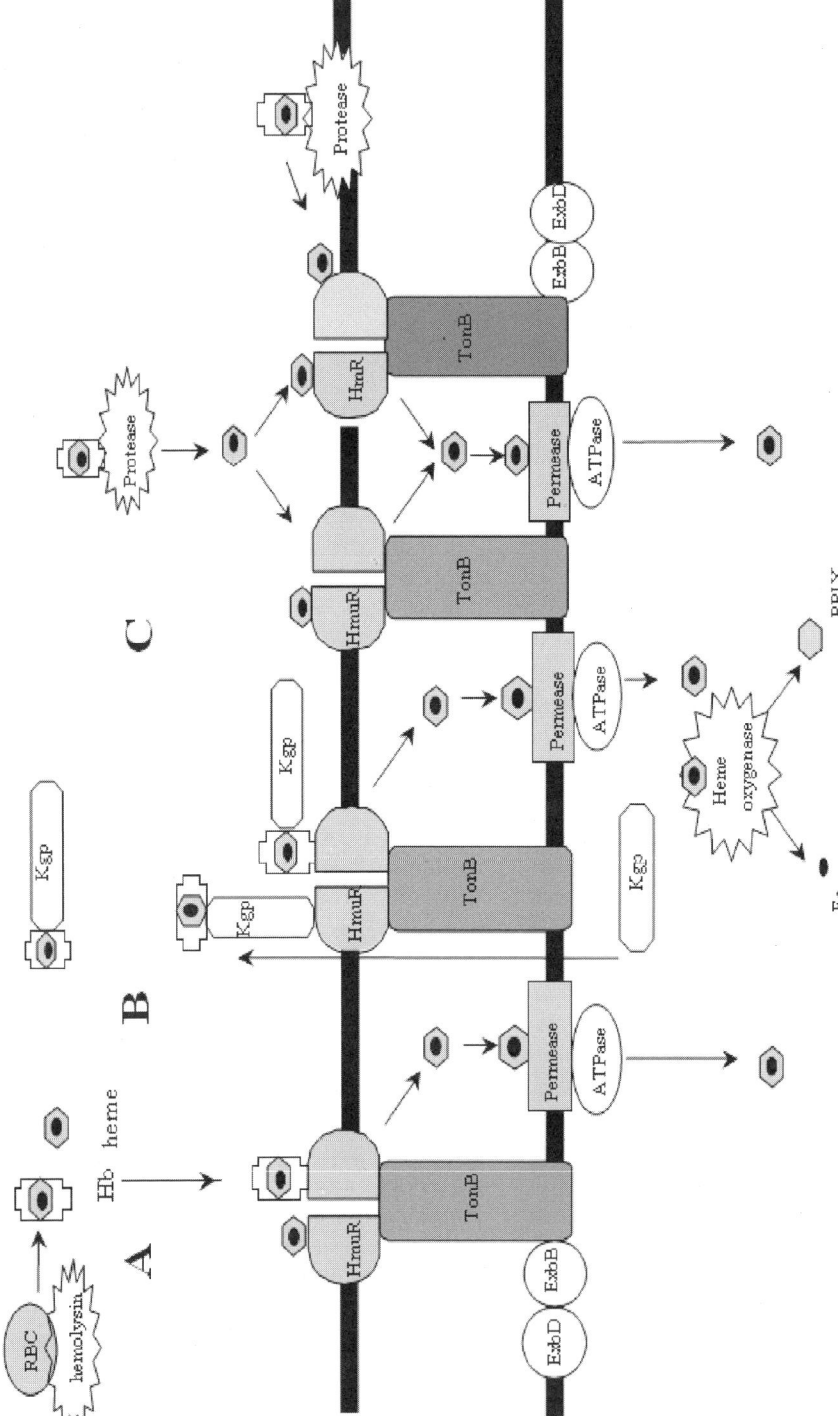

FIGURE 2 Three proposed mechanisms of heme transport in *P. gingivalis*. (A) The degradation of a red blood cell (RBC) by a hemolysin liberates hemoglobin (Hb) and possibly heme. These iron sources may bind directly to HmuR working in cooperation with the putative TonB, ExbB, and ExbD proteins. (B) Soluble Kgp, acting as a hemophore-like protein, binds free hemoglobin and delivers it to HmuR. Either Kgp or hemoglobin may directly interact with HmuR. (C) Membrane-bound Kgp degrades hemoglobin, liberating free heme. This free heme may bind to HmuR or to a separate heme receptor (HmR). In all three scenarios, the transport of heme into the cytoplasm is facilitated by the use of a cytoplasmic membrane-bound permease working in cooperation with an ATPase. Once internalized, a putative degradative protein, such as heme oxygenase, may degrade heme.

nisms. Recent studies in our laboratory have led to the identification of a *P. gingivalis fur* homolog. The *P. gingivalis fur* gene consists of 501 bp and encodes a 19.2-kDa protein. The *fur* homolog is present within various *P. gingivalis* strains. The *P. gingivalis fur* homolog partially complemented the *fur* mutation in *E. coli* strain H1780. As has been demonstrated with Fur proteins of other bacteria, repressor activity was dependent on the addition of iron to the medium; this indicates that iron functions as a corepressor of the *P. gingivalis* Fur protein.

Studies of other gram-negative organisms have documented that heme or hemoglobin may also regulate the expression of heme and hemoglobin transport genes. Heme can displace iron from the *E. coli* Fur protein, leading to the possibility that heme could serve as a Fur corepressor in this system, and positive regulation by heme has been observed in *Bordetella* spp. (see chapter 20). Heme regulation of gene expression has also been observed in the gram-positive pathogen *Corynebacterium diphtheriae* (see chapter 22). Transcription of *hmuO*, which encodes a heme oxygenase, is repressed by DtxR (a diphtheria toxin repressor protein that is functionally similar to Fur) and iron and activated by hemin and hemoglobin. Activation of *hmuO* transcription in response to heme involves a two-component regulatory system. A specific sensor protein senses the heme in the environment and transduces a signal to an activator protein, which functions to activate the transcription of the *hmuO* promoter.

Additional studies in our laboratory and in others have indicated that hemin binding and transport in *P. gingivalis* are induced by hemin. This suggests that a hemin-inducible component(s) may be involved in *P. gingivalis* hemin binding and utilization. Binding of hemin in *P. gingivalis* may occur through both high- and low-affinity receptors. *P. gingivalis* hemin-repressible proteins may represent low-affinity binding receptors and may function as components of a hemin uptake system that is present only under low environmental levels of hemin. In contrast, the hemin-inducible proteins may be involved in high-affinity binding of hemin.

These results suggest that in *P. gingivalis*, the expression of proteins involved in hemin transport may be regulated at several different levels.

P. GINGIVALIS IRON TRANSPORT SYSTEMS AND THE LINK TO PATHOGENESIS

The influence of iron on the virulence of *P. gingivalis* has been documented. However, the role of iron in the regulation of specific virulence genes has not been defined. We, as well as others, have demonstrated that hemin-limited *P. gingivalis* cells exhibit greater lethality in a mouse model of infection. Growth of *P. gingivalis* under hemin limitation results in a reduction in the number of fimbriae on the cell surface, in reduced expression of the major fimbrillin gene *fimA*, and in an increase in the number of extracellular vesicles. Hemin restriction has also been reported to influence the hemolytic activity, hemagglutination, and trypsin-like protease activity of *P. gingivalis*. Studies in our laboratory with a *P. gingivalis kgp* mutant (MSM-3), which exhibits an increased-virulence phenotype in mice, showed that this strain expresses increased levels of *rgpA* and *rgpB* transcripts. This correlates with an increased arginine-specific cysteine proteinase activity. The increased virulence of *P. gingivalis* grown under hemin limitation may be due to the increased expression of the *rgpA* and *rgpB* genes, which may contribute to the pathogenic potential of *P. gingivalis*. Despite these recent observations, the regulation by iron of virulence factors in *P. gingivalis* remains a poorly investigated area. It is anticipated that with the recent construction of a *P. gingivalis*-specific DNA microarray, the role of iron in global gene expression in *P. gingivalis* will be addressed in the near future.

SUMMARY

The establishment of a pathogen within a specific niche requires that the pathogen sense the environmental conditions of the host. The environmental conditions in the periodontal pocket during infection are not precisely known. However, inflammation and tissue damage can result in an altered pattern of nutrients; when gingival fluid flow is increased and bleeding occurs, the availability of hemin may increase, resulting in the enrichment of *P. gingivalis*. *P. gingivalis* is highly adapted as a periodontopathic organism, since under environmental conditions likely to prevail during the initial stages of infection (iron-limiting conditions), it maximally produces enzymes that have tissue-damaging potential. Like many other pathogens, one mechanism used by *P. gingivalis* for the regulation of iron transport genes and possibly virulence genes is regulation via the transcriptional regulator Fur. Due to the obligate requirement for iron by *P. gingivalis* the expression of specific iron uptake systems is an important survival mechanism. The development of effective control strategies for *P. gingivalis* infection requires an understanding of the iron transport systems required for in vivo growth of this organism. *P. gingivalis* does not produce siderophores as do many other pathogens. Recent investigations of the acquisition of heme by *P. gingivalis* have led to the identification of several outer membrane receptors. While the function of many of these putative receptors remains to be determined by mutational analysis, the initial analyses have provided us with our primary insight into how *P. gingivalis* acquires heme. The involvement of the gingipains, particularly Kgp, indicates that in addition to specific outer membrane proteins, *P. gingivalis* may use a hemophore-like delivery system.

SUGGESTED READING

Aduse-Opoku, J., J. Slaney, M. Rangarajan, J. Muir, K. Young, and M. A. Curtis. 1997. The Tla protein of *Porphyromonas gingivalis* W50: a homolog of the R1 protease precursor (PrpR1) is an outer membrane receptor required for growth on low levels of hemin. *J. Bacteriol.* **179:**4778–4788.

Amano, A., M. Kuboniwa, K. Kataoka, K. Tazaki, E. Inoshita, H. Nagata, H. Tamagawa, and S. Shizukuishi. 1995. Binding of hemoglobin by *Porphyromonas gingivalis*. *FEMS Microbiol. Lett.* **134:**63–67.

Bramanti, T. E., and S. C. Holt. 1992. Localization of a *Porphyromonas gingivalis* 26—kilodalton heat-modifiable, hemin-regulated surface protein which translocates across the outer membrane. *J. Bacteriol.* **174:**5827–5839.

Dashper, S. G., A. Hendtlass, N. Slakeski, C. Jackson, K. J. Cross, L. Brownfield, R. Hamilton, I. Barr, and E. C. Reynolds. 2000. Characterization of a novel outer membrane hemin binding protein of *Porphyromonas gingivalis*. *J. Bacteriol.* **182:**6456–6462.

Genco, C. A., and D. W. Dixon. 2001. Emerging strategies in microbial heme capture. *Mol. Microbiol.* **391:**1–11.

Genco, C. A., B. M. Odusanya, and G. Brown. 1994. Binding and accumulation of hemin in *Porphyromonas gingivalis* are induced by hemin. *Infect. Immun.* **62:**2885–2892.

Karunakaran, T., T. Madden, and K. Kuramitsu. 1997. Isolation and characterization of a hemin-regulated gene, *hemR*, from *Porphyromonas gingivalis*. *J. Bacteriol.* **179:**1898–1908.

Letoffe, S., F. Nato, M. E. Goldberg, and C. Wandersman. 1999. Interactions of HasA, a bacterial hemophore, with hemoglobin and with its outer membrane receptor HasR. *Mol. Microbiol.* **33:**546–555.

Lewis, J. P., J. A. Dawson, J. C. Hannis, D. Muddiman, and F. L. Macrina. 1999. Hemoglobinase activity of the lysine gingipain protease (Kgp) of *Porphyromonas gingivalis* W83. *J. Bacteriol.* **181:**4905–4913.

Nakayama, K., D. B. Ratnayake, T. Tsukuba, T. Kadowaki, K. Yamamoto, and S. Fujimura. 1998. Haemoglobin receptor is intragenically encoded by the cysteine proteinase-encoding genes and the haemagglutinin-encoding gene of *Porphyromonas gingivalis*. *Mol. Microbiol.* **27:**51–61.

Olczak, T., D. W. Dixon, and C. A. Genco. 2001. Binding specificity of the *Porphyromonas gingivalis* heme and hemoglobin receptor HmuR, gingipain K, and gingipain R1 for heme, porphyrins, and metalloporphyrins. *J. Bacteriol.* **183:**5599–5608.

Paramaesvaran, M., K. A. Nguyen, E. Caldon, J. A. McDonald, S. Najdi, G. Gonzaga, D. B. Langley, A. DeCarlo, M. J. Crossley, N. Hunter, and C. A. Collyer. 2003. Porphyrin-mediated cell surface heme capture from hemoglobin by *Porphyromonas gingivalis*. *J. Bacteriol.* **185:**2528–2537.

Pike, R. N., W. McGraw, J. Potempa, and J. Travis. 1994. Lysine- and arginine-specific proteinases from *Porphyromonas gingivalis*. Isolation, characterization and evidence for the existence of complexes with hemagglutinins. *J. Biol. Chem.* **269:**406–411.

Potempa, J., A. Banbula, and J. Travis. 2000. Role of bacterial proteinases in matrix destruction and modulation of host responses. *Periodontology* **24:**153–192.

Shi, Y., D. B. Ratnayake, K. Okamoto, N. Abe, K. Yamamoto, and K. Nakayama. 1999. Genetic analyses of proteolysis, hemoglobin binding, and hemagglutination of *Porphyromonas gingivalis*. *J. Biol. Chem.* **274:**17955–17960.

Shibata, Y., K. Hiratsuka, M. Hayakawa, T. Shiroza, H. Takiguchi, Y. Nagatsuka, and Y. Abiko. 2003. A 35-kDa co-aggregation factor is a hemin binding protein in *Porphyromonas gingivalis*. *Biochem. Biophys. Res. Commun.* **300:**351–356.

Simpson, W., T. Olczak, and C. A. Genco. 2000. Characterization and expression of HmuR, a TonB-dependent hemoglobin receptor of *Porphyromonas gingivalis*. *J. Bacteriol.* **182:**5737–5748.

Slakeski, N., S. G. Dashper, P. Cook, C. Poon, C. Moore, and E. C. Reynolds. 2000. A *Porphyromonas gingivalis* genetic locus encoding a heme transport system. *Oral Microbiol. Immunol.* **15:**388–392.

Smalley, J. W., J. Silver, P. J. Marsh, and A. J. Birss. 1998. The periodontopathogen *Porphyromonas gingivalis* binds iron protoporphyrin in the mu-oxo dimeric form: an oxidative buffer and possible pathogenic mechanism. *Biochem J.* **331:**681–685.

Sroka, A. E., M. Sztukowska, J. Potempa, J. Travis, and C. A. Genco. 2001. Degradation of host heme proteins by the lysine- and arginine-specific cysteine proteinases (gingipains) of *Porphyromonas gingivalis*. *J. Bacteriol.* **183:**5609–5616.

CORYNEBACTERIUM DIPHTHERIAE

Michael P. Schmitt

22

The genus *Corynebacterium* encompasses a diverse group of gram-positive, nonsporulating bacteria and includes both pathogenic and nonpathogenic species. *Corynebacterium diphtheriae* is the predominant human pathogen in this group, and almost all of the research that has examined iron transport and iron-regulated systems in the corynebacteria was done with this species. Most of the interest in iron transport and iron-regulated genes in *C. diphtheriae* originated from decades of research studying the expression and production of diphtheria toxin.

In 1884, Loeffler first demonstrated that the diphtheria bacillus was the causative agent of diphtheria. This finding was followed in 1888 by the discovery by Roux and Yersin that diphtheria toxin isolated from cell-free filtrates could alone reproduce the predominant systemic symptoms associated with *C. diphtheriae* infection in an animal model. Studies from the 1930s demonstrated that high levels of iron in the growth medium repressed the production of diphtheria toxin and that enhanced synthesis of the toxin was observed only after growth of *C. diphtheriae* in an iron-depleted medium.

Diphtheria toxin, the primary virulence determinant of *C. diphtheriae*, is a toxic protein secreted by virulent strains of *C. diphtheriae* that are lysogenized by a related group of bacteriophages that harbor the structural gene for diphtheria toxin, *tox*. Diphtheria toxin is recognized as the first reported iron-regulated virulence factor and is the only iron-regulated factor used as a licensed human vaccine. The iron-repressible phenotype of diphtheria toxin that was observed by these early investigators is similar to the iron-dependent repression that was found years later for other bacterial factors, including siderophores, toxins, and iron-regulated outer membrane proteins.

Diphtheria is typically manifested in two predominant forms: either as an infection of the upper respiratory tract with associated systemic symptoms due to the disseminated diphtheria toxin or as a less severe cutaneous infection that is caused by either toxinogenic or nontoxinogenic *C. diphtheriae* strains. The more severe respiratory disease is caused by toxin-producing strains of *C. diphtheriae* that colonize the mucosal membranes of the upper respiratory tract. The organism is rarely found outside of the local area of infection, and lesions caused by the organisms occur predominantly on the surfaces of the pharyngeal mucosa and adjacent regions of the respiratory tract. The infection

---

Michael P. Schmitt, Center for Biologics Evaluation and Research, Food and Drug Administration, Bethesda, MD 20892.

Iron Transport in Bacteria, Edited by Jorge H. Crosa, Alexandra R. Mey, and Shelley M. Payne
© 2004 ASM Press, Washington, D.C.

is characterized by edema, necrosis of the epithelium, and formation of a pseudomembrane. The pseudomembrane, a diagnostic characteristic of diphtheria, is an inflamed area that consists of layers of necrotic tissue and bacteria and is susceptible to bleeding. In severe cases of diphtheria, the swelling associated with the larynx and trachea, known as "bull neck," can result in asphyxiation. Moreover, the secreted diphtheria toxin is absorbed into the circulatory system and causes severe systemic lesions. The primary systems affected by the toxin include cardiac and neurological functions, although numerous cell types are susceptible to the toxin. The mucous membranes that are colonized by C. diphtheriae are predicted to be relatively low in available iron since diphtheria toxin, which in vitro is produced only under low-iron conditions, is expressed in respiratory diphtheria. The associated edema, bleeding, and cell lysis at the site of infection generates an environmental milieu that is potentially rich in host iron sources. Iron sources available to C. diphtheriae during an infection may include transferrin that is present in serum, lactoferrin found on mucous membranes, ferritin, and various heme-containing proteins.

Diphtheria toxin, the prototype for bacterial toxins and a model for the A-B-type toxins specifically, is a 58-kDa protein that is capable of entering and intoxicating a wide variety of eukaryotic cells. The toxin is synthesized as a single polypeptide and is subsequently cleaved by the cellular trypsin-like protease, furin, into a 21-kDa fragment A and a 37-kDa fragment B. The two fragments remain linked by a disulfide bond. Fragment A, which includes the N-terminal domain, possesses an ADP-ribosyltransferase activity that results in the inhibition of protein synthesis through the ADP-ribosylation of a modified histidine residue known as diphthamide, which is located on elongation factor 2. Fragment B is required for binding to the cell surface and for facilitating the transfer of the enzymatically active fragment A into the cell cytosol.

Within 20 years after the findings by Roux and Yersin that diphtheria toxin was the primary virulence determinant for C. diphtheriae, a diphtheria equine antitoxin became widely accepted as a treatment for diphtheria, and by the 1920s it was known that formaldehyde-inactivated diphtheria toxin, referred to as toxoid, was protective against the disease. The currently licensed toxoid vaccine neutralizes the severe symptoms associated with toxin activity, and its widespread use is responsible for the reduction of disease in much of the world. The vaccine is prepared from culture filtrates of the PW8 strain of C. diphtheriae, a strain that naturally overproduces diphtheria toxin.

The recent diphtheria epidemic in the newly independent states of the former Soviet Union, the largest outbreak in the world in over 40 years, has resulted in a renewed interest in the disease and the organism. Additional attention to C. diphtheriae has come from recent reports showing that a majority of the adult population in the United States and Europe lack adequate immunity to diphtheria. This enhanced interest has resulted in the completion of the genome sequence from a clinical isolate of C. diphtheriae obtained from the epidemic in Eastern Europe. The complete genome of C. diphtheriae, along with the recent development of new genetic tools for this organism, will facilitate the identification and characterization of new genes, including those involved in iron transport and iron regulation.

IRON TRANSPORT SYSTEMS IN C. DIPHTHERIAE

Siderophore-Dependent Transport

Since much of the research on C. diphtheriae is directed toward diphtheria toxin, it is perhaps not surprising that the first report to provide evidence of an iron transport system was made while investigating diphtheria toxin regulation. In 1983, Cryz et al. identified mutants in C. diphtheriae C7, designated HC1, HC3, HC4, and HC5 (HC mutants), in which diphtheria toxin production was no longer sensitive to iron repression. The HC mutants showed reduced growth in low-iron medium relative to the wild-type strain and produced elevated levels of toxin during growth in both high- and low-iron medium. ^{59}Fe uptake experiments revealed that the HC mutants were strongly defective in iron transport, and it was proposed that these mutants lacked components of an iron transport system. The overproduction of toxin observed with the HC mutants in high-iron medium was thought to be an indirect effect of reduced iron uptake. It was postulated that the defect in iron transport resulted in lower intracellular iron concentrations, even in high-iron medium, which caused the overexpression of toxin.

A few years earlier, Kanei et al. had also isolated *C. diphtheriae* mutants in which toxin synthesis was no longer sensitive to iron repression and proposed that these strains harbored a mutation in a chromosomal gene involved in iron-dependent repression of the *tox* gene. It was later shown that one of these regulatory mutants, C7(β)hm723, contained a point mutation in the *dtxR* gene, the diphtheria toxin repressor (*dtxR* is discussed in more detail below). Like the HC mutants, C7(β)hm723 produced high levels of toxin constitutively, but it differed from the HC mutants in that its growth was not impaired in low-iron medium. Surprisingly, C7(β)hm723 was moderately deficient in iron uptake, a finding which suggests that the mutation in the regulatory locus in C7(β)hm723 may also affect iron transport; however, since this mutant was derived by chemical mutagenesis, the possibility exists that this strain may carry multiple mutations.

Following the identification of iron transport mutants, Russell and Holmes reported the existence of a high-affinity iron transport system in *C. diphtheriae*. They showed that iron transport was an energy-dependent process which could be disrupted by metabolic inhibitors that affected the electrochemical proton gradient. A subsequent study demonstrated that an iron-chelating factor was secreted into the extracellular medium by C7 during growth under low-iron conditions, thus providing the first definitive evidence for a corynebacterial siderophore, designated corynebactin. A siderophore-biosynthetic mutant of C7, HC6, was identified based on its inability to produce any detectable siderophore and its poor growth in low-iron medium. HC6 also showed somewhat reduced growth relative to the wild-type C7 strain in high-iron medium. Partially purified corynebactin stimulated the growth of HC6 in a bioassay; however, the siderophore failed to stimulate growth of the HC mutants in low-iron medium, an observation consistent with the iron transport defect in these mutants. A variety of bacterial and fungal siderophores, including enterobactin, DHBA, rhodotorulic acid, ferrichrome A, desferrioxamine B, and vibriobactin, failed to stimulate growth of the HC6 mutant. Only aerobactin, a hydroxamate-type siderophore, and corynebactin allowed the growth of HC6 in low-iron medium. It was also shown that corynebactin failed to stimulate the growth of various gram-negative bacteria that were defective in siderophore production.

The *C. diphtheriae* PW8 strain, which is used in the production of diphtheria toxin for vaccine development, is also defective for siderophore synthesis. The siderophore biosynthesis defect in PW8 appears to be spontaneous, and, like HC6, the PW8 strain grows poorly in both high- and low-iron medium, although both strains show a more severe growth defect in low-iron conditions than in high-iron medium. Growth of PW8 in low-iron conditions is enhanced in the presence of corynebactin, which supports the hypothesis that the low-iron growth defect is due to the inability of this strain to produce siderophore. Although it is not known if the defect in siderophore synthesis in PW8 and HC6 is the cause of the reduced growth in iron-rich medium, it was reported that spontaneous revertants of HC6 regained wild-type characteristics with regard to both growth and siderophore production, suggesting that a single mutation is the cause of both phenotypes. The mutations that are responsible for the defect in biosynthesis of the siderophore in HC6 and PW8 have not been mapped, and the genes required for siderophore synthesis have not been identified.

Although the corynebactin siderophore was partially purified from C7 culture supernatants, the structure of the siderophore has not been determined. While corynebactin does not produce a positive reaction in the assays used for the detection of either catechol (Arnow) or hydroxamate type (Csaky) siderophores, certain biochemical tests suggest that the siderophore has some catechol characteristics. It is not known if the structure of the *C. diphtheriae* siderophore is similar to the structure of a siderophore recently reported for *C. glutamicum*, also termed corynebactin.

More recent studies have demonstrated that siderophore synthesis and diphtheria toxin production are coordinately regulated by iron. HC iron transport mutants overexpress both toxin and siderophore under high-iron conditions, suggesting that diphtheria toxin and the high-affinity siderophore transport system are part of an iron regulon in *C. diphtheriae* in which their expression is controlled by a shared regulatory mechanism.

In a subsequent study, two *dtxR*- and iron-regulated promoters, designated IRP1 and IRP2, were cloned from the chromosome of *C. diphtheriae* C7. The irp1 gene, located downstream from the IRP1 promoter, encodes a lipoprotein that is homologous to the periplasmic binding component of ABC-type iron transporters found in gram-negative bacteria. It is proposed that in gram-positive bacteria, homologs of the periplasmic component of ABC siderophore transporters are lipoproteins that are anchored to the cell surface by their N-terminal lipid moiety and function as cell surface siderophore receptors. In comparison to gram-positive bacterial sequences, Irp1 has the highest sequence homology to the *Bacillus subtilis* FhuD protein, the putative ferrichrome receptor. Clones containing the *irp1* gene and downstream sequences, which were subsequently shown to encode additional components of the *irp1* ABC transport system, failed to complement the iron transport defect in any of the HC mutants. This finding suggests that the ABC transport system encoded by *irp1* is not involved in the transport of corynebactin.

The iron-chelate or siderophore that is transported by the *irp1* system has not been identified. No mutants with mutations in this system have been constructed, and it seems likely that the *irp1* ABC transporter is involved in the uptake of a siderophore or iron-chelating compound that is not synthesized by *C. diphtheriae*. The predicted amino acid sequence of an open reading frame downstream from the *irp2* promoter had no match to any sequence in various protein databases.

SELEX (the systematic evolution of ligands by exponential enrichment) is a PCR-based procedure that was used to amplify and enrich for DNA fragments capable of binding to DtxR. A DNA fragment containing a novel DtxR-binding site from *C. diphtheriae* C7 was isolated by this technique and cloned into a promoter-probe plasmid, where it exhibited iron- and *dtxR*-regulated transcriptional activity in *Escherichia coli*. Downstream from this iron-regulated promoter element, which was termed IRP6, is an operon that encodes an ABC-type iron transport system composed of three genes, designated *irp6A*, *irp6B*, and *irp6C* (Fig. 1). The product of *irp6A* is predicted to be a lipoprotein with homology to the periplasmic binding component of gram-negative ABC transporters. The *irp6B* and *irp6C* genes are predicted to encode a permease and an ATP-binding protein, respectively. A clone containing the entire *irp6* operon complemented the siderophore overexpression phenotype of the previously identified iron transport mutants HC1, HC4, and HC5. The HC3 mutant was not complemented by the *irp6* operon. Subsequent complementation studies with subclones harboring the individual genes encoded by the *irp6* operon were done to identify the defective genes in the various HC mutants, and sequence analysis of certain *irp6* alleles was used to identify the specific genetic defects. These studies revealed that the HC1 and HC4 mutants each contained single missense mutations in *irp6A* and *irp6C*, respectively. HC5 carried a chain-terminating mutation in the *irp6B* gene. Although HC3 was not complemented by the *irp6* operon, this mutant displayed enhanced

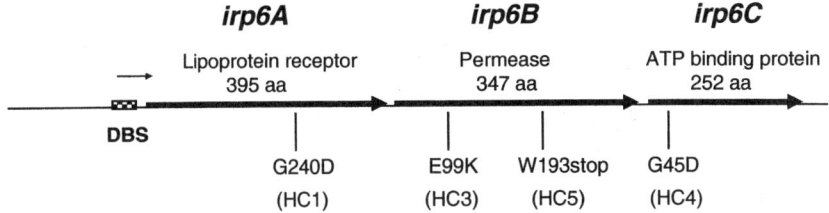

FIGURE 1 Genetic map of the *C. diphtheriae irp6* operon and locations of mutations. The function of the predicted product encoded by each gene is indicated below the gene designation, and the size of the product is indicated in amino acids (aa). A DtxR-binding site (DBS) overlaps the promoter for the operon, and an arrow indicates the direction of transcription. The locations of the various point mutations in the HC mutants and the resulting amino acid changes caused by the mutations are indicated below the genetic map.

sensitivity to heme, a property shared by C7(β)hm723, a *dtxR* mutant of C7. Sequence analysis of the *dtxR* allele in HC3 revealed a stop codon in place of Trp104, a mutation predicted to produce a defective DtxR protein. HC3 was also found to contain a missense mutation in *irp6B*. As predicted, the HC3 strain that harbored both the cloned *dtxR* gene and *irp6* operon exhibited wild-type characteristics.

The findings from this study indicate that the *irp6* operon encodes an iron-regulated ABC transport system required for the uptake of the *C. diphtheriae* siderophore (Fig. 1). This study also indicates that the HC3 mutant contains at least two mutations, and it is possible that the other HC mutants also contain multiple mutations. Recent advances in transposon mutagenesis and in the construction of site-specific mutations in the chromosome of C7 should facilitate the creation of defined mutations in the *irp6* operon. The construction of defined chromosomal mutations in the *irp6* genes in the wild-type C7 background will provide definitive confirmation for the role of the *irp6* operon in corynebactin transport. The genes required for the synthesis of the siderophore have not been identified, and they do not appear to be linked to the *irp6* transport operon. Since siderophore production is coordinately regulated with diphtheria toxin, the genes encoding the biosynthetic enzymes are likely to be under the control of an iron- and *dtxR*-regulated promoter. Furthermore, the coordinate regulation of toxin and siderophore also suggests that the siderophore is produced during *C. diphtheriae* infection. Animal models are not available to evaluate the capacity of *C. diphtheriae* to colonize and survive on mucosal surfaces. The development of an animal model system is needed to determine if the high-affinity siderophore transport system and other iron acquisition mechanisms in *C. diphtheriae* are required for colonization and virulence of this pathogen.

Heme-Iron Transport Systems

Both toxinogenic and nontoxinogenic strains of *C. diphtheriae* are able to colonize the upper respiratory tract of humans, an area presumed to be low in available iron. The ability of *C. diphtheriae* strains to colonize and persist in the human host for months and even years in the absence of toxin production suggests that factors encoded on the bacterial genome are important for survival and probably include systems involved in the acquisition of host iron sources. In 1997, studies were initiated to determine the capacity of *C. diphtheriae* to use various host compounds as iron sources. These studies revealed that *C. diphtheriae* and the related species *C. ulcerans* were able to use heme, hemoglobin, and transferrin as iron sources for growth in iron-depleted environments. The *C. diphtheriae* siderophore transport mutant, HC1, which contains a defect in Irp6A, the putative siderophore receptor, was unable to use transferrin as an iron source but could utilize both heme and hemoglobin. These findings suggest that the ability of *C. diphtheriae* to use iron bound to transferrin is dependent on the siderophore, while the acquisition of iron from heme and hemoglobin does not require a functional siderophore transport system and may instead involve a novel iron acquisition mechanism.

In earlier studies, it was shown that certain gram-positive bacteria could use heme compounds as iron sources, but the factors required for heme transport in these organisms were not identified. The molecular mechanisms involved in heme transport have been most extensively studied in various gram-negative pathogens, including *Yersinia* species, *Vibrio cholerae*, *Shigella dysenteriae*, and *E. coli* O157:H7 (see chapters 5, and 14 though 16). The components of gram-negative heme transport systems include an outer membrane heme receptor that actively transports heme across the outer membrane by a TonB-dependent mechanism. An ABC transport system is involved in the active transport of heme through the cytoplasmic membrane and includes a periplasmic binding protein, a plasma membrane permease, and an ATPase. The mechanism of iron removal from the heme transported into the cytosol was not known for these gram-negative organisms when these systems were initially identified, and only recently has this mechanism been elucidated for some gram-negative bacteria (see discussion below).

Studies to determine the mechanism of heme transport in *C. diphtheriae* and in *C. ulcerans* involved the isolation of mutants that were defective in the utilization of heme and hemoglobin as iron sources. Chemical mutagenesis was used to introduce mutations into the chromosome of *C. diphtheriae* and *C. ulcerans*, and mutants that were specifically defective in the utilization of heme and hemoglobin as iron sources were isolated based on their resistance to streptonigrin, an antibiotic that has been commonly used to identify iron transport mutants. Streptonigrin, which is readily internalized by bacteria, interacts with intracellular iron pools, resulting in the formation of reactive oxygen species which have deleterious effects on macromolecules and can result in cell death. Bacteria that are defective in iron transport are more resistant to the lethal effects of streptonigrin, since they accumulate less intracellular iron. The streptonigrin enrichment method could not be used with the wild-type C7 strain for the isolation of heme utilization mutants, presumably due to interference caused by the activity of the siderophore. This difficulty was not encountered with the *C. diphtheriae* siderophore transport mutant, HC1, or with a wild-type strain of *C. ulcerans*. Several mutants of *C. ulcerans* and from the *C. diphtheriae* HC1 strain were identified that exhibited a diminished ability to use both heme and hemoglobin as iron sources; however, all of the mutants maintained some capacity to utilize heme and hemoglobin as iron sources if the iron source was supplied at sufficiently high levels. Mutants that could use heme but not hemoglobin (or vice versa) were not found, suggesting that a common mechanism is involved in the acquisition of iron from these sources.

To identify genes involved in heme utilization, a plasmid library carrying chromosomal DNA fragments from *C. diphtheriae* C7 was used to transform several of the *C. ulcerans* heme utilization mutants. *C. ulcerans*, which is more amenable to genetic manipulation than is *C. diphtheriae*, was used as a surrogate host to perform these initial complementation studies. *C. ulcerans* is also a good model system for *C. diphtheriae* infection and colonization, since occasional isolates of *C. ulcerans* produce diphtheria toxin and cause a diphtheria-like illness in humans that is virtually indistinguishable from the disease caused by toxinogenic *C. diphtheriae*. Clones that were able to restore the ability to use both heme and hemoglobin to the various *C. ulcerans* mutants were identified. Based on the results of these complementation studies, the *C. ulcerans* heme utilization mutants could be divided into two distinct groups: those that lack heme oxygenase function and those that are defective in heme transport. The first group of mutants, represented by strain CU29, was complemented by plasmid pCD293, while the second group, represented by CU84, was complemented by plasmid pCD842. All of the *C. diphtheriae* HC1 heme utilization mutants were strongly complemented by clones carrying pCD293 and, surprisingly, were also weakly complemented by pCD842.

C. diphtheriae Heme Oxygenase

A single gene, designated *hmuO*, carried on pCD293 is responsible for the complementation of the CU29 mutant. HmuO, a 24.1-kDa cytosolic protein, has significant homology to heme oxygenases, a well-characterized family of eukaryotic proteins not previously found in bacteria. HmuO has 33% identity and 70% similarity to the N-terminal 221 residues from the human heme oxygenase protein, HO-1. The high degree of relatedness is quite surprising, considering the wide evolutionary divergence between these species. Specific sequences important for the activity of heme oxygenases are conserved between HmuO and the human HO-1. Most notably, histidine residues at position 20 and 128 in HmuO, which correspond to residues H25 and H132, respectively, in HO-1, are important for the enzymatic activity of HO-1, and the H25 residue in HO-1 is the primary heme-iron ligand (the corresponding residue in HmuO has a similar function). Another area of similarity between these proteins is a 24-amino-acid region in the central portion of the protein that is highly conserved among the HO-1 class of heme oxygenases.

Heme oxygenases degrade heme through the oxidative cleavage of the heme porphyrin backbone and the subsequent liberation of biliverdin, carbon monoxide, and iron. While heme oxygenases had not been found in bacteria prior to the identification of HmuO in *C. diphtheriae*, a role for a heme oxygenase-like enzyme was proposed as part of the mechanism for the utilization of heme as an iron source in bacteria. It is proposed that HmuO functions in the final step of heme utilization in *C. diphtheriae* and *C. ulcerans* once heme has traversed the cytoplasmic membrane and entered the cytosol, HmuO degrades the heme, which liberates the heme-bound iron.

In a subsequent study, the HmuO protein was overproduced in an *E. coli* expression system and the native protein was purified. HmuO was shown to have a catalytic activity similar to that previously observed for eukaryotic heme oxygenases. HmuO was able to bind heme in a 1:1 ratio, and the heme-HmuO complex was shown to yield biliverdin, carbon monoxide, and iron as the final reaction products. These studies with the purified HmuO protein confirmed its function as a bacterial heme oxygenase and provided support for the role of this protein in the utilization of heme as an iron source through its ability to degrade heme and release the iron.

Pathogenic species of *Neisseria* and *Pseudomonas* were recently shown to encode heme oxygenases involved in the removal of iron from heme. Other bacterial pathogens, such as *V. cholerae*, *E. coli*, and various *Shigella* species, which utilize heme compounds as iron sources, do not appear to possess heme oxygenases, and the mechanisms used by these organisms to extract the heme-bound iron remains unknown. One of the genes within the heme transport locus of *V. cholerae* is able to complement the heme utilization defect in *Corynebacterium hmuO* mutants. The function of this gene has not been determined, and the predicted product has minimal sequence homology to any known protein, but the protein does not appear to have typical heme oxygenase activity.

hmuTUV Heme Transport Locus in *C. diphtheriae*

A second distinct class of mutants in *C. ulcerans* with diminished capacity to utilize heme and hemoglobin as iron sources was identified from the initial mutant analysis. This group of mutants was not complemented by clones carrying *hmuO* but was complemented by plasmids represented by pCD842. Sequence analysis of the chromosomal insert in plasmid pCD842 identified four genes, designated *hmuT*, *hmuU*, *hmuV*, and *orfX* (Fig. 2). The predicted products of the *hmuTUV* genes are homologous to the ABC heme transporters of gram-negative bacteria. The *hmu* genes are closely linked, suggesting that they may be organized as an operon. No promoter elements were identified upstream of *hmuT*, the first gene in this putative operon, suggesting that an additional gene(s) may be present upstream of *hmuT*. The *orfX* gene, which is located downstream from the *hmu* operon, does not appear to play a role in heme transport and is not required for complementation of the heme utilization mutants.

The 4.5-kb chromosomal insert present on plasmid pCD842 terminates approximately 200 bp upstream of the start codon for *hmuT*, and a small segment containing the 3′ end of an open reading frame, designated *htaA*, was identified immediately upstream of *hmuT*. When the determination of the entire *C. diphtheriae* genome sequence was finished earlier this year, the complete *htaA* open reading frame was identified. The full-length *htaA* gene is predicted to encode a product of 591 amino acids that contains a signal sequence and a putative membrane-spanning region at its C terminus. The sequence of HtaA suggests that it is anchored to the *C. diphtheriae* cell surface or possibly secreted into the extracellular environment. HtaA does not possess an LPXTG anchoring motif at its C terminus. A recent GenBank search revealed that HtaA has homology to putative membrane proteins of unknown function from *Streptomyces coelicolor*. Examination of the sequence upstream of HtaA has identified a putative promoter element and a possible DtxR binding site, suggesting that HtaA is iron regulated (Fig. 2). Furthermore, sequence analysis suggests that *htaA* may be part of the *hmu* operon, since only a 9-bp gap is present between the stop codon of *htaA* and the start codon of *hmuT*. A role for HtaA in the transport or utilization of heme or hemoglobin as iron sources has not been demonstrated.

HmuT has homology to the periplasmic binding component of ABC transport systems and, like other protein of this class in gram-positive bacteria, is predicted to be a lipoprotein anchored to the cell surface. HmuT from *C. diphtheriae* exhibits the highest homology to the periplasmic binding component, HmuT, from the *Yersinia pestis* ABC heme transport system (25% identity and 39% similarity). HmuU and HmuV are homologous to the plasma membrane permease and the ATPase components of ABC heme transport systems, respectively. HmuU and HmuV also show the highest homology to their counterparts in the *Y. pestis* ABC heme transport system (36% identity and 62% similarity for HmuU, 40% identity and 56% similarity for HmuV).

Experimental evidence provided strong support that the *C. diphtheriae* HmuT protein functions as a cell surface-binding protein for heme and possibly hemoglobin. The native HmuT protein was overexpressed in both *E. coli* and *C. diphtheriae*, and fractionation studies showed that HmuT is localized to the cytoplasmic membrane, as would be predicted for

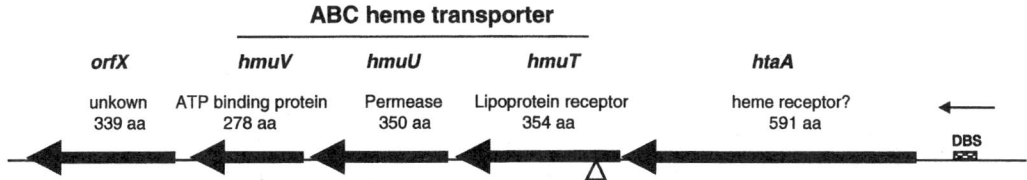

FIGURE 2 Genetic map of the *C. diphtheriae hmu TUV* heme transport locus. The products of the *hmuT*, *hmuU*, and *hmuV* genes are homologous to ABC heme transport proteins. A function for the proteins encoded by *htaA* or *orfX* has not been demonstrated. The size of the predicted products encoded by each gene in the operon is indicted in amino acids (aa). A DtxR-binding site (DBS) overlaps the promoter for the operon, and an arrow indicates the direction of transcription. The vector integration mutation in the *hmuT* gene is indicated by a triangle below the genetic map.

a cell surface lipoprotein. The HmuT protein present in the cytoplasmic membrane fractions from *E. coli* bound both hemin agarose and hemoglobin agarose, with HmuT recovered at greater than 90% purity after incubation with either of these heme-containing agarose resins. Unbound hemoglobin effectively competed with the hemoglobin agarose resin for binding to the HmuT protein, demonstrating a specificity of HmuT for hemoglobin binding. High-level expression of HmuT in the presence of [^{3}H] palmitate, which specifically labels the lipid moiety on lipoproteins, or in the presence of globomycin, a specific inhibitor of signal peptidase II, confirmed that the mature form of HmuT is in fact a lipoprotein (Fig. 3 shows a model of bacterial heme transport).

A surprising observation from the complementation studies with the *C. diphtheriae hmuTUV* and *hmuO* genes was the capability of either pCD293 or pCD842 to at least partially complement the *C. diphtheriae* HC1 heme utilization mutants. Sequence analysis of the *hmuO* allele from three of the HC1 heme utilization mutants revealed that they all carried point mutations in the *hmuO* gene. While it is

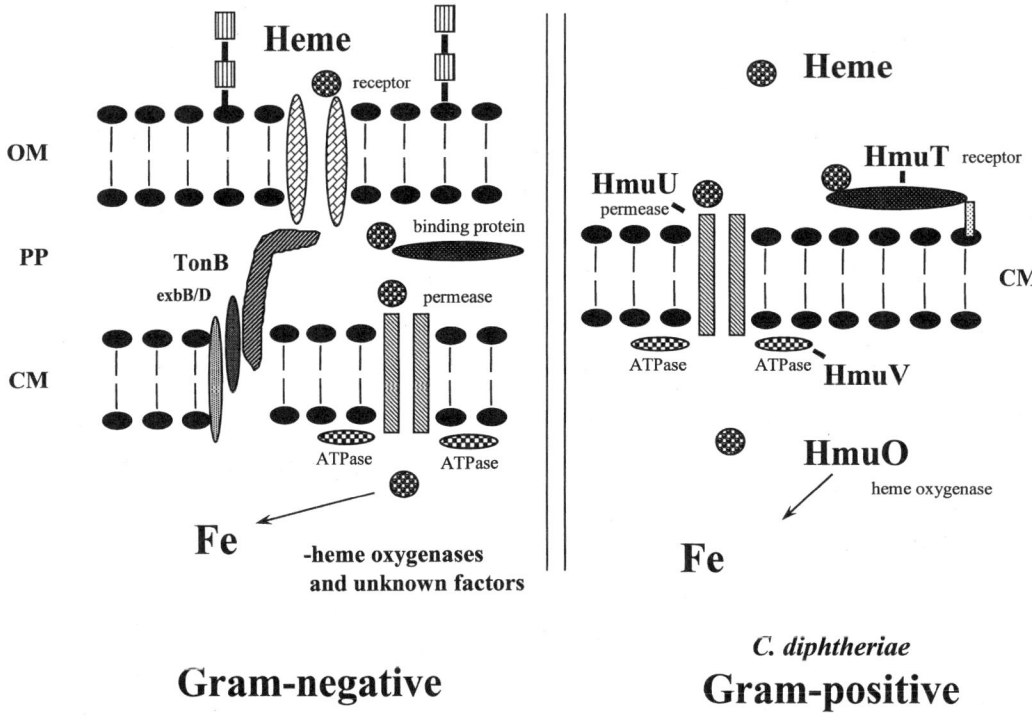

FIGURE 3 Comparison of heme utilization between gram-negative and gram-positive bacteria (*C. diphtheriae*). In gram-negative bacteria, heme is transported across the outer membrane by a TonB-dependent outer membrane heme receptor. In the periplasm, heme binds to a periplasmic binding protein, which delivers heme to a cytoplasmic membrane permease. The permease facilitates the transport of heme into the cytosol in an energy-requiring process that involves the hydrolysis of ATP by an ATPase located on the cytosolic side of the plasma membrane. The mechanism of iron extraction from heme is thought to involve a heme oxygenase in *Neisseria* and *Pseudomonas* species; however, many other gram-negative bacteria do not appear to possess heme oxygenases, and the mechanism of iron removal for these organisms is not known. In gram-positive bacteria such as *C. diphtheriae*, heme is proposed to bind to the HmuT lipoprotein at the cell surface. Heme is then transported into the cytosol by the HmuU permease, which utilizes energy from the hydrolysis of ATP provided by the ATPase activity of the HmuV protein. Once inside the cell, the heme oxygenase activity of HmuO degrades the heme, which results in the release of iron. OM, outer membrane; PP, periplasm; CM, cytoplasmic membrane.

unclear how the cloned *hmuTUV* genes can partially complement these *hmuO* mutations in *C. diphtheriae*, it is known that high levels of heme stimulate the growth of the *Corynebacterium* heme utilization mutants in low-iron medium. It is possible that the higher copy number of the cloned *hmuTUV* genes allows increased transport of heme into these strains and thus permits limited growth in the presence of heme under low-iron conditions.

Analysis of the *C. diphtheriae* HC1 heme utilization mutants suggests that the heme utilization defect in these strains is caused by mutations only in the *hmuO* gene and that mutations in the *hmuTUV* genes are not present in any of these *C. diphtheriae* mutants. The reason for the lack of mutations in the *hmuTUV* locus in *C. diphtheriae* became clear once defined mutations in the *hmuT* genes in *C. diphtheriae* and *C. ulcerans* were constructed. A vector integration method was developed to generate targeted mutations in the chromosome in both of these species. Integration mutations in the *hmuT* gene in *C. ulcerans* resulted in a defect in heme utilization, whereas a similar mutation in the *hmuT* gene in *C. diphtheriae* C7 had no effect on the ability of that strain to use heme or hemoglobin as iron sources. This finding suggests that alternate mechanisms for the acquisition of heme exist in *C. diphtheriae* and also provides an explanation of why chemically derived mutations in the *hmuTUV* locus were not identified in *C. diphtheriae*. A defined mutation in the *hmuO* gene in wild-type *C. diphtheriae* C7 was recently constructed in this laboratory and, like the *hmuT* mutation, had no effect on the ability of C7 to utilize heme as an iron source. When the identical *hmuO* mutation was constructed in the siderophore transport mutant, HC1, a defect in heme utilization was observed that was indistinguishable from the chemically derived HC1 heme utilization mutants. It is unclear why an *hmuO* mutation in HC1 results in a heme utilization defect while the identical mutation in the wild-type C7 background has no obvious phenotype. It is possible that the *hmuO* mutation in HC1, a strain that is already iron starved due to an iron transport defect, may further reduce the iron availability to HC1 when the strain is grown with heme as the sole iron source. The findings from the previous studies suggest that alternate mechanisms for both transport (*hmuTUV*) and the extraction of iron from heme (*hmuO*) exist in wild-type *C. diphtheriae*. Similarly, *C. ulcerans* may also encode additional systems for heme utilization, since this organism still maintains some ability to use heme and hemoglobin as iron sources when carrying mutations in either the *hmuTUV* or *hmuO* genes.

Heme Iron Regulation of *hmuO*

Sequence analysis of the *hmuO* promoter region identified an area that matched the consensus DtxR-binding site at 15 of 19 bases. DNase I footprinting experiments confirmed that DtxR binds this sequence and suggested that, like the *tox* promoter, the *hmuO* promoter is iron and DtxR regulated. Analysis of an *hmuO-lacZ* reporter fusion in *C. diphtheriae* indicated that expression from the *hmuO* promoter was repressed under high-iron conditions. However, unlike expression at the *tox* promoter, the *hmuO* promoter was poorly expressed under low-iron conditions. Surprisingly, when heme or hemoglobin was added to low-iron medium, a 20-fold increase in transcription from the *hmuO* promoter was observed and the addition of iron to the heme-containing medium resulted in partial repression from the *hmuO* promoter. The dual regulation, activation by heme and repression by iron, observed with the *hmuO* promoter was the first finding of a heme-activated gene in bacteria. Heme activation had been seen previously with eukaryotic genes, and recent reports indicate that the heme transport genes in *Bordetella* species are heme activated. Northern analysis confirmed the heme activation at the *hmuO* promoter and indicated that *hmuO* was transcribed as a single cistron. A gene located upstream of *hmuO* but divergently transcribed was also regulated by DtxR and iron but was not heme activated. The sequence of this upstream gene has no homologs in the database, and its function is not known.

The mechanism involved in heme activation at the *hmuO* promoter was investigated in a subsequent study. Deletion of sequences upstream of the *hmuO* promoter resulted in a reduction in heme activation, suggesting that a binding site for a *trans*-acting factor involved in the heme-dependent activation of *hmuO* expression may be located in this upstream region. In addition, it was observed that the *hmuO* promoter is poorly expressed in *E. coli*, presumably because *E. coli* does not encode the factor(s) required for heme activation. This lack of heme activation of the *hmuO* promoter in *E. coli* was then used to identify *C. diphtheriae* genes involved in the heme-dependent expression of *hmuO*. Two independent clones, designated pTSB2 and pTSB5, obtained from a *C. diphtheriae* plasmid library were shown to activate the expression of the *hmuO-lacZ* fusion in *E. coli*. For each clone, a single gene, termed *chrA* in pTSB2 and *cstA* in pTSB5, was responsible for the enhanced expression; however, the activation of the *hmuO-lacZ* fusion in *E. coli* by either *chrA* or *cstA* was very weak and was not heme inducible. The predicted products of both genes show striking homology to re-

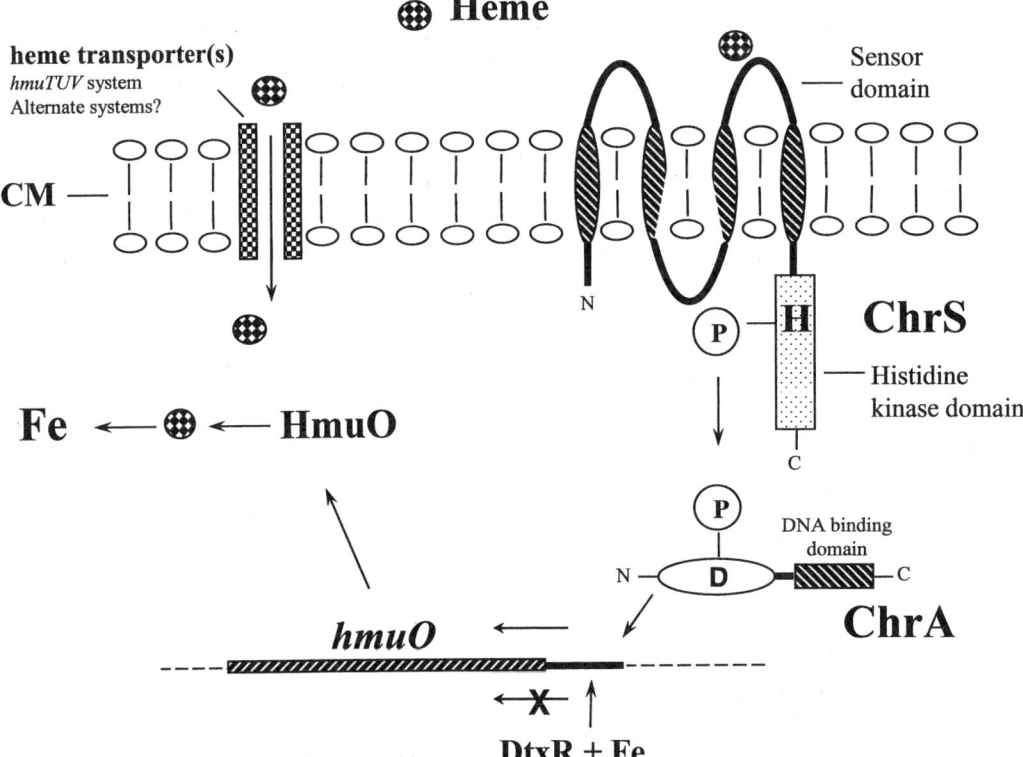

FIGURE 4 Heme-dependent activation of *hmuO*. It is proposed that the sensor kinase protein, ChrS, detects extracellular heme through its N-terminal region, which is predicted to contain at least four transmembrane helices (indicated by diagonally striped ovals) and two extracellular loop regions. The detection of heme or hemoglobin is thought to result in autophosphorylation of ChrS at a conserved His residue (H) located within the cytosolic histidine kinase domain (boxed region). The phosphoryl group (P) is then transferred to a conserved Asp residue (D) on ChrA. Phosphorylation activates the DNA-binding function of ChrA and allows ChrA to bind upstream of the *hmuO* promoter and activate transcription. Under high-iron conditions, DtxR represses the transcription of the *hmuO* promoter. The *hmuO* gene, therefore, is optimally expressed in low-iron environments in the presence of heme. The HmuO protein is proposed to degrade the cytosolic heme and liberate the heme-bound iron.

sponse regulators, the activating factor of two-component signal transduction systems. ChrA and CstA are approximately 40% identical to each other and show similar relatedness to other members of the NarL family of response regulators. Upstream of both *chrA* and *cstA* were partial open reading frames encoding the C-terminal end of the presumed cognate sensor kinases for each of these systems. The genes encoding the sensor kinases for *chrA* and *cstA* were cloned from the *C. diphtheriae* chromosome and designated *chrS* and *cstS*, respectively. The complete two-component system composed of *chrA-chrS* was cloned onto a low-copy-number plasmid and was shown to strongly activate the *hmuO* promoter in *E. coli* in the presence of either heme or hemoglobin. The *cstA-cstS* system, however, failed to activate the *hmuO* promoter in the presence of heme, and the function of this system remains to be determined.

The heme activation of the *hmuO* promoter in *E. coli* requires the sensor kinase component, ChrS, which is proposed to be involved in the detection of heme or a signal generated in response to heme. The heme activation in *E. coli* also requires the presence of a gram-negative heme uptake system that is needed to transport heme through the outer membrane. The ChrS protein, like other sensor kinases, has high homology to histidine kinases in its C-terminal region but has minimal homology to any proteins in its N terminus, the region of the protein that is proposed to interact with or "sense" the environmental signal. The sensor kinase portions of many two-component systems have loop regions in their N-terminal sequences that extend outside of the plasma membrane. These loop regions in gram-negative bacteria are located in the periplasm, but in gram-positive organisms such as *C. diphtheriae*, these extensions are exposed to the extracellular environment. It is proposed that the loop regions in the N-terminal portion of the ChrS protein interact with an extracellular signal, either heme or hemoglobin, that causes a conformational shift in ChrS, which results in autophosphorylation at a conserved histidine residue in the C-terminal histidine kinase domain.

Through a phosphotransfer mechanism, ChrS then activates ChrA by phosphorylation, allowing ChrA to bind upstream of the *hmuO* promoter and turn on transcription (Fig. 4).

More recent results from our laboratory have revealed that site-directed mutations in the chromosomal copies of *chrS* and *chrA* result in a heme-sensitive phenotype for *C. diphtheriae* C7. The reason for this heme sensitivity is not known; however, reduced expression of the *hmuO* gene, which would be an expected result of a *chrA* or *chrS* mutation, is not the reason for the heme-sensitive phenotype. It is likely that the heme-sensitive phenotype caused by mutations in *chrS* or *chrA* may be the result of reduced expression of other *chrA-chrS*-regulated genes.

IRON REGULATION OF DIPHTHERIA TOXIN

Most of what is known regarding iron regulation in *C. diphtheriae* is derived from studies that have examined the expression of the *tox* gene. It was first reported in the 1930s that the iron levels in growth medium controlled the expression of diphtheria toxin. Over the next several decades, numerous investigators confirmed and extended this finding, but the mechanism by which iron exerted its effect remained unresolved. It was not until the 1970s that a series of studies provided the first significant clues to explain the iron-dependent regulation of toxin synthesis. In 1974, Murphy and coworkers showed that toxin was produced when bacteriophage DNA encoding the *tox* gene was applied to an *E. coli* S-30 extract. It was further noted that the addition of cellular extracts from the nonlysogenic *C. diphtheriae* C7(−) strain to the *E. coli* S-30 system inhibited toxin production. These experiments provided the first evidence that a factor encoded on the bacterial chromosome repressed toxin synthesis. In subsequent studies, mutants of *C. diphtheriae* were isolated in which the production of toxin was no longer sensitive to iron repression. These mutations were found to occur either in the phage genome, later mapped to the operator sequence for the *tox* gene, or on the chromosome of *C. diphtheriae*, where it was postulated that a gene required for the repression of toxin expression was inactivated. From these studies, it was proposed that *C. diphtheriae* produces a chromosomally encoded repressor protein that, under iron-replete conditions, interacts with iron and inhibits *tox* expression by binding to a specific operator sequence located upstream of the *tox* gene.

The Diphtheria Toxin Repressor DtxR

In 1990, Boyd et al. cloned the diphtheria toxin repressor gene, *dtxR*, from *C. diphtheriae* and showed that it repressed the expression of a *tox* promoter-*lacZ* fusion construct in an iron-dependent manner in *E. coli*. The *dtxR* gene was shown to encode a predicted product of 226 amino acids that had little if any amino acid similarity to any known proteins, including the functionally similar Fur protein (see chapter 13). The *dtxR* gene did not affect the expression of Fur-regulated proteins in *E. coli*, and the *E. coli fur* gene had no effect on the expression of the *dtxR*-regulated *tox* promoter. The cloned *dtxR* gene was shown to restore iron-dependent regulation to toxin and siderophore production in the toxin-regulatory mutant C7(β)hm723. Sequence analysis of the *dtxR* allele from C7(β)hm723 showed that it carried a single point mutation, located in a putative helix-turn-helix motif, and, as predicted, the cloned mutant allele from C7(β)hm723 had greatly reduced ability to repress the expression of a *tox*-*lacZ* reporter fusion in *E. coli*. Moreover, a recently described *C. diphtheriae* C7 mutant with a *dtxR* transposon insertion mutation was shown to have a similar phenotype to that of C7(β)hm723 with respect to toxin and siderophore expression. It was also demonstrated that DtxR is required for the protection of *C. diphtheriae* against oxidative stress induced by H_2O_2.

Footprinting experiments showed that DtxR protects an approximately 30-bp region that overlaps the *tox* promoter from DNase I digestion. In addition, previous in vivo studies demonstrated that various metals, in addition to iron, could repress toxin production in *C. diphtheriae*. Several different divalent metals activate the binding of DtxR to the *tox* promoter in vitro, including Fe^{2+}, Mn^{2+}, Co^{2+}, Ni^{2+}, Cd^{2+}, and Zn^{2+}. Several of these metals also activate the binding of Fur to its cognate operators. In 1994, two additional DtxR-regulated promoters, designated IRP1 and IRP2, were cloned from *C. diphtheriae* and shown to have *dtxR*- and iron-regulated promoter activity in *E. coli*. The DtxR-binding sites that overlap the IRP1 and IRP2 promoters have significant similarity to the binding site at the *tox* operator. A 19-bp consensus DtxR-binding site was derived from an alignment of the DtxR-binding sites at the *tox*, IRP1, and IRP2 promoters as well as from a putative binding site upstream from an iron-regulated gene in *Streptomyces pilosus*. The DtxR consensus binding site (TTAGGTTAGCCTAACCTAA), which shows no sequence similarity to the Fur-binding site in *E. coli*, contains a 9-bp inverted repeat with a 1-bp gap. An almost identical consensus DtxR-binding site was derived by an in vitro affinity selection technique. At present, a total of eight DtxR-binding sites have been identified in *C. diphtheriae*. These include the previously discussed binding sites at *tox*, IRP1, IRP2, IRP6, and *hmuO*. Three additional promoters, designated IRP3, IRP4, and IRP5, were identified using the same promoter screen that was used to identify IRP1 and IRP2. The functions of the predicted products encoded downstream from IRP2, IRP4, and IRP5 are not known; however, an open reading frame encoding a product of 15 kDa downstream of IRP3 has similarity to transcriptional activators.

The ability of DtxR to regulate numerous genes in *C. diphtheriae* indicates that DtxR, like Fur, is a global iron-dependent repressor. The lack of significant amino acid similarity between DtxR and Fur suggests that these repressors represent the prototypes of two distinct classes of iron-dependent regulatory proteins.

Structural Analysis of DtxR

Analysis of the amino acid sequence of DtxR suggests that it contains a helix-turn-helix DNA-binding motif at its N-terminal region.

DtxR also possesses a single cysteine, as well as several histidine residues that may play roles in metal binding. The functional regions of DtxR and the residues associated with metal binding are depicted in Fig. 5. Mutagenesis studies identified residues in DtxR that are required for repressor function. Analysis of the single Cys residue (C102) suggests that this amino acid is important for repressor function, since saturation mutagenesis has shown that repressor activity can be maintained only if an Asp residue is substituted for the Cys. In a separate study, random mutagenesis of *dtxR* showed that most of the mutations that reduced or abolished repressor activity reside in the N-terminal half of the protein, suggesting that this region contains most of the elements required for function.

In 1995, two groups independently solved the crystal structure of DtxR. Qiu et al. reported the structure of DtxR in the presence of six different divalent cations. These studies revealed that the repressor contains three distinct domains that include an N-terminal helix-turn-helix DNA-binding region (domain 1); a central domain involved in dimerization, which also contains most of the metal-binding residues (domain 2); and a poorly resolved C-terminal domain that was subsequently shown to have homology to Src homology 3-like (SH3-like) motifs (domain 3). Two metal-binding sites, termed site 1 and site 2, were identified. Site 2 includes the unique Cys residue (C102) as well three other residues:

M10 (from domain 1), E105, and H106. In most of the crystal structures of the wild-type DtxR, metal-binding site 2 was often not bound by metal or was poorly occupied; this was thought to be due to oxidation of the unique Cys residue, which resulted in the formation of inactive disulfide-linked dimers. Schiering et al. reported the structure of both apo- and metal (Ni^{2+})-bound DtxR and identified two domains within the repressor: an N-terminal domain that includes the previously identified regions involved in DNA binding, metal binding, and dimerization, and a poorly resolved C-terminal domain. Although Schiering et al. found that Ni^{2+} was bound only to site 1 in their structure, they proposed that metal-binding site 2 is the primary metal-binding site required for repressor activity. This conclusion was based on the arrangement of the residues within site 2 and on previous mutagenesis analysis. Further support for this proposal came from site-directed mutagenesis studies demonstrating that mutations in each of the residues in site 2 strongly diminished repressor function while residue substitutions in site 1 had only a modest effect on activity. To avoid the problems associated with oxidation of C102 in the wild-type DtxR protein, the crystal structure of a functional variant of DtxR, termed DtxR (C102D), was resolved. In this structure, both metal-binding sites are occupied, and it was proposed that metal-binding site 1 is an ancillary site that makes little if any contribution to the activation of DtxR and

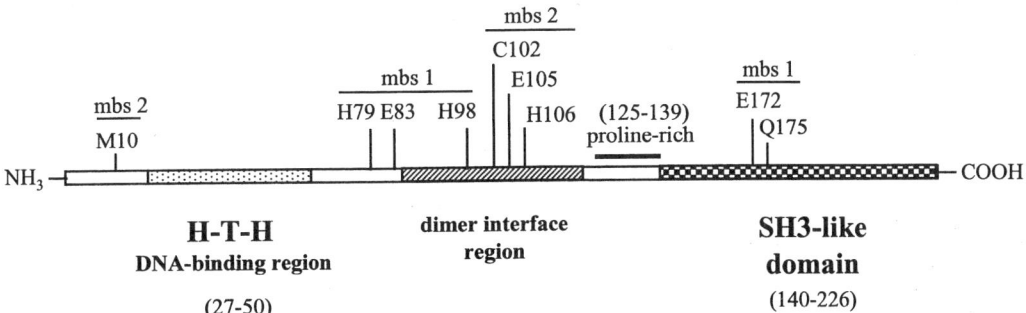

FIGURE 5 Functional domains of DtxR and amino acid residues associated with metal-binding sites 1 and 2 (mbs 1 and mbs 2). H-T-H, helix-turn-helix.

that site 2 is the primary metal-binding site required for repressor function.

The structures of Ni^{2+}-DtxR (C102D) and Co^{2+}-DtxR (wild type) bound to DNA have recently been solved. The structure of Ni^{2+}-DtxR (C102D) was determined in the presence of the *tox* operator sequence, while the wild-type Co^{2+}-DtxR structure was resolved bound to a DNA fragment carrying the DtxR consensus binding site. Both structures show that two dimers of DtxR bind to DNA, with one dimer binding on the opposite face of the DNA helix from the other dimer. A novel feature of this structure is that there appears to be no interaction between the pair of dimers. Several residues within the helix-turn-helix domain of DtxR interact with the phosphate backbone of the DNA, but only three residues from this region make base-specific contacts. Two different models have been proposed to describe the metal activation of DtxR bound to DNA. In one model, the metal-dependent activation of DtxR results in a shift in the DNA-binding helices of DtxR such that these regions have an optimal fit in the major groove of the DNA-binding site. It was proposed that this shift of the DNA-binding helices is due to a hinge-like motion of the DNA-binding domain relative to the dimerization domain. In the second model, it was proposed that metal activation of the holo-repressor dimer brings the N-terminal DNA-binding helices closer together through a caliper-like motion, which places the helices into an optimal alignment with the major groove of the DNA-binding site. In this model, the six N-terminal residues undergo a metal-dependent helix-to-coil transition that further enhances the interaction between DtxR and its operator sequence.

The function of the C-terminal SH3-like domain has not been determined and remains an area of significant interest. Recent structural data indicate that specific conserved residues in the C-terminal region of DtxR not only are involved in metal binding at site 1 (E172 and Q175) (Fig. 5) but also interact at the dimer interface and at the DNA-binding helices. From these observations, it was proposed that the C-terminal region might play a role in the activation of DtxR by helping the N-terminal DNA-binding domain to attain the proper orientation for binding to its cognate operators. In a separate study, the C-terminal domain of DtxR was found to interact with a proline-rich region near the junction of the N- and C-terminal domains. Residues in this region also bind to the dimer interface in domain 2. These findings have led to the proposed model in which the C-terminal domain functions as a "molecular switch." This model predicts that in the absence of metal, the C-terminal region stabilizes the monomeric form of the protein by binding to the proline-rich region. In the presence of metals, polar residues within the proline-rich region would interact at the dimer interface and stabilize the active repressor.

Family of DtxR-Like Repressors

DtxR is the prototype of a family of metal-dependent repressors. Because of the recent completion of numerous bacterial genomes, homologs of DtxR have now been identified in many gram-positive and gram-negative bacteria. The crystal structure of IdeR, the DtxR homolog from *Mycobacterium tuberculosis*, was recently determined and has many similarities to that of DtxR. DtxR-like proteins vary in size and in their metal selectivity, and members of this family of repressors appear to be responsive primarily to Fe^{2+} or Mn^{2+}. A second DtxR-like protein from *C. diphtheriae*, MntR, has recently been described. The MntR repressor is 223 amino acids long and shows 27% identity and 41% similarity to DtxR. MntR regulates the expression of a promoter located upstream of an ABC metal transport operon by an Mn^{2+}-dependent mechanism. In *C. diphtheriae*, only Mn^{2+} activates MntR in vivo whereas Fe^{2+} and Mn^{2+} both activate DtxR (although in vivo, Fe^{2+} is a more effective co-repressor for DtxR than is Mn^{2+}). In contrast to in vivo analysis, in vitro studies show that the same metals that activate DtxR also activate MntR; interestingly, DNase I footprint analysis revealed that MntR protects an unusually large 73-bp region. The size of this MntR-binding site suggests that as many as three dimer pairs may bind to this region, if it is assumed that MntR interacts with DNA in a manner similar to the interaction of DtxR. Mn^{2+}-activated DtxR-like repressors present in *B. subtilis* and in various gram-negative bacteria lack the SH3-like C-terminal domain, which suggests that this domain plays no role in repressor activation among this group of regulatory proteins. MntR from *C. diphtheriae* contains the C-terminal region, although the function for this domain in MntR is not known.

SUGGESTED READING

Boyd, J. M., O. N. Manish, and J. R. Murphy. 1990. Molecular cloning and DNA sequence analysis of a diphtheria *tox* iron-dependent regulatory element (*dtxR*) from *Corynebacterium diphtheriae*. *Proc. Natl. Acad. Sci. USA* **87:**5968–5972.

Collier, R. J. 2001. Understanding the mode of action of diphtheria toxin: a perspective on progress during the 20th century. *Toxicon* **39:**1793–1803.

Cryz, S. J., L. M. Russell, and R. K. Holmes. 1983. Regulation of toxinogenesis in *Corynebacterium diphtheriae*: mutations in the bacterial genome that alter the effects of iron on toxin production. *J. Bacteriol.* **154:**245–252.

Drazek, E. S., C. A. Hammack, and M. P. Schmitt. 2000. *Corynebacterium diphtheriae* genes required for acquisition of iron from hemin and hemoglobin are homologous to ABC hemin transporters. *Mol. Microbiol.* **36:**68–84.

Kanei, C., T. Uchida, and M. Yoneda. 1977. Isolation from *Corynebacterium diphtheriae* C7(β) of bacterial mutants that produce toxin in medium with excess iron. *Infect. Immun.* **18:**203–209.

Murphy, J. R., A. M. Pappenheimer, and S. T. De Borms. 1974. Synthesis of diphtheria *tox*-gene products in *Escherichia coli* extracts. *Proc. Natl. Acad. Sci. USA* **71:**11–15.

Pohl, E., R. K. Holmes, and W. G. Hol. 1999. Crystal structure of a cobalt-activated diphtheria toxin repressor-DNA complex reveals a metal-binding SH3-like domain. *J. Mol. Biol.* **292:**653–667.

Qian, Y., J. H. Lee, and R. K. Holmes. 2002. Identification of a DtxR-regulated operon that is essential for siderophore-dependent iron uptake in *Corynebacterium diphtheriae*. *J. Bacteriol.* **184:**4846–4856.

Qiu, X., L. M. Christophe, S. Zhang, M. P. Schmitt, R. K. Holmes, and W. G. Hol. 1995. Three-dimensional structure of the diphtheria repressor in complex with divalent cation co-repressor. *Structure* **3:**87–100.

Russell, L. M., S. J. Cryz, and R. K. Holmes. 1984. Genetic and biochemical evidence for a siderophore-dependent iron transport system in *Corynebacterium diphtheriae*. *Infect. Immun.* **45:**143–149.

Schiering, N., X. Tao, H. Zeng, J. R. Murphy, G. A. Petsko, and D. Ringe. 1995. Structures of the apo- and the metal ion-activated forms of the diphtheria *tox* repressor from *Corynebacterium diphtheriae*. *Proc. Natl. Acad. Sci. USA* **92:**9843–9850.

Schmitt, M. P. 1997. Utilization of host iron sources by *Corynebacterium diphtheriae*: identification of a gene whose product is homologous to eukaryotic heme oxygenase and is required for acquisition of iron from heme and hemoglobin. *J. Bacteriol.* **179:**838–845.

Schmitt, M. P. 1999. Identification of a two-component signal transduction system from *Corynebacterium diphtheriae* that activates gene expression in response to the presence of heme and hemoglobin. *J. Bacteriol.* **181:**5330–5340.

Schmitt, M. P. 2002. Analysis of a DtxR-like metalloregulatory protein, MntR, from *Corynebacterium diphtheriae* that controls the expression of an ABC-metal transporter by a Mn^{2+}-dependent mechanism. *J. Bacteriol.* **184:**6882–6892.

Schmitt, M. P., and R. K. Holmes. 1991. Iron-dependent regulation of diphtheria toxin and siderophore expression by the cloned *Corynebacterium diphtheriae* repressor gene *dtxR* in *C. diphtheriae* C7 strains. *Infect. Immun.* **59:**1899–1904.

Schmitt, M. P., and R. K. Holmes. 1994. Cloning, sequence and footprint analysis of two promoter/operators from *Corynebacterium diphtheriae* that are regulated by the diphtheria toxin repressor and iron. *J. Bacteriol.* **176:**1141–1149.

Schmitt, M. P., E. M. Twiddy, and R. K. Holmes. 1992. Purification and characterization of the diphtheria toxin repressor. *Proc. Natl. Acad. Sci. USA* **89:**7576–7580.

Tao, X., and J. R. Murphy. 1992. Binding of the metalloregulatory protein DtxR to the diphtheria *tox* operator requires a divalent heavy metal ion and protects the palindromic sequence from DNase I digestion. *J. Biol. Chem.* **267:**21761–21764.

White, A., X. Ding, J. C. VanderSpeck, J. R. Murphy, and D. Ringe. 1998. Structure of the metal-ion-activated diphtheria toxin repressor/*tox* operator complex. *Nature* **394:**502–506.

Wilks, A., and M. P. Schmitt. 1998. Expression and characterization of a heme oxygenase (HmuO) from *Corynebacterium diphtheriae*. *J. Biol. Chem.* **273:**837–841.

PATHOGENIC MYCOBACTERIA

G. Marcela Rodriguez and Issar Smith

23

Like most living cells, pathogenic mycobacteria require iron for growth, but the concentration of free iron in the host is far below the level required to support bacterial growth. The levels of free iron present in both blood and tissue fluids of host organisms are further reduced during infection as a mean of defense against invading microorganisms. This hypoferremia is a component of the acute-phase response to infection and has been detected in individuals affected by tuberculosis. Despite these iron-withholding mechanisms, pathogens multiply and cause disease. Therefore, they must be able to compete for iron in the host. Indeed, pathogens have evolved a variety of strategies to scavenge iron, as discussed elsewhere in this book. Some secrete reducing agents that convert ferric iron bound to transferrin into the ferrous form, which is released from the protein. Others utilize citrate as a low-affinity iron carrier. A few bacterial species can directly obtain iron from ferri-transferrin after binding of this molecule to specific receptors expressed on their cell surface. Other bacteria can produce cytolytic toxins and hemolysins to lyse cells and obtain iron from heme-containing proteins. Synthesis of such toxins is often regulated by iron availability, with increased expression of their structural genes when iron is limiting. Mycobacterial pathogens belong to the group of bacteria that acquire iron by producing and secreting low-molecular-weight, high-affinity iron chelators or siderophores, which can compete for iron with host iron binding proteins. A mutant strain of *Mycobacterium tuberculosis* unable to produce siderophores grows poorly in low-iron conditions in vitro, as well as in human macrophages. This observation confirmed the long-suspected essential role of siderophores for the in vivo growth of *M. tuberculosis* and underlines the importance of iron acquisition systems as potential targets for therapeutic intervention.

Although many studies of iron metabolism have been conducted with nonpathogenic mycobacteria, particularly *M. smegmatis*, our current knowledge about iron acquisition and transport in pathogenic mycobacteria is more limited. This chapter focuses on the information presently available about iron metabolism and its regulation in pathogenic mycobacteria; this information is derived mainly from experiments conducted with *M. tuberculosis*.

IRON ACQUISITION IN MYCOBACTERIA

Production of siderophores is the main mechanism used by mycobacteria to acquire iron. These iron-solubilizing molecules are pro-

---

G. Marcela Rodriguez and Issar Smith, TB Center, Public Health Research Institute at the International Center for Public Health, 225 Warren St., Newark, NJ 07103–3535.

duced in large amounts when the bacterium encounters iron limitation. Mycobacteria produce two types of siderophores: a membrane-associated molecule termed mycobactin and secreted molecules that vary in composition depending whether they are derived from a pathogenic or a saprophytic mycobacterium. The term "exochelin" was initially used collectively in reference to all secreted mycobacterial siderophores. However, soon after their discovery, it was recognized that secreted siderophores produced by pathogenic mycobacteria were chloroform soluble whereas those from nonpathogens were not. Later, when the structure of those molecules was elucidated, it became clear that the secreted siderophores of pathogens and saprophytes were very different. Since then, the term "exochelin" has been used by most researchers to refer to the peptidic, non-salicylate-containing molecule produced by avirulent mycobacteria. Indeed, exochelins are short, linear peptides in which hydroxamate groups form the major iron binding centers. The secreted siderophores synthesized by pathogenic mycobacteria are generally known as carboxymycobactins. Pathogenic mycobacteria do not produce exochelins, and for at least two slow-growing mycobacteria, *M. bovis* and *M. intracellulare*, it has been shown that they are unable to take up iron provided in the form of ferri-exochelin, suggesting that slow-growing mycobacteria do not possess the machinery to transport iron delivered by this siderophore. *M. leprae* is an exception since it has lost the *mbt* genes conferring the ability to produce mycobactin and carboxymycobactin. It is not known if *M. leprae* produces any siderophore, but when isolated from armadillo livers this pathogen takes up iron from the ferri-exochelin derived from *M. neoaurum* but not from those derived from *M. vaccae* and *M. smegmatis* or from the ferri-carboxymycobactin from *M. bovis*. The basis of this specificity for exochelin of *M. neoaurum* is unknown, but this molecule is unusual in containing a β-hydroxyhistidine residue not present in the other exochelins.

Mycobactins

Mycobactin was first described as a growth factor for *M. paratuberculosis*, the causative agent of Johne's disease in cattle. This mycobacterium and some strains of *M. avium* cannot grow in laboratory media without mycobactin or carboxymycobactin supplementation. The way in which these pathogens might acquire iron in vivo is discussed in a later section. Mycobactins have been detected in almost all mycobacteria examined. The first reported structure of a mycobactin was that of *M. phlei* mycobactin, elucidated by Snow through chemical degradation. Since then, properties and structures of mycobactins derived from many mycobacteria have been described. The general structure of this molecule is shown in Fig. 1. It is composed of a hydroxyaromatic acid, an oxazoline moiety, a β-hydroxy acid and two ε-N-hydroxylysine residues. The two hydroxamic acids of the ε-N-hydroxylysines, the nitrogen of the oxazoline ring, and the phenolate oxygen atom form the Fe^{3+} binding center. A methyl substitution can be found in some mycobactins at position 6 of the phenolic ring or at the 5′ position of the oxazoline (Fig. 1). Both types of oxazoline are found in *M. tuberculosis*. Mycobactins produced by different species vary mainly in the alkyl substitutions of the hydroxy acid and in the length of the acyl molecule, which varies between 10 and 21 carbons. Even in a single mycobacterial species, a variety of mycobactin molecules differing only in the length of their long alkyl chain can be identified. The presence of this long alkyl chain makes mycobactin insoluble in water and ensures its retention within or in close proximity to the cytoplasmic membrane.

Carboxymycobactins

Carboxymycobactins were identified as the chloroform-soluble extracellular siderophores produced by pathogenic mycobacteria. Carboxymycobactin has been found in clinical and laboratory strains of *M. tuberculosis* as well as in isolates of *M. africanum*, *M. scrofulaceum*, *M. triviale*, *M. xenopi*, *M. avium*, and *M. paratuberculosis*. The structures of carboxymycobactins

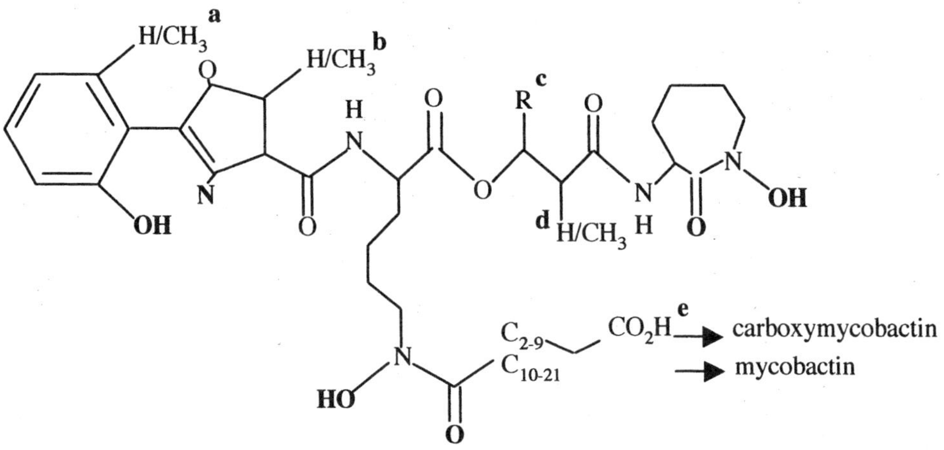

FIGURE 1 Core structure of mycobactin and carboxymycobactin. Substituents: (a) H can be found in carboxymycobactins and in some mycobactins, whereas in others the group is a CH_3. (b) Both of these substitutions can be found in mycobactins and in carboxymycobactins. (c) CH_3 is found in the *M. tuberculosis* carboxymycobactin and in mycobactins from *M. fortuitum*, *M. thermoresistibile*, *M. avium*, *M. smegmatis*, and *M. tuberculosis*; C_2H_5 is found in carboxymycobactin of *M. avium* and in mycobactins from *M. avium*, *M. phlei*, and *M. terrae*; C_{15-18} is found in mycobactin from *M. marinum*; and C_3H_9 is found in mycobactin from *M. paratuberculosis*. (d) Both of these substitutions can be found in mycobactins and carboxymycobactins. (e) A long alkyl substitution, C_{10-21}, is found in mycobactins, whereas a shorter form, C_{2-9}, is found in carboxymycobactin. Molecules carrying the methyl ester have been found in *M. tuberculosis* and *M. avium*.

from *M. tuberculosis*, *M. avium*, and *M. bovis* have been elucidated. They are variants of the mycobactin molecule in which the long alkyl chain is replaced by a shorter chain (two to nine carbons) terminating in a carboxylic acid (Fig. 1). As a consequence, these molecules are less lipophilic than mycobactin and, instead of being retained in the membrane, are released into the medium, where they remain soluble. The name "carboxymycobactin" was used in the first publication describing this structure because of the presence of the terminal carboxyl group. However, since forms of this molecule in which the carboxylic acid is a methyl ester were found in *M. tuberculosis* and *M. avium*, some authors prefer to use the term "exomycobactin" instead of "carboxymycobactins". While exochelins have not been detected in pathogenic mycobacteria, the carboxymycobactins have been found in the saprophyte *M. smegmatis*, although at very low concentrations (1 to 10% of the total iron binding capacity). The relevance of this small amount of carboxymycobactin to *M. smegmatis* iron uptake is unknown.

Mycobactin and Carboxymycobactin Biosynthesis

Early precursor incorporation studies provided significant information about the building blocks used for mycobactin biosynthesis. The aromatic nucleus is formed by salicylic acid in most mycobactins or by 6-methylsalicylate in the mycobactins produced by *M. phlei*, *M. aurum*, and *M. thermoresistibile*. L-Lysine is the precursor of both ε-N-hydroxylysines, while the β-hydroxy acid molecule is derived from the condensation of two propionate molecules, as in mycobactin P (from *M. phlei*), or two acetates, as in mycobactin S (from *M. smegmatis*). Serine and threonine are thought to be the precursors of the unsubstituted and the methyl forms of the oxazoline molecule, respectively.

More recent investigations of mycobacterial siderophore biosynthesis have been greatly aided by the availability of the complete genome sequence of *M. tuberculosis*. A cluster of ten genes, *mbtA* to *mbtJ*, and two unlinked genes, *pptT* and *ascP*, encoding the appropriate enzymes for mycobactin and carboxymycobactin core synthesis, were identified in the chromosome of this mycobacterium. Several of these genes encode proteins highly homologous to nonribosomal peptide synthases (NRPSs), which are involved in the synthesis of many other siderophores (see chapter 2). NRPSs are large modular proteins that produce peptidic molecules in a sequential process that involves activation of the amino acid to its acyl adenylate in the activation domain, linkage of the activated amino acid to the enzyme at the peptidyl carrier protein domain (through a phosphopantetheinyl arm attached to the enzyme posttranslationally), and condensation to another similarly activated amino acid in a reaction catalyzed by the condensation domain.

The proteins encoded by the *mbt* cluster show significant homology to proteins involved in the biosynthesis of other structurally related siderophores such as yersiniabactin from *Yersinia pestis* and pyochelin from *Pseudomonas aeruginosa*. Based on those homologies, putative functional roles for the Mbt proteins were proposed: three NRPS peptide synthetases (MbtB, MbtE, and MbtF), two polyketide synthetases (MbtD and MbtC), an isochorismate synthetase (MbtI), an adenylating and salicyl-AMP ligase enzyme (MbtA), and a hydroxylase (MbtG). Two additional genes, *pptT* and *ascP*, were suggested to encode phosphopantetheinyl transferases required for the attachment of the phosphopantetheinyl arm at the peptidyl carrier protein domain. The proposed activity was confirmed for PptT and MbtA by overexpression and purification of these proteins from *E. coli* followed by experimental evaluation of their activity in vitro.

Based on the postulated activities for the Mbt proteins and the information from precursor incorporation assays, a pathway for mycobactin and carboxymycobactin synthesis has been proposed (Fig. 2). Salicylate, the first building block, can be provided by MbtI, which is homologous to anthranilate synthetases and salicylate-forming enzymes. Salicylate is then activated as its acyl adenylate by MbtA and attached to MbtB. MbtB can activate serine or threonine, condense it with the salicylate moiety, and cyclize this product to an hydroxyphenyloxazoline. The activation and condensation of the two lysine-derived moieties can be performed by either MbtE or MbtF, while both polyketide synthethases, MbtC and MbtD, have the appropriate modules to assemble the β-hydroxy acyl moiety. MbtG has homology to lysine- and ornithine-N-oxygenases; therefore, it could catalyze the hydroxylation of both lysine residues. No biological role has been assigned to the proteins encoded by the remaining two genes in this cluster, *mbtH* and *mbtJ*.

Disruption of the *mbtB* gene in *M. tuberculosis* results in loss of production of both mycobactin and carboxymycobactin, indicating a direct link between the *mbt* genes and siderophore synthesis. This further demonstrates that there is a common biosynthetic pathway for the production of mycobactin and carboxymycobactin. However, the enzymatic pathways responsible for generating carboxymycobactin and mycobactin by transferring the different-length alkyl substitutions to the first N-OH-lysine have not yet been identified.

SIDEROPHORES AND IRON TRANSPORT

The *mbtB* mutant, which is defective in the biosynthesis of mycobactin and carboxymycobactin, is impaired for growth in low-iron media and is attenuated for growth in human macrophages. This confirms the essential role of these siderophores for iron uptake in vitro and in host cells. The reduced growth in host cells is consistent with the induction of siderophore genes observed in infected macrophages and mouse lungs. However, the specific contribution of each siderophore to iron uptake and the events involved in transport of the metal inside the bacterial cell remain to be determined. Several studies have described the ability of carboxymycobactin and mycobactin to compete with host iron sources. Because it is

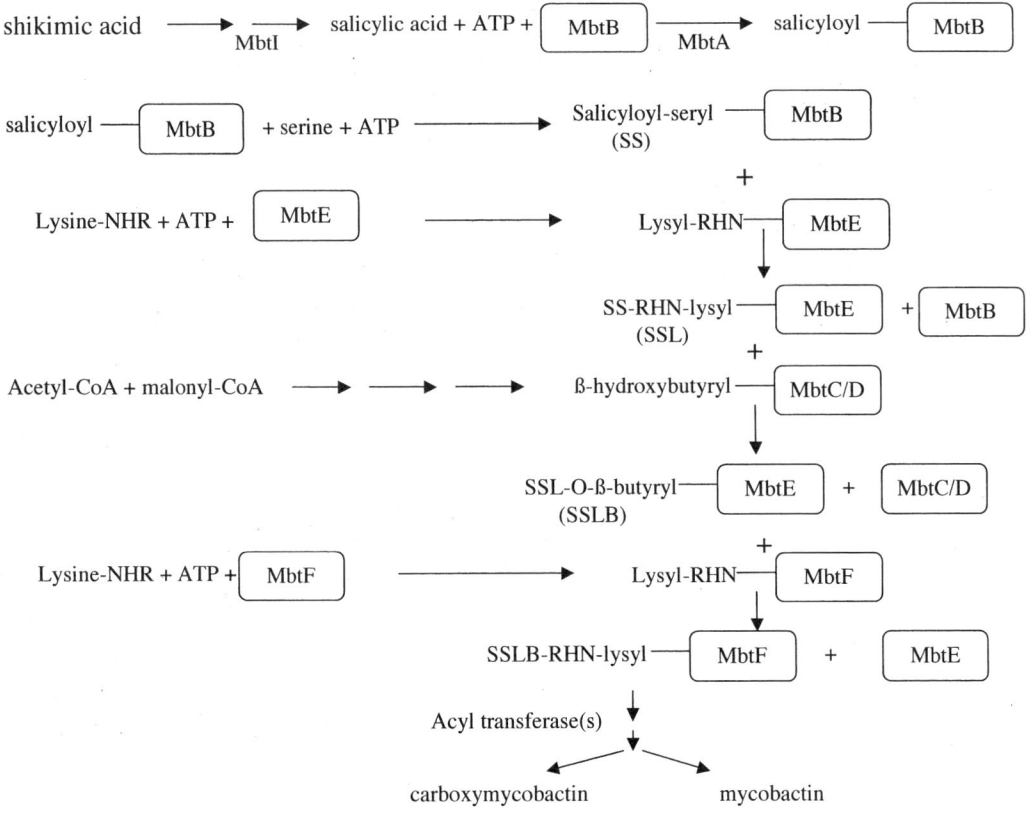

FIGURE 2 Proposed scheme for the biosynthesis of mycobactin and carboxymycobactin from *M. tuberculosis*. See the text for a detailed description of the predicted reactions. Confirmation of this model awaits the characterization of the proposed enzymatic activity in vitro for the known enzymes as well as the identification of the acyltransferase(s), the nature of its specificity, and the precise substrate used for these reactions.

secreted, carboxymycobactin has greater access to environmental iron than does mycobactin. Purified carboxymycobactins from *M. tuberculosis* can chelate iron from transferrin whether it is 95 or 40% iron saturated (the latter figure being its approximate percent saturation in human serum). This effect is consistent with the observation that carboxymycobactin from *M. bovis* BCG reversed the tuberculostatic effect of serum, which is thought to result from transferrin chelation of free iron. Transferrin is likely to be a significant iron source for intracellular mycobacteria. Ferri-transferrin binds to a specific receptor (TfR) on the eukaryotic cell surface and is internalized by receptor-mediated endocytosis. In infected macrophages, transferrin can traffic to the *M. tuberculosis* phagosome and can be used by the bacteria as a source of iron. Carboxymycobactin is probably responsible for sequestering iron from transferrin in the phagosome, since normal release of iron from transferrin triggered by the acidic pH of the early endosome is reduced as a result of mycobacterium inhibition of phagosome acidification.

Carboxymycobactin can also capture iron from lactoferrin and ferritin in vitro. Mycobactin, on the other hand, being located intracellularly, cannot obtain iron from extracellular sources such as transferrin. It can acquire iron directly only from low-molecular-weight iron compounds that penetrate the cell envelope, such as ferric salicylate and ferric ammonium citrate. However, mycobactin accepts iron from ferri-carboxymycobactin. Even though these two molecules have equivalent affinities for iron, the higher concentration of mycobactin at the cell surface could favor transfer of the metal from incoming ferri-carboxymycobactin to mycobactin. Thus, an interplay of the two siderophores might take place by which mycobactin functions as an intracellular acceptor molecule for iron captured by carboxymycobactin. Mycobactin could then act as an ionophore for the transport of iron across the lipid envelope of the cell. *M. tuberculosis* can also acquire iron from an intracellular pool in iron-preloaded macrophages, but the mechanism involved in obtaining this iron is unknown.

It has also been postulated that mycobactin could serve as an iron-holding molecule that prevents an influx of excess iron if the metal suddenly becomes available after a period of iron limitation. It has been shown that *M. smegmatis* represses the biosynthesis of porphyrins and some iron-containing proteins when grown in a low-iron medium; thus, when iron becomes available, the cell might not be prepared to incorporate it immediately into heme and iron-containing proteins. This iron-holding role of mycobactin would require a mechanism for the controlled release of iron from the siderophore, according to the anabolic rate of the cell. As discussed below, a molecule better suited for that job could be the iron storage protein bacterioferritin.

Iron from ferri-carboxymycobactin could also be transported inside the cell without using mycobactin as an intermediate. In gram-negative as well as gram-positive bacteria, ferri-siderophores bind directly to highly specific receptor proteins. The binding proteins then deliver the ferri-siderophore to ATP binding cassette (ABC) importers in the cytoplasmic membrane. Initial reports that incorporation of iron from ferri-carboxymycobactin into *M. intracellulare* and *M. bovis* was insensitive to energy poisons and uncouplers of ATP biosynthesis suggested that it was unlikely that acquisition of iron from ferri-carboxymycobactin was an active transport process. Rather, it was concluded that iron from ferri-carboxymycobactin would be taken up into pathogenic mycobacteria by facilitated diffusion, perhaps in a porin-mediated manner. However, since it is now known that carboxymycobactin can transfer iron to mycobactin in the cell envelope, it is quite possible that the cell-associated iron measured in those experiments over a short period was mainly bound to mycobactin. Iron transfer from carboxymycobactin to mycobactin most probably does not involve input of energy and therefore is expected to be insensitive to inhibitors of ATP generation. This would be similar to the transfer of iron from ferri-salicylate to mycobactin observed in iron-deficient *M. smegmatis*, which has been also shown to be ATP independent. Thus, there is insufficient evidence to rule out the possibility that the transport of iron through the plasma membrane of pathogenic mycobacteria is an active, energy-dependent process as in other bacteria.

A common characteristic of components of iron acquisition systems in prokaryotes and in higher organisms is that they are expressed only under conditions of iron limitation. Therefore, identification of iron-regulated proteins has been used to find protein involved in iron uptake. Two proteins, Irp10 and MtaA, were identified in *M. tuberculosis* grown under low-iron conditions. Because MtaA has homology to metal-transporting P-type ATPases, it has been suggested to be involved in the uptake of free iron. Recently, genes induced in *M. tuberculosis* in response to iron deficiency were identified by transcriptional profiling using DNA microarrays. Genes detected by microarray analysis included the *mbt* genes for siderophore synthesis and Rv1348 and Rv1349, genes encoding proteins with homology (46% similar) to YbtP and YbtQ from *Y. pestis*. YbtP and YbtQ are ABC transporters involved in iron uptake, probably functioning in the transport of ferri-yersiniabactin across the inner membrane of *Y. pestis* (see chapter 15). YbtP, YbtQ, and the Rv1348 and Rv1349 proteins are a distinct subfamily of iron ABC transporters and contain the membrane-spanning domain and the ATPase fused in one protein. The amino terminus of the protein contains the membrane-spanning domain, while the ATPase is located at the carboxy terminus. The structures of yersiniabactin and mycobactin are very similar, and so it is possible that the proteins encoded by Rv 1348 and Rv 1349 are similar to their *Yersinia* homologs and are involved in iron transport from ferri-siderophores in *M. tuberculosis*. Rv1348 and Rv1349 are regulated by the iron-dependent regulator IdeR (discussed in the last section of this chapter), further supporting their role in iron acquisition.

Another iron-regulated gene, Rv2039c, together with Rv2038c and Rv2040c, comprises a putative ABC transporter. These genes are annotated in the genome sequence as encoding a sugar transporter; however, the protein encoded by Rv2039c also has 40% similarity to the FutA protein from *Synechocystis* spp. FutA is part of the ferric iron uptake transporter family (www-biology.ucsd.edu/~msaier/transport) and is similar to many ferric binding periplasmic protein components of ABC iron transporters. Demonstration of the role of these and other iron-regulated proteins in *M. tuberculosis* iron transport awaits genetic and biochemical characterization.

Intracellular iron is probably released from siderophores by reduction, since ferrous iron is much less effectively coordinated than ferric iron by siderophores. Extracts of *M. smegmatis* grown anaerobically contain a NADH/NADPH-dependent ferri-mycobactin reductase activity. This activity is not specific for ferri-mycobactin since it could also reduce iron in ferri-exochelin, ferri-ferrioxamine B, and ferric ammonium citrate, suggesting that it may be associated with an enzyme whose primary role need not be that of a siderophore reductase. In addition, ferri-mycobactin was reduced by extracts from other microorganisms, including *Escherichia coli*, indicating that ferri-mycobactin reductase activity may be catalyzed by enzymes such as dehydrogenases of broad specificity, as opposed to a specific mycobacterial enzyme. Although this has not been tested, it is expected that ferri-carboxymycobactin would be as readily reducible as ferri-mycobactin.

SIDEROPHORE-INDEPENDENT IRON UPTAKE

Given the importance of iron, pathogenic bacteria often have multiple pathways to obtain this nutrient. While solubilization of iron by siderophores is probably the principal way in which mycobacteria acquire iron, it is not the only mechanism for iron uptake. A ferric citrate uptake system that works independently of the exochelin route, and is not affected by metabolic inhibitors, has been reported to function in *M. smegmatis*. A similar pathway may occur in pathogenic mycobacteria, since two genes equivalent to the *E. coli* ferric-dicitrate transporter-encoding genes, *fecB* and *fecB2*, are present in the *M. tuberculosis* genome. In fact, the *M. tuberculosis mbtB* siderophore mutant can grow normally at low concentrations of ferric citrate (5 μM), indicating a siderophore-independent pathway for uptake of this form of iron. An extracellular ferric reductase was identified in *M. paratuberculosis* growing in culture and in infected macrophages. This enzyme was able to remove iron from ferric ammonium citrate, ferritin, and transferrin by reduction of the metal, suggesting that it may be an important mechanism for iron acquisition by this mycobacterium. Reduction might be a mechanism by which *M. paratuberculosis* acquires iron in vivo; however, the presence of the reductase does not overcome the siderophore dependence exhibited by this mycobacterium in vitro, a phenomenon that remains unexplained. If a similar reductase is produced by other pathogenic mycobacteria, it could be an alternative and direct way to acquire iron from host tissue sources.

Uptake of xenosiderophores, i.e., siderophores produced by nonmycobacterial microorganisms, has been examined by studying some nonpathogenic mycobacteria. *M. smegmatis* can obtain iron from a variety of xenosiderophores, some of which transfer iron to exochelins or mycobactin. However, others, including aerobactin, arthrobactin, enterobactin, ferrichrome, myxochelin C, omibactin, rizoferrin, and triacetylfusarine, can be used as iron sources by *M. smegmatis* mutants that are deficient in exochelin and mycobactin biosynthesis. Thus, these xenosiderophores can supply iron in a way that is independent of mycobactin and exochelin. If similar pathways occur in pathogenic mycobacteria, they would have significant relevance as possible pathways to deliver antimycobacterial drugs. The use of siderophores as drug carriers is a potentially useful approach to specifically deliver antibiotics via an essential transport system (see chapter 27).

IRON STORAGE

Iron storage proteins can bind internalized iron that is not immediately utilized. This prevents the production of toxic oxygen radicals catalyzed by free iron, as discussed below. Two types of iron storage proteins have been described: ferritins, found in both eukaryotes and prokaryotes, and bacterioferritins (Bfr proteins) found in bacteria and fungi. These two proteins have similar quaternary structures but differ in that ferritins bind nonheme iron while Bfr proteins are heme proteins containing protoporphyrin IX as the prosthetic group. The *M. leprae* Bfr is a 380-kDa iron-rich protein containing 1,000 to 4,000 atoms of iron per molecule. Bfr has been detected in tissue-derived *M. leprae* and is a major antigen of *M. paratuberculosis*, recognized by the serum of infected animals. Sequences homologous to that of *bfr* have also been found in *M. avium*, *M. intracellulare*, *M. scrofulaceum*, and *M. tuberculosis*. Although it was initially found in membrane extracts, Bfr can be isolated as a soluble protein from *M. leprae* sonicates. Its sequence contains short stretches of hydrophobic amino acids consistent with a soluble or peripheral membrane protein.

M. tuberculosis possesses genes encoding a bacterioferritin-like protein (BfrA) and a ferritin (BfrB). Although the roles of these proteins in iron metabolism has not been demonstrated, it has been shown that expression of the genes is enhanced by high iron levels in the medium, consistent with a role in handling excess iron. Regulation of *bfrA* is complex. The gene is transcribed from three independent promoters, one active under high-iron conditions and two active under low-iron conditions. The high-iron promoter is activated by iron and the iron-dependent regulator IdeR, whereas the low-iron promoters are repressed by those same conditions. The high-iron promoter is 5 to 10 times more active than the low-iron ones, resulting in more efficient transcription of this gene when the bacteria are cultivated in iron-rich medium. However, significant amounts of transcripts originating from the low-iron promoters are detected. This points to a role for the BfrA produced during iron deficiency. This protein could function as an iron-holding molecule during transitions from low- to high-iron environments, similarly to Bfr in *Synechocystis* spp. In *Synechocystis* spp., Bfr is also expressed under low- and high-iron conditions, and it seems to play a role as an intermediate in iron acquisition, acting as a temporary depository. As iron availability increases, the cell needs to sequester iron until it can incorporate it into newly synthesized proteins. While such a role has been postulated for mycobactin, Bfr might be better suited for a temporary storage function, given its high iron binding capacity and its localization. Unlike mycobactin, Bfr is located within the cytoplasm of the mycobacterial cell, where it could hold incoming iron that the cell is not prepared to use and can donate it to iron-requiring proteins as they are synthesized. Interestingly, Mössbauer spectroscopy analyses show that the transfer of iron from ferric citrate to mycobactin and then slowly to Bfr occurs in *M. smegmatis* and *M. fortuitum*. Taken together, this result and the expression of *bfrA* under low-iron conditions support a role for mycobactin as an intermediate in iron acquisition and for Bfr as a repository.

Expression of *bfrB* is regulated by iron, similarly to that of other ferritin-encoding genes. High levels of intracellular iron activate its transcription, and no transcription is detected in cells growing under low-iron conditions, consistent with a role in metal storage during growth in iron-rich environments.

IRON REGULATION

In aerobic organisms, iron acts as a central catalyst in the generation of toxic oxygen radicals from hydrogen peroxide and superoxide via the Haber-Weiss and Fenton reactions. For this reason, intracellular iron levels must be strictly controlled. Iron homeostasis is the result of coordinated assimilation, utilization, and storage of this metal. In mycobacteria, as in other prokaryotes, iron metabolism is generally regulated at the transcriptional level in response to extracellular iron levels. More than 100 genes were found by transcriptional profiling to be differentially expressed in *M. tuberculosis,* according to iron availability in the medium. Iron deficiency induced about two-thirds of those genes, while the remainder were up regulated in iron-rich medium. The response to iron deficiency included up regulation of siderophore biosynthetic enzymes and other proteins possibly involved in iron uptake, and changes in metabolism to meet the challenges of iron deficiency.

Members of two families of iron-dependent transcriptional regulators, the Fur and DtxR families, regulate the bacterial iron-responsive genes. Fur (for "ferric uptake regulator") homologs are widespread among prokaryotes including gram-negative and gram-positive bacteria (see chapter 13). In general, Fur and related proteins act as transcriptional repressors since they bind to Fe^{2+} and to a specific sequence in iron-regulated promoters inhibiting the transcription of downstream genes. DtxR, the prototype of the second family of iron-dependent regulators, was identified in *Corynebacterium diphtheriae*, where it regulates the synthesis of the diphtheria toxin and of iron acquisition systems, as well as the protective response against oxidative stress (see chapter 22). DtxR-like proteins are also found in other actinomycetes, including mycobacteria, and in several gram-negative and gram-positive bacteria. The DtxR-like proteins from *Streptococcus gordonii*, *Treponema pallidum*, *Bacillus subtilis*, and *E. coli* control genes encoding Mn^{2+} transport systems.

Representatives of both metal regulator families, Fur and DtxR, are found in mycobacteria. *M. tuberculosis* possesses two Fur-like proteins, FurA and FurB, and two DtxR homologs, IdeR and SirR. The *M. leprae* genome sequence has genes coding for IdeR, FurB, FurA, and SirR, but the last two are pseudogenes. Nothing is currently known about the role of FurB and SirR, while the available information about FurA suggests that this regulator is involved in the control of the oxidative stress response, since it modulates the expression of the catalase-peroxidase-encoding gene, *katG*.

The major role of regulating iron metabolism in *M. tuberculosis* is performed by the iron-dependent regulator IdeR, a close homolog of DtxR. IdeR and DtxR are structurally (90% identity in the first 180 amino acids) and functionally related, and IdeR interacts with the same DNA sequences as does DtxR in vitro and in cells. Using iron as a cofactor, this regulator binds to a 19-bp inverted repeat consensus sequence or iron box (TTAGGTTAGGCTAACCTAA) at the promoters of iron-regulated genes and modulates their transcription. The structure of IdeR shows the presence of two binding sites for divalent metals and three distinct functional domains: the amino-terminal domain containing a helix-turn-helix DNA binding motif, a dimerization domain that also bears most of the metal binding residues, and the carboxy-terminal domain characterized by having an SH3 (Src homology domain 3)-like folding, suggesting possible interactions with other proteins.

IdeR is indispensable in *M. tuberculosis*, since disruption of the gene encoding this protein is lethal in this mycobacterium. IdeR is responsible for the differential expression of one-third of the *M. tuberculosis* genes whose expression is affected by iron availability. IdeR-regulated genes include the siderophore synthesis genes (*mbtA* to *mbtJ*), genes encoding potential iron transporters, and the genes encoding bacterioferritin and ferritin. Under low-iron conditions, genes involved in iron uptake such as the *mbt* cluster are transcribed, but when the intracellular levels of iron increase, a complex of Fe(II)-IdeR is formed which has DNA binding activity. Iron boxes on IdeR-repressed genes are generally located in the −10 region; consequently, binding of IdeR to the promoter prevents access of the RNA polymerase, thereby inhibiting the transcription of those genes (Fig. 3). Taking advantage of the similarity of IdeR and DtxR, Manabe et al. overexpressed in *M. tuberculosis* a mutant *C. diphtheriae* DtxR protein, E175K, that behaves as an iron-independent repressor. The resulting *M. tuberculosis* strain was attenuated for growth in mice. This effect is thought to be due to constitutive repression of iron acquisition systems controlled by IdeR, although this was not directly demonstrated. In contrast, the absence of IdeR in *M. tuberculosis* and *M. smegmatis* results in an inability to repress siderophore production.

While transcription of genes involved in iron uptake is repressed by iron and IdeR, that of genes involved in iron storage, such as *bfrA* and *bfrB*, is induced. In the promoters of those genes, tandem iron boxes are located farther upstream (100 to 106 bp) from the transcriptional start site, suggesting a mechanism of activation by which Fe(II)-IdeR bound to these upstream sites contacts the RNA polymerase and favors the initiation of transcription (Fig. 3). Thus, in the presence of iron, IdeR can both negatively and positively regulate transcription. In the latter case, IdeR acts as an activator.

Like many other bacteria, mycobacteria couple the control of iron homeostasis and the protective response to oxidative stress. This is evidenced by the requirement for IdeR in order to mount an efficient oxidative stress response, since *M. smegmatis* and *M. tuberculosis* *ideR* mutant strains show increased sensitivity

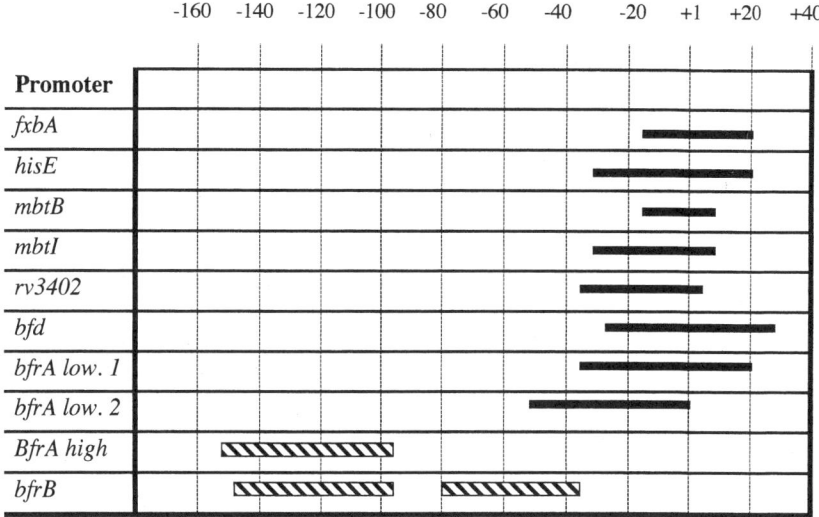

FIGURE 3 Genes regulated by the iron-dependent regulator IdeR. The genes shown have been characterized as direct targets of IdeR regulation by DNA binding and DNase footprinting analyses; all are from *M. tuberculosis* except for *fxbA*, which is from *M. smegmatis*. The transcriptional start sites in the promoters, indicated by +1, were mapped by primer extension analysis. The boxes indicate IdeR binding sites, defined as protected regions in footprinting analysis. In some cases there is more than one iron box in this sequence. The black boxes indicate IdeR binding sites in the promoters of genes that are repressed by iron and IdeR. The striped boxes indicate the sites required for IdeR binding in the promoters of genes whose transcription is activated by iron and IdeR.

to hydrogen peroxide and superoxide. In *M. smegmatis*, this effect may be due to decreased levels of catalase and the superoxide dismutase SodA, since IdeR is required for full expression of *katG* and *sodA*. No decrease in the activity of these enzymes was detected in the *M. tuberculosis ideR* mutant, but the decreased expression of bacterioferritin and ferritin in the absence of IdeR is probably a predisposing factor for iron-mediated oxidative damage, as has been observed in other systems.

CONCLUSIONS

Solid evidence has accumulated that supports the importance of iron acquisition and regulation of iron metabolism to the pathogenicity of mycobacterial species. Advances had been made in understanding the biochemistry, genetics, and biological role of mycobacterial siderophores, and significant progress has also been achieved in elucidating the adaptive response to iron availability and the mechanisms regulating that response. The future challenge consists of integrating this information to decipher the molecular events involved in iron uptake and to identify the players in addition to siderophores that are part of the iron acquisition and utilization machinery. This will allow the development of strategies to intervene in this process, using them as new weapons to battle the persistent problem of tuberculosis and other mycobacterial diseases.

ACKNOWLEDGMENTS

We thank the colleagues at the TB center of the Public Health Research Institute for helpful discussions.

Work from our laboratory discussed in this chapter was supported by a Parker B. Francis postdoctoral fellowship (awarded to G. M. R.) and NIH research grants AI-44856 and HL-64544 (awarded to I. S.).

SUGGESTED READING

Braun, V., and H. Killmann. 1999. Bacterial solutions to the iron-supply problem. *Trends Biochem. Sci.* **24:**104–109.

Clemens, D. L., and M. A. Horwitz. 1996. The *Mycobacterium tuberculosis* phagosome interacts with early endosomes and is accessible to exogenously administered transferrin. *J. Exp. Med.* **184:**1349–1355.

Cole, S. T., R. Brosch, J. Parkhill, T. Garnier, C. Churcher, D. Harris, S. V. Gordon, K. Eiglmeier, S. Gas, C. E. Barry III, F. Tekaia, K. Badcock, D. Basham, D. Brown, T. Chillingworth, R. Connor, R. Davies, K. Devlin, T. Feltwell, S. Gentles, N. Hamlin, S. Holroyd, T. Hornsby, K. Jagels, A. Krosh, J. McLean, S. Moule, L. Murphy, K. Oliver, J. Osborne, M. A. Quail, M.-A. Rajandream, J. Rogers, S. Rutter, K. Seeger, J. Skelton, R. Squares, S. Squares, J. E. Sulston, K. Taylor, S. Whitehead, and B. G. Barrell. 1998. Deciphering the biology of *Mycobacterium tuberculosis* from the complete genome sequence. *Nature* **393:**537–544.

Cole, S. T., K. Eiglmeier, J. Parkhill, K. D. James, N. R. Thomson, P. R. Wheeler, N. Honore, T. Garnier, C. Churcher, D. Harris, K. Mungall, D. Basham, D. Brown, T. Chillingworth, R. Connor, R. M. Davies, K. Devlin, S. Duthoy, T. Feltwell, A. Fraser, N. Hamlin, S. Holroyd, T. Hornsby, K. Jagels, C. Lacroix, J. Maclean, S. Moule, L. Murphy, K. Oliver, M. A. Quail, M. A. Rajandream, K. M. Rutherford, S. Rutter, K. Seeger, S. Simon, M. Simmonds, J. Skelton, R. Squares, S. Squares, K. Stevens, K. Taylor, S. Whitehead, J. R. Woodward, and B. G. Barrell. 2001. Massive gene decay in the leprosy bacillus. *Nature* **409:**1007–1011.

De Voss, J. J., K. Rutter, B. G. Schroeder, and C. E. Barry III. 1999. Iron acquisition and metabolism by mycobacteria. *J. Bacteriol.* **181:**4443–4451.

De Voss, J. J., K. Rutter, B. G. Schroeder, H. Su, Y. Zhu, and C. E. Barry III. 2000. The salicylate-derived mycobactin siderophores of *Mycobacterium tuberculosis* are essential for growth in macrophages. *Proc. Natl. Acad. Sci. USA* **97:**1252–1257.

Feese, M. D., B. P. Ingason, J. Goranson-Siekierke, R. K. Holmes, and W. G. Hol. 2001. Crystal structure of the iron-dependent regulator from *Mycobacterium tuberculosis* at 2.0-Å resolution reveals the Src homology domain 3-like fold and metal binding function of the third domain. *J. Biol. Chem.* **276:**5959–5966.

Gobin, J., and M. Horwitz. 1996. Exochelins of *Mycobacterium tuberculosis* remove iron from human iron-binding proteins and donate iron to mycobactins in the *M. tuberculosis* cell wall. *J. Exp. Med.* **183:**1527–1532.

Gobin, J., C. H. Moore, J. J. R. Reeve, D. K. Wong, B. W. Gibson, and M. A. Horwitz. 1995. Iron acquisition by *Mycobacterium tuberculosis*: isolation and characterization of a family of iron-binding exochelins. *Proc. Natl. Acad. Sci. USA* **92:**5189–5193.

Gold, B., G. M. Rodriguez, M. P. Marras, M. Pentecost, and I. Smith. 2001. The *Mycobacterium tuberculosis* IdeR is a dual functional regulator that controls transcription of genes involved in iron acquisition, iron storage and survival in macrophages. *Mol. Microbiol.* **42:**851–865.

Hantke, K. 2001. Iron and metal regulation in bacteria. *Curr. Opin. Microbiol.* **4:**172–177.

Homuth, M., P. Valentin-Weigand, M. Rhode, and G. F. Gerlach. 1998. Identification and characterization of a novel extracellular ferric reductase from *Mycobacterium paratuberculosis*. *Infect. Immun.* **66:** 710–716.

Litwin, C. M., and S. B. Calderwood. 1993. Role of iron in regulation of virulence genes. *Clin. Microbiol. Rev.* **6:**137–149.

Matzanke, B. F., R. Bohnke, U. Mollmann, R. Reissbrodt, V. Schunemann, and A. X. Tratwein. 1997. Iron uptake and intracellular metal transfer in mycobacteria mediated by xenosiderophores. *BioMetals* **10:**193–203.

Oram, D. M., A. Avdalovic, and R. K. Holmes. 2002. Construction and characterization of transposon insertion mutations in *Corynebacterium diphtheriae* that affect expression of the diphtheria toxin repressor (DtxR). *J. Bacteriol.* **184:**5723–5732.

Quadri, L. E. N., J. Sello, T. A. Keating, P. H. Weinreb, and C. T. Walsh. 1998. Identification of a *Mycobacterium tuberculosis* gene cluster encoding the biosynthetic enzymes for assembly of the virulence-conferring siderophore mycobactin. *Chem. Biol.* **5:**631–645.

Ratledge, C. 1999. Iron metabolism, p. 260–286. *In* C. Ratledge and J. Dale (ed.), *Mycobacteria: Molecular Biology and Virulence*. Blackwell Science Publishing, Oxford, United Kingdom.

Rodriguez, G. M., and I. Smith. 2003. Mechanisms of iron regulation in mycobacteria: role in physiology and virulence. *Mol. Microbiol.* **47:**1485–1494.

Rodriguez, G. M., M. I. Voskuil, B. Gold, G. K. Schoolnik, and I. Smith. 2002. *ideR*, an essential gene in *Mycobacterium tuberculosis*: role of IdeR in iron-dependent gene expression, iron metabolism, and oxidative stress response. *Infect. Immun.* **70:** 3371–3381.

Weinberg, E. D. 1999. Iron loading and disease surveillance. *Emerg. Infect. Dis.* **5:**346–352.

Wong, D. K., J. Gobin, M. A. Horwitz, and B. W. Gibson. 1996. Characterization of exochelins of *Mycobacterium avium*: evidence for saturated and unsaturated and for acid and ester forms. *J. Bacteriol.* **178:**6394–6398.

LEGIONELLA

Nicholas P. Cianciotto

24

INTRODUCTION

The *Legionella* Organism

In the summer of 1976, a mysterious outbreak of respiratory illness occurred among attendees of a bicentennial celebration of the American Legion in Philadelphia. Intensive examination of patient specimens ultimately led to the identification of a new bacterial genus, *Legionella*. The acute pneumonia associated with *Legionella* infection was named Legionnaires disease. In the last 25 years, the *Legionella* genus has grown to include 49 species, comprising 70 serogroups. Furthermore, it is now recognized that the legionellae are common causes of community-acquired and nosocomial pneumonia. Approximately one-half of the *Legionella* species have been implicated as agents of human disease. However, *L. pneumophila*, the species first isolated from the Philadelphia outbreak, is responsible for 85 to 90% of Legionnaires' disease cases. Within the *L. pneumophila* species, serogroup 1 strains represent the largest group of clinical isolates. *L. pneumophila*, as well as several other *Legionella* species, can also elicit a flu-like illness known as Pontiac fever.

The gram-negative legionellae are common inhabitants of natural and man-made aquatic environments, surviving free, in biofilms, and as intracellular parasites of protozoa. Indeed, *L. pneumophila* has been isolated from lakes, streams, rivers, and wet soil throughout the world and may be present within approximately 60% of large-building plumbing systems. Protozoan hosts for *L. pneumophila* include amoebae and ciliates, belonging to such genera as *Acanthamoeba*, *Hartmannella*, *Naegleria*, and *Tetrahymena*. The legionellae are thin, catalase-positive, motile rods that can exist within a temperature range of 4 to 63°C and a pH range of 5.4 to 8.1. They perform aerobic respiration and obtain carbon and energy from amino acids. They are routinely cultured on buffered charcoal yeast extract (BCYE) agar, which contains L-cysteine and an iron supplement. Based on rRNA analysis, the legionellae are most closely related to *Coxiella burnetii*, the agent of Q fever. Given the greater clinical significance of *L. pneumophila*, our understanding of the legionellae, including their mechanisms of iron transport, derives mainly from studies of *L. pneumophila*. Although the remainder of this section provides an overview of *L. pneumophila* pathogenesis, there are a number of recent review articles

---

Nicholas P. Cianciotto, Department of Microbiology-Immunology, Northwestern University Medical School, Chicago, IL 60611.

Pathogenesis of Legionnaires' Disease

L. pneumophila infection follows either the inhalation of contaminated aerosols generated by air conditioners and other devices or the aspiration of potable water. In the lungs the bacterium grows within the resident macrophages that line the alveoli. Following this intracellular multiplication, polymorphonucleocytes, additional macrophages, and erythrocytes infiltrate the alveoli, and capillary leakage results in edema. Chemokines and inflammatory cytokines released by infected macrophages help trigger the host defense responses. Effective clearance of *Legionella* organisms is thought to require the Th1 T-cell response and its associated cytokines, including gamma interferon, which activates macrophages, rendering them nonpermissive for the parasite. Thus, legionellosis is largely a disease of immunocompromised individuals such as transplant recipients. Other risk factors for contracting Legionnaires' disease infections include smoking, the male sex, advanced age, and alcoholism.

The ability of *L. pneumophila* to grow within macrophages is central to pathogenesis. Indeed, the majority of legionellae seen in lung samples are associated with alveolar macrophages. Furthermore, the susceptibility of an animal species correlates with the ability of *L. pneumophila* to infect its macrophages, and bacterial mutants that are impaired for in vitro infection of macrophages have reduced virulence. It is widely thought that the adaptation of *Legionella* to protozoan niches in nature endowed it with the ability to infect mammalian phagocytes. *L. pneumophila* enters the macrophage by conventional or coiling phagocytosis. Opsonization with the complement component C3 can promote phagocytosis, but entry by this pathway limits the oxidative burst and thereby may enhance bacterial intracellular survival. However, opsonin-independent phagocytosis also appears to be important. Although this second mode of entry may elicit an oxidative burst, *L. pneumophila* strains can display innate resistance to hydrogen peroxide, superoxide anions, and hydroxyl radicals. After entry, legionellae reside within a nascent phagosome that does not fuse with endosomes or lysosomes, thereby avoiding acidification and degradative enzymes. The phagosome soon associates with the rough endoplasmic reticulum, and replication of the bacteria by binary fission ensues within this compartment. Late in the intracellular cycle, the *Legionella* phagosome appears to fuse with acidic lysosomal compartments, but bacterial grow continues. Ultimately, the *L. pneumophila* phagosome fills the host cell. Macrophage death involves an early induction of caspase-3-mediated apoptosis and a late necrosis that appears to be triggered by a pore-forming activity.

Processes in addition to macrophage infection probably contribute to *L. pneumophila* disease. The bacterium may replicate or, at a minimum, must survive extracellularly in the alveoli. The ability of *L. pneumophila* strains to resist complement and cationic peptides may be especially relevant for extracellular survival, particularly following the onset of inflammation. The examination of necropsy material suggests that *L. pneumophila* may also grow within the alveolar epithelium. Indeed, in vitro, the microbe grows within alveolar type I and type II cells. The importance of extra-macrophage processes is also implied by the fact that some mutants are not defective for macrophage infection but are impaired for animal virulence.

Virulence Factors of *L. pneumophila*

A variety of *L. pneumophila* surface structures have been implicated in pathogenesis. *Legionella* lipopolysaccharide contains some endotoxic activity, and changes in lipopolysaccharide are associated with increases in serum resistance, intracellular growth, and virulence. In addition, the loss of *rcp*, a gene that appears to modify lipid A, reduces the ability of *L. pneumophila* to resist cationic peptides and to infect host cells. Whereas type IV pili modestly promote bacterial attachment to host cells, flagella promote invasion independently of adherence. The major outer membrane porin is a binding site for complement components and thus mediates opsonophagocytosis. The Mip protein is a surface-exposed propyl-proline isomerase that is required for the early stages of intracellular infection and for full virulence in animals, and the 60-kDa heat shock protein enhances epithelial cell invasion. The *rtxA* gene promotes adherence and virulence, although the structure and localization of its protein product are unclear. *L. pneumophila* secretes a pigment, a variety of degradative enzymes, and putative toxins, and it has both type II and type IV protein secretion systems. Acid phosphatases, an RNase(s), a zinc metalloprotease, mono-, di-, and triacylglycerol lipases, a phospholipase A, a lysophospholipase A, and a phospholipase C are secreted via the type II system, which is itself dependent on the type IV prepilin peptidase. Mutations within the genes encoding the type II secretion system diminish infectivity for macrophages, protozoa, and animals. The *Legionella* type IV secretion system, known as Dot/Icm, is absolutely critical for the ability of *L. pneumophila* to inhibit phagosome-endo/lysosomal fusions and to establish its replicative niche. A pore-forming activity is associated with Dot/Icm; however, it does not alone account for the aberrant phagosome trafficking seen in infected cells. Thus, effector proteins secreted through the Dot/Icm apparatus/pore are probably responsible for altering host cell function. Several infectivity factors have been localized to the *L. pneumophila* periplasm or cytoplasm. A Cu-Zn superoxide dismutase resides in the periplasm, affording resistance to toxic superoxide anions, and the KatB catalase-peroxidase is needed for optimal intracellular infection.

The growth phase is important to *Legionella* pathogenesis. Overall, stationary-phase bacteria are more infective for macrophages than are exponential-phase organisms. This phenotype is initiated by amino acid depletion and involves the bacterial stringent response. It is not known whether this response results in increased expression of one or more of the known virulence factors or whether it involves additional, uncharacterized factors.

IMPORTANCE OF IRON FOR *L. PNEUMOPHILA*

Iron is a key requirement for *L. pneumophila* replication. Indeed, legionellae are most readily isolated from water systems that contain elevated iron levels. As determined in experiments with bacteria on bacteriologic media, the iron requirement of *L. pneumophila* was originally estimated to be 3 to 13 μM for minimal growth and >20 μM for optimal growth. However, recent studies using chemically defined media indicate that the *Legionella* iron requirement is <1 μM. In standard BCYE agar, iron is added in the form of ferric pyrophosphate, although ferric chloride, ferric nitrate, and ferrous sulfate have also been used successfully in artificial media. On incubation with ^{55}FeCl$_3$, virulent *L. pneumophila* strains take up significant amounts of radiolabeled iron in an energy-dependent process that is resistant to trypsin and pronase. We have demonstrated that *L. pneumophila* can also bind and utilize hemin as a source of iron. As in other bacteria, iron is required by *L. pneumophila* for its role as a cofactor in enzymes such as superoxide dismutase and aconitase. Iron also appears to catalyze the formation of homogentisic acid melanin, a brownish pigment evident in *L. pneumophila* stationary-phase supernatants.

Iron acquisition is particularly relevant for *Legionella* pathogenesis. The administration of supplemental iron to experimental animals increases their susceptibility to *Legionella* infection, and *L. pneumophila* grown under iron-depleted conditions exhibits a reduced ability to cause disease. A strain derived by plate passage of a wild-type strain that has an elevated iron requirement is avirulent, although the parent and mutant strains accumulate equal amounts of iron. Four lines of evidence indicate that the ability of *L. pneumophila* to replicate within mammalian cells is dependent on iron. First, human monocytes and macrophages treated with iron chelators such as desferrioxamine, apotransferrin, and apolactoferrin do not support *Legionella* replication. Second, the cytokine gamma interferon inhibits bacterial growth by reducing the amount of iron in the host cell. Third, macrophages from A/J mice become permissive for intracellular infection following the addition of iron to the macrophages. Finally, decreased levels of the receptor for transferrin, the primary iron binding and transport protein of the host cell, are correlated with a reduced ability of legionellae to establish successful intracellular infection. Although the relationship between the *L. pneumophila* phagosome and the host cell endosomes and lysosomes is well understood, the precise source of intracellular iron is still unclear but presumably involves active access to the labile-iron pool.

The first genetic evidence for the importance of iron in *Legionella* physiology was the identification of *L. pneumophila fur*. In 1994, we cloned *Legionella fur* through complementation analysis of an *Escherichia coli fur* mutant. The *L. pneumophila fur* gene encodes a 15.0-kDa protein that cross-reacts with anti-*E. coli* Fur antibodies and whose repressive activity is, as expected, highest in legionellae grown in iron-rich media. Sequence analysis indicated that *L. pneumophila* Fur has an amino acid identity of over 54% and a similarity of over 72% to Fur from *E. coli*, *Pseudomonas aeruginosa*, and others. The promoter region of *L. pneumophila fur* contains sequences homologous to the Fur binding site, suggesting that *fur* is autoregulated in *Legionella*. Southern analysis indicates that *fur* is conserved among *Legionella* species. *L. pneumophila fur* cannot be insertionally inactivated, suggesting that Fur is essential for *Legionella* aerobic growth. However, mutations in *fur* could be obtained by selecting for resistance to manganese. The importance of *fur* is substantiated by the identification of multiple iron- and Fur-regulated genes in *L. pneumophila*.

Although the importance of iron for *Legionella* has always been clear, the genes and mechanisms used by the bacterium to acquire and transport iron have been rather elusive. However, recent investigations, which are described in detail in the following section, have begun to yield significant insight. For example, continued examination of culture supernatants has finally yielded an *L. pneumophila* siderophore activity. In addition, potential components of *Legionella* iron acquisition systems have been uncovered through the identification of iron-regulated genes, the characterization of mutants defective for growth under low-iron conditions, the screening of recombinant *E. coli* strains for iron-related phenotypes, and the examination of the unfinished *L. pneumophila* genome database for potential iron acquisition genes. Although more study is required for a precise understanding of how the recently identified factors contribute to iron acquisition, growth, and virulence, current data already indicate that the legionellae have evolved multiple iron assimilation pathways, once again highlighting the importance of iron to bacterial survival.

SIDEROPHORE-MEDIATED IRON TRANSPORT

Initial Negative Results

In 1983, it was reported that *L. pneumophila* does not make siderophores. This conclusion was based on negative results both in a bioassay and in the Arnow and Csáky assays, which detect catecholate and hydroxamate structures, respectively. In 1991, the issue of *Legionella* siderophores was revisited using the chrome azurol S (CAS) assay, which detects iron chelators independently of structure. That study identified CAS reactivity in supernatants taken from statically grown *L. pneumophila* cultures, suggesting the existence of a noncatecholate, nonhydroxamate siderophore. However, in 1996, we determined that such CAS reactivity is due to the high level of cysteine in the chemically defined medium used to grow *Legionella*. When we repeated the CAS assay using supernatants from cultures generated with cysteine-free media, we detected no siderophore activity. A 1997 study, which used chemostat cultures, also failed to identify any CAS-, Arnow-, or Csáky-reactive substance in iron-starved *L. pneumophila* cultures. Taken together, these data, along with the belief that the legionellae survive much of the time as intracellular parasites, had promoted the notion that *L. pneumophila* does not produce siderophores.

Discovery of Legiobactin

The story of *Legionella* siderophores changed significantly in 2000, when we demonstrated that *L. pneumophila* strains grown under specific conditions produced a high-affinity iron chelator. Indeed, when grown with shaking at 37°C in a low-iron chemically defined medium (CDM), *L. pneumophila* strains secrete a substance that is highly reactive in the CAS assay. Interestingly, the siderophore-like activity is observed only when the cultures are inoculated with bacteria that had been grown to log or early stationary phase. Inocula derived from late-stationary-phase cultures, despite ultimately growing in the CDM, fail to elicit CAS reactivity. The production of siderophore-like activity is not strictly dependent on the medium used to derive the inoculum; i.e., both CDM and buffered yeast extract (BYE) broth are satisfactory. However, it is influenced, to a relatively minor degree, by the amount of iron in that medium. The size of the inoculum also influences the ability of *L. pneumophila* to grow in the iron-deplete CDM and to ultimately produce CAS reactivity. The *L. pneumophila* CAS-reactive compound is made by virulent serogroup 1 strains 130b (ATCC BAA-74), Philadelphia-1 (ATCC 33217), and Oxford-4032E (ATCC 43110), as well as by clinical and environmental isolates representing all nine of the other *L. pneumophila* serogroups tested.

The CAS-reactive compound in *Legionella* cultures has the characteristics of a siderophore. The presence of CAS-reactive material correlates with enhanced aerobic growth in an iron-depleted defined medium, and peaks in reactivity occur during the late log to early stationary phases of growth. The chelating activity is subject to iron repression; i.e., adding iron to the CDM decreases CAS reactivity, whereas reducing the amount of iron in the CDM cultures increases CAS reactivity. The CAS-reactive compound is less than 1 kDa in size and is resistant to heat and proteases, and the CAS reactivity is not due to the pH or constituents (i.e., cysteine or phosphate) of the medium or the tyrosine-based pigment that is produced by growing legionellae. Further, the observed CAS reactions are rapid and intense; i.e., supernatants routinely display $\geq 1,000$ μM desferrioxamine equivalents within 2 min, suggesting significant amounts of a high-affinity iron chelator. Importantly, CAS-positive supernatants facilitate the growth of wild-type legionellae in BCYE agar containing otherwise inhibitory concentrations of the iron chelator 2,2′-dipyridyl (DIP), indicating biological activity. We have called the iron-chelating activity in *L. pneumophila* supernatants legiobactin, although it is formally possible that the CAS reactivity is due to the action of multiple siderophores.

CAS-reactive *L. pneumophila* supernatants remain negative in assays for catechols and hydroxamates. This is not due to *L. pneumophila* elaborating a substance that interferes with siderophore recognition, since mixtures of supernatants and known siderophores retain positivity in the assays for these chemical structures. Together, these data suggest that legiobactin is not a typical catecholate or hydroxamate. In support of this notion, ethyl acetate, dichloromethane, benzyl alcohol, butanol, and phenol chloroform do not extract the *Legionella* siderophore activity. Specific detection assays indicate that the CAS reactivity is also not salicylate or citrate. Preliminary studies using Sephadex A-50 (anionic), Sephadex C-25 (cationic), and XAD-16 (uncharged) resins suggest that legiobactin has both hydrophilic and hydrophobic characteristics. Legiobactin is stable on long-term storage at 4 to 37°C but is sensitive to acid treatment.

Representatives of most other *Legionella* species tested, including clinical and environmental isolates, also secrete a compound with CAS reactivity. Indeed, *L. birminghamensis, L. bozemanii, L. brunensis, L. erythra, L. feeleii, L. israelensis, L. jamestowniensis, L. londiniensis, L. moravica, L. parisiensis, L. rubrilucens,* and *L. wadsworthii* all yield significant desferrioxamine equivalents when grown in the iron-depleted

CDM. Although it cannot be concluded these species are producing legiobactin, their supernatants, like that of the *L. pneumophila* strains, are negative in the Arnow and Csáky assays. *L. cherrii*, *L. cincinnatiensis*, *L. gratiana*, *L. hackeliae*, *L. longbeachae*, *L. maceachernii*, and *L. oakridgensis* do not display significant CAS reactivity. However, these bacteria grow poorly in the CDM, because the medium is lacking an undefined growth factor other than iron. Thus, it is not clear whether these legionellae lack a siderophore, since the absence of CAS reactivity could have been an indirect effect of minimal growth. *L. micdadei* is a special case in that some of its strains do not exhibit CAS reactivity despite showing good growth in the CDM. The ability of certain *L. micdadei* strains, as well as *L. pneumophila* strains inoculated from late stationary phase, to achieve growth without exhibiting CAS reactivity suggests that the legionellae can employ alternative types of iron uptake in the absence of legiobactin.

Genetics and Regulation of Siderophore Production

Not surprisingly, our understanding of the genetic basis of *Legionella* siderophore expression is in its infancy. One *L. pneumophila* gene that is involved in legiobactin production has been identified. This gene, designated *lbtA* (for "legiobactin gene A"), encodes a predicted product that is homologous to multiple hydroxamate synthetases. The 66-kDa LbtA is 36% identical and 56% similar to another *Legionella* iron-regulated protein, FrgA (see below), and is 23 to 26% identical and 39 to 46% similar to known siderophore biosynthesis genes IucA and IucC, two of the enzymes involved in aerobactin biosynthesis by *E. coli* and *Shigella* species (see chapter 14), *Bordetella bronchiseptica* alcaligin AlcC, and *Sinorhizobium meliloti* rhizobactin RhbF. More importantly, several independently derived *lbtA* mutants have 40 to 60% less CAS-reactive material when they are grown in iron-depleted CDM, indicating that LbtA is required for optimal legiobactin production. Although this is not proven, we suspect that LbtA is involved in the biosynthesis of legiobactin rather than in siderophore export. Because several Fur boxes precede *lbtA*, we think that *Legionella* Fur mediates, at least in part, the iron regulation associated with legiobactin production. Southern blot hybridizations indicate that *lbtA* is widely distributed among *Legionella* strains, a result compatible with the fact that many species produce CAS reactivity.

Another gene potentially involved in siderophore production is *frgA* (for "ferric regulated gene A") the gene originally identified in our screen for iron-regulated loci in strain 130b. When bacteria are grown on media containing various amounts of free iron, *frgA* is repressed up to 60-fold by high-iron conditions and is Fur regulated, based on its derepression when introduced into manganese-resistant *fur* mutants of strain 130b. The presence of two Fur boxes in the promoter region of *frgA* supports the observed iron regulation. The predicted amino acid sequence of the 63-kDa FrgA has significant homology to the *lbtA* product and to several hydroxymate synthetases. This was somewhat surprising, given that *Legionella* does not appear to produce a hydroxamate siderophore. We observed an overall 22 to 23% identity and 36 to 40% similarity between FrgA and the aerobactin biosynthesis proteins IucA and IucC, and FrgA also has homology to proteins involved in the biosynthesis of *B. bronchiseptica* alcaligin (i.e., 22% identity and 40% similarity to AlcC) and *S. meliloti* rhizobactin (i.e., 20% identity and 40% similarity to RhbF) (see chapters 20 and 30). The relationship between FrgA and the known siderophore biosynthesis enzymes is significant for reasons in addition to the overall homologies. FrgA aligns with and is comparable in size to the hydroxymate synthetases; e.g., it is predicted to have the same number of amino acids as IucA. FrgA and the others have the greatest homology (i.e., ca. 50 to 75% identity) in regions thought to be enzyme active sites. Additionally, the only known proteins to which FrgA has significant homology are enzymes involved in siderophore production. With the subsequent discovery of legiobactin, we assessed whether *frgA* contributes to the siderophore biosynthesis seen in CDM broth cultures. However, the *frgA* mutant (i.e., strain NU229) proved not to be defective for CAS reactivity. Furthermore, the mutant is not impaired for growth in low-iron bacteriological media, including iron-depleted CDM and Chelex-treated BYE broth. Taken together, these data indicate that *frgA* is not required for optimal legiobactin production or for extracellular growth in low-iron conditions. However, since *frgA* is clearly expressed during growth in low-iron broth cultures, it remains possible that *frgA* is involved in siderophore biosynthesis but that another CAS-reactive iron chelator can compensate for the loss of FrgA and its resultant siderophore. Alternatively, *frgA* may be most critical for the production of a siderophore that is expressed under different growth conditions, such as during intracellular infection (see below).

Role of *lbtA* and *frgA* in Intracellular Infection

To begin to assess the role of siderophores in *Legionella* intracellular infection, we determined the relative ability of the *lbtA* mutant to infect human U937 cell macrophages and *Hartmannella vermiformis* amoebae. However, the use of standard infection protocols showed that all *lbtA* mutants were unimpaired in their abilities to infect host cells, including iron-depleted amoebae, suggesting that LbtA and legiobactin are not required for optimal intracellular infection. These data do not, however, demonstrate that legiobactin is not expressed or relevant intracellularly, since it is possible that the loss of *lbtA* is compensated for by another siderophore. Interestingly, *frgA* mutants are defective for intracellular growth in macrophages. For example, quantitative infection assays demonstrate that an *frgA* mutant is impaired approximately 80-fold for growth in U937 cells. Reconstruction of the mutant by allelic exchange proves that the defect is due to the inactivation of *frgA* and is not a spontaneous, second-site mutation. Finally, *trans*-complementation of the mutation demonstrates that the infectivity defect is directly due to the loss of FrgA. The *frgA* mutant grows normally in cocultures with the *Hartmannella* amoebae. Taken together, these data indicate that *frgA*, unlike *lbtA*, is required for at least some forms of intracellular infection. Furthermore, they raise the intriguing possibility that *L. pneumophila* encodes a siderophore that, unlike legiobactin, is necessary for optimal intracellular replication. Since Southern blots indicate that *frgA* is absent from non-*L. pneumophila* strains, it is tempting to speculate that such a siderophore would be specific to *L. pneumophila*. The presence and function of a second siderophore remain to be shown experimentally.

Identification of *pvc*-Like Genes in *L. pneumophila*

We have recently identified *L. pneumophila* genes that are analogous to the *pvc* genes of *P. aeruginosa*, which are involved in the biosynthesis of the siderophore pyoverdine (see chapter 19). More specifically, in the genome of strains 130b and Philadelphia-1, there is an open reading frame whose predicted product has large regions with 41% identity and 61% similarity to *Pseudomonas* PvcA. Immediately downstream of that gene, there is an open reading frame whose predicted product has segments that are 48% identical and 64% similar to *Pseudomonas* PvcB. Mutations in these *pvc*-like genes do not diminish the CAS reactivity of *L. pneumophila* CDM cultures, indicating that these genes are not required for legiobactin production. However, it is possible that *pvc*-like genes may be involved in the synthesis of another *L. pneumophila* siderophore.

Utilization of Ferric Citrate

In some bacteria, endogenously or exogenously produced ferric citrate serves as an iron transporter (see chapter 11). We do not detect citrate in CAS-reactive supernatants derived from low-iron CDM cultures of *L. pneumophila*, and the *L. pneumophila* genome database, which is about 90% complete, does not reveal a ferric citrate transport system that would be analogous to the Fec system of *E. coli*. Significantly, the addition of (iron-free) citrate to media inhibits the growth of legionellae that have not been induced to produce legiobactin. Thus, at present, we do not think that ferric citrate operates as a *Legionella* siderophore. However, the addition of ferric citrate to low-iron BCYE agar can promote the growth of *L. pneumophila*, suggesting that a factor (siderophore) produced by *Legionella* can extract the iron from the citrate chelator or that conditions that induce legiobactin production also induce ferric dicitrate utilization.

Summary and Current Models

With the discovery of legiobactin and *lbtA*, we now have biochemical and genetic evidence for the existence of a *Legionella* siderophore. CAS assays and Southern hybridizations suggest that this siderophore is produced by all *L. pneumophila* strains and perhaps by other *Legionella* species as well. It is likely that the influence of the bacterial inoculum on siderophore production is the main reason that we and others failed to detect legiobactin in the past. The discovery of legiobactin and its promotion of growth in media lacking iron also indicate that the *L. pneumophila* requirement for iron is not as great as had been intimated. In retrospect, it is not surprising that *L. pneumophila* produces a siderophore, since many other aquatic bacteria, when examined, are found to produce this type of iron scavenger. Based on the behavior of *lbtA* mutants, we suspect that legiobactin is more critical for extracellular than for intracellular growth. Thus, it is conceivable that the siderophore is vital for bacterial growth within biofilms in aquatic environments. Although *lbtA* has homology to enzymes involved in the biosynthesis of several well-studied hydroxamates, legiobactin may prove to be novel in structure, for several reasons. First, the observed CAS reactivity is Csáky (as well as Arnow and Rioux) negative and is not extractable with the typical solvents. Second, an examination of the *lbtA* locus as well as the unfinished *L. pneumophila* genome database has not revealed additional genes (other than *frgA*) that would be predicted to promote the biosynthesis of a known hydroxamate. Ongoing studies are aimed at clarifying the function, structure, complete genetic basis, and regulation of legiobactin.

Although we have not yet detected a biochemical activity associated with the *L. pneumophila frgA* and *pvcAB* genes, we hypothesize, based on sequence data, that one or both of them has the potential to be involved in siderophore production. At present, the argument for a role for *frgA* in iron acquisition is bolstered by the fact that the gene is iron regulated and is required for optimal intracellular infection of macrophages, an iron-stressed environment. Since both FrgA and LbtA have homology to IucA and IucC, it is formally possible that they, like the Iuc or Rhb proteins, catalyze separate but similar reactions in the biosynthesis of a single siderophore. However, the fact that *frgA* and *lbtA* mutants have different phenotypes under extracellular and intracellular growth conditions indicate that such a scenario is unlikely. In addition, the *frgA* and *lbtA* genes are rarely present together in the different species of *Legionella*: i.e., *frgA* appears to be limited largely to *L. pneumophila*. Thus, we strongly suspect that *L. pneumophila* produces multiple siderophores. Whereas legiobactin is clearly associated with extracellular replication, a siderophore associated with *frgA* would be predicted to be critical for intracellular infection. Ongoing studies need to address the roles of *frgA* and *pvcAB* in iron uptake and intracellular growth. The notion that *L. pneumophila* has evolved multiple siderophores to flourish in distinct intra- and extracellular niches is appealing.

NONSIDEROPHORE IRON TRANSPORT SYSTEMS

L. pneumophila FeoB and Ferrous Iron Transport

Several reports indicate that Fe^{2+} is present within a eukaryotic cytosolic pool, the presumed source of iron for intracellular legionellae. Although the *L. pneumophila* phagosome does not fuse with endosomes and lysosomes and has near neutral pH during early stages of the intracellular life cycle, it appears to fuse with low-pH cellular compartments during later stages of the infection. If there were a concomitant decrease in phagosomal pH, the amount of soluble ferrous iron available to the bacteria may increase. Previous reports suggest that L. pneumophila can use ferrous iron during replication in bacteriological media, since wild-type bacteria show reduced growth in the presence of the ferrous iron chelator dipyridyl. Therefore, we have begun to consider the means by which L. pneumophila acquires and/or utilizes ferrous iron.

To assess the role of ferrous iron transport in Legionella physiology and pathogenesis, we identified and mutated the *feoB* gene in virulent L. pneumophila strain 130b. As in E. coli (see chapter 12), the L. pneumophila *feoB* gene is contained within a *feoAB* operon. L. pneumophila FeoB has 44% identity and 61% similarity and is similar in size to E. coli FeoB. Like its E. coli counterpart, Legionella FeoB is predicted to have 10 transmembrane domains, consistent with an inner membrane location of this protein. L. pneumophila FeoA has 52% identity and 70% similarity and is identical in size to E. coli FeoA. Upstream of the *feoA* coding region is a putative Fur binding site, suggesting that L. pneumophila *feoAB*, like its E. coli counterpart, is subject to regulation by bacterial intracellular Fe^{2+} levels. L. pneumophila *feoB* insertion mutants exhibit decreased uptake of ferrous, but not ferric, radiolabeled iron. Whereas wild-type legionellae show five fold increases in $^{55}Fe^{2+}$ accumulation over a 20-min assay period, a *feoB* mutant does not exhibit significant iron uptake on incubation. The mutant's defect in iron assimilation is complemented by the addition of *feoB* in *trans*, confirming that FeoB is required for optimal ferrous iron transport by L. pneumophila.

Growth of L. pneumophila on standard BCYE agar or in BYE broth is unaffected by the loss of FeoB. However, the *feoB* mutant has a reduced ability to grow on low-iron BCYE agar and in low-iron BYE broth. The mutant takes 13 to 15 days to form colonies on BCYE agar that lacks the usual iron supplement, as opposed to 3 days for wild-type legionellae. Consistent with its ferrous iron uptake defect, the mutant shows accelerated colonial growth on low-iron agar when plated around wells containing ferric but not ferrous iron salts. Similarly, the defect in growth in low-iron BYE broth is exacerbated by the addition of dipyridyl, a result that can be reversed by additional supplementation with ferric chloride. Interestingly, the *feoB* mutant forms colonies on low-iron BCYE agar in only 4 to 5 days when it is plated next to confluent lawns of either itself or wild-type bacteria, suggesting that a secreted factor of L. pneumophila can reverse iron starvation in ferrous transport mutants. In support of this notion, the mutant shows accelerated colony growth when plated around wells containing CAS-positive supernatants that are derived from either itself or wild-type strains, indicating, once again, that Legionella CAS reactivity involves a biologically active material. Supernatants obtained from the *lbtA* siderophore mutant are unable to rescue the *feoB* mutant from low-iron starvation. The *feoB* mutant is also 2.5 log units more resistant to streptonigrin, an antibiotic that requires intracellular iron for its activity, than is the wild type, confirming its decreased ability to acquire iron during extracellular growth. The reduced ability of the mutant to grow in low-iron conditions is complemented by *feoB* in *trans*. Taken together, these data indicate that L. pneumophila FeoB and Fe^{2+} transport are required for extracellular replication under low-iron conditions. Although aerobic growth conditions typically result in a predominance of ferric iron, these data suggest that ferrous iron is present within Legionella cultures grown in the presence of oxygen. Using a ferrozine-based assay to measure ferrous iron, we have confirmed that L. pneumophila culture supernatants contain significant amounts of both ferrous and ferric iron. The FeoB inner membrane permease may also transport Fe^{2+} that has been generated in the periplasm, following reduction of acquired Fe^{3+} by a periplasmic ferric reductase (see below).

The role of FeoB in L. pneumophila intracellular infection has been assessed using U937 cells and *H. vermiformis* amoebae. In the human macrophage cell line, the *feoB* mutant exhibits a 10-fold reduced recovery compared to the wild type at 24 h, increasing to 15-fold at 48 and 72 h postinoculation. Further evidence of the role of *L. pneumophila* FeoB in infections of macrophages was obtained using a cytopathicity assay, which monitors the reduction in host cell viability that is associated with intracellular infection. Whereas >95% of the host cells are destroyed 72 h after inoculation with wild-type bacteria, the viability of macrophages infected with FeoB-deficient bacteria does not decrease significantly over the assay period. The reduced intracellular infectivity of the mutant can be complemented by the introduction of a plasmid containing *feoB*, confirming that *L. pneumophila* FeoB is required for optimal intracellular infection of macrophages. A correlation between iron levels in macrophages and the growth defect observed in the *feoB* mutant has been obtained from infection assays using U937 cells that were depleted of iron by treatment with dipyridyl. Indeed, the 15-fold growth defect seen with untreated U937 cells at 48 h postinoculation increases to 10^5-fold in the presence of dipyridyl. In monolayers treated with dipyridyl, fewer mutant bacteria are recovered than were originally inoculated, suggesting that some of the infecting bacteria had died. When amoebal cocultures were performed in the presence of dipyridyl, the *feoB* mutant exhibited a 150-fold reduction in growth relative to that of the wild type. Together, these results indicate that *feoB* and ferrous transport are important for the growth and survival of *L. pneumophila* in macrophages and amoebae. The role of *L. pneumophila feoB* in disease was examined using a competition assay and the A/J mouse model of pneumonia. At 3 days after intratracheal inoculation of the lungs of mice with equal numbers of wild-type and mutant bacteria, the wild type had outgrown the mutant in the lungs by three-fold, indicating that FeoB promotes in vivo growth.

In summary, *L. pneumophila* FeoB is a ferrous iron transporter that is important for extracellular and intracellular growth. The infectivity defect of the *feoB* mutant further suggests that ferrous iron is present and accessible in the *L. pneumophila* intracellular niche. The fact that the *feoB* mutant is not completely defective for intracellular growth in macrophages or amoebae and exhibits only a modest virulence defect in mice implies that other uptake systems and/or iron sources can overcome decreased ferrous iron uptake. These data represent the first evidence for the importance of ferrous iron transport to intracellular replication by a human pathogen.

L. pneumophila IraAB, a Promoter of Intracellular Infection and a Possible Transporter of Iron-Loaded Peptides

Following random mutagenesis with mini-Tn*10*, we isolated 17 *L. pneumophila* mutants that are hypersensitive to the iron chelator EDDA and/or resistant to streptonigrin. These strains are designated *ira* mutants (for "defective iron acquisition/assimilation"). None of the *ira* mutants are defective for legiobactin production. Six of the *ira* mutants are defective for infection of U937 cells; i.e., they display prolonged lag phases and in some cases replicate at a lower rate. Relative to wild-type strains, these *ira* mutants show reduced recoveries whose reductions range from 3- to 1,000-fold. Along with the *feoB* mutant, the *ira* mutants offer genetic evidence that iron acquisition is critical for intracellular infection by *L. pneumophila*.

The NU216 *ira* mutant is hypersensitive to EDDA and displays a severe intracellular lag phase and a low rate of replication, such that macrophage monolayers infected with the mutant yield 1,000-fold fewer bacteria relative to the wild type at 48 h postinoculation. It is also impaired, albeit modestly, for growth within *Hartmannella* amoebae. Importantly, NU216 and its allelic equivalent, mutant NU216R, are approximately 100-fold more sensitive than the wild type to treatment with the iron chelator desferrioxamine, confirming that they are defective for intracellular iron acquisition. The NU216R mutant has also been examined for its ability to cause disease in guinea pigs following intratracheal inoculation. Fewer bacteria were recovered from the lungs and spleens of animals infected with the mutant than were recovered from these organs of wild-type-infected animals. Moreover, animals infected with the mutant clear the bacteria, and the infection does not elicit the high fever or weight loss characteristic of wild-type infection. Thus, the *ira* locus defective in NU216R appears critical for both intracellular growth and virulence. However, it is not clear how IraA promotes infection or serves as a dispensable facilitator of iron acquisition.

Sequence analysis revealed that the mutation in NU216R lies in the first gene of a two-gene operon. This gene (*iraA*) encodes a 272-amino-acid protein that shows sequence similarity to methyltransferases. Indeed, IraA has the S-adenosylmethionine (SAM)- binding motif present in several prokaryotic and eukaryotic small-molecule methyltransferases, which use SAM as a donor to methylate a wide variety of acceptors. IraA has significant homology to both the *E. coli trans*-aconitate methyltransferase and the *Acetobacter aceti* phosphatidylethanolamine methyltransferase, an enzyme involved in the production of phosphatidylcholine, a molecule known to be present in unusually high levels in *L. pneumophila*. The second gene (*iraB*) encodes a 501-amino-acid protein that is highly similar to the PTR2 family of di/tripeptide transporters present in both prokaryotes and eukaryotes. The best-studied member of this class of proteins is the 463-amino-acid DtpT protein from *Lactococcus lactis*. IraB, like DtpT, is predicted to be an inner membrane protein with 12 membrane-spanning domains. A mutant (NU244) containing a disruption in *iraB* consistently show reduced growth in iron-depleted BYE broth. In contrast to the *iraA* mutant, the *iraB* mutant does not have a defect in macrophages, including iron-stressed macrophages, suggesting that *iraB* is more important for growth under low-iron extracellular than intracellular conditions. Based on sequence data, we hypothesize that IraB is involved in a novel pathway that imports iron-loaded peptides as an iron source. Although peptides can be components of siderophores, there have been no reports of free peptides involved in iron transport in bacteria. However, cysteine-containing N-terminal peptides enhance iron uptake in Caco-2 intestinal cells. *L. pneumophila* secretes multiple proteases, and it is possible that some of them generate peptides that bind iron. Alternatively, iron-loaded peptides might exist in the bacterium's environment. The *iraAB* genes are conserved among strains of *L. pneumophila* but are largely absent from other *Legionella* species.

Cytochrome *c* Maturation System and *L. pneumophila* Iron Acquisition

The EDDA-hypersensitive mutant NU208 has a reduced ability to grow on low-iron media; i.e., unlike the wild type, it produces 1,000-fold fewer colonies on BCYE agar lacking added iron than on fully iron-replete BCYE plates. It is also dramatically impaired for replication in U937 cell monolayers. The numbers of mutant bacteria do not increase for at least 24 h. Subsequent intracellular replication is at a lower rate than that of the wild type, such that mutant-infected monolayers yield at least 1,000-fold fewer bacteria at 48 h postinoculation. Correlating with the observed replication defect, the mutant fails to elicit a cytopathic effect on the U937 cells. The mutant's infectivity defect is exacerbated by treatment of the macrophages with desferrioxamine, indicating that the mutant is impaired for intracellular iron acquisition. Reconstruction of the NU208 mutation confirms that the iron acquisition and infectivity defects are due to the mini-Tn*10* insertion and not a spontaneous second-site mutation. The transposon disruption lies within a gene that is highly similar to the cytochrome *c* maturation gene, *ccmC*. CcmC is generally recognized for its role in the heme export step of cytochrome biogenesis. Indeed, NU208 lacks cytochrome *c*. Three additional *L. pneumophila* mutants that had been first identified for their diminished cytopathicity toward U937 cells contain mutations in *ccmC* as well as a second cytochrome *c* biogenesis gene, *ccmF*. All *ccm* mutants are highly defective for growth within *Hartmannella* and *Acanthamoeba* amoebae. Like the *ccmC* mutants, the *ccmF* mutant is impaired for growth in media lacking iron supplements but grows normally in iron-supplemented media. Thus, the *Legionella ccm* locus promotes both intracellular infection and iron acquisition. Complete sequence analysis of the *ccm* locus from strain 130b identified *ccmA* through *ccmH*, which are analogs of each of the *ccm* genes found in other bacteria. Our observations represent the fourth example of a linkage between *ccm* genes and bacterial iron acquisition, complementing the results of recent studies with *Paracoccus*, *Pseudomonas*, and *Rhizobium* species. The manner in which *ccm* promotes iron acquisition is unclear, although, in the other bacteria, *ccm* mutants have altered siderophore expression. Since the *L. pneumophila ccm* mutants are not defective for legiobactin production, it is possible that the *Legionella ccm* locus promotes the expression of a second siderophore or facilitates growth under low-iron conditions by a novel mechanism. The infectivity defects of our mutants indicate, for the first time, that *ccm* genes can be necessary for bacterial growth within an intracellular niche.

Hemin Binding

Hemin can replace ferric pyrophosphate as an iron source for *L. pneumophila* grown in CDM broth or on YP agar. In addition, *L pneumophila* binds hemin in a liquid hemin binding assay. Taken together, these data indicate the existence of a *Legionella* hemin utilization system and the presence of surface structures that promote hemin uptake. As one genetic approach toward addressing this possibility, we screened a genomic library of *L. pneumophila* strain 130b for a *Legionella* locus that could confer hemin binding on *E. coli*. A single gene, designated *hbp* (for "hemin binding promotion" was identified. The cloned *hbp* gene also confers on *E. coli* the ability to bind Congo red dye, a property that correlates with hemin binding. The monocistronic *hbp* gene is predicted to encode a secreted 15.5-kDa protein, which does not have homology to any known protein. There are two potential Fur binding sites upstream of the *hbp* open reading frame, suggesting that iron levels regulate the gene. An *hbp* mutant of strain 130b displays a 42% reduction in hemin binding, confirming that *hbp* potentiates the binding of hemin by *L. pneumophila*. The mutant, however, is unaltered in its ability to grow within U937 cells and *Hartmannella* amoebae, indicating that *hbp* is not required for intracellular infection. These data suggest that heme binding is not required during intracellular infection. They do not, however, prove that heme acquisition is inoperative during *L. pneumophila* infection of the mammalian host. The *hbp* gene is conserved among strains of *L. pneumophila* but is present within a minority of the other *Legionella* species.

Interactions with Transferrin and Lactoferrin

L. pneumophila does not bind or directly take up iron from transferrin. However, the addition of iron-loaded transferrin, but not apotransferrin, to *L. pneumophila* cultures can stimulate bacterial multiplication. The zinc metalloprotease secreted by *L. pneumophila* is able to degrade transferrin, and it has been proposed that such degradation can provide an iron source to growing legionellae. However, this form of iron acquisition is not likely to be critical for macrophage infection, since the *Legionella*-containing phagosome lacks transferrin and since *Legionella* protease mutants display wild-type levels of intracellular growth. Iron released from transferrin might enhance extracellular bacterial growth in the mammalian lungs. Apolactoferrin is inhibitory to *L. pneumophila* growth, and under certain culture conditions, apolactoferrin binds to *Legionella* bacteria, resulting in a bactericidal activity. Iron-loaded lactoferrin has no effect on *L. pneumophila* cultures and is not cleaved by the *Legionella* protease.

Ferric Reductases

L. pneumophila possesses a 25-kDa cytoplasmic and a 38-kDa periplasmic ferric reductase. The purified enzymes readily reduce the iron within ferric citrate, with the periplasmic enzyme having the greater specific activity. Glutathione is the favored reductant for the periplasmic enzyme, whereas NADPH serves best for the cytoplasmic reductase. Both enzyme activities are present within iron-limited and iron-replete cultures and within virulent and avirulent organisms. There are no reports of surface or secreted ferric reductases produced by *L. pneumophila*. Thus, *Legionella* ferric reductases promote iron assimilation rather than acquisition, processing Fe^{3+} that has been transported into the periplasm or cytoplasm. Although ferric citrate can serve as a substance for the purified reductases, it may not be physiologically relevant for iron transport across *Legionella* membranes, for reasons stated above. An iron delivery system that could be a substrate for the *Legionella* reductase(s) is a ferrisiderophore, perhaps legiobactin. The discovery of *L. pneumophila* FeoB provides a mechanism by which ferrous iron produced in the periplasm by the ferric reductase can be carried into the bacterial cytoplasm.

CONCLUDING REMARKS

In summary, *L. pneumophila* appears to possess multiple pathways for iron acquisition (Fig. 1). The organism may produce several siderophores; i.e., there is biochemical and genetic evidence for legiobactin, and there are DNA sequence data for the existence of enzymes that may be involved in synthesis of hydroxamate- and pyoverdine-like iron chelators. The membrane-associated Ccm complex is clearly required for growth under low-iron conditions; however, additional work is needed to determine how it facilitates iron acquisition or assimilation. Based on observations with other bacteria, it is possible that a *ccm* gene product(s) influences *Legionella* siderophore production. Once internalized, the ferrisiderophores may be acted on by the *Legionella* periplasmic ferric reductase (Pfr), yielding ferrous iron that would be transported across the inner membrane by FeoB. Alternatively, an ABC-type transporter might deliver the ferrisiderophore across the inner membrane, whereupon the cytoplasmic ferric reductase (Cfr) would generate ferrous iron. *L. pneumophila* extra- and intracellular environments contain significant amounts of ferrous iron, and so, a Fe^{2+} uptake pathway may also feed into FeoB. The *L. pneumophila iraAB* locus promotes growth under low-iron intra- and extracellular conditions and is predicted to encode an inner membrane peptide trans-

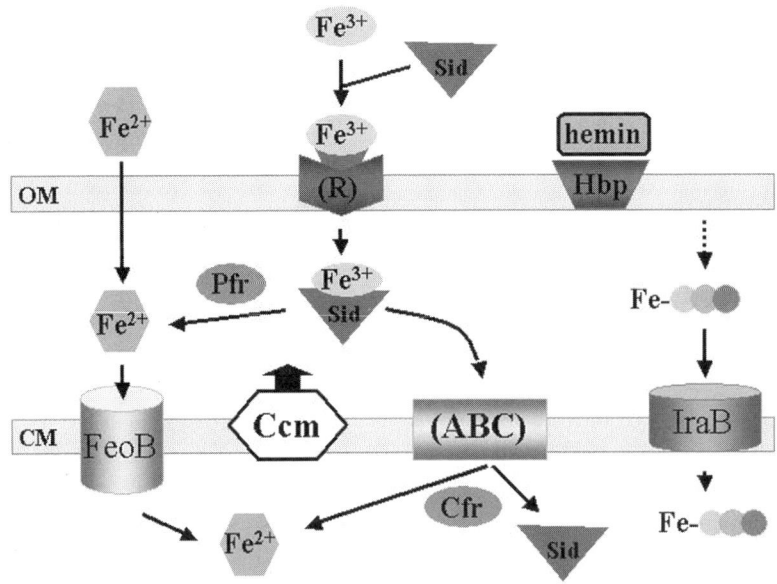

FIGURE 1 Model of *L. pneumophila* iron acquisition. See the text for details.

porter. We speculate that IraB imports iron-loaded peptides that serve as a source of iron, although it is possible that *iraAB* facilitates iron acquisition in a less direct manner such as by promoting the expression of a peptidic siderophore. In addition to utilizing ferric and ferrous iron, *L. pneumophila* can bind and use hemin, in part through the action of the Hbp protein. This suggests that the *Legionella* parasite, like a number of other pathogenic bacteria, may utilize heme-containing proteins as iron sources. The existence of multiple iron uptake systems in *L. pneumophila* is quite compatible with the fact that the bacterium resides within such a variety of environments. It is likely that several iron acquisition pathways will prove to be relevant for *Legionella* intracellular infection and pathogenesis. Clearly, *L. pneumophila* possesses some of the same iron uptake factors as do other gram-negative bacteria. However, studies of *Legionella* are still providing new insight into iron acquisition. For example, the legiobactin siderophore appears to be novel in terms of its structure and regulation of expression. Similarly, molecules such as IraA and IraB have never before been implicated in an iron-related activity. Furthermore, observations with *L. pneumophila frgA*, *ccmC*, and *feoB* indicate that siderophores, cytochrome *c* maturation systems, and ferrous iron (transport) need to be examined more extensively as components of intracellular parasitism. Finally, some *L. pneumophila* factors (e.g., *hbp*) as yet have no homologs in the database. Thus, there is a strong foundation for continued research on *L. pneumophila* iron acquisition.

ACKNOWLEDGMENTS

I thank past and present members of my laboratory for their dedication, creativity, and hard work. Our studies on *Legionella* iron acquisition are supported by NIH grant AI34937.

SUGGESTED READING

Byrd, T. F., and M. A. Horwitz. 2000. Aberrantly low transferrin receptor expression on human monocytes is associated with nonpermissiveness for *Legionella pneumophila* growth. *J. Infect. Dis.* **181:** 1394–1400.

Fields, B. S., R. F. Benson, and R. E. Besser. 2002. *Legionella* and Legionnaires disease: 25 years of investigation. *Clin. Microbiol. Rev.* **15:**506–526.

Gebran, S. J., C. Newton, Y. Yamamoto, R. Widen, T. W. Klein, and H. Friedman. 1994. Macrophage permissiveness for *Legionella pneumophila* growth modulated by iron. *Infect. Immun.* **62:** 564–568.

Goldoni, P., P. Visca, M. C. Pastoris, P. Valenti, and N. Orsi. 1991. Growth of *Legionella* spp. under conditions of iron restriction. *J. Med. Microbiol.* **34:** 113–118.

Hickey, E. K., and N. P. Cianciotto. 1994. Cloning and sequencing of the *Legionella pneumophila fur* gene. *Gene* **143:**117–121.

Hickey, E. K., and N. P. Cianciotto. 1997. An iron- and Fur-repressed *Legionella pneumophila* gene that promotes intracellular infection and encodes a protein with similarity to the *Escherichia coli* aerobactin synthetases. *Infect. Immun.* **65:**133–143.

James, B. W., W. S. Mauchline, P. J. Dennis, and C. W. Keevil. 1997. A study of iron acquisition mechanisms of *Legionella pneumophila* grown in chemostat culture. *Curr. Microbiol.* **34:**238–243.

James, B. W., W. S. Mauchline, R. B. Fitzgeorge, P. J. Dennis, and C. W. Keevil. 1995. Influence of iron-limited continuous culture on physiology and virulence of *Legionella pneumophila*. *Infect. Immun.* **63:**4224–4230.

Johnson, W., L. Varner, and M. Poch. 1991. Acquisition of iron by *Legionella pneumophila*: role of iron reductase. *Infect. Immun.* **59:**2376–2381.

Liles, M. R., T. Aber Scheel, and N. P. Cianciotto. 2000. Discovery of a nonclassical siderophore, legiobactin, produced by strains of *Legionella pneumophila*. *J. Bacteriol.* **182:** 749–757.

Liles, M. R., and N. P. Cianciotto. 1996. Absence of siderophore-like activity in *Legionella pneumophila* supernatants. *Infect. Immun.* **64:**1873–1875.

Mengaud, J. M., and M. A. Horwitz. 1993. The major iron-containing protein of *Legionella pneumophila* is an aconitase homologous with the human iron-responsive element-binding protein. *J. Bacteriol.* **175:**5666–5676.

O'Connell, W. A., E. K. Hickey, and N. P. Cianciotto. 1996. A *Legionella pneumophila* gene that promotes hemin binding. *Infect. Immun.* **64:** 842–848.

Poch, M. T., and W. Johnson. 1993. Ferric reductases of *Legionella pneumophila*. *Biometals* **6:** 107–114.

Pope, C. D., W. O'Connell, and N. P. Cianciotto. 1996. *Legionella pneumophila* mutants that are defective for iron acquisition and assimilation and intracellular infection. *Infect. Immun.* **64:**629–636.

Reeves, M. W., L. Pine, S. H. Hutner, J. R. George, and W. K. Harrell. 1981. Metal requirements of *Legionella pneumophila*. *J. Clin. Microbiol.* **13:**688–695.

Reeves, M. W., L. Pine, J. B. Neilands, and A. Balows. 1983. Absence of siderophore activity in *Legionella* species grown in iron-deficient media. *J. Bacteriol.* **154:**324–329.

Robey, M., and N. Cianciotto. 2002. *Legionella pneumophila feoAB* promotes ferrous iron uptake and intracellular infection. *Infect. Immun.* **70:**5659–5669.

Viswanathan, V. K., P. H. Edelstein, C. D. Pope, and N. P. Cianciotto. 2000. The *Legionella pneumophila iraAB* locus is required for iron assimilation, intracellular infection, and virulence. *Infect. Immun.* **68:**1069–1079.

Viswanathan, V. K., S. Kurtz, L. L. Pedersen, Y. Abu-Kwaik, K. Krcmarik, S. Mody, and N. P. Cianciotto. 2002. The cytochrome c maturation locus of *Legionella pneumophila* promotes iron assimilation and intracellular infection and contains a strain-specific insertion sequence element. *Infect. Immun.* **70:**1842–1852.

STAPHYLOCOCCUS, STREPTOCOCCUS, AND *BACILLUS*

David E. Heinrichs, Andrea Rahn, Suzanne E. Dale, and Michael Tom Sebulsky

25

In contrast to the information available for iron uptake systems in gram-negative bacteria, there is a relative paucity of information in the literature describing iron transport in gram-positive bacteria. However, over the last several years, our knowledge about iron uptake systems in gram-positive bacteria has increased steadily, and with the availability of genome sequence information, there should be a substantial rise in the number of publications over the next few years. This chapter focuses on our existing knowledge of the molecular basis for iron transport in the genera *Staphylococcus*, *Streptococcus*, and *Bacillus*.

STAPHYLOCOCCUS

The staphylococci are nonmotile, facultatively aerobic, gram-positive cocci that are distinguished by their growth in irregularly shaped clusters and by the presence of pentaglycine cross-bridges within their peptidoglycan. Most species are natural inhabitants of mucous membranes and the skin of mammals. The staphylococci, however, are probably most notable for their ability to cause disease in humans and other mammals. Although infections are especially frequent in the hospital setting, staphylococci are also among the most common causes of bacterial infections in the community. Infections range from rather mild skin infections (e.g., boils, furuncles, and minor wound infections) to those of a more serious nature (e.g., endocarditis, osteomyelitis, and septic shock), illustrating the pathogenic diversity of the staphylococci. Staphylococci can be divided into two major groups: coagulase positive and coagulase negative. While coagulase-negative staphylococci (CoNS), notably *Staphylococcus epidermidis*, have come into prominence due to their opportunistic ability to colonize foreign medical devices, by far the most extensively studied staphylococcal species is coagulase-positive *Staphylococcus aureus*, owing to it being both a frequent and a highly versatile pathogen.

Staphylococcal Siderophores

Siderophore production has been detected in iron-starved culture supernatants of the majority of staphylococcal species. The relative amounts of siderophore produced among various species and strains of *Staphylococcus* seem to vary remarkably and depends largely on culture conditions. Two staphylococcal siderophores, staphyloferrin A and staphyloferrin B, have

---

David E. Heinrichs, Andrea Rahn, Suzanne E. Dale, and Michael Tom Sebulsky, Department of Microbiology and Immunology, University of Western Ontario, London, Ontario, Canada N6A 5C1.

Iron Transport in Bacteria, Edited by Jorge H. Crosa, Alexandra R. Mey, and Shelley M. Payne
© 2004 ASM Press, Washington, D.C.

been chemically characterized. A third siderophore, termed aureochelin, was identified by its positive test with Swain reagent, suggesting that it may contain one or more catechol groups. The structure of aureochelin remains uncharacterized. Finally, a fourth siderophore, termed staphylobactin, was recently identified, and some genetic information is now available concerning its biosynthesis (see below).

Staphyloferrin A (Fig. 1) was first identified in iron-starved culture supernatants of a veterinary pathogen, *S. hyicus* DSM 20459. This highly hydrophilic siderophore, with an iron-free molecular mass of 480 Da, consists of two molecules of citric acid, each amide-linked to D-ornithine. Staphyloferrin A, which is produced by the majority of CoNS and by several strains of *S. aureus*, can be produced in greater quantities if the culture medium is supplemented with D-ornithine.

The chemical structure of staphyloferrin B (Fig. 1) has also been determined. Staphyloferrin B is composed of L-2,3-diaminopropionic acid, succinic semialdehyde, ethylenediamine, and citrate. The siderophore has a molecular mass of 448 Da and lacks the symmetry of staphyloferrin A. Structurally, staphyloferrin B exhibits a high degree of similarity to vibrioferrin, a siderophore produced by *Vibrio parahaemolyticus*. The genetics underlying staphyloferrin A and staphyloferrin B production remain uncharacterized, and no staphyloferrin A- or B-deficient knockout strains have yet been generated. Knockout strains are needed to elucidate the importance of staphyloferrin A and staphyloferrin B production to staphylococcal pathogenicity and to investigate the importance of these siderophores to the iron-restricted growth of staphylococci either in vitro or in vivo. These studies would be especially important for staphylococci that produce both staphyloferrin A and staphyloferrin B or alternative siderophores and may shed light onto which, if either, is the preferred siderophore under different iron-restricted growth conditions.

The first molecular-genetic characterization of siderophore production in the staphylococci has recently been reported. An iron-regulated nine-gene operon (Fig. 2), termed *sbn* (for "siderophore biosynthesis"), was demonstrated to contain some genes involved in the production of a siderophore termed staphylobactin, a siderophore significantly larger than either of the staphyloferrins. Several predicted protein products of this operon are similar to siderophore biosynthetic enzymes and amino acid-modifying enzymes present in other bacteria, suggesting that they are involved in siderophore production in *S. aureus*. A Fur box (see chapter 13) is located immediately upstream of the first open reading frame, and *lacZ* fusions constructed throughout the operon show that its expression is iron regulated through the ninth coding region. Insertional inactivation of the fifth coding region, *sbnE*, resulted in the loss of detectable staphylobactin in culture supernatants. The *sbnE* mutant also demonstrated debilitated growth in iron-restricted laboratory culture and was attenuated in a murine kidney abscess model of staphylococcal infection. These data illustrate the importance of this siderophore to the iron-restricted growth of *S. aureus*. The *sbn* genes were not detected in

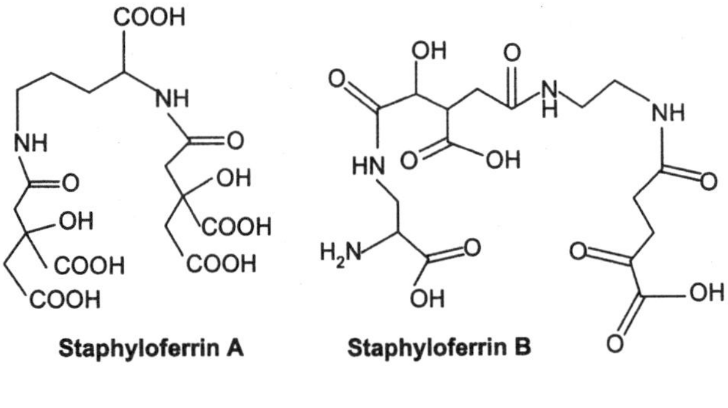

FIGURE 1 Chemical structures of the staphylococcal siderophores staphyloferrin A (left) and staphyloferrin B (right).

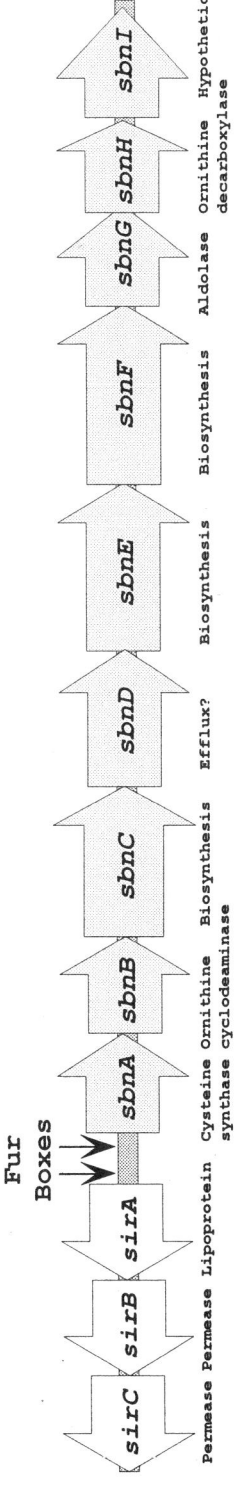

FIGURE 2 Physical map of the *sirABC* genes and the siderophore biosynthetic locus (*sbn*) present in the *S. aureus* genome. Expression of each of the operons is iron regulated and controlled by the activity of the Fur protein. Shown below the open reading frames are the predicted functions of the encoded proteins

several CoNS species, therefore, the presence of these genes may potentially contribute to differences in the pathogenicity observed between S. aureus and the CoNS.

Iron-Siderophore Transport in Staphylococci

Several different known and putative ferric siderophore transport systems have been found in S. aureus (Table 1), and genome sequences indicate the potential for additional transporters, although little is known about the substrates for the majority of these transporters or their importance to the iron-restricted growth of S. aureus.

Transport of ferric hydroxamates in S. aureus is accomplished through the activity of several Fhu proteins (see chapter 11 for a review of the E. coli Fhu transport system). In S. aureus RN6390, the Fhu system is composed of five polypeptides: FhuC, FhuB, FhuG, FhuD1, and FhuD2. A three-gene operon, *fhuCBG*, whose expression is controlled by exogenous iron levels via the Fur protein, encodes a classical ABC transporter that is required for uptake of all hydroxamate siderophores in S. aureus. The nomenclature of the genes is based on their homology to ferric hydroxamate transport proteins in Escherichia coli and Bacillus subtilis. S. aureus FhuC has high levels of similarity to

TABLE 1 Known and putative genes involved in iron uptake in Staphylococcus, Streptococcus, and Bacillus

Organism	Gene(s)	Description of the product(s)
S. aureus	fhuCBG	ABC transporter for Fe^{3+} hydroxamates
	fhuD1	Lipoprotein receptor for Fe^{3+} hydroxamates
	fhuD2	Lipoprotein receptor for Fe^{3+} hydroxamates
	sirABC	ABC transporter; similar to Fe^{3+} siderophore transporters
	sstABCD	ABC transporter; similar to Fe^{3+} siderophore transporters
	isdA or stbA or frpA	Transferrin and/or heme binding
	isdBCDEF	Heme binding and transport?
	N315: SA0112–SA0120 Mu50: SAV0116–SAV0214	Siderophore biosynthesis
	N315: SA1983–SA1980 Mu50: SAV2181–SAV2178	Siderophore biosynthesis and transport
	Unknown	Staphyloferrin A biosynthesis
	Unknown	Staphyloferrin B biosynthesis
	fur	Transcriptional repressor of iron uptake
S. epidermidis	sitABC	Iron regulated; probable Mn^{2+} transporter
	sirR	Iron-dependent repressor, probably of Mn^{2+} transporters
S. pneumoniae	piaABCD	Iron uptake ABC transporter; required for virulence; encoded on PAI and probably the dominant iron transporter
	piuBCDA	Iron uptake ABC transporter; required for virulence
	pitABCD	Iron uptake ABC transporter; required for full virulence
S. pyogenes	shp (formerly spy 1796)	Iron-regulated, cell surface heme-associated protein
	spy1795–1787	Present downstream of, and contranscribed with, shp
B. subtilis	fhuBGC, fhuD	ABC transporter for Fe^{3+} hydroxamates
	dhbACEBF	Bacillibactin biosynthesis
	ycdNOPQ	Similar to ferrichrome ABC transporter
	yfmCDEF	Similar to ferrichrome ABC transporter
	yfiY, yfiZ, yfhA, yusV	Similar to Fe^{3+} dicitrate transport permease
	yhfQ	Similar to Fe^{3+} dicitrate binding protein
	yvrABC	Similar to iron transport proteins
	feuABC	Iron uptake ABC transporter
	ytoA	Similar to ferripyochelin binding protein
	fur	Transcriptional repressor of iron uptake
	foxD	Receptor for ferrioxamine hydroxamates
B. cereus	Unknown	Hemoglobin and heme uptake

ATP binding proteins and possesses both Walker A ^{36}GPNGCGKS^{43} and Walker B ^{160}IIFLDEPTTYLD^{171} motifs. The FhuB and FhuG proteins display significant hydrophobic characteristics, and each is predicted to contain eight or nine transmembrane domains. These proteins are thought to constitute a membrane-embedded permease specific for ferric hydroxamate complexes. Mutations in either FhuB or FhuG eliminate ferric hydroxamate uptake in S. aureus.

The FhuD1 and FhuD2 proteins, each encoded by monocistronic mRNAs at different locations within the S. aureus genome, are ferric hydroxamate binding lipoproteins present at the external face of the cytoplasmic membrane. They are members of a large family of binding proteins known, or thought, to be involved in the binding and transport of ferric siderophores (see chapter 8). FhuD1 and FhuD2 both possess a distinguishable prolipoprotein signal peptide with a characteristic lipoprotein modification motif (FhuD1, ^{15}LTAC^{18}; FhuD2,^{15}LAAC^{18}), and both fractionate exclusively to the detergent phase of Triton X-114 cell extracts, indicating that they are amphipathic. While they have only 28.9% identity and 51.3% similarity to each other, they possess overlapping ligand specificities. However, significant differences are seen in the ligand affinities between the two proteins, with the FhuD2 protein showing significantly higher affinity than FhuD1 for several ferric hydroxamates (ferrichrome: FhuD1 K_d = 0.2 μM, FhuD2 K_d = 20nM; Desferal, FhuD1 K_d = 1.0 μM, FhuD2 K_d = 50 nM; aerobactin: FhuD1 K_d = 40 μM, FhuD2 K_d = 1.5 μM; rhodotorulic acid: FhuD1 K_d = 6 μM, FhuD2 K_d = 9 μM). Given the higher affinity of FhuD2 (versus FhuD1) for several ferric hydroxamates, the FhuD2 protein may be the preferred, or natural, binding protein for the S. aureus hydroxamate uptake system. In support of this, an examination of seven strains of S. aureus whose genomes have been sequenced reveals that *fhuD2* is present in all seven genomes while *fhuD1* is present in only five (absent in N315 and MRSA252).

Even with the limited sequence similarity between the S. aureus FhuD proteins and the E. coli homolog (approximately 10 to 15% identity and 22 to 28% total similarity), the occurrence of conserved residues at comparable locations provides enough similarity to thread the amino acid sequence of the S. aureus FhuD homologs to the high-resolution structure of the E. coli FhuD protein. Doing so indicates that the S. aureus FhuD structures probably adopt the same overall fold as that of their E. coli counterpart. This correlates with data from our group that identified mutations in highly conserved glutamic acid residues in FhuD2 (E97A and E231A [these residues are highly conserved within a large family of ferric siderophore binding proteins]) that abrogate the protein's ability to transport hydroxamate siderophores in S. aureus but do not alter its ligand binding properties. The mutations map to regions predicted to be exposed at the outer surface of each of the two lobes of the protein and outside of the predicted binding pocket in FhuD2. These residues may therefore play a central role in interactions with the permease components FhuB and FhuG.

The *sirABC* operon (for "staphylococcal iron regulated") was described by Heinrichs et al. in 1999. Predicted products of the operon have significant similarity to iron-siderophore uptake systems in several bacteria, the most similar being the Cbr proteins, which are involved in iron acquisition in the plant pathogen *Erwinia chrysanthemi* (see chapter 26). The first gene (*sirA*) encodes an iron-regulated 37-kDa lipoprotein, while the predicted products of *sirB* and *sirC* are highly hydrophobic, implying residence within the membrane, and probably act as components of a heterodimeric permease. Absent from this operon is a gene that would encode the ATPase component of an ABC transporter. Immunoblotting, using anti-SirA antisera, was used to show iron-regulated expression of SirA in S. aureus and to show that several strains of S. epidermidis expressed a SirA homolog. Expression of the SirA homolog in S. epidermidis, however, did not appear to be as dependent on growth under iron starvation conditions as it was in *S. aureus*. The exact nature of the substrate for the Sir transport system remains unresolved but is potentially staphylobactin, the siderophore produced by the activity of products of the *sbn* operon, which is transcribed in the opposite direction to the *sirABC* genes (Fig.2).

The *sstABCD* (for "staphylococcal siderophore transporter") operon encodes another putative ferric siderophore transporter in *S. aureus*. *sstA* and *sstB* are predicted to encode integral membrane proteins, while *sstC* encodes a predicted ATPase and *sstD* encodes a putative 38-kDa lipoprotein. The Sst proteins have significant similarities to siderophore transport proteins in other bacteria, including the Fat proteins in *Vibrio anguillarum* (see chapter 16), involved in the uptake of anguibactin. Expression of antisense RNA to *sstD* was used to decrease the expression of the Sst transport system. Although the antisense RNA decreased the observable expression of SstD in vitro, there was no growth reduction in vivo in a rat chamber implant model of *S. aureus* infection. Several possibilities could explain the failure to observe a growth defect, including residual SstD expression, the potential use of multiple different siderophores in vivo, and a redundancy in the lipoprotein receptor that may associate with the Sst transporter. Indeed, the last situation is observed for the ferric hydroxamate uptake system (described above) in that either one of two lipoproteins, FhuD1 or FhuD2, may be utilized for the uptake of ferric hydroxamates. At present, the substrate for the Sst transport system is unknown.

Nonsiderophore Iron Transport Systems in Staphylococci

The Sit (for "staphylococcal iron transport") system in *S. epidermidis* was investigated in studies that identified an iron-regulated 32-kDa lipoprotein (SitC). SitC is encoded from the third gene of an operon (*sitABC*) in *S. epidermidis*, and although expression of the operon is iron regulated, SitC has similarity to bacterial adhesins (FimA of *Streptococcus parasanguis* and ScaA of *Streptococcus gordonii*) and SitA and SitB have similarities to manganese transporters. Expression of the *sitABC* genes is controlled by the product of the *sirR* gene, divergently transcribed from the *sitABC* genes. SirR was shown to bind to a region of dyad symmetry, termed the Sir box, which overlaps the transcriptional start site of *sitABC*. Binding of SirR to the Sir box was dependent on the presence of Fe^{2+} or Mn^{2+}. The role of the Sit proteins in metal ion uptake requires further investigation to elucidate their exact role in the biology of the staphylococci.

Significant effort has been devoted to studying the role of transferrin in iron acquisition by *S. aureus* and *S. epidermidis*. Early studies indicated that staphylococcal growth can be enhanced by human transferrin under iron-restricted conditions and that *S. aureus*, *S. epidermidis*, *Staphylococcus capitis*, *Staphylococcus haemolyticus*, and *Staphylococcus hominis*, but not *Staphylococcus warneri* or *Staphylococcus saprophyticus*, are capable of binding transferrin to the cell surface. The association of human ^{125}I-transferrin with *S. aureus* and *S. epidermidis* was time and concentration dependent and could be inhibited by the presence of unlabeled ligand, either apotransferrin or iron-saturated transferrin. The interaction was effectively disrupted by the presence of human, rat, and rabbit transferrins but not by bovine and porcine transferrins, ovotransferrin, or lactoferrins. Receptor activity was partially iron regulated in *S. epidermidis* but appeared to be constitutive in *S. aureus*. In addition to binding transferrin, *S. aureus* and *S. epidermidis* are capable of converting diferric human transferrin into apotransferrin via an energy-requring process. It is not known how the released iron is transported into the cytoplasm.

The *S. aureus* receptor for transferrin has not been definitively identified. Initial reports suggested that transferrin binding was mediated by a 42-kDa protein located in the *S. aureus* cell wall, which was later identified through N-terminal sequencing as being a glyceraldehyde-3-phosphate dehydrogenase (GAPDH). Subsequent studies, however, showed that insertional inactivation of the *gap* gene (encoding

S. aureus GAPDH) did not abolish transferrin binding. Instead, a cell wall-anchored protein, StbA (for "Staphylococcal Transferrin Binding"), was shown to be responsible for the observed binding of *S. aureus* RN6390 to horseradish peroxidase (HRP)-conjugated transferrin. StbA possesses a C-terminal cell-wall sorting signal (LPKTG) that would allow the protein to be recognized by the *S. aureus* sortase enzyme for anchoring to the cell wall. StbA is encoded from within a locus of iron-regulated genes that were reported virtually simultaneously by three groups, leading to several designations (*stbA*, *sirD* to *sirH*; *isdA* to *isdF* and *srtB*; and *frpA* to *frpF*) (Fig. 3) for the genes in this locus. For the purpose of simplicity, we refer to genes within this locus by the *isd* designation from this point forward. Two of the predicted products from this locus, a putative lipoprotein (IsdE) and a putative membrane permease (IsdF), have similarity to binding proteins and membrane components of ABC transporters, respectively, involved in iron transport processes in other bacteria. Expression of these genes is tightly coordinated to iron levels via the Fur repressor, as is typical of iron uptake systems. Further studies showed that the IsdA protein (also designated FrpA and StbA) bound some peroxidase-conjugated proteins in a nonspecific manner, bringing into question the validity of the identification of the protein as a transferrin binding protein, since HRP-transferrin was used in those experiments. HRP is a glycoprotein with a hemin prosthetic group, and hemin alone is detectable by chemiluminescence detection methods. A complicating observation is that the IsdA protein seems not to bind unconjugated HRP. Subsequent studies by O. Schneewind's group reported that IsdA, in addition to several other Isd proteins (see below), binds heme. They hypothesize that the extracytoplasmic membrane Isd proteins may participate in the shuttling of heme-iron across the staphylococcal cell envelope as a means of acquiring host iron sources. In agreement with this, heme and hemoglobin can enhance the iron-restricted growth of *S. aureus* strains. Two lines of evidence have shown that the IsdA protein is expressed in *S. aureus* grown in vivo. Cell wall extracts of *S. aureus* grown in rat peritoneal implants expressed an iron-regulated protein with an equivalent molecular mass to that of IsdA, and convalescent-phase sera from patients with *S. aureus* septicemia reacted with recombinant IsdA expressed in *E. coli*. Efforts are under way to verify the true function of the IsdA protein and to clarify the nature of transferrin binding to the *S. aureus* cell surface.

The IsdB protein, more than twice the size of IsdA, also possesses a C-terminal LPXTG motif, indicating that it, too, is cell wall anchored in *S. aureus*. The expression of this protein, like that of IsdA, is iron regulated. Together, these two proteins represent the predominant iron-regulated cell wall-anchored

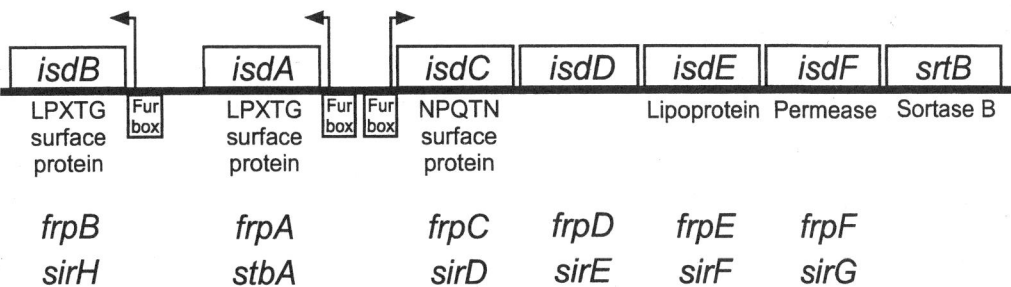

FIGURE 3 Physical map of the *isd* region in the *S. aureus* genome. Also shown are alternate designations given to the genes based on publications that appeared in the literature almost simultaneously. Expression of each of the three transcripts within the locus is iron regulated as a result of Fur binding sites present within the promoter regions.

proteins expressed at the surface of *S. aureus*. Sequence analysis of the two cell surface proteins indicates that the N-terminal region of IsdA has sequence similarity to the C-terminal region of IsdB, indicating that these proteins may interact with the same or similar substrates. Interestingly, the IsdB protein has no observable affinity for HRP-transferrin. The IsdA, IsdB, and IsdC proteins all share a conserved domain, referred to as the NEAT domain (for "near Fe^{3+} transporter"). This domain, approximately 125 amino acids in length, is found in proteins on the extracellular side of the cytoplasmic membrane in organisms including *S. aureus*, *Listeria innocua*, *Listeria monocytogenes*, *Streptococcus pyogenes*, and *Clostridium perfringens* and is probably playing an integral role in interactions with Fe ligands. Interestingly, two copies of this domain are found within the IsdB protein.

Work from O. Schneewind's laboratory identified the *isd* gene cluster through searches for a second sortase gene in *S. aureus*. The second sortase gene, termed *srtB*, was found to reside within the *isdC* operon (Fig. 3), and its expression was found to be iron regulated. Whereas SrtA recognizes and cleaves secreted proteins with the C-terminal LPXTG sorting signal motif (present in both IsdA and IsdB), IsdC was shown to be a substrate for SrtB activity due to the presence of the C-terminal sorting signal NPQTN. Further analyses, using a *srtB* mutant derivative of *S. aureus* strain Newman, showed that although SrtB activity was not required for the establishment of *S. aureus* infection in a murine kidney abscess model (there was no discernable defect in pathogenicity between the mutant and wild type at 5 days postinoculation), *srtB* mutants were significantly impaired in their ability to persist in kidneys beyond 5 days, in contrast to the wild type. These data, together with the observed ability of the IsdA protein to bind heme, suggest that gene products from this locus may be important for acquisition of host iron sources and may aid in the ability of the bacterium to survive in vivo for extended periods. Data suggesting that siderophores are also important for prolonged survival in the same mouse model of *S. aureus* infection (see above) indicate that a combination of the two iron acquisition systems is probably required for full fitness of the bacteria in vivo.

STREPTOCOCCUS

The streptococci are a diverse group of organisms, which includes commensal species as well as obligate human pathogens. They have been relatively well studied, since members of this group of bacteria are associated with infections in a variety of sites within humans. For the most part, investigations into the ability of streptococci to transport iron have been limited to *Streptococcus pneumoniae* and *Streptococcus pyogenes*.

S. pneumoniae is commonly found in the human respiratory tract, where it resides as a transient commensal. However, it is known to also cause serious infections, including bacterial pneumonia, otitis media, sepsis, and meningitis. Groups of individuals most susceptible to these infections are young children and the elderly. In fact, *S. pneumoniae* is the leading cause of invasive bacterial infections in these populations. *S. pneumoniae* infections are also becoming a serious concern for people who are immunocompromised due to an underlying disease.

S. pyogenes, commonly known as group A streptococcus, or GAS, is associated with nasopharyngeal infections and impetigo but is also capable of causing serious invasive infections such as bacteremia, necrotizing fasciitis, and streptococcal toxic shock syndrome. Untreated infection can lead to the development of rheumatic fever and rheumatic heart disease. Interestingly, humans appear to be the only reservoir of *S. pyogenes*.

Streptococcal Iron Acquisition

Analysis of the iron-limited growth of *S. pneumoniae*, using either a chemical assay (chrome azurol S) or a bioassay, has failed to detect the presence of siderophores in culture supernatants. In addition, a survey of the *S. pneumoniae* genome sequence does not identify any genes encoding putative siderophore biosynthetic enzymes. Similarly, there is no evidence to suggest that *S. pyogenes* has the capacity to produce siderophores. However, the expression of transport systems that facilitate iron uptake from heterologous siderophores (or xenosiderophores) in their environment is a common bacterial iron acquisition strategy, and the ability of streptococci to scavenge iron from siderophores produced by other bacteria remains to be examined. The genomes of both *S. pneumoniae* and *S. pyogenes* appear to encode ABC transporters with sequence similarity to known iron-siderophore uptake systems. However, the substrates for these putative transporters remain to be determined. Interestingly, it has been noted that *S. pneumoniae* causes virtually the same range of infections as *Neisseria meningitidis* and *Haemophilus influenzae*, neither of which produces siderophores.

Growth of *S. pneumoniae* in iron-limited media, although not enhanced by transferrin or lactoferrin, was enhanced in the presence of hemin and hemoglobin, and an undefined mutant defective in hemin utilization was less virulent in animal models than was its parental counterpart. Although early reports indicated that *S. pneumoniae* cells could bind lactoferrin, it would appear that this interaction does not result in iron uptake. A closer examination of the ability of *S. pneumoniae* to utilize hemin and hemoglobin as sources of iron showed that the cells bound hemin via both protein and nonprotein, possibly carbohydrate, components in a process that does not require metabolic activity. A 43-kDa hemin binding protein was identified that localized to the pneumococcal cell surface and was conserved among all pneumococcal strains tested. It is postulated that hemin is bound to the cell surface by both the protein and nonprotein components and that it acts as an iron reservoir there: The additional machinery required to transport heme-iron into the cytoplasm has not been identified.

More recent studies of the iron acquisition processes of *S. pneumoniae* have focused on three operons that code for putative protein products with significant similarity to ABC transporters involved in iron uptake. The operons have been termed *piuBCDA* (previously *pit1BCDA*), *piaABCD* (previously *pit2ABCD*), and *pitABCD*. Each operon encodes a putative ATPase, a putative lipoprotein, and two putative transmembrane permease components. The *pia* operon is contained within a 27-kb segment of the chromosome with divergent G + C content. This region has been termed pneumococcal pathogenicity island 1. Mutagenesis of the operons indicates that all three are involved in iron uptake but that the *piaABCD* operon appears to play a more prominent role than the others. This may be explained by the fact that the *pia* genes are transcribed at higher levels than the *piu* or *pit* genes. Mixed-infection experiments showed that the *pia* mutant strain was consistently attenuated in animal models, whereas the *pit* or *piu* mutant strains showed little or no attenuation, depending on the route of infection. It would appear that these operons encode redundant iron uptake mechanisms. Although the natural substrate for these operons has not been confirmed, growth defects in iron-restricted media as a result of mutation of either the *piu* or *pia* operon can be partially restored by the addition of hemoglobin but not by the addition of lactoferrin or transferrin. Growth of the *piu pia* double mutant could not be enhanced by hemoglobin, consistent with a role for these transporters in scavenging iron from this protein. Such a phenotype has not been confirmed for the *pit* mutant. Further research is required to establish the true role of these transporters in the biology of *S. pneumoniae*.

Like *S. pneumoniae*, the growth of iron-starved *S. pyogenes* can be promoted by the presence of hemin and heme-containing compounds such as hemoglobin, myoglobin, and catalase but not transferrin or lactoferrin. The ability of GAS to use heme-containing proteins as an iron source appears to be limited to proteins in which the heme component is associated through hydrogen bonding rather than a covalent attachment, since the growth of *S.*

pyogenes is not enhanced by cytochrome *c*. This limitation may be imposed by the as yet uncharacterized mechanism responsible for iron release and uptake.

Although the mechanism for heme utilization has not been fully elucidated, a nine-gene operon has recently been identified that may play an important role in this process. The first gene encodes a protein, termed Shp, which is located on the cell surface and can interact with heme. Convalescent-phase sera from humans and experimentally infected mice contain antibodies against Shp, indicating that Shp is expressed in vivo and elicits an immune response. A comparison of *shp* transcript levels from *S. pyogenes* grown in iron-replete and iron-limited media shows that *shp* is expressed at higher levels under iron-restricted than under iron-replete conditions. The other open reading frames within the operon encode proteins that include a putative lipoprotein, a putative membrane permease, two putative ATP binding proteins, and two putative transporter proteins. All of these components are consistent with a role for this operon in iron acquisition.

While the majority of studies of the iron acquisition processes in streptococci have been carried out with *S. pneumoniae* and *S. pyogenes*, the ability of *Streptococcus intermedius* to acquire iron has also been examined. *S. intermedius* is part of the normal flora of humans but has been associated with abscesses and infective endocarditis. Chemical tests failed to detect the presence of siderophores in culture supernatants, consistent with the findings for pneumococci. However, growth of the organism could be supported by the addition of holotransferrin to iron-restricted medium. This is the first report of transferrin utilization by a member of the streptococci. It is hypothesized that iron may be released from the transferrin molecule by significant localized decreases in pH that are mediated by growth of the organism.

BACILLUS

Bacillus subtilis is second only to *E. coli* as the most extensively studied prokaryote and serves as an important model gram-positive bacterium with which to study molecular aspects of iron uptake systems. *B. subtilis* is genetically very well characterized, and its genome sequence contains numerous predicted iron uptake systems. *B. subtilis* is not pathogenic, but the *Bacillus cereus* group of *Bacillus* species contains members (notably *Bacillus anthracis* and *B. cereus*) capable of causing disease in humans. *B. anthracis* is the causative agent of anthrax and has come into prominence recently due to its use as a biological weapon. *B. anthracis* can infect all mammals and can cause two forms of anthrax, cutaneous and systemic. The cutaneous form results from entry of cells or spores into the skin via a cut or abrasion and leads to the formation of a characteristic black eschar. Cutaneous infections generally remain localized, but a small percentage of untreated cases progress to fatal septicemia. Systemic anthrax generally occurs as a result of infection via the respiratory or gastrointestinal route. Gastrointestinal anthrax symptoms include nausea, vomiting, abdominal pain, and, in some cases, hemorrhagic diarrhea. Inhalation anthrax progresses within days from early flulike symptoms to sudden onset of respiratory distress and death results from respiratory failure, sepsis, and shock. *B. cereus* is capable of causing food poisoning as well as acute infections including meningitis, pneumonia, endophthalmitis, and septicemia. Moreover, *B. cereus* can become an opportunistic pathogen in immunocompromised patients such as those suffering from AIDS or undergoing cancer treatment.

Siderophore Production in *Bacillus*

One of the first siderophores to be characterized in bacteria was schizokinen (Fig. 4), a monohydroxamate mixed-ligand siderophore produced by *Bacillus megaterium* and some strains of *B. subtilis*, in addition to *Ralstonia solanacearum* and *Anabaena* spp. Schizokinen contains a citrate backbone that is substituted with 1-amino-3-(*N*-acetyl-*N*-hydroxylamino)propane. Seminal studies of *Bacillus* by Byers, Lankford, and coworkers identified a component present in culture supernatants that would

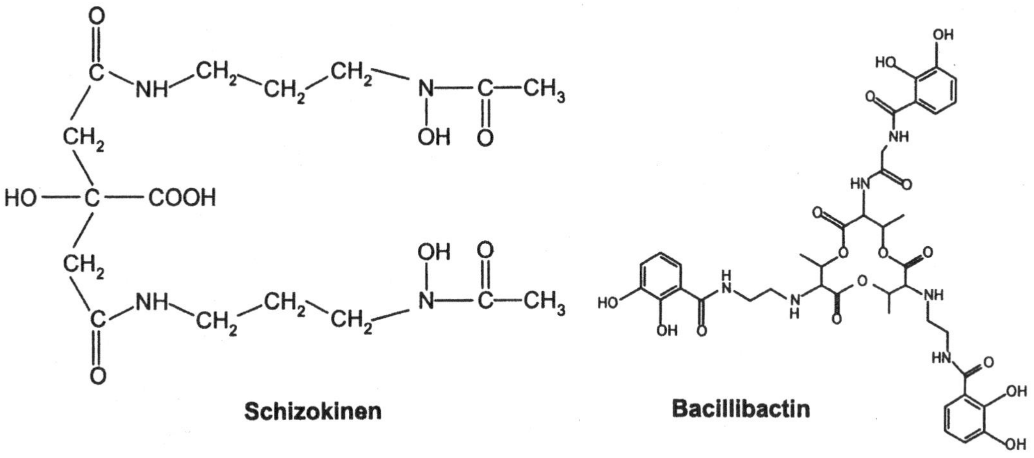

FIGURE 4 Chemical structures of the *Bacillus* siderophores schizokinen (left) and bacillibactin (right).

decrease the inoculum-dependent lag in growth in chemically defined media, and the components were termed "schizokinens," from the Greek for "growth stimulating." Strategies for generating and selecting siderophore-deficient mutations and complementing mutant growth deficiencies were also developed soon thereafter. Of course, this work stimulated many similar studies, leading to the identification of additional siderophores and the characterization of their transport in many different bacterial species.

When confronted with iron limitation, *B. subtilis* produces bacillibactin (dihydroxybenzoate-Gly-Thr)$_3$, an 882-Da trilactone siderophore (Fig. 4). The structure of bacillibactin is identical to that of corynebactin, a siderophore produced by *Corynebacterium glutamicum*. The 12-kb *dhbACEBF* operon required for bacillibactin biosynthesis, named on the basis of similarities to homologs in the *E. coli* enterobactin (*ent*) gene cluster (see chapter 9), is located at 291° on the *B. subtilis* chromosome. Unlike the enterobactin assembly and uptake locus, where *ent* biosynthetic genes are interspersed with *fep* genes and the *fes* gene (required for enterobactin uptake and processing, respectively), the genes required for bacillibactin biosynthesis are located within a single operon. Figure 5 illustrates the genetic organization of the *dhb* operon and the biosynthetic steps involved in the production of bacillibactin from chorismate.

Transcription of the *dhb* operon is iron regulated, and this regulation is controlled by the activity of the Fur protein. Transcription of the operon occurs during periods of iron starvation from a σ^A-dependent promoter and is inhibited by the binding of Fur protein to a Fur binding consensus sequence found in the operator region. Using a two-dimensional proteome analysis approach, Hoffman et al. identified a surprising result in that conditions of high salinity, even in the presence of moderate iron concentrations, induce an iron starvation response in *B. subtilis*, and they showed that expression of Dhb proteins was induced under high-salt conditions. The affect of salt stress on derepression of *dhb* genes is probably mediated through an effect on the Fur protein, since salt stress did not result in higher derepression of iron-regulated genes in a *fur* mutant background.

Synthesis of bacillibactin begins with chorismate and proceeds through the intermediate 2,3-dihydroxybenzoate (DHB) (Fig. 5). DhbC is an isochorismate synthase that converts chorismate to isochorismate. Interestingly, *B.*

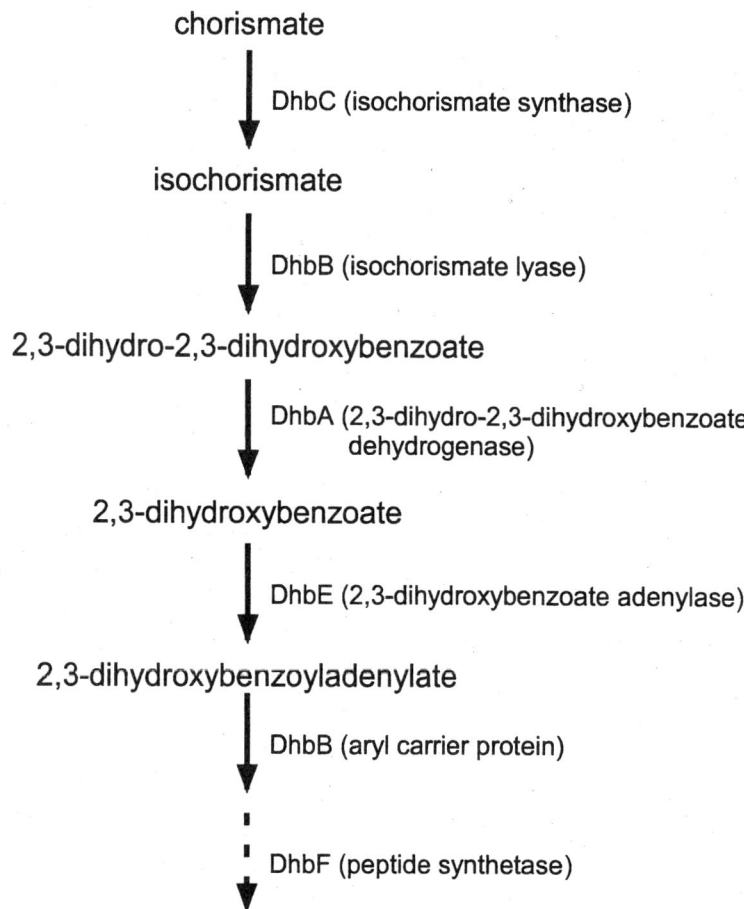

FIGURE 5 Physical map of the *dhbACEBF* operon in *B. subtilis*, and the biosynthetic pathway of bacillibactin.

subtilis possesses a gene, *menF*, encoding a second isochorismate synthase for the synthesis of the respiratory chain component, menaquinone. Unlike *dhbC*, iron concentrations do not control the transcription of *menF*. Moreover, although *dhbC* could compensate for the loss of *menF* and synthesize menaquinone in a *menF* mutant, *menF* could not substitute for the DHB biosynthetic function of *dhbC*.

Based on significant (80%) similarity to EntB, it is likely that DhbB, encoding a 2,3-dihydro-2,3-dihydroxybenzoate synthetase, performs the first dedicated step in the biosynthesis of DHB (see chapter 2 for a discussion of the mechanism of enterobactin synthesis). This enzyme performs a hydrolytic cleavage of the enolpyruvyl group on isochorismate to produce 2,3-dihydro-2,3-dihydroxybenzoate, the substrate for DhbA. Relatively little is known about the role of DhbA in bacillibactin biosynthesis aside from what may be inferred from the function of EntA. EntA and DhbA have 41% identity and 80% total similarity. Therefore, DhbA is likely to encode the 2,3-dihydro-2,3-dihydroxybenzoate dehydrogenase required for an NAD^+-dependent oxidation to produce DHB.

The cyclic amino core for bacillibactin assembly is formed by the activity of DhbB, DhbE, and DhbF, which, together, form a large multidomain enzyme complex termed a nonribosomal peptide synthetase. Together, DhbB, DhbE, and DhbF form the modular architecture required for precursor activation and modification to produce the completed siderophore and function in an "assembly-line" fashion. DhbE contains a single adenylation domain that recognizes DHB and activates it in an ATP-PP_i exchange reaction. DhbB, which, as mentioned above, possesses 2,3-dihydro-2,3-dihydroxybenzoate synthetase activity, also contains an aryl carrier protein domain. Thus, activated DHB can probably be transferred to DhbB from DhbE.

DhbF is a large (265-kDa) dimodular protein responsible for the addition of glycine and threonine into the DHB-Gly-Thr units. Each module within DhbF consists of a condensation domain, an adenylation domain, and a peptidyl carrier protein. The extreme C-terminal part of DhbF contains a thioesterase domain that is responsible for cyclization and release of the completed bacillibactin structure from the nonribosomal peptide synthetase complex.

Iron Transport in *Bacillus*

There is limited information in the literature describing the transport of iron or iron complexes in *Bacillus* species. The iron-restricted growth of *B. cereus*, a human pathogen, is inhibited by transferrin and lactoferrin but enhanced by hemoglobin, heme, and heme-albumin complexes. At this time, however, no molecular information is available regarding the nature of the protein(s) responsible for these observed activities.

Although *B. subtilis* produces bacillibactin as discussed above, the proteins required for its uptake are, at present, uncharacterized, but may be among the putative ferric siderophore uptake systems listed in Table 1. *B. subtilis* can also utilize ferric hydroxamates for growth under iron-restricted conditions. The uptake of ferric hydroxamates requires the products of the *fhuBGC* and *fhuD* genes, transcribed from two divergently arranged, iron-regulated promoters. While *fhuBGC* encode the two components of a heterodimeric permease and an ATPase, respectively, *fhuD* encodes the solute binding lipoprotein. *B. subtilis* FhuD is a 32-kDa lipoprotein whose processing to the mature lipidated form is blocked by exposure to globomycin. Similar to the scenario for *S. aureus*, FhuD does not appear to be the only binding protein for ferric hydroxamates in *B. subtilis*, since a strain bearing a *fhuD* mutation, although incapable of utilizing ferrichrome, ferrichrysin, and ferricrocin, was still capable of using ferrioxamines. The *foxD* gene product probably serves as the receptor for the ferrioxamine hydroxamates. Analysis of the completed *B. subtilis* genome reveals several other putative iron transporters (Table 1), suggesting that *B. subtilis* may be capable of utilizing a wide range of iron complexes. The Helmann laboratory has used microarray technology to show that many of these putative transporters exist within a *B. subtilis* global iron-regulatory network that is controlled by the *B. subtilis* Fur homolog (see below).

REGULATION

The *E. coli* Fur protein (see chapter 13) is the prototype for a large family of metallo regulatory proteins. In *B. subtilis*, three Fur homologs, Fur, PerR, and Zur, have been found. Zur mediates zinc homeostasis by regulating zinc uptake transporters, while PerR, the peroxide regulon repressor, is the central regulator of the inducible peroxide stress response. *B.*

subtilis Fur, approximately 33% similar to *E. coli* Fur, controls iron uptake. In an interesting contrast to *E. coli* Fur, *B. subtilis* Fur does not respond to Mn(II) in vivo. Fur regulation in *B. subtilis* has been demonstrated for a number of genes and operons, many of which are known or thought to encode iron uptake systems. The use of DNA microarrays has indicated that the *B. subtilis* Fur regulon consists of at least 20 operons and 39 total genes. DNase I footprinting showed that each of these 20 operons was a target for Fur regulation. Transcription from the *dhb* operon promoter, which possesses a perfect (19 of 19) Fur box, was shown to be extremely sensitive to iron concentrations. Indeed, transcription from this promoter was significantly repressed by iron concentrations as low as 100 nM and completely repressed by 1 μM iron. Studies of additional Fur-regulated genes, where less perfect matches to the Fur consensus were found, indicated that their repression levels were not as sensitive to iron concentrations but that all were repressed by iron concentrations as low as 5 μM. The *B. subtilis fur* gene is not iron regulated but is controlled by PerR.

Like *B. subtilis*, *S. aureus* also contains three Fur homologs, Fur, PerR, and Zur, and Fur controls iron transport in *S. aureus*. Expression of the *fhu*, *sir*, and *sst* transport systems, as well as the *isd* genes, is regulated by exogenous iron concentrations. The Fur protein was shown to interact with an *S. aureus* Fur box consensus sequence present in the *fhuCBG* promoter region, and expression of several iron-regulated genes in *S. aureus* is constitutive in a *fur*-null strain. In *S. aureus*, mutations in *fur* result in oxidative stress. Evidence has been obtained that loss of Fur activity results in constitutive siderophore production, expression of ferric siderophore transporters, and decreased *katA* (encoding catalase) expression. Not only did the *fur* mutant show a growth defect in rich and in metal ion-depleted media, but also it was attenuated in a murine skin abscess model of infection, demonstrating the importance of iron homeostasis to the fitness of *S. aureus*.

CONCLUDING REMARKS

It is clear that molecular studies of iron transport systems in gram-positive bacteria, especially members of the genera *Staphylococcus*, *Streptococcus*, and *Bacillus*, are in their infancy. However, the completion over the last several years of genome sequences from multiple species of these bacteria will no doubt correlate with a rise in the interest in studying iron uptake in these important groups of bacteria. It is expected that many initial discoveries of uptake systems will be based on homologies to well-characterized systems in gram-negative bacteria but that important novel and interesting advances will be made upon subsequent detailed analyses. A thorough understanding of the mechanistic details of iron transport in gram-positive bacteria, coupled with their relationship to pathogenesis, may help identify novel strategies and targets that will aid in our struggle to deal with drug resistance, which is so prominent in many gram-positive bacteria.

ACKNOWLEDGMENTS

Studies in our laboratory are funded by operating grants (to D.E.H.) kindly provided by the Canadian Institutes of Health Research (CIHR) and the Natural Sciences and Engineering Research Council (NSERC). D.E.H. holds a CIHR New Investigator award, while S.E.D. and M.T.S. are recipients of NSERC postgraduate and CIHR doctoral awards, respectively.

SUGGESTED READING

Andrade, M. A., F. D. Ciccarelli, C. Perez-Iratxeta, and P. Bork. 2002. NEAT: a domain duplicated in genes near the components of a putative Fe^{3+} siderophore transporter from Gram-positive pathogenic bacteria. *Genome Biol.* **3:**0047.1–0047.5.

Baichoo, N., T. Wang, R. Ye, and J. D. Helmann. 2002. Global analysis of the *Bacillus subtilis* Fur regulon and the iron starvation stimulon. *Mol. Microbiol.* **45:**1613–1629.

Brown, J. S., S. M. Gilliland, and D. W. Holden. 2001. A *Streptococcus pneumoniae* pathogenicity island encoding an ABC transporter involved in iron uptake and virulence. *Mol. Microbiol.* **40:**572–585.

Bsat, N., A. Herbig, L. Casillas-Martinez, P. Setlow, and J. D. Helmann. 1998. *Bacillus subtilis* contains multiple Fur homologues: identification of the iron uptake (Fur) and peroxide regulon (PerR) repressors. *Mol. Microbiol.* **29:**189–198.

Clarke, T. E., V. Braun, G. Winkelmann, L. W. Tari, and H. J. Vogel. 2002. X-ray crystallographic structures of the *Escherichia coli* periplasmic protein FhuD bound to hydroxamate-type siderophores and the antibiotic albomycin. *J. Biol. Chem.* **277:**13966–13972.

Cockayne, A., P. J. Hill, N. B. L. Powell, K. Bishop, C. Sims, and P. Williams. 1998. Molecular cloning of a 32-kilodalton lipoprotein component of a novel iron-regulated *Staphylococcus epidermidis* ABC transporter. *Infect. Immun.* **66:**3767–3774.

Dale, S. E., A. Doherty-Kirby, G. Lajoie, and D. E. Heinrichs. 2004. Role of siderophore biosynthesis in the virulence of *Staphylococcus aureus*: identification and characterization of genes involved in the production of a siderophore. *Infect. Immun.* **72:**29–37.

Haag, H., H. P. Fiedler, J. Meiwes, H. Drechsel, G. Jung, and H. Zähner. 1994. Isolation and biological characterization of staphyloferrin B, a compound with siderophore activity from staphylococci. *FEMS Microbiol. Lett.* **115:**125–130.

Heinrichs, J. H., L. E. Gatlin, C. Kunsch, G. H. Choi, and M. S. Hanson. 1999. Identification and characterization of SirA, an iron-regulated protein from *Staphylococcus aureus*. *J. Bacteriol.* **181:**1436–1443.

Horsburgh, M. J., E. Ingham, and S. J. Foster. 2001. In *Staphylococcus aureus*, Fur is an interactive regulator with PerR, contributes to virulence, and is necessary for oxidative stress resistance through positive regulation of catalase and iron homeostasis. *J. Bacteriol.* **183:**468–475.

May, J. J., T. M. Wendrich, and M. A. Marahiel. 2001. The *dhb* operon of *Bacillus subtilis* encodes the biosynthetic template for the catecholic siderophore 2,3-dihydroxybenzoate-glycine-threonine trimeric ester bacillibactin. *J. Biol. Chem.* **276:**7209–7217.

Mazmanian, S. K., H. Ton-That, K. Su, and O. Schneewind. 2002. An iron-regulated sortase anchors a class of surface protein during *Staphylococcus aureus* pathogenesis. *Proc. Natl. Acad. Sci. USA* **99:**2293–2298.

Mazmanian, S. K., E. P. Skaar, A. H. Gaspar, M. Humayun, P. Gornicki, J. Jelenska, A. Joachmiak, D. M. Missiakas, and O. Schneewind. 2003. Passage of heme-iron across the envelope of *Staphylococcus aureus*. *Science* **299:**906–909.

Meiwes, J., H.-P. Fiedler, H. Haag, H. Zähner, S. Konetschny-Rapp, and G. Jung. 1990. Isolation and characterization of staphyloferrin A, a compound with siderophore activity from *Staphylococcus hyicus* DSM 20459. *FEMS Microbiol. Lett.* **67:**201–206.

Modun, B., R. W. Evans, C. L. Joannou, and P. Williams. 1998. Receptor-mediated recognition and uptake of iron and human transferrin by *Staphylococcus aureus* and *Staphylococcus aureus* and *Staphylococcus epidermidis*. *Infect. Immun.* **66:**3591–3596.

Modun, B., and P. Williams. 1999. The staphylococcal transferrin-binding protein is a cell wall glyceraldehyde-3-phosphate dehydrogenase. *Infect. Immun.* **67:**1086–1092.

Morrissey, J. A., A. Cockayne, J. Hammacott, K. Bishop, A. Denman-Johnson, P. J. Hill, and P. Williams. 2002. Conservation, surface exposure, and in vivo expression of the Frp family of iron-regulated cell wall proteins in *Staphyloccus aureus*. *Infect. Immun.* **70:**2399–2407.

Morrissey, J. A., A. Cockayne, P. J. Hill, and P. Williams. 2000. Molecular cloning and analysis of a putative siderophore ABC transporter from *Staphylococcus aureus*. *Infect. Immun.* **68:**6281–6288.

Rowland, B. M., T. H. Grossman, M. S. Osburne, and H. W. Taber. 1996. Sequence and genetic organization of a *Bacillus subtilis* operon encoding 2,3-dihydroxybenzoate biosynthetic enzymes. *Gene* **178:**119–123.

Sato, N., S. Ikeda, T. Mikami, and T. Matsumoto. 1999. *Bacillus cereus* dissociates hemoglobin and uses released heme as an iron source. *Biol. Pharm. Bull.* **22:**1118–1121.

Sebulsky, M. T., and D. E. Heinrichs. 2001. Identification and characterization of *fhuD1* and *fhuD2*, two genes involved in iron-hydroxamate uptake in *Staphylococcus aureus*. *J. Bacteriol.* **183:**4994–5000.

Sebulsky, M. T., D. Hohnstein, M. D. Hunter, and D. E. Heinrichs. 2000. Identification and characterization of a membrane permease involved in iron-hydroxamate transport in *Staphylococcus aureus*. *J. Bacteriol.* **182:**4394–4400.

Sebulsky, M. T., B. H. Shilton, C. D. Speziali, and D. E. Heinrichs. 2003. The role of FhuD2 in iron(III)-hydroxamate transport in *Staphylococcus aureus*. Demonstration that FhuD2 binds iron(III)-hydroxamates but with minimal conformational change and implication of mutations on transport. *J. Biol. Chem.* **278:**49890–49900.

Tai, S. S., T. R. Wang, and C. J. Lee. 1997. Characterization of hemin binding activity of *Streptococcus pneumoniae*. *Infect. Immun.* **65:**1083–1087.

Taylor, J. M., and D. E. Heinrichs. 2002. Transferrin binding in *Staphylococcus aureus*: involvement of a cell wall anchored protein. *Mol. Microbiol.* **43:**1603–1614.

Xiong, A., V. K. Singh, G. Cabrera, and R. K. Jayaswal. 2000. Molecular characterization of the ferric-uptake regulator, Fur, from *Staphylococcus aureus* *Microbiology* **146:**659–668.

ERWINIA, A PLANT PATHOGEN

Dominique Expert, Lise Rauscher, and Thierry Franza

26

SOFT ROT DISEASE

The enterobacterial *Erwinia chrysanthemi* and *E. carotovora* species belong to the soft rot *Erwinia* group. Both species have recently been reclassified in the genus *Pectobacterium* as *P. chrysanthemi* and *P. carotovorum*, respectively, but this nomenclature, based on ribosomal DNA sequences analysis, has not yet been widely accepted. These bacteria are characterized by the production of large quantities of extracellular pectolytic enzymes as well as other plant cell wall-degrading enzymes such as cellulases, xylanases, and proteases. The ability to produce a wider range of enzymes or isoenzymes more rapidly and in larger quantities than that of pectolytic saprophytic microorganisms enables *Erwinia* species to invade living plants. Pectin is the major component of primary plant cell walls. Maceration of parenchymatous tissues, leading to pectin breakdown and cell wall disorganization, is the main symptom associated with soft rot diseases caused by these bacteria. Cell wall deconstruction can weaken plant cells, cause cell death, and thus allow bacteria free access to cellular nutrients. Bacterial attack affects a wide range of plants, including vegetables (potato, chicory, and maize) and ornamentals (African violet, carnation, and orchid). Soft rot can develop on growing plants or on the harvested crop, either in storage or in transit. The infection may be systemic or latent when environmental conditions for development of the disease are not met. Latent infection of potato tubers by the soft rot *Erwinia* species is widespread, and because disease tends to develop only when host resistance is impaired, the *Erwinia* species have often been described as opportunistic pathogens. Besides soft rot itself, symptoms such as wilting and dwarfing have been observed on growing plants. Thus, the economic importance of losses caused by these bacteria can be great, depending on the value of the crop and the severity of the attack. The extent of losses is influenced by the climate as well as the conditions of growth and storage.

Epidemiological studies of soft rot diseases have revealed that pectolytic *Erwinia* species are generally not endemic in soil. Temperature and humidity are important factors affecting their survival in soils. In common with many other plant-pathogenic bacteria, the soft rot *Erwinia* species can overwinter in contaminated plant residues remaining in the soil after harvest. The bacteria persist, albeit in decreased numbers, as long as the plant material is not completely

---

Dominique Expert, Lise Rauscher, and Thierry Franza, Laboratory of Plant Pathology, UMR 217 INRA/INA P-G/Université Paris 6, 16 rue Claude Bernard, 75005 Paris, France.

decomposed. Diseased plants can act as foci for the spread of the bacteria to uncontaminated plants. Soil water (rainfall and irrigation), airborne insects, and aerosols all play a role in dissemination. Recurrent distribution in infected planting material remains a major cause of infection. Soft rot *Erwinia* species are usually absent in seed. Seed collected from diseased plants can be contaminated externally, but the bacteria tend to die out rapidly as the seed dry. In contrast, soft rot bacteria introduced on the planting material can extensively infect vegetatively reproduced crops. Therefore, the development of molecular tools has been a major challenge for diagnosis and control of soft rot diseases.

E. chrysanthemi strain 3937, isolated from *Saintpaulia ionantha* H. Wendl. (African violet), has been recognized by plant pathologists and microbiologists as a valuable model for the analysis of phytopathogenicity determinants at the molecular level. Strain 3937 causes a systemic disease characterized by the progressive spread of bacterial cells within plant tissues of this host. *E. chrysanthemi* 3937 is the subject of an ongoing genome annotation project, and analysis of this phytopathogenic coliform has benefited from the development of a wide range of genetic tools that were successfully used for the identification of structures and functions involved in virulence of other coliforms. Virulence factors identified in *E. chrysanthemi* include a variety of secreted proteins. Among the various pectinases secreted by strain 3937, three pectate lyase isoenzymes with a broad pI range (PelA, PelD, and PelE) and one pectin methyl esterase (PemA) play a major role in bacterial progression from the leaf to the petiole. Many studies have shown that the bacterium has acquired the means of modulating its pectolytic machinery via complex regulatory pathways, which allow fine-tuning of gene expression. Besides cell wall-degrading enzymes, other factors proved to be important for disease development. Among them, a manganese superoxide dismutase, a methionine sulfoxide reductase and on [Fe-S] assembly machinery confer on the bacterium the ability to survive in planta and to cope with damage caused by oxidative stress during infection. Two high-affinity iron acquisition systems mediated by siderophores are required for full virulence. This chapter focuses on the one of the siderophores, chrysobactin, and the role of iron in this disease. Since chrysobactin is structurally related to enterobactin, a comparison of these two iron acquisition systems is presented when appropriate.

SIDEROPHORE-MEDIATED IRON TRANSPORT SYSTEMS

Siderophore Production

E. chrysanthemi 3937 synthesizes two structurally unrelated siderophores, achromobactin and chrysobactin, which are produced in a sequential manner in culture supernatants of bacterial cells grown under iron limitation (Fig. 1). In low-iron cultures, achromobactin can be detected during the mid-exponential phase of growth whereas chysobactin appears much later. Chrysobactin, the first siderophore of strain 3937 to be characterized, is a catechol, whereas achromobactin is a hydroxycarboxylate siderophore that does not contain any catecholate or hydroxamate groups. Achromobactin was identified in chrysobactin-deficient mutants that still had siderophore activity. Chrysobactin-deficient mutants fail to grow in the presence of the ferric iron chelator EDDHA [ethylenediamine-N,N'-bis(2-hydroxyphenylacetic acid)], but the production of achromobactin enables them to thrive on a medium containing the ferrous iron chelator 2,2'-dipyridyl. In this respect, achromobactin resembles a preventive system that helps bacterial cells to cope transiently with a loss of iron until chrysobactin is fully induced.

Structural Properties of Chrysobactin

Chrysobactin, identified as α-N-[N-(2,3-dihydroxybenzoyl)-D-lysyl]-L-serine, belongs to a class of siderophores that are basically dihydroxybenzoic acid (DHB)-modified amino acids or peptides. Unlike the tricatecholate siderophore enterobactin and other hexadentate ligands, which are strong iron binders,

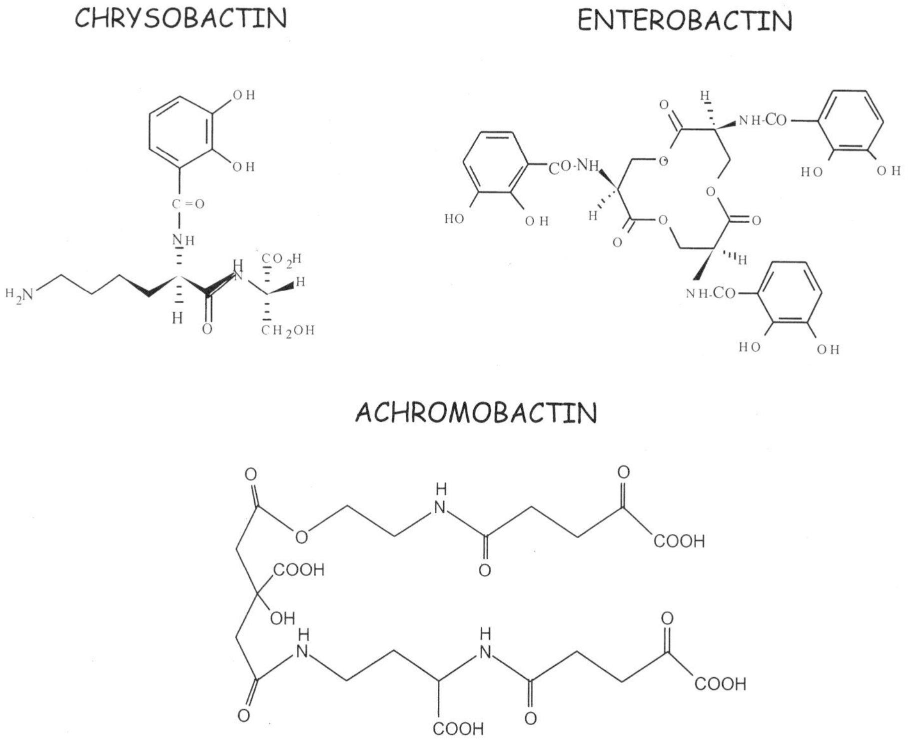

FIGURE 1 Structures of the siderophores chrysobactin and achromobactin produced by *E. chrysanthemi* 3937. Enterobactin produced by *E. coli* K-12 is used by *E. chrysanthemi* as a xenosiderophore.

chrysobactin possesses only three potential coordination sites for complexing ferric iron: two hydroxyl groups on the catechol moiety and the terminal carboxylate group of serine. Persmark and Neilands have shown that only catecholate hydroxyl groups are involved in the chelation, suggesting that chrysobactin is a bidentate and thus relatively weak ligand. However, depending on the pH and metal/ligand concentration ratio, chrysobactin was found to form ferric complexes of different stoichiometries, from 1:1 to 1:3 (Fe^{3+} to chrysobactin). When iron is in four fold or greater excess, there is a mixture of bis and tris complexes in solution at physiological pH. To further understand the biological function of chrysobactin, A.-M. Albrecht-Gary and coworkers have investigated the coordination properties of the different ferric complexes of this siderophore. They found that chrysobactin is a less effective ferric chelator than hexadentate siderophores such as enterobactin or ferrioxamine B. However, chrysobactin has a higher pFe value than citrate or malate (pFe of Cb and citrate are 17.1 and 14.8, respectively), which are known to be the major ferric carriers in plants, and it can effectively sequester the iron from their ferric complexes.

Transport of Ferric Chrysobactin

As in *E. coli* and other gram-negative species, the passage of ferric chrysobactin through the bacterial outer membrane involves a receptor, Fct (for "ferric chrysobactin transport"), energized by the TonB protein. The receptor was identified as a protein of ca. 81,000 Da by sodium dodecyl sulfate-polyacrylamide gel electrophoresis and sequence analysis of the *fct* gene. The kinetics of ferric chrysobactin uptake have apparent values for K_m and V_{max} of about 30 nM and 90 pmol/mg/min, respectively. Isomers of chrysobactin and analogues with progressively shorter side chains are taken up as efficiently as the native siderophore, indicating that this receptor primarily recognizes the catechol-iron center. The Fct receptor does not mediate the uptake of ferric enterobactin, which is used by *E. chrysanthemi* cells as a xenosiderophore. Analysis of the genome sequence of strain 3937 revealed the presence of an open reading frame encoding a protein 65% identical to the ferric enterobactin receptor precursor FepA from *Escherichia coli* that may be responsible for enterobactin uptake. The passage of ferric chrysobactin and ferric enterobactin through the inner membrane of *E. chrysanthemi* cells involves the same permease, CbuBCDG (for "chrysobactin uptake"), whose components form an ABC transporter with homology to the *E. coli* ferric enterobactin permease FepBCDG. Cross-feeding assays and transport experiments have shown that *E. chrysanthemi cbu* mutants altered in this permease do not obtain iron from either of these siderophores.

Biosynthesis and Degradation of Chrysobactin

Using a collection of insertion mutants unable to grow in the presence of EDDHA, it was possible to characterize the different stages required for biosynthesis of chrysobactin and the transport back to the cell of its ferric complex. A part of the corresponding genetic region was cloned by functional complementation of *E. coli* mutants defective in the formation of DHB, which is a common intermediate in enterobactin and chrysobactin biosynthetic pathways. The map of the genes and the list of chrysobactin biosynthesis and transport proteins are shown in Fig. 2 and Table 1, respectively. The first operon identified, *fct-cbsCEBA*, encodes the receptor protein Fct and functional homologues of the EntC, EntE, EntB, and EntA enzymes that lead to the formation of DHB-AMP, a common intermediate in enterobactin and chysobactin biosynthesis. Analysis of the *fct* gene sequence revealed a strong resemblance of its promoter region to the bidirectional promoter controlling the expression of the *fepA-entD* and *fes-entF* operon in *E. coli*. A second operon, *cbsHF*, transcribed divergently with respect to *fct-cbsCEBA*, was identified in the 5′-upstream region of the *fct* gene. The two genes of this operon, *cbsH* and *cbsF*, have identity to the *E. coli fes* and *entF* genes, respectively. The *fes* and *entF* genes encode ferric enterobactin esterase and enterobactin synthase, respectively. Polar mutations generated in the *cbsH* gene interrupt the chrysobactin biosynthetic pathway, confirming that the *cbsF* gene encodes the enzyme CbsF, required for chrysobactin synthesis. In both organisms, the genes involved in synthesis and transport of enterobactin and chrysobactin, respectively, are clustered into six transcriptional units that are controlled by three divergent promoters (Fig. 2). It is interesting that the *fct-chuB*

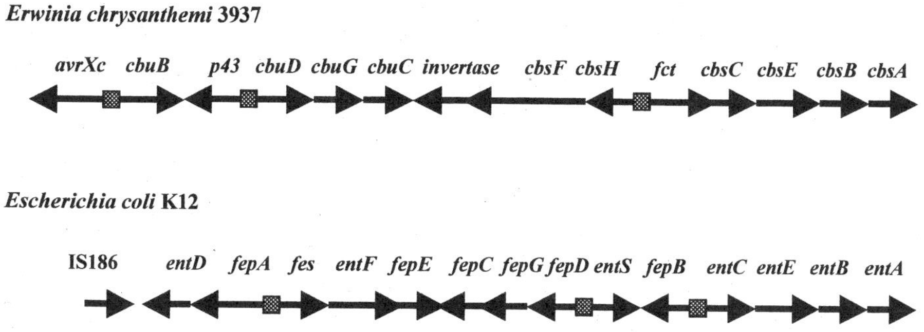

Erwinia chrysanthemi 3937

avrXc cbuB   p43 cbuD cbuG cbuC invertase   cbsF cbsH   fct   cbsC cbsE cbsB cbsA

Escherichia coli K12

IS186   entD fepA fes entF fepE fepC fepG fepD entS fepB   entC entE entB entA

FIGURE 2 Comparison of the genetic organization of the loci involved in enterobactin (*E. coli* K-12) and chrysobactin (*E. chrysanthemi* 3937) biosynthesis and transport. Squares and arrows represent bidirectional promoters and genes or operons, respectively. In *E. chrysanthemi* 3937, only the *fct-cbsH* bidirectional promoter has been determined experimentally.

TABLE 1 Proteins involved in the biosynthesis of chrysobactin and transport of its ferric complex

Name	No. of amino acids	Homology
Chrysobactin biosynthetic proteins		
CbsA	252	66% identical to the 2,3-dihydro-2,3-dihydroxybenzoate dehydrogenase EntA from *E. coli* K-12
CbsB	291	66% identical to the 2,3-dihydro-2,3-dihydroxybenzoate synthetase, isochorismatase EntB from *E. coli* K-12
CbsC	395	54% identical to the isochorismate hydroxymutase 2, synthase EntC from *E. coli* K-12
CbsD	273	46% identical to the yersiniabactin biosynthetic component YbtD (phosphopantetheinyl transferase) from *Yersinia pestis* KIM
CbsE	545	66% identical to the 2,3-dihydroxybenzoate–AMP ligase EntE from *E. coli* K-12
CbsF	2,864	44% identical to the enterobactin synthetase EntF from *E. coli* K-12
Chrysobactin transport proteins		
Fct	735	Ferrichrysobactin outer membrane receptor precursor; 37% identical to the ferrichrome iron receptor FhuA from *E. coli* K-12
CbuB	363	55% identical to the ferric enterobactin-binding periplasmic precursor FepB from *E. coli* K-12
CbuC	268	66% identical to the ATP-binding component of the ferric vibriobactin ABC transporter from *Vibrio cholerae*
CbuD	358	58% identical to the ferric enterobactin transport system permease protein FepD from *E. coli* K-12
CbuG	354	56% identical to the ferric enterobactin transport system permease protein FepG from *E. coli* K-12
Other functions		
CbsH	434	Chrysobactin peptidase; 46% identical to the ferric enterobactin esterase from *E. coli* K-12
P43	426	50% identical to the enterobactin secretion machinery protein EntS (P43) from *E. coli* K-12

segment is inverted relative to the corresponding *fepA-fepB* region of *E. coli*. No *fepE* gene homologue was found in *E. chrysanthemi*. In addition, the *cbsD* gene does not appear to map within the chrysobactin genetic region.

The *cbsF* gene, of 8,595 bp, can potentially encode a polypeptide of 2,864 amino acid residues with a molecular mass of 312 kDa and a pI of 5.71. Since chrysobactin contains one amide bond and one peptide bond, the CbsF protein is predicted to be a nonribosomal peptide synthetase with a multimodular structure allowing the assemblage of the three components, DHB, D-lysine, and L-serine. On the basis of the data described in the literature concerning nonribosomal peptide synthetase functional domains and enterobactin biosynthesis, we identified two functional modules of CbsF (using the nomenclature of Marahiel et al.) (Fig. 3). The first module (amino acids 1 to 1540), with its condensation, adenylation, thiolation, and epimerization domains, contains the information for epimerization of the L-lysine substrate and formation of the DHB-lysine bond. The second module (amino acids 1541 to 2864), with condensation, adenylation, thiolation, and thioesterase domains, is probably involved in the formation of the Lys-Ser peptide linkage and the cleavage of chysobactin from the protein template (see chapter 2). In this scheme, the CbsB product must be a functional homologue of EntB with both isochorismate lyase and aryl

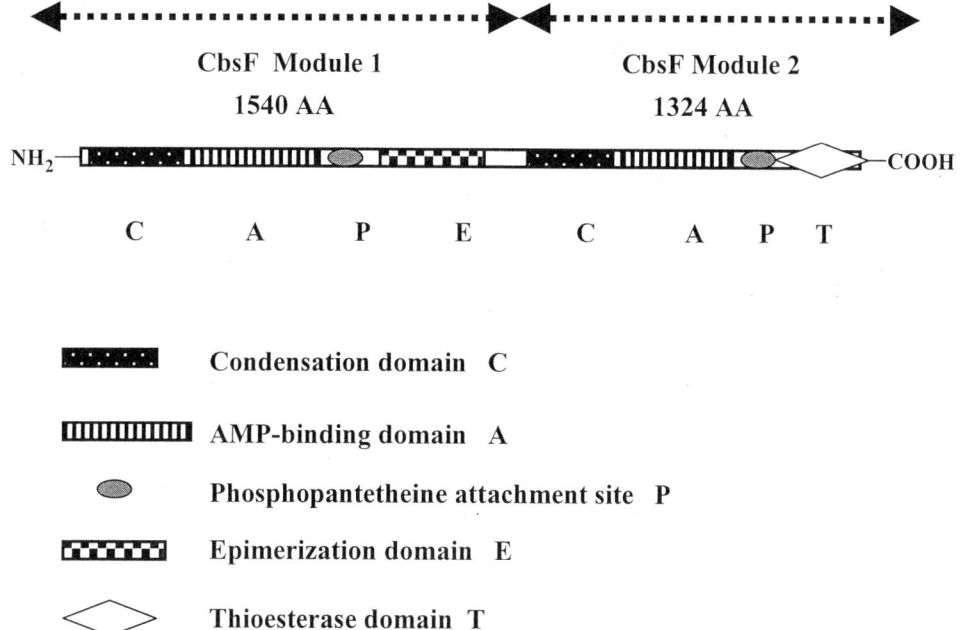

FIGURE 3 Modular structure of the chrysobactin synthetase CbsF protein. The different catalytic domains involved in the last steps of chrysobactin biosynthesis are shown.

carrier protein domains. An EntD homologue with a phosphopantetheinyl transferase activity must also occur in *E. chrysanthemi*. Analysis of the genome sequence revealed the existence of a gene (*cbsD*) encoding a protein (CbsD) with identity to the yersiniabactin biosynthetic component YbtD from *Yersinia pestis* KIM, predicted to be a phosphopantetheinyl transferase. The homologies between the CbsC, CbsE, CbsB, and CbsA gene products and their *E. coli* counterparts, EntC, EntE, EntB, and EntA, respectively, are presented in Table 1.

The sequence analysis of the *cbsH* gene product revealed characteristics of the S9 prolyl oligopeptidase family. In agreement with sequence predictions, it was demonstrated that the CbsH protein is an enzyme that degrades chrysobactin in the bacterial cytosol. Given the chrysobactin structure, it is expected that the CbsH protein catalyzes the cleavage of the lysyl bond, generating DHB-lysine and serine. The lack of chrysobactin hydrolysis in a nonpolar *cbsH*-negative mutant results in growth inhibition caused by intracellular ferric iron chelation. The CbsH protein has been proposed to be a peptidase that prevents the bacterial cells from being iron depleted by intracellular chrysobactin. Mössbauer spectroscopy of whole cells at various states of [^{57}Fe] chrysobactin uptake showed that this enzyme is not required for intracellular iron removal. Nor is it required for iron removal from enterobactin in *E. chrysanthemi*. We did not identify any other *E. coli fes* gene homologue in the *E. chrysanthemi* genome, indicating that the Fes enzymatic activity is not essential for ferric enterobactin utilization in this bacterium.

Fur-DEPENDENT IRON REGULATION

The regulatory role of iron in *E. chrysanthemi* was first investigated for the *fct-cbsCEBA* operon. An mRNA transcript of ca. 8,400 bp, corresponding to the size expected for this operon, was identified only in bacterial cells grown under iron limitation. Studies with various *lacZ* gene fusions and promoter-mapping experiments showed that the *fct-cbsCEBA* and

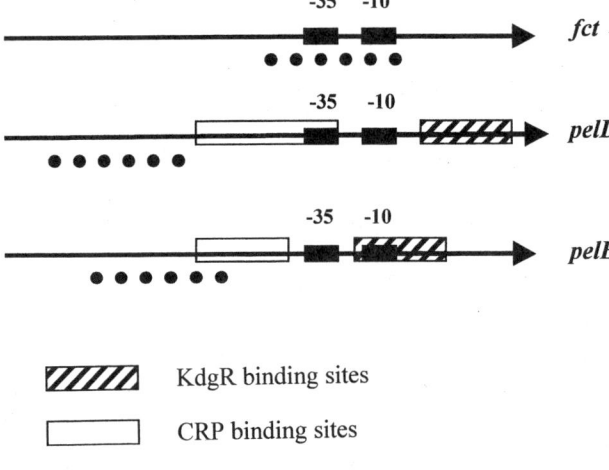

FIGURE 4 Schematic view of the promoter regions of the *fct*, *pelD*, and *pelE*, genes from *E. chrysanthemi* 3937. The −35 and −10 promoter elements are represented by black rectangles. KdgR- and CRP-binding sites identified by footprinting experiments are indicated by hatched and white boxes, respectively. The confirmed Fur-binding regions are underlined with circles.

cbsHF operons are divergently transcribed. Two promoter elements, P and P′, were determined, and the presence of two potential Fur-binding sites overlapping these promoters indicated the occurrence of a *fur*-like gene in *E. chrysanthemi*. Iron availability was also shown to regulate the transcription of genes (*pelB* to *pelE*) encoding pectate lyases (PelB to PelE): moderate iron limitation derepresses transcription of the genes *pelB*, *pelC*, and *pelE*, whereas greater iron deficiency is necessary to derepress the *pelD* gene. The isoenzymes PelD and PelE play a major role in the virulence of strain 3937 for African violets. Inspection of the promoter region of the *pelD* and *pelE* genes revealed two conserved Fur boxes located, respectively, 30 and 50 bp upstream from the −35 promoter site of these two genes. In addition to their unusual position, these Fur boxes overlap the binding sites for the cyclic AMP (cAMP)-cAMP receptor protein (CRP) complex characterized by DNase I footprinting experiments.

The *E. chrysanthemi fur* gene was cloned by functional complementation of an *E. coli fur* mutant. It encodes a protein of 148 amino acid residues exhibiting 89% identity to *E. coli* Fur. A *fur* null mutant was constructed. Analysis of transcriptional fusions to different iron-regulated genes established that Fur negatively controls iron transport and the genes encoding PelD and PelE. To further analyze the molecular mechanisms of this control, the *E. chrysanthemi* Fur protein was purified. Band shift assays showed that Fur specifically binds in vitro to the regulatory region of the *fct*, *pelD*, and *pelE* genes. The Fur-binding sites were determined by DNase I footprinting experiments (Fig. 4). For the *fct* promoter, the protection patterns covers the Fur boxes and includes the −35 and −10 elements of the P promoter. This indicates that the process involving direct competition between the RNA polymerase and Fur, characterized in *E. coli*, probably occurs for the control of the *fct* gene. On the other hand, for the *pelD* and *pelE* genes, the sequence protected by Fur includes a part of the binding site of CRP. In this case, Fur would act as an anti-activator of transcription by blocking the action of CRP. The affinity with which Fur binds the *pelD* and *pelE* regulatory regions is lower than that observed for the promoter of the *fct* gene.

IRON HOMEOSTASIS AND PLANT PATHOGENESIS

The molecular data reported above indicate that low iron availability is a triggering signal for coordinated expression of the genes encoding pectate lyases and those involved in chrysobactin-mediated iron uptake. Pectate lyases provide bacteria with carbon sources derived from the degraded pectin. Pectin catabolism leads to the formation of 2-keto-3-deoxygluconate (KDG), which is the main inducer of this metabolic pathway: transcription of the genes involved in pectinolysis is repressed by a common regulator, KdgR. In the presence of pectic inducers such as KDG and iron, the repressor KdgR is inactivated and Fur is active and can thus prevent the activation of the *pelD* and *pelE* genes by CRP. In addition, the negative metalloregulation by Fur of iron transport genes was found to be partially relieved during pectinolysis, regardless of the iron level. To further investigate this question, an insertional mutagenesis of the chromosome was carried out to find mutations conferring increased expression of both the *pelD* and *fct* genes. One insertion giving rise to this phenotype appeared to inactivate the expression of the *kdgK* gene encoding the KDG kinase. The KDG phosphorylation is the last step of the pectin or galacturonate catabolic pathway before this compound is converted into 3-phosphoglyceraldehyde and pyruvate. A mutation in the *kdgK* gene, which blocks this metabolic conversion, leads to the intracellular accumulation of KDG and derepresses genes involved in pectinolysis. On the other hand, the effect of the *kdgK* mutation on derepression of the *fct* gene in the presence of iron was intriguing. This effect can be reproduced by the addition of KDG to the culture medium but is not mediated by the regulator KdgR. These data indicate that intracellular accumulation of KDG modulates not only the expression of the pectate lyase-encoding genes but also that of iron transport genes. Of the possible interpretations of these results, the authors favored the idea of a direct action of this compound on the Fur regulatory activity. It thus seems that the two pathogenicity determinants, iron acquisition and pectinolysis, are regulated in a coordinate manner by responding to common signals: iron availability and pectin. This metabolic coupling should confer an important advantage on the bacteria during pathogenesis (Fig. 5).

Which condition do bacteria encounter in planta? After leaf inoculation of *Saintpaulia* plants, bacterial cells can be visualized in the intercellular spaces, attacking wall material around the cell intersects, and moving within the cell wall. During the first 24 h, the growth rate of a chrysobactin-deficient mutant is the same as that of the parental strain, but after 36 h the growth of the mutant decreases significantly. There is a good correlation between bacterial growth rate and symptom evolution. Lesions initiated with the mutant become dry within a few days. Thus, leaf intercellular fluid represents the environmental conditions encountered by bacteria on infection. Bacterial cells grown in intercellular fluid from healthy plants behave as if they were iron restricted: they induce their high-affinity transport systems. Chrysobactin was detected in fluids of diseased plants 24 h postinfection. Derepression of the chrysobactin system is caused by the presence of strong iron ligands in plant fluids. Plants are known to produce a variety of phenolics in response to various stresses, and some of these compounds might be involved in iron scavenging. Competition for nutritional iron was also studied by using a plant-bacterium system: iron incorporation into plant ferritin appeared to be considerably reduced in bacterially treated suspension soybean cells. The same effect was visualized during treatment of soybean cells with intercellular fluids from diseased leaves or with pure chrysobactin. Therefore, chrysobactin can act as a toxic compound for the plant by reducing the amount of iron present in tissues colonized or inoculated with *E. chrysanthemi*. The temporal pattern of expression of the pectate lyase-encoding genes was compared to that of the chrysobactin operon in bacteria grown in planta. Induction of the chrysobactin operon can be detected 10 h after inoculation, concomitantly with induction of the pectate lyase-encoding genes, except for the *pelD* gene, which is induced earlier, i.e., after 6 h. A double mutant impaired in the production of both the PelD isoenzyme and chrysobactin does not survive in planta. These data indicate that the production of pectinases

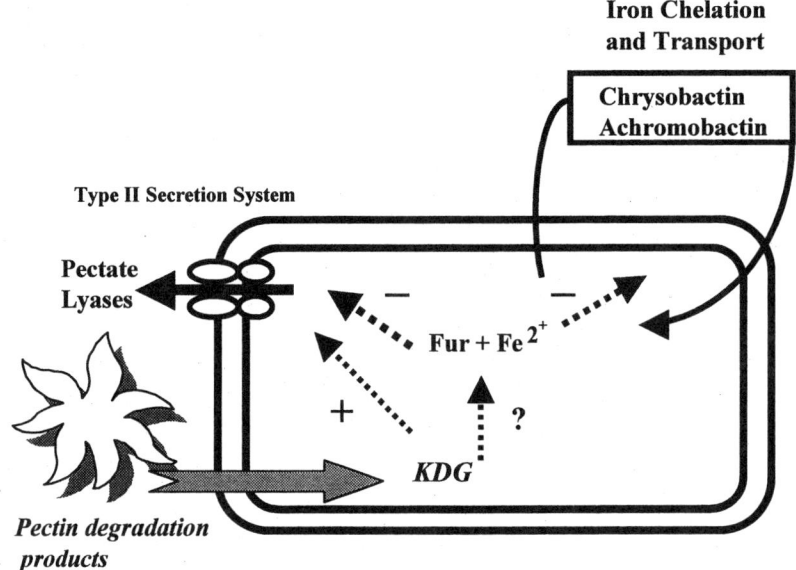

FIGURE 5 Coordinated regulation of pectinolysis and iron transport in *E. chrysanthemi* 3937. Under low-iron conditions, the Fur-mediated transcriptional repression of the genes involved in iron transport and pectinolysis is relieved. In the presence of iron, intracellular accumulation of the pectin degradation product KDG turns on the transcription of the pectinase-encoding genes by inactivation of the transcriptional repressor KdgR, as well as that of the genes involved in iron acquisition by an unknown mechanism.

is tightly associated with the ability of the bacterium to acquire iron via chrysobactin. In initiating the maceration symptom, the PelD isoenzyme may, via the chrysobactin-dependent transport route, enhance the release of plant iron-containing nutrients used by the pathogen. When the bacteria are starved for iron prior to inoculation, the maceration process is accelerated. In this bacterium-plant interaction, the appearance of symptoms appears to depend on the initial iron status of the pathogen and iron availability in the host. These physiopathological data corroborate the molecular model depicting the central role of the Fur protein in iron sensing.

CONCLUSION

Studying the role of chrysobactin in the pathogenicity of *E. chrysanthemi* 3937 has contributed to an appreciation of the regulatory mechanisms involved in the expression of virulence by plant-pathogenic bacteria. Unlike typical phytopathogens capable of inducing a cell death reaction called the hypersensitive response on nonhost plants, *E. chrysanthemi* has a broad host range and does not display the basic characteristics of compatible and incompatible reactions mediated by type III secretion systems. A type III secretion system exists in strain 3937, however. The ability of this strain to infect a large number of hosts is probably related to the performance of its pectolytic equipment. A rapid destruction of plant tissues must allow the bacteria to overcome the host defenses and be somehow independent of the type III secretion system. The existence of a fine-tuning mechanism such as iron sensing to ensure a balanced production of pectate lyases during the infection process is probably a selective advantage for strain 3937. To control its iron homeostasis, this bacterium produces two different siderophores, whose contributions are probably modulated by the iron fluctuations encountered during the infection process. Mutants affected in the biosynthesis of both chrysobactin and achromobactin are less virulent than the simple mutants. Mutants with impaired ferric achromobactin permease are derepressed for the production of chrysobactin. Thus, the role in pathogenicity of achromobactin compared to chrysobactin must be explored. Another important question is whether plants control and modify their iron storage during infection, possibly by developing reactions similar to the iron-withholding response induced by animal and human pathogens. It has recently been shown that *E. chrysanthemi* 3937 induces systemic symptoms in the plant model *Arabidopsis thaliana*. Such a pathosystem should help to identify plant iron mobilization reactions that may occur during pathogenesis.

ACKNOWLEDGMENTS

We thank all the colleagues and friends who have been involved in the development of the chrysobactin story. We are particularly grateful to A.-M. Albrecht-Gary, J. S. Buyer, J.-P. Laulhère, M. A. McIntosh, B. F. Matzanke, J. B. Neilands, and M. Persmark for their support and contribution to this work. We also thank N. Perna and her collaborators for giving us the opportunity to participate in the annotation of the *E. chrysanthemi* 3937 genome.

REFERENCES

El Hassouni, M., J.-P. Chambost, D. Expert, F. van Gigsegem, and F. Barras. 1999. The minimal gene set member *mrsA*, encoding peptide methionine sulfoxide reductase, is a virulence determinant of the plant pathogen *Erwinia chrysanthemi*. *Proc. Natl. Acad. Sci. USA* **96:**887–892.

Enard, C., A. Diolez, and D. Expert. 1988. Systemic virulence of *Erwinia chrysanthemi* 3937 requires a functional iron assimilation system. *J. Bacteriol.* **170:**2419–2426.

Enard, C., and D. Expert. 2000. Characterization of a *tonB* mutation in *Erwinia chrysanthemi* 3937: TonB$_{Ech}$ is a member of the enterobacterial TonB family. *Microbiology* **146:**2051–2058.

Franza, T., and D. Expert. 1991. The virulence-associated chysobactin iron uptake system of *Erwinia chrysanthemi* 3937 involves an operon encoding transport and biosynthetic functions. *J. Bacteriol.* **173:**6874–6881.

Franza, T., I. Michaud-Soret, P. Piquerel, and D. Expert. 2002. Coupling of iron assimilation and pectinolysis in *Erwinia chrysanthemi* 3937. *Mol. Plant-Microbe Interact.* **15:**1181–1191.

Franza, T., C. Sauvage, and D. Expert. 1999. Iron regulation and pathogenicity in *Erwinia chrysanthemi* strain 3937: role of the Fur repressor protein. *Mol. Plant-Microbe Interact.* **12:**119–128.

Glasner, J. D., P. Liss, G. Plunket III, A. Darling, T. Prasad, M. Rusch, A. Bynes, M. Gilson, B. Biehl, F. R. Blattner, and N. T. Perna. 2003. ASAP, a systematic annotation package for community analysis of genomes. *Nucleic Acids Res.* **31:**147–151.

Lu, C., J. S. Buyer, J. F. Okonya, and M. J. Miller. 1996. Synthesis of optically pure chrysobactin and immunoassay development. *BioMetals* **9:**377–383.

Marahiel, M. A., T. Stachelhaus, and H. D. Mootz. 1997. Modular peptide synthetases involved in nonribosomal peptide synthesis. *Chem. Rev.* **97:**2651–2673.

Masclaux, C., and D. Expert. 1995. Signaling potential of iron in plant-microbe interactions: the pathogenic switch of iron transport in *Erwinia chrysanthemi Plant J* **7:**121–128.

Masclaux, C., N. Hugouvieux Cotte-Pattat, and D. Expert. 1996. Iron is a triggering factor for differential expression of *Erwinia chrysanthemi* 3937 pectate lyases during pathogenesis of African violets. *Mol. Plant-Microbe Interact.* **9:**198–205.

Münzinger, M., H. Budzikiewicz, D. Expert, C. Enard, and J.-M. Meyer. 2000. Achromobactin, a new citrate siderophore of *Erwinia chrysanthemi*. *Z. Naturforsch.* **55C:**328–332.

Nachin, L., M. El Hassouni, L. Loiseau, D. Expert, and F. Barras. 2001. SoxR-dependent response to oxidative stress and virulence of *Erwinia chrysanthemi*: the key role of SufC, an orphan ABC ATPase. *Mol. Microbiol.* **39:**969–972.

Neema, C., J.-P. Laulhère, and D. Expert. 1993. Iron deficiency induced by chrysosbactin in *Saintpaulia* leaves inoculated with *Erwinia chrysanthemi*. *Plant Physiol.* **102:**967–973.

Perombelon, M. C. M. 2002. Potato diseases caused by soft rot erwinias: an overview of pathogenesis. *Plant Pathol.* **51:**1–12.

Persmark, M., D. Expert, and J. B. Neilands. 1989. Isolation, characterization and synthesis of chrysobactin, a compound with siderophore activity from *Erwinia chrysanthemi*. *J. Biol. Chem.* **264:**3187–3196.

Persmark, M., D. Expert, and J. B. Neilands. 1992. Ferric iron uptake in *Erwinia chrysanthemi* mediated by chrysobactin and related catechol type compounds. *J. Bacteriol.* **174:**4783–4789.

Rauscher, L., D. Expert, B. F. Matzanke, and A. X. Trautwein. 2002. Chrysobactin-dependent iron acquisition in *Erwinia chrysanthemi*: functional study of an homologue of the *Escherichia coli* ferric enterobactin esterase. *J. Biol. Chem.* **277:**2385–2395.

Santos, R., T. Franza, M.-L. Laporte, C. Sauvage, D. Touati, and D. Expert. 2001. Essential role of superoxide dismutase on the pathogenicity of *Erwinia chrysanthemi* strain 3937. *Mol. Plant-Microbe Interact.* **14:**758–757.

Sauvage, C., and D. Expert. 1994. Differential regulation by iron of *Erwinia chrysanthemi* pectate lyases: pathogenicity of iron transport regulatory mutants. *Mol. Plant-Microbe Interact.* **7:**71–77.

Sauvage, C., T. Franza, and D. Expert. 1996. Analysis of the *Erwinia chrysanthemi* ferrichrysobactin receptor gene: resemblance to the *Escherichia coli fepA-fes* bidirectional promoter region with hydroxamate receptors. *J. Bacteriol.* **178:**1227–1231.

THERAPEUTIC USES OF IRON(III) CHELATORS AND THEIR ANTIMICROBIAL CONJUGATES

Vinay Girijavallabhan and Marvin J. Miller

27

IRON ASSIMILATION IN BACTERIA

As indicated throughout this book, iron plays an essential role in nearly all life-forms on Earth. It is required for the proper function of enzymes that facilitate electron transport, oxygen transport, and other life-sustaining processes. Microorganisms are known to use iron for multiple functions including reduction of oxygen for the synthesis of ATP and for the formation of heme. Since iron is essential to cellular function, the competition for iron between host and bacteria is one of the most important factors governing the course of bacterial infection. Details of microbial iron assimilation are provided in other chapters and are only briefly summarized here to set the stage for our discussion of the therapeutic potential of exploiting microbial iron sequestration.

Although iron is one of most abundant metals on this planet, the assimilation of ionic iron, especially iron (III) is made very difficult by its insolubility at physiological pH [the K_{sp} of Fe(OH)$_3$ is $< 10^{-18}$ at pH 7]. Many microorganisms circumvent this problem of low iron bioavailability by synthesizing and excreting siderophores, iron chelators that solubilize and sequester iron. Under iron-stressed conditions, these iron complexes can be recognized by the outer membrane receptors of the microorganism and then are transported through the cellular membrane via active transport.

The several hundred siderophores that have been isolated and characterized have been found to vary greatly in composition and topology. Some examples of siderophores with common iron binding moieties are shown in Fig. 1. Siderophores can be classified by either their iron binding moieties or their backbone structure. Iron binding moieties such as hydroxamic acids, catechols, and α-hydroxy carboxylic acids are found in most siderophores. Mixed-ligand siderophores that contain combinations of iron binding groups have also been isolated.

As indicated by the generalized trihydroxamate in Fig. 2, the most effective siderophores contain multiple ligands that bind iron(III) in an octahedral complex with the iron binding sites in a particular orientation to minimize the entropic effects of separate ligands. These octahedral complexes are recognized by outer membrane receptors such as the ferrichrome (FhuA) and enterobactin (FepA) receptors in

---

Vinay Girijavallabhan and Marvin J. Miller, Department of Chemistry and Biochemistry, University of Notre Dame, Notre Dame, IN 46556.

Iron Transport in Bacteria, Edited by Jorge H. Crosa, Alexandra R. Mey, and Shelley M. Payne
© 2004 ASM Press, Washington, D.C.

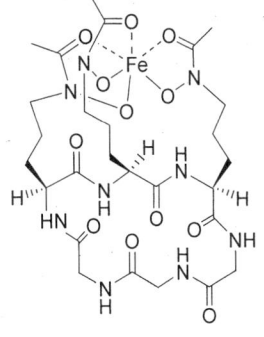

FIGURE 1 Representative siderophores.

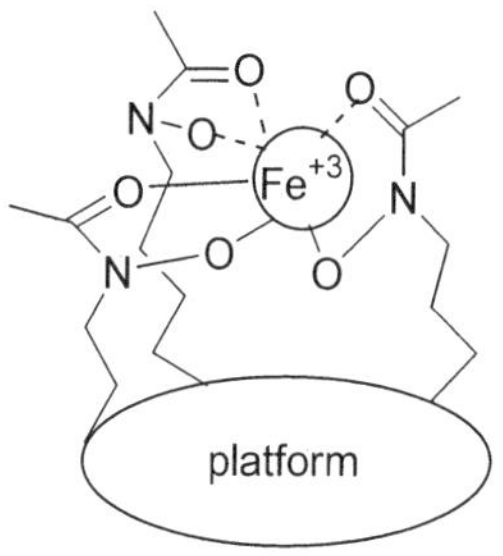

FIGURE 2 Generalized trihydroxamate siderophore analog.

Escherichia coli (see chapter 4). The FhuA receptor is a multifunctional protein that recognizes the hydroxamate siderophore ferrichrome (structure 3 in Fig. 1) and many other low-molecular-weight compounds and biopolymers. FepA transports the catechol siderophore enterobactin (structure 1 in Fig. 1) across the outer membrane. Outer membrane receptors in gram-negative bacteria provide a pathway for ferric siderophore complexes to enter the cell via the active transport system (Fig. 3). Since there is no energy source on the outside of the cytoplasmic membrane, the ability of these receptors to transport siderophores is dependent on TonB to transduce energy provided by the proton motive force of the cytoplasmic membrane (see chapter 7). While the most detailed studies of siderophore transport have been done with *E. coli*, similar transport systems have also been found in other bacterial and fungal species.

Chirality can play a critical role in the recognition of siderophores by outer membrane receptors. Chelation of the siderophore to iron (III) creates a new chiral center at the metal, so that either a left-handed or right-handed octahedral complex can form. Although these two complexes are in equilibrium due to kinetic lability, the chirality of the metal is often essential to receptor recognition. The difference in binding constants of the two isomers leads to stereoselective recognition by the outer membrane receptor proteins. Transport of the desired isomer shifts the equilibrium that would lead to the maximum utilization of iron (III) available in the environment. The stereochemistry of the non-iron binding part of the siderophore can also be essential for recognition. Changing the stereochemistry can alter

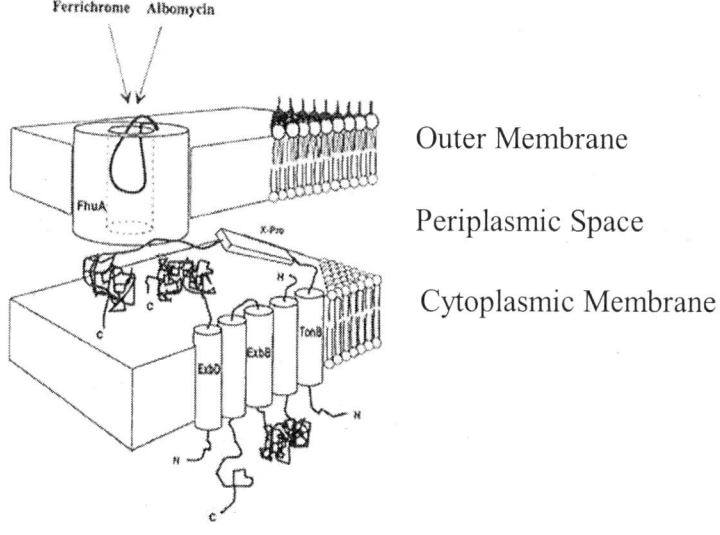

FIGURE 3 Siderophore transport system. Reprinted from Braun et al. (1998) with permission from the publisher.

the conformation of the iron binding ligands, disrupting the binding to iron. These subtle changes can have large adverse affects on recognition and transport.

Once inside the cell, the organism must have a method of releasing the iron from the siderophore. The strong affinity of oxygen for "hard" ferric ions allows for a possible iron release via reduction of the iron (III) inside the cell. Siderophore decomposition is another means of removal of iron from the siderophore. Therefore, four parameters are necessary for efficient iron (III) uptake: (i) iron (III) binding to the siderophore, (ii) outer membrane recognition of the siderophore-iron complex, (iii) efficient transport of the iron complex across the cellular membrane, and (iv) iron release.

NATURAL SIDEROPHORE ANTIBIOTICS

Numerous siderophores possess antibiotic properties. These antibiotic siderophores, or sideromycins, defend against iron assimilation by foreign organisms. Some extremely efficient iron chelators may deprive pathogenic microbes of iron that is essential for growth by blocking the iron transport receptors with nonfunctional siderophores. Iron starvation can also be achieved by creating siderophores bound to metabolically less useful metals. This can prevent the microorganism from acquiring the iron it needs since it recognizes and transports these siderophores.

Study of the mycobactins, a family of natural siderophores produced by mycobacteria, has

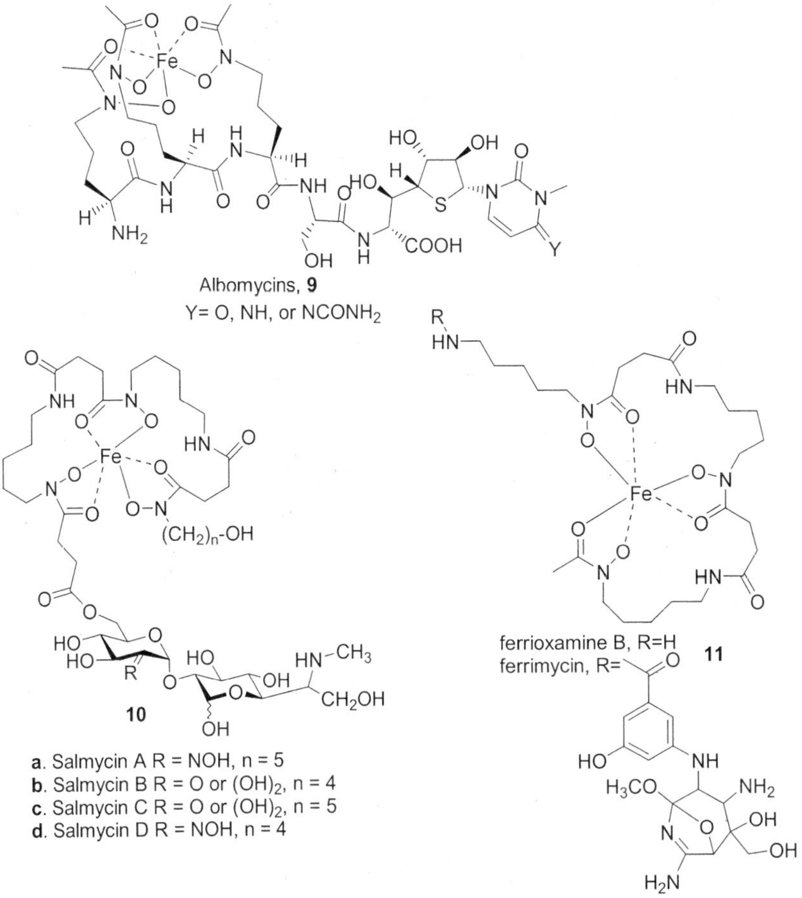

FIGURE 4 Natural siderophore antibiotics.

shown that the synthesis of analogs with similar structures can inhibit the growth of several mycobacteria, such as *Mycobacterium tuberculosis*. Other natural siderophores, such as albomycins and salmycins, are conjugated to toxic compounds that are released on entry into the cell. For iron chelation, albomycins utilize a tripeptide of δ-*N*-hydroxy-δ-*N*-acetyl-L-ornithine residues that is found in ferrichrome, but unlike ferrichrome, which forms a cyclic hexapeptide, albomycin contains a toxic moiety attached via a serine spacer (structure 9 in Fig. 4). Salmycins A to D consist of the natural siderophore danoxamine ($n = 5$) conjugated to aminoglycoside moieties and are known to be active against staphylococci and streptococci (structure 10 in Fig. 4). Ferrimycin mimics the transport properties of ferrioxamine B (structure 11 in Fig. 4).

While nature has demonstrated the use of sideromycins as iron transport-mediated drug delivery agents, these compounds have not found extensive use as therapeutic agents because the exact modes of action are unknown and organisms reportedly develop rapid resistance to these natural drug conjugates. It should be noted that these resistant strains are often nonpathogenic, since the resistance may result in bacteria that have altered iron transport systems and can no longer compete for iron.

The finding of natural siderophore antibiotics, such as salmycins and albomycin, prompted the search for novel synthetic siderophores that could be utilized in a variety of therapies. These siderophore analogs can contain drugs with known modes of action and could combat problems of drug resistance. Several examples of therapeutically useful synthetic siderophores are described in the following sections.

USE OF SIDEROPHORES AS THERAPEUTIC AGENTS

Iron Overload
Although iron is essential to most life-forms, large amounts of iron are toxic. Iron overload in patients is often caused by repeated blood transfusions, as in patients affected by Cooley's anemia, or the misuse of iron vitamins. The most common drug to treat iron overload is the siderophore desferrioxamine B (Desferal) (structure 12 in Fig. 5), but drawbacks to the use of this siderophore are its poor oral activity, short retention time in the body, and slow iron chelation. Daily subcutaneous or intravenous treatments are required to remove excess iron. Therefore, new iron chelators with better pharmacological effects are being sought. It has been discovered that iron removal from transferrin or lactoferrin is accelerated by mediator anions that modify the protein structure and hence make the metal more accessible. Catechol groups are known to bind to proteins and change their conformation, increasing the rate of iron binding. Bazzicalupi and coworkers synthesized a series of siderophore analogs that contained several catechols (four to six) and studied their binding to iron (structures 13 and 14 in Fig. 5). These siderophores removed iron from transferrin quickly in vitro and the rates of abstraction were faster than that of Desferal, suggesting that this novel siderophore could be useful in treating iron overload cases. Further in vivo studies are required to determine if these siderophore analogs have improved oral activity and low toxicity.

Mycobacterium Species
The resurgence of tuberculosis infections worldwide has reemphasized the constant need for new antimycobacterial agents. Tuberculosis causes approximately 3 million deaths every year, and a total of nearly 2 billion people are infected worldwide. One of the major problems in treating tuberculosis is the inability of many therapeutic agents to traverse the mycobacterial cell envelope and reach their intended targets. One possible weakness of mycobacteria is in the complex mechanism of iron metabolism. Snow found that antagonists of the growth of one type of mycobacteria might be found in another species of mycobacteria and demonstrated that natural mycobactins M or N could suppress the growth of *M. tuberculosis*, *M. paratuberculosis*, and *M. kansasii* (structures

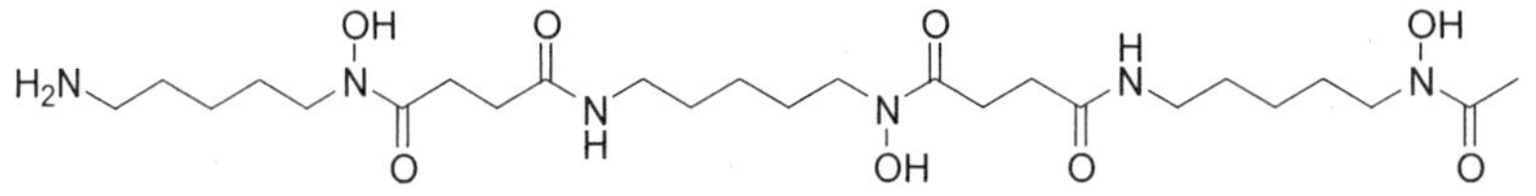

desferrioxamine, **12**

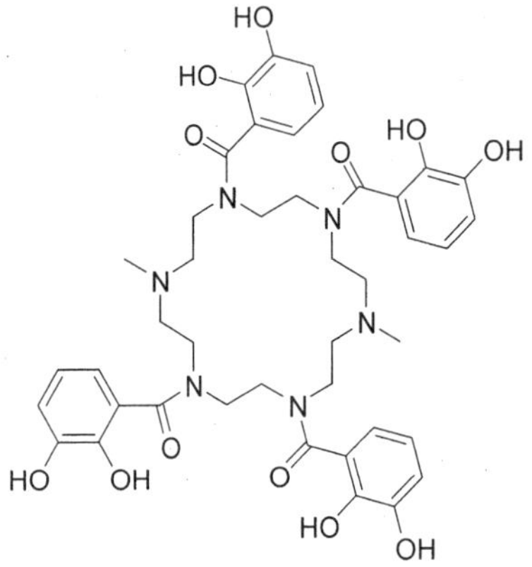

13

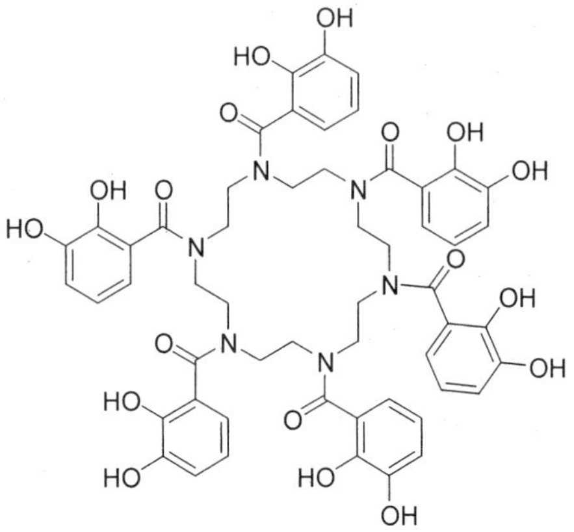

14

FIGURE 5 Desferroxamine and synthetic chelators for iron overload treatment.

15 and 16 in Fig. 6). This finding was further supported by Miller and Hu, who found that synthetic mycobactin S inhibited the growth of *M. tuberculosis* (99% inhibition when given at 12.5 μg/ml) (structure 17 in Fig. 6). These results suggest that mycobactin S is competing with the natural mycobactin T (structure 18 in Fig. 6) in the iron uptake pathway, inhibiting siderophore recognition. Several other mycobactin analogs were synthesized by the Miller

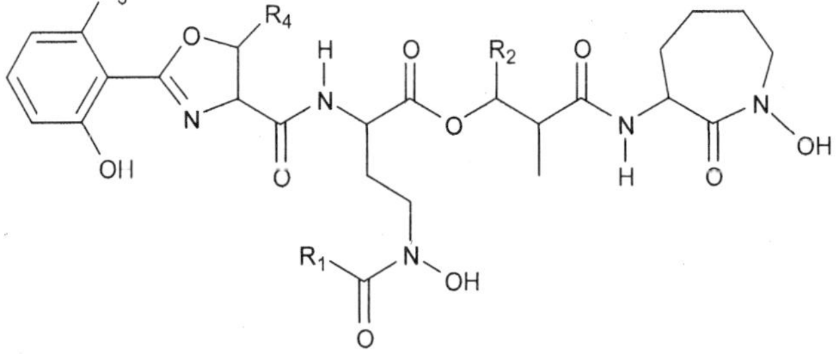

15 Mycobactin M, R_1=Me, R_2= $C_{15\text{-}18}$, R_3=H, R_4=Me

16 Mycobactin N, R_1=Et, R_2=$C_{15\text{-}18}$, R_3=H, R_4=Me

17 Mycobactin S, R_1=$C_{13\text{-}19}$, R_2=Me, R_3=H, R_4=H

18 Mycobactin T, R_1=$C_{17\text{-}20}$, R_2=Me, R_3=H, R_4=H

FIGURE 6 Natural mycobactins.

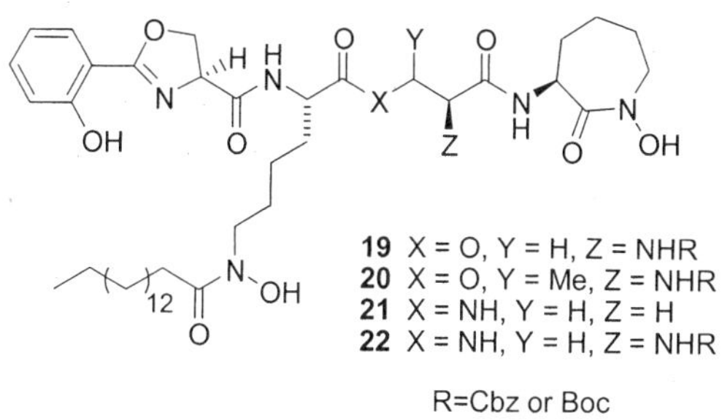

FIGURE 7 Synthetic mycobactin analogs.

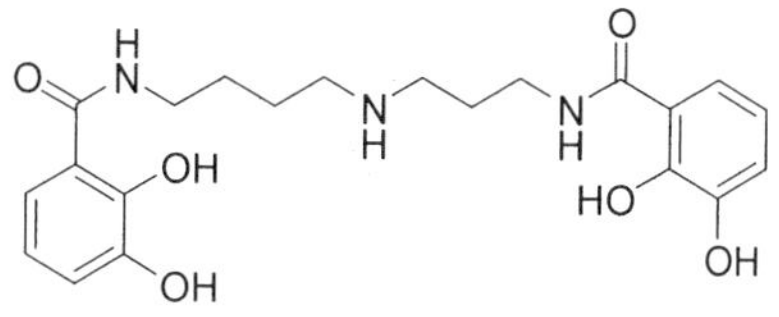

FIGURE 8 FR401, a bis-catecholspermidine.

group and found to have high inhibitory activity against *M. tuberculosis* H37Rv when given at a concentration of 12.5 μg/ml (structures 19 to 22 in Fig. 7). The most active analog, structure 22 (MIC, <0.2 μg/ml), is thought to have increased hydrophobic interactions with the mycobacterium cell wall due to the bulky Boc protecting group, preventing the delivery of iron into the cell. Enhancement of the iron binding ability of the molecule due to stabilization of the iron complex by the large Boc group can also cause iron starvation and thus explain the excellent activity. The syntheses of mycobactin analogs that bind other metals have also been investigated in order to disrupt the iron uptake system of mycobacteria. Mycobactins bound to chromium inhibit the growth of *M. tuberculosis*, but Ratledge and coworkers do not think that this type of inhibition will be therapeutically useful due to ligand preference for iron and the toxic effects of many metals.

Plasmodium Species

Malaria is one of the major causes of mortality in tropical regions due to the prevalence of strains of *Plasmodium falciparum* that are resistant to common antimalarial drugs such as chloroquine. The impermeability of the membrane of this microorganism necessitates the use of new methods of drug delivery. Iron chelation therapy was found to be a suitable treatment of malaria. The siderophore desferrioxamine has activity against strains of malaria, but the activity is slow to develop. Many other, more lipophilic iron chelators were synthesized to increase the intracellular accessibility of the drug and improve the speed of action. Therapy was also found to improve when iron chelators with differing chelating abilities were used in various combinations. Kunesch and coworkers have accomplished the conjugation of macrolide antibiotics to catecholspermidines (structure 23 in Fig. 8). These compounds have activity comparable to that of the macrolides against antibiotic-resistant mycobacteria and also inhibit the growth of *P. falciparum* by sequestering iron that the organism requires for survival. To find a potent antimalarial cocktail, this bis-catecholate siderophore was used in combination therapies with antimalarial drugs to which resistance has developed. The siderophore and tetracyclines were found to have synergistic or additive anti-*P. falciparum* effects that are due to their similar mechanisms of actions against both ribonucleotide reductase and dihydroorotate dehydrogenase. The ribonucleotide reductase and dihydroorotate dehydrogenase enzymes are iron-dependent enzymes that are known to be sensitive to the intracellular concentration of iron and therefore are inhibited by siderophores. These combination drug therapies are required due to the slow activity of antimalarial drugs. New methods of drug delivery may eliminate the need for this type of treatment.

Various Bacteria

Siderophore-metal complexes that cleave DNA have been investigated as potential drugs. Iron complexes of EDTA and copper complexes of Desferal are known to oxidatively attack the sugar residues of the DNA backbone. Cleavage at the bond between C-4 and H is

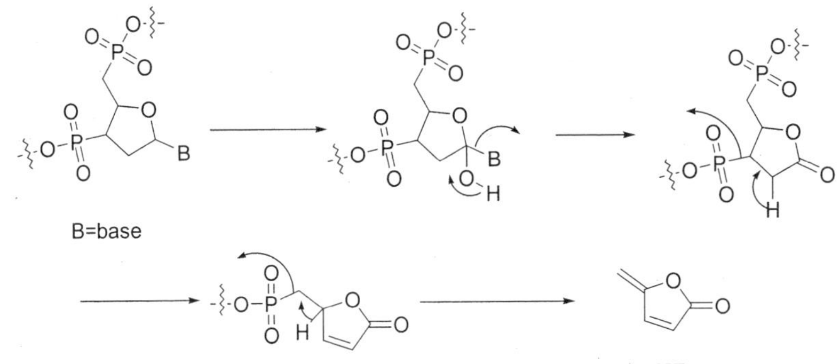

FIGURE 9 Mechanism for the generation of 5-methylene furanone (MF).

known for EDTA, but Ganesh and coworkers have used high-performance liquid chromatography and ^{31}P nuclear magnetic resonance analysis of cleavage products to determine that copper complexes of Desferal attack at C-1 in the minor groove. It was suggested that the copper-Desferal complex binds to DNA and then causes cleavage through redox chemistry at the metal center. This cleavage was suppressed when radical scavengers were added, suggesting that copper created hydroxy radicals responsible for DNA damage. The isolation of 5-methylene furanone (Fig. 9) by high-performance liquid chromatography analysis further established this type of cleavage mechanism. Copper-siderophore complexes could be used as drug delivery agents or antimalarial agents due to their ability to cause DNA damage.

Pyridine-2,6-dithiocarboxylic acid (PDTC) (structure 24 in Fig. 10) is a metal chelator synthesized by *Pseudomonas* spp. It is known to have strong affinities for many metals such as cobalt, copper, zinc, iron, and nickel. Crawford and coworkers have used radioactive ^{59}Fe studies to show that this metal chelator does not act as the cell's primary siderophore; rather, PDTC is an antibacterial agent active against other organisms. Studies demonstrated that PDTC inhibits the growth of *E. coli* under iron-limiting conditions. The growth was found to resume when iron, cobalt, or copper was added to the cultures. This indicates that the mode of inhibition by PDTC is through chelation to metals. In addition to its antimicrobial properties, the copper complex of PDTC was found to have redox activity. Previous work by Lewis has shown that the copper complex can degrade carbon tetrachloride. This suggests that PDTC could perform other extracellular chemical transformations. The addition of zinc to the cultures resulted in a dramatic increase in the activity of PDTC, indicating that the zinc complex of PDTC is highly toxic or that PDTC can enhance the delivery of zinc into the cell.

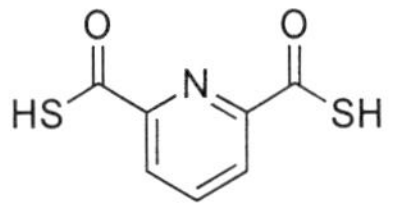

Immunomodulators

The search for novel immunomodulators led to the discovery that desferrioxamine can inhibit the ribonucleoside reductase, suppressing DNA synthesis. This reductase is a non-heme-iron-containing protein that requires ATP and Mg^{2+} as cofactors. Another siderophore, IC202C (structure 25 in Fig. 11), produced by *Streptoalloteichus* sp. strain 1454–19, was also found to significantly inhibit this reductase and

pyridine-2,6-thiocarboxylic acid, **24**

FIGURE 10 Structure of 2,6-dithiocarboxylic acid.

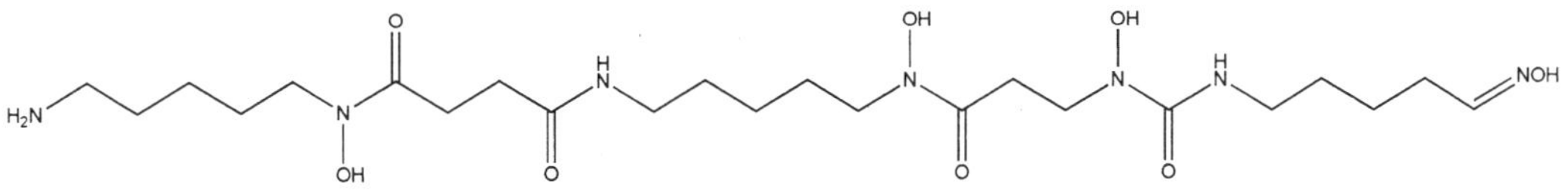

FIGURE 11 Structure of IC202C.

show strong immunosuppressive activity. The exact mechanism of this inhibition is unknown, but it is thought that the siderophore inactivates the enzyme due to removal of bound iron from the reductase. This hypothesis was proposed due to the inactivity of the preformed siderophore iron complex.

Cancer

Neoplasia is the cause of 5 million deaths each year worldwide, and hepatocellular carcinomas, in particular, are one of the most common malignancies in the world. Resistance to chemotherapy treatments, as well as the asymptomatic nature of the disease, renders it very difficult to treat. Kicic and coworkers discovered that the siderophore desferrioxamine mesylate displays antineoplastic activity in hepatocellular carcinoma cells. This siderophore was found to inhibit iron uptake from transferrin, depending on the concentration and incubation time. Due to the poor membrane permeability of desferrioxamine mesylate, incubation times of over 24 hs were required to efficiently chelate intracellular iron. This finding prompted the testing of several other siderophores that differed in iron binding affinity, iron chelation (ferrous versus ferric), molecular size, and membrane permeability. These studies demonstrated that the impermeable strong ferric chelator diethylenetriaminepentaacetic acid (DTPA) (structure 26 in Fig. 12) was a very effective antineoplastic agent. It was suggested that this activity is due to the binding of ferric ions at the surface of the cell or within the endocytotic vesicle, thereby preventing iron uptake. Since iron on the cell surface is reported to be involved in *trans*-plasma membrane electron transport that is essential for the oxidation of NADH and cell proliferation, the chelation of iron could possibly disrupt cellular function.

Another useful iron chelator was the permeable ferrous chelator dipyridine (structure 27 in Fig. 12), which was shown to be a strong inhibitor of cell proliferation and iron uptake. Dipyridine was also found to be an effective iron chelator in reticulocytes and erythroleukemic cells, in which it inhibited transferrin recycling and receptor-mediated endocytosis. Its potential mode of action was suggested to be the chelation of iron (II) after

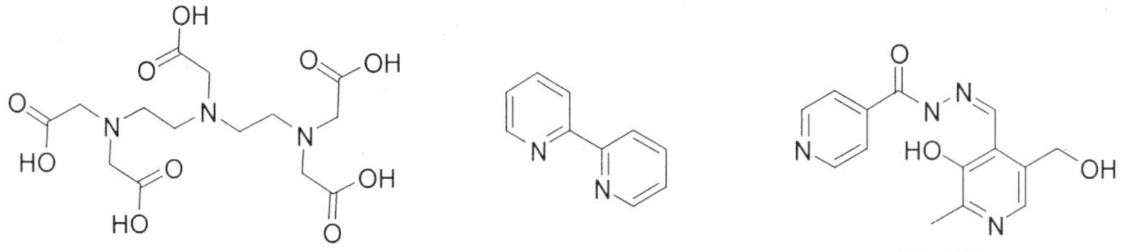

FIGURE 12 Structure of DTPA (structure 26) and other iron chelators (structures 27 and 28).

it was released from transferrin. Permeable ferric chelators, such as pyridoxal isonicotinoyl hydrazone (structure 28 in Fig. 12) and its analogs, were also found to be excellent inhibitors of tumor cell proliferation due to their ability to acquire iron from transferrin. All these iron chelators were found to have relatively low toxicity and excellent selectivity for tumorigenic cells.

DESIGN OF SIDEROPHORE-DRUG CONJUGATES

The existence of sideromycins such as albomycins and salmycins prompted attempts to synthesize mimics of natural siderophores that could be conjugated to antimicrobial agents and targeted specifically to microorganisms. These antibiotic siderophores led to a new form of drug delivery that utilizes the pathogen's own iron transport system. Siderophore-drug conjugates could allow the active transport of drugs that are normally therapeutically ineffective because they cannot permeate the cellular membrane by themselves. It may also be possible to transport drugs that cannot be used in therapy due to toxicity, since deleterious effects might be masked by the siderophores. Efficient release of these toxic drugs within the cell by a microbe-triggered drug release would be an excellent example of directed drug delivery.

The generalized siderophore-drug conjugate consists of four components: iron, ligand (the siderophore), linker, and a drug. The ability to use siderophores as drug delivery agents is dependent on several factors. First, it is essential that the synthetic iron binding domain of the siderophore be able to bind iron(III) in a specific conformation for recognition and transportation by outer membrane receptors. The Shanzer group has synthesized several nonnatural siderophores to map the surface of ferrichrome receptors. This work demonstrated that iron chelators have two domains of siderophore-receptor interactions. The first domain is the iron binding domain, which is very sensitive to structural modification, whereas the second domain consists of the linker to the iron binding moiety, which is more tolerant of modification (Fig. 13). The first domain is sensitive to manipulation since it is in direct contact with the recognition sites on the outer membrane receptors. The second domain can be manipulated as long as it does not interfere with the orientation of the iron binding ligands. Conformational studies by Shanzer have shown that the length of the linker is very important for recognition of the siderophore by outer membrane receptors. Synthetic siderophores with different linker lengths were studied for their ability to affect the growth of bacteria. Only siderophores with the longer linker provided either promotion or inhibition of the performance of ferrichrome by competing with its membrane receptors. Molecular modeling of the iron(III) octahedral complexes of the two siderophores showed that the linkers provided different conformations, implying that siderophore-iron complexes adopt a specific conformation that is recognized by outer membrane receptors.

Second, it is important that linkers and drugs do not influence the ability of the siderophore to be recognized and transported by the outer membrane receptors. Recent studies by Shanzer demonstrated that the attachment of fluorescent markers via a linker to iron binding ligands does not interfere with iron chelation and receptor recognition. The fluorescent probes were attached to the second domain of the siderophore, therefore, supporting the

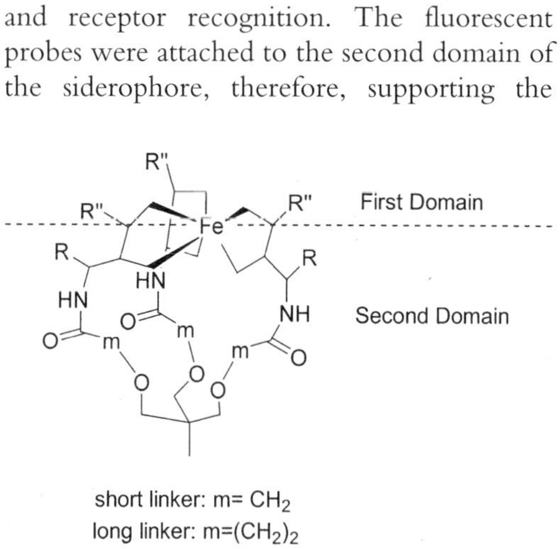

short linker: m= CH_2
long linker: m= $(CH_2)_2$

FIGURE 13 Generalized structure of Shanzer's ferrichrome analogs.

two-domain hypothesis and also allowing the detection and identification of pathogenic organisms. This work also proved that it is possible to add fairly bulky substituents to siderophores without compromising recognition and transport.

A third factor essential to binding and transport of siderophore antibiotics is the ability of synthetic siderophores to compete with natural siderophores for iron and transport. The iron binding ability of a siderophore alone does not accurately reflect the ability of the siderophore to compete for iron under deficient conditions. Studies by Raymond and coworkers have shown that the mode of siderophore-iron transport, solubility, stability, and accessibility of siderophores for iron uptake are important factors in iron acquisition by siderophores. For example, even though enterobactin (structure 1 in Fig. 1) has a thermodynamic stability constant of the order of 10^{52}, aerobactin (10^{23}) (structure 7 in Fig. 1) may provide a selective advantage for the growth of *E. coli* strains under certain iron-stressed conditions, due to the poor solubility of enterobactin and its absorption to albumin in the presence of serum at physiological pH. The strong binding of iron by the catechols in enterobactin also requires that the siderophore be destroyed in order to access the iron. Siderophore antibiotics must have good solubility, stability, and iron release in order to compete with other siderophores and be effectively recognized and transported by receptors.

The fourth factor involves the ability of the siderophore to transport the drug to its site of action. The activity of siderophore antibiotics could be due to the intact conjugate, the released drug, or possibly even the siderophore. Therefore, studies must determine where the activity of the drug conjugates arises. Growth promotion studies of the siderophore can help to determine if it is recognized by the organism and if it is toxic. Susceptibility studies of the siderophore-drug conjugates can be used to determine the MICs of the conjugate, but they might not differentiate between the activity of the intact conjugate and the released drug. If a siderophore is conjugated to antibiotics, it may be essential that the agent be released on entry into the cell to provide the desired activity. Many drugs may have diminished or no activity when the siderophore is attached; therefore, the choice of linker is extremely important. The linker must be stable enough to survive the extracellular media but must also be labile enough to be released at the site of action. The following examples demonstrate how the study of siderophore-drug conjugates has developed in the past few years.

ANTIMICROBIAL STUDIES OF SIDEROPHORE-DRUG CONJUGATES

Iron transport-mediated drug delivery has been demonstrated against several strains of bacteria and fungi. This "Trojan horse" mechanism of drug delivery utilizes the pathogen's own iron transport system to act as a delivery agent of therapeutic agents. The earliest example of antibiotic-siderophore conjugates was a study by Diddens et al. in 1977, in which sulfonamide drugs were coupled to the siderophore desferrioxamine (structure 29 in Fig. 14). Many of these conjugates that contained nicotinic acid linkers were found to have promising activity in assays.

Monodentate Siderophore Antibiotics

Many groups synthesized simple iron chelators such as catechols and hydroxamic acids that were conjugated to β-lactam antibiotics at the C-6 or C-7 position (structures 30 to 32 in Fig. 14). Unlike natural siderophores which usually contain two or three bidentate ligands, these were the simplest siderophores, with only one bidentate ligand. These conjugates were found to have excellent activities against gram-negative bacteria, even in cases where the initial drug could not permeate the cell membrane alone. This suggested that the activity was due to active transport via the iron transport system. The growth of bacterial colonies was delayed significantly by these conjugates, and the bacteria that eventually grew lacked the outer membrane receptors necessary to assimilate iron and

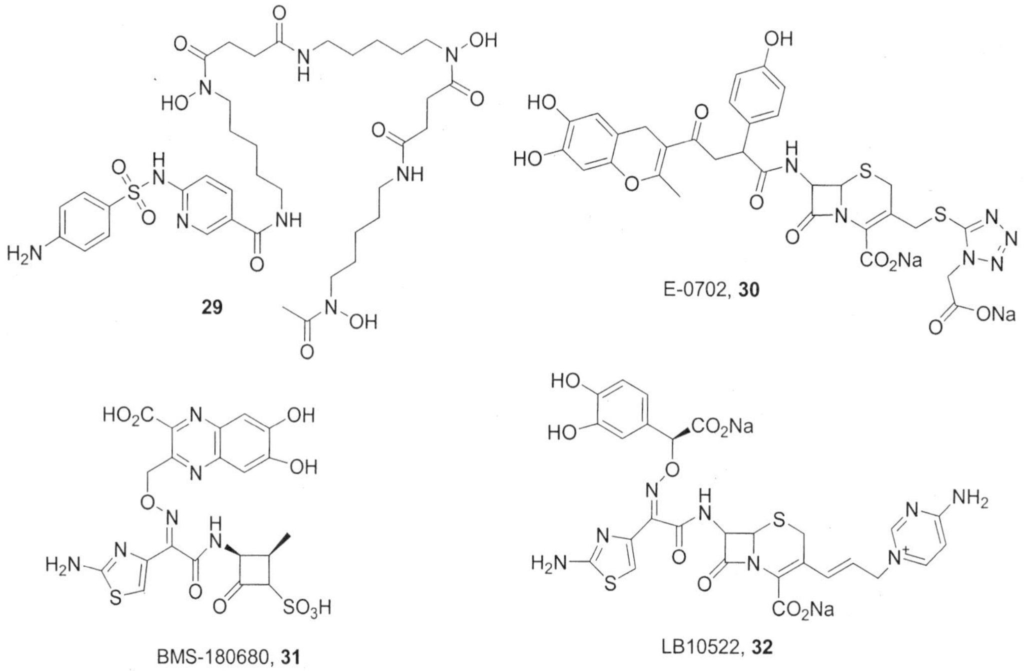

FIGURE 14 Synthetic siderophore antibiotics.

therefore could not survive well under iron-deficient conditions.

In 2000, Kline and coworkers also synthesized novel siderophores conjugated to β-lactam antibiotics. Since siderophore receptors often lack specificity for the type of iron binders that they recognize, the group designed a panel of small iron binding molecules to measure the competition between siderophore antibiotics and different siderophores. A screen of 300 iron binding compounds demonstrated that many of these salicylimine-containing compounds could function as siderophores in gram-negative bacteria. Compounds that were found to promote growth were converted to antibiotic conjugates and tested against both *E. coli* and *Pseudomonas aeruginosa* (Fig. 15). Some siderophores lost their ability to transport iron when antibiotics were covalently linked, but most displayed good activity, suggesting that the siderophores did not impair the ability of the antibiotic to interact with its target. Therefore, they could not predict which sidero-phores would be good transporters of antibiotics. This indicated that the ability of siderophores to transport iron is not the sole factor in determining which siderophores would be viable drug conjugates. One interesting finding was that many of the antibiotic conjugates retained or improved their activity against *tonB* mutants, which lack receptor-mediated active transport, in both *E. coli* and *P. aeruginosa*. Since the bactericidal activity could not be attributable to active transport by a siderophore transport system, the MexAB/OprM efflux system may be contributing to the activity of these conjugates.

Bi- and Tridentate Siderophore Antibiotics

The success of single bidentate siderophore-like components as antibiotic carriers was complemented by the synthesis of conjugates with two or three bidentate ligands similar to the natural siderophore antibiotics albomycin and salmycin (structures 9 and 10 in Fig. 4). It was

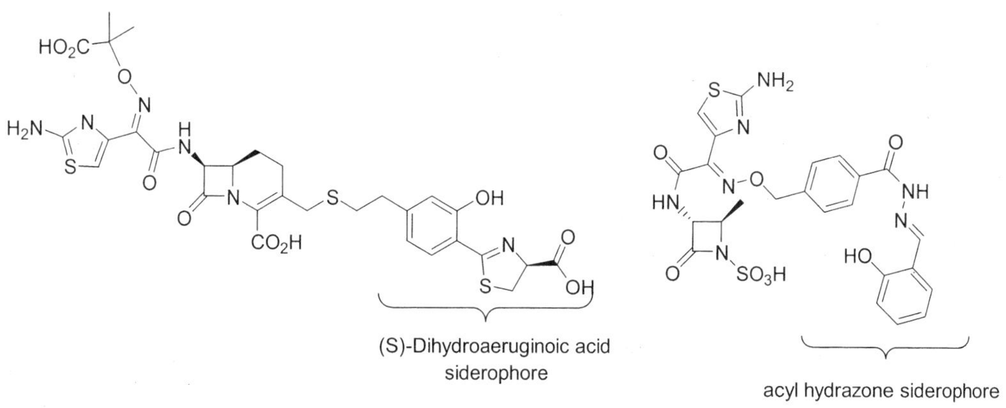

FIGURE 15 Antibiotics containing siderophore components.

hypothesized that these iron complexes would have lower entropy and could be more easily recognized by outer membrane receptors. The synthesis and study of these conjugates demonstrated that they provided more stable iron complexes and improved recognition and assimilation. Several examples are provided to display the utility of these synthetic siderophores as drug carriers.

The Miller group has synthesized numerous siderophore-drug conjugates that have excellent activity against bacteria. One of the first of these conjugates was based on the natural siderophore found in the antibiotic albomycin. This siderophore was synthesized and conjugated to the β-lactam antibiotic Lorabid (structure 33 in Fig. 16). These conjugates were found to significantly delay the growth of *E. coli* X580. Similar to the observation of mono bidentate siderophore antibiotics, mutant strains resistant to this antibiotic conjugate lacked the outer membrane receptor FhuA and showed impaired growth in vivo when implanted in the peritoneal cavities of rats. Studies of siderophore recognition requirements were done by manipulating the stereochemistry of the phenylglycine chain. These changes did not provide any alteration of the activity against bacteria, proving that the stereochemistry in this second domain was not essential to recognition and activity. Catechol spermidine-based siderophore antibiotics (structure 34 in Fig. 16) were also synthesized and found to have excellent activity against *E. coli* X580. Similar to the

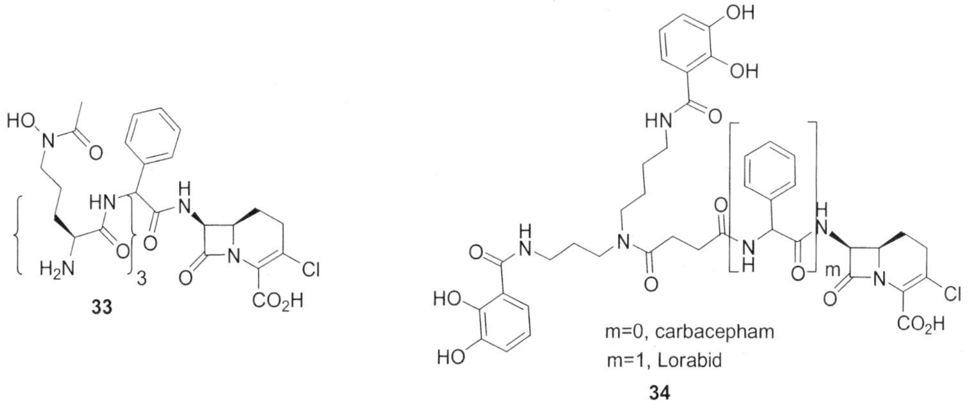

FIGURE 16 Carbacephalosporin conjugates of siderophore components.

hydroxamate siderophore in albomycins, these catechol siderophores were found to induce the formation of mutant strains that did not possess the Cir outer membrane protein associated with iron assimilation. Treatment with cocktails of both catechol and hydroxamate siderophore antibiotics generated bacterial mutants that were found to be missing both hydroxamate and catechol outer membrane receptors and therefore could not assimilate iron through either pathway.

The Heinisch group has synthesized numerous bis- and tris-catecholates based on diamino acids and found that the tris-catecholates displayed the highest siderophore activity because they had the strongest chelation ability (structures 35 and 36 in Fig. 17). These synthetic siderophores were conjugated to β-lactams such as ampicillin, amoxicillin, cephalexin, and ceflacor, and most were found to have excellent activity against gram-negative bacteria. This group also synthesized mixed bis-catecholate mono hydroxamates and found that conjugates of these siderophores also displayed good activity. In these hydroxamates, the NOH and CO groups are exchanged (reversed hydroxamate) with respect to natural hydroxamates. Recently, this group synthesized aminoacyl penicillin conjugates with acylated bis-catecholate siderophores based on secondary diamino acids (structures 37 and 38 in Fig. 18). They used acylated catecholates and the corresponding benzoxazine derivatives as siderophore components to facilitate the synthesis and to also decrease the pharmacological side effects caused by free catecholates. Several of these conjugates displayed excellent activities against several strains of bacteria including *P. aeruginosa*, which is resistant to penicillin-based antibacterial agents. This suggested that the siderophore drug conjugates were being recognized by the catechol outer membrane receptors and actively transported across the cell envelope.

Siderophore Therapeutics for Highly Resistant Targets

Pyoverdines (structures 39 and 40 in Fig. 19) are siderophores synthesized by *Pseudomonas aeruginosa*, a gram-negative bacterium that primarily infects cystic fibrosis patients and other compromised hosts. Antibiotic resistance in *P. aeruginosa* is due to the acquisition of genes encoding enzymes that attack drugs (i.e., β-lactamases), changes in the drug target, or reduced membrane permeability. The Kinzel group synthesized analogs of pyoverdine that contained the antibiotic ampicillin in order to circumvent ampicillin resistance due to problems with membrane permeability. These conju-

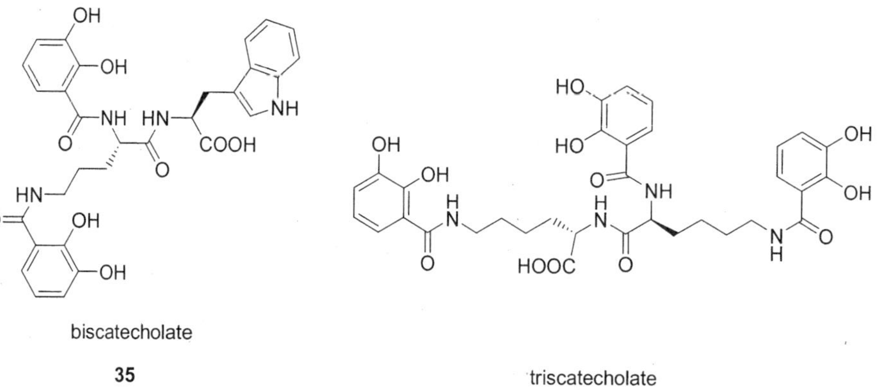

FIGURE 17 Diamino acid-based bis- and tris-catecholates.

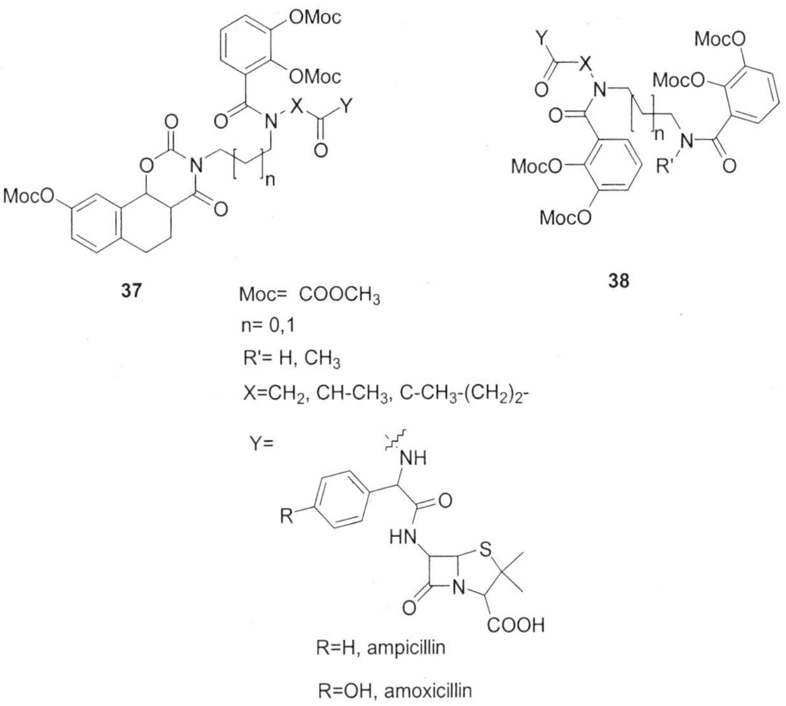

FIGURE 18 Penicillin conjugates of bis-catecholates.

gates exhibited excellent antimicrobial activities but were inactive against *P. aeruginosa* mutants that did not contain outer membrane receptors for pyoverdinas. Pyoverdine transport mutants are nonpathogenic.

The Abdallah group has synthesized pyoverdine-quinolone adducts as a potential specific therapy against *P. aeruginosa*. Pyoverdine was conjugated to quinolones that are known to target the DNA gyrase. Two different spacer arms were used between the siderophore and the drug. The first spacer was a labile methylenedioxy group, whereas the other was a stable amide linkage (structures 41 and 42 in Fig. 20). The pyoverdine antibiotic analogs transported ^{55}Fe even though there are large structural modifications in the siderophore. These drug conjugates were tested in inhibition assays with the gyrase of *E. coli* and found to have a less efficient inhibitory activity than that of the quinolone alone, probably due to the steric hindrance of the bulky siderophore. The MICs against strains of *P. aeruginosa* showed poor antibiotic activity. The activity seemed to be better with more labile linkages, and it was therefore hypothesized that the release of the amide-linked drug in the organism was not efficient and the antibiotic is partly excreted bound to the siderophore. The methylenedioxy spacer also had the disadvantage of premature hydrolysis before entry into the cell; therefore, a slightly more stable spacer is required for efficient function.

Because Desferal (desferrioxamine) is an effective antimalarial agent, the Miller group synthesized analogs of Desferal that were conjugated to the DNA-intercalating drugs nalidixic acid and anthraquinone carboxylic acid (structures 43 and 44 in Fig. 21). It was shown that these conjugates could bind to DNA via the intercalating drugs and that the iron could cause redox damage to DNA via Fenton-type processes. The nalidixic acid conjugate was also found to have excellent activity against *Plasmodium falciparum* D6 and *P. falciparum* W2 (50%

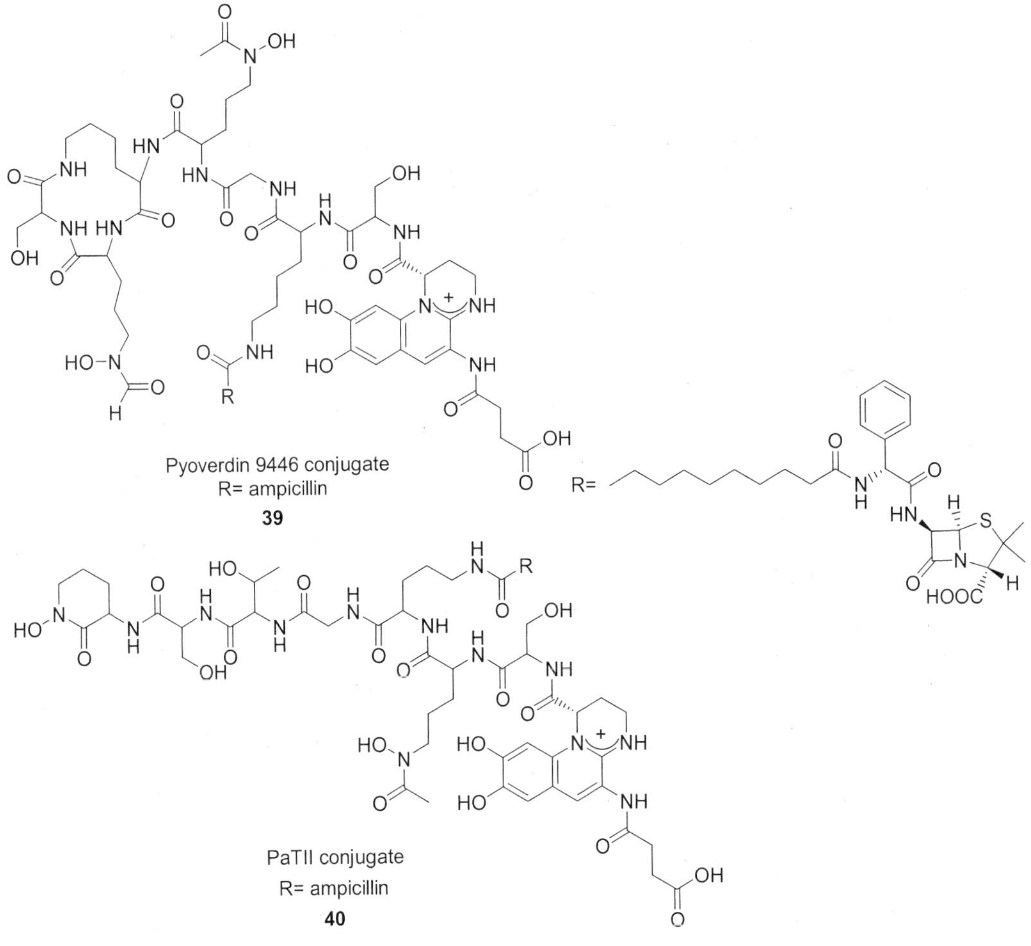

FIGURE 19 Pyoverdine–ampicillin conjugates.

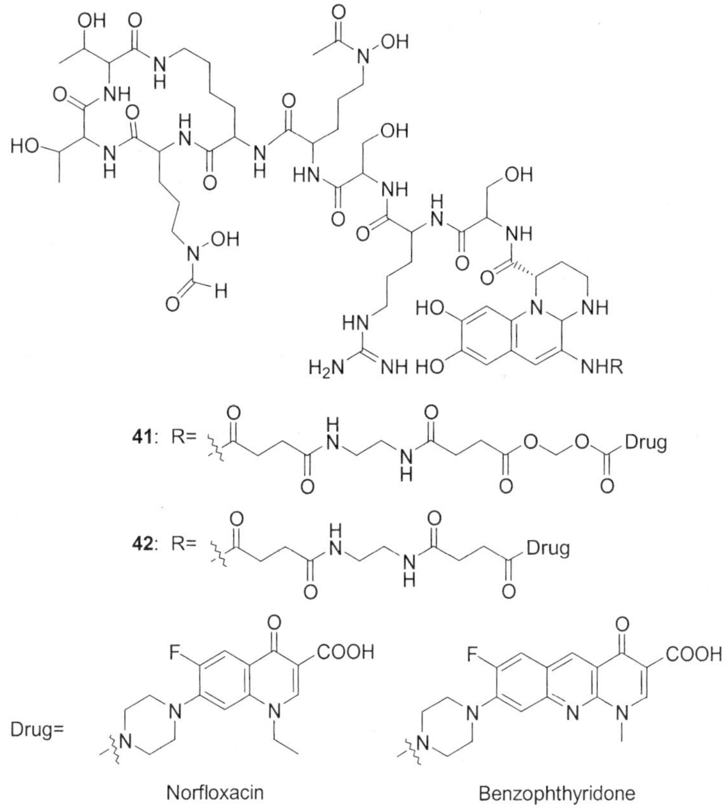

FIGURE 20 Pyoverdine-quinoline antibiotic conjugates.

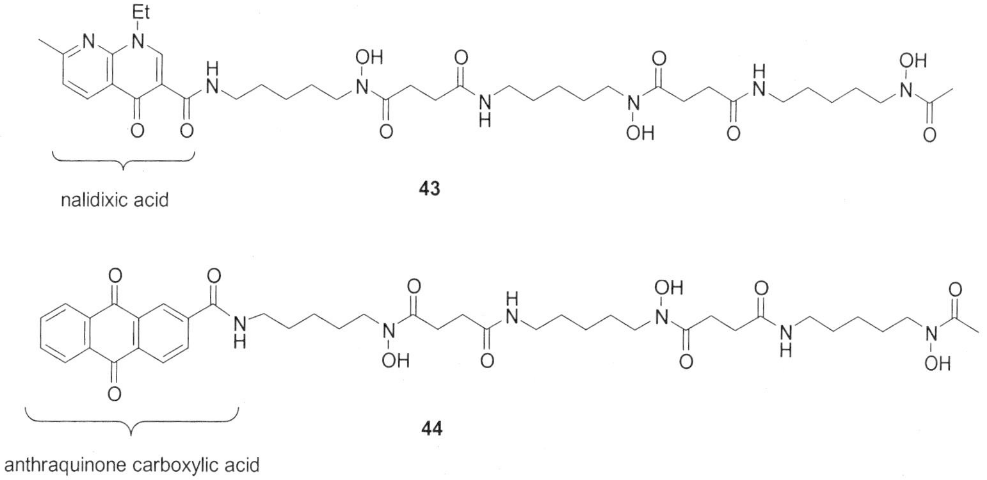

FIGURE 21 Conjugates of desferroxamine and nalidixic acid (structure acid 43) and anthraquinone carboxylic acid (structure 44).

inhibitory concentration of ~0.6 μg/ml for both strains), both of which are known to be resistant to common antimalarial drugs.

Multiwarhead Siderophore Antibiotics

Most of the antibacterial siderophores synthesized have been found to inhibit the growth of bacteria, but only a few of them have shown activity that was stronger than that of the unconjugated drug itself. A potential solution to this problem developed by the Miller group was to create siderophore conjugates that contained multiple drugs attached to one siderophore. These "multiwarhead" siderophore conjugates would enable the transport of drugs with similar or dissimilar actions. A trihydroxamate cyanuric acid siderophore (structure 45 in Fig. 22) based on the natural dihydroxamate siderophore rhodotorulic acid (structure 46 in Fig. 22) was chosen as the core. Rhodotorulic acid is a dihydroxamate-containing siderophore; therefore, it binds iron in a 3:2 ratio. It was once considered for the treatment of thalassemia, but pain at injection sites and poor oral bioavailability made it ineffective to use as a drug. The synthetic trihydroxamate siderophore analog, structure 45, was shown to bind iron in a 1:1 ratio while having little or no toxicity to mammals. The iron binding ability of the cyanuric acid siderophore was also expected to be better than that of rhodotorulic acid based on computational studies. A succinyl linker was incorporated to allow potential microbe-triggered drug release on deferration of the siderophore (Fig. 23). This type of linker is seen in the natural siderophore antibiotic salmycin. Growth promotion studies demonstrated the ability of the siderophores to promote the growth of bacteria under iron-deficient conditions. Monoerythromycylamine conjugates promoted the formation of mutant strains under iron-stressed conditions similarly to several other siderophore drug conjugates. These mutant strains are assumed to be nonpathogenic, similar to those found previously. The multiwarhead capability of these siderophores was tested by the synthesis of mono, bis, and tris 5-fluorouridine (5-FU) siderophore analogs (structures 47 to 49 in Fig. 24). 5-FU is an active metabolite of the commonly used antibacterial and antifungal agent 5-fluorocytosine. It is known to covalently inactivate thymidylate synthase and therefore inhibit the synthesis of DNA. The ability to inhibit the growth of *E. coli* X580 was found to increase with each 5-FU addition even when the total 5-FU concentration was kept constant. The tris 5-FU conjugate was found to have stronger inhibition activity than 5-FU under iron-deficient conditions. Currently, many antifungal analogs of this

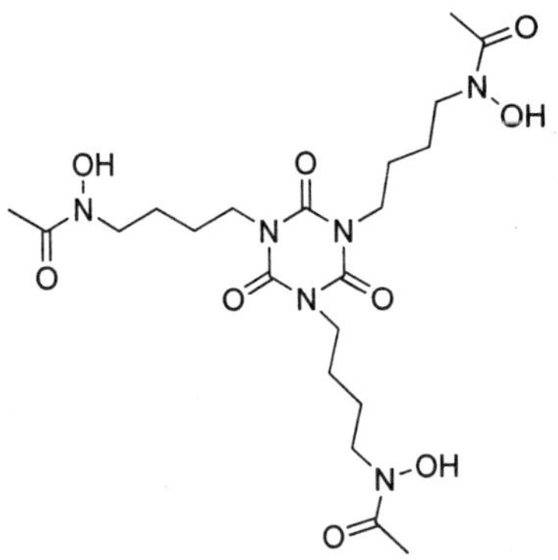

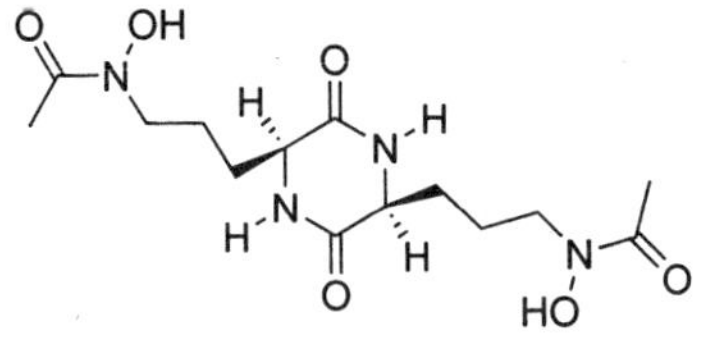

FIGURE 22 Trihydroxamic acid analog of rhodotorulic acid (structure 45) and rhodotorulic acid (structure 46).

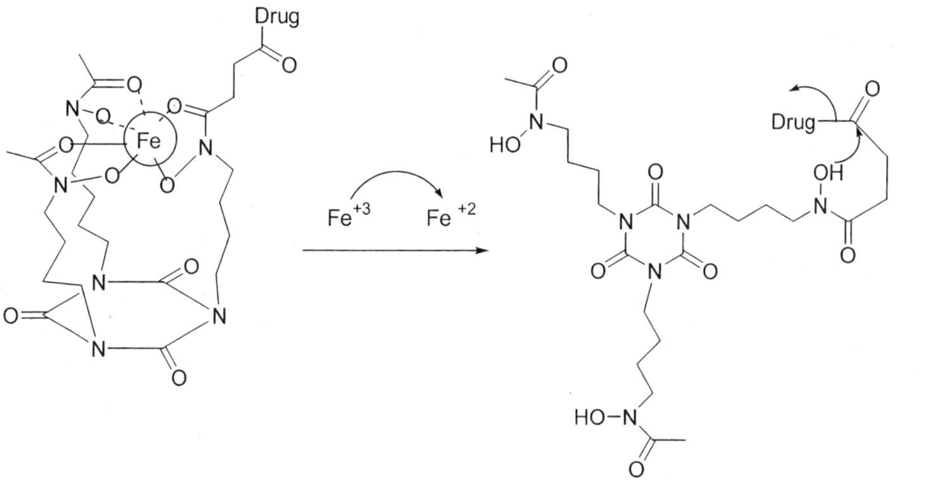
Drug
Fe
Fe+3
Fe +2
HO
HO–N
OH
Drug

FIGURE 23 Proposed redox-based drug release.

siderophore are being synthesized and tested against a variety of fungal strains.

SUMMARY

Iron is essential to the survival of most organisms, but due to its insolubility, many microorganisms utilize siderophores in order to assimilate the iron that is required for survival. These siderophores may be essential to the growth and proliferation of microorganisms and are targets in several therapeutic applications. Determination of the structures of outer membrane siderophore receptors such as FhuA and FepA by X-ray crystallography is a significant achievement and should help to further our understanding of the mechanism of iron trans-

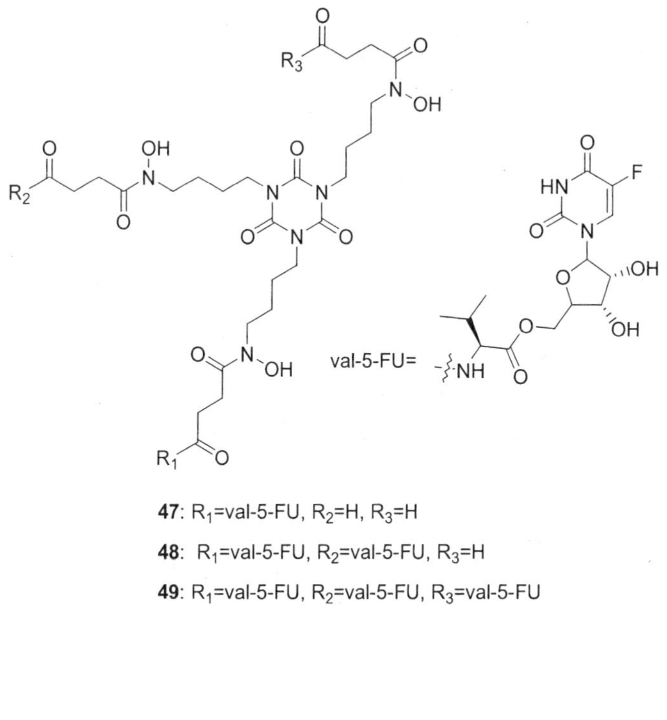
R3
R2
R1
val-5-FU=
HN
F
OH
NH
47: R1=val-5-FU, R2=H, R3=H
48: R1=val-5-FU, R2=val-5-FU, R3=H
49: R1=val-5-FU, R2=val-5-FU, R3=val-5-FU

FIGURE 24 Drug conjugates of isocyanurate-based trihydroxamate.

port. Determination of the active sites in the receptors will provide valuable information in understanding the receptor-ligand recognition on the molecular level. Further studies of siderophores will facilitate a more detailed understanding of the essential role of iron in microbial virulence as well as the potential for the efficient design and synthesis of siderophores that can be used as therapeutic agents.

An especially exciting aspect of siderophore research is the ability to use siderophores as transporters of drugs. By covalently attaching a drug to a synthetic siderophore, it is possible to achieve active transport of the drug via the iron transport system of the organism. This has allowed the development a new form of microbe-selective drug delivery with the ability to deliver drugs that cannot permeate the cell membrane by themselves. This "Trojan horse" method of drug delivery can also allow the development of alternative strategies for the treatment of multidrug-resistant microbial infections. Several siderophore-drug conjugates have been synthesized and found to have excellent potential as both antibacterial and antifungal agents. Further research in the area of siderophore-drug conjugates will address concerns about drug release for activity and novel methods to release drugs from their siderophore carriers.

SUGGESTED READING

Bazzicalupi, C., A. Bencini, A. Bianchi, V. Fusi, C. Giorgi, L. Messori, M. Migliorini, P. Paoletti, and B. Valtancoli. 1998. Synthesis and characterisation of two new catechol-based iron (III) ion-sequestering agents. *J. Chem. Soc. Dalton Trans.* **1998**:359–367.

BosneDavid, S., L. Bricard, F. Ramiandrosia, A. DeRoussent, G. Kunesch, and A. Andremont. 1997. Evaluation of growth promotion and inhibition from mycobactins and nonmycobacterial siderophores (desferrioxamine and FR160) in *Mycobacterium aurum*. *Antimicrob. Agents Chemother.* **41**:1837–1839.

Braun, V., K. Hantke, and W. Koster. 1998. Bacterial iron transport: mechanisms, genetics, and regulation. *Metal Ions Biol. Syst.* **35**:67–145.

FungTomc, J., K. Bush, B. Minassian, B. Kolek, R. Flamm, E. Gradelski, and D. Bonner. 1997. Antibacterial activity of BMS-180680, a new catechol-containing monobactam. *Antimicrob. Agents Chemother.* **41**:1010–1016.

Ghosh, M., L. J. Lambert, P. W. Huber, and M. J. Miller. 1995. Synthesis, bioactivity, and DNA-cleavage ability of desferrioxamine B-nalidixic acid and anthraquinone-carboxylic acid conjugates. *Biorg. Med. Chem. Lett.* **5**:2337–2340.

Heinisch, L., S. Wittmann, T. Stoiber, A. Berg, D. Ankel-Fuchs, and U. Möllmann. 2002. Highly antibacterial active aminoacyl penicillin conjugates with acylated bis-catecholate siderophores based on secondary diamino acids and related compounds. *J. Med. Chem.* **45**:3032–3040.

Hennard, C., Q. C. Truong, J. F. Desnotted, J. M. Paris, N. J. Moreau, and M. A. Abdallah. 2001. Synthesis and activity of pyoverdin-quinoline adducts: a prospective approach to a specific therapy against *Pseudomonas aeruginosa*. *J. Med. Chem.* **44**:2139–2151.

Hu, J., and M. J. Miller. 1997. Total synthesis of a mycobactin S, a siderophore and growth promoter of *Mycobacterium smegmatis*, and determination of its growth inhibitory activity against *Mycobacterium tuberculosis*. *J. Am. Chem. Soc.* **119**:3462–3468.

Joshi, R. R., S. M. Likhite, K. Kumar and K. N. Ganesh. 1994. DNA cleavage by Cu(II)-Desferal: identification of C1-hydroxylation as the initial event for DNA damage. *Biochim. Biophys. Acta* **1199**:285–292.

Kicic, A., A. C. Chua, and E. Baker. 2001. Effect of iron chelators on proliferation and iron uptake in hepatoma cells. *Cancer* **92**:3093–3110.

Kim, Y. Z., J. C. Lim, J. H. Yeo, C. S. Bang, S. S. Kim, T. H. Lee, S. H. Oh, Y. C. Moon, and C. S. Lee. 1996. Synthesis and antibacterial activities of novel C(7)-catechol-substituted cephalosporins. *J. Antibiot.* **49**:496–498.

Kinzel, O., R. Tappe, I. Gerus, and H. Budzikiewicz. 1998. The synthesis and antibacterial activity of two pyoverdin-ampicillin conjugates, entering *Pseudomonas aeruginosa* via the pyoverdin iron uptake pathway. *J. Antibiot.* **51**:499–507.

Kline, T., M. Fromhold, T. E. McKennon, S. Cai, J. Treiberg, N. Ihle, D. Sherman, W. Schwan, M. J. Hickey, P. Warrener, R. Witte, L. L. Brody, L. Goltry, L. M. Barker, S. U. Anderson, S. K. Tanaka, R. M. Shawar, L. Y. Nguyen, M. Langthorne, A. Bigelow, L. Embruscado, and E. Naeemi. 2000. Antimicrobial effects of novel siderophores linked to β-lactam antibiotics. *Bioorg. Med. Chem.* **8**:73–93.

Liu, Z. D., and R. C. Hider. 2002. Design of clinically useful iron (III)-selective chelators. *Med. Res. Rev.* **22**:26–64.

Lu, Y., and M. J. Miller. 1999. Synthesis and studies of multiwarhead siderophore-5-fluorouridine conjugates. *Bioorg. Med. Chem.* **7**:3025–3038.

Miller, M. J., F. Malouin, E. K. Dolence, C. M. Gasparski, M. Ghosh, P. R. Guzzo, B. T. Lotz, J. A. McKee, A. A. Minnick, and M. Teng. 1993. Iron transport-mediated drug delivery, p. 135–159. *In* P. H. Bentley and R. Ponsford (ed.), *Recent Advances in the Chemistry of Anti-infective Agents*. Royal Society of Chemistry special publication no. 119. The Royal Society of Chemistry, London, United Kingdom.

Minnick, A. A., J. A. McKee, E. K. Dolence, and M. J. Miller. 1992. Iron transport-mediated antibacterial activity of and development of resistance to hydroxamate and catechol siderophore-carbacephalosporin conjugates. *Antimicrob. Agents Chemother.* **36:**840–850.

Neilands, J.B. 1995. Siderophores—structure and function of microbial iron transport compounds. *J. Biol. Chem.* **270:**26723–26726.

Nudelmann, R., O. Ardon, Y. Hadar, Y. Chen, J. Libman, and A. Shanzer. 1998. Modular fluorescent-labeled siderophore analogues. *J. Med. Chem.* **41:**1671–1678.

Ratledge, C., and L. G. Dover. 2000. Iron metabolism in pathogenic bacteria. *Annu. Rev. Microbiol.* **54:**881–941.

Raymond, K. N. 1994. Recognition and transport of natural and synthetic siderophores by microbes. *Pure Appl. Chem.* **66:**773–781.

Roosenberg, J. M., Jr., Y. M. Lin, Y. Lu, and M. Miller. 2000. Studies and syntheses of siderophores, microbial iron chelators, and analogs as potential drug delivery agents. *Curr. Med. Chem.* **7:**159–197.

Snow, G. A. 1970. Mycobactins: iron-chelating growth factors from mycobacteria. *Bacteriol. Rev.* **34:**99–125.

Vergne, A. F., A. J. Walz, and M. J. Miller. 2000. Iron chelators from mycobacteria (1954–1999) and potential therapeutic applications. *Nat. Prod. Rep.* **17:**99–116.

Wittman, S., M. Schnabelrauch, I. Scherlitz-Hofmann, U. Möllmann, D. Ankel-Fuchs, and L. Heinisch. 2002. New synthetic siderophores and their β-lactam conjugates based on diamino acids and dipeptides. *Biorg. Med. Chem.* **10:**1659–1670.

IRON TRANSPORT AND ECOLOGY

ECOLOGY OF SIDEROPHORES

Günther Winkelmann

28

When examining siderophores at a chemical or molecular level, we sometimes forget to appreciate the biology of siderophores at the level of the environment, where they play a crucial role in maintaining active ecosystems, even at very low concentrations of soluble iron. Ecology describes the relationship between an organism and the environmental factors that allow its growth, development, and, finally, evolution. However, environments are highly variable and consist of a large number of interacting factors such as temperature, pH, redox potential, osmotic pressure, water activity, salinity, and general and essential nutrients. Most often, simply the presence or absence of certain nutrients defines an ecological niche where certain organisms can grow and others are prevented from colonizing. This chapter attempts to illustrate the occurrence of siderophores within different bacterial genera and explains the relationship between environmental factors and the biological function of siderophores, which may be summarized under the term "ecology of siderophores."

The ecology of siderophores may be considered in the context of an ecological niche defined by low iron availability and by the expression of high-affinity siderophore-mediated iron transport systems by the organisms that occupy this niche. Iron is one of the most crucial growth-limiting factors for virtually all aerobic microorganisms. Only a few exceptions are known, including certain lactobacilli, streptococci, and *Borrelia* species that have to some extent replaced iron with manganese. Despite the importance of iron, this element is not readily available due to its low solubility in the prevailing aerobic environment. We now know that the appearance of oxygen and the subsequent formation of insoluble iron hydroxides in the oceans, which occurred when cyanobacteria evolved oxygenic photosynthesis about 3 billion years ago, represents the most challenging event during evolution. In fact, many of the microorganisms that existed in the previous reducing atmosphere are living today hidden in water containing sulfide, anoxic zones of soils and lakes, deep marine areas, or even in the enteric tracts of higher animals. Since ferrous iron is readily soluble in a reducing environment, there is no need to assume the presence or synthesis of Fe^{3+}-binding siderophores in anaerobes, although specific transport systems for ferrous iron do exist.

In addition to being scarce, ferric iron can be highly toxic to aerobic microorganisms due

Günther Winkelmann, Institute of Microbiology, University of Tübingen, Tübingen, Germany.

Iron Transport in Bacteria, Edited by Jorge H. Crosa, Alexandra R. Mey, and Shelley M. Payne
© 2004 ASM Press, Washington, D.C.

to the Haber-Weiss-Fenton reaction, which yields peroxides and hydroxyl radicals leading to the degradation of enzymes, DNA, RNA, and membrane lipids. To prevent cellular damage, most aerobic microorganisms contain catalase and superoxide dismutase. Nevertheless, Fe^{3+} is dangerous as well as essential to most living cells, and the balance between these two properties is maintained by the iron-binding, transport, and storage events of cell metabolism, in which siderophores play a major role. The focus of this chapter is on selected aerobic, chemoorganotrophic bacterial groups, which represent only a part of the currently known bacteria.

SIDEROPHORE PRODUCTION IN SOIL

Soil is a rich source of different bacterial genera, of which the streptomycetes (*Actinomycetales*), a group of gram-positive, GC-rich, spore-forming bacteria that show a characteristic mycelium-like growth, have attracted special attention because of their ability to synthesize important secondary metabolites and antibiotics. The mesophilic phase of degradation of organic matter in soil is dominated by the autochthonous microbial population of the streptomycetes, which are widespread in the upper zones of all kinds of terrestrial and marine sediments, although other spore-forming rods, such as *Bacillus*, and various non-spore-forming gram-positive genera, such as *Arthrobacter*, *Rhodococcus*, and *Nocardia*, are present as well.

Streptomycetes are known to utilize polymeric carbohydrates such as starch, chitin, and cellulose, which are abundant in the upper layers of soils. Thus, rotting plant parts and fungal and insect material are the natural substrates of most streptomycetes. In certain cases when compost has accumulated, the mesophilic microbial population changes to a thermophilic population (>45°C) dominated by the genera *Bacillus* and *Thermus*. However, strains of *Streptomyces* and *Thermoactinomyces* may also be present.

While the presence of easily degradable organic substrates is a precondition for rapid growth of streptomyces in soil, iron availability and siderophore production seems to be of equal importance. Rapid outgrowth from spores requires the presence of highly effective iron acquisition systems. Although not yet experimentally proven, the production of siderophores by *Streptomyces* species and fungi probably occurs in the upper, aerobic zones of soils, where organic matter is degraded and iron is bound to humic compounds as illustrated in Fig. 1. The production of siderophores leads to growth promotion of the producing strains as well as of some other microbial populations that live in the same habitat and are able to utilize exogenous siderophores. This gives siderophores a much wider impact in the environment.

The detection of siderophores from *Streptomyces pilosus* dates back to 1960, when a series of structurally different compounds containing iron-binding oxygenated amines, called ferrioxamines, were isolated. The ferrioxamines are powerful ferric chelators ($K \approx 10^{30}$), forming red-brown complexes due to the typical ligand-to-metal charge transfer band at 430 to 440

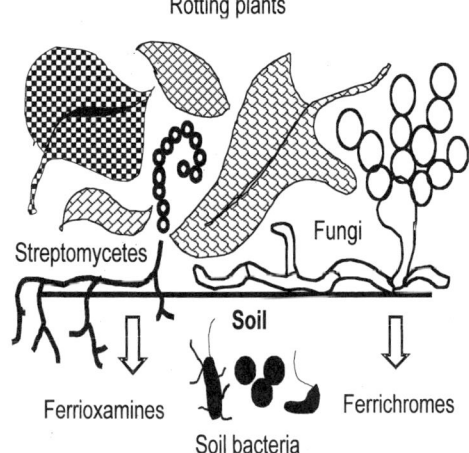

FIGURE 1 Siderophore production in humic soil. Streptomycetes and fungi are the predominant microorganisms in the upper layer of humic soil, producing desferrioxamines and desferrichromes, respectively, under iron limitation. Siderophores solubilize iron bound to humic matter and subsequently support the growth of the producing strains as well as other soil microorganisms.

nm ($\varepsilon \approx 2800$ M^{-1} liter^{-1}). Besides the linear ferrioxamine B, the cyclic ferrioxamine E ($K = 10^{32}$) is produced in considerable amounts in some strains. Besides *S. pilosus*, several other species, including *S. glaucescens* and *S. olivaceus*, synthesize ferrioxamines. *S. pilosus* produces a number of desferrioxamines (A through H), of which desferrioxamine B is produced industrially and marketed as the mesylate salt (Desferal) by Novartis as a drug for the treatment of acute and chronic iron overlaod, as well as for removing dissolved aluminum. Thus, iron overload due to frequent transfusions of HbE/β-thalassemia patients is prevented by chelation therapy with Desferal.

Ferrioxamines are not restricted to the genus *Streptomyces* but also occur in the related genera *Micromonospora* and *Nocardia*. Although the exact classification of the *Nocardia* species was not certain at the time, the first name given to the isolated compound was nocardamine, which turned out later to be identical to ferrioxamine E, isolated from *Streptomyces* strains. Ferrioxamine E was subsequently found in *Pseudomonas stutzeri*, a soil saprophyte member of the nonfluorescent pseudomonads. This is surprising since the fluorescent pseudomonads produce predominantly the fluorescent peptide siderophores pyoverdines, which have recently been used as a taxonomic marker (see chapter 29). It is interesting that ferrioxamine E is also produced in large amounts among the enterobacterial genera *Pantoea*, *Erwinia*, and *Enterobacter*, which are common in soil and plant material.

Transport studies of *S. pilosus* revealed that ferrioxamine B and its N-acetyl derivative, ferrioxamine D$_1$, are superior to the cyclic ferrioxamines E and D$_2$. Crystal structures of ferrioxamines B and E showed that the molecules crystallize as low-energy racemic mixtures of Λ-N-cis,cis and Δ-N-cis,cis coordination isomers. In addition, kinetically stable chromic *cis* and *trans* isomers have been prepared. The chromic isomers acted as inhibitors of uptake of all ferrioxamines tested, excluding any geometrical preference during recognition and transport in *S. pilosus*. Since ferrioxamines in solution are mixtures of geometric (*cis* and *trans*) and optical (λ and Δ) isomers, the observed preference for ferrioxamine B transport in *S. pilosus* has been explained by different shielding of the carbonyl and oxime faces, which correlate well with the shielded oxime faces in ferrichromes. However, as long as the involved transport proteins in the gram-positive streptomycetes are unknown, this explanation remains speculative. A comparative study with cloned FoxA receptors from the gram-negative species *Erwinia herbicola* and *Yersinia enterocolitica* did not demonstrate preferred uptake of any ferrioxamine, arguing against a mechanism where the oxime versus carbonyl faces play a role in recognition in enterobacteria.

Although ferrioxamines are well known for solubilization and transport of ferric iron into microbial cells, their intracellular fate and iron-releasing mechanisms are largely unknown. The pH-independent redox potential of ferrioxamine E has been determined as -477 mV versus Nernst hydrogen electrode, which initially was considered too negative to be reduced by biological reductants such as NADPH, NADH, or FADH$_2$. However, it was suggested that the redox equilibrium within the cell can be altered by the presence of ferrous iron-specific ligands by, pH reduction, or by a hydrophobic environment such as that found within membranes. Thus, a biological reduction of ferrioxamines may well be explained under nonequilibrium conditions by a coupling process.

Although exact measurements are still lacking, ferrioxamines appear to be widespread in many soils due to the ubiquitous occurrence of the strepomycetes in decaying materials. However, fungi of the Ascomycetes and Basidiomycetes might likewise contribute to the total hydroxamate content in soil by producing ferrichromes and other environmentally resistant hydroxamate siderophores. This has been documented by growth promotion tests using the hydroxamate-dependent *Microbacterium flavescens* JG9 strain. Fungi might dominate the upper layers of soil, provided that there is enough organic material available. Indeed, fungi are the prevailing organisms in soils containing plenty of decaying plant material leading to a pH of <5 which is generally unfavorable for streptomycetes. Thus, the pH finally determines the ecology and distribution of various siderophores in soils.

Ferrioxamine biosynthesis and transport systems are not restricted to the streptomycetes. As mentioned above, certain enterobacterial genera (*Enterobacter, Erwinia, Pantoea,* and *Hafnia*) are able to synthesize and utilize ferrioxamines under iron limitation, suggesting that enterobacteria not occurring in the intestine have maintained the ability to produce ferrioxamines while enterobacteria, such as *Escherichia coli, Shigella, Salmonella,* and *Yersinia,* living in the gut seem to have lost this ability. An outer membrane transport protein (FoxA) specifically designed for the uptake of ferrioxamines in enterobacteria was first found in *Erwinia herbicola,* a plant-associated enterobacterial species that is now designated *Pantoea agglomerans.*

SIDEROPHORES AND PLANTS

While siderophore uptake systems may be superior to reductive mechanisms in freshwater, marine, and calcareous soil ecosystems, iron acquisition by dicotyledonous plants can occur via reductive mechanisms and proton release by the root epidermis. As is the case in well-buffered alkaline soil, proton release mechanisms for iron acquisition in alkaline aqueous environments seem to be difficult. To overcome such an environmental stress, monocotyledonous graminaceous plants have developed so-called phytosiderophores, which may have helped to colonize unfavorable terrestrial areas worldwide. Although plants are generally not able to use fungal and bacterial siderophores as iron sources, there are reports showing optimal utilization by mono- and dicotyledonous plants of some fungal mono- and dihydroxamates, which originate from partial degradation of fungal trihydroxamate siderophores.

SIDEROPHORES IN THE AQUATIC ENVIRONMENT

The microbial population of freshwater is to some degree very similar to that of soil, since there is no strict separation between the two habitats and many bacteria are simply washed out from the soil (transient population). In addition, freshwater lakes and surface water contain large numbers of resident or indigenous species, such as *Pseudomonas, Azomonas, Aeromonas,* and *Alcaligenes* species. In contrast, soil is rich in spore-forming *Bacillus, Nocardia,* and *Streptomyces* and non-spore-forming irregular *Arthrobacter, Rhodococcus,* and *Mycobacterium* species. The aquatic habitat is often dominated by photosynthetic bacteria of the nonsulfur and sulfur purple groups. Zones of different redox potentials give rise to specific chemolithotropic bacterial populations using inorganic compounds as electron donors. Thus, the freshwater ecosystem is a rich source of microorganisms. However, their source of iron nutrition is largely unknown. The marine offshore environment, containing species of *Alteromonas, Pseudoalteromonas, Shewanella,* and *Vibrio,* is quite distinct due to its higher salinity and different nutritional composition. There are several reports of the occurrence of novel siderophores from marine bacteria, which are treated in chapter 16. Some of the opportunistic pathogenic marine vibrios seem to profit from high-affinity hydroxamate siderophore uptake systems, which result in increased virulence. Interestingly, the freshwater cyanobacteria, which perform oxygenic photosynthesis, produce siderophores. Thus, *Anabaena* strains synthesize schizokinen and anachelins. Due to their continued oxygen production under illumination, iron utilization by reductive transport systems is difficult to achieve and chelation of ferric iron by siderophores seems to be an effective means of sequestering iron for growth in oxygen-containing aquatic environments.

SIDEROPHORE PRODUCTION IN THE INTESTINAL TRACT

Many bacteria that inhabit the intestinal tracts of vertebrate hosts produce siderophores, although the question whether ferric siderophore transport is required in the gut still remains. The lower part of the intestine is largely an anaerobic environment, and *E. coli* contains a ferrous iron transport system, encoded by the *feoABC* genes, of which the FeoB protein is located in the cytoplamic membrane and is required for ferrous iron transport into the cell (see chapter 12). The ecological significance of such a system has been investigated by using enterochelin-producing and enterochelin-nonproducing *E. coli* K-12 *feoB* mutants. Only the parent strain (*fepA*$^+$ *feoB*$^+$) was able to survive for a longer time in the mouse intestine. The *feoB* mutants were unable to survive in the mouse intestine, indicating that ferrous iron uptake is an important route for iron acquisition in the anaerobic intestine. Another experiment using *feoB* and *tonB* mutants of *Salmonella enterica* serovar Typhimurium revealed that *feoB* mutants were outcompeted by the wild type during mixed colonization of the mouse intestine, confirming the view that ferrous iron uptake is essential for *Salmonella* and *E. coli* in the anaerobic intestine. Mutations in TonB, the energy-transferring protein for all the high-affinity outer membrane receptors involved in iron transport, attenuated the infection by the intragastric route, indicating that exogenous hydroxamates and/or other TonB-dependent iron sources may contribute to a successful intestinal colonization. Nevertheless, a *tonB feoB* double mutant was still able to infect the liver and spleen when given intraperitoneally, implying that additional iron acquisition systems are present in *Salmonella* (see chapter 14).

The presence in *Salmonella* of a ferrioxamine uptake system in addition to the FepA, Fiu, and Cir systems may enable *S. enterica* to colonize a wider area in the gut and particularly the ileum, which is generally more acidic than the colon. The ferrioxamines are more stable than the catecholates in acidic environments. A ferrioxamine receptor, FoxA, is present in several enteric bacteria, although not in *E. coli*, and has been characterized in *Y. enterocolitica* and *S. enterica*. The occurrence and origin of ferrioxamines in the gut are still unclear, since the predominant members of the gut flora do not produce these compounds, but it is likely that ferrioxamines in the gut originate from the food intake. Ferrioxamines originating from soil streptomycetes, which can overcome the acidic barrier of the stomach, may be utilized by *S. enterica* in the proximal small intestine, a region with a lower pH. Intake of dietary ferrioxamines might be low in humans but significant in other animals, like cattle, swine, ducks, and hens, which take up large amounts of mud together with nutrients.

Ferrioxamine uptake systems also may support the survival of *Salmonella* and other enterics living in the environment where ferrioxamines are abundant. Hence, some enterobacteria may survive when their habitats changes between the soil and the gut. As shown in a previous section, a number of ferrioxamine-producing enterobacterial genera are still plant associated and are not found in the gut, suggesting that this is an evolutionarily old habitat.

Aerobactin, another siderophore widely produced in the family *Enterobacteriaceae*, is regarded as a better iron scavenger in serum than is enterobactin, which may be bound up by albumin, and may also be more effective in the more acidic areas of the gut. Besides iron uptake via ferrioxamines, enterobactin, or aerobactin, *Salmonella* species are able to produce salmochelins, a newly characterized class of catecholate siderophores taken up by the IroN outer membrane receptor.

DEGRADATION OF SIDEROPHORES

The bacterial ferrioxamines and the fungal ferrichromes represent environmentally highly stable iron complexes. Although down regulation of siderophore biosynthesis keeps the siderophore content of iron-containing soils low, early measurements using a biological assay based on growth promotion of a ferric hydroxamate-requiring strain of *Microbacterium flavescens* JG9 (previously erroneously classified as *Arthrobacter* or *Aureobacterium*) showed the presence of detectable amounts of siderophores in certain soils. However, reports of the microbial degradation of ferric siderophores are scarce and were initially focused on ferrichrome-type siderophores by Neilands' group. These early studies showed that *Pseudomonas* species are able to degrade the peptide backbone of ferrichrome and ferrichrome A. Recently, studies of the degradation of the iron-free desferrioxamines have been reported. Although the occurrence of desferrisiderophores in a natural environment seems to be unusual, a bacterial isolate, DFBC5, tentatively identified as a *Rhizobium loti*-like bacterium, has been isolated that can degrade desferrioxamine B via monohydroxamates by a hydroxamate hydrolase characterized as a microbial serine protease. Another isolate (ASP-1), identified as *Azospirillum irakense*, was able to degrade desferrioxamine B and several other linear and cyclic desferrioxamines. Bacteria of the genus *Azospirillum* are nitrogen-fixing, microaerophilic soil bacteria that are commonly found associated with roots of grasses. The degradation activity was not detected in related species such as *A. lipoferum*, *A. brasilense*, or *A. amazonense* and seem to be a characteristic feature of *A. irakense* species. The ecological significance of desferrioxamine-degrading bacteria is unclear, since desferrioxamines may be rapidly converted to the ferric complexes in most natural soils containing ferric hydroxides. Ferrioxamines are known to function as siderophores in *Erwinia*, *Pantoea*, and *Enterobacter* species, which have been isolated from the rhizosphere of grasses, wheat, and barley. Desferrioxamines may occur in large amounts in several different situations. First, the increased utilization of iron at the roots of grasses may produce an ecological niche where ferrioxamines might be present in larger amounts as iron-free molecules. The resulting desferrioxamines may subsequently be utilized by *A. irakense* as an additional carbon and nitrogen source, as illustrated in Fig. 2. Another possibility is that accompanying rhizosphere bacteria, such as *Pseudomonas stutzeri* a strong denitrifyer in soil and a known producer of desferrioxamines under iron deficiency, coexist with *A. irakense* in this habitat. A third possibility is that the *Azospirillum* enzymes that split the desferrioxamines may have other, as yet unknown, substrates. The amide bonds in desferrioxamines differ from the corresponding amide bonds in peptides by their achiral nature. Therefore, the desferrioxamine-

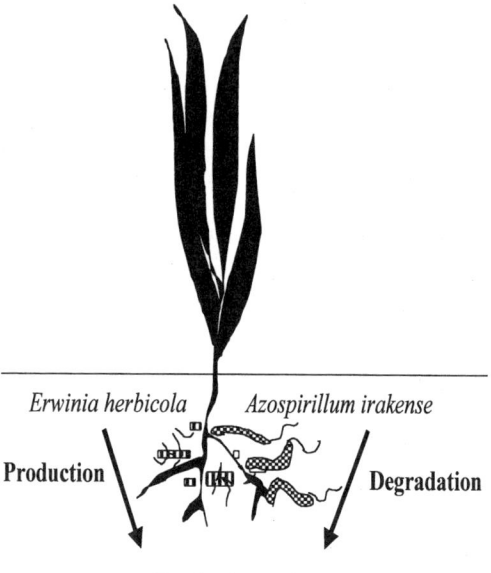

FIGURE 2 Proposed model for the production and degradation of desferrioxamines by *E. herbicola* and *A. irakense* on the root surface. Species of both microorganisms have been isolated from the root surface of grasses, wheat, and barley.

degrading activity might represent an amidase-like hydrolytic enzyme that may play other roles in the metabolism of *A. irakense*. Determination of the precise role of desferrioxamine-degrading *A. irakense* strains in microbial habitats on root surfaces requires further analysis.

ECOLOGICAL IMPACT OF SIDEROPHORES

The ecological impact of siderophore production and utilization is not completely understood, and much more needs to be discovered in order to comprehensively describe the ecology of siderophores in soil, surface waters, or marine environments. Siderophores do not accumulate in large amounts in nature, and their occurrence is restricted to environments of low iron availability. In laboratory cultures, growth of microorganisms cannot be increased simply by the addition of iron if the pH of the medium is alkaline. Even in an acidic environment, microorganisms may suffer from a shortage of iron due to the lack of iron-binding siderophores. Microorganisms first need to sense iron limitation for subsequent expression of the biosynthesis of siderophores and their cognate transport systems. Siderophore production is not observed in rich nutrient media because of the tight regulation and the rapid consumption of siderophores during growth. However, in low-iron laboratory media, overproduction of siderophores is a general phenomenon among microorganisms. The amount of siderophores produced exceeds the amount that is really needed for the growth of the producing culture. From this observation, it may be assumed that microbial pioneers start to colonize a low-iron habitat by excretion of siderophores, which in turn leads to a succession of further microbial populations that do not necessarily synthesize siderophores in large amounts but have already expressed a variety of siderophore transport systems within their membranes. For example, low-iron spent media that have been used to grow fungi subsequently support the growth of various bacteria due to the presence of fungal siderophores such as ferrichromes. The reverse is often observed with yeasts that are unable to synthesize siderophores but grow well aerobically in low-iron media when enterobactin is present. Thus, siderophore-nonproducing organisms may profit from the preceding settlement of siderophore producers. This kind of mutualism, where siderophores from one species improve the growth of other species, may be either facultative, when siderophores are not absolutely required but improve the growth, or obligate, when siderophores are absolutely required for growth.

Due to their ability to utilize the prevailing organic matter and to synthesize and excrete siderophores, several bacterial groups seem to dominate certain habitats such as soil, water, plants, or animal bodies. However, iron solubilization and products excreted by pioneer species may allow new species to colonize low-iron environments. There is every reason to assume that the buildup of siderophores and other nutrients allows successive colonization by different bacterial and fungal species, ultimately resulting in a stable indigenous microbial population. Although the bacterial population size may vary due to inadequate carbon and nitrogen supplies, the contribution of siderophores to maintenance of homeostasis helps ensure that microbial communities remain stable. Bacteria possessing inefficient siderophore transport systems are outcompeted, as has been shown by numerous experiments with pathogens in the plant rhizosphere and in the animal intestine. The currently known bacterial siderophores and their producing organisms are listed in Table 1. An assignment to their main ecological habitat is also made, although the borderlines between these habitats may not be sharp.

From an ecological point of view, the biosynthesis of hydroxamate-type siderophores in soil seems advantageous over the synthesis of catecholates, because trihydroxamates are generally stable within a pH range of 2 to 8 while the ferric complexes of catecholates are unstable below pH 5. Therefore, hydroxamate biosynthesis is common among bacteria and fungi living in soil or marine environments where pHs of less than 5 may be encountered or where oxidation of catecholates may prevent efficient iron acquisition. As shown in Fig. 3, the biosynthesis of most hydroxamate siderophores of fungal and bacterial origin starts with N^{δ}-hydroxyornithine and results in various linear and cyclic hydroxamate siderophores of high environmental stability. Ornibactins are found in the nonfluorescent pseudomonad-like *Burkholderia cepacia* strains, colonizing low-pH plant rhizospheres, while exochelins have been detected in various mycobacterial species known to survive environmentally unfavorable conditions and acid-containing phagosomes. The dihydroxamate siderophores alcaligin, bisucaberin, and putrebactin have been detected in *Alcaligenes*, *Vibrio* and *Alteromonas*, and *Shewanella* species, respectively, living in marine or freshwater environments, indicating that hydroxamates also play an important role in aqueous environments. However, *Vibrio* strains are also known to grow in alkaline environments, and this

TABLE 1 Siderophores and producing organisms grouped according to their main ecological habitat

Siderophore and habitat	Organism
Soil and surface water	
Agrobactin	*Agrobacterium tumefaciens*
Aminochelin	*Azotobacter vinelandii*
Amonabactins	*Aeromonas hydrophila*
Amycolachrom	*Amycolatopsis orientalis, A. azurea, A. mediterranei*
Anachelins	*Anabaena cylindrica*
Arthrobactin	*Arthrobacter pascens*
Azotobactin	*Azotobacter vinelandii*
Azotochelin	*Azotobacter vinelandii*
Azoverdin	*Azomonas macrocytogenes*
Carboxymycobactin	*Mycobacterium* spp.
Cepabactin	*Burkholderia cepacia*
Cepachelin	*Burkholderia cepacia*
Chryseomonin	*Chryseomonas luteola*
Corrugatin	*Pseudomonas corrugata*
Corynebactin	*Corynebacterium glutamicum, Bacillus subtilis*
Enantio-Rhizoferrin	*Ralstonia pickettii*
Enterobactin	*Klebsiella* spp., *Enterobacter* spp., *Erwinia* spp.
Exochelin	*Mycobacterium* spp.
Ferribactin	*Pseudomonas fluorescens*
Ferrioxamines (A to H)	*Streptomyces pilosus, S. griseus, S. griseoflavus, S. aureofaciens, S. coelicolor, S. lavendulae, S. olivaceus*, etc.
Ferrioxamines (D_2, E, and G)	*Enterobacter agglomerans, E. cloacae, E. intermedium, Ewingella americana, Pantoea agglomerans, P. dispersa, Hafnia alvei*
Heterobactins (A and B)	*Rhodococcus erythropolis*
Itoic acid	*Bacillus subtilis*
Maduraferrin	*Actinomadura madurae*
Mycobactin	*Mycobacterium* spp.
Myxochelins	*Angiococcus disciformis*
Nannochelins	*Nannocystis excedens*
Nocobactins	*Nocardia* spp.
Ornibactins (C_4, C_6, and C_8)	*Burkholderia cepacia, B. vietnamiensis*
Parabactin	*Paracoccus denitrificans*
Protochelin	*Azotobacter vinelandii*
Pyochelin	*Pseudomonas* spp.
Pyoverdines	*Pseudomonas fluorescens, P. putida, P. aeruginosa*, etc.
Quinolobactin	*Pseudomonas fluorescens*
Schizokinen	*Bacillus megaterium, Anabaena* spp., *Ralstonia solanacearum*
Serratiochelin	*Serratia marcescens*
Spirillobactin	*Azospirillum brasilense*
Plant pathogens and plant-associated bacteria	
Achromobactin	*Erwinia chrysanthemi*
Chrysobactin	*Erwinia chrysanthemi, E. carotovora* subsp. *carotovora*
Ferrorosamine A	*Erwinia rapontici*
Ferrioxamines (D_2, E, and G)	*Erwinia herbicola, E. ananas, E. stewartii, E. uredovora, E. amylovora*
Pyoverdines	*Pseudomonas* spp. (rhizosphere)
Rhizobactin 1021	*Rhizobium meliloti*
Vicibactin	*Rhizobium leguminosarum*

(Continued)

TABLE 1 *Continued*

Siderophore and habitat	Organism
Marine bacteria	
Aerobactin	*Vibrio hollisae, V. mimicus*
Alterobactins (A and B)	*Alteromonas luteoviolacea*
Anguibactin	*Vibrio anguillarum*
Aquachelins (A to D)	*Halomonas aquamarina, Halomonas* spp.
Bisucaberin	*Alteromonas haloplanktis, Vibrio salmonicida*
Ferrioxamine G	*Vibrio* spp.
Fluvibactin	*Vibrio fluvialis*
Marinobactins (A to E)	*Marinobacter* spp.
Vibrioferrin	*V. parahaemolyticus*
Vulnibactin	*V. vulnificus*
Human and animal pathogens	
Acinetobactin	*Acinetobacter baumannii, A. haemolyticus*
Acinetoferrin	*A. haemolyticus*
Aerobactin	*Escherichia coli, Shigella flexneri, S. boydii, Salmonella* spp., *Yersinia enterocolitica*
Alcaligin	*Bordetella pertussis, B. bronchiseptica*
Bisucaberin	*Vibrio salmonicida*
Carboxymycobactins	*Mycobacterium tuberculosis, M. avium, M. bovis,* etc.
Enterobactin	*Escherichia coli, Klebsiella* spp., *Salmonella enterica* serotype Typhimurium, *Salmonella* spp., *Serratia marcescens, Shigella dysenteriae, S. sonnei*
Exochelins	*Mycobacterium tuberculosis, Mycobacterium* spp.
Mycobactins	*Mycobacterium tuberculosis, M. avium, M. bovis,* etc.
Ornibactins	*Burkholderia cepacia*
Pyoverdines	*Pseudomonas aeruginosa*
Salmochelins	*Salmonella enterica, Escherichia coli* (uropathogenic)
Staphyloferrin (A,B)	*Staphylococcus aureus, S. hyicus,* etc.
Vibriobactin	*Vibrio cholerae*
Yersiniabactin	*Yersinia enterocolitica, Y. pestis*

property has been used for enrichment procedures (TCBS medium) developed for vibrios, although the production of catecholates in TCBS medium has not been demonstrated.

Recently, a novel class of siderophores named heterobactins (A and B) was isolated from *Rhodococcus erythropolis* strain IGTS8, as well as from other strains. These siderophores have both catecholate and hydroxamate iron donor groups. Although both functional groups are present, heterobactin A was originally identified as being transported by a catecholate receptor and heterobactin B was identified as being transported by a hydroxamate receptor. This suggests that, depending on the prevailing environmental conditions, iron can be taken up via the hydroxamate or the catecholate receptor without changing the biosynthetic route. Thus, *Rhodococcus* is well equipped to sequester iron under different environmental conditions by using only one biosynthetic route. Indeed, the production of two siderophores, a catecholate and a hydroxamate, by a single strain is a common strategy to overcome different environmental conditions. Thus, many *E. coli* strains produce the plasmid-encoded siderophore aerobactin in addition to enterobactin. Likewise, strains of *Pantoea, Erwinia,* and *Enterobacter* synthesize both enterobactin and ferrioxamines, suggesting an ecological adaptation to variable environmental conditions.

In soils containing decaying plant material, a drop in pH to values of 3 to 4 is not uncom-

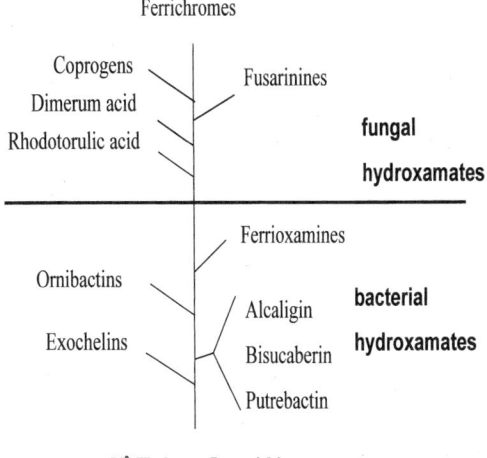

N⁶-Hydroxy-L-ornithine

FIGURE 3 Major hydroxamate siderophores produced by bacteria and fungi originating from oxygenated ornithine.

is to enable iron binding in a neutral to alkaline environment rather than to resist acidic conditions. For example, soil bacteria such as *Azotobacter* and *Agrobacterium* produce the catecholate-type siderophores azotochelin, protochelin, and aminochelin, which restricts their growth to alkaline or neutral soils. The ability to also synthesize hydroxamate siderophores greatly enhances the environmental range and activity for other soil bacteria. An example of an absolute requirement for hydroxamate siderophores is the soil bacterium *Microbacterium flavescens* JG9 (previously classified as *Arthrobacter* or *Aureobacterium*), which is often used as a bioassay organism. The requirement for hydroxamate siderophores is even more important when a large amount of plant material is present, which can be degraded by any acid-producing *Enterobacter*, *Listeria*, or *Lactobacillus* species present. Except for certain calcarous soils that may have buffering properties, the degradation of organic material originating from plants and woods may finally result in a humic acid matrix whose components have different molecular sizes. Soils represent mixtures of sand, silt, clay, and humic matter arranged into a microcosmos for bacteria, fungi,

mon. Therefore, several enterobacterial genera profit from the ability to biosynthesize ferrioxamines in addition to enterobactin. However, as shown in Fig. 4, many bacteria living in soil or aqueous environments produce only catecholates. This suggests that their main function

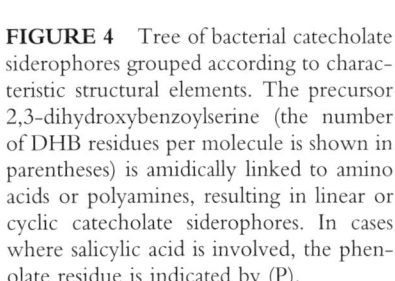

FIGURE 4 Tree of bacterial catecholate siderophores grouped according to characteristic structural elements. The precursor 2,3-dihydroxybenzoylserine (the number of DHB residues per molecule is shown in parentheses) is amidically linked to amino acids or polyamines, resulting in linear or cyclic catecholate siderophores. In cases where salicylic acid is involved, the phenolate residue is indicated by (P).

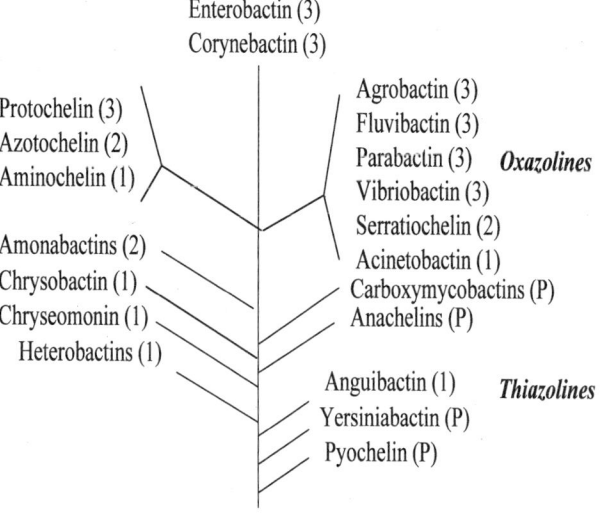

2,3-Dihydroxybenzoic acid / Salicylic acid

plants, and animals. Most often, soil particles consist of biomass fragments of humic matter and sand or clay connected by adhesive exopolymers produced by microorganisms. Since humic soils behave like cation exchangers, a variety of metal ions are bound, of which iron is of the utmost importance. Although iron may be abundant in humic matter, it is not easily available to microorganisms, and it seems reasonable that the formation constants of iron-complexing siderophores must be high enough to compete with those of iron bound to humic matter. Ferrioxamines are optimally adapted to extract iron from soils containing large amounts of humic acids under slightly acidic conditions.

INTERRELATIONSHIP BETWEEN BACTERIAL AND FUNGAL SIDEROPHORES

It is an interesting fact that enzymes for siderophore biosynthesis are found among bacteria as well fungi. For example, the key enzymes, such as amino acid-N^8-oxygenases, which provide the building blocks for hydroxamate siderophores, are present in both prokaryotes and eukaryotes. Fungi of the Ascomycetes and Basidiomycetes synthesize a number of hydroxamate siderophores, among which ferrichromes, fusarinines, and coprogens predominate. The ability of enterobacteria, bacilli, and staphylococci to utilize iron from fungal hydroxamates is well known and has been amply documented elsewhere in this book. Bacterial outer membrane receptors for ferrichromes and coprogen may have evolved as an adaption to the exogenously available fungal siderophores by mutational alteration of an ancestral receptor such as the enterobactin receptor FepA. Similarly, the periplasmic binding protein-dependent ABC transporters for hydroxamates FhuB/C and FhuD may have been adapted to recognize a variety of fungal hydroxamates in gram-negative and gram-positive bacteria. Thus, evolution of siderophore transport systems may have an ecological basis and can be regarded as an adaptation to the prevailing iron source.

While the adaptation theory may be valid for the evolution of siderophore transport systems, the biosynthetic genes for siderophores seem to be highly conserved and specific to either bacterial or fungal genera. Thus, lysine-N^6- and ornithine-N^5-oxygenases are found only in bacteria and fungi, respectively, and are not present in higher organisms. Specific siderophore biosynthetic pathways are also found in the polycarboxylate class. R,R-Rhizoferrin, a fungal polycarboxylate siderophore, has been isolated from *Rhizopus microsporus* and other, members of the Mucorales, while an enantiomer of rhizoferrin, possessing an S,S-configuration of citryl residues, was identified as a siderophore in *Ralstonia pickettii*. This suggests that although the final products appear similar, the biosynthetic pathways for the fungal and bacterial rhizoferrin might be different. As shown in Fig. 5, two types of citrate-based siderophores exist, the polycarboxylate type and the citrate-hydroxamate type. While iron is bound exclusively to carboxy and α-hydroxycarboxy residues in the polycarboxylates, the citrate-hydroxamates also possess two hydroxamate groups. The latter compounds represent a combination of two ligand types in one molecule, which considerably increases the efficiency of iron binding. The iron-binding properties of the citrate-hydroxamate siderophores are dominated largely by the hydroxamate donor groups, resulting in a higher acid stability than that of the polycarboxylates, which are stable in a pH range of 4 to 9. Citrate-containing siderophores may be regarded as the simplest microbial iron chelates which are in contrast to the fungal ferrichromes, which represent cyclic hexapeptides.

There are some exceptions to the general rule regarding the distribution of siderophore biosynthesis genes among bacteria and fungi. Ferrichromes are fungal siderophores, but a corresponding open-chain ferrichrome analogue, linked to an antibiotic residue in albomycin, has been isolated from *Streptomyces* strains, indicating the existence of ferrichrome biosynthetic genes in bacteria. A ferrichrome-like dihydroxamate, named amycolachrome,

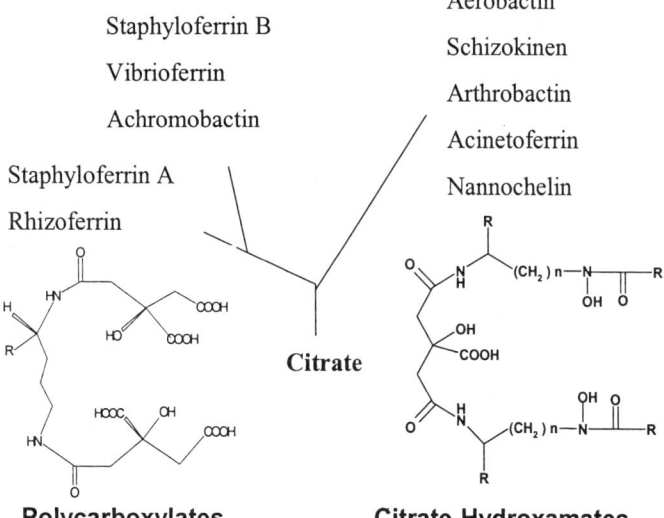

FIGURE 5 Citrate-containing siderophores comprising polycarboxylates and citrate-hydroxamates. While Fe^{3+} is coordinated exclusively by carboxy and α-hydroxycarboxy ligands in polycarboxylates, the citrate-hydroxamates have a mixed coordination dominated by the hydroxamate donor groups.

has also been isolated from low-iron cultures of the related *Amycolatopsis orientalis*. These few examples show that fungal biosynthetic genes for peptides of the ferrichrome class can also be found among certain bacteria. Responsible for the highly developed cyclic peptides are ferrichrome-specific peptide synthetases, which may not exist in bacteria.

While utilization of fungal siderophores by bacteria is very often observed, the reverse situation, utilization of bacterial siderophores by fungi, is not well documented. Although desferrioxamine therapy has been reported to cause a fungal infection called mucormycosis (Zygomycetes), many ascomycetous fungi such as *Aspergillus*, *Penicillium*, and *Chrysonilia* (Anamorph: *Neurospora*) are unable to utilize ferrioxamines and other siderophores produced by bacteria because of insufficient recognition by the fungal hydroxamate transport systems located in the cytoplasmic membrane. However, the ascomycetous yeast *Saccharomyces cerevisiae* is well adapted to utilize a variety of siderophores via low and high-affinity transport systems (Fig. 6). The yeasts possess effective ferric reductases (Fre1 to Fre7), which are located on the outside of the cytoplasmic membrane and enable the reduction of a variety of soluble ferric chelators such as ferric citrate and ferrioxamine B. In addition, a high-affinity ferrioxamine-utilizing system involving a membrane transporter (Sit1p) of the major facilitator superfamily (MFS) has been identified. By using a functional genomic strategy to identify sequences of previously uncharacterized MFS transporter genes, further siderophore transporters have been identified in *S. cerevisiae*. Another strategy of cDNA microarrays of the genome of *S. cerevisiae* identified these transporters (called ARN1 to ARN4) as being under the control of the major iron-dependent transcription factor AFT1. Among these, a transporter specific for enterobactin (Enb1p) could be identified, indicating that fungi have successfully adapted to enterobacterial siderophores (Fig. 6). This suggests that *S. cerevisiae*, which is unable to synthesize siderophores, may profit from the presence of certain enterobacteria living in the same habitat. Moreover, members of the aspergilli that are known to synthesize their own hydroxamate siderophores are also able to express additional transporters for the bacterial siderophore enterobactin. Thus, in *A. nidulans*, the *mirB* gene encodes

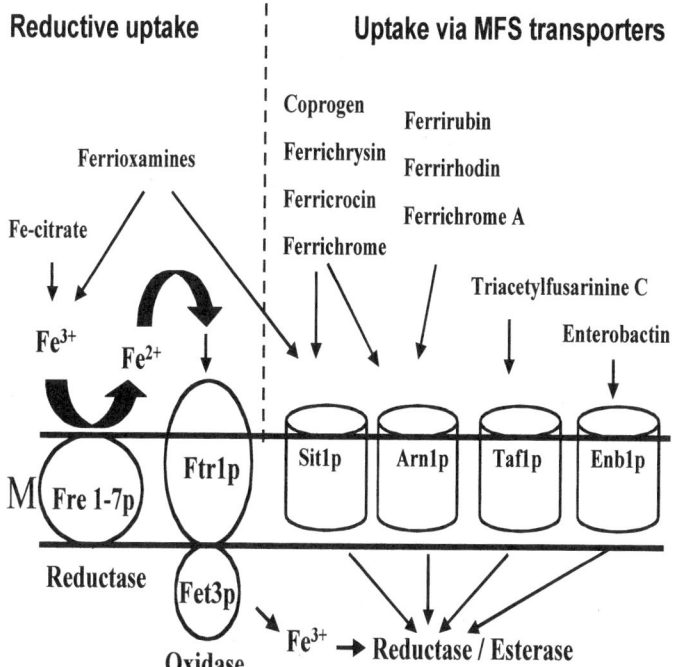

FIGURE 6 Major routes of bacterial and fungal siderophore uptake in the fungus *S. cerevisiae*. Ferric citrate is taken up via the reduction route (Fre, Ftr1, and Fet3), while ferrioxamines may enter the cells either via reduction or, like all other siderophores, via specific (Sit1p) membrane transporters of the MFS. M, cytoplasmic membrane; Fre 1–7p, membrane reductases; Ftr1p, high-affinity Fe^{2+}/Fe^{3+} transporter; Fet3p, multicopper oxidase; Si1p, Arn1p, Taf1p, Enb1 (also called Arn1–4p), MSF transporters exhibiting specificity for different siderophores of the hydroxamate and catecholate classes.

a transporter (MIRB) for the fungal siderophore triacetylfusarinine C while *mirA* expresses a transporter (MIRA) specific for enterobactin. Sequence analysis suggested that the transporters for enterobactin in *S. cerevisiae* and *A. nidulans* evolved independently after the two species split from each other. These examples show that transporters for bacterial siderophores in fungi have developed in an ecological niche where enterobacteria and fungi may have lived together for a long evolutionary period.

SUGGESTED READING

Bister, B., D. Bischoff, G. J. Nicholson, M. Valdebenito, K. Schneider, G. Winkelmann, K. Hantke, and R. D. Süssmuth. 2004. The structure of salmochelins: C-glucosylated enterobactins of *Salmonella enterica*. *BioMetals* **17:**471–481.

Boukhalfa, H., and A. L. Crumbliss. 2002. Chemical aspects of siderophore mediated iron transport. *BioMetals* **15:**325–339.

Carrano, C. J., M. Jordan, H. Drechsel, D. G. Schmid, and G. Winkelmann. 2001. Heterobactins: a new class of siderophores from *Rhodococcus erythropolis* IGTS8 containing both hydroxamate and catecholate donor groups. *BioMetals* **14:**119–125.

Crosa, J. H., and C. T. Walsh. 2002. Genetics and assembly line enzymology of siderophore biosynthesis in bacteria. *Microbiol. Mol. Biol. Rev.* **66:**223–249.

Drechsel, H., and G. Winkelmann. 1997. Iron chelation and siderophores, p. 1–49. *In* G. Winkelmann and C. J. Carrano (ed.), *Transition Metals in Microbial Metabolism*. Harwood Academic Publishers, Amsterdam, The Netherlands.

Hantke, K., G. Nicholson, W. Rabsch, and G. Winkelmann. 2003. Salmochelins, siderophores of *Salmonella enterica* and uropathogenic *Escherichia coli* strains, are recognized by the outer membrane receptor IroN. *Proc. Natl. Acad. Sci. USA* **100:**3677–3682.

Leong, S. A., and G. Winkelmann. 1998. Molecular biology of iron transport in fungi, p. 147–186. *In* A. Sigel and H. Sigel (ed.), *Metal Ions in Biological Systems*, vol. 35. Marcel Dekker, Inc., New York, N.Y.

Martinez, J. S., J. N. Carter-Franklin, E. L. Mann, J. D. Martin, M. G. Haygood, and A. Butler. 2003. Structure and membrane affinity of a suite of amphiphilic siderophores produced by a marine bacterium. *Proc. Natl. Acad. Sci. USA* **100:**3754–3759.

Meyer, J.-M. 2000. Pyoverdines: pigments, siderophores and potential taxonomic markers of fluorescent *Pseudomonas* species. *Arch. Microbiol.* **174:** 135–142.

Ratledge, C., and L. G. Dover. 2000. Iron metabolism in pathogenic bacteria. *Annu. Rev. Microbiol.* **54:**881–941.

Raymond, K. N., E. A. Dertz, and S. S. Kim. 2003. Enterobactin: an archetype for microbial iron transport. *Proc. Natl. Acad. Sci. USA* **100:** 3584–3588.

Winkelmann, G. (ed.). 2001. *Microbial Transport Systems.* Wiley-VCH, Weinheim, Germany.

Winkelmann, G., and C. J. Carrano (ed.). 1997. *Transition Metals in Microbial Metabolism.* Harwood Academic Publishers, Amsterdam, The Netherlands.

ENVIRONMENTAL FLUORESCENT *PSEUDOMONAS* AND PYOVERDINE DIVERSITY: HOW SIDEROPHORES COULD HELP MICROBIOLOGISTS IN BACTERIAL IDENTIFICATION AND TAXONOMY

Jean-Marie Meyer and Valérie A. Geoffroy

29

The genus *Pseudomonas* represents a group of bacteria of fundamental, medical, and environmental interest. Widely distributed in nature and present in most environments, these bacteria are remarkable in their metabolic diversity, which allows them to degrade an impressive variety of chemicals including xenobiotic compounds. Although some species are recognized as phytopathogens, e.g., the *Pseudomonas syringae* group, many plant-related pseudomonads have been described as potent biocontrol agents with protecting or stimulating plant growth effects. Most of the pseudomonads are nonpathogenic for humans or animals. The ones with virulence factors are usually considered as opportunistic pathogens and include a few well-characterized species, e.g., the former *Pseudomonas mallei* and *Pseudomonas pseudomallei* (recently reclassified in the *Burkholderia* genus) and the best-known species, *Pseudomonas aeruginosa* (see chapter 19).

The genus *Pseudomonas* originally comprised bacteria exhibiting just a few general phenotypic characteristics and included gram-negative rod-shaped motile cells with one or a few polar flagella, presenting a strict aerobic metabolism. With such a broad definition, numerous bacterial species were included in the genus, and it is not surprising that the group is presently considered to be very heterogeneous, as indicated by the most recent taxonomical investigations. The classical division of the genus into five main RNA clusters according to RNA-RNA and RNA-DNA hybridization relationships has now been abandoned, and numerous species have been ranked in novel genera such as *Burkholderia* or *Ralstonia*. Bacteria described in this chapter belong to the so-called *Pseudomonas* sensu stricto genus, which is the former RNA group I of Palleroni's classification. This genus can be divided into two subgroups according to the ability of the bacteria to produce the yellow-green, water-soluble, and brightly fluorescent pigment pyoverdine, characteristic of the fluorescent *Pseudomonas* species. Such a subdivision into fluorescent and nonfluorescent *Pseudomonas* species, based on a very easily distinguishable phenotypic characteristic, is common, although it is not officially recognized from a taxonomic point of view. As it turns out, all species ranked in the genus have very high homology at the 16S RNA gene level, whether or not they produce the fluorescent pigment. Thus, the recent finding that the structural diversity of pyoverdines cor-

---

Jean Marie Meyer and Valérie A. Geoffroy, Laboratoire de Microbiologie et Génétique, Université Louis-Pasteur-CNRS, FRE 2326, 67000 Strasbourg, France.

Iron Transport in Bacteria, Edited by Jorge H. Crosa, Alexandra R. Mey, and Shelley M. Payne
© 2004 ASM Press, Washington, D.C.

relates well with bacterial taxonomic diversity at the species level suggests that pyoverdine molecules could be considered powerful taxonomical markers for fluorescent *Pseudomonas* species. As detailed in the present chapter, siderophore-based methods could be successfully used for bacterial identification. Moreover, preliminary data for other bacteria (nonfluorescent *Pseudomonas* species, *Burkholderia* species, rhizobia, and bradyrhizobia, as well as enterobacteria, mycobacteria, and *Aeromonas* species) strongly indicate that any siderophore could be used as a taxonomic marker for its producing bacteria, which means that the emerging concept of siderotyping as a taxonomic tool could be applied to a large part of the microbial world.

SIDEROPHORES OF FLUORESCENT *PSEUDOMONAS* SPECIES

Pseudomonads have particularly strong nutritional requirements for iron, which is necessary for numerous important biochemical processes including the synthesis of energy via cytochrome-driven electron transport. As strict aerobic bacteria, they possess many enzymes or compounds that, like the cytochromes, require iron to be biologically active. One could say that there is no possible life for pseudomonads without iron since DNA synthesis is known to require iron for the ribotide reductase activity. The biological importance of iron comes from its particular facility to move from the reductive Fe^{2+} form to the oxidized Fe^{3+} form by the loss of one electron, and vice versa. However, like most of the free-living microorganisms, pseudomonads are faced with the problem of iron nonavailability in aerated environments. In the presence of oxygen, and especially at pH values close to neutrality, most of the iron does not exist as free Fe^{3+} ions but as particularly insoluble Fe^{3+} hydroxides. The common strategy of the pseudomonads to solubilize and internalize iron involves the production of siderophores, which is the classical method used by most of the free-living bacteria.

When the bacterial iron demand is not satisfied and, consequently, growth becomes limited by the iron supply, bacteria start to excrete large amounts of specific Fe^{3+}-scavenging molecules named siderophores. Siderophore production by the fluorescent *Pseudomonas* species is particularly easy to follow since the main siderophore of these bacteria is the yellow-green fluorescent pigment pyoverdine. The siderophore-producing bacterial culture starts to turn yellow-green during the exponential phase of growth and reaches its maximal intensity of color (indicating maximal production of siderophore) at the beginning of the stationary phase. The color of pyoverdine in solution is very different from the dark green color attributable to phenazines, the other type of pigment produced by certain species including *P. aeruginosa*, *Pseudomonas chlororaphis*, and *Pseudomonas aureofaciens*.

Besides pyoverdine, fluorescent *Pseudomonas* species produce supplementary siderophores:

- Pyochelin, made by many, if not all, *P. aeruginosa* strains but never purified from cultures of other fluorescent *Pseudomonas* species.
- Salicylic acid, detected in one strain of *Pseudomonas fluorescens* (strain CHA0). Both salicylic acid, part of the pyochelin molecule, and pyochelin have been described as siderophores produced by strains belonging to the *Burkholderia* (formely *Pseudomonas*) *cepacia* complex.
- Pseudomonin, produced by two *P. fluorescens* strains, one being a fish isolate and the other being a plant-related strain. This siderophore is a salicylate-based compound resulting from the condensation of salicylate with threonine and histamine.
- Quinolobactin, produced by *P. fluorescens* ATCC 17400. This siderophore corresponds to an original quinoleinic structure not biosynthetically related to pyochelin or pyoverdine, although a quinoleinic cycle is part of the chromophore of pyoverdine.
- Pyridine-2,6-bis(monothiocarboxylic acid), recently described as a secondary siderophore produced by two *P. putida* strains and one strain of *Pseudomonas stutzeri*.

These supplementary siderophores, detected together with pyoverdine in culture supernatants of some fluorescent *Pseudomonas* species, are usually produced in very small amounts compared to pyoverdine. Interestingly, pyoverdine-deficient mutants may overproduce some of these compounds. Since the affinity of these supplementary siderophores for Fe^{3+} is much lower than the affinity of pyoverdine, they could be considered rescue systems useful when the pyoverdine-mediated iron uptake is deficient. However, to assess the relative importance of such supplementary systems, a systematic characterization of siderophores other than pyoverdines within numerous representatives of the main species found in natural environments (e.g., *P. fluorescens* and *P. putida*), as well as comparative studies between mono- and multisiderophore-producing strains, must be carried out.

PHYSIOLOGICAL AND BIOCHEMICAL FEATURES OF PYOVERDINE

Pyoverdine is specifically synthesized by fluorescent *Pseudomonas* species growing under iron deficiency. Such conditions are easily attained in vitro by using synthetic media such as succinate medium or Casamino acids (CAA). The iron content of these media is usually less than 2 μM and comes from the contaminating iron supplied by chemicals, water, and glassware. Most environmental isolates are good pyoverdine producers in both media, with ca. 100 to 200 mg of pyoverdine produced per liter of medium. A few individual strains, however, produce good amounts of pyoverdine in CAA medium while demonstrating no production in succinate medium. These strains probably have defects in pyoverdine biosynthesis at the level of amino acid metabolism and therefore are of particular interest for studying the pyoverdine biosynthesis pathway.

As already mentioned, pyoverdine was first described as a yellow-green pigment, its color being due to the absorbance of light in the visible range, at around 400 nm. Therefore, detection and quantification of pyoverdine production are particularly easy compared to all other siderophores. Because of the sharpness of the maximal absorption peak and the high molar absorbance value of pyoverdine ($\varepsilon_M = 20,000$ at pH 7.0), a spectrophotometric measurement of the absorbance of a pyoverdine-containing culture supernatant can result in a precise determination of the amount of siderophore produced by the bacteria. The maximal absorbance wavelength varies according to the pH of the pyoverdine solution, from 385 nm at acidic pH (<pH 5) to 415 nm at pH >9. Thus, precise comparative studies require careful examination of the pH value reached by each culture at the time of measurement.

Most of the studies described in the following sections require the purification of pyoverdine from the culture supernatant. This can be achieved very easily by using a convenient and rapid chromatographic method based on the fixation of pyoverdine on an XAD-Amberlite resin followed by its elution from the column with a 50% methanol-water solution. The gold-yellow powder obtained by evaporation and lyophilization of the methanolic solution usually yields close to 100% of the pyoverdine material excreted by the bacteria, as estimated by spectrophotometry. The material can be considered biologically pure since, as far as we have experienced, it contains only pyoverdine as a siderophore (pyochelin, quinolobactin, and salicylic acid are either not retained by or not released from the resin). However, from a chemical point of view, the XAD-purified material is typically composed of a mixture of several pyoverdine molecules with an identical biological efficiency, as detailed below.

Another convenient and rapid purification procedure for mini-scale preparation is the use of C_{18}-Sep-Pack cartridges (Waters), which retain pyoverdines as well as ferripyoverdines.

STRUCTURAL FEATURES OF PYOVERDINE

As illustrated in Fig. 1, which represents the pyoverdine produced by a strain of the recently described species *Pseudomonas costantinii*, three different structural parts are distinguishable within the pyoverdine molecule: (i) a chromo-

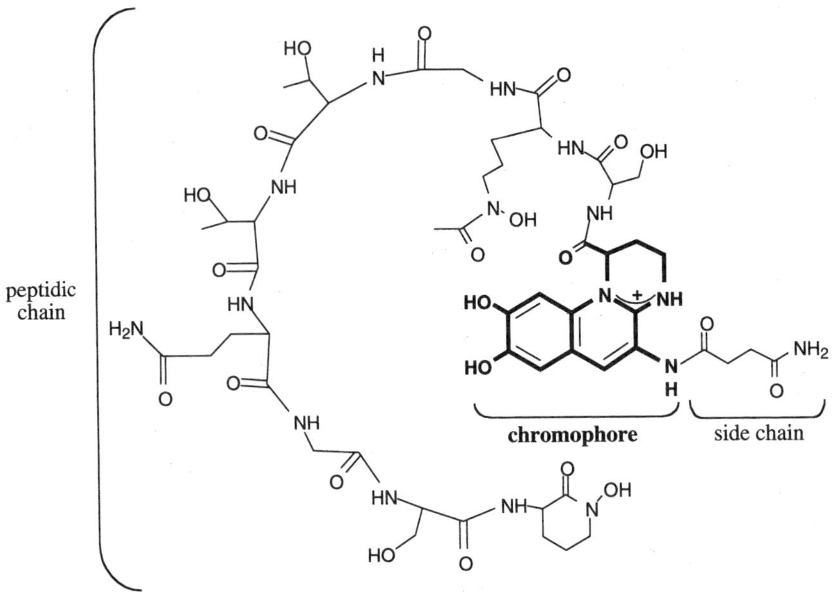

FIGURE 1 Structure of the *P. costantinii* pyoverdine. Reprinted from Fernandez et al., 2001, with permission from the publisher.

phore based on a quinoline heterocycle, responsible for the yellow-green color and bright fluorescence of pyoverdine; (ii) a side chain formed by a succinamide residue attached to the NH$_2$ group of the chromophore; and (iii) a peptidic chain branched by its N-terminal amino acid to the carboxylic group of the chromophore. This molecule is only one example among hundreds of pyoverdine structures already described in the literature or identifiable by the siderotyping methods described below. Structural differences are seen in each of the three parts of pyoverdine. Differences among pyoverdines originating from different strains are located mainly within the peptide part of the molecules, while differences between pyoverdine molecules purified from a single culture supernatant involve mostly one or both of the other parts of the molecule.

Structure Diversity at the Dicarboxylic Side Chain Level

Together with the molecule described in Fig. 1, another closely related molecule has been found as a minor compound in the XAD-purified extract isolated from the *P. costantinii* culture. Its structure differed only by the nature of the dicarboxylic side chain branching from the chromophore, with a succinic acid group replacing the succinamide residue. The two compounds (isoforms) differ by their respective electric charge and can be well separated by electrophoresis (see below).

The co-occurrence of pyoverdine isoforms in a culture supernatant is very common. Several different side chains have been described so far, corresponding to components of the citric acid cycle and their amide derivatives (Table 1). The most commonly encountered are the succinic acid and succinamide derivatives, which are usually found simultaneously in culture supernatants, as described for the *P. costantinii* pyoverdine. Isoelectric focusing IEF profiles determined at different times during bacterial growth clearly showed an inverse evolution between the two isoform bands. Thus, the succinamide form appeared as the most intense band during the exponential growth phase and progressively declined in intensity in favor of the succinic form during the stationary phase.

TABLE 1 Possible side chains of pyoverdine isoforms

Nature of the side chain	Chemical structure[a]
Succinic acid	(Chr-NH)-CO-CH_2-CH_2-COOH
Succinamide	(Chr-NH)-CO-CH_2-CH_2-$CONH_2$
Malic acid	(Chr-NH)-CO-CH_2-CHOH-COOH
Malamide	(Chr-NH)-CO-CH_2-CHOH-$CONH_2$
Glutamic acid	(Chr-NH)-CO-CH_2-CH_2-$CHNH_2$-COOH
α-Ketoglutaric acid	(Chr-NH)-CO-CH_2-CH_2-CO-COOH

[a] Presented with the acyl residue bound to the NH_2 group of the chromophore (Chr; see Fig. 1).

The succinic form has been shown by in vitro degradation to be a spontaneous hydrolysis product: incubation of an aqueous pyoverdine solution buffered at pH 9 (which is the pH value reached at the plateau during growth in a succinate medium) for 24 h resulted in a 90% transformation of the succinamide to the succinic acid form. Moreover, monitoring a culture at a stabilized neutral pH favoured the accumulation of the succinamide isoform. Thus, the succinamide form, and not the succinic acid form, should be the molecule synthesized by the bacteria. The same explanation is probably valid for the pyoverdines presenting the malic acid/malamide isoforms. As discussed in the section on the pyoverdine biosynthesis pathway (below), these multiple isoforms are all chemically or enzymatically derived from a unique precursor, which is glutamic acid. Many bacterial strains are characterized by isoelectrofocusing profiles presenting up to five pyoverdine isoform bands, suggesting the accumulation of most, if not all, possible isoforms during growth of the bacterial culture. Co-occurrence has been chemically demonstrated in many cases. There is as yet no rational explanation for such a diversity at the side chain level of pyoverdines, except that the differences are related to the chemical instability of certain structural functions, e.g., the amide groups of the side chains. Biologically, no differences have been observed between the various isoforms: they were able to mediate iron uptake with an equal efficiency.

Structure Diversity at the Chromophore Level

Pyoverdine-related compounds differing from the major pyoverdines by modifications affecting the chromophore have occasionally been described. Together with the pyoverdine molecules harboring the usual (1S)-5-amino-2,3-dihydro-8,9-dihydroxy-1H-pyrimido-[1,2-a]quinoline-1-carboxylic acid chromophore (Fig. 2a), some minor compounds may coexist in the growth medium supernatant. They have been named 5,6-dihydropyoverdine, 5,6-dihydropyoverdine-7-sulfonic acid, or succinopyoverdine (Fig. 2b to d), according to the structural change at the chromophore level. The chromophore shown in Fig. 2e was first described as the chromophore of azotobactin, the fluorescent pigment and siderophore of *Azotobacter* spp., suggesting that the *Pseudomonas* and *Azotobacter* genera are as closely related at the siderophore level as they are at the taxonomical level.

Other pyoverdine-related compounds, named isopyoverdines, have been defined as molecules containing a chromophore with the carboxylic group located at the C-3 instead of the C-1 of the heterocycle (Fig. 2f). The modification could be seen as the consequence of an inverted cyclization reaction of 2,4-diaminobutanoic acid during the tetrahydropyrimidine ring formation. Interestingly, the isopyoverdines so far described in the literature are produced exclusively by strains belonging to the *P. putida* complex.

The first iron-complexing compound isolated from a fluorescent *Pseudomonas* strain and

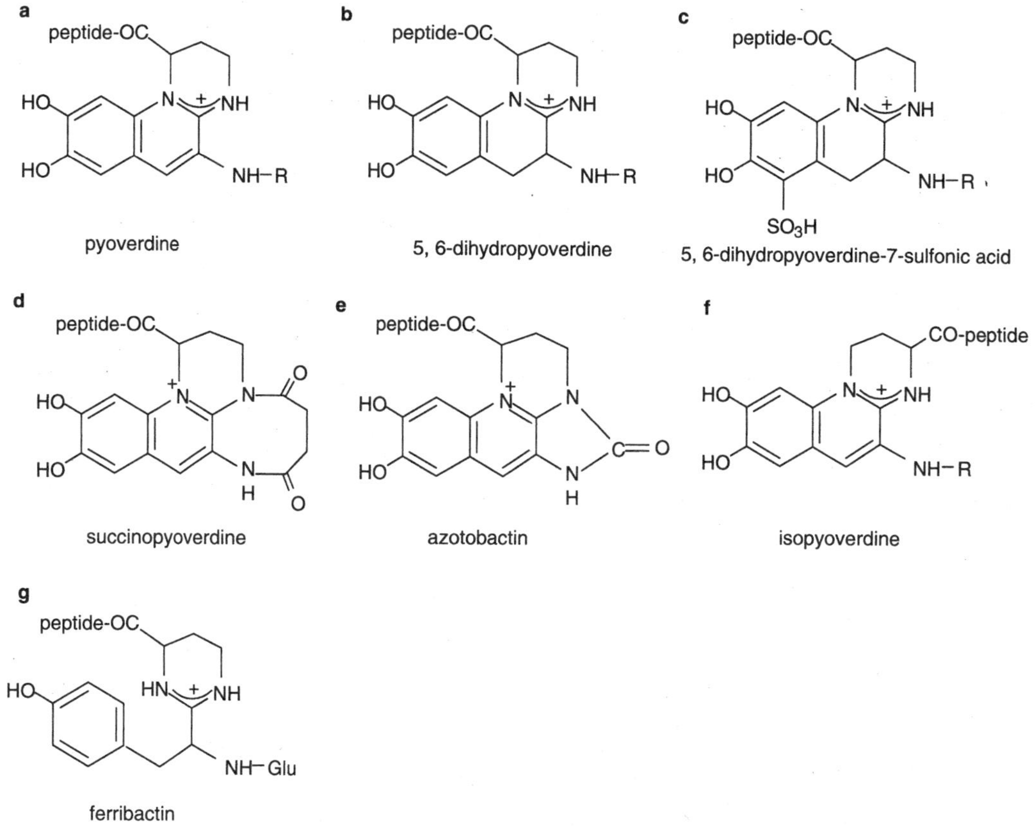

a pyoverdine

b 5,6-dihydropyoverdine

c 5,6-dihydropyoverdine-7-sulfonic acid

d succinopyoverdine

e azotobactin

f isopyoverdine

g ferribactin

FIGURE 2 Variations at the chromophore level as found in different pyoverdine-related compounds.

tentatively assigned to a siderophore function was ferribactin. This compound was neither colored nor fluorescent and, in fact, corresponded to a pyoverdine-like molecule containing instead of the chromophore a tyrosine bound to a diaminobutanoic acid and to a glutamyl residue (Fig. 2g). Since then, it has been shown that ferribactin should be considered a biosynthetic precursor of pyoverdine. According to the amino acid composition ascribed to the peptide of ferribactin, it could be assumed that the strain was related to *P. fluorescens* ATCC 13525, whose pyoverdine has the same amino acid composition. Ferribactins have also been found in other *Pseudomonas* strains and in some cases have been described as being coproduced with pyoverdines.

It is well established that pyoverdine chromophore-like residues are generated as precursors or side products during the general biosynthetic pathway of pyoverdine. Thus, the existence of these minor compounds should be viewed as the result of certain side reactions during the sophisticated process of pyoverdine biosynthesis.

Structure Diversity at the Peptide Level

The pyoverdine produced by *P. costantinii* (Fig. 1) has a peptide part composed of 9 amino acid residues, namely, two glycine, two D-serine, one L-threonine, one D-*allo*-threonine, one L-glutamine, and two D-δ-N-hydroxyornithine residues. Concerning these two last residues, the first is acylated on the δ-N with an acetyl group and the second is internally cyclized. Thus, these two form a hydroxamate group [-CO-NOH-] that, together with the catecholate group of the chromophore, can participate in the complexing of Fe^{3+}. The peptide part of the pyoverdine molecule is remarkable because of the presence of unusual amino acids never found in natural peptides and also by the presence of D-amino acids, a feature which is particularly uncommon in natural compounds, except in some peptide antibiotics of microbial origin, e.g., gramicidin or streptomycin.

The detailed structure of the pyoverdine produced by *P. aeruginosa* ATCC 15692 (PAO1 strain) is described in chapter 19. When comparing CAA cultures of the *P. aeruginosa* and *P. costantinii* strains, no straight phenotypic differences concerning color and fluorescence due to pyoverdine could be observed. The pyoverdines produced by the two strains have the same chromophore and thus show the same or very similar spectral properties. A major difference, however, exists between the two compounds at the level of their respective peptide chains. The *P. aeruginosa* compound contains 8 amino acid residues, with some being identical (serine, threonine, N^{δ}-hydroxyornithine) to the ones present in the *P. costantinii* pyoverdine and others being original (arginine, lysine). Another feature differentiating the two compounds is the linearity of the *P. costantinii* peptide, whereas the one of *P. aeruginosa* contains an internal cycle involving half of the amino acyl residues. Such a cyclic peptide substructure is unusual in natural compounds, except in certain peptide antibiotics such as actinomycins, bacitracins, and cyclosporins.

A total of 44 pyoverdine structures, differing by their respective peptide chains, have already been determined. The amino acid residues or derivatives which form the peptide chains of each of the 44 structurally different pyoverdines are given in Table 2. The most highly represented amino acid residues among these peptides are, in decreasing order of occurrence, serine, ornithine, threonine, lysine, glycine, alanine, aspartic acid, and glutamine. Less frequent amino acids such as arginine, hydroxyhistidine, 2,4-diaminobutanoic acid, and valine are found in some pyoverdines. Sulfur-containing amino acids have never been found.

The same is true for cyclic or aromatic amino acids (with the exception of one histidine-containing pyoverdine) and for the aliphatic residues leucine and isoleucine. Thus, the peptide chain is characterized by a predominance of hydrophilic amino acid residues conferring water solubility to the pyoverdine molecule, a property which is a prerequisite for its siderophore function.

A particular feature of the pyoverdine molecules is the high percentage of D-amino acid residues, reaching close to 37% of the total number of amino acyl residues in some of the 44 pyoverdines listed in Table 2. This feature may render the molecules highly resistant to common peptidases but may also be way to increase the variety among pyoverdines while limiting the number of amino acids.

As a general feature, the peptide part of pyoverdines contains two amino acid-derived residues involved in the complexing of iron. They can be either β-hydroxy amino acids (usually L- or D-*threo*-β-hydroxyaspartic acid) and/or hydroxamic acids derived from L- or D-ornithine. Different ways are used to build up the hydroxamate functions: (i) the δ-amino group of ornithine is hydroxylated and acylated, with the acyl group typically being formyl or acetyl (β-hydroxybutanoyl for the pyoverdine of *P. putida* strain C); or (ii) the δ-*N*-hydroxyornithine is internally cyclized to created a piperidone ring, usually at the C terminus of the peptide chain.

As shown in Table 2, a majority of pyoverdines (24 of 44) carry two δ-*N*-hydroxyornithine derivatives as the iron-complexing moieties, while most of the others (18 of 44) have a combination of one δ-*N*-hydroxyornithine derivative and one β-hydroxyaspartic acid residue (or, in one case, β-hydroxyhistidine). Only two pyoverdines complex iron without using δ-*N*-hydroxyornithine, using instead two β-hydroxyaspartic acid residues.

Another factor in the diversity among pyoverdines is the length of the peptide part, which can vary from 6 to 12 amino acid residues, with a majority of pyoverdines (14 of 44) carrying 7 amino acids. The pyoverdine groups con-

TABLE 2 Amino acid composition of the peptidic part of 44 pyoverdines with fully determined structures[a]

Nature and no. of amino acyl residues in the pyoverdine peptide produced by strain (columns, left to right): P.fl. G173; P.a. 15692; P.sp. 96-318; P.sp. 96-312; P.sp. 95-275; P.ch. PL16; P.fl. CHAO; P.fl. SR8.3; P.sp. G24; P.tol. 2192; P.sp. 3b; P.fl. PL8; P.fl. PL7; P.co. PS3a; P.p. 90-40; P.fl. A6; P.p. 90-44; P.fl. 13525; P.fl. 12; P.fl. 18.1; P.sp. 96-188; P.a. 6; P.a. R'; P.a. 27853; P.sp. A214; P.fl. PL9; P.fl. 9AW; P.fl. 1.3; P.fl. 17400; P.sp. 2908; P.fl. BTP2; P.p. Gwose; P.p. G4R; P.fl. 51W; P.fl. ii; P.fl. W; P.p. G176; P.p. 12633; P.p. WCS358; P.p. 90-33; P.sp. B10; P.p. C; P.p. 90-51; P.syr. 19310.

Amino acid residue[b]	values across 44 strains
L-aOHOrn	2; 2; ; ; ; ; 2; ; 1; ;
D-aOHOrn	; ; 2; 2; 2; ;
L-fOHOrn	; 2; ; ; ; ; ; ; 1; ; ; 1; 1; 1; 1; 1; 1; ; ; ; ; ; ; ; 1; 1; ; ; 1; 1; ; ; ; ; ; ; ; ; ; ; ; ; ;
D-fOHOrn	; ; ; ; ; 2; ; ; ; ; ; ; ; ; ; ; ; 1; 1; 1; 1; 1; 1; 1; ; ; ; ; ; ; ; ; ; ; ; ; ; ; ; ; ; ; ;
L-cOHOrn	; 1; 1; 1; 1; ; ; ;
D-cOHOrn	; ; ; ; ; ; ; ; ; ; ; ; ; ; ; 1; 1; ; ; ; ; ; ; ; ; ; ; 1; ; ; ; ; ; ; ; ; ; ; ; ; ; ; ;
D-bOHOrn	; 1; ; ; ; ; ; ; ; ; ; ; ; ; ; 1; 1; ;
OHHis	; ;
L-OHAsp	; 1; 1; ; ; 1; 1; 1; 1; 1; 1; ; 1; 1; 1; 1; 1; ; 1; 1;
D-OHAsp	; 2; ; ; ; ; ; ; ; ; ; ; ; ; ; ; ; ; ; ;
L-Ala	; ; ; ; ; 2; 2; 1; ; ; ; 1; 1; ; 1; ; 2; 1; 2; 2; ; ; ; ; ; ; ; 1; ; ; ; ; ; 1; 1; 1; 1; 1; 1; ; ; 1; 2
D-Ala	1; ; ; ; ; 1; 1; ; 3; 1; ; ; ; ; 3; 3; 1; ; ; ; ; ; ; ;
Arg	; 1; 1; 2; ; ; ; ; ; ; ; ; ; ; ; ; ;
L-Asp	; ; ; ; ; ; 1; 1; ; ; ; 1; ; ; ; ; ; ; ; ; 2; ; ; ; ; ; 1; ; ; ; ; ; ; ; ; ; ; ; ; ; ; ; ;
D-Asp	1; ; ; ; ; ; ; ; ; ; ; ; ; ; 2; ;
L-DAB	; 1; 1; ; ; ; ; ; 1; 1; ; ; ; ; ; ; ; ; ; ; ; ; ;
D-DAB	; 1; ; ; ; ; ; ; ; ;
D-Glu	; 1; ; ; ; ; ; 1; ; ; ; ;
L-GluNH₂	; ; ; ; ; ; ; ; ; ; ; ; ; ; ; ; 1; 1; ; ; ; ; ; ; ; ; ; 1; 1; ; ; ; ; ; ; ; ; ; ; ; ; ; ;
D-GluNH₂	; 3; ; ; ; ; ; 1; ; ; ; ; ; ; ; ; ;
Gly	; ; ; ; ; ; ; ; ; ; ; ; 1; ; 2; 1; 1; 2; 2; 2; ; ; ; ; 1; 1; ; ; 2; ; ; ; 1; 3; 3; ; ; ; 1; ; ; ; ;
L-Lys	1; 1; 1; 2; 2; 1; 2; 1; 2; 1; 1; ; ; ; ; ; 1; 2; 1; 2; 2; ; ; 1; ; ; 1; ; ; ; 1; 1; ; ; 2; ; ; ; ; ; ; 2; 1
D-Lys	; ; ; ; ; ; ; ; ; ; ; 1; 1; 1; ; ; ; ; ; ; ; ; 1; ; ; ; ; ; ; 1; 1; ; ; ; ; ; ; ; ; ; ; ; ;
Orn	1; 1; ; ; ; ; ; ; ; 1; ; ; ; ; ; ; ; ; ; ; ;
L-Ser	1; 2; 2; 3; 1; ; 1; 1; 1; 1; ; 1; 1; 2; 1; 1; 1; 2; ; ; ; ; ; ; ; 1; 1; ; ; 2; 2; ; ; ; ; ; ; ; 1; ; ; ; 2
D-Ser	1; 2; 2; 1; 2; ; ; ; 2; ; ; ; ; ; ; ; ; ; ; ; ; ; ; ; ; 2; 2; ; ; ; ; ; ; ; ; ; ; ; ; ; ; ; ;
L-Thr	1; 2; ; ; ; 1; 1; 1; 1; 2; ; ; ; ; ; 1; 1; ; ; ; ; ; ; ; ; 1; 1; ; ; 2; 1; ; ; ; ; ; ; 2; ; ; ; 2; 2; 2
L-allo Thr	; ; ; ; ; ; ; ; ; ; 1; ;
D-Thr	; ; ; ; ; ; ; ; ; ; 2; ;
D-allo Thr	; ; ; ; ; ; ; ; ; ; ; ; ; ; 1; 1; 1; 1; ; ; 1;
L-Val	; 1; ; ; ; ; ; ;

[a] Abbreviations: aOHOrn, δN-acetyl-δN-hydroxyornithine; fOHOrn, δN-formyl-δN-hydroxyornithine; bOHOrn, δN-β-hydroxy-butyryl-δN-hydroxy-ornithine; cOHOrn, cyclo-hydroxyornithine (3-amino-1-hydroxypiperidone-2); OHHis, threo-β-hydroxyhistidine; OHAsp, three-β-hydroxyaspartic acid; DAB, diaminobutanoic acid; P.fl., Pseudomonas fluorescens; P.a., Pseudomonas aeruginosa; P.sp., Pseudomonas species; P.ch., Pseudomonas chlororaphis; P.tol., Pseudomonas tolaasii; P.co., Pseudomonas costantinii; P.p., Pseudomonas putida; P.syr., Pseudomonas syringae.

[b] The amino acid residues involved in iron complexation are indicated in bold type.

taining 8, 9, or 10 residues each have seven members. There are five pyoverdines carrying 6 amino acids, two pyoverdines with 11 residues, and two with 12 residues. The number of amino acid residues determines the spatial conformation necessary for iron binding, as well as the interaction site for the membrane receptor. Most of the amino acids are linked by a peptide bond, except when lysine is involved, as in the pyoverdines of strains B10, G176, 90–51, *P. syringae*, *P. putida* ATCC 12633, and *P. putida* WCS358 (Table 2), in which case an ε-NH$_2$ binding is used.

In conclusion, numerous variations occurring at the level of the peptide chain, as well as in the chromophore or in the chromophore side chain, result in the existence of hundreds of molecules. Indeed, the two major questions that can be raised are (i) why there is such a diversity among molecules having an identical biological function and (ii) how such diversity is possible. The first question is probably related to the strict specificity of *Pseudomonas* strains toward their respective pyoverdines, as can be seen in iron uptake studies. The methods used by bacteria to introduce so many modifications into an otherwise similar structure are related mainly to a biosynthetic pathway based on peptide synthetases, as with the synthesis of peptide antibiotics.

Structure Diversity and Biosynthetic Pathway of Pyoverdines

Because of their short length, their unusual amino acid composition, and the possibility of internal cyclic substructures, the peptide chains of pyoverdines appear to be closely related to peptidic antibiotics of microbial origin such as the gramicidins or tyrocidin produced by *Bacillus* spp., actinomycin from *Streptomyces* spp., and syringomycin and syringotoxin produced by strains of *P. syringae*. Thus, the hypothesis that pyoverdines might be synthesized through a multienzyme thiotemplate mechanism involving peptide synthetases, such as the mechanism used for peptidic antibiotics, was favored over that of synthesis through a classical ribosomal pathway. A hypothetical pathway has been proposed for the biosynthesis of the *P. aeruginosa* ATCC 15692 pyoverdine that integrates the different possible biosynthetic steps and the various pyoverdine-related compounds already recognized as by-products or biosynthetic precursors of pyoverdine. One of the first pieces of indirect biochemical evidence in favor of the nonribosomal pathway was the characterization of iron-regulated cytoplasmic proteins (IRCPs) detectable in iron-starved cells of all fluorescent *Pseudomonas* strains tested. These proteins, which varied in number (two to five) depending on the bacterial strains, were characterized by particularly high molecular masses ranging from 180 to 600 kDa, as tentatively determined by sodium dodecyl sulfate-polyacrylamide gel electrophoresis. Such values are in the same range as the molecular masses of peptide synthetases involved in peptidic antibiotic synthesis. Moreover, comparisons of IRCP profiles showed that these proteins strongly correlated with the type of pyoverdine produced: 10 strains producing structurally different pyoverdines each had a specific IRCP pattern, while strains producing identical pyoverdines demonstrated identical IRCP profiles. It was also shown that some pyoverdine-deficient mutants exhibited a perturbed IRCP profile compared to the wild type. Direct evidence for the existence of pyoverdine-related peptide synthetases came from the characterization of specific genes, mainly in *P. aeruginosa* ATCC 15692 (PAO1 strain). The encoded proteins had all the characteristics defined for antibiotic peptide synthetase components, i.e., adenylation, thiolation, and condensation domains, forming well-differentiated biosynthetic modules. Besides *pvdD* and *pvdI*, another recently described peptide synthetase gene within the PAO1 chromosome is *pvdL*. Interestingly, the modules found in PvdL appear to bind and activate precisely those amino acyl residues, i.e., diaminobutanoic acid, D-tyrosine, and L-glutamic acid, predicted to be involved in the biosynthesis of the chromophore. Thus, if the assumptions concerning the biochemical function of PvdL are correct, its corresponding gene should be present in each pyoverdine-producing strain, since the chromophore part is identical in all pyoverdines. In silico DNA sequence comparisons, as well as IRCP pattern comparisons, favor such a general distribution of *pvdL*: for instance, the 480-kDa IRCP detected in PAO1 is the protein which corresponds best to PvdL (4,363 estimated amino acid residues versus 4,342 residues for PvdL). A protein with an apparent molecular mass identical or close to the experimental average value (480 ± 22 kDa) was found in 10 of the 11 IRCP patterns. The other IRCPs presented a great, strain-dependent variability and should therefore correspond to peptide synthetases involved in the biosynthesis of the peptide part of the respective pyoverdines. They probably harbor specific biosynthetic modules responsible for synthesizing the variable peptide chains. The present view concerning peptide synthetases is that these modules are functionally and structurally independent and can be rearranged in different orders and combinations to generate different products. Pyoverdines, by their huge structural diversity, thus represent a particularly attractive field of investigation, namely, to precisely define the different modules necessary for the fluorescent *Pseudomonas* species to generate so many different structures.

SIDEROTYPING

With the increasing number of structurally known pyoverdines, the probability of finding an already published structure when studying an unknown pyoverdine is increasing. Some of the structures in Table 2 have been described twice in the literature, and the pyoverdine of strain B10, a North American isolate, has been found independently in two other strains, originating from Italy and Ireland. Thus, a method by which a pyoverdine under study is easily identified as having a novel structure is greatly needed. Such a typing method must be highly discriminative and easy to carry out. It could be an analytical method, focused on the pyoverdine molecule. However, in order to remain attractive, the method should not require a highly pure pyoverdine preparation involving many purification steps. The typing method could also be a biological method focused on the pyoverdine-producing strain, its goal being to discriminate between bacterial strains and to ensure that the producing isolate is not identical to any of the known pyoverdine producers. Indeed, strain identification by the classical taxonomic method would be time-consuming and is not considered here. Instead, the high level of specificity of the pyoverdine-mediated iron uptake system exhibited by a fluorescent pseudomonad could be used as an efficient bacterial discriminative tool.

Several analytical and biological methods have been tested in our laboratory. Two of them, namely, pyoverdine IEF and pyoverdine-mediated iron uptake assays, are presently used routinely, but other methods have been described.

IEF as a Powerful Analytical Siderotyping Method

The method of choice for a rapid characterization of pyoverdines is IEF, a method of electophoresis conventionally used for discriminating proteins on the basis of their isoelectric pH values (pH_i). Figure 3 shows the profiles of several of the different pyoverdine isoforms that are often found in a culture supernatant of a given strain (succinate, succinamide, malate, malamide, and α-ketoglutarate [see above]). They are usually well separated on a gel and contribute to several fluorescent bands characterized by different pH_i values. Strains producing identical pyoverdines usually have the same IEF profile (Fig. 3, lanes 5 and 6), whereas strains producing structurally different pyoverdines typically demonstrate different IEF patterns (lanes 1, 2, 3, 5, 7, 8, and 9). Nonfluorescent siderophores produced by fluorescent pseudomonads, together with pyoverdines, e.g., pyochelin, salicylic acid, or quinolobactin, and other compounds such as those produced by nonfluorescent pseudomonads, e.g., ornibactins, cepabactin, or desferrioxamines, are revealed by an overlay of 1% melted agarose in CAS reagent, which reveals siderophores as yellow to pink spots appearing at the surface

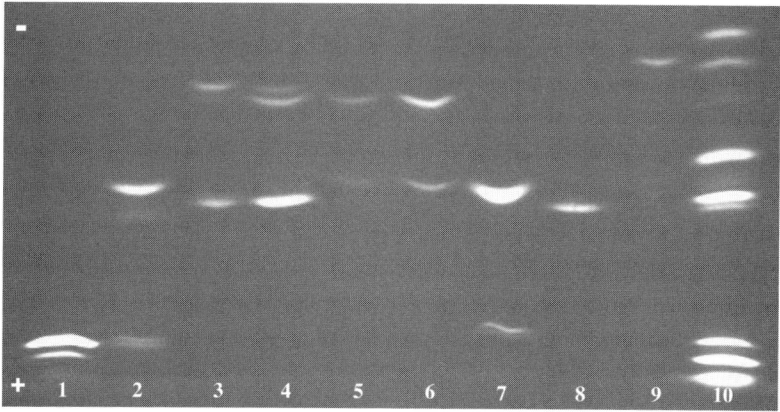

FIGURE 3 IEF patterns of pyoverdines produced by *Pseudomonas* sp. strain LBSA1 (lane 1), *P. fluorescens* CIP 59.27 (lane 2), *P. fluorescens* ATCC 13525 (lane 3), *P. fluorescens* CIP 73.25 (lane 4), *P. chlororaphis* CIP 75.23 (lane 5), *P. chlororaphis* CIP 103295 (lane 6), *P. fluorescens* ATCC 17559 (lane 7), *P. fluorescens* CIP 104377 (lane 8), *P. tolaasii* CIP 106735 (lane 9), and the internal pH$_i$ standard (lane 10). A convenient device for electrofocusing is the mini-IEF gel apparatus from Bio-Rad. Following the manufacturer's recommendations, 5% polyacrylamide gels (10 by 6.5 cm; 0.4 mm thick) containing commercially available ampholines (the large pH 3 to 10 range is the most useful) are freshly cast (within 1 h), loaded with 10 to 20 samples (usually 1 μl of a 20 fold-concentrated CAA culture supernatant or 1 μl of a 5 mM purified siderophore aqueous solution), and electrophoresed for 15 min at 100 V, 15 min at 200 V, and 1 h at 450 V. The bright fluorescent pyoverdine bands are detected during exposure of the samples to UV light (350 nm).

of the gel. The pH$_i$ is determined from an experimental calibration curve constructed by using commercially available standards or by slicing the electrophoresed gel into 0.5-cm bands which are incubated in 2 ml of 10 mM KCl for 30 min before measurement of the pH. This general procedure has been established with a small collection of pyoverdines of different bacterial origin that were selected for their full coverage of the pH gradient. A mixture of these pyoverdines is now used systematically as an internal standard for pH$_i$ determination (Fig. 3, lane 10).

With the increasing number of structurally different pyoverdines, their discrimination by IEF may become a less reliable siderotyping test. Certain compounds differing only slightly in structure can have almost identical IEF profiles (e.g., pyoverdines of *P. fluorescens* 1.3 and *P. putida* ATCC 17400, which differ by a single neutral amino acid [Table 2]). On the other hand, structurally identical pyoverdines have been observed to have somewhat modified IEF patterns, differing by the number of isoform bands (Fig. 3, lanes 3 and 4). Therefore, a second siderotyping method based on the biological function of pyoverdines should be used together with the IEF method.

Pyoverdine-Mediated Iron Uptake as a Powerful Biological Siderotyping Method

The siderophore-mediated iron uptake capacity of a given strain can be easily measured by incubating iron-starved cells in a nonproliferation incubation medium in the presence of a ferrisiderophore labeled with ^{59}Fe. The cells are then rapidly filtered and counted for radioactivity. The incubation medium is the same as the one used for cell growth, except that it does not contain one element essential for growth, e.g., nitrogen. The labeled ferrisiderophore is prepared by mixing a ^{59}FeCl$_3$ commercial solution (5 μl) with an excess of the purified ligand (i.e., 50 μl of a 1 mM pyoverdine solution) to ensure a total solubilization of the iron through complexation. Since some components of the incubation medium (i.e., the phosphates) could interfere by slowing the siderophore-iron complex formation, controls without bacteria are included to verify the iron solubility. Cells are carefully washed to remove the native siderophore and resuspended in nitrogen-free medium to reach a cell suspension of 0.3 optical density unit at 600 nm units. The ^{59}Fe-pyoverdine complex is added to the cell suspension at time zero, and the mixture is incubated at 30°C with gentle shaking. Samples (1 ml) are withdrawn at 2, 5, 10, 15, and 20 min and rapidly filtered on an ultrafiltration membrane (0.45-μm pore size). The membranes are then washed twice with 2 ml of fresh incubation medium and wrapped in aluminum foil, and the radioactivity is measured. The total radioactivity per milliliter of the incubation mixture, corresponding to a known amount of labeled iron, is estimated by counting 1 ml of unfiltered bacterial suspension. The amount of labeled iron associated with the cells is determined accordingly.

In the context of siderotyping, the protocol could be simplified by doing a single sampling after 20 min of incubation. Kinetics studies showed that maximal incorporation is usually reached at that time. Thus, the volume of the incubation medium could be limited to 2 ml, making it feasible to test many different pyoverdines in one experiment. A diagram like the one depicted in Fig. 4, showing the pyoverdine-mediated iron uptake capacity of the *P. costantinii* type strain toward its own pyoverdine and 34 other pyoverdines of various bacterial origin, could be obtained within 3 h of starting the cell harvest. Thus, the method allows a rapid investigation of many different pyoverdines within a reasonable length of time and therefore could be used for analyzing collections of strains containing a large number of isolates.

Figure 4 shows that the *P. costantinii* strain is able to recognize, and use at close to 100% efficiency, the pyoverdine produced by a strain tentatively described as a *P. aureofaciens* strain. It is well established that the *P. aureofaciens* strain efficiently incorporates the pyoverdine of *P. costantinii* and that the two pyoverdines have

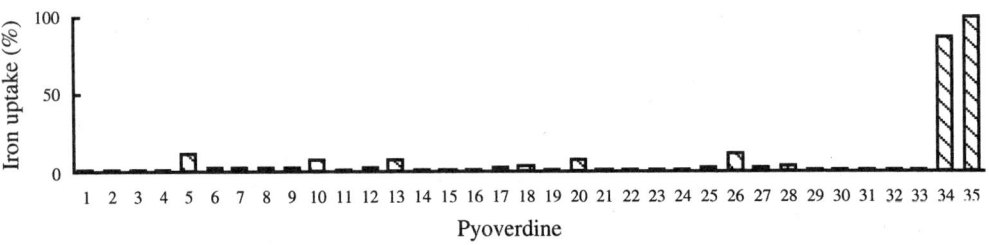

FIGURE 4 Heterologous pyoverdine-mediated ^{59}Fe incorporation in *P. costantinii* CFBP 5705. Values on the ordinate correspond to the percentage of ^{59}Fe incorporation after 20 min of incubation, compared to the homologous system. Numbers 1 to 35 on the abscissa correspond to the different pyoverdines tested, originating from the following bacterial strains: 1, *Pseudomonas* sp. strain E8; 2, *P. syringae* ATCC 19310; 3, *P. fluorescens* 9AW; 4, *P. putida* ATCC 12633; 5, *P. fluorescens* 51W; 6, *P. aeruginosa*, Pa6; 7, *P. fluorescens* CCM 2798; 8, *P. fluorescens* CHA0; 9, *P. tolaasii* LMG 2342; 10, *P. aeruginosa* ATCC 27853; 11, *P. fluorescens* ii; 12, *P. fluorescens* SB8.3; 13, *P. fluorescens* ATCC 17400; 14, *P. fluorescens* 1.3; 15, *Pseudomonas* sp. strain 267; 16, *P. fluorescens* ATCC 13525; 17, *P. aeruginosa* ATCC 15692; 18, *P. fluorescens* 18.1; 19, *P. fluorescens* 12; 20, *P. fluorescens* CFBP 2392; 21, *P. putida* CFBP 2461; 22, *Pseudomonas* sp. strain ATCC 15915; 23, *P. monteilii* CFML 90–54; 24, *P. mosselii* CFML 90–77; 25, *P. rhodesiae* CFML 92–104; 26, *P. putida* CFML 90–33; 27, *Pseudomonas* sp. strain CFML 90–40; 28, *Pseudomonas* sp. strain CFML 90–42; 29, *Pseudomonas* sp. strain CFML 90–51; 30, *Pseudomonas* sp. strain CFML 90.52; 31, *Pseudomonas* sp. strain 7SR1; 32, *Pseudomonas* sp. strain 2908; 33, *Pseudomonas* sp. strain A214; 34, *P. aureofaciens*; 35, *P. costantinii* CFBP 5705.

identical IEF patterns. Thus, the two siderotyping methods suggested an identity between the two pyoverdines, a conclusion that was confirmed by structural studies.

Pyoverdine-mediated iron uptake studies also allowed the grouping of some strains demonstrating efficient cross-incorporations, although their respective pyoverdines had different IEF patterns. This is illustrated in Table 3 for two groups of two and three pyoverdines, respectively. The difference in structure between the pyoverdines of *P. chlororaphis* PL16 and *P. fluorescens* CHA0 concerns a unique amino acyl residue (an alanine in PL16, and a lysine in CHA0). Indeed, the change has a strong effect on the overall electric charge of the two molecules but apparently does not have a drastic effect on iron uptake. It can be concluded that the ferripyoverdine receptor recognition site in these two pyoverdines must be located in the other, common part of the peptide chains. The recognition site could be more precisely defined for the three cross-reacting *P. fluorescens* PL7, *P. fluorescens* PL8, and *P. aeruginosa* ATCC 27853 pyoverdines (Table 3), since the common peptide part in these three compounds is limited to the dipeptidyl motif Gly-D-*allo*-Thr. Interestingly, the same motif is present in some other pyoverdines, but at different positions in the peptide chains. These pyoverdines did not cross-react with the PL7-PL8-ATCC 27853 group, demonstrating that the spatial position of the motif is an important prerequisite for the binding between the Fe^{3+}-pyoverdine complex and its outer membrane receptor.

Siderotyping, Siderotypes, and Siderovars

The main goal of siderotyping was initially to detect novel structures of pyoverdines by analyzing strains with a particular siderotype, i.e., a specific pyoverdine IEF pattern and a strict specificity of uptake of the pyoverdine it produces. Strains showing an identical pyoverdine IEF pattern and cross-reacting at about 100% efficiency in pyoverdine-mediated iron uptake studies were classified in the same bacterial group named a siderovar. In most cases, structural studies confirmed that the different siderotypes, as defined by IEF and iron uptake, were each characterized by a particular pyoverdine structure. Conversely, these studies demonstrated identical pyoverdine structures for strains with an identical siderotype. An example is the analysis of fluorescent strains isolated from Antarctic soils. In that study, the siderotyping patterns of two of the isolates were identical to those of the type strain *P. fluorescens* ATCC 13525, for which the pyoverdine structure was already known. Two other isolates shared an apparently new siderotype, resulting in the identification of an original pyoverdine structure (the pyoverdine of strain 9AW in Table 2). Another strain with unique siderotyping features was also found to synthesize an original pyoverdine (the pyoverdine of strain 51W in Table 2). It should be stated that no

TABLE 3 Examples of structurally related pyoverdines demonstrating cross-reactivity in pyoverdine-mediated iron uptake[a]

Strain	Pyoverdine peptide structure[b]	Pyoverdine isoform pH$_i$
P. chlororaphis PL16	**Asp-fOHOrn-Lys-(Thr-Ala-Ala-fOHOrn-Ala)**	7.5–5.4–4.1
P. fluorescens CHA0	**Asp-fOHOrn-Lys-(Thr-Ala-Ala-fOHOrn**-Lys)	8.5–7.5–5.3
P. aeruginosa ATCC 27853	Ser-fOHOrn-Orn-**Gly-aThr**-Ser-cOHOrn	8.9–8.7–7.6–7.4
P. fluorescens PL8	Lys-aOHOrn-Ala-**Gly-aThr**-Ser-cOHOrn	9.0–8.9–7.6
P. fluorescens PL7	Ser-aOHOrn-Ala-**Gly-aThr**-Ala-cOHOrn	7.7–5.2

[a] The common parts in the peptide chains are in bold type.
[b] Abbreviations: aOHOrn, δN-acetyl-δN-hydroxyornithine; fOHOrn, δN-formyl-δN-hydroxyornithine; cOHOrn, cylohydroxyornithine (3-amino-1-hydroxypiperidone-2); a Thr, allo-Thr. The D-amino acids are underlined. A dotted underline in the first two pyoverdine peptides means that one Ala is an L-form and the other is a D-form.

endemic pyoverdines specific to Antarctica were revealed during that study: according to data gathered in our laboratory, the *P. fluorescens* ATCC 13525 and the *P. fluorescens* 9AW siderotypes are very common and widely distributed in nature (see below). The pyoverdine of strain 51W, however, appeared to be particularly rare, since it was found only in the corresponding strain from the Antarctic and in one strain isolated in Finland.

At present, the two siderotyping methods have allowed us to recognize 106 different siderotypes among nearly 2,000 fluorescent pseudomonads analyzed. The uniqueness of about two-thirds of these pyoverdines has been confirmed by structural studies. They correspond to the 44 well-defined structures described in Table 2, and to 35 other pyoverdines for which partial structure information has already been obtained in the laboratory of H. Budzikiewicz, Köln University, Cologne, Germany. Thus, 27 pyoverdine structures remain to be determined.

Siderotyping and Taxonomy

The taxonomy of *Pseudomonas* has undergone many changes during the last 10 years, mostly thanks to genomic tools which, together with the phenotypic methods, were used for bacterial identification on the basis of polyphasic taxonomy. New genera were created, and many new species in the *Pseudomonas* sensu stricto genus were recognized. Some of these corresponded to strains first identified by a limited set of phenotypic characteristics as *P. fluorescens* or *P. putida* (e.g., the species *Pseudomonas monteilii* and *Pseudomonas mosselii*, which were derived from the *P. putida* group). From siderotyping studies done with numerous strains belonging to taxonomically well-defined new taxa (13 species and 26 clusters or genomospecies), the following general rules were defined: (i) all strains belonging to a given species have an identical siderotype, i.e., produce an identical pyoverdine; and (ii) each species is characterized by an original pyoverdine.

It became clear from these general conclusions that siderotyping could be very useful for bacterial identification and *Pseudomonas* taxonomy: following the recognition of the type of pyoverdine it produces, a taxonomically undefined fluorescent pseudomonad is classified in a corresponding siderovar. Then a species name can be assigned if the siderovar already contains bacteria belonging to a taxonomically well-defined species. Such an approach has been successfully used to classify phenotypic clusters into defined species, in agreement with DNA hybridization experiments which confirmed that the corresponding taxa hybridized at 70% or more. Indeed, any siderovar that presently remains unrelated to a well-defined species could be considered to represent a putative new species. This has been recently demonstrated for the newly described *P. costantinii* species, which fulfills all the criteria required for a species and whose strains produce the original pyoverdine shown in Fig. 1.

Table 4 gives an overview of the data collected to date in our laboratory, showing that a majority of well-defined fluorescent *Pseudomonas* species (13 of 22) can be perfectly correlated to a unique siderovar. It also shows, however, that a few species (so far, *P. aeruginosa*, *P. grimontii*, and *P. lini*) do not fit with the general rule of "one species = one siderovar" but are composed of strains belonging to two or three different siderovars. Another exception is that a unique siderotype may be distributed within several species. This was shown for *Pseudomonas brenneri* and *Pseudomonas gessardii*, with both species containing the siderotype of strain *P. fluorescens* W. Another example is given by the unique pyoverdine found within strains belonging to the different genomospecies of the *P. syringae* complex. So far, the most widely distributed siderotype among fluorescent *Pseudomonas* species is that of the *P. fluorescens* ATCC 13525 pyoverdine, which, according to electrospray mass spectrometry, appears to be produced also by representatives of the *Pseudomonas orientalis*, *Pseudomonas cedrella*, *Pseudomonas palleroniana*, and *Pseudomonas veronii* species.

In contrast, strains clustered according to a limited number of phenotypic characteristics

TABLE 4 Correlation between siderotype and species

Species	No. of isolates	No. of siderotypes
P. cedrella	5	
P. orientalis	6	1
P. palleroniana	11	
P. veronii	8	
P. brenneri	12	1
P. gessardii	13	
P. jessenii	8	1
P. migulae	10	
P. brassicacearum	10	1
P. costantinii	6	1
P. fuscovaginae	13	1
P. kilonensis	12	1
P. libanensis	6	1
P. mandelii	25	1
P. monteilii	9	1
P. mosselii	11	1
P. rhodesiae	7	1
P. salomonii	15	1
P. syringae	26	1
P. thivervalensis	6	1
P. tolaasii	8	1
P. lini	15	2
P. grimontii	34	3
P. aeruginosa	88	3
P. fluorescens biovar I	28	10
P. fluorescens biovar II	45	13
P. fluorescens biovar III	39	20
P. fluorescens biovar IV	19	8
P. fluorescens biovar V	48	20
P. putida biovars A + B	70	33

characteristics are taxonomically heterogeneous (a feature presently well recognized by the taxonomists) but also may be used to propose a new strain allocation within a polyphasic defined species. Hence, by significantly limiting the panel of type strains to be investigated for comparison purposes, siderotyping information can be of great value to the taxonomist.

Siderotyping and Phylogeny

Since the peptide part of pyoverdines is under the control of well-defined enzymatic blocks programmed by genetic modules (see chapters 2 and 19), it is not astonishing to find some evidence for a link between pyoverdine structure and bacterial phylogeny. Such evidence can be obtained by correlating the structures of pyoverdines produced by taxonomically well-defined strains with the positions of these strains in the phylogenetic tree derived from 16S rDNA sequences. For instance, most of the phylogenetically related strains forming the "P. syringae group" produce the same pyoverdine. Among two exceptions, P. meliae strains were not typeable because of the absence of a pyoverdine system while P. cichorii strains demonstrated some heterogeneity at the pyoverdine level with, noteworthyingly, some of them producing a pyoverdine of the P. syringae type. Another example concerns the four species sharing the siderotype of P. fluorescens ATCC 13525, namely, P. orientalis, P. cedrella, P. palleroniana, and P. veronii, which were all shown recently to be phylogenetically related to P. fluorescens ATCC 13525.

Better knowledge about the pyoverdine structures and the precise taxonomic positions of the pyoverdine-producing strains is necessary before we can establish additional correlations. It remains intriguing that bacteria producing isopyoverdines instead of pyoverdines belong exclusively to the P. putida complex and that most of the P. putida pyoverdine peptides are linked to the chromophore by an aspartyl residue, while alanine, serine and lysine are usually found in the pyoverdines of other species. Thus, a species-specific distribution of

within the different biovars of the P. fluorescens or P. putida taxonomic groups demonstrate a great heterogeneity at the sidcrotyping level. A minimum of 8 different siderotypes have been recognized among 19 strains belonging to the P. fluorescens biovar IV, while 33 siderotypes were found among the 70 strains classified in the two biovars A and B of the P. putida group (Table 4). Moreover, as shown above, some of the 10 siderovars found in P. fluorescens biovar I corresponded to siderotypes characterizing the newly described P. jessenii, P. mandelii, or P. rhodesiae species. Thus, siderotyping not only suggests that strains recognized as P. fluorescens or P. putida by a restricted panel of phenotypic peptide synthetases involved in pyoverdine biosynthesis could be postulated.

Siderotyping and Environmental Microbiology

The methods of siderotyping are very fast and easy to perform. They require limited amounts of biological material (at the most a few milliliters of a bacterial culture). Thus, they are particularly convenient for studies requiring simultaneous analyses, for comparison purposes, of a great number of strains, as is usually the case in epidemiology, bacterial taxonomy, or any study relevant to bacterial populations. The method has already proved its efficiency in the bacterial identification of fluorescent *Pseudomonas* species and in the detection of new species.

So far, our experience has extended to nearly 2,000 isolates obtained within more than 30 collections from different geographical origins and from different natural environments worldwide. To date, they have allowed the recognition of 106 different siderotypes. Approximately one-third have been correlated with well-defined species, while the rest have yet to be classified. Other aspects, besides taxonomy, could be easily developed using siderotyping. Biodiversity could be evaluated by detecting siderotypes of natural environmental isolates. Among the collections analyzed, the *P. fluorescens* ATCC 13525 siderotype is the most widely distributed (Table 5). As described above, this siderotype is characteristic of several species (Table 4). This peculiarity could explain its prevalence. *P. mandelii* and *P. rhodesiae* siderotypes are also well distributed among mineral water isolates and very often are detected by siderotyping in collections originating from natural sources, rivers, or soils. Interestingly, the pyoverdine of *Pseudomonas* sp. strain B10, which has been described several times (see above), appears to be widespread, since it was found in 62 strains within 11 collections, with a majority of these strains being isolated from plants or cultivated soil.

Another benefit of siderotyping might be the clarification of the clonal status of isolates among a bacterial population. In cases where the IEF patterns are well differentiated, the single IEF method might be sufficient to successfully and very quickly discriminate between isolates of clonal origin or those belonging to the same siderovar (which demonstrate a unique IEF pattern) and isolates very probably belonging to different species (which are characterized by different IEF patterns). Thus, an answer could be reached within a day for as many as 100 isolates. Another application of

TABLE 5 The most common siderotypes and their corresponding siderovars found in 28 bacterial collections of diverse environmental origins representing 1,340 fluorescent *Pseudomonas* isolates[a]

Siderotype	No. of isolates	No. of collections	Putative siderovar species
13525	155	18	*Pseudomonas fluorescens* sv. 13525, *Pseudomonas chlororaphis* sv. 9446, *Pseudomonas orientalis*, *Pseudomonas cedrella*, *Pseudomonas grimontii* sv. 13525, *Pseudomonas veronii*
96–318	118	7	*Pseudomonas lomagnae*
SB8.3	95	12	*Pseudomonas mandelii*
syr	79	2	*Pseudomonas syringae*
W	70	7	*Pseudomonas gessardii*
B10	62	11	*Pseudomonas* sp.
9AW	55	9	*Pseudomonas jessenii*
PL9	37	3	*Pseudomonas lini*
Lille25	37	11	*Pseudomonas rhodesiae*
18.1	34	9	*Pseudomonas grimontii* sv. 18.1
PL8	33	7	*Pseudomonas* sp.

[a] *P. aeruginosa* strains of clinical origin are not included. sv., serovar.

the method could be to trace and measure the survival of field-released bacteria, as long as the siderotype of the experimental strain is not represented within the natural endemic strains. Finally, the method could provide an easy way to search for new isolates of a given bacterial species, by selecting for strains able to specifically incorporate the corresponding pyoverdine. Meanwhile, it should be possible to select poorly represented fluorescent pseudomonads within an heterogenous bacterial population. Providing that the medium is supplemented with pyoverdine, growth of the pyoverdine-utilizing bacteria should be greatly favored, resulting in a specific enrichment of the pseudomonads belonging to the pyoverdine-corresponding siderovar. Depending on the specificity level expressed by the pyoverdine, it should thus be possible to achieve, during a single experimental step, both isolation and identification of a fluorescent *Pseudomonas* strain.

ACKNOWLEDGMENTS

We are grateful to H. Budzikiewicz and his team for the enthusiastic and fruitful collaboration maintained for many years, allowing the identification of many pyoverdines and siderophores and thus actively supporting siderotyping since its beginning.

SUGGESTED READING

Anzai, Y., H. Kim, J.-Y. Park, H. Wakabayashi, and H. Oyaizu. 2000. Phylogenic affiliation of the pseudomonads based on 16S rRNA sequence. *Int. J. Syst. Evol. Microbiol.* **50:**1563–1589.

Böckmann, M., K. Taraz, and H. Budzikiewicz. 1997. Biogenesis of the pyoverdine chromophore. *Z. Naturforsch.* **52C:**319–324.

Bossis, E., P. Lemanceau, X. Latour, and L. Gardan. 2000. The taxonomy of *Pseudomonas fluorescens* and *Pseudomonas putida*: current status and need for revision. *Agronomie* **20:**51–63.

Budzikiewicz, H. 2004. Siderophores of the Pseudomonadaceae *sensu stricto* (fluorescent and non-fluorescent *Pseudomonas* spp.). *Prog. Chem. Org. Nat. Prod.* **87:**81–237.

Fernandez, D. U., R. Fuchs, K. Taraz, H. Budzikiewicz, P. Munsch, and J.-M. Meyer. 2001. The structure of a pyoverdine produced by a *Pseudomonas tolaasii*-like isolate. *BioMetals* **14:**81–84.

Fuchs, R., M. Schäfer, V. Geoffroy, and J.-M. Meyer. 2001. Siderotyping: a powerful tool for the characterization of pyoverdines. *Curr. Top. Med. Chem.* **1:**31–35.

Gardan, L., P. Bella, J.-M. Meyer, R. Christen, P. Rott, W. Achouak, and R. Samson. 2002. *Pseudomonas salomonii* sp. nov., pathogenic on garlic, and *Pseudomonas palleroniana* sp. nov., isolated from rice. *Int. J. Syst. Env. Microbiol.* **52:**2065–2074.

Georges, C., and J.-M. Meyer. 1995. High-molecular-mass, iron-repressed cytoplasmic proteins in fluorescent *Pseudomonas*: potential peptide-synthetases for pyoverdine biosynthesis. *FEMS Microbiol. Lett.* **132:**9–15.

Hohnadel, D., and J.-M. Meyer. 1988. Specificity of pyoverdine-mediated iron uptake among fluorescent *Pseudomonas* strains. *J. Bacteriol.* **170:** 4865–4873.

Kersters, K., W. Ludwig, M. Vancanneyt, P. Devos, M. Gillis, and K. H. Schleifer. 1996. Recent changes in the classification of pseudomonads: an overview. *Syst. Appl. Microbiol.* **19:**465–477.

Kleinkauf, H., and H. von Dörhen. 1996. A nonribosomal system of peptide biosynthesis. *Eur. J. Biochem.* **236:**335–351.

Koedam, N., E. Wittouck, A. Gaballa, A. Gillis, M. Höfte, and P. Cornelis. 1994. Detection and differentiation of microbial siderophores by isoelectric focusing and chrome azurol S overlay. *BioMetals* **7:**287–291.

Konz, D., and M. A. Marahiel. 1999. How do peptide synthetases generate structural diversity? *Chem. Biol.* **6:**R39–R48.

Lehoux, D.E., F. Sanschagrin, and R.C. Levesque. 2000. Genomics of the 35-kb *pvd* locus and analysis of novel *pvdIJK* genes implicated in pyoverdine biosynthesis in *Pseudomonas aeruginosa*. *FEMS Microbiol. Lett.* **190:**141–146.

Maurer, B., A. Müller, W. Keller-Schierlein, and H. Zähner. 1968. Ferribactin, ein Siderochrom aus *Pseudomonas fluorescens* Migula. *Arch. Microbiol.* **60:** 326–329.

Merriman, T.R., M.E. Merriman, and I.L. Lamont. 1995. Nucleotide sequence of *pvdD*, a pyoverdine biosynthetic gene from *Pseudomonas aeruginosa*: PvdD has similarity to peptide synthetases. *J. Bacteriol.* **177:**252–258.

Meyer, J.-M. 2000. Pyoverdines: pigments, siderophores and potential taxonomic markers of fluorescent *Pseudomonas* species. *Arch. Microbiol.* **174:** 2745–2753.

Meyer, J.-M., A. Stintzi, V. Coulanges, S. Shivaji, J.A. Voss, K. Taraz, and H. Budzikiewicz. 1998. Siderotyping of fluorescent pseudomonads: characterization of pyoverdines of *Pseudomonas fluorescens* and *Pseudomonas putida* strains from Antarctica. *Microbiology* **144:**3119–3126.

Meyer, J.-M., V.A. Geoffroy, N. Baida, L. Gardan, D. Izard, P. Lemanceau, W. Achouak, and N. Palleroni. 2002. Siderophore typing, a powerful tool for the identification of fluorescent and nonfluorescent pseudomonads. *Appl. Environ. Microbiol.* **68:**2745–2753.

Meyer, J.-M., V. A. Geoffroy, C. Baysse, P. Cornelis, I. Barelmann, K. Taraz, and H. Budzikiewicz. 2002. Siderophore-mediated iron uptake in fluorescent *Pseudomonas*: characterization of the pyoverdine-receptor binding site of three cross-reacting pyoverdines. *Arch. Biochem. Biophys.* **397:**179–183.

Mossialos, D., U. Ochsner, C. Baysse, P. Chablain, J.-P. Pirnay, N. Koedam, H. Budzikiewicz, D. U. Fernandez, M. Schäfer, J. Revel, and P. Cornelis. 2002. Identification of new, conserved, non-ribosomal peptide synthetases from fluorescent pseudomonads involved in the biosynthesis of the siderophore pyoverdine. *Mol. Microbiol.* **45:**1673–1685.

Munsch, P., T. Alatossava, N. Marttinen, J.-M. Meyer, R. Christen, and L. Gardan. 2002. *Pseudomonas costantinii* sp. nov., another causal agent of brown blotch disease, isolated from cultivated mushroom sporophores in Finland. *Int. J. Syst. Evol. Microbiol.* **52:**1973–1983.

Palleroni, N. J. 1984. *Pseudomonas*, p. 141–199. *In* N. R. Krieg and J. G. Holt (ed.), *Bergey's Manual of Systematic bacteriology*, vol. 1, The Williams & Wilkins Co., Baltimore, Md.

Quadri, L. E. 2000. Assembly of aryl-capped siderophores by modular peptide synthetases and polyketide synthases. *Mol. Microbiol.* **37:**1–12.

Ravel, J., and P. Cornelis. 2003. Genomics of pyoverdine-mediated iron uptake in pseudomonads. *Trends Microbiol.* **11:**195–200.

Vandamme, P., B. Pot, M. Gillis, P. De Vos, K. Kersters, and J. Swings. 1996. Polyphasic taxonomy, a consensus approach to bacterial systematics. *Microbiol. Rev.* **60:**407–438.

MECHANISMS AND REGULATION OF IRON UPTAKE IN THE RHIZOBIA

Andrew W. B. Johnston

30

RHIZOBIA AND THEIR NEED FOR IRON

The rhizobia probably comprise the single most important beneficial group of eubacteria known to humans. This is because they form and occupy nodules on the roots (and, rarely, the stems) of leguminous plants, within which they reduce or "fix" atmospheric N_2 to NH_3, a form of N that the plants assimilate into nitrogenous compounds. Legumes include some of the world's major crops, such as clover, beans, soybeans, and alfalfa, which require no nitrogenous fertilizers. The benefits of legumes to soil fertility and human nutrition have been known since the time of the Roman Empire, the name "Fabian" being derived from the "grower of Faba beans." Importantly, different legumes are nodulated by different strains, species, or genera of rhizobia. Nearly all of these are α-proteobacteria, although they are widely dispersed within that taxon (those that are included in this chapter are listed in Table 1). At the 16S sequence level of taxon allocation, *Sinorhizobium* and *Rhizobium* are closely related and, indeed, closely resemble the phytopathogen *Agrobacterium* and the mammalian pathogen *Brucella*, the causative agent of brucellosis and other diseases. More distant is *Mesorhizobium*; even less similar is the microsymbiont of soybeans, *Bradyrhizobium*, which is nearly identical to the photosynthetic bacterium *Rhodopseudomonas*.

Over and above its agronomic and ecological importance, the interaction between rhizobia and legumes is an exquisite example of coupled differentiation and development in two very different organisms. The nodule is no mere tumor but, instead, is a complex organ with various cell types that differ markedly in structure, biochemistry, and patterns of gene expression from those in "normal" root tissues. Likewise, the bacteria undergo dramatic changes during nodule development as they differentiate into the N_2-fixing bacteroids, which are amply supplied with nutrients from the host plant. Once the nodules senesce, many millions of rhizobia are released into the soil to replenish the free-living populations of these bacteria.

In both the oligotrophic soils and the nutritionally sumptuous nodule, the ability of rhizobia to acquire iron is an important component of their fitness, but for different reasons in the two environments. In the soil, they must compete with other members of the flora and fauna for the metal, whereas in nodules there is a

---

Andrew W. B. Johnston, School of Biological Sciences, University of East Anglia, Norwich NR4 7TJ, United Kingdom.

TABLE 1 Species of rhizobia referred to in this chapter

Species name	Host plant(s)	Genome size and source[a]
Bradyrhizobium japonicum	Soybeans	ca. 9.1 Mb (http://www.kazusa.or.jp/rhizobase/Bradyrhizobium/index.html)
Mesorhizobium loti	Lotus	7.6 Mb (http://www.kazusa.or.jp/rhizobase/)
Rhizobium leguminosarum		ca. 7.8 Mb (http://www.sanger.ac.uk/Projects/R_leguminosarum/)
bv. phaseoli	*Phaseolus* beans	
bv. trifolii	Clovers	
bv. viciae	Peas, lentils, vetches	
Sinorhizobium meliloti	Alfalfa	6.5 Mb (http://sequence.toulouse.inra.fr/meliloti.html)

[a] Websites of the currently available rhizobial genomic sequences are shown.

particular demand for iron, since many polypeptides required for N_2 fixation are iron-containing proteins. The characteristic pink color of root nodules is due to leghemoglobin, an abundant protein made by the plant that delivers O_2 to the metabolically active rhizobia. Bacteroids, too, have major iron-containing proteins, most notably the polypeptides of the nitrogenase complex. This is a heteromultimer comprising iron-containing nitrogenase reductase plus nitrogenase itself, which contains an iron-molybdenum cofactor at which N_2 reduction occurs. Bacteroids also have abundant cytochromes and other electron donors, all of which contain iron.

Although there have been detailed genetic analyses of several properties of the rhizobia, studies of their iron metabolism have been somewhat neglected despite the importance of this metal. This has begun to be rectified in the last few years, but most studies have been concerned with the behavior of free-living rhizobia, not those in the nodule. However, the recently available sequences of *Bradyrhizobium japonicum*, *Mesorhizobium loti*, *Sinorhizobium meliloti*, and *Rhizobium leguminosarum* (Table 1) have already been of value in the identification of likely genes for iron uptake and, without doubt, will continue to be so.

One lesson from much of the earlier work on rhizobial iron uptake is that measurement of siderophore synthesis or uptake solely on the basis of the sizes of the halos formed by rhizobia on chrome azural sulfonate (CAS) siderophore indicator plates can be misleading. The reduced halos produced on CAS agar by several different classes of mutants were shown to be due to indirect, sometimes trivial effects. These mutants are not considered here, nor is there a list of all the siderophores and/or other iron sources that have been identified for the rhizobia (see chapter 28 for a more detailed list of rhizobial siderophores). This chapter considers only recent experimental and in silico analyses that directly apply to the iron biology of rhizobia.

At least two main sets of observations have emerged from these studies and are recurring themes in this review. The first is that the rhizobia have great flexibility in their use of iron sources compared to many other bacteria. The second is that some paradigms of iron-responsive gene regulation do not operate in the rhizobia or, at least, not all rhizobia. In the following, discussion, the evidence supporting these statements is presented.

MECHANISMS OF IRON UPTAKE BY FREE-LIVING RHIZOBIA

Siderophores

Most strains of the fast-growing rhizobial genera (e.g., *Rhizobium* and *Sinorhizobium*) make at least one siderophore in free-living culture, these being catechols, carboxylates, or di- and trihydroxamates. Siderophore production is much less common in strains of slow-growing rhizobia, notably *B. japonicum*.

The first chemically characterized rhizobial siderophore, rhizobactin, is made by a strain of *S. meliloti* and has, unusually, two iron-ligating carboxyl groups (Fig. 1). Its structure was elucidated in J. B. Neilands' laboratory, but there has been no genetic analysis of this transport system. There have been detailed biochemical and genetic studies of only two rhizobial siderophores. These are the trihydroxamate vicibactin (VB), made by most strains of *R. leguminosarum* (Fig. 1), and the chemically distinct dihydroxamate rhizobactin 1021 (RB1021), synthesized by different strains of *S. meliloti*, including the well-characterized strain 1021—hence the name RB1021 (Fig. 1). The following presents a description of the molecular genetic analyses of the genes involved in the synthesis and uptake of VB and RB1021.

VB Synthesis and Transport

VB is a "hybrid" between a polyketide and a nonribosomal peptide, with a ring structure comprising three residues each of N-2-acetyl-N-5-hydroxy-D-ornithine and hydroxybutanoic acid arranged in alternate ester and peptide bonds (Fig. 1). *R. leguminosarum* has a cluster of eight *vbs* genes arranged in four operons, all of which are involved in VB synthesis (Fig. 2a). A pathway for VB synthesis was proposed by isolating mutants defective in each of these genes and identifying the chemical nature of the VB-like molecules that were formed (Fig. 3; Table 2). Although it remains to be ratified, this scheme accounts for the phenotypes of individual *vbs* mutants and the likely functions of the corresponding protein products, as predicted from their amino acid sequences. Although some of the steps resemble those in other siderophore biosynthetic pathways, there are also some unusual features. It is proposed that the first building block is hydroxylated ornithine (made by VbsO) joined to and then activated by VbsS. VbsS is a large, multicomponent protein similar to many nonribosomal peptide synthases, which is modified, via VbsP, by the addition of pantothenate cofactor. The remaining steps occur on this multifunctional VbsS enzyme scaffold, with different Vbs gene products supplying the enzymatic modifications needed for certain steps (Table 2; Figure 3). Unlike other peptide synthases, VbsS has no epimerization domain, and it was proposed that VbsL is the cognate epimerase for this step. Remarkably, VbsL closely resembles the MetB cystathione-γ-lyase of *Escherichia coli* and other bacteria, the epimerase activity of such lyases being subsumed for this step in the siderophore biosynthesis. VbsL$^-$ mutants make a functional, Fe^{3+}-binding siderophore, but the ornithine is not acetylated. Thus, the D form of the siderophore still binds Fe^{3+}, but epimerization must be needed for the final, acetylation step. This final modification (thought to be catalyzed by the acetylase VbsC) is also not required for iron-binding activity but may stabilize or otherwise "improve" the activity of the mature VB molecule. The *vbsG* gene, although absolutely needed for VB synthesis, has no identified function. However, VbsG homologues occur in other bacteria, where they are also required to synthesize other polyketides and non ribosomal peptides. VbsD resembles a family of export proteins, suggesting a role in siderophore export, a topic that, in general, receives little attention. The *vbs* gene cluster is sufficient for VB synthesis, since its transfer on a wide-host-range plasmid to other α-proteobacteria, such as *Paracoccus* or *Rhodobacter*, conferred on them the ability to make authentic VB.

Turning to the uptake of VB, *R. leguminosarum* appears to have a fairly conventional Fhu-like uptake ABC transporter, similar to that described for *E. coli* (see chapter 11). In the 8401(pRL1JI) strain of *R. leguminosarum*, the *fhuA1* gene that specifies the outer membrane VB-Fe^{3+} receptor is adjacent to the *vbs* genes (Fig. 2a) and is followed, in the same operon, by a gene whose product resembles FhuF, an *E. coli* ferrireductase that probably removes Fe^{3+} from hydroxamates. The other three genes, *fhuCDB*, which specify the periplasmic protein (FhuD), the inner membrane transporter (FhuB), and the cognate ATPase

Vicibactin (*Rhizobium*)

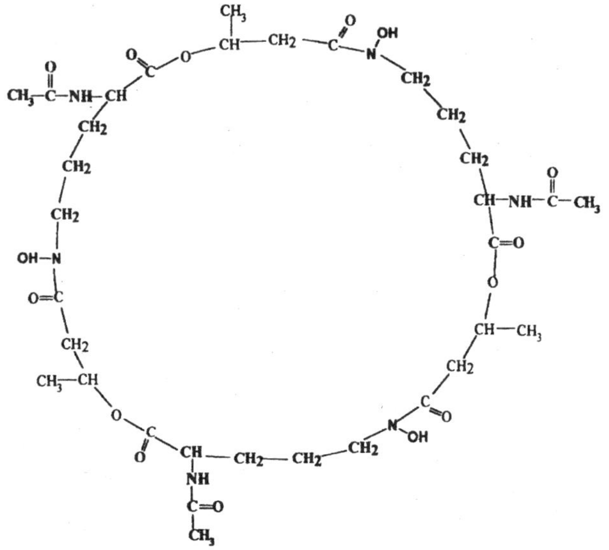

Rhizobactin 1021 (*Sinorhizobium*)

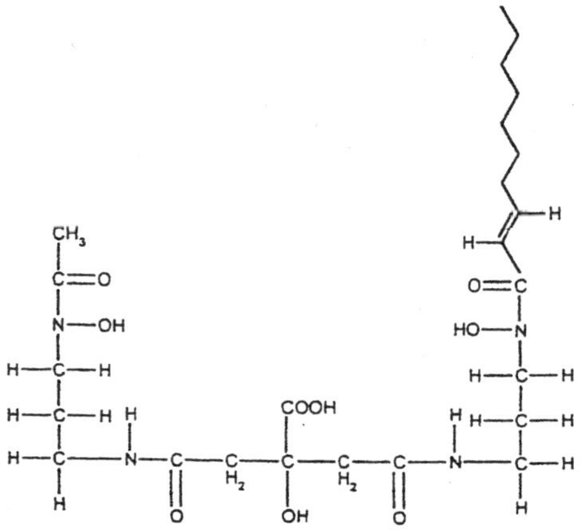

Rhizobactin (*Sinorhizobium*)

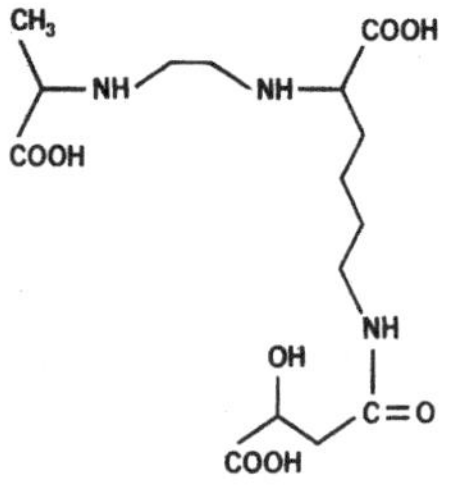

FIGURE 1 Structures of three chemically characterized siderophores made by rhizobia.

(a) Genes for VB biosynthesis and uptake in *Rhizobium leguminosarum*

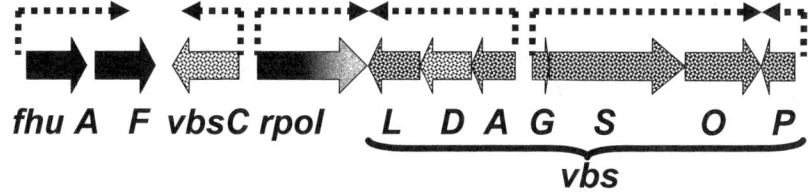

(b) Genes for RB1021 biosynthesis and uptake in *Sinorhizobium meliloti*

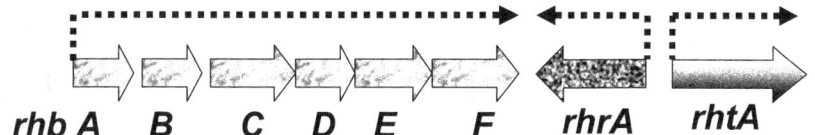

FIGURE 2 Organization of gene clusters involved in siderophore synthesis and uptake in *R. leguminosarum* and *S. meliloti*. Relative locations of genes for the synthesis and uptake of the siderophores VB by *R. leguminosarum* (a) and RB1021 by *S. meliloti* (b) are shown. Dotted lines indicate transcriptional organization of the different operons. Gene functions are given in the text, Table 2, and Fig. 3 and 4. Data from Carter et al. (2002) for VB and from Lynch et al. (2001) for RB1021.

(FhuC), are in another operon, not closely linked to the functional *fhuA1* gene. Next to *fhuCDB* in strain 8401(pRL1JI) is a pseudogene version of *fhuA* (ψ*fhuA*), strewn with stop codons and lacking a promoter. The surface location of the FhuA protein makes it susceptible to attack by some bacteriophages, bacteriocins, and antibiotics. Therefore, such receptors are prime candidates for rapid evolutionary loss through the generation of pseudogenes, thus avoiding the attentions of local outbreaks of bacteriophage that recognize sensitive forms of FhuA.

A different field isolate of *R. leguminosarum* (strain 3841), whose genomic sequence has been determined (http://www.sanger.ac.uk/Projects/R_leguminosarum/), also contains *fhuA1* but has another, intact and functional, *fhuA* gene (*fhuA2*) and lacks the pseudogene ψ*fhuA*. The DNA sequence of *fhuA2* is very similar to that of ψ*fhuA* but not to that of the functional *fhuA1* version in either strain 3841 or strain 8401(pRL1JI). Thus, *fhuA1* and *fhuA2* of strain 3841 did not arise by recent duplications, but ψ*fhuA* of strain 8401(pRL1JI) may be very recently derived from a *fhuA2* gene. Therefore, there is significant strain-to-strain variation in the genes involved in iron uptake, even in the same species.

Synthesis and Uptake of RB1021

RB1021 is the most widespread siderophore among strains of *S. meliloti* and has a similar structure to aerobactin, made by some enterics, and to schizokinen, made by *Bacillus* (Fig. 1). RB1021 is distinguished, though, by a lipid moiety that may facilitate the formation of micelles, protecting it from extracellular damage.

A gene cluster involved in the synthesis, uptake, and regulation of RB1021 was identified on a large, so-called "symbiotic plasmid" which also contains genes for nodulation and N_2 fixation (Fig. 2b). The receptor for RB1021 is specified by *rhtA*, whose product, not surprisingly, resembles the IucA aerobactin receptor. The *rhbABCDEF* operon specifies RB1021 synthesis, the likely chemical starting

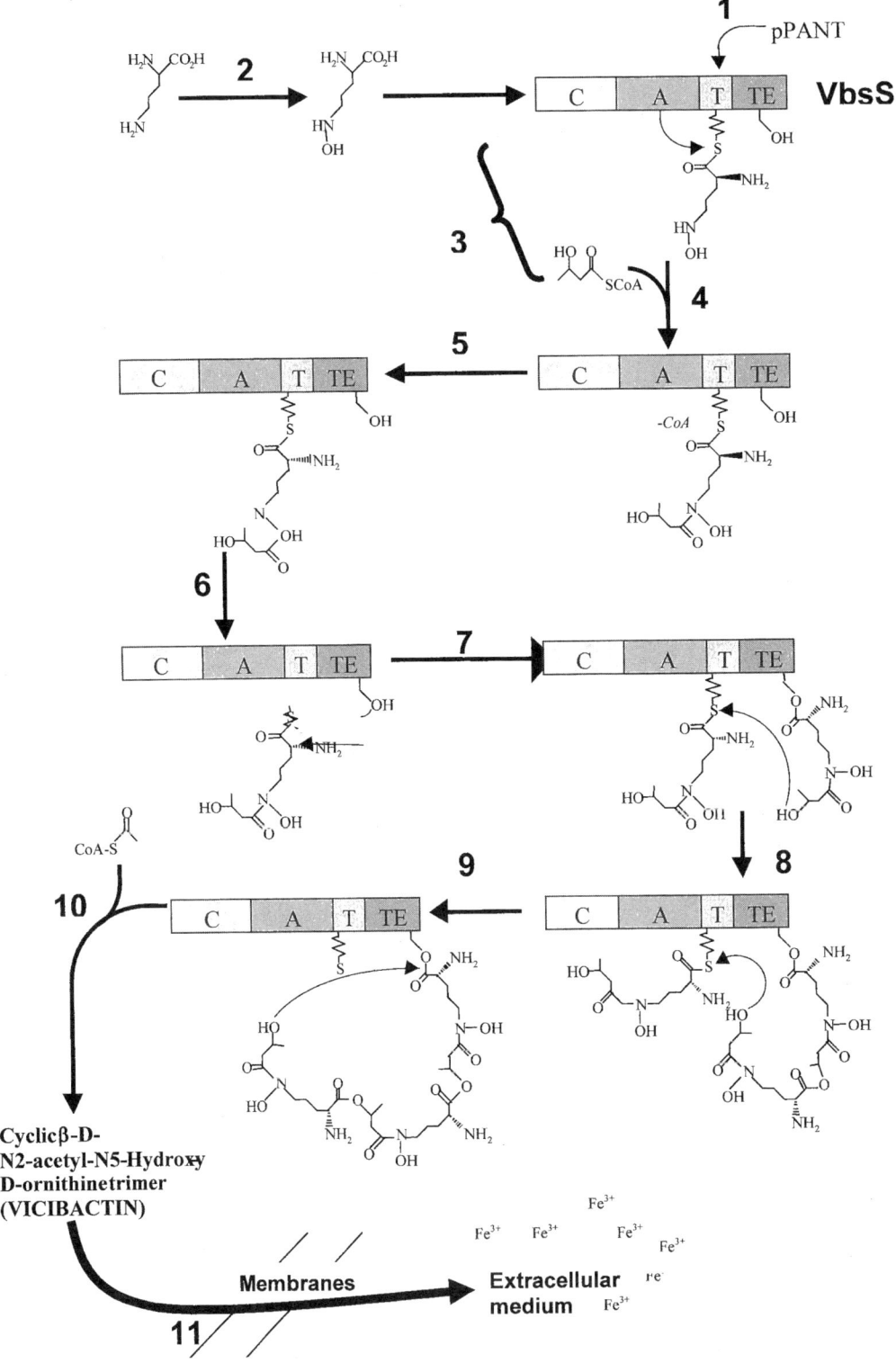

FIGURE 3 Pathway of VB synthesis. The proposed pathway for VB synthesis in *R. leguminosarum* is shown. The numbered steps correspond to those in Table 2. Adapted from Carter et al. (2002).

TABLE 2 Proposed functions of individual *vbs* genes in the VB biosynthetic gene cluster of *R. leguminosarum*

Pathway step (Fig. 3)	Mediated by Vbs product	Proposed enzyme activity	Comments on Vbs homologue(s)
1	VbsP	Addition of phosphopantotheinate to VbsS	Resembles many phosphopantothemate transferases
2	VbsO	Hydroxylation of L-ornithine	Resembles hydroxylases, including RhbE, involved in RHB1021 synthesis
3	VbsS	Multifunctional nonribosomal peptide synthase	Resembles multifunctional peptide synthases, including those for various antibiotics and other siderophores
4	VbsA	Addition of butyryl coenzyme A to hydroxy-L-ornithine on VbsS scaffold	Resembles several acetylases, including RhbD, involved in RHB1021 synthesis
5	VbsL	Epimerization of L-ornithine	Similar to cystathione γ-lyases, including MetB, in the methionine biosynthetic pathway
6	VbsS	Interdomain chain transfer	Provides the scaffold, comprising several domains needed for steps 6 through 9
7	VbsS	Dimerization by ester bond formation	See above
8	VbsS	Trimerization	See above
9	VbsS	Cyclization and release by ester bond formation	See above
10	VbsC	Acetylation of D-VB	Similar to RibTD, an acetylase involved in riboflavin synthesis
11	VbsD	Possible export of VB from the cell	Similar to the "NorM" class of efflux proteins

point being 1,3-diaminopropane (Fig. 4) rather than lysine, the progenitor for aerobactin. The diaminopropane is probably made by the RhbA and RhbB proteins, which respectively resemble Dta and Dcc, which synthesize this molecule in *Acinetobacter baumannii*. Consistent with the similarities of aerobactin and RB1021, RhbB, RhbC, RhbD, and RhbE all resemble individual Iuc polypeptides involved in the aerobactin synthesis. A closely linked *S. meliloti* open reading frame (ORF) has a deduced product with similarity to a malonyl coenzyme, which may add the lipid "tail" to RB1021, but a mutation in this ORF did not affect siderophore activity (measured on CAS plates). However nonlipolated RB1021 is identical to schizokinen, a siderophore found in nature, and so this decoration may act only to enhance the activity or stabilize RB1021 in the environment.

Other Rhizobial Siderophore Transporters—"Real" and "Virtual"

S. meliloti strain 1021 also has two homologues of *fhuA*, termed *fhuA1* and *fhuA2*, both of which are located on its chromosome, but the products of these genes are not highly homologous to either FhuA1 or FhuA2 of the closely related *R. leguminosarum*. Given that RB1021, the only siderophore known to be made by strain 1021, is not imported by a FhuA-type protein, these *S. meliloti* FhuA proteins may be receptors for hydroxamates made by other bacteria rather than being "home-grown" ones. Further, there are no obvious homologues of FhuB, FhuC, or FhuD in *S. meliloti* and so any hydroxamates that may bind to FhuA1 or FhuA2 of *S. meliloti* must be internalized by some other importer system.

In contrast to *S. meliloti* and *R. leguminosarum*, the proteome of *Mesorhizobium* has no

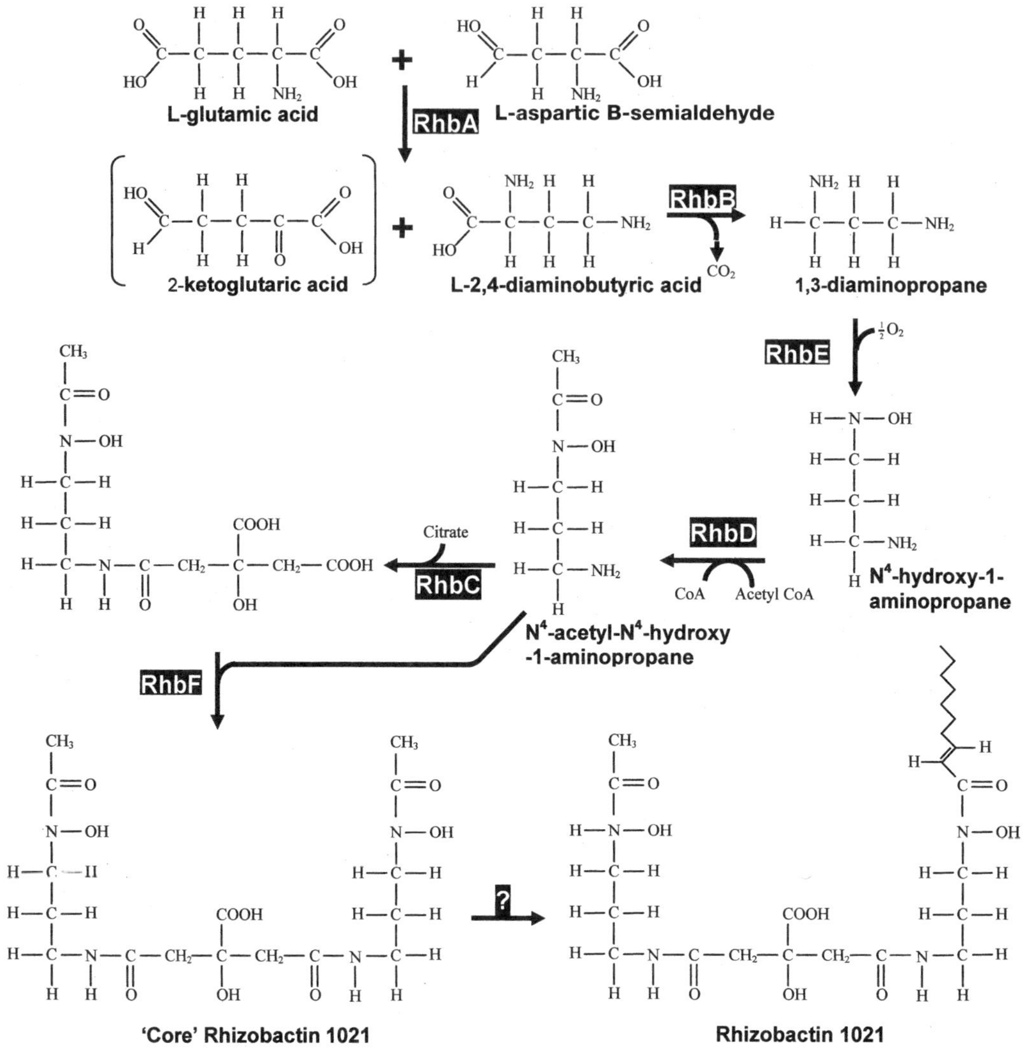

FIGURE 4 Pathway of RB1021 synthesis. The proposed pathway for RB1021 synthesis in *S. meliloti* is shown. Adapted from Lynch et al. (2001).

Fhu-like receptors or any of the other machinery needed to import hydroxamates. To date, there are no published papers on iron uptake in this rhizobial species. In *B. japonicum*, FegA is an outer membrane protein that has homology to Fe^{3+}-hydrozamate receptors and accumulates in iron-depleted cells. Given that there are no reports of hydroxamate synthesis by this species, the finding of this iron stress protein is surprising, particularly since FegA plays a role in symbiotic N_2 fixation (see below).

Phenotypes of Siderophore-Defective Mutants in Free-Living Culture and in Symbiosis

Mutants defective in producing VB in *R. leguminosarum* grew only slightly less well than wild-type strains on iron-depleted medium, pointing to the presence of other iron uptake systems in these bacteria. The failure to make or import VB or RB1021 also had no detectable effect on symbiotic N_2 fixation of *R. leguminosarum* on peas or *S. meliloti* on alfalfa. Surprisingly, several of the relevant genes (*rhb*, *vbs*, *rbtA*, and *fhu*) are not even expressed in bacteroids; this is a real paradox, given the high iron demands of these cells. Thus, these siderophores are not the only, and perhaps not even the major, sources of iron in N_2-fixing bacteroids.

There is one oddity in the literature concerning defects in iron uptake systems and symbiotic N_2 fixation. A FegA$^-$ mutant of *B. japonicum* produced nonfixing nodules on soybeans. It is not clear why FegA is required for functional bacteroids in soybeans whereas the loss of a hydroxamate receptor (FhuA) in *R. leguminosarum* has no effect in pea nodules. The morphology and physiology of soybean nodules differ from those on peas; it may be that FegA, on the *B. japonicum* cell surface, acts as a recognition marker required for normal nodule development quite independent from iron uptake per se.

Heme Utilization in Rhizobia: Identification of Hmu Transporters

Unlike siderophores, which bacteria make themselves, other iron sources come "ready-made" and tend to be used only by bacteria inhabiting specialized niches. One such iron source is heme, which is locally abundant, especially in higher organisms. Not surprisingly, the use of heme as an iron source is common among pathogens of animals, which may have one or more transport systems dedicated to its uptake. However, it was noted that rhizobia could grow with heme as the sole iron source. The *hmu* genes responsible for heme uptake have been characterized in *B. japonicum* and *R. leguminosarum*. In both organisms, Hmu$^-$ mutants were affected in their ability to use heme as the sole iron source. The *R. leguminosarum hmuPSTUV* operon specifies an ABC-type heme transporter similar to that found in *Yersinia* and *Serratia*. The exact functions of HmuP and HmuS are unknown. The other gene products, HmuT, HmuU, and HmuV, comprise the "standard" components of ABC transporters. The arrangement of *hmu* genes in *Bradyrhizobium* is similar, with *hmuTUV* divergently transcribed from *hmuR*, whose product resembles the HasR outer membrane heme-hemophore receptor of *Pseudomonas aeruginosa*. *Mesorhizobium* also has an *hmuR* gene divergently transcribed from *hmuTUV*, while *R. leguminosarum* and the closely related *S. meliloti* lack HmuR. The last two species presumably bind heme to the cell surface by another mechanism.

Hmu$^-$ mutants of *B. japonicum* cannot use heme as a source of iron, but *R. leguminosarum hmu* mutants undergo only reduced growth in heme; therefore, they may have another, unidentified, heme transporter. Hmu$^-$ mutants of both species induce apparently normal nodules on their respective hosts, suggesting that heme is not a major iron source for bacteroids.

TonB and Iron Uptake in Rhizobia

In many bacteria, the TonB-ExbB-ExbD protein complex transduces energy to transport various molecules, including siderophores and heme (see chapter 7). In *B. japonicum*, *R. leguminosarum*, and *S. meliloti* (but not *M. loti*), the *hmu* genes are next to a *tonB*-like gene, and in *M. loti* and *B. japonicum*, *tonB* is adjacent to *exbBD*, just as in several other bacteria. In *R. leguminosarum* and *B. japonicum*, *tonB* is absolutely required for the use of heme as an iron source and, in the former, for VB import. Thus, rhizobial TonB proteins have similar functions in iron nutrition to the TonB proteins in other bacteria. However, mutations in the *tonB*-like gene *bll7071* of *B. japonicum*, although totally blocked for heme utilization, can use exogenous siderophores (ferrichrome) as iron sources. In fact, *B. japonicum* has a second gene (*bll7153*) whose product resembles TonB, perhaps accounting for the lack of any defect in siderophore uptake in *bll7071* mutants (http://www.kazusa.or.jp/rhizobase/Bradyrhizobium/index.html).

R. leguminosarum TonB$^-$ mutants and *B. japonicum* mutants defective in the *bll7071 tonB*-like gene induce functional nodules on their respective hosts. Since TonB is required to transport more than one iron source, this shows that none of these TonB-dependent transporters is crucial for iron nutrition in bacteroids.

Other ABC-Type Transporters

In addition to using heme as the sole iron source, some of the rhizobia encode periplasmic binding protein-dependent ABC transporters that have homologs in many bacterial pathogens. These can be divided into two types. The Fbp homologs are members of the FeT cluster of ABC transporters, which normally have a cognate membrane receptor for an iron-carrying molecule such as transferrin or lactoferrin. The second group, of which Sit is the prototype, transport manganese or, less effectively, ferrous iron and have no known outer membrane receptor associated with transport.

Inspection of the genomes of *Mesorhizobium*, *Rhizobium*, and *Sinorhizobium* reveals at least three notable features that relate to these types of metal transporters (Table 3).

- *R. leguminosarum* and *S. meliloti* each contain not just one but at least two separate operons, one of which appears to specify an Fbp-like transporter and one which specifies a Sit-like transporter.
- The levels of identity between these ABC transporters of the rhizobia and homologues in distantly related bacteria are remarkably high. Table 3 presents several examples of >65% identity among the amino acid sequences of the periplasmic binding proteins.
- The particular portfolio of transporters is different in the three rhizobial genera. Thus, only *S. meliloti* has a close homologue of the *Pasteurella* FbpA system, but it lacks the "YfuA"-like transporter that is found in both *Mesorhizobium* and *Rhizobium*. Conversely, *Mesorhizobium* lacks the SitABCD system possessed by the other two rhizobia (see also below).

Taken together, these observations strongly suggest that these species of rhizobia acquired

TABLE 3 Similarities among the Fbp-like proteins of rhizobia and selected homologues in other taxa.

Rhizobial species	ORF or contig[a]	Homologues in other bacteria[b]	% Amino acid identity[c]
Mesorhizobium loti	mlr3495	*Neisseria gonorrhoeae* FbpA	35
		Haemophilus influenzae HitA	36
		Yersinia enterocolitica YfuA	69
		R. leguminosarum 111b05	69
Rhizobium leguminosarum	rhiz395e07	*Salmonella* SitA	77
		Haemophilus influenzae YfeA	70
		Yersinia pestis YfeA	59
		S. meliloti SMc02509	74
	rhiz111b05	*Neisseria gonorrhoeae* FbpA	35
		Haemophilus influenzae HitA	40
		Yersinia enterocolitica YfuA	60
		M. loti mlr3485	69
Sinorhizobium meliloti	SMc00784	*Pasteurella haemolytica*[a] FbpA	51
	SMc02509	*Salmonella* SitA	73
		Haemophilus influenzae YfeA	70
		Yersinia pestis YfeA	63
		R. leguminosarum 395e07	74

[a] The ORF designations for *M. loti* and *S. meliloti* are from http://www.kazusa.or.jp/rhizobase/ and http://sequence.toulouse.inra.fr/meliloti.html, respectively. The *R. leguminosarum* contigs refer to those in http://www.sanger.ac.uk/Projects/R_leguminosarum/ on 18 November 2002.
[b] Examples of very similar proteins in more than one of the rhizobial species are shown in bold type.
[c] Percent identities are taken from BLAST comparisons.

these transporters independently and that these genes were almost certainly obtained by lateral gene transfer from different "donor" bacterial species. The very striking similarities of the gene products to those of distantly related bacteria indicate that these events occurred relatively recently in evolutionary time.

Do any rhizobial Fbp systems have cognate outer membrane receptors, analogous to those in bacterial pathogens, and, if so, what form(s) of complexed iron do they recognize? Are the various transporters optimally active (or expressed?) in different environments, with some, perhaps, being more effective at (for example) a particular pH or O_2 tension—or in the nodulation process? Alternatively, the multiple iron transport system genes may simply reflect the mosaic-like genomes of rhizobia, having been acquired from other bacteria almost by chance. The characterization of single or perhaps multiple mutants, defective in more than one such transporter, is required before at least some of those questions can be answered. Perhaps ironically, the one system that has been analyzed in detail, namely, the *sitABCD* operon, has recently been shown to be involved in the uptake not of iron but of manganese (see below).

In contrast to the fast-growing rhizobia discussed above, the deduced proteome of *Bradyrhizobium* does not appear to have Fbp-like or Sit-like proteins with close counterparts in bacterial pathogens. *B. japonicum*, like the other rhizobia, has many other ABC transporters, nearly all of which have no known specific function, and it remains to be seen if any of these are dedicated to iron import.

In Silico Analyses and "Missing Genes"

The availability of genome sequences allows one to search in silico for genes that are likely to specify iron uptake systems. Not without interest, though, are the genes that are absent but might have been predicted to be present. In the rhizobia, there are at least two cases of such "dogs that do not bark," as discussed below.

CITRATE UPTAKE

Ferric iron, complexed to citrate, is an effective iron source for many bacteria (see chapter 11), including rhizobia. However, *Bradyrhizobium*, *Mesorhizobium*, *Rhizobium*, and *Sinorhizobium* have no close homologues of the FecA receptor or any other components of the TonB-dependent FecABC transporter that imports citrate in *E. coli*. Another difference between rhizobia and *E. coli* is that in *R. leguminosarum*, *tonB* mutations do not affect the use of citrate as an iron source. Since no genetic studies of this process have been done, it is unclear how rhizobia sequester Fe-citrate.

FERROUS IRON UPTAKE

Bacteroids exist in a very low pO_2 environment, and the enzyme ferrireductase occurs in the nodules on soybean plants. Therefore, a priori, it might be thought that soluble, ferrous iron, which would be available under those conditions, would be a likely iron source for bacteriods. Indeed, Fe^{2+} is more readily transported across bacteroid membranes than is the ferric form. As discussed in chapter 12, enteric bacteria import Fe^{2+} via a dedicated transport system involving FeoB. The genome sequence of *B. japonicum* reveals a gene (*blr6523*) whose product is 31% identical to *E. coli* FeoB and which has a greater similarity to putative ferrous iron transporters of *Rhodobacter* and *Caulobacter* (64 and 62% identical, respectively). Genetic analyses are needed to elucidate the role (if any) of this *B. japonicum* gene in symbiotic N_2 fixation on soybeans. Interestingly, in contrast to *Bradyrhizobium*, the genomes of *Mesorhizobium*, *Rhizobium*, and *Sinorhizobium* have no genes that specify a FeoB-like protein (nor do those of the closely related *Brucella* and *Agrobacterium*). To date, there have been no physiological studies of the ability of the bacteroids of the fast-growing rhizobial genera to import ferrous iron. It is important to establish if the FeoB-like transporter in *Bradyrhizobium* allows this species to import Fe^{2+} and, if so, whether this reflects a fundamental difference in the iron nutrition in the nodules of soybeans compared to the various hosts of *Mesorhizobium*, *Rhizobium*, and *Sinorhizobium*.

IRON-RESPONSIVE GENE REGULATION IN RHIZOBIA

As shown repeatedly in this book, the expression of bacterial genes involved in iron uptake and other aspects of "iron biology" responds to iron availability, and the rhizobia are no exception. Studies with gene reporters and regulatory mutants show that there are similarities between the rhizobial iron regulons and those in other bacteria but that there are also important differences.

Range of Iron-Responsive Genes in Rhizobia

Most of the genes involved in iron uptake that were mentioned above are expressed at reduced levels under iron-replete, conditions compared to iron-depleted conditions. Thus, the *R. leguminosarum fhuAF*, *fhuCDB*, *hmuPSTUV*, and *tonB* operons are all expressed at ca. 8- to 10-fold higher levels under iron deprivation, as are at least two operons that specify Fbp-like transporters. Of the four *vbs* operons involved in VB synthesis, three (*vbsGSO*, *vbsADL*, and *vbsC*) behave in a similar fashion, but *vbsP* is an exception, being expressed at the same level irrespective of iron availability. Perhaps the VbsP pantetheine transferase modifies other (unidentified) peptide synthases that are not involved in iron metabolism and are required under iron-replete conditions.

Similarly, in *S. meliloti*, the *rhbABCDEF* and *rhtA* operons involved in RB1021 synthesis and uptake are expressed at elevated levels in iron-depleted cells. In *B. japonicum*, transcription of the *hmu* and *tonB* genes is reduced by exogenous iron, as is that of the *fhuA* homologue, *fegA*. In this species, genes involved in heme biosynthesis are also subject to complex, iron-dependent control (see below).

Two very different "local" regulators that affect just a few cognate genes for siderophore synthesis and/or uptake have been identified in the rhizobia. One of these, RpoI, is an RNA polymerase σ factor found in *R. leguminosarum*; the other, termed RhrA, is an *S. meliloti* transcriptional regulator.

RpoI, an ECF σ Factor for VB Synthesis

Within the *vbs* gene cluster of *R. leguminosarum* is *rpoI*, a gene that is absolutely required for VB synthesis (Fig. 2). RpoI is closely related to the large family of extracytoplasmic function (ECF) σ factors for RNA polymerase, which recognize genes with products located at the cell surfaces. All known ECF σ factors that are specifically involved in iron uptake are in a subgroup, even among diverse bacterial taxa. RpoI is most similar to PvdS, a σ factor for several genes including those for siderophore synthesis in *Pseudomonas*.

In RpoI$^-$ mutants, transcription of *vbsGSO* and *vbsADL* was undetectable even under low-iron conditions, indicating that RpoI is the σ factor for these two VB synthesis operons. Subsequent in vitro transcription runoff experiments with purified RpoI protein confirmed this. RpoI recognized sequences close to the promoters of *vbsGSO* and *vbsADL* that resemble sequences important for the activity of the *Pseudomonas* PvdS σ factor. In contrast to *vbsGSO* and *vbsADL*, transcription of two other *vbs* genes, *vbsP* and *vbsC*, is normal in RpoI$^-$ mutants. This seems surprising, especially for *vbsC*, whose transcription responds to iron availability. Likewise, RpoI is not needed to transcribe *fhuAF*, *fhuCDB*, *tonB*, and *hmuPSTUV*, although they are also iron responsive. It is not known if RpoI recognizes promoters other than those for *vbsGSO* and *vbsADL* or if the other iron-responsive promoters in *R. leguminosarum* have their own "specialized" σ factors.

Transcription of *rpoI* itself is reduced but not abolished in iron-replete medium. The mechanism that governs this partial repression is unknown; however, site-directed mutagenesis of the *rpoI* promoter region generated constitutive mutants that expressed *rpoI* at very high levels. These may have lost an iron-responsive, repressor-binding site, but the identity of the putative *trans*-acting repressor is unknown, although the regulatory RirA protein is a possible candidate.

Many σ factors are subject to modulation by anti-σ factors. For example, the function of PvdS of *Pseudomonas* is inhibited in a process resembling that between the σ factor FecI and its cognate anti-σ factor FecR in the Fe^{3+} citrate uptake mechanism (see chapter 11). Through an analogous signaling cascade, PvdS fails to work properly in *Pseudomonas* mutants that do not make the cognate siderophore pyoverdine. This is because the anti-σ factor FpvR is locked into its inhibitory mode since it cannot respond to the signal mediated by the outer membrane receptor, FpvA, bound to the cognate siderophore. This is an adaptive response in which bacteria make the siderophore only if it is likely to be returned to the cells (see chapter 19). In contrast to PvdS, RpoI-dependent transcription of *vbsGSO* and *vbsADL* is unaffected in *vbs* mutants that are completely defective in VB synthesis. Thus, RpoI has no anti-σ factor whose action depends on VB. This tallies with the idea that the form of control seen for PvdS is particularly well suited for bacteria (including *Pseudomonas*) that make more than one siderophore, since it directly links the extracellular levels of a given siderophore with the activity of the genes for its synthesis. To our knowledge, *R. leguminosarum* synthezises only VB, and so its need for this type of quorum-sensing loop is less pressing. It remains to be seen if there is any anti-RpoI σ factor, but there is no candidate anti-σ factor gene near *rpoI*.

RhrA, an AraC-Like Regulator for RB1021 Synthesis

Located between the *S. meliloti rhbABCDEF* biosynthetic operon and the *rhtA* "transport" gene is *rhrA*, which is also required to synthesize RB1021. In *rhrA* mutants, neither *rhbABCDEF* nor *rhtA* is expressed, indicating that RhrA is a positively acting transcriptional regulator for these genes. Indeed, the RhrA sequence places it in the AraC-like family of regulators, including PchR and YbtA, which regulate the synthesis of the siderophores pyochelin and yersiniabactin in *P. aeruginosa* and *Y. pestis*, respectively.

It was not established if RhrA is the sole regulator of *rhtA* and *rhbABCDEF*. There are no genes near the *rhb-rht* gene cluster that specify a σ factor, but this does not preclude a specialized σ factor, analogous to RpoI, that recognizes the *rhb* and/or *rht* promoters. Potential *fur* boxes were said to occur upstream of both *rhtA* and *rhbABCDEF*, but these differ markedly from canonical sequences, and, to my knowledge, it has not been reported if these respond to Fur-mediated control.

Rhizobial Regulators with a Broader Regulatory Spectrum

In addition to these cases of "local" gene control, recent studies of the more "global" regulation of iron-responsive genes in the rhizobia have been done. Although some rhizobia have a gene whose deduced product resembles Fur of other bacteria, it plays, at best, only a minor role in iron-dependent gene regulation. Instead, in at least one species, the regulator for iron sensing is a newly discovered protein, RirA, whose activities may be confined to some of the rhizobia and very close relatives. Another unusual feature of rhizobia and closely related α-proteobacteria is that they possess another member of the Fur superfamily, termed Irr, which mediates iron-dependent gene control of at least one gene in a novel way.

RirA, a Novel Transcriptional Regulator?

In *R. leguminosarum*, mutations were identified in a chromosomal gene, *rirA*, which caused deregulated, iron-independent expression of all the iron-responsive operons mentioned above, namely, *vbsADL*, *vbsGSO*, *vbsC*, *hmuPSTUV*, *tonB*, *fhuAF*, and *fhuCDB*. Also, two operons that specify Fbp-like Fe^{2+} ABC transporters and an operon that probably encodes an iron-sulfur cluster are transcribed at high levels in RirA⁻ mutants in iron-replete medium. Transcription of *rirA* itself is enhanced ca. twofold in iron-replete medium compared to iron-deficient medium.

The sequence of RirA has no detectable similarity to that of Fur. Among the available bacterial genome sequences, there are some very close homologues of RirA (>70% identity), but these are confined to just two of the rhizobial genera (*Mesorhizobium* and *Sinorhizobium*) and their very close relatives (*Brucella* and *Agrobacterium*). Interestingly, with the exception of *Rhizobium*, the *rirA* gene is adjacent to or very close to genes involved in iron uptake or metabolism (heme uptake in *Mesorhizobium* Fbp-like transporter in *Sinorhizobium* and *Agrobacterium*, and bacterioferritin in *Brucella*). Other α-proteobacteria, even quite close relatives (e.g., *Rhodobacter*), have no near homologues of RirA but do encode proteins with ca. 30% identity to RirA. Significantly, *B. japonicum* has no close homologue of RirA, showing that this protein is not a signature of the rhizobia, per se.

Low-level (ca. 30% identical) homologues also occur in many, less closely related eubacterial taxa. Of these homologues, only two, Rrf2 of *Desulfovibrio vulgaris* and IscR of *E. coli*, have been studied in any detail. The former is implicated in regulating cytochrome synthesis; IscR, too, is a transcriptional regulator, repressing the *isc* structural genes that are responsible for the synthesis of FeS clusters. Interestingly, IscR itself contains FeS clusters and acts as a repressor only when it is charged with these. A conserved feature of these RirA homologues (even the more divergent ones) is a triad of cysteines near the C terminus, which are candidates for binding of metal-containing ligands such as FeS clusters.

Molecular modeling of the N-terminal end of RirA of *R. leguminosarum* predicted a similar overall structure to the iron-dependent transcriptional regulator DtxR, although the two proteins have no sequence similarity. These analyses suggest that RirA may be a DNA-binding metalloprotein. However, there are no highly conserved promoter-proximal motifs common to all the genes that are deregulated in *rirA* mutants, suggesting that RirA does not simply bind to conserved, *cis*-acting DNA sequences near iron-responsive promoters.

Fur Superfamily in Rhizobia

As described in chapter 13, Fur is widely perceived as the major controller of iron-responsive gene expression in many gram-negative bacteria, since it is a negative regulator of genes involved directly or tangentially in iron acquisition, utilization, or signaling. *E. coli* Fur can also repress a small RNA (*ryhB*) that negatively regulates the expression of some genes, leading to the Fur-dependent expression of some genes, such as *sod* and *bfr*, that were originally thought to be positively regulated by Fur. Fur is just one member of a wider superfamily that includes the Zn-responsive regulator Zur, the oxygen stress protein PerR, and a fourth member, Irr, which, to date, appears to be confined to the rhizobia and their near relatives. (The tree in Fig. 5 shows the relatedness of available members of the rhizobial Fur superfamily to a selection of such proteins in other bacteria.)

Bradyrhizobium, *Sinorhizobium*, and *Rhizobium* each have one close homologue of the conventional Fur protein, whose structure in *R. leguminosarum* is similar to that in *Pseudomonas aeruginosa*. In contrast, *M. loti* has no *fur* gene, although it does have close homologues of Zur and Irr (Fig. 5). This observation, in itself, indicates that at least some rhizobia may be unusual with respect to the importance of Fur. Genome scanning also shows that Fur boxes are conspicuously absent from the promoter regions of rhizobial iron-regulated genes.

Mutations in the *fur*-like gene of both *B. japonicum* and *R. leguminosarum* had no effect on the iron-dependent control of expression of the *hmu* and *tonB* genes in either species. More extensive studies of *R. leguminosarum* showed that the iron-responsive *fhu*, *vbs*, and *fbp*-like genes also exhibit normal, iron-responsive expression in Fur⁻ mutants. For *S. meliloti*, too, transcriptomic studies showed that Fur does not play a global regulatory role in Fe-responsive gene regulation.

The first experimental signs that rhizobial Fur was unusual came from O'Brian's studies of the role of Fur in heme biosynthesis in *B. japonicum*. The first, dedicated step in this path-

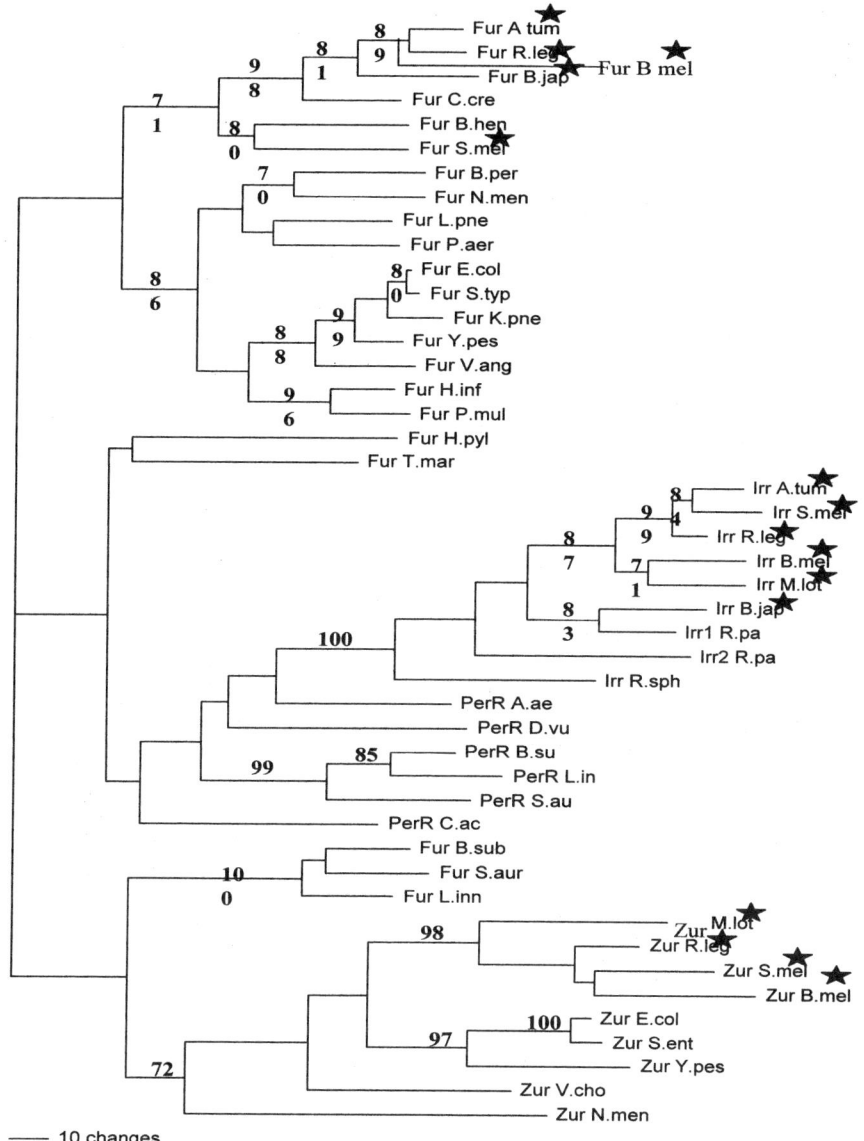

FIGURE 5 Phylogenic tree of the Fur superfamily in the rhizobia, compared with homologues in other bacterial taxa. A maximum-parsimony phylogenetic tree was constructed from amino acid sequences of Fur and Fur-like sequences by using PAUP★ version 4.0b10. The tree is rooted at the midpoints, and bootstrap values are shown for nodes with values of >70. Branch lengths are proportional to the degree of amino acid change. Proteins were obtained from the GenBank database. A.ae, *Aquifex aeolicus*; A.tum, *Agrobacterium tumefaciens*; B.jap, *Bradyrhizobium japonicum*; B.sub, *Bacillus subtilis*; B.hen, *Bartonella henselae*; B.mel, *Brucella melintensis*; B.per, *Bordetella pertussis*; C.ac, *Clostridium acetobutylicum*; C.cre, *Caulobacter crescentus*; D.vu, *Desulfovibrio vulgaris*; E.col, *Escherichia coli* K-12; H.inf, *Haemophilus influenzae*; H.pyl, *Helicobacter pylori*; K.pne, *Klebsiella pneumoniae*; L.pne, *Legionella pneumophila*; L.inn, *Listeria innocua*; M.lot, *Mesorhizobium loti*; N.men, *Neisseria meningitidis*; P.aer, *Pseudomonas aeruginosa*; P.mul, *Pasteurella multocida*; R.leg, *Rhizobium leguminosarum*, R.sph, *Rhodobacter sphaeroides*; R.pa, *Rhodopseudomonas palustris*; S.ent, *Salmonella enterica*; S.mel, *Sinorhizobium meliloti*, S.typ, *Salmonella enterica* serovar Typhimurium; S.aur, *Staphylococcus aureus*; T.mar, *Thermotoga maritima*; V.ang, *Vibrio anguillarum*; V.cho, *Vibrio cholerae*; Y.pes, *Yersinia pestis*. Proteins from *Agrobacterium*, *Brucella*, *Mesorhizobium*, *Rhizobium*, and *Sinorhizobium* are marked with stars.

way in these (and other) rhizobia is mediated by aminolevulinic acid synthase, specified by *hemA*. Normally, *hemA* expression is enhanced by growth at high levels of iron, but this did not occur in a Fur⁻ mutant. This suggests that Fur may be a positive regulator of *hemA*, but the mechanism(s) involved is not known.

A second *B. japonicum* gene, termed *irr*, was identified whose transcription responded to Fur, but, again, in a way that differs from Fur in "model" bacteria. Irr itself is a member of the Fur superfamily and is an iron-responsive regulator in its own right (see below). Transcription of *irr* is moderately repressed by iron in the wild type but not in a Fur⁻ mutant, superficially resembling the normal mode of Fur-dependent gene control in other bacterial taxa. Importantly, though, there is no Fur box near the *irr* promoter, yet *B. japonicum* Fur can bind to this region of DNA in vitro. Intriguingly, purified *B. japonicum* Fur not only binds to this novel "*irr* box" but also retains the ability to bind to conventional *E. coli* Fur boxes. In contrast, *E. coli* Fur does not bind to DNA spanning the *irr* promoter.

Purified Fur of *R. leguminosarum* can also bind to canonical Fur boxes, and its cloned *fur* gene partially corrects the regulatory defect of an *E. coli fur* mutant. Thus, even when *fur* boxes do not occur in the promoter regions of iron-responsive genes, *R. leguminosarum* Fur may retain the ability to bind to such canonical motifs. Put another way, these observations show that the Fur proteins of *B. japonicum* and *R. leguminosarum* are functional and that the terminology "Fur" was not based solely on sequence alignments.

Irr, a Rhizobium-Specific Member of the Fur Superfamily

The Irr subgroup of the Fur superfamily is confined to the rhizobia and its close relatives (Fig. 5). Irr is a transcriptional repressor of *hemB*, which is normally repressed in iron-depleted media and which specifies aminolevulinic dehydratase, the second enzymatic step in heme synthesis. In contrast to the classical Fur proteins, the repressive properties of Irr are reduced in cells grown at high levels of iron, due to two very different mechanisms. First, *irr* transcription is reduced in a Fur-dependent fashion under high-iron conditions. More importantly, the Irr protein is very unstable in cells grown in iron-rich media, being subject to an unusual system of posttranslational modification that does not involve Fur. This iron-dependent instability is not direct but is mediated via intracellular heme, which binds to a heme-regulatory motif at the Irr N-terminal region. This was elegantly demonstrated by the greatly increased stability of Irr in either of two different classes of *B. japonicum* mutants. In one case, a cysteine in the heme-binding site in Irr was replaced by alanine, eliminating the heme "target"; the other had a mutation in a heme biosynthetic gene, thereby entirely removing this effector molecule from the scene. Even this is not the whole story. It was also found that the transfer of heme to Irr was facilitated when the heme was bound to ferrochelatase, which inserts the iron into the porphyrin ring, the final step in heme synthesis. Although this heme-dependent instability of Irr provides a sophisticated link between the final step in heme biosynthesis and the transcriptional control of an early step, it may not apply to all bacterial Irr proteins. Those of other rhizobia, *Agrobacterium*, and *Brucella* lack the characteristic heme-regulatory motif (an exception is one of the Irr polypeptides of *Rhodopseudomonas palustris*, a close relative of *B. japonicum* that has two *irr* genes in its genome).

It may be instructive to see if and how the Irr proteins of these other rhizobial species affect *hemB* expression in the absence of this heme-docking site. At present, the full range of proteins whose transcription is affected by Irr is unknown. Nor do we know why this Fur-like regulator is confined, to date, to such a small group of bacteria. We do know, however, that in *R. leguminosarum*, at least, the lack of a "classical" phenotype of Fur⁻ mutants is not due to functional redundancy between Fur and Irr. Double mutants that lack both these proteins also show normal iron-responsive regulation of genes that are involved in iron uptake. The *B.*

japonicum genome sequence reveals two *irr*-like genes, the originally described *bll0768* and a second homologue *(blr1216)* of unknown function. Clearly, the role of this "new" member of the Fur family in *Bradyrhizobium* requires further investigation.

Rhizobium "Fur" Is a "Mur"

The reason why the "Fur" of *Rhizobium* and *Sinorhizobium* plays no apparent role in Fe-responsive gene regulation in these bacteria has now become clear. It was found that in *S. meliloti*, as in other bacteria, the SitABCD transporter, despite its name (for "*Salmonella* iron transporter") is primarily a transporter of Mn^{2+} rather than of Fe^{2+}. Since *sitABCD* of *S. meliloti* is adjacent to its *fur*-like gene and since rhizobial Fur was a regulator with no known function, it was tempting to think that the *Rhizobium* Fur might be involved in regulating *sitABCD* in response to Mn rather than to Fe. This turns out to be true. In both *S. meliloti* and *R. leguminosarum*, the expression of *sitABCD* was greatly increased in cells starved of Mn. However, Fe depletion had no effect on its transcription. Importantly, this regulation was totally abolished in mutants that had insertions in the *fur* gene of these two species. Thus their "Fur" is a "Mur" (manganese-responsive regulator). In *R. leguminosarum*, the *sitABCD* operon has two promoters, both of which are Mn repressible. In gel shift experiments, purified Mur bound to regions near both these promoter regions, even though neither had any similarity to canonical *fur* boxes. Thus, Mur of *R. leguminosarum* recognizes canonical *fur* boxes in response to Fe^{2+} but can also bind to very different sequences in *Rhizobium* itself.

As mentioned above, *M. loti* has no close homologue of Fur. It is perhaps no coincidence that *M. loti* also does not have an operon that corresponds to *sitABCD*. The *sitABCD* genes of *S. meliloti* and *R. leguminosarum* may have been recently acquired, together with their cognate *fur*-like regulatory gene. This is particularly plausible for *S. meliloti*, where *mur* and *sitABCD* are contiguous. In contrast, the sequenced strain (at least) of *M. loti* did not acquire *sitABCD* and hence would have no need for their *fur*-like gene, *mur*.

CONCLUSIONS

It is clear that the rhizobia have great flexibility for iron uptake and show real differences in their iron-responsive gene regulation compared to other gram-negative bacteria. Why should this be so?

Rhizobial genomes are relatively large, those of *Rhizobium* and *Bradyrhizobium*, for example, being ca. 8 and ca. 9 Mb, respectively. Only a small part (ca. 150 kb) of these genomes is occupied by the "symbiotic" genes for nodulation and N_2 fixation. These genomes are larger, by several megabases, than those of the close relatives *Brucella* (3.3 Mb), *Agrobacterium* (4.8 Mb), and the benthic *Caulobacter* (4.1 Mb). Importantly, it is now evident that as much as 50% of the genome of one rhizobial species does not resemble those of other, even closely related species (e.g., *S. meliloti* and *R. leguminosarum*). It seems, then, that there is a core, "rhizo-genome" into which has been introduced subgenomic chunks from other genera by lateral gene transfer. Given the species-specific nature of these "extra" genes, they must have been acquired independently by different rhizobia from different donors. In some cases, including those for some iron uptake systems, the extra genes are remarkably similar in sequence to homologues in distantly related taxa, indicating that they were obtained very recently. These mosaic rhizobial genomes confer increased metabolic flexibility. For example, rhizobia have more than 100 different ABC transporters, the ligands for most of which are unknown but whose presence must extend the range of available nutrients in the soils and other environments.

The evolution of this flexibility may have two driving forces. One of these may be the patchy, usually low-level availability of nutrients in the soil, encouraging the rhizobia to "snack" on whatever happens to be available at a particular time and place. Such an environment is not, of course, confined to the rhizobia—why, then, does this group adapt by enlarging its genome? Speculatively, the symbiotic life-style of rhizobia, in which one rhizobial cell can induce one nodule that contains millions of its progeny, may be especially well suited to the "large-genome" state. Within a few weeks of initiating nodule formation, the direct progeny of the initially infective individual are liberated back to the soil, massively augmenting the numbers of that species. The luxurious life that the rhizobia spend in the nodule, hugely increasing their numbers courtesy of the host, may allow them to cope with the extra cost of replicating and generally "managing" the large amounts of ancillary DNA. Other, nonsymbiotic, bacteria that also face nutritional shortages (e.g., *Caulobacter*) do not have the occasional nutritional boost enjoyed by rhizobia and so must be genomically more "lean and mean."

Focusing on mechanisms of iron uptake, these lines of argument may explain why rhizobia not only use siderophores but also contain at least two transport systems more frequently associated with pathogens. Concerning the use of one of these atypical substrates, namely, heme, this ability may be hardwired into the "rhizo-genome," since the HmuTUV protein sequences predicted from the available rhizobial genomes are more similar to each other than to those of other bacterial groups. Thus, the *hmuTUV* genes are likely to have been acquired by rhizobia long ago. This is in sharp contrast to genes encoding the Fbp-like and Sit-like transporters.

Concerning the mechanisms of Fe-responsive gene regulation, it is clear that at least some of the rhizobia do not use Fur as their global regulator but that, instead, the very different RirA protein plays the same role as Fur (or a very similar role). Although some rhizobia have Fur-like proteins, in all cases examined so far, these proteins differ markedly from classical Fur of *E. coli* or *P. aeruginosa*, in terms of either the *cis*-acting regulatory DNA sequences that they recognize or the identity of their metallic corepressor, or both.

It is not clear why, specifically, one very small bacterial clade should have evolved a RirA-mediated system of gene control, assuming that *R. leguminosarum* exemplifies what applies to the other species with very close homologues of RirA. The fact that *Brucella* is also among this small clade indicates that it is not the particular symbiotic and soil-dwelling life-style per se that is the selective force which caused RirA to be the functional counterpart of Fur. Rather, RirA may have evolved from an ancestor of the less similar RirA progenitor, which is found throughout the eubacteria, and this may have occurred prior to the divergence of *Rhizobium*, *Mesorhizobium*, *Sinorhizobium*, *Agrobacterium*, and *Brucella* from the other α-proteobacteria.

The fact that *B. japonicum* has no near homologue of RirA protein means that no single model can describe the iron regulon for all rhizobial species. It also shows that RirA is not a key part of the ability to nodulate and to fix nitrogen; rather, it appears that RirA evolved into its present role as an iron-responsive regulator very recently (in evolutionary terms), after the separation of the more distant *Bradyrhizobium* from its faster growing relatives.

We have no explanation of why RirA should be the Fe regulator of choice for these bacteria, nor do we know, at this stage, how extensive is the range of bacterial species that have an RirA regulon. This is largely because, perhaps surprisingly, there have been very few direct studies of Fe-responsive gene regulation in the α-proteobacteria. There are very close homologues of RirA only in the very close relatives of *Rhizobium* and it may be that yet other regulators are responsible for Fe-responsive gene regulation in more distant relatives among the α-proteobacteria. What is required, therefore, is direct experimental analysis of the Fe-regulated genes in genetically amenable α-proteobacteria such as *Caulobacter* or *Rhodobacter*.

Finally, it must be admitted that the results obtained in the last few years have been disappointing in terms of increasing our understanding of what happens in the nodule, perhaps the more important and interesting aspect of iron nutrition in the rhizobia. This is largely because the majority of studies have been done with the more amenable free-living forms of rhizobia. These, however, have been of value in showing which processes are not important for iron uptake by bacteroids. Thus, mutants that are defective in the synthesis and/or uptake of heme, of siderophores, and of the TonB-dependent transport systems can induce N_2-fixing nodules, showing that none of these mechanisms is the sole source of iron for bacteroids. Initial in silico information, too, has not pointed to any obvious nodule-specific means of iron acquisition. Indeed, genomic analyses actually appear to refute the possibility of some potentially likely candidates, such as the Fec system of citrate import, being involved. Likewise, the FeoB Fe^{2+} transporter, a good a priori candidate for iron nutrition in bacteroids, has not been found in the faster-growing rhizobial species, although a possible role for this transporter in *Bradyrhizobium* is suggested from the genome of this genus. Further, there do not appear to be any novel siderophore synthesis systems that are quiescent in the free-living state but might be involved in making bacteroid-specific siderophores.

The very flexibility and multiplicity of the iron uptake systems in the rhizobia may mean that no single, predominant uptake system occurs in the symbiotic state. To address this, further work may well require the phenotypic characterization of multiple mutants defective in more than one transport system, coupled to postgenomic analyses to identify all the genes whose expression is subject to iron availability and/or which are especially active in the root bacteroids.

Clearly, much remains to be done in coupling biochemistry, physiology, and genetics to unravel this question; most probably, studies of different rhizobium-legume symbioses are needed to obtain the full picture.

ACKNOWLEDGMENTS

Work from my laboratory was funded by the BBSRC of the United Kingdom. My thanks go to Jon Todd, Margaret Wexler, Kay Yeoman, Andrew Hemmings, Edith Mireles-Diaz, Andy Curson, Ola Kolade, Stefan Weidner, and Alf Puhler for sharing unpublished data.

SUGGESTED READING

Carter, R. A., P. S. Worsley, G. Sawers, G. L. Challis, M. J. Dilworth, K. C. Carson, J. A. Lawrence, M. Wexler, A. W. B. Johnston, and K. H. Yeoman. 2002. The *vbs* genes that direct synthesis of the siderophore vicibactin in *Rhizobium leguminosarum*: their expression in other genera requires ECF σ factor RpoI. *Mol. Microbiol.* **44:** 1153–1166.

Dilworth, M. J., K. C. Carson, R. G. F. Giles, L. T. Byrne, and A. R. Glenn. 1998. *Rhizobium leguminosarum* bv. *viciae* produces a novel cyclic trihydroxamate siderophore, vicibactin. *Microbiology* **144:**781–791.

Hamza, I., S. Chauhan, R. Hassett, and M. R. O'Brian. 1998. The bacterial Irr protein is required for coordination of heme biosynthesis with iron availability. *J. Biol. Chem.* **273:**21669–21674.

Hamza, I., R. Hassett, and M. R. O'Brian. 1999. Identification of a functional *fur* gene in *Bradyrhizobium japonicum*. *J. Bacteriol.* **181:**5843–5846.

Hamza, I., Z. H. Qi, N. D. King, and M. R. O'Brian. 2000. Fur-independent regulation of iron metabolism by Irr in *Bradyrhizobium japonicum*. *Microbiology* **146:**669–676.

Hirsch, A. M., M. R. Lum, and J. A. Downie. 2001. What makes the rhizobia-legume symbiosis so special? *Plant Physiol.* **127:**1484–1492.

Johnston, A. W. B., K. H. Yeoman, and M. Wexler. 2001. Metals and the rhizobial-legume symbiosis—uptake, utilization and signalling. *Adv. Microb. Physiol.* **45:**114–156.

Kolade O. O., P. Bellini, M. Wexler, A. W. B. Johnston, J. G. Grossmann, and A. M. Hemmings. 2002. Structural studies of the Fur protein from *Rhizobium leguminosarum*. *Biochem. Soc. Trans.* **30:**771–774.

LeVier, K., D. A. Day, and M. L. Guerinot. 1996. Iron uptake by symbiosomes from soybean nodules. *Plant Physiol.* **111:**893–900.

LeVier, K., and M. L. Guerinot. 1996. The *Bradyrhizobium japonicum fegA* gene encodes an iron-regulated membrane protein with similarity to hydroxamate-type siderophore receptors. *J. Bacteriol.* **178:** 7265–7275.

Lynch, D., J. O'Brien, T. Welch, J. H. Crosa, and M. O'Connell. 2001. Genetic organization of the region encoding regulation, biosynthesis and transport of rhizobactin 1021, a siderophore produced by *Sinorhizobium meliloti*. *J. Bacteriol.* **183:** 2576–2585.

Moreau, S., D. A. Day, and A. Puppo. 1998. Ferrous iron is transported across the peribacteroid membrane of soybean nodules. *Planta* **207:**83–87.

Nienaber, A., H. Hennecke, and H. M. Fischer. 2001. Discovery of a haem uptake system in the soil bacterium *Bradyrhizobium japonicum*. *Mol. Microbiol.* **41:**787–800.

Noya, F., A. Arias, and E. Fabiano. 1997. Heme compounds as iron sources for nonpathogenic *Rhizobium* species. *J. Bacteriol.* **179:**3076–3078.

Platero, R. A., M. Jaureguy, F. J. Battistoni, and E. R. Fabiano. 2003. Mutations in *sitB* and *sitD* genes affect manganese-growth requirements in *Sinorhizobium meliloti*. *FEMS Microbiol. Lett.* **218:**65–70.

Pohl, E., J. C. Haller, A. Mijovilovich, W. Meyer-Klaucke, E. Garman, and M. L. Vasil. 2003. Architecture of a protein central to iron homeostasis: crystal structure and spectroscopic analysis of the ferric uptake regulator. *Mol. Microbiol.* **47:**903–915.

Qi, Z. H., I. Hamza, and M. R. O'Brian. 1999. Heme is an effector molecule for iron dependent degradation of the bacterial iron response regulator (Irr) protein. *Proc. Natl. Acad. Sci. USA* **96:**13056–13061.

Qi, Z., and M. R. O'Brian. 2002. Interaction between the bacterial iron response regulator and ferrochelatase mediates genetic control of heme biosynthesis. *Mol. Cell* **9:**155–162.

Schwartz, C. J., J. L. Giel, T. Patschkowski, C. F. Luther, J. Ruzicka, H. Beinert, and P. J. Kiley. 2001. IscR, an Fe-S cluster-containing transcription factor, represses expression of *Escherichia coli* genes encoding Fe-S cluster assembly proteins. *Proc. Natl. Acad. Sci. USA* **98:**14895–14900.

Stevens, J. B., R. A. Carter, H. Hussain, K. C. Carson, M. J. Dilworth, and A. W. B. Johnston. 1999. The *fhu* genes of *Rhizobium leguminosarum*, specifying siderophore uptake proteins: *fhuDCB* are adjacent to a pseudogene version of *fhuA*. *Microbiology* **145:**593–601.

Todd, J. D., M. Wexler, G. Sawers, K. H. Yeoman, P. S. Poole, and A. W. B. Johnston. 2002. RirA, an iron-responsive regulator in the symbiotic bacterium *Rhizobium leguminosarum*. *Microbiology* **148:**4059–4071.

Visca, P., L. Leoni, M. J. Wilson, and I. L. Lamont. 2002. Iron transport and regulation, cell signalling and genomics: lessons from *Escherichia coli* and *Pseudomonas*. *Mol. Microbiol.* **45:**1177–1190.

Wexler, M., J. D. Todd, O. Kolade, A. M. Hemmings, G. Sawers, and A. W. B. Johnston. 2003. Fur is not the global regulator of iron uptake genes in *Rhizobium leguminosarum*, *Microbiology* **149:**1357–1365.

Wexler, M., K. H. Yeoman, J. B. Stevens, N. G. de Luca, G. Sawers, and A. W. B. Johnston. 2001. The *Rhizobium leguminosarum tonB* gene is required for the uptake of siderophores and haem as sources of iron. *Mol. Microbiol.* **41:**801–816.

Yeoman, K. H., A. G. May, N. G. deLuca, D. B. Stuckey, and A. W. B. Johnston. 1999. A putative ECF sigma actor gene, *rpoI*, regulates siderophore production in *Rhizobium leguminosarum*. *Mol. Plant-Microbiol. Interact.* **12:**994–999.

Yeoman K. H., S. Mitelheiser, G. Sawers, and A. W. B. Johnston. 2003. The ECF sigma factor RpoI of *R. leguminosarum* initiates transcription of the *vbsGSO* and *vbsADL* siderophore biosynthetic genes *in vitro*. *FEMS Microbiol. Lett.* **223:**239–244.

Yeoman, K. H., F. Wisniewski-Dye, C. Timony, J. B. Stevens, N. G. deLuca, J. A. Dowine, and A. W. B. Johnston. 2000. Analysis of the *Rhizobium leguminosarum* siderophore-uptake gene *fhuA*: differential expression in free-living bacteria and nitrogen-fixing bacteroids and distribution of an *fhuA* pseudogene in different strains. *Microbiology* **146:**829–837.

INDEX[a]

ABC proteins, 40–42
ABC transport systems, 78–80
ABC transporters, 113
Achromobactin, 403
  structure of, 404
Actinobacillus, 72
Aerobactin, 6, 9–10, 203–205, 423–424
  structure of, 204
  synthesis of, 29, 203, 204
Albomycin, 122, 126–127, 162, 163, 416, 426, CP12
  structure of, 158, 159
Alcaligin, 5, 8, 312–313, CP2
  AlcS permease and, 316
  Bordetella, biosynthesis genes, 314–316
    siderophore system, genetic organization of, 314–315
    system proteins and homologs, 315–316
  ferric, transport mutants, AlcR-positive regulator and *alcABCDER* operon, 316
  molecular structure of, 313–314
  sensing, 318
  siderophore utilization by, 313
Alcaligin receptor, FauA ferric, 316–317
Alcaligin system genes, transcriptional activation of, 317–318
  transcriptional repression of, 317
Amoxicillin, 426
Ampicillin, 426, 428
Anguibactin
  biosynthesis of, 243–244
  regulation of, 245
  transport of, 244–245

Anthraquinone carboxylic acid, conjugates of, 430
Antibiotics, siderophore, activity of, 423
  bi- and tridentate, 424–426
  catechol spermidine-based, 416, 426
  for highly resistant targets, 426–430
  monodentate, 423–424
  multiwarhead, 430–431
  natural, 416–417
  redox-based drug release of, 431
  synthetic, 425, 426–427
  transport of, by FhuA, 163–164
  Trojan horse, 126–128
  with siderophore components, 425
Aryl carrier protein domain, 136–137
ATPases, heme-specific, 79
Azospirillum, 442
Azospirillum irakense, 442

Bacillibactin, 397–399
Bacillus, 396–400
  Fur homolog regulation in, 399–400
  iron transport in, 399
  iron uptake in, genes involved in, 390
  siderophores of, production in, 396–399
    structures of, 396, 397
Bacillus anthracis, 396
Bacillus cereus, 396, 399
Bacillus megaterium, 396
Bacillus subtilis, 20–21, 178, 396, 397–398
Bacteria, gram-negative, iron uptake in, 178, 179
  gram-positive, heme transport across outer membrane in, 80–81
  iron assimilation in, 413–416
  pathogenic, iron transport systems in, 197–433
  siderophore biosynthesis in, 18–37

[a] CP, color plate.

Bordetella, 311–328
  bhu genes, regulation of, 324–325
  ferrimone-inducible iron acquisition in vivo by, 325–327
  heme utilization by, 322–325
  heme utilization genes, 322–324
  regulatory genes of, 77
  xenosiderophore utilization by, 318–319
Bordetella avium, 83
Bordetella bronchiseptica, 311, 319, 325–327
Bordetella parapertussis, 311
Bordetella pertussis, 311, 319, 325–327
  in vivo growth of, 311–312
  iron acquisition by, 312–313
Bradyrhizobium, 469
Bradyrhizobium japonicum, iron uptake in, 477–478
Brucella abortus, 139
BtuB and TonB proteins, structures of, in absence and presence of vitamin B_{12}, 153, CP15
  N-domain of, 148–149, CP13
BtuB structure(s), 61–63, 155
  β5-β6 loop in, 63
  N-terminal plug domain and 22-strand β-barrel and, 62
BtuC protein, FhuB and, sequence homology between, 167, 168–169
Burkholderia, 451
Burkholderia cepacia, 83, 293–294
  siderophore-mediated iron uptake in, 307–308

Campylobacter jejuni, 106
Cancer, iron chelators for tumor inhibition in, 421
Carboxymycobactin(s), 361–362
  biosynthesis of, 362–363, 364
  capture of iron by, 364
  structure of, 362
Cefaclor, 426
Cephalexin, 426
Chorismate, 135
Chrysobactin, 403
  biosynthesis of, and degradation of, 405–407
    proteins involved in, 405, 406
  ferric, transport of, 404–405
  in plants, 409
  structural properties of, 403–404
  synthetase CbsF protein, structure of, 406, 407
Citrate, and iron uptake in *P. aeruginosa*, 295, 306
  ferric. *See* Ferric citrate
  uptake of, by rhizobia, 479
Citrate-hydroxamates, 447, 448
Citrate-mediated Fe^{3+} transport, 169–170
Cobalamin substrate, BtuB structures and, 61–63
  interactions of, 62
Colibacillosis, 217
Colicin M, 162
Colicins, 103, 106–108
Coprogen, 122, 125, CP12

Corynebacterium diphtheriae, 77–78, 80–81, 87, 89, 90, 92, 186, 344–359, 368
  genetic map of *irp6* operon of, 347, 348
  heme iron regulation of *hmuO* and, 353–355
  heme-iron transport systems in, 348–350
  heme oxygenase in, 350
  heme utilization in gram-negative and gram-positive bacteria, compared, 352
  hmuTUV heme transport locus in, 350–353
  iron transport systems in, 345–355
  siderophore-dependent transport in, 345–348
Corynebacterium glutamicum, 178, 397
Corynebacterium ulcerans, 80
Corynebactin, ferric, 7–8, CP1
Coxiella burnetii, 372
Cytochrome *c* maturation system and *L. pneumophila* iron acquisition, 382–383
Cytoplasmic membrane, transport of ferrichrome across, 164–168

Deinococcus radiodurans, 88–89
Desferal, 122, 125, 420, 439, 442, CP12
  conjugates of, 429, 430
  in iron overload syndrome, 417, 418
Desferrioxamine(s). *See* Desferal
Diethylenetriaminepentaacetic acid, 422
2,3-Dihydroxybenzoylserine, 134, 138
Diphtheria, characteristics of, 344–345
  forms of, 344
Diphtheria toxin, 345
  iron regulation of, 355–359
Diphtheria toxin repressor, 356
  functional domains of, 357, 358
  structural analysis of, 356–358
Diphtheria toxin repressor-like repressors, family of, 358–359
Dipyridine, 422
2,6-Dithiocarboxylic acid, 420
DtxR, structure of, 189–190

Enterobacteriaceae, 104, 105, 181
Enterobactin, 6, 20–21, 199–202, 295, 304–306, 319, 423–424
  biosynthesis of, 24, 25, 134–138
  Bordetella, utilization genes, 319–321
    utilization system proteins and homologs, 320, 321
  circular dichroism spectrum of, 14, 15
  cluster genes, two-component regulatory systems of, 142
  ferric, 7–8, CP1
  gene cluster, genes in, 135
  iron transport system, components of, 228–229
  structure of, 133, 134
  utilization genes, regulation of, 321–322
Enterochelin. *See* Enterobactin

Erwinia, 402–412
  siderophore-mediated iron transport systems of, 403–407
Erwinia carotovora, 402
Erwinia chrysanthemi, 402, 403, 405, 407–411
  Fur-dependent iron regulation in, 407–410
  siderophore production by, 403
Erwinia chrysanthemi 3937, regulation of pectinolysis and iron transport in, 409, 410
Erwinia herbicola, 440
Escherichia coli, 5, 20–21, 40–42, 51, 73, 82, 83, 97, 99, 104, 106, 116, 158, 169, 178, 182, 185, 214–217, 405
  animal-pathogenic, 216–217
  enterobactin locus in, 200
  extraintestinal isolates of, 215–216
  Fur protein of, 399–400
  Fur target sequences in, 193–194
  heme sources and, 278
  intestinal isolates of, 215
  K-12, iron transport, energetics, and regulation in, 131–196
  pathogenic, 199–218
ExbB proteins, 99
  TonB and ExbD proteins, 96–112
  topology of, 97
  transmembrane domains, conservation in, 108, 109
ExbD proteins, 99
  TonB and ExbB proteins, 96–112
  topology of, 97
  transmembrane domains, conservation in, 110
Exochelin(s), 361
  biosynthesis of, 29–30
Exomycobactins. *See* Carboxymycobactin(s)
Extragenic palindrome sequences, 144

Fe^{3+} ion, 3–4
Fe^{3+} transport, citrate-mediated, 169–170
Fe^{2+} transport systems, 182–183
Fe-TRENCAM, 152
FecA, 51–52, 143–144, CP5–7
  apices and switch helices in, 55, 56–57, CP6–7
  crystal structures of, 155, 160, 161
  globular domain and lock region in, 58
    formation of transient channel in, 59–60, CP6, CP8
  N-domain of, 148–149, CP13
  N-terminal globular domain in, structure of, 53–54, CP5–7
  second site of interaction with TonB, 61, CP9
  signaling, and signaling pathway, 172–173
  structure of, reference points in, 58–59, CP8
FecI σ factor, 173–174
FeEnt, adsorption and desorption of, 151
  binding of, conformational motion in L7 during, 151–152

binding to FepA, 152
interactions with TonB, 152–154
loosely and tightly bound, 151
transport through FepA, 152, CP14
FeEnt uptake, binding stage of, 149–152
  internalization stage of, 152–155, CP14
  site-directed mutagenesis in, 149–150
  two-site binding model of, 149–150
Feo genes, in bacteria, distribution of, 181–182
  regulation of expression of, 181
FeoB, and pathogenicity, 182
  G protein in, 178–180
  in *L. pneumophila* intracellular infection, 380–381
  transport of, by *L. pneumophila*, 379–381
FepA, 51, 52, CP5–7
  affinity for ligands, 151
  apices and switch helices in, 55, 56–57, CP7
  binding by, 140
  C-domain of, 149
  crystal structure of, 140, 155
  electron spin resonance spectroscopy of, 150
  FeEnt binding to, 152
  FeEnt transport through, 152, CP14
  fluorescence spectroscopy of, 150–151
  globular domain and lock region in, 58
  homology region identified with, 60–61
  lock region in, mutants of central residues of, 59
  N-domain of, 148–149, 156, CP13
  second site of interaction with TonB, 60–61
  selective permeability and, 155–156
  selectivity for ligands, 152
  transport biochemistry of, 147–157
FepB, 140, 143, 145
FepC, 144
FepD, 144
FepG, 144
Ferribactin, 455–456
Ferric citrate, and *L. pneumophila*, 378–379
  dinuclear, crystal structure of, 170
  induction by, 170–172
  transcription regulation, mechanism of, 174–175
  transport of, and signaling pathway, 172–173
  transport system, regulation by iron, 174
Ferric dicitrate, 208
Ferric iron binding protein, 117–118
Ferric reductases, *L. pneumophila* and, 383–384
Ferric uptake regulation gene. *See* Fur gene
Ferrichrome, 29–30, 126, 138–139, 152, 158, 163
  Pseudomonas and, 441
  structure of, 158, 159
  transport of, 158–159
    across cytoplasmic membrane, 164–168
    across outer membrane, 160–164
  uptake of, interaction of components in, 168–169
Ferrichrome A, 8, CP4
Ferrichrome analogs, 423
Ferrienterobactin, uptake of, 139–141
Ferrienterobactin esterase, 138

Ferrihemes, 83–84
Ferrioxamine B, 5, 8, CP3
    and iron uptake in *P. aeruginosa*, 295, 306
Ferrioxamines, 438–439, 440
Ferrisiderophore receptor family(ies), 51, 53, CP5–7
Ferrous iron, transport of, 178–184, 208–209
        by *L. pneumophila*, 379–381
        uptake of, by rhizobia, 479–480
        sit operon for, 209, 210
Ferrous iron genes, in *E. coli* K-12, 178–180
Ferrous iron transport system, 178–182, 234
FetA, as functional siderophore receptor, 269–270
Fhu iron transport system, components of, 228–229
FhuA, 51, 52, 60, 158–159, CP5–7
    affinity for ligands, 151
    and FhuD, interaction of, 168
    apices and switch helices in, 55, 56–57, CP7
    crystal structures of, 155, 160, 161
    globular domain and lock region in, 58
    N-domain of, 148–149, CP13
    opening of channel in, 162
    second site of interaction with TonB, 60–61
    transport of antibiotics by, 163–164
    wild-type, 162–163
FhuB and BtuC protein, sequence homology
        between, 167, 168–169
    and FhuC, interaction of, 169
    and FhuD, interaction of, 168–169
    integral membrane protein, 165–168
    topology of, 165
FhuC and FhuB, interaction of, 169
FhuD, 140–141, 164–165
    and FhuA, interaction of, 168
    and FhuB, interaction of, 168–169
    crystal structure of, 165
Fit ABC iron transport system, 233
Fiu ABC iron transport system, 233–234
5-Fluorouridine, 431, 432
FpvA, 299–300
    in pyoverdine biosynthesis, 300–302
    receptor-mediated gene expression of, 300
Freshwater environment, siderophores in, 440
Fungal siderophores, 6
Fur, 189–190
    binding sequence, 190–192
    functions of, 141–142
    interaction with target sequences, consequences
        of, 192–193
    regulation by, 142, 187
    structure of, 189–190
    target sequences, in *E. coli*, 193–194
    target sites, 190–192
Fur boxes, and other regulatory elements, 194–195
Fur superfamily, in rhizobia, 482–484
        as manganese-responsive regulator, 485
        Irr subgroup of, 483, 484–485

G protein, in FeoB, 180
Gingipains, of *P. gingivalis*, functions of, 337
    of *Porphyromonas gingivalis*, role in heme
        acquisition, 335–339
    structures of, 336
Gly residues, 60
Glycine, 60
Gonorrhea, 256

Haemophilus, 67, 273–292
    heme and iron requirement and acquisition by,
        273–276
    hemoglobin of, 280–283
Haemophilus ducreyi, 83
    heme sources and, 279
Haemophilus influenzae, 38–40, 43, 71, 72, 78, 273,
        280–281
    accessory proteins and, 289
    genomic sequences of, 290
    heme acquisition by, 284–285, 286, 287
    heme and iron acquisition proteins of, 275–276
    heme sources and, 278–279
    hemoglobin/hemoglobin-haptoglobin binding
        proteins of, 281, 282
    hemophore system of, 43
    lactoferrin and, 288
    regulation of iron and heme acquisition
        mechanisms by, 289–290
    sources of iron for, 285–289
    transferrin and, 288
    utilization of hemoglobin-haptoglobin complexes
        by, 283–284
Haptoglobin, 67
Has hemophore system, 237–238
has operon, of *S. marcescens*, 44–45
Helicobacter pylori, 82–83, 181–182, 188
Heme, 206–208
    acquisition mechanisms, regulation by *H.
        influenzae*, 289–290
    acquisition of, by *H. influenzae*, 284–285, 286,
        287
        by *P. aeruginosa*, 308
        by *P. gingivalis*, 331–332
        gingipains of *P. gingivalis* in, 335–339
    and iron acquisition proteins, of *H. influenzae*,
        275–276
    roles of, in vivo, 277
    and iron requirement and acquisition, by
        Haemophilus, 273–276
    as source of iron, pathogens using, 86
    assimilation systems, 71
    availability of, bacteria influencing, 76–77
    bacteria binding or utilizing, 71
    bacteria sensing presence of, 77–78
    bacterial, and hemoprotein receptors, 66–85
        uptake systems, 66, 68–70
    binding and utilization, in *P. gingivalis*, 330–332

binding of, by *Legionella*, 383
cytoplasmic fate of, 81–82
degradation of, chemical steps in, 88
hemoglobin-bound, as iron source, 267–268
intercalation within DNA, 87
iron regulation of *hmuO* by, in *C. diphtheriae*, 353–355
membrane transport of, periplasmic and cytoplasmic, 78–80
methods of bacteria to acquire, 76–78
molecular weight of, 67
permeases and ATPases of, 79
regiospecificity of, 91–92
regulation of utilization of, in *P. gingivalis*, 339–341
sources of, 276–285
structure of, 67
transport of, in gram-negative bacteria, 71
   in *P. gingivalis*, 339, 340
   in virulence, 83–84
   TonB proteins and, 82–83
transport systems, 67
   of *Yersinia*, 234–238
unbound, 276–279
utilization of, by *Bordetella*, 322–325
   in *C. diphtheriae*, 352
   in rhizobia, 477
Heme acquisition pathways, 86–87
Heme acquisition systems, hemophore-dependent, 38–47
Heme-albumin complexes, 280
Heme assimilation systems, 74
Heme-binding proteins, in *P. gingivalis*, 333
Heme-hemopexin complexes, 279–280
Heme-iron transport systems, in *C. diphtheriae*, 348–350
Heme-mediated iron uptake systems, of *Vibrio*, 250–251
Heme oxygenase-like proteins, 90
Heme oxygenase proteins, amino acid sequence alignment of, 92, 93
Heme oxygenases, 87–89
   bacterial, 86–95
   bacterial phytochrome, 88–89
   distribution among bacterial species, 92–93
   in *C. diphtheriae*, 350
   mammalian and bacterial, structural similarities in, 89–92
Heme periplasmic binding proteins, 79
Heme pocket, 90–91
Heme receptors, general, 72–76
Heme-specific ABC transporters, 78–80
Heme transport locus, *hmuTUV*, in *C. diphtheriae*, 350–353
Heme transport systems, phase-variable expression of, 78
   regulated expression by bacteria, 77
   Y. enterocolitica and, 79–80

Heme transporters, and nonhomologous receptors, 76
Heme uptake systems, 80–81
Heme utilization genes, *Bordetella*, 322–324
Hemin. *See* Heme
Hemin utilization receptor, of *P. gingivalis*, 334–335
HemO, 88–89
Hemoglobin, 66, 67, 206–208
   binding and utilization of, in *P. gingivalis*, 334
   heme moieties in, 86
   of *Haemophilus*, 280–283
   receptors for, 267
Hemoglobin-haptoglobin complexes, utilization by *H. influenzae*, 283–284
Hemoglobin/hemoglobin-haptoglobin binding proteins, of *Haemophilus influenzae*, 281, 282
Hemolysins, 67
   of *Vibrio*, 252–254
Hemophore-dependent receptors, 72
Hemophore-receptor interactions, 43–44
   heme delivery in, 44
Hemophore receptors, outer membrane, 43
Hemophores, 38–40, 87
   α-helical C-terminal secretion signal of, 42
   and siderophores, iron acquisition pathways of, 39
   functions of, 46
   HasA, 40, 41
   heme acquisition systems of, 38–47
   secretion of, 40–42
Hemoprotein-specific receptors, 72–76
HemR receptor, 333
Heterobactins, 445
High-molecular-weight protein 2, 35
HmbR, as receptor for hemoglobin, 267
Hmu/Hem hemoprotein transport and regulatory system, 235
Hmu/Hem uptake system, 234–237
Hmu transporters, identification of, 477
HmuR receptor, of *P. gingivalis*, 334–335
HpuA/HpuB receptor, 267–268
HpuAB, as receptor for hemoglobin, 267
   function in humans, 268
   genetics and regulation of, 268
α-Hydroxyacids, 208

ibtA, and *frgA*, in *Legionella* intracellular infection, 378
   as *Legionella* siderophore, 379
IC202C, 420–422
IdeR, genes regulated by, 369
   in *M. tuberculosis*, 368–369
Immunomodulators, 420–422
Intestinal tract, siderophore production in, 440–441
IraAb, 381–382
Iron, acquisition mechanisms, regulation by *H. influenzae*, 289–290
   acquisition of, by *B. pertussis*, 312–313

Iron (continued)
    by *L. pneumophila*, 384–385
    cytochrome *c* maturation system and, 382–383
    by *P. gingivalis*, 331–332
    by streptococci, 394–396
    in mycobacteria, 360–363
   acquisition systems, in enteric pathogens, 200
   and virulence of *P. gingivalis*, 341–342
   assimilation of, in bacteria, 413–416
   bacterial uptake of, periplasmic binding proteins in, 113–129
   ferric, 18–19
    octahedral configuration of, 18, 19
   ferrimone-inducible acquisition of, by *Bordetella*, 325–327
   ferrous. *See* Ferrous iron
   free, for *H. influenzae*, 285–288
   in hemoglobin, 86
   in regulation of diphtheria toxin, 355–359
   lactoferrin as source of, 263–267
   regulation of, evolving complexity of, 188–189
    Fur-dependent, in *E. chrysanthemi*, 407–410
    genetics of, 187–188
    in mycobacteria, 368–370
   required by *P. gingivalis*, 330
   source(s) of, for *H. influenzae*, 285–289
    for pathogens, heme and heme-bound proteins as, 86
    hemoglobin-bound heme as, 267–268
   steady-state distribution in humans, 87
   storage of, by mycobacteria, 367–368
   transport of, and ecology, 436–488
    ferrichrome- and citrate-mediated, 158–177
    in *Bacillus*, 399
    in pathogenic bacteria, 197–433
    nonsiderophore, in *Staphylococcus*, 392–394
    siderophore-mediated, by *P. aeruginosa*, 294–304
    in *Legionella*, 375–379
    in *Staphylococcus*, 392–394
   transport systems, enterobactin-dependent, of *Yersinia*, 227
    FhuBCD siderophore-dependent, of *Yersinia*, 226–227
    in biology of enteric pathogens, 209–217
    nonsiderophore, 206–209, 211
    of *Legionella*, 379–384
    of *Neisseria*, 259–263
    of *P. gingivalis*, 341–342
    siderophore-dependent, of *Yersinia*, 221–229
    siderophore-independent, of *Yersinia*, 230–234
    siderophore-mediated, of *Erwinia*, 403–407
   uptake of, from transferrin, 262–263
    in *Bacillus*, genes involved in, 390
    in gram-negative bacteria, 178, 179
    in *P. aeruginosa*, 305–306, 307
    in rhizobia, 469–488
    in *Staphylococcus*, genes involved in, 390–391
    in *Streptococcus*, genes involved in, 390
    pyoverdine-mediated, 463
    siderophore-independent, by mycobacteria, 366–367
    siderophore-mediated, by *Neisseria*, 268–270
    by *Vibrio*, 242–250
    in *B. cepacia*, 307–308
   uptake systems, redundancy in *Neisseria*, 270–271
    TonB-dependent, 270
Iron acquisition systems, of *Neisseria*, 258–259, 260
Iron heme transport protein, 333–334
Iron(III) chelators, therapeutic uses of, 413–433
Iron overload, siderophores in, 417
Iron-protoporphyrin IX. *See* Heme
Iron-responsive gene, regulation of, in rhizobia, 480–485
Iron transport systems, in *C. diphtheriae*, 345–348
irp6 operon, genetic map of, in *C. diphtheriae*, 347, 348
Irr, of Fur superfamily, 483, 484–485
Isopyoverdines, 455, 456

Johnne's disease, 361

α-Ketoacids, 208
Klebsiella pneumoniae, 169

L-Lysine, 362
Lactoferrin, as iron source, 263–267
   H. influenzae and, 288
   interactions with *L. pneumophila*, 383
Lactoferrin LbpBA receptor, 263–265
Lactoferrin receptor, function of, in humans, 266–267
   genetics and regulation of, 265–266
Legiobactin, 376–377, 379
Legionella, 372–386
   genetics and regulation of siderophore production by, 377–378
   intracellular infection, 385
    FeoB in, 381
    ibtA and *frgA* in, 378
   nonsiderophore iron transport systems of, 379–384
   siderophore-mediated iron transport in, 375–379
Legionella pneumophila, 372–373
   FeoB and ferrous iron transport by, 379–381
   ferric citrate and, 378–379
   ferric reductases and, 383–384
   hemin binding by, 383
   importance of iron for, 374–375
   interactions with transferrin and lactoferrin, 383
   IraAB, functions of, 381–382
   iron acquisition by, 384–385
    cytochrome *c* maturation system and, 382–383
   pvc-like genes in, 378
   siderophore-mediated iron transport and, 375–379
   virulence factors of, 373–374

Legionnaires' disease, 372
  pathogenesis of, 373
  risk factors for, 373
Ligand competition studies, 10, 11–12
Ligand internalization, in FeEnt uptake, 154–155
Lipopolysaccharide O-antigen, 151
Lysine-N^6-hydroxylation, 28

Malaria, 419
Malleobactin, 307
Maltose binding protein, 116–117
Manganese, 437
  iron transport system of *Y. pestis*, 230–232
  uptake of, *sit* operon for, 209, 210
Marine offshore environment, siderophores in, 440
Meningitis, bacterial, 256
Mesorhizobium, 469
Metal transporters, in outer membrane, 147–148
  N-domains of, 156
  structural features of, 148–149, CP13
Metallo-regulated genes, Fur-mediated repression of, 185, 186
5-Methylene furanone, generation of, 420
Micacocidin, chelation of ferric iron by, 30, 31
Microbacterium flavescens, 441
Microcin, 162
Mucormycosis, 448
Mycobacteria, as therapeutic agents, 417–419
  iron acquisition in, 360–363
  iron regulation in, 368–370
  iron storage by, 367–368
  pathogenic, 360–371
  siderophore-independent iron uptake by, 366–367
Mycobacterium avium, 361
Mycobacterium bovis, 361
Mycobacterium fortuitum, 367
Mycobacterium intracellulare, 361
Mycobacterium leprae, 361
Mycobacterium neoaurum, 361
Mycobacterium paratuberculosis, 361, 366
Mycobacterium smegmatis, 28–29, 360, 361, 362, 365, 366, 367
Mycobacterium tuberculosis, 360, 361, 363, 365, 366, 367, 368, 417–419
  biosynthesis of mycobactin and carboxymycobactin from, 363, 364
  IdeR in, 368–369
Mycobacterium vaccae, 361
Mycobactin synthetic analogs, 418–419
Mycobactin(s), 361, 416–417
  biosynthesis of, 362–363, 364
  capture of iron by, 364–365
  natural, 417–418
  structure of, 362, 366
Myoglobin, 66
Myxochelins, biosynthesis of, 26

Nalidixic acid, conjugates of, 430
Neisseria, 256–272
  adherence-related virulence factors of, 258
  do not produce siderophores, 268–269
  HpuAB of, 75–76
  iron acquisition systems of, 258–259, 260
  nonsiderophore iron transport systems of, 259–263
  pathogenic, similar lifestyles of, 257
  siderophore-mediated iron uptake by, 268–270
  surface adhesins and, 257–258
  virulence factors in, 256–257
Neisseria gonorrhoeae, 256
  transferrin-iron acquisition system in, 263
Neisseria lactamica, 256
Neisseria meningitidis, 72, 78, 81, 82, 83, 88, 89, CP10, 256
  HmubR of, 76
NGAL, 5
Nonribosomal peptide synthetases, 19–20, 21
NU216R, 381–382

Oregon Green maleimide, 101
Ornibactins, 306–307, 443
Ornithine, 28
Outer membrane, metal transporters in, 147–148
  transport of ferrichrome across, 160–164
Outer membrane metal transport systems, 149
Outer membrane receptors, 148
Outer membrane transporters, non-hemophore-dependent, 71

Pantoea agglomerans, 440
Pathogens, enteric, biology of, iron transport systems in, 209–217
  gram-negative, and diseases caused by, 200
  iron acquisition systems in, 200
Pectobacterium carotovorum, 402
Pectobacterium chrysanthemi, 402
Penicillin conjugates, 427
Peptides, nonribosomal, biosynthesis of, 19–20, 21
Periodontal diseases, 329
Periodontitis, 330
Periplasmic binding proteins, class 9, 118–119, 120
  classification by clusters, 114
  clustering of, 121
  functions of, 113
  in bacterial iron uptake, 113–129
  structural features of, 114–116
  topological arrangement of, 114–115
Plants, siderophores and, 440
Plasma, human, contents of, 66–67
Plasmodia, as therapeutic agents, 418–419
Plasmodium falciparum, 429
  as therapeutic agents, 419
Plesiomonas shigelloides, 78
  heme sources and, 279

Polycarboxylates, 447, 448
Porins, 51, 113
   structure of, 155–156
Porphyromonas gingivalis, 72, 77, 78, 83, 329–343
   acquisition of iron and heme by, 331–332
   colonization by, 329
   gingipains of, functions of, 337
      role in heme acquisition, 335–339
      structures of, 336
   heme-binding proteins in, 333
   heme transport in, 339, 340
   heme utilization in, regulation of, 339–341
   hemin binding and utilization in, 330–332
   hemin utilization receptor and, 334
   hemoglobin binding and utilization in, 334
   HemR receptor and, 333
   HMUR of, 75
   IhtB protein and, 333–334
   iron requirements of, 330
   iron transport systems of, and link to pathogenesis, 341–342
   TonB-linked receptor and, 333
   virulence of, influence of iron on, 341–342
Proteins, accessory, *H. influenzae* and, 289
   as nonhomologous receptors, 76
   heme periplasmic binding, 79
   iron acquisition, and heme, roles of, in vivo, 277
   iron transport, structural studies of, 49–129
   outer membrane receptor, formation of transient channel in, 59–60
      globular domains in, topology and structure of, 53–54
      simultaneous sequence alignment of, 57–61
      structure of, 51–65, CP5
      bipartite gating, 52–53
      two domains of, 51–52
   periplasmic binding. *See* Periplasmic binding proteins
Proton motive force, 96–112
   cytoplasmic membrane, protein shuttling and, 103–104
Protonophore carbonyl cyanide *m*-chlorophenyl hydratone, 103
Pseudobactin. *See* Pyoverdine(s)
Pseudomonas, 293–310
   environmental fluorescent, diversity of, 451–468
   ferrichrome and, 441
   putative iron receptors and regulators in, 300, 301, 302
   taxonomy of, 464–465
   uptake post-outer membrane in, 306–307
Pseudomonas aeruginosa, 71, 72, 81, 82, 89, 105, 106, 293, 294
   ferric pyoverdine receptor of, 299–300
   Fur protein of, structure of, 189–190
   heme acquisition by, 308
   siderophore-mediated iron transport and, 294–304
   siderophores of, 294–302
   siderophores synthesized by, 428
Pseudomonas aureofaciens, 462–463
Pseudomonas costantinii, pyoverdine of, 453–454, 456, 462
Pseudomonas fluorescens, 72
Pseudomonas putida, 104–111, 319
Pseudomonas stutzeri, 439
Pseudomonin, 452
Pyochelin, 295, 302–304, 307, 452
   biosynthesis of, and transport of, 302–303
      genetics of, 303
   regulation of production of, 303–304
Pyoverdine-quinoline antibiotic conjugates, 429
Pyoverdine(s), 428–429, 452
   biosynthesis of, genetics of, 296, 297
   biosynthetic pathway of, and structure diversity of, 459–460
   chromophore level of, structure diversity at, 455–456
   dicarboxylic side chain level of, structure diversity at, 454–455
   environmental fluorescent, diversity of, 451–468
   heterologous, uptake of, 302
   iron uptake mediated by, 463
   peptide level of, structure diversity at, 456–459
   peptidic part of, amino acid composition of, 457, 458
   physiological and biochemical features of, 453
   regulation of production of, 297–299
   side chains of, 454, 455
   siderotyping of, 460–467
      and environmental microbiology, 466–467
      and phylogeny, 465–466
      and taxonomy of, 464–465
      benefits of, 466
      IEF electrophoresis for, 460–461
      pyoverdine-mediated iron uptake as method of, 461–463
      siderotypes, and siderovars of, 463–464
   structural features of, 453–460
   structure of, 294–296
   transport of, 299–300
Pyridine-2,6-bis(monothiocarboxylic acid), 452
Pyridoxal isonicotinoyl hydrazone, 422

Q fever, 372
Quinolobactin, 452

Ralstonia, 451
Rhizobactin, structure of, 472
Rhizobactin 1021, structure of, 472
Rhizobia, 77
   ABC-type transporters of, 478–479
   citrate uptake by, 479
   Fbp-like proteins of, 478–479

ferrous iron uptake by, 479–480
free-living, iron uptake by, mechanisms of, 470–480
Fur superfamily in, 482–484
heme utilization in, 477
in silico analyses and "missing genes," 479–480
iron-responsive genes in, range of, 480
  regulation of, 480–485
iron uptake in, mechanisms of, and regulation of, 469–488
regulators of, with broader regulatory spectrum, 481
Rpol of, 480–481
siderophore transporters of, 475–476
TonB and iron uptake in, 477–478
Rhizobium, 469
Rhizobium leguminosarum, 185, 471–473
  iron uptake in, 477–478
  RirA of, 481–482
  siderophore synthesis and uptake in, 471–472
  vicibactin biosynthetic gene cluster of, functions of, 474
Rhizobium loti-like bacterium, 442
Rhodobacterium sphaeroides, 92–93
Rhodococcus, 445
Rhodotorulic acid, 5, 8, 430, CP2
RhrA, regulator for RB1021 synthesis, 481
Rifamycin, 162
  structure of, 158, 159
RirA, 481–482
Rpol, of *R. leguminosarum*, 480–481

Saccharomyces cerevisiae, 448–449
Salicylic acid, 452
Salmochelins, 202–203, 441
  structure of, 202
Salmonella, 441
  enterobactin locus in, 200
  environments harboring, 211
  iron transport and pathogenesis in, 211–212
  pathogenic, 199–218
Salmonella enterica, 169, 178, 211–213
  heme sources and, 278
Salmycin, 416, 426
Schizokinen(s), 396–397
Serine, 362
Serratia marcescens, 40, 41, 72, 82, 104
  has operon of, 44–45
  hemophore of, crystal structure of, 42
    receptor of, 43
  TonB-like proteins in, 44
Shigella, 213–214
  enterobactin locus in, 200
  pathogenic, 199–218
Shigella dysenteriae, 80, 81–82
  heme transport locus of, 207
Shigella flexneri, 213–214

Siderophore antibiotics. *See* Antibiotics, siderophore
Siderophore binding proteins, 119–125
Siderophore-defective mutants, phenotypes of, 476–477
Siderophore-dependent iron acquisition pathways, 71
Siderophore-dependent transport, in *C. diphtheriae*, 345–348
Siderophore-drug conjugates, antimicrobial studies of, 422–431
  components of, 423
  design of, 422–423
Siderophore-mediated iron transport systems, 199–206
Siderophore transport system, 415
Siderophores, α-hydroxy carboxylate-based, 414
  and plants, 440
  antibiotic properties of, 416–417, 424
  as therapeutic agents, 416–417
  bacterial and fungal, interrelationship between, 447–449
    uptake in fungus, 448–449
  bacterial catecholate, grouped according to structural elements, 446
  binding of, 54–56, CP5–7
  biochemical and physical properties of, 3–17
  biosynthesis in bacteria, 18–37
  carboxylate, 7
  catechol-based, 414
  catecholate, 6, 15, 16, 20–25
    biosynthesis of, 22, 23
  citrate-hydroxamate, 447, 448
  classification of, 413
  containing ligands binding iron(III), 413–415
  degradation of, 441–442
  denticity of, 12–14
  description of, 242
  ecological impact of, 442–447
  ecology of, 437–450
  electronic structure and spectra of, 14–15
  five-member heterocyclic ring-containing, 30–35
  fluorescent, 7
  fungal, 6
  grouped according to ecological habitat, 444–445
  hydroxamate, 6–7, 25–30
    biosynthesis in soil, 443, 446
    scaffolds of, 28
    structures of, 27
  hydroxamate-based, 414
  hydroxamate-type, binding sites in, 125–126, CP12
    space-filling models of, 123
    structures of, 122, 123
  in aquatic environment, 440
  internalization by, 154–155
  mixed ligand, 414
  of fluorescent *Pseudomonas* species, 452–453

Siderophores (*continued*)
  phenolate, 20–25
    biosynthesis of, 22, 23
  pM values of, 9–11, 13
  production of, in intestinal tract, 440–441
    in soil, 437–440
  protonation and iron formation constants of, 10, 11–12
  recognition of, 125–126
  secondary, 138–139
  solution thermodynamics of, 8–11
  stability of, conformational effects and, 14, CP2
  structure(s) of, 4–6, 7–8, CP1–4
  tetradentate, 5, 8, CP2
  types of, 4, CP1
  with common iron binding moieties, 413, 414
Sinorhizobium, 469
Sinorhizobium meliloti, 28
  RB1021 siderophore of, synthesis and uptake of, 473–475, 476
Soft rot disease, 402–403
Soil, siderophore production in, 437–440
Staphylococcal siderophore transporter, 392
Staphylococci, 387–394
  iron-siderophore transport in, 390–392
  iron uptake in, genes involved in, 390–391
  nonsiderophore iron transport in, 392–394
  siderophores produced by, 387–390
Staphylococcus aureus, 67, 80, 387, 392–394
  Fur homologs of, 400
  genome, *isd* region of, 393
  heme transport system in, 81
  siderophore biosynthetic locus of, 388, 389
Staphylococcus epidermidis, 387, 392
Staphyloferrin A, 388
Staphyloferrin B, 388
Streptococci, 394–396
  iron acquisition by, 394–396
  iron uptake in, genes involved in, 390
Streptococcus intermedius, 394–396
Streptococcus pneumoniae, 394, 395, 396
Streptococcus pyogenes, 67, 80, 394, 395, 396
  heme transporters of, 76
Streptomyces pilosus, 438–439
Streptomycetes, 437–440
Synechocystis, 367

TbpA, 259–260
TbpB, 259–260
Thermococcales, 92–93
Threonine, 362
TonA mutant phenotypes, 96–97
TonB, in rhizobia, 477–478
TonB-like proteins, 44, 56
  sites of FecA interaction with, 61, CP9
  sites of FhuA interaction with, 60

TonB-linked receptor, *P. gingivalis* and, 333
TonB mutant phenotypes, 96–97
TonB proteins, 97–98
  and BtuB, structures of, in absence and presence of vitamin B_{12}, 153, CP15
  and ExbB and ExbD proteins, 96–112
  energy transduction, model for, 102–103, CP11
  "extra," C-terminal region of, 107, 108
  heme transport and, 82–83
  in various species, 104–111
  interactions to FeEnt with, 152–154
  iron uptake systems dependent on, 270
  phenotypic activity assays, compared, 99–100
  protein synthesis inhibitors and, 104
  shuttling between membranes, 100–101
  topology of, 97
  transmembrane domain, conservation in, 105
Transferrin, as iron source of *Neisseria*, 259–263
  H. influenzae and, 288
  interactions with *L. pneumophila*, 383
  iron uptake from, 262–263
Transferrin-binding proteins, 259–260
Transferrin-iron acquisition system, in *N. gonorrhoeae*, 263
Transferrin-iron internalization, proteins in, 260–261
Transmembrane signaling device, induction of ferric citrate transport system by, 170–175
Treponema pallidum, 118
Trihydroxamate, isocyanurate-based, drug conjugates of, 431, 432
Tuberculosis, 417–419

Ustilago sphaerogena, 158

Vibrio, 241–255
  energy-transducing complexes of, 252, 253
  siderophore-mediated iron uptake by, 242–250
  siderophores of, structures of, 243
  transporters in, 251–252
Vibrio anguillarum, 32, 83, 185, 188, 193, 241, 242, 243–245, 251–252
Vibrio cholerae, 30–35, 71, 79, 82, 106, 241, 242, 245–247, 251
Vibrio harveyii, 241, 242
Vibrio parahaemolyticus, 241, 242, 248–250
Vibrio vulnificus, 76, 77, 241, 242, 247–248, 252
Vibriobactin, biosynthesis of, 32–35, 245–247
  location of, 245
  regulation of, 247
  transport of, 247
Vibrioferrin, 249–250
Vicibactin, structure of, 472
  synthesis of, and transport of, 471–473
  pathway of, 474
Virulence, bacterial, heme transport in, 83–84

Vitamin B$_{12}$, structures of TonB and BtuB in absence and presence of, 153, CP15
Vitamin B$_{12}$ transporter, structure of, 165–167
Vulnibactin, 248

Xanthomonas campestris, 106
Xenosiderophore, utilization by *Bordetella*, 318–319
Xenosiderophore transport systems, 205–206, 304
Xylella fastidiosa, 106

Yersinia, 219–240
    HemR and HmuR of, 73–75
    iron regulation and storage by, 238
    pathogenesis of, iron and hemoprotein transport systems and, 224
Yersinia enterocolitica, 72, 73, 74, 81, 86–87, 219, 223, 226–227, 236, 237, 238
    heme-specific ABC transport by, 79–80, 234–235

Yersinia pestis, 32, 72, 78, 219–222, 226–227, 236–237, 238, 407
    derivatives of, gene fusion in, 225
    manganese ABC iron transport system of, 230–232
    Yfe iron transport system of, 230–232
    Ysu and Ynp siderophore biosynthesis systems, 227–229
Yersinia pseudotuberculosis, 219, 223, 226–227
Yersiniabactin, 32–35, 205
    iron transport system, 221–226
        model of, 222
    structure of, 221, 366
Yfu iron transport system, 232–233
Yiu iron transport system, 232–233
Ynp iron transport system, components of, 228–229

Zur protein, 186–187